Lecture Notes in Electrical Engineering

Volume 661

The book series *Lecture Notes in Electrical Engineering* (LNEE) publishes the latest developments in Electrical Engineering—quickly, informally and in high quality. While original research reported in proceedings and monographs has traditionally formed the core of LNEE, we also encourage authors to submit books devoted to supporting student education and professional training in the various fields and applications areas of electrical engineering. The series cover classical and emerging topics concerning:

- Communication Engineering, Information Theory and Networks
- Electronics Engineering and Microelectronics
- Signal, Image and Speech Processing
- Wireless and Mobile Communication
- Circuits and Systems
- Energy Systems, Power Electronics and Electrical Machines
- Electro-optical Engineering
- Instrumentation Engineering
- Avionics Engineering
- Control Systems
- Internet-of-Things and Cybersecurity
- Biomedical Devices, MEMS and NEMS

For general information about this book series, comments or suggestions, please contact leontina.dicecco@springer.com.

To submit a proposal or request further information, please contact the Publishing Editor in your country:

China

Jasmine Dou, Associate Editor (jasmine.dou@springer.com)

India, Japan, Rest of Asia

Swati Meherishi, Executive Editor (Swati.Meherishi@springer.com)

Southeast Asia, Australia, New Zealand

Ramesh Nath Premnath, Editor (ramesh.premnath@springernature.com)

USA, Canada:

Michael Luby, Senior Editor (michael.luby@springer.com)

All other Countries:

Leontina Di Cecco, Senior Editor (leontina.dicecco@springer.com)

**** Indexing: The books of this series are submitted to ISI Proceedings, EI-Compendex, SCOPUS, MetaPress, Web of Science and Springerlink ****

More information about this series at http://www.springer.com/series/7818

Margarita N. Favorskaya · Saad Mekhilef ·
Rajendra Kumar Pandey · Nitin Singh
Editors

Innovations in Electrical and Electronic Engineering

Proceedings of ICEEE 2020

Springer

Editors
Margarita N. Favorskaya
Department of ICT
Reshetnev Siberian State University
of Science and Technology
Krasnoyarsk, Russia

Rajendra Kumar Pandey
National Power Training Institute (NPTI)
Ministry of Power, Government of India
Varanasi, India

Department of Electrical Engineering
Indian Institute of Technology (BHU)
Varanasi, India

Saad Mekhilef
Department of Electrical Engineering
University of Malaya
Kuala Lumpur, Malaysia

Nitin Singh
Department of Electrical Engineering
MNNIT Allahabad
Allahabad, India

ISSN 1876-1100 ISSN 1876-1119 (electronic)
Lecture Notes in Electrical Engineering
ISBN 978-981-15-4691-4 ISBN 978-981-15-4692-1 (eBook)
https://doi.org/10.1007/978-981-15-4692-1

This Springer imprint is published by the registered company Springer Nature Singapore Pte Ltd.
The registered company address is: 152 Beach Road, #21-01/04 Gateway East, Singapore 189721, Singapore

ICEEE 2020 Organization

Chief Patrons

Mr. Suneel Galgotia, Chancellor, Galgotias University, India
Mr. Dhruv Galgotia, CEO, Galgotias University, India
Prof. R. K. Pandey, Director General NPTI Government of India

Patrons

Prof. Preeti Bajaj, Vice Chancellor, Galgotias University.
Dr. Manju Mam, Director, NPTI Faridabad
Prof. Pardeep Kumar, Pro Vice Chancellor, Galgotias University

General Chairs

Prof. B. Mohapatra, Dean SECE Galgotias University, India
Dr. N. V. Kumar, Director R&D, NPTI Faridabad

Conference Chair and Chairman, Oversight Committee

Prof. Rabindra Nath Shaw, Galgotias University, India

Conference Secretaries

Er. Himanshu Karn, Galgotias University, India
Prof. D. Saravanan, Galgotias University, India

Technical Chairs

Prof. Valentina Balas, University of Arad, Romania
Prof. Rohit Tripathi, Galgotias University, India
Prof. Sheetla Prasad, Galgotias University, India

Publication Chairs

Prof. Saad Mekhilef, University of Malaya, Malaysia
Prof. Pratima Walde, Galgotias University, India
Prof. Rashid Ansari, Galgotias University, India

Springer/ICEEE Liaison

Aninda Bose, Senior Editor, Springer Nature

International Advisory Board

Prof. Vincenzo Piuri, University of Milan, Italy
Prof. Georges Zissis, President, IEEE IAS
Prof. Lakhmi C. Jain, University of Technology, Sydney
Dr. Tamas Ruzsanyi, Ganz-Skoda Electric Ltd., Hungary
Prof. Valentina Balas, University of Arad, Romania
Prof. N. R. Pal, President, IEEE CIS
Prof. T. George, University of Piraeus, Greece
Prof. Yen-Wei Chen, Ritsumeikan University, Japan
Prof. Milan Simic, RMIT University, Australia
Prof. M. Paprzycki, Polish Academy of Sciences
Prof. Maria Virvou, University of Piraeus, Greece
Prof. D. P. Kothari, Ex Director IIT Delhi, India
Prof. B. K. Panigrahi, Professor, IIT Delhi, India

Prof. C. Boccaletti, Sapienza University, Italy
Dr. Nishad Mendis, Det Norske Veritas, Australia
Dr. Akshay Kumar, Concordia University, Canada

Technical Program Committee and Reviewers

Dr. A. R. Abhyankar
Dr. Aditi Sharan
Dr. Ajay Mittal
Dr. Sudhir Kumar Sharma
Dr. Ajai Jain
Dr. Alok Kushwaha
Dr. Amit Agarwal
Dr. Amalendu Patnaik
Dr. Anil K. Ahlawat
Dr. Anil K. Singh
Dr. Anuradha
Dr. Arun Kumar Verma
Dr. Aseem Chandel
Dr. Asheesh K. Singh
Dr. Ashutosh Dixit
Dr. Asif Ekbal
Dr. B. Dushmanta Kumar Patro
Dr. Baij Nath Kaushik
Dr. Bhaskar Biswas
Dr. Bharat Singh Rajpurohit
Dr. C. Patvardhan
Dr. C. Rama Krishna
Dr. C. K. Nagpal
Dr. Chandra Sekaran
Dr. Chiranjeev Kumar
Dr. Chittaranjan Hota
Dr. D. Bhagwan Das
Dr. D. A. Mehta
Dr. D. S. Kushwaha
Dr. D. S. Yadav
Dr. Desh Deepak Sharma
Dr. Dhram Singh
Dr. Dimple J. Gupta
Dr. Diwakar Bhardwaj
Dr. Girish Patnaik
Dr. Jai Govind Singh
Dr. Joy Deep Mitra

Dr. K. V. Arya
Dr. Kiran Kumar Pattanaik
Dr. Kishor K. Bhoyar
Dr. Komal Kumar Bhatia
Dr. Lalit Kumar Awasthi
Dr. M. K. Dutta
Dr. M. P. Singh
Dr. Madhavi Sinha
Dr. Manisha Sharma
Dr. Mohd. Rihan
Dr. Mayank Pandey
Dr. Munesh C. Trivedi
Dr. N. Badal, KNIT
Dr. Nanhay Singh
Dr. Narendra Kohli
Dr. Naresh Chauhan
Dr. Naveen Kumar
Dr. Neelam Duhan
Dr. Neeraj Tyagi
Dr. O. P. Verma
Dr. Pooja Jain
Dr. Pooja Pathak
Dr. Prabhat Ranjan
Dr. Prabhakar Tiwari
Dr. Prabin Panigrahi
Dr. Pragya Dwivedi
Dr. Pradeep Sharma
Dr. Pramod Kumar
Dr. Pramod Kumar Singh
Dr. Punam Bedi
Dr. R. K. Singh
Dr. R. S. Yadav
Dr. R. S. Rao
Dr. Rahul Rishi
Dr. Rajesh Prasad
Dr. Reena Dadhich
Dr. Ruchika Malhotra
Dr. S. P. Tripathi
Dr. Sapna Gambhir
Dr. Suneeta Agarwal
Dr. Sujoy Das
Dr. Sukomal Pal
Dr. Sunil Kumar Khatri
Dr. Tanveer Siddiqui
Dr. Tarun Shrimali

Dr. Vasudha Bhatnagar
Dr. Vishal Bhatnagar
Dr. Yashpal Singh
Prof. Herbert H. C. Lu
Dr. Senthilrajan Agni
Dr. Abhineet Anand
Dr. Anurag Baghel
Dr. Balamurugan Balusamy
Dr. Priti Bansal
Dr. Sonia Bansal
Dr. Annappa Basava
Dr. Rohit Beniwal
Dr. Vandana Bhasin
Dr. Rodrigo Bortoletto
Dr. John Moses Cyril
Dr. Pinaki Chakraborty
Dr. Sansar Chauhan
Dr. Rahul Chaurasiya
Dr. Surya Deo Choudhary
Dr. Anurag Dixit
Dr. Nripendra Narayan Das
Dr. Indrani Das
Dr. Aparna Datt
Dr. Parneeta Dhaliwal
Dr. Chandrakant Divate
Dr. Rajesh Dubey
Dr. Arman Faridi
Dr. Ankush Ghosh
Dr. Utkarsh Goel
Dr. Pallavi Goel
Dr. Amit Goel
Dr. Priyanka Goyal
Dr. Deepak Gupta
Dr. Suneet Gupta
Dr. Raza Haidri
Dr. Syed Shabih Hasan
Dr. Manas Hati
Dr. Brijesh Iyer
Dr. Manisha Jailia
Dr. Prashant Johri
Dr. Jegathesh Amalraj Joseph
Dr. Sandeep K. Singh
Dr. Vinay Kumar
Dr. Amita Kapoor
Dr. Sandhya Katiyar

Dr. Anvesha Katti
Dr. Ruqaiya Khanam
Dr. Aanchal Khatri
Dr. Shrawan Kumar
Dr. Devendra Kumar
Dr. Avneesh Kumar
Dr. Arun Kumar
Dr. Sanjeev Kumar
Dr. Vipin Kumar
Dr. Sanjay Kumar
Dr. Bhavnesh Kumar
Dr. Sandeep Kumar
Dr. Neetesh Kumar
Dr. M. Mohanraj
Dr. Ramakrishnan Malaichamy
Dr. Manas Kumar Mishra
Dr. Baibaswata Mohapatra
Dr. Thiyagarajan Muthunatesan
Dr. Rashid Mahmood
Dr. Yogendra Meena
Dr. Gitanjali Mehta
Dr. A. K. Mishra
Dr. Keshav Niranjan
Dr. Manoj Panda
Dr. Sanjeev Pippal
Dr. V. A. Sankar Ponnapalli
Dr. Shiv Prakash
Dr. Sheetla Prasad
Dr. Mohammed Abdul Qadeer
Dr. R. Gunasundari Ranganathan
Dr. Ranjeet Kumar Ranjan
Dr. Rohit Raja
Dr. Bharti Rana
Dr. Mukesh Rawat
Dr. Navaid Zafar Rizvi
Dr. S. Pravinth Raja
Dr. Anil Kumar Sagar
Dr. Rajeev Sharma
Dr. Birendra Kumar Sharma
Dr. Shrddha Sagar
Dr. Jyoti Sahni
Dr. Mohd. Saifuzzaman
Dr. Kavita Saini
Dr. Kamalesh Sethuramalingam
Dr. Priestly Shan

Dr. Gavaskar Shanmugam
Dr. Dilip Kumar Sharma
Dr. R. P. Sharma
Dr. Mayank Sharma
Dr. Sudhir Sharma
Dr. Lokesh Kumar Sharma
Dr. Vishnusharma
Dr. Jitendra Singh
Dr. Girish Singh
Dr. Karan Singh
Dr. Harikesh Singh
Dr. Prashant Singh
Dr. Neetay Singh
Dr. Ajay Shanker Singh
Dr. Arun Solanki
Dr. Subhranil Som
Dr. Ritesh Srivastava
Dr. Vijayalakshmi Subramanian
Dr. Hardeo Kumar Thakur
Dr. Pradeep Tomar
Dr. Shashi Kant Verma
Dr. Sohan Kumar Yadav
Dr. Vinod Yadav
Dr. Dileep Yadav
Dr. Chandra Yadav
Dr. Emre Yay
Dr. Aasim Zafar
Dr. Usha Chauhan
Dr. Chetna Dabas
Dr. Sanjoy Das
Dr. Sumithra Gavaskar
Dr. Vimal Kumar

Preface

The book constitutes selected high-quality papers presented in International Conference on Electrical and Electronics Engineering (ICEEE 2020) jointly organized by School of Electrical, Electronics and Communication Engineering, Galgotias University, India, and National Power Training Institute Faridabad, India in February 2020. It discusses issues in electrical, computer, and electronics engineering and technologies. The selected papers are organized into three sections—Power and Energy, Control system and power electronics, drives and renewable energy. In-depth discussions on various issues under topics provide an interesting compilation for researchers, engineers, and students.

We are thankful to all the authors that have submitted papers for keeping the quality of the ICEEE 2020 at high levels. The editors of this book would like to acknowledge all the authors for their contributions and the reviewers. We have received an invaluable help from the members of the International Program Committee and the chairs responsible for different aspects of the workshop. We appreciate also the role of Special Sessions Organizers. Thanks to all of them, we had been able to collect many papers on interesting topics, and during the conference, we had very interesting presentations and stimulating discussions.

Our special thanks go to Leopoldo Angrisani, (Editor in Chief, Springer, Lecture Notes in Electrical Engineering Series) for the opportunity to organize this guest-edited volume.

We are grateful to Springer, especially to Aninda Bose (Senior Editor, Hard Sciences Publishing), for the excellent collaboration, patience, and help during the evolvement of this volume.

We hope that the volume will provide useful information to professors, researchers, and graduated students in the area of power, control and renewable energy and applications, and all will find this collection of papers inspiring, informative, and useful. We also hope to see you at a future ICEEE event.

Krasnoyarsk, Russia — Margarita N. Favorskaya
Kuala Lumpur, Malaysia — Saad Mekhilef
Varanasi, India — Rajendra Kumar Pandey
Allahabad, India — Nitin Singh

Contents

About the Editors

Prof. (Dr.) Margarita N. Favorskaya is a Professor and Head of Department of ICT at Reshetnev Siberian State University of Science and Technology, Russian Federation. Professor Favorskaya is a member of KES organization since 2010, the IPC member and the Chair of invited sessions of over 30 international conferences. She serves as a reviewer in many international journals the Reviewer, Guest Editor, and Book Editor (Springer). She is the author or the co-author of 200 publications and 20 educational manuals in computer science and Electrical Engineering. She co-authored/co-edited books for Springer recently. She supervised nine Ph.D. candidates and presently supervising four Ph.D. students. Her main research interests are in the allocations of artificial intelligence and information technologies in Electrical Engineering, Electronic Engineering, Computer Engineering, Bio-Medical Engineering, Communication and related disciplines.

Prof. (Dr.) Saad Mekhilef is an IET Fellow and IEEE senior member. He is the associate editor of IEEE Transaction on Power Electronics and Journal of Power Electronics. He is a Professor at the Department of Electrical Engineering, University of Malaya since June 1999. He is currently the Dean of the faculty of Engineering and the Director of Power Electronics and Renewable Energy Research Laboratory-PEARL-He is the author and coauthor of more than 400 publications in international journals and proceedings (253 ISI journal papers) and five books with more than 17000 citations and 64 H-index, 116 Ph.D. and master students have graduated under his supervision. He has six patents. He frequently invited to give keynote lectures at international conferences. Prof. Saad listed by Thomson Reuters (Clarivate Analytics) as one of the Highly Cited (Hi Ci) engineering researcher in the world, and included in the Thomson Reuters' The World's Most Influential Scientific Minds: 2018. He is actively involved in industrial consultancy for major corporations in the power electronics projects. His research interests include power conversion techniques, control of power converters, renewable energy, and energy efficiency.

Prof. (Dr.) Rajendra Kumar Pandey is a Professor at the Department of Electrical Engineering, Indian Institute of Technology (BHU), Varanasi, and is the Director General, National Power Training Institute (NPTI). He has over 36 years' experience in the fields of high-voltage direct current (HVDC) transmission technology and flexible AC transmission systems (FACTS) device control, intelligent power control along with the operation of power systems in open access. He has published more than 150 peer-reviewed papers in both national and international journals and at leading conferences.

Dr. Nitin Singh is currently working as an Associate Professor in Department of Electrical Engineering, MNNIT Allahabad, Prayagraj, India. He is an alumnus of Motilal Nehru National Institute of Technology Allahabad, India. His primary areas of research are Power System Planning, Electricity markets, Artificial Intelligence, implementation of optimization algorithms and machine learning techniques for power system optimization. He has published more than 45 publications in International Journals and in Proceedings of International Conferences of repute. He is serving as reviewer of several Scopus, SCI and E-SCI-indexed journals. He has organized several international conferences under the banner of IEEE and Springer. He is Member of IEEE, member of IEEE Computational Intelligence Society, Power Energy Society, and Industrial Application society. He is currently serving as Secretary, IEEE Joint Chapter of Industrial Electronics/Power Electronics/Control System at MNNIT Allahabad.

An Improved Simplex based Particle Swarm Optimization for Environmentally Constrained Economic Dispatch Problem in Thermal Power Plants

Namarta Chopra, Yadwinder Singh Brar, and Jaspreet Singh Dhillon

Abstract Though the particle swarm optimization method is successfully used optimization approach in many of the engineering problems, it has some of the drawbacks like premature convergence, entrapment in the local optima and stagnation in the solution when number of iterations are increased. In this paper, a novel hybrid method is proposed by doing the integration between stochastic and deterministic methods. The conventional particle swarm optimization is integrated with simplex search method to overcome the drawbacks and to improve the results. The improvement in the results is justified by testing the proposed algorithm in a practically environment constrained economic dispatch problem comprising losses and the results obtained are compared with conventional PSO and other optimization methods reported in the literature. The proposed algorithm is also tested on various benchmark functions to prove its validity in different engineering applications. The outcome reveals that the method is effective and robust in practical applications.

Keywords Simplex search method · Particle swarm optimization · Multiobjective problem · Constraints

1 Introduction

In the category of evolutionary algorithms, conventional particle swarm optimization (PSO) is a popular nature-based stochastic optimization approach, inspired by the natural behavior of swarm of birds, fishes or taken as bird flocking and fish schooling. PSO was proposed by Kennedy and Eberhart in 1995 from the social behavior of birds. Since then it was applied on various engineering applications like in power systems, neural networks engineering design and many more and found

N. Chopra (✉) · Y. S. Brar
I.K.G. Punjab Technical University, Kapurthala, India
e-mail: namartachopra@yahoo.co.in

J. S. Dhillon
Sant Longowal Institute of Engineering and Technology, Longowal, India

M. N. Favorskaya et al. (eds.), *Innovations in Electrical and Electronic Engineering*, Lecture Notes in Electrical Engineering 661, https://doi.org/10.1007/978-981-15-4692-1_1

to be successful as alone or as hybrid with other optimization algorithms. The two main search operators for PSO are *pbest* and *gbest* in the vast search space. These search operators depend on the previous experience of the velocity and position of individual particles in swarm so far. It was done to improve the observations as the algorithm proceeds. Mathematically also, the working concept of PSO is very simple and easily understandable, which motivates the research scholars to work upon it. Thus, from the last two decades, there is an incredible use of PSO algorithm both in the conventional as well as modified form upon the constrained and unconstrained single and multiobjective problems.

Owing to certain drawbacks in PSO like premature convergence, local entrapment, stagnation problem, etc., led to the extensive research and advances in the basic PSO algorithm and resulted in significant developments over the last decade. The working of PSO algorithm mainly depends upon the parameters associated with it like inertia weight, social and cognitive parameters. In recent studies, [1] presents the adaptive and fuzzy-based parameter control methods, in which adaptive type is used for inertia weight control and fuzzy logic control is applied on social and cognitive parameters. This prevents the problem of local optima entrapment to some extent. In [2, 3], a chaotic PSO approach is used in the evolution phase and implicit filtering method is then used in the learning phase. Later, they used the crossover operation along with the chaotic sequences. The hybridization of PSO and genetic algorithm (GA) is done by Fergougui et al. [4] which led to improvement in results in terms of execution time as compared to basic PSO and GA alone. Gaing [5] solves the economic dispatch having several constraints like prohibited operating zones and ramp rate limits using PSO method and compared their results with GA method that showed the improved quality of results using PSO. After solving for single objective problem, many researchers started working with PSO on multiobjective problem and again proved its superiority as compared to many other algorithms. Economic dispatch problem was also solved by Jadoun et al. [6] using modified PSO incorporating new constriction function, which regulates and controls the velocity of particles for better exploration. A new multiobjective PSO was proposed in [7] and was successfully applied on economic emission dispatch problem. A bare bones multiobjective PSO was proposed by Zhang et al. [8] which included updation of particles without requiring the tuning of any control parameters, secondly to expand the search capability, a mutation operator with action range varying over time is required, and to update the global particle leaders particle diversity approach is added. A quantum mechanics theory was also incorporated with basic PSO in [9] and quantum behaved PSO (QPSO) was proposed; it also uses selective probability behavior and named it as SQPSO. Selvakumar and Thanushkodi [10] presented civilized swarm optimization (CSO) by integrating society civilization algorithm (SCA) with conventional PSO. Another crisscross search PSO (CSPSO) [11] is an integrated version of PSO and crisscross search optimization. Two search operators are incorporated, i.e., horizontal crossover which improves the global convergence ability and vertical crossover which improves the swarm diversity. A new global PSO (NGPSO) was also proposed and applied on economic emission dispatch problem [12]. It improves its global search by new position updating equation for all particles and uses randomness in their distribution by slightly disturbing the flight trajectories. In [13], a levy flight-based voltage PSO, two main parameters position and velocity are taken as analogous to voltage and change in voltage. Some of the derivative-free methods [14] are also integrated with conventional PSO, in order to improve the results and to eliminate some of the drawbacks of the later. A simplex-based chaotic PSO was proposed by Gaing and Lin [15], which includes the chaotic map integrated with downhill simplex method in the conventional PSO algorithm.

In this paper, the properties of conventional PSO are hybridized with a derivative-free simplex search method. As PSO is a stochastic method and simplex method is deterministic type, so in this type of hybridization, stochastic methods are used for base level search in order to find global optima, whereas the results are further refined using deterministic methods as local level search. Thus, an improved simplex-based particle swarm optimization method (ISPSO) is proposed and tested on several benchmark functions. The results obtained using ISPSO are compared with conventional PSO, and then the proposed method is applied on the application of multiobjective environmental constrained economic dispatch (MECED) problem in thermal power plants.

2 Mathematical Modeling for MECED Problem

2.1 *Economic and Emission Objective*

The basic MECED problem is to minimize the environmental emissions and overall fuel cost from thermal power units while subjecting to various constraints. The total cost of fuel for generating units with valve point loading effect is represented as piecewise quadratic function with the supposition that the incremental cost curves of the units are monotonically growing. Quadratic fuel cost equation can be given as follows:

$$F_C = \sum_{i=1}^{\mathrm{nu}} \left(a_i P_i^2 + b_i P_i + c_i + \left| d_i \sin\left\{ e_i \left(P_i^{\min} - P_i \right) \right\} \right| \right) \tag{1}$$

where F_C represents the overall fuel cost, P_i is the power generated by ith generator, and a_i, b_i, c_i, d_i and e_i are the fuel cost coefficients.

Mathematical equation for pollutants' emissions with valve point loading effect is represented as follows:

$$E_T = \sum_{i=1}^{nu} \left(\alpha_i P_i^2 + \beta_i P_i + \gamma_i + \eta_i \mathrm{e}^{(\delta_i P_i)} \right) \; (i = 1, 2, \ldots, \mathrm{nu}) \tag{2}$$

where E_T is the total emission, α_i, β_i, γ_i, η_i, δ_i are the emission coefficients of ith unit, and nu is the total number of power units as given in Eqs. 1 and 2.

2.2 Constraint Handling

Fuel cost equation must satisfy the inequality and equality constraints as follows:

$$\sum_{i=1}^{\text{nu}} P_i = P_D + P_L \tag{3}$$

where P_L is the total transmission loss and P_D is the total power demand. Kron's loss formula is used to express transmission line losses as follows:

$$P_L = \sum_{i=1}^{\text{nu}} \sum_{j=1}^{\text{nu}} P_i B_{ij} P_j + \sum_{i=1}^{\text{nu}} B_{0i} P_i + B_{00} \tag{4}$$

P_j and P_i are powers for generating units j and i. B_{0j}, B_{ii} *and* B_{00} are the loss coefficients.

The inequality constraint imposed on the outputs of the generator units is given as follows:

$$P_i^{\min} \le P_i \le P_i^{\max} \quad (i = 1, 2, \ldots, \text{nu}) \tag{5}$$

where $P_i^{\min}$ and $P_i^{\max}$ are the minimum and maximum power assigned at unit i [16–18].

2.3 Optimization Problem

To solve the multiobjective optimization problem subjecting to various constraints requires a feasible decision making due to the presence of conflicting objectives. In the case of MECED problem, the best result must be such that both fuel cost and pollutants' emissions are minimized simultaneously. The two objectives to be minimized can be converted into a single objective using method of price penalty factor (PPF). Collectively, the multiobjective problem can be formulated as follows:

$$F_{Tj} = F_C + h_j E_T \quad (j = 1, 2, 3, 4) \tag{6}$$

where h_j is the price penalty factor and can be calculated by taking the ratio between the total fuel cost at maximum or minimum power to total emissions at minimum or maximum power of the particular generators as max–min, min–max, max–max and min–min penalty factors.

$$h_1 = \frac{F_C\left(P_i^{\min}\right)}{E_T\left(P_i^{\max}\right)};\ h_2 = \frac{F_C\left(P_i^{\min}\right)}{E_T\left(P_i^{\min}\right)};\ h_3 = \frac{F_C\left(P_i^{\max}\right)}{E_T\left(P_i^{\max}\right)};\quad h_4 = \frac{F_C\left(P_i^{\max}\right)}{E_T\left(P_i^{\min}\right)} \tag{7}$$

The optimization problem is to minimize Eq. (8) by using different values of price penalty factors as given in Eq. (7).

$$\text{MECED} = \min\{F_{T1}, F_{T2}, F_{T3}, F_{T4}\} \tag{8}$$

Here, the value of Eq. (8) will become the optimal solution for the MECED problem.

2.4 *Convergence and Quality Checking*

The proposed ISPSO is applied to find the optimal solution of considered case in the test systems undertaken in further sections. The quality assurance of the proposed algorithm in different test systems is also checked by comparing their best value, worst value, mean and standard deviation with the conventional PSO through thirty independent runs with different random variables. The quality indexes as mean value (μ) and standard deviation (σ) are applied to check the effectiveness of the proposed method. They are calculated using Eqs. (9) and (10) as follows:

$$\mu = \frac{1}{t}\sum_{i=1}^{t} f_i \tag{9}$$

$$\sigma = \sqrt{\frac{1}{t}\sum_{i=1}^{t}(f_i - \mu)^2} \tag{10}$$

where f is the evaluated value at each run and t is the number of trials.

The results obtained for each test system are compared with other optimization methods cited in the literature survey including both deterministic and heuristic methods.

3 Proposed Hybridization of PSO with Simplex Search Method

3.1 Conventional PSO

PSO is initialized with an array of random members in the predefined space for search. These random members adjust their velocity and position in the complete process of optimization in order to attain the optimum result. After each iteration, the swarms are modified by two values, *gbest* and *pbest*. *gbest* is the best value accomplished from the total population, and *pbest* is the best result achieved so far in between all the participated members [19, 20].

Initialization of swarm particle's velocity and position is done as follows:

$$\text{Velocity vector, } v_i(0) = r\left(v_i^{\max} - v_i^{\min}\right) + v_i^{\min} \tag{11}$$

$$\text{Position vector, } Pi(0) = P_i^{\min} + r\left(P_i^{\max} - P_i^{\min}\right) \tag{12}$$

$v_i^{\min}$, $v_i^{\max}$, $P_i^{\min}$ and $P_i^{\max}$ are minimum and maximum limits of respective vectors.

Upgradation of initial velocity and position of ith particles is done as follows:

$$\begin{aligned} v_i(m+1) = wv_i(m) + c_1r_1(m)(pbest_i(m) - P_i(m)) \\ + c_2r_2(m)(gbest_i(m) - P_i(m)) \end{aligned} \tag{13}$$

$$P_i(m+1) = P_i(m) + v_i(m+1) \tag{14}$$

$$\begin{aligned} w = w^{\max} - \left(\frac{w^{\max} - w^{\min}}{\text{iter}^{\max}}\right)\text{iter}; \\ with \quad w^{\max} = 0.9 \text{ and } w^{\min} = 0.4 \end{aligned} \tag{15}$$

$P_i(m)$ and $v_i(m)$ are position and velocity of ith particle during mth movement, $c_1 = c_2 = 2$ are acceleration constants, iter is the count of iterations, $\text{iter}^{\max}$ is the maximum iterations count, w is the weight of inertia, and r_1, r_2 are the random numbers between 1 and 0.

3.2 Simplex Search Method

This technique was originally suggested by Spendley in 1962 and after that improved by Nelder and Mead for getting local minima from a set of several variables. To get an optimal point, a sequence of simplexes is generated iteratively. In nu dimensions (it is equal to the number of generators units under consideration), only nu + 1 points are used in the initial simplex. This algorithm performs operational calculations as reflection, contraction and expansion for improving the inferior points iteratively [14]. Various steps to be followed are as follows:

- Find the best minimum point (x_l), the worst maximum point (x_h) and next to worst point (x_g), from first simplex variables.
- Compute centroid (x_{cj}) from all starting points leaving the worst point as follows:

$$x_{cj} = \frac{1}{\mathrm{nu}} \sum_{j=1, j \neq h}^{\mathrm{nu}+1} x_j \quad (j = 1, 2, \ldots, \mathrm{nu}) \tag{16}$$

- Calculate the new reflected point (x_{rj}) as follows:

$$\text{Reflected point, } x_{rj} = 2x_{cj} - x_{hj} \tag{17}$$

- If $F_T(x_{rj}) < F_T(x_{lj})$, perform expansion operation as follows:

$$x_{\mathrm{new},j} = (1 + \gamma)x_{cj} - \gamma x_{hj} \tag{18}$$

where γ is the expansion controlling factor.

- If $F_T(x_{rj}) \geq F_T(x_{hj})$, perform the inside contraction as follows:

$$x_{\mathrm{new},j} = (1 - \beta)x_{cj} + \beta x_{hj} \tag{19}$$

- If $F_T(x_{gj}) < F_T(x_{rj}) < F_T(x_{hj})$, perform the outside contraction as follows:

$$x_{\mathrm{new},j} = (1 + \beta)x_{cj} - \beta x_{hj} \tag{20}$$

where β is the contraction controlling factor.

- Change x_{hj} with $x_{\mathrm{new},j}$ and repeat again the steps from Eqs. (16)–(20) and form new simplex.
- Check whether the termination criterion is satisfied by Eq. (21), otherwise continue the iterations.

$$\left[\sum_{j=1}^{\mathrm{nu}+1} \frac{\left(F_T(x_j) - F_\tau(x_{cj})\right)^2}{\mathrm{nu} + 1} \right]^{1/2} \leq \in \tag{21}$$

where $\in$ is the termination parameter.

The approximated values for the parameters are $\gamma \approx 2.0$, $\beta \approx 0.5$ and $\in \approx 0.001$.

3.3 The Proposed ISPSO Algorithm

In the hybridized proposed algorithm, PSO being a stochastic method is used for finding the global optimum solution as a base level search, whereas simplex being a deterministic method and is hybridized for further refining the solution by local level search. Different steps involved in the functioning of the proposed algorithm as shown in Fig. 1 on the MECED problem are as follows:

Step1: Input the generator unit data, and initialize the PSO and simplex parameters.
Step2: Set the movement counter, $k = 0$.

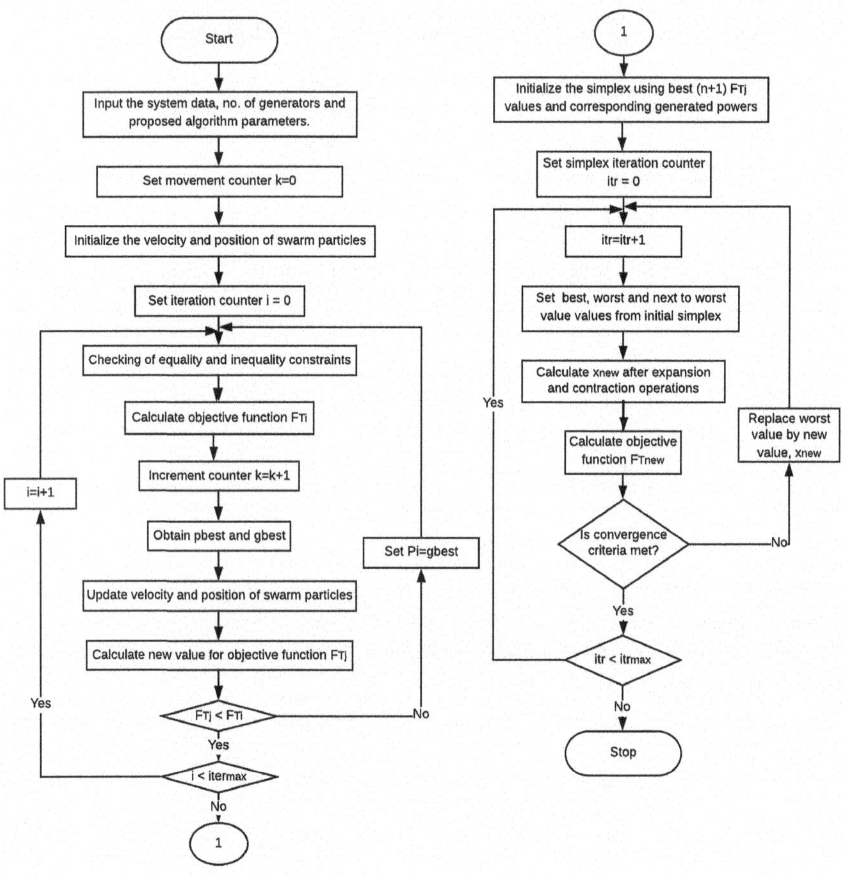

Fig. 1 Flowchart for the proposed ISPSO

Step3: Calculate initial velocity and position of particles using Eqs. (11)–(12) and check the equality and inequality constraints from the corresponding generator data.
Step4: Calculate objective function and increment the movement counter, $k = k + 1$.
Step5: Find pbest and gbest.
Step6: Calculation of inertia weight, new velocity and position using Eqs. (13)–(15) and checking the velocity constraints for their maximum and minimum values.
Step7: Calculate new solution of objective function and check whether $k \leq \text{iter}^{\max}$, otherwise repeat from Step 4 to get the optimum solution.
Step8: Input the nu + 1 best solutions from PSO to simplex algorithm.
Step9: Set iteration counter, $t = 1$.
Step10: Set best point (x_l), worst point (x_h) and next to worst point (x_g) and form the first starting simplex.
Step11: Compute the centroid and reflected points using Eqs. (16)–(17). After this perform the expansion and contraction operations to get the new optimum points using Eqs. (18)–(20).
Step12: Check equality and inequality constraints for the new points and increment the counter $t = t + 1$.
Step13: Repeat the algorithm from step 11 to get the best solution until Eq. (21) is satisfied.
Step14: Else calculate the final objective function.
Step15: Stop.

4 Experimental Results and Analysis

4.1 Validity Using Benchmark Functions

The validity of the proposed improved simplex-based PSO is checked by considering several benchmark functions, and the fitness results are compared with conventional PSO. The code is formulated in MATLAB 7 and executed on core i3 with 3 GB RAM. The parameters undertaken are as follows: total particles in swarm = 10, $w^{\min} = 0.4$, $w^{\max} = 0.9$, $c_1 = c_2 = 2$, $\gamma = 2.0$, $\beta = 0.5$ and $\in = 0.001$.

Eight benchmark test functions are studied to check the convergence speed, efficiency and robustness of the proposed algorithm. These are non-separable, separable, differentiable, unimodal, multimodal, scalable, non-scalable, discontinuous and continuous in nature [21]. Results using conventional PSO and ISPSO for best, mean and worst fitness after random thirty trials are shown in Table 1.

From this comparison, it is observed that the proposed ISPSO shows better performance over the conventional PSO. The minimum, mean and maximum results for the

Table 1 Quality comparison for benchmark functions using PSO and ISPSO

S. No.	Benchmark test function	PSO—Fitness			ISPSO—Fitness		
		Best	Mean	Worst	Best	Mean	Worst
1	Booth	3.56	5.26	9.81	1	2.25	4.28
2	Sphere	1.88×10^{-07}	2.89×10^{-05}	9.50×10^{-05}	6.56×10^{-13}	8.55×10^{-11}	9.56×10^{-10}
3	Step	7.51×10^{-08}	3.54×10^{-06}	9.94×10^{-06}	5.75×10^{-20}	2.54×10^{-18}	2.47×10^{-17}
4	Ackley	3.85×10^{-10}	6.55×10^{-09}	4.49×10^{-08}	1.85×10^{-16}	1.44×10^{-14}	9.90×10^{-14}
5	Rastrigin	2.89×10^{-03}	3.01×10^{-02}	9.82×10^{-02}	3.40×10^{-10}	2.63×10^{-09}	8.15×10^{-09}
6	Schwefel	7.04×10^{-06}	2.97×10^{-05}	9.50×10^{-05}	9.56×10^{-10}	1.98×10^{-08}	8.43×10^{-08}
7	Rosenbrock	12.49	13.75	15.79	1	1.91	2.94
8	Griewank	8.03×10^{-03}	3.30×10^{-02}	9.35×10^{-02}	7.18×10^{-09}	7.06×10^{-08}	9.46×10^{-07}

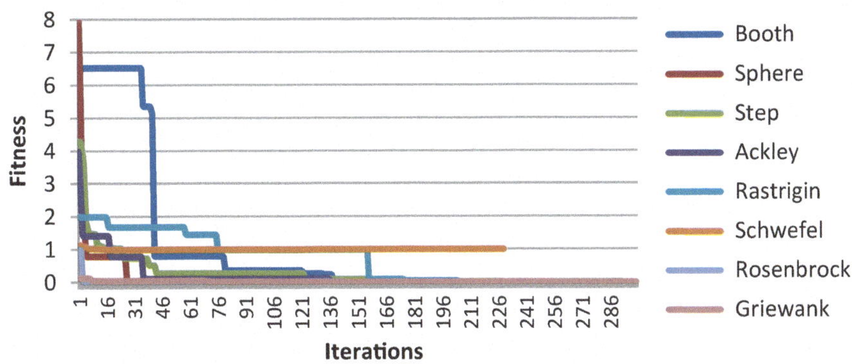

Fig. 2 Convergence for benchmark functions using ISPSO

test functions are improved when the conventional PSO is hybridized with simplex method. The convergence characteristics using ISPSO for these eight benchmark test functions are shown in Fig. 2, and it is observed that Booth, Step and Rastrigin functions converge slowly as compared to Sphere, Ackley, Schwefel, Griewank and Rosenbrock functions but still achieves the optimal solution within the specified iterations.

4.2 Application on MECED Problem

In this section, two cases are undertaken to study the effectiveness of the proposed algorithm on the problem of MECED.

Case 1(Small-scale system): This case study consists of ten generator systems with valve point loading effect along with transmission losses [9]. The results obtained at 2000 MW with different PPF as decision making using ISPSO are shown in Table 2. It shows that the best solution is obtained using max–max price penalty factor as decision-making technique. MECED solutions with ISPSO and max–max PPF for different power demands are given in Table 3. The MECED is solved with max–max PPF at power demands of 1700, 1800, 1900, 2000, 2100, 2200 and 2300 MW.

From Table 3, it is observed that as the demand power increases, the overall fuel cost, pollutants' emissions, transmission losses and final MECED also increase. The comparison analysis with valve point loading at 2000 MW for fuel cost and pollutants' emissions with other methods is presented in Table 4. It is found that

Table 2 MECED at different PPF at 2000 MW (case 1)

PPF (*h*)	Max–Min	Max–Max	Min–Max	Min–Min
MECED (Rs./h)	181153.78	161433.02	180518.58	179962.52

Table 3 MECED solutions using ISPSO for different power demands (case 1)

Unit power output (MW)	Power demand (MW)						
	1700	1800	1900	2000	2100	2200	2300
P1	24.99	25	32.34	54.79	55	55	55
P2	59.99	60	60	62.39	75.88	80	80
P3	79.99	80	80	81.94	95.43	111.09	120
P4	79.99	80	80	81.94	95.43	111.09	130
P5	109.99	110	110	111.94	125.43	141.09	160
P6	179.99	180	180	181.94	195.43	211.09	240
P7	249.99	250	253.48	275.94	289.43	300	300
P8	299.99	300	300	310.65	324.14	339.81	340
P9	275.48	383.70	432.85	455.31	468.79	484.46	518.08
P10	399.99	400.34	449.48	470	470	470	470
Total P	1760.49	1869.04	1978.16	2086.85	2194.98	2303.66	2413.08
P_L	60.47	69.03	78.15	86.86	94.99	103.67	113.09
Total $F_C \times 10^5$ (Rs./h)	9.5799	1.01212	1.07088	1.13207	1.20108	1.27437	1.35591
Emissions, E_T (Kg/h)	3101.68	3397.09	3778.31	4106.11	4449.56	4862.25	5328.24
MECED $\times 10^5$ (Rs./h)	1.32	1.39	1.49	1.61	1.77	1.96	2.19

Table 4 Comparison of MECED solutions using ISPSO with other methods for load demand of 2000 MW (case 1)

Unit power output (MW)	ISPSO	ABC_PSO [22]	DE [21]	NSGA-II [21]	SPEA-2 [21]
P1	54.79	55	54.95	51.95	52.98
P2	62.39	80	74.58	67.26	72.81
P3	81.94	81.14	79.43	73.68	78.11
P4	81.94	84.22	80.69	91.35	83.61
P5	111.94	138.34	136.85	134.05	137.24
P6	181.94	167.51	172.64	174.95	172.92
P7	275.94	296.83	283.82	289.44	287.20
P8	310.65	311.58	316.34	314.05	326.40
P9	455.31	420.34	448.59	455.69	448.88
P10	470	449.16	436.43	431.81	423.90
P_L	86.86	84.17	NA	NA	NA
Total $F_C \times 10^5$ (Rs./h)	1.13207	1.13420	1.13480	1.13540	1.13520
Emissions, E_T (Kg/h)	4106.10	4120.10	4124.90	4130.20	4109.10
Execution time (s)	5.36	NA	NA	NA	NA

NA—Not available in the cited literature

using the proposed algorithm for 2000 MW load demand, fuel cost obtained is 113207.648 Rs./h and emissions are 4106.104 kg/h which are comparatively lower when compared with other methods given in [22, 23]. The quality comparison of the proposed algorithm ISPSO with conventional PSO is shown in Table 5. From these results, it is observed that the results are improved for the hybrid combination of PSO with simplex method as compared to conventional PSO. The fuel cost convergence characteristics are also shown in Fig. 3.

Case 2 (Large-scale system): This case study consists of eighty generator systems along with losses. It is replicated data of forty unit systems from [24]. This system consists of valve point loading effect along with the transmission losses having total

Table 5 Quality comparison for case 1 with conventional PSO

Parameter	Units	Min. value	Mean value	Max. value	Std dev
Total F_C (Rs./h)	PSO	113433.68	113574.61	113759.71	84.00
	ISPSO	113207.65	113431.78	113563.22	83.63
Total E_T (Kg/h)	PSO	4125.54	4237.18	4430.38	91.16
	ISPSO	4106.10	4231.09	4367.51	72.92
MECED (Rs./h)	PSO	173102.91	173687.97	174327.10	387.54
	ISPSO	161433.01	162136.74	162904.97	380.19

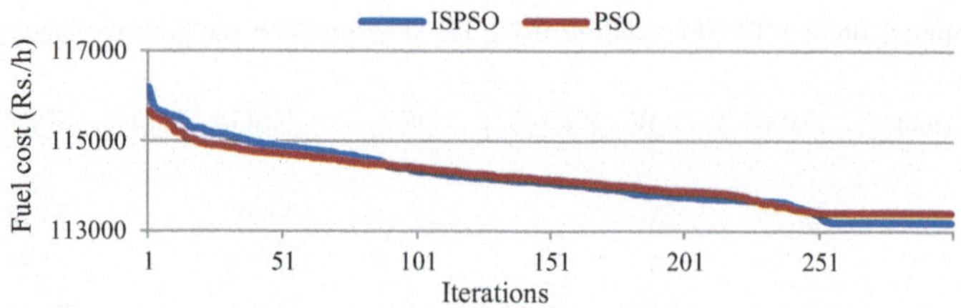

Fig. 3 Convergence of fuel cost with ISPSO and PSO for case 1

demand of 21,000 MW. The best results obtained are reported in Table 6 and are compared with other methods available in the literature in Table 7. This comparison is done on the basis of total fuel cost obtained from different methods, and it is

Table 6 Results of eighty generator test systems using ISPSO (case 2)

Power unit	Output power (MW)	Power unit	Output power (MW)	Power unit	Output power (MW)	Power unit	Output power (MW)
P1	595.74	P21	422	P41	408.39	P61	510.00
P2	88.72	P22	492.84	P42	36	P62	476.56
P3	64.76	P23	520.90	P43	60.01	P63	433.13
P4	60	P24	523.60	P44	60	P64	422
P5	136.53	P25	477.86	P45	120	P65	422
P6	60	P26	422	P46	75.17	P66	422
P7	68	P27	478.41	P47	77.80	P67	486.61
P8	229.73	P28	145.47	P48	248.57	P68	110
P9	253.68	P29	136.63	P49	264.71	P69	110
P10	201.03	P30	125.20	P50	200	P70	110
P11	213.15	P31	69.57	P51	271.37	P71	67
P12	274.41	P32	155.82	P52	356.04	P72	170.91
P13	220.08	P33	168.51	P53	340.49	P73	168.46
P14	443.40	P34	159.73	P54	474.92	P74	185.74
P15	441.42	P35	168.60	P55	484.40	P75	168.51
P16	412.04	P36	195.74	P56	402.62	P76	138.99
P17	401.82	P37	195.74	P57	422.75	P77	130
P18	484.77	P38	74.25	P58	490.66	P78	81.17
P19	409.50	P39	97.53	P59	485.85	P79	55.16
P20	422	P40	75.62	P60	477.06	P80	65.75
Net power		21077.61	Power losses	77.61		Total F_C(Rs./h)	242859.59
Emissions (Kg/h)		761820.38		MECED (Rs./h)		6790.34	

Table 7 Comparison of results with other methods (case 2)

Method	Min. fuel cost	Mean fuel cost	Max. fuel cost	Std dev	Time (s)
FAPSO [1]	244273.5	NA	NA	NA	NA
NAPSO [1]	242844.1	NA	NA	NA	NA
MSSA [25]	242909.3	243037.3	243229.9	53.8	8.7
SSA-II [25]	243398.2	243575.4	243760.9	68.9	6.7
SSA-I [25]	243173.6	243306.1	243490.8	72.7	8.5
SCA [10]	250864.1	254579.8	261099.6	NA	848.3
CSO [10]	243195.4	243546.6	244038.7	NA	27.8
THS(t=8) [25]	243192.7	243457.4	NA	120.9	NA
ISPSO	242859.6	242879.1	242899.6	12.4	38.7

observed that ISPSO shows better results in terms of mean, standard deviation and computation time. Only NAPSO [1] shows better results in terms of best fuel cost, as this does not include emissions in its objective function. The computation time of ISPSO is also better than CSO [10] and SCA [10]. Thus, it shows the effectiveness of the proposed method for large-scale systems also. The quality comparison and convergence with conventional PSO are shown in Table 8 and Fig. 4.

Table 8 Quality comparison for case 2 with conventional PSO

Parameter	Units	Min. value	Mean value	Max. value	Std dev
Total fuel cost, (Rs./h)	PSO	242865.83	242885.44	242923.8	13.518
	ISPSO	242859.59	242879.14	242899.58	12.418
Total E_T (Kg/h)	PSO	761827.6	761860.1	761905.34	21.503
	ISPSO	761820.38	761855.07	761898.11	22.164
MECED (Rs./h)	PSO	702813.01	702836.63	702870.4	18.056
	ISPSO	702795.02	702816.78	702845.77	16.376

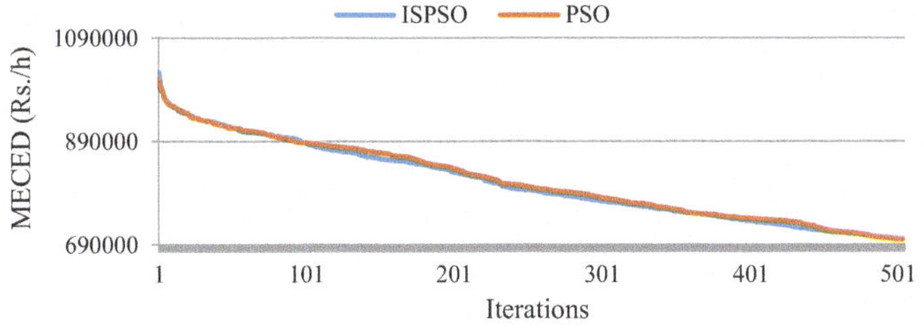

Fig. 4 Fuel cost convergence characteristics of ISPSO and PSO for case 2

5 Conclusions

This paper proposes an improved simplex-based particle swarm optimization as a novel evolutionary algorithm, which integrates the stochastic search method with deterministic search method. In this combination, PSO is used as a global search mechanism, and simplex is then further used to improve the results by again performing the local search on it. Thus, the integration of these two helps in modifying the results quality as compared to conventional PSO. Several benchmark functions are considered to show the validity and superiority of the proposed algorithm. From the practical point of view, it is also tested on small-scale and large-scale multi-objective environmentally constrained economic dispatch problem in thermal power plant, which again shows its superiority and robustness as compared to other methods discussed in the literature.

Acknowledgements The authors are indebted to I.K.G. Punjab Technical University, Kapurthala, and Amritsar College of Engineering and Technology, Amritsar, for providing the advanced research facilities while preparing this research paper.

References

1. Niknam T, Mojarrad HD, Meymand HZ (2011) Non-smooth economic dispatch computation by fuzzy and self adaptive particle swarm optimization. Appl Soft Comput 11:2805–2817. https://doi.org/10.1016/j.asoc.2010.11.010
2. Dos Santos Coelho L, Mariani VC (2007) Economic dispatch optimization using hybrid chaotic particle swarm optimizer. In: IEEE international conference on systems, man and cybernetics, 2007. ISIC. IEEE, pp 1963–1968
3. Park JB, Jeong YW, Shin JR, Lee KY (2010) An improved particle swarm optimization for nonconvex economic dispatch problems. IEEE Trans Power Syst 25:156–166. https://doi.org/10.1109/TPWRS.2009.2030293
4. Fergougui AE, Ladjici AA, Benseddik A, Amrane Y (2018) Dynamic economic dispatch using genetic and particle swarm optimization algorithm. In: 2018 5th international conference on control, decision and information technologies (CoDIT), pp 1001–1005. https://doi.org/10.1109/CoDIT.2018.8394962
5. Gaing ZL (2003) Particle swarm optimization to solving the economic dispatch considering the generator constraints. IEEE Trans Power Syst 18:1187–1195. https://doi.org/10.1109/TPWRS.2003.814889
6. Jadoun VK, Gupta N, Niazi KR, Swarnkar A (2014) Nonconvex economic dispatch using particle swarm optimization with time varying operators. Adv Electr Eng 2014:1–13. https://doi.org/10.1155/2014/301615
7. Abido MA (2009) Multiobjective particle swarm optimization for environmental/economic dispatch problem. Electr Power Syst Res 79:1105–1113. https://doi.org/10.1016/j.epsr.2009.02.005
8. Zhang Y, Gong DW, Ding Z (2012) A bare-bones multi-objective particle swarm optimization algorithm for environmental/economic dispatch. Inf Sci 192:213–227. https://doi.org/10.1016/j.ins.2011.06.004
9. Niu Q, Zhou Z, Zhang HY, Deng J (2012) An improved quantum-behaved particle swarm optimization method for economic dispatch problems with multiple fuel options and valve-points effects. Energies 5:3655–3673. https://doi.org/10.3390/en5093655

10. Selvakumar AI, Thanushkodi K (2009) Optimization using civilized swarm: solution to economic dispatch with multiple minima. Electr Power Syst Res 79:8–16. https://doi.org/10.1016/j.epsr.2008.05.001
11. Meng A, Li Z, Yin H, Chen S, Guo Z (2016) Accelerating particle swarm optimization using crisscross search. Inf Sci 329:52–72. https://doi.org/10.1016/j.ins.2015.08.018
12. Zou D, Li S, Li Z, Kong X (2017) A new global particle swarm optimization for the economic emission dispatch with or without transmission losses. Energy Convers Manag 139:45–70. https://doi.org/10.1016/j.enconman.2017.02.035
13. DeoBodha K, Mukherjee V, KumarYadav V, Saurabh K, Anium S (2018) A levy flight based voltage particle swarm optimization for multiple-objective mixed cost-effective emission dispatch. In: 2018 8th international conference on cloud computing, data science engineering (Confluence), pp 1–6. https://doi.org/10.1109/CONFLUENCE.2018.8442919
14. Deb K (2012) Optimization for engineering design: algorithms and examples. PHI Learning Pvt. Ltd.
15. Gaing ZL, Lin CH (2011) Contingency-constrained optimal power flow using simplex-based chaotic-PSO algorithm. Appl Comput Intell Soft Comput 2011:1–13. https://doi.org/10.1155/2011/942672
16. Chopra N, Brar YS, Dhillon JS (2017) Modified particle swarm optimization using simplex search method for multiobjective economic emission dispatch problem. In: 2017 3rd international conference on condition assessment techniques in electrical systems (CATCON), pp 171–176. IEEE
17. Chopra N, Brar YS, Dhillon JS (2020) Hybridized particle swarm optimization on constrained economic dispatch problem. J Comput Theor Nanosci 17:322–328
18. Singh D, Dhillon JS (2019) Ameliorated grey wolf optimization for economic load dispatch problem. Energy 169:398–419. https://doi.org/10.1016/j.energy.2018.11.034
19. Rezaie H, Abedi M, Rastegar S, Rastegar H (2019) Economic emission dispatch using an advanced particle swarm optimization technique. World J Eng. https://doi.org/10.1108/WJE-04-2018-0126
20. Chopra N, Brar YS, Dhillon JS (2017) Dispatch of multiarea system with tie line constraints using modified particle swarm optimization. Int J Electr Electron Comput Sci Eng 4:129–137
21. Jamil M, Yang XS (2013) A literature survey of benchmark functions for global optimization problems. Int J Math Model Numer Optim 4:150. https://doi.org/10.1504/IJMMNO.2013.055204
22. Basu M (2011) Economic environmental dispatch using multi-objective differential evolution. Appl Soft Comput 11:2845–2853. https://doi.org/10.1016/j.asoc.2010.11.014
23. Manteaw ED, Odero NA (2012) Combined economic and emission dispatch solution using ABC_PSO hybrid algorithm with valve point loading effect. Int J Sci Res Publ 2:1–9
24. Sinha N, Chakrabarti R, Chattopadhyay PK (2003) Evolutionary programming techniques for economic load dispatch. IEEE Trans Evol Comput 7:83–94. https://doi.org/10.1109/TEVC.2002.806788
25. Elsayed WT, Hegazy YG, Bendary FM, El-bages MS (2016) Modified social spider algorithm for solving the economic dispatch problem. Int J Eng Sci Technol 19:1672–1681. https://doi.org/10.1016/j.jestch.2016.09.002

FC-TBSR Compensator for Reactive Power Compensation and Voltage Swell Mitigation

Swapnil Patil, Suraj Pawar, Anwar Mulla, and Dadaso Patil

Abstract This paper introduces FC-TBSR, which is one of the controllers in flexible AC transmission system. The thyristor binary switched reactor contains thyristors which are connected in antiparallel which acts as bidirectional switch in series with a reactor and fixed capacitor connected across. The reactor banks are arranged in binary steps. Usually, FACTS devices are used for voltage regulation and reactive power compensation but it has some drawbacks like harmonics injection and it will not give fine voltage control because it has fixed values. To avoid and overcome these drawbacks, a new FC-TBSR compensator has been proposed.

Keywords FACTS devices · FC-TBSR · Reactive power compensation · Voltage swell · Binary current

1 Introduction

In the network of the power system, starting from the generating station to the load end, the power is transmitted through the transmission line. Nowadays large load and many different regions over a very long distance are interconnected to each other. Large attention is given toward the voltage stability and reactive power compensation, which are some of the problems of power system network.

S. Patil (✉) · S. Pawar
Annasaheb Dange College of Engineering and Technology, Ashta, Maharastra, India
e-mail: patilsd4017@gmail.com

S. Pawar
e-mail: pawarsuraj57@gmail.com

A. Mulla
Daulatrao Aher College of Engineering, Karad, Maharastra, India
e-mail: ammaitp@rediffmail.com

D. Patil
Walchand College of Engineering, Sangli, Maharastra, India
e-mail: dadasorpatil@gmail.com

M. N. Favorskaya et al. (eds.), *Innovations in Electrical and Electronic Engineering*, Lecture Notes in Electrical Engineering 661, https://doi.org/10.1007/978-981-15-4692-1_2

In the flexible AC transmission system, different controllers are used to compensate these problems. Recently, TSR-based SVC and TCR-based SVC have been used for the compensation. The TSR-based SVC system injects less harmonics in the system [1]. So the filters are not required for this system, which is one of the advantages of TSR system. TSR is a shunt compensator that can absorb reactive power. The TSR controlling technique is in step only, either fully ON or fully OFF. The TSR is the most suitable solution for the reactive power compensation. But sometimes, there is need of reactive power in step for precise control. To overcome this problem, TBSR is introduced. This reactor splits into binary order to allow variation reactive power that is fine control [2, 3].

For a transmission line, the capacitance of line impacts the sending-end and receiving-end voltage and current. During the light-load condition, the line produces more reactive power than required. So, there is excess reactive power in the system. A device that absorbs reactive power must be added to the system. We know the reactive power is absorbed by the inductor. The reactive power generated by capacitor is consumed by shunt reactor connected in parallel with the line.

Recently, as the power system is very huge, number of problems occurs in it. Some of them are related to system stability, voltage stability, and steady-state stability. Number of solutions is available to overcome these problems, from which compensation of the reactive power is one of the solution. With the FC-TBSR, the reactive power is compensated by switching the TBSR bank as per the requirement [4–8].

2 Desirable Features

- Cycle-by-cycle compensation of reactive power (Fig. 3).
- Transient-free switching [4].
- Reactor banks are arranged in binary sequential manner to achieve almost step less control (Fig. 4).
- It mitigates the voltage swell [9].
- FC-TBSR compensator is mainly used for mitigation of Ferranti effect (Fig. 7).
- It generates less or negligible harmonics as compared with TCR.
- It will increase the efficiency of a system.
- FC-TBSR compensator has fast response time.

3 Proposed Topology of FC-TBSR

The FC-TBSR consists of antiparallel connected thyristors in series with a reactor and fixed capacitor shown in Fig. 1. In the reactor, transients can occur during switching to avoid that following conditions are satisfied [10–12]:

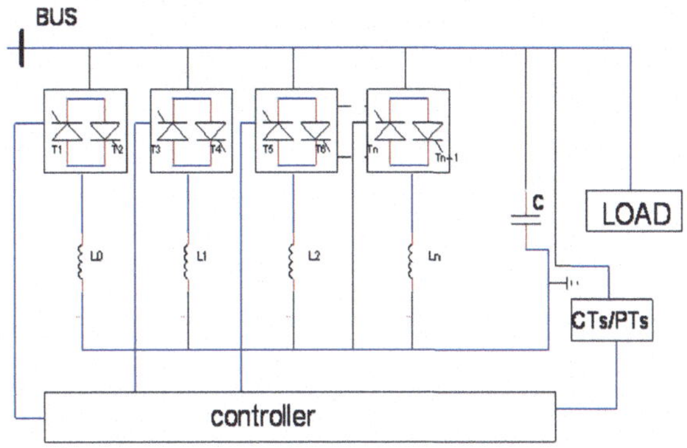

Fig. 1 Proposed topology of FC-TBSR compensator

1. At zero crossing of current OR.
2. At positive or negative peak of supply voltage.

The objectives of this project are to compensate reactive power and voltage swell mitigation by using TBSR compensator. The voltage regulation is achieved by compensation of reactive power.

Binary conversion will be carried out by using equation [13]:

$$Q_L = 2^n L + 2^{n-1} L + \cdots + 2^2 L + 2^1 L + 2^0 L \tag{1}$$

Figure 1 shows the hardware design for the TBSR compensator. The reactive power produced in a system is compensated with the help of this system. The TSR compensator produces less harmonics in system, and so the filter circuit is not required in this system. For fine voltage regulation, we have added steps in binary format in the TSR compensator.

In the TBSR system, banks are arranged as L1, L2, L3, and L4 H with a fixed capacitor. They make the switching operation transient free and binary voltage generation for all combination. The voltage is sensed and fed to controller to perform necessary calculations and this will enable close matching of compensated reactive power.

The sequence used for switching operation is as follows:-

Step-1: Reactor L0 is switching ON for one cycle, and switch OFF for one cycle.
Step-2: Reactor L1 is switching ON for two cycles, and switch OFF for two cycles.
Step-3: Reactor L2 is switching ON for four cycles, and switch OFF for four cycles.
Step-4: Reactor L3 is switching ON for eight cycles, and switch OFF for eight cycles.

In this way, we get step less control of reactive power.

Table 1 Control and switching strategy

S. No.	L3 (800 VAR)	L2 (400 VAR)	L1 (200 VAR)	L0 (100 VAR)	(QL) Load demand
1.	OFF	OFF	OFF	OFF	0
2	OFF	OFF	OFF	ON	100
3	OFF	OFF	ON	OFF	200
4.	OFF	OFF	ON	ON	300
5.	OFF	ON	OFF	OFF	400
6.	OFF	ON	OFF	ON	500
7.	OFF	ON	ON	OFF	600
8.	OFF	ON	ON	ON	700
9.	ON	OFF	OFF	OFF	800
10.	ON	OFF	OFF	ON	900
11.	ON	OFF	ON	OFF	1000
12.	ON	OFF	ON	ON	1100
13.	ON	ON	OFF	OFF	1200
14.	ON	ON	OFF	ON	1300
15.	ON	ON	ON	OFF	1400
16.	ON	ON	ON	ON	1500

3.1 Control Strategy

In this paper, the reactor bank is chosen in binary sequence and switching operation is also in binary sequence. Control strategy of FC-TBSR compensator (Table 1).

3.2 Closed-Loop Control Circuit

The closed-loop control system consists of PI controller, CT, PT, ADC, transient-free switching, and FC-TBSR bank as shown in Fig. 2. The controller of a compensator is the most important part of a circuit. Reactive power is compensated by using controller. In the closed-loop system, the voltage v is sensed by potential transformer (PT) and current I is sensed by current transformer (CT). The sensed values are given to the controller as a input along with the reference. The controller used here is PI controller. It determines the values of the reactive power of the system and generates necessary control signal, i.e., firing signal to achieve desirable output.

The block diagram of FC-TBSR compensator operating in closed loop is shown in Fig. 2. The reactive power $Q_{\text{ref},}$ i.e., reference reactive power is calculated from $Q_{\text{demand},}$ i.e., reactive power demand of the load. Actual reactive power Q_{actual} is calculated by sensing voltage and current.

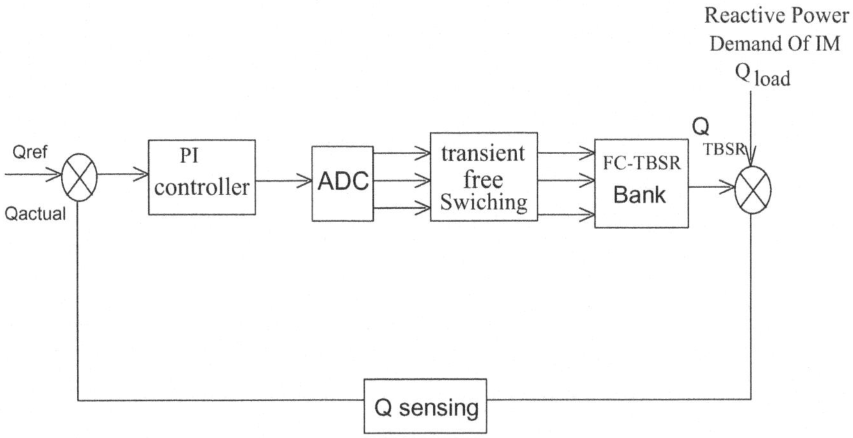

Fig. 2 Closed-loop control

The generated error in the $Q_{\text{reference}}$ and Q_{actual} is given to PI controller as input. The output of PI controller is given to analog-to-digital converter. The output of ADC is given to TBSR bank in such a way that no transient occurs. The switching of the TBSR bank is controlled according to the reactive power demand.

4 Simulation Results

4.1 Binary Current Generation

Binary current generated by TBSR is demonstrated in the result shown in Fig. 3. The total current is being increased step by step. The reactor current from branches is IL0, IL1, IL2, and IL3, respectively. Figure 3 shows the compensating current for the load and Fig. 4 shows the total compensating current.

4.2 Resultant Binary Current Generation

The waveform of Fig. 4 illustrated the total compensating current required for compensation of 100 VAR reactive power to 1500 VAR reactive power by achieving step less control. Current magnitude increases from 0 to 5.8 A as the reactive power increases.

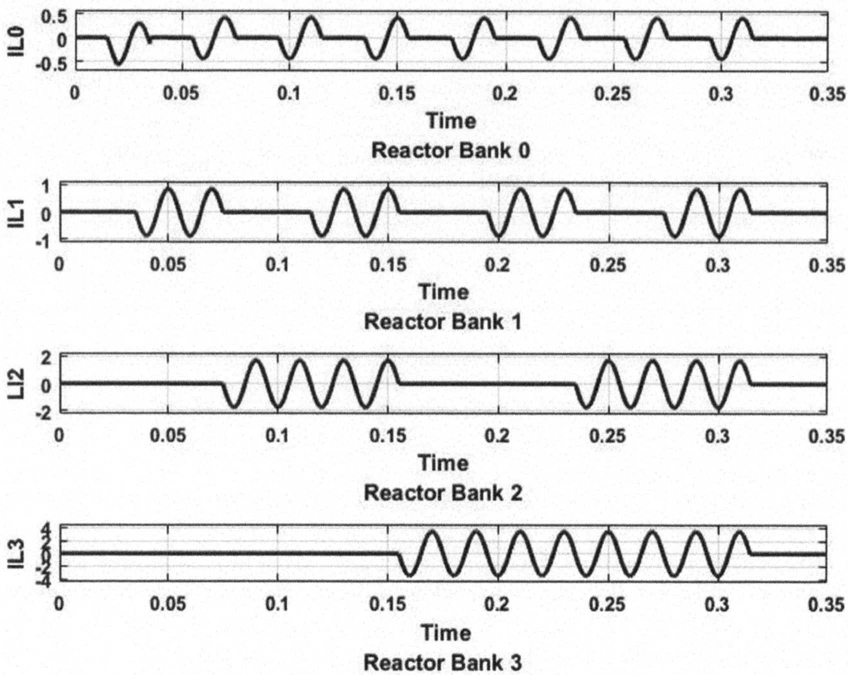

Fig. 3 Current through each reactor

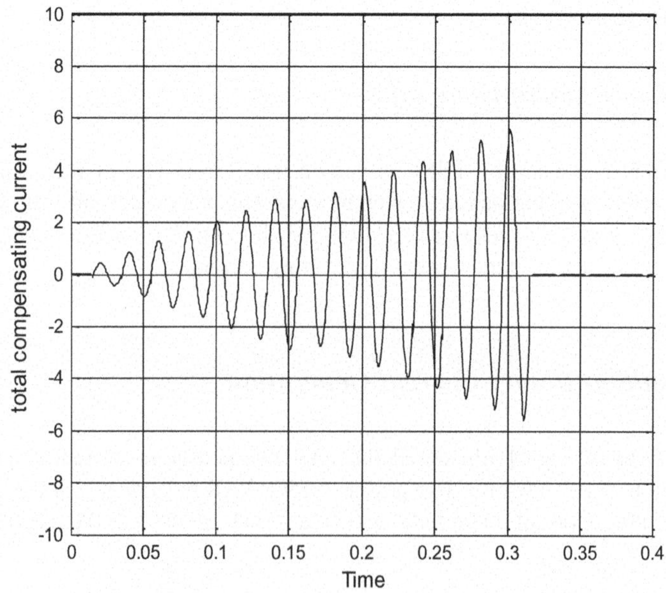

Fig. 4 Total compensating (Binary) current

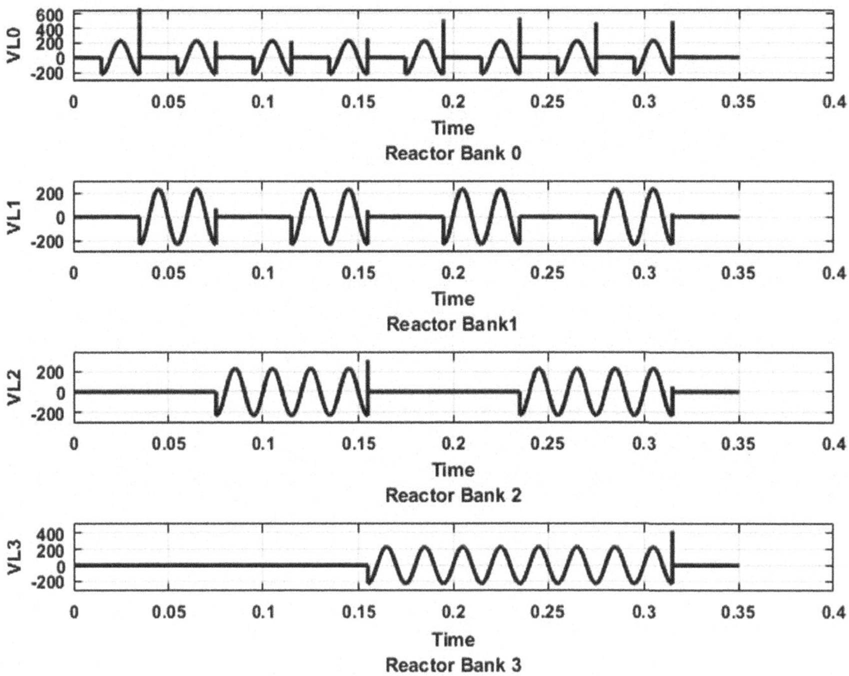

Fig. 5 voltage across each reactor

4.3 Voltage Across Reactors

The result of simulation of Fig. 5 shows the voltage across reactor generated by TBSR, by using four reactor banks. The voltages across the reactors are V_{L0}, V_{L1}, V_{L2}, and V_{L3}, respectively.

4.4 Voltage Swell and Mitigation

The above waveform shows the voltage swell occurred in the system. The voltage swell occurs due the Ferranti effect and the voltage of the system increases up to 330 V shown in Fig. 6. To achieve such type of result, we used pi type transmission line.

Figure 7 shows the voltage swell compensation of pi transmission line with the help of FC-TBSR Compensator. Due to reactive power compensation by TBSR compensator, it maintains voltage level up to 230 V.

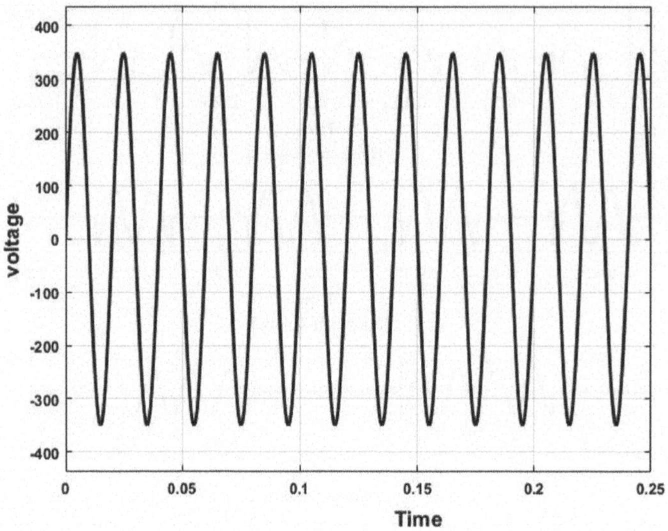

Fig. 6 Voltage swell

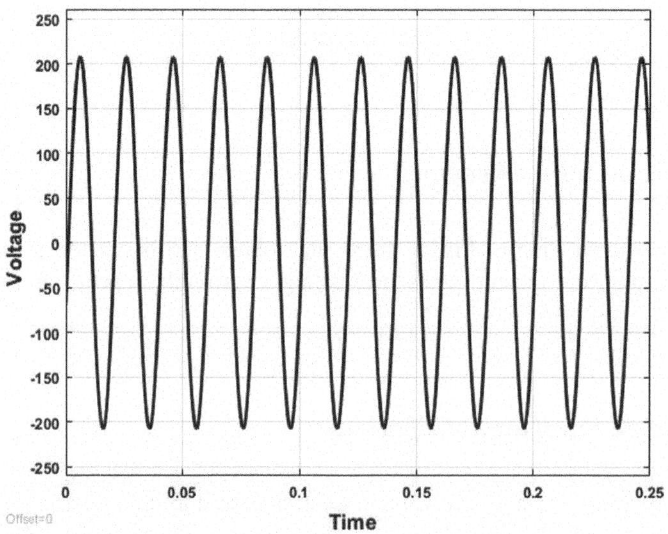

Fig. 7 Voltage swell compensation

4.5 FFT Analysis

Figure 8 shows the FFT analysis of compensator current. It is 1.39% of fundamental component. It is achieved due to switching manner of reactors.

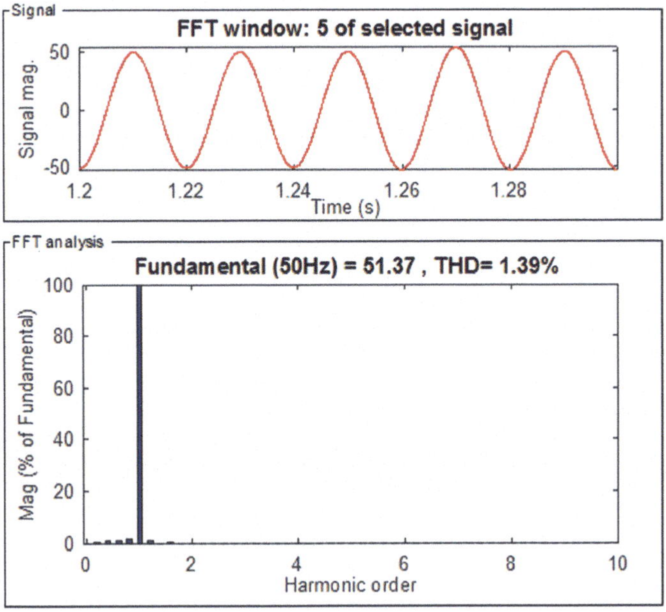

Fig. 8 FFT (THD %) analysis of compensator

5 Conclusions

In this paper, we have presented the modeling and simulation of the FC-TBSR compensator in both open loop and close loop. The simulation is verified by using MATLAB simulation. The simulation result shows the cycle-by-cycle compensation of reactive power (Figs. 3 and 4). Also the voltage swell generated is mitigated by compensator (Fig. 7). After installation of the compensator in the power system, FC-TBSR gives almost step less reactive power compensation (Fig. 4). As compared to the TCR and other compensators, the harmonics (Fig. 8) generation in this compensator is less. So this system does not need filter circuit for the harmonic reduction which is one of the advantages of this system.

Acknowledgements This work was supported by Shivaji University, Kolhapuer under Research Initiation Scheme and carried out at the Electrical Engineering Department, Annasaheb Dange College of Engineering and Technology, Ashta, Sangli, India.

References

1. Dewan SB (1969) A DC controlled reactor using thyristor. Paper 11.6. In: INTERMAG Conference Amsterdam, The Netherlands, April 15–18
2. Gelen A, Yalcinoz T (2007) Analysis of TSR-based SVC for a three-phase system with static and dynamic loads. In: Proceedings of the IEEE international conference on electrical engineering (ICEE), Lahore, Pakistan, 11–12 April 2017, pp 1–6
3. Gelen A, Yalcinoz T (2006) Modeling and simulation of TSR based SVC on voltage regulation for three-bus system. In: 3rd international symposium and exhibition on electrical, electronic and computer engineering, (ISEECE-2006). Near East University, Nicosia, TRNC, 23–25 Nov 2006, pp 167–171
4. Patil SD, Shende KV, Mulla AM, Patil DR (2016) Microcontroller based TBSC compensator with transient free switching for reactive power compensation. In: International conference on signal processing, communication, power and embedded system (SCOPES)-2016, 978-1-5090-4620-1/16/$31.00©2016. IEEE, pp 216–220
5. Patil S, Shende K, Mulla A, Patil D (2017) TBSC Compensator. Chapter No. 22, Transactions on engineering technologies. Springer, Singapore. https://doi.org/10.1007/978-981-10-2717-8_22. Print ISBN 978-981-10-2716-1, Online ISBN 978-981-10-2717-8. Springer, Date: 07 February 2017, pp 305–318
6. Patil S, Kumbhar N, Mulla AM, Patil DR (2015) Binary current generation with transient free switching for TBSC compensator. In: Proceedings of WCECS 2015, vol I, San Francisco, USA, 21–23 Oct 2015
7. Patil SD, Mulla AM, Gudaru U, Patil DR (2014) An innovative transient free TBSC compensator with closed loop control for fast varying dynamic load. In: Proceedings of the world congress on engineering and computer science, WCECS 2014, vol I, San Francisco, USA, 22–24 Oct 2014
8. Patil S, Shende K, Mulla A, Patil D (2017) TBSC Compensator. Chapter No. 22, Transactions on engineering technologies. Springer Singapore. https://doi.org/10.1007/978-981-10-2717-8_22, Print ISBN 978-981-10-2716-1, Online ISBN 978-981-10-2717-8, Date: 07 February 2017, pp 305–318
9. Lokhande S, Patil S, Shende K, Patil D, Mulla A (2016) Introduction to FC-TBSR based SVC for voltage regulation and reactive power compensation. In: IEEE international conference on power electronics, drives and energy systems (PEDES). IEEE Conference Publications 978-1-4673-8888-7/16/$31.00©2016. IEEE, pp 1–5. https://doi.org/10.1109/pedes.2016.7914416
10. Patil S, Mulla A, Patil D (2015) TBSC compensator: application and simulation results for starting and voltage sag mitigation of induction motor. Am Int J Res Sci Technol Eng Math:1–11. AIJRSTEM 15–106; © 2015, AIJRSTEM All Rights Reserved
11. Patil S, Shinde Y, Gudaru KSU, Patil DR (2013) Transient free TBSC compensator for dynamic reactive load with closed loop control. In: Proceedings of the WCECS 2013, vol I, San Francisco, USA, 23–25 Oct 2013
12. Patil SD, Mujawar II, Gudaru U, Patil DR (2013) A closed loop TBSC compensator for direct online starting of induction motors with voltage sag mitigation. In: Proceedings of the world congress on engineering and computer science, WCECS 2013, vol 1, San Francisco, USA, 23–25 Oct 2013
13. Patil SD, Mujawar I, Mujawar I, Patil DR, Gudaru U (2014) TBSC-TCR compensator simulation: a new approach in closed loop reactive power compensation of dynamic loads. In: Proceedings IMECS 2014, vol II, Hong Kong, 12–14 Mar 2014

A Unique Multiple DGs Allocation Technique for Loss Minimization in Distribution System

Ankush Tandon and Sarfaraz Nawaz

Abstract Nowadays, distributed generation units are used in distribution systems for active power loss minimization. It is pre-requisite to compute the appropriate size and location of DGs (solar PV module) and capacitors to achieve maximum loss reduction. The prominent aim of this paper is to maximize percentage loss reduction of distribution systems. An innovative mathematical term, Loss Constant (LC), is anticipated in this paper. The LC determines rating and position of multiple DG and Capacitors units separately. The above method is experienced on IEEE 69 bus standard distribution system. The efficacy of above methodology is proved by considering three different load levels for aforesaid test system. Results are found encouraging and optimistic for both test systems.

Keywords Real power loss (RPL) · Loss constant (LC) · Distributed generation (DG) · Distribution systems (DS)

1 Introduction

In India, the RPL of distribution network are on higher side (around 25%). In distribution systems, DG is incorporated for active power loss reduction. In DG technology, both conservative and non-conservative energy sources are adopted. In conventional sources gas turbine, fuel cells, reciprocating engines, micro-turbine are utilized while biomass, cogeneration, wind power plant, solar PV array falls in the category of non-conventional sources. Placing of DG units is the predicament of determining suitable location and its rating, while satisfying the cost constraints. Optimal sizing of isolated hydropower plant is calculated by tabu search in [13]. The DG allocation problem is resolved by analytical method in RDS [3, 9]. Genetic algorithm technique has been utilized in [16, 17] to evaluate optimal location and rating of DG units. In [14], conventional linear programming (LP) method is adopted to solve DG place-

A. Tandon (✉) · S. Nawaz
Swami Keshvanand Institute of Technology M & G, Jagatpura, Jaipur, India
e-mail: eeankush.1986@gmail.com

M. N. Favorskaya et al. (eds.), *Innovations in Electrical and Electronic Engineering*, Lecture Notes in Electrical Engineering 661, https://doi.org/10.1007/978-981-15-4692-1_3

ment problem. Classical Kalman filter approach is adopted in [18]. Tabu Search (TS) methodology is adopted for RPL in [20]. In [4] genetic algorithm is utilized to solve the allocation problem. ABC algorithm [2] has been applied to determine the best location and size of multiple DGs. Zhang et.al [28] adopted integrated energy model for allocating DG units in. In [11], Injeti and Kumar applied new method to diminish the RPL in distribution system. Evolutionary technique such as simulated annealing are used to decide both position and rating of DG units. In [21], sensitivity analysis technique is presented to resolve OPDG problem. In [27], an analytical approach is projected to resolve OPDG problem in balanced DS.

Reactive power compensation is habitually done by shunt capacitors. It is desirable to allocate capacitors at optimal location with appropriate size for reduction of RPL and for enhancement of voltages. The rating and position of capacitors are calculated by particle swarm optimization in [23] , simulated annealing [5], cuckoo search algorithm [6], optimization technique based on teaching learning algorithm [25], based on flower pollination [1], heuristic algorithm [10], genetic algorithm [26] are applied in literature to solve capacitor placement problem.

A new approach has been presented here for finding optimal position and rating of DG and capacitor units. A simple and new mathematical expression, line constant (LC), is formulated to resolve DG and capacitor units allocation problem. Rating and location of both units are identified by line constant separately. The method is tested on conventional IEEE 69 bus system. The results for same systems are found promising and optimistic. The same results have outperformed other latest techniques proposed in the literature.

2 Problem Statement

The aim of this work is to scale down real power losses which occurred in distribution system. For minimizing power losses, it is essential to install the DG and capacitor units of suitable rating at best possible places. A simple line diagram of bus a and b is shown in Fig. 1. Capacitor units or DG are placed at bus b. The RPL for n-bus system is calculated by using:

$$RPL_n = \sum_{a=1}^{n} \sum_{b=1}^{n} \frac{|V_a|^2 + |V_b|^2 - 2|V_a||V_b| \cos\theta_{ab}}{Z^2} R \tag{1}$$

It is mandatory for optimal placement of DG or capacitor units that the ratio of real power losses (after and before shunt compensation) is to be minimum.

Hence, the objective function of the problem is :

$$Min.(f) = RPL_n \tag{2}$$

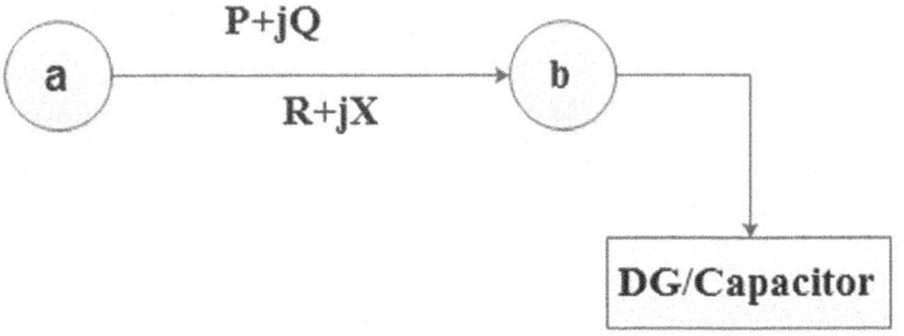

Fig. 1 Line diagram of bus a and b

The operational constraints are as follows:

1. Power balance rule should not be violated.
2. The total generated power of DG and capacitor units should be less than the load of the system.
3. Voltage profile of each bus should maintain Indian standards (±5%).
4. The line current should remain between its minimum and maximum values.

3 Proposed Approach

In this paper, a latest method is adopted to minimize RPL in DS. The RPL are reduced by inserting DG or capacitor units at appropriate positions. The ratio of P_2 and P_1 is termed as "Loss Constant (LC)" and is expressed as:

$$\mathrm{LC} = \frac{P_2}{P_1} \tag{3}$$

where,

P_1: RPL for base case before compensation. P_2: RPL after shunt compensation.

The value of P_2 is required to be minimum in order to identify the position as well as rating of DG or capacitor. Loss Constant (LC) value is determined at each bus with explicit DG size for calculating the location and range of DGs. Candidate bus position of DG is identified by determining LC values, and the bus which have minimum LC value will be the candidate bus for appointment of DG unit.

The steps involved in computational procedure are mentioned below:

1. Calculate real power losses for base case (P_1).
2. Initiate with 1% DG or capacitor value of total load of the system (PD and QD).
3. Compute P_2 and "LC" of the system bus using equation no 3.
4. Increase DG/ capacitor size in minute step and calculate P_2 and "LC."
5. Minimum amount of "LC" value yields optimal DG/ capacitor size.
6. Stop the program when there is change in bus number.
7. The meticulous bus will be the optimal location.
8. To determine more sites of DG or capacitor units repeat Steps 4–7.

4 Results

Standard IEEE 69 bus system [23] is used to examine the above approach. The projected technique is tested by considering three different loading levels also. The code for the proposed methodology is written in MATLAB software.

4.1 Test System I (Standard IEEE 69 Bus System)

Figure 2 exhibits line diagram of test system-I. Total real and reactive load of the system is 3802 kW and 2694 kVAr [23], respectively. The different cases are as follows:

I: Placement of DG only
II: Placement of Capacitor only
III: Placement of DGs & capacitors concurrently.

All the three different cases mentioned above are tested on three different loading conditions which are as follows: Light load (50%), nominal load (100%) and heavy load (160%) (Figs. 3 and 4).

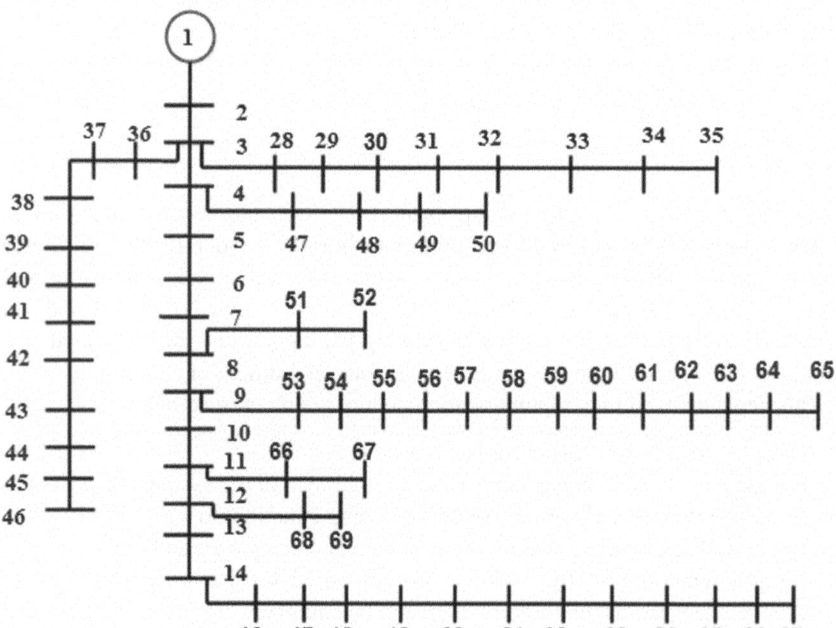

Fig. 2 IEEE-69 bus distribution system

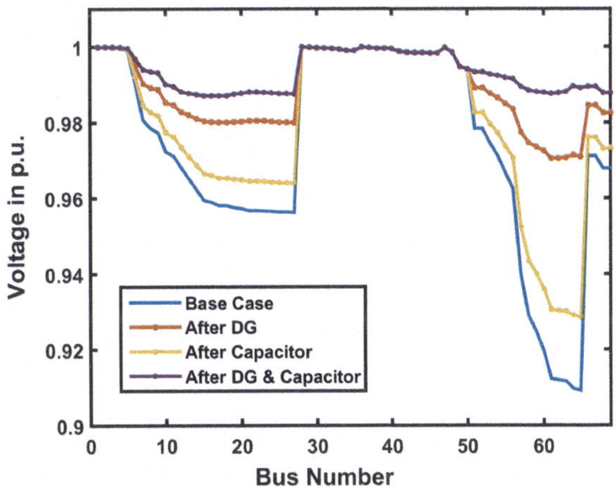

Fig. 3 Voltage outline of 69 bus system (Nominal loading level for CASE-III)

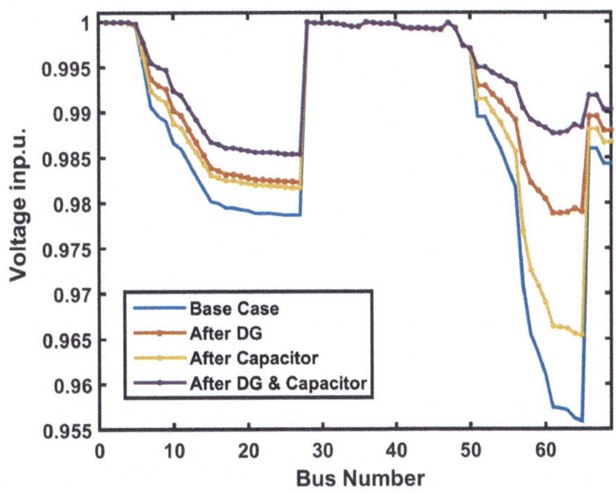

Fig. 4 Voltage outline of 69 bus system (light loading level for CASE-III)

1. **Case-I: DGs (Solar PV Module) placement only**:
   By using proposed method, the size and position of DG units are identified. The candidate buses are selected while taking into account a ceiling of 50% DG penetration level. Table 1 exhibits the results of standard 69 bus system. The most favorable place for DG is determined at bus no. 21, 61 and 64. The real power losses after DG allotment are 77 kW (nominal load), 25 kW (light load) and 210 kW (heavy load). Table 2 justifies that percentage RPL reduction is remarkable of the projected technique as compared to other techniques proposed in the literature. The minimum voltage profile after applying proposed technique is 0.9708 pu which is appreciably better as the DG size is 1.80 MW. The basis of comparison is total size of DG in MW, minimum bus voltage in per unit and percentage loss reduction.

Table 1 Result of CASE-I for 69 bus system

Cases	Parameter	Load levels		
		50%	100%	160%
Before compensation	P_{loss} (kW)	53.31	225	643
	V_{min} (pu)	0.956	0.909	0.846
After compensation with DG	Rating of DG (location)	319(61) 281(64)	300(21) 949(61) 551(64)	382(21) 1569(61) 749(64)
	DG rating (kW)	600	1800	2700
	P_{loss} (kW)	25	77	210
	V_{min} (per unit)	0.98	0.97	0.95
	Percentage RPL	54	66	67

Table 2 Results of comparison for CASE-I

Method	DG rating (kW)	V_{min} (per unit)	Percentage RPL reduction
Bacterial foraging technique (BFT) [22]	2020	0.98	62.2
Modified BFT [22]	1870	0.97	63.2
Combined GA and PSO [23]	3000	0.99	64
Harmony search [25]	1780	0.96	62
Analytical [26]	1840	0.97	64
Proposed	1800	0.97	66

2. **Case-II: Shunt Capacitor placement only**:
   In this case, shunt capacitors are placed for RPL reduction. Capacitors optimum location and its size are determined by the same technique. Table 3 shows results of three different loading conditions. Similarly, the best location of capacitor banks is found at bus no. 21, 61 and 64. The real power losses are reduced to 150 kW from 225 kW at nominal load level after installation of 1300 kVAr capacitors. The percentage loss reduction at all three loading levels is remarkable. The results after capacitor allocation are compared with latest optimization technique like DSA [24], FPA [1]. Table 4 exhibits the comparison of results for Case-II on various parameters such as real power loss, percentage loss reduction, minimum voltage profile, rating and size of capacitors with other proposed techniques in the literature. It is quite evident to say that percentage loss reduction for the proposed technique accounts for 33.34% which is better than FPA [1] which yields 32.44% of RPL reduction. Analogously the voltage profile is also enhanced considerably after placement of capacitor units. It is clearly reported from results reported

Table 3 Result of Case-II for 69 bus system

Load levels		50%	100%	160%
After capacitor allotment only	Rating of capacitor (kVAr) with position	420(61)	720(61)	1550(61)
		70(21)	230(21)	220(21)
		160(64)	350(64)	620(64)
	Total size (kVAr)	640	1300	2390
	Prealloss (kW)	37	150	400
	V_{min} (pu)	0.97	0.931	0.88
Power loss reduction in %		30.6	33.34	37.8

Table 4 Results compared with other techniques for test system-I (CASE-II)

Parameters	Without capacitor	DSA [24]	FPA [19]	Proposed
Real power loss (kW)	225	147	152	150
% loss reduction	–	34.66	32.44	33.34
V_{min} (pu)	0.909	0.93	0.93	0.93
Rating of capacitor (location) (kVAr)		900(61)		720(61)
	–	450(15)	1450(61)	230(21)
		450(60)		350(64)
Total size in kVAr	–	1800	1450	1300

above that losses reduce significantly with simultaneously decrease in capacitor size as compared to other contemporary approaches.

3. **Case-III: Placement of DGs and capacitor concurrently**:
Table 5 exhibits brief picture of results for concurrent allotment of DG and capacitor of 69 bus system. At nominal load level, a reduction of 94.05% is observed for real power losses in Table 5. The size identified by the proposed technique for both DG and capacitor are 1800 kW and 1300 kVAr, respectively, as shown in Table 6. It can be observed from the table; the size of DG units investigated by PSO, IMDE, BBO, ICA, IPSO techniques is more than the proposed technique for nearly same loss reduction (Fig. 5).

Table 5 Result of CASE-III

Case	Items	Loading levels		
		Light loading (50%)	Nominal loading (100%)	Heavy loading (160%)
Base case	P_{loss} (kW)	51.61	225	652.47
	Reactive power loss (kVAr)	24.28	102.15	290.25
	V_{min} (pu)	0.956	0.9092	0.845
Allocation of DG and capacitor	DGs size (kW)	600	1800	2700
	Capacitors size (kVAr)	660	1300	2340
	Ploss (kW)	8.60	13.05	38.9
	% real power loss reduction	83.33%	94.05%	94%
	Reactive power loss	5.04	10.35	30.23
	% Reactive power loss reduction	79.4%	90%	89.5%
	V_{min} (pu)	0.985	0.99	0.974

Table 6 Comparative results for CASE-III

Parameters	PSO [12]	IMDE [15]	BBO [7]	ICA [22]	IPSO [8]	Proposed
DGs size in MW	1.82	2.21	2.32	1.83	1.81	1.8
Capacitors size in MVAr	1.3	1.3	2.7	1.3	2.1	1.3
RPL in kW	23.2	13.9	55	23.21	17.32	13.05
% loss reduction (%)	89.70	93.85	75.6	89.68	92.3	94.05
V_{min} (pu)	0.98	0.99	0.97	0.972	0.9775	0.99

5 Conclusions

A novel and simple technique is presented here to minimize RPL in DS. The prominent objective of power loss has been attained by appropriate allotment of DG and capacitors. A new mathematical formulation, Loss Constant (LC), is formulated for finding candidate bus position and rating. The efficacy of projected method is examined on standard 69 bus system at three different loading conditions. Results of

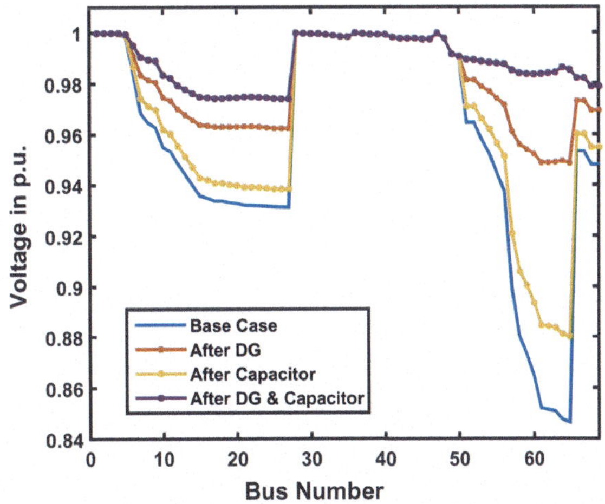

Fig. 5 Voltage outline of 69 bus system (heavy loading level for CASE-III)

standard test system are compared with latest optimization approaches and achieved optimal for all load levels. The proposed technique yields significant upgrading in bus voltages and RPL reduction after placement of DG and capacitor unit simultaneously.

Acknowledgements The authors would like to thank the management of Swami Keshvanand Institute of Technology, Management & Gramothan. Also, the authors wish to express their very sincere thanks to the peer reviewers, for their helpful suggestions and valuable comments.

References

1. Abdelaziz AY, Ali ES, Elazim SA (2016) Flower pollination algorithm and loss sensitivity factors for optimal sizing and placement of capacitors in radial distribution systems. Int J Electr Power Energy Syst 78:207–214
2. Abu-Mouti FS, El-Hawary M (2011) Optimal distributed generation allocation and sizing in distribution systems via artificial bee colony algorithm. IEEE Trans Power Delivery 26(4):2090–2101
3. Acharya N, Mahat P, Mithulananthan N (2006) An analytical approach for dg allocation in primary distribution network. Int J Electr Power Energy Syst 28(10):669–678
4. Borges CL, Falcao DM (2006) Optimal distributed generation allocation for reliability, losses, and voltage improvement. Int J Electr Power Energy Syst 28(6):413–420
5. Chiang HD, Wang JC, Tong J, Darling G (1994) Optimal capacitor placement, replacement and control in large-scale unbalanced distribution systems: system modeling and a new formulation. In: Proceedings of IEEE/PES transmission and distribution conference. IEEE, pp 173–179
6. Das P, Banerjee S (2013) Placement of capacitor in a radial distribution system using loss sensitivity factor and cuckoo search algorithm. Int J Sci Res Manage 2(4):751–757
7. Ghaffarzadeh N, Sadeghi H (2016) A new efficient bbo based method for simultaneous placement of inverter-based dg units and capacitors considering harmonic limits. Int J Electr Power Energy Syst 80:37–45

8. Ghanegaonkar SP, Pande V (2017) Optimal hourly scheduling of distributed generation and capacitors for minimisation of energy loss and reduction in capacitors switching operations. IET Gener Transm Distrib 11(9):2244–2250
9. Gözel T, Hocaoglu MH (2009) An analytical method for the sizing and siting of distributed generators in radial systems. Electric Power Syst Res 79(6):912–918
10. Hamouda A, Sayah S (2013) Optimal capacitors sizing in distribution feeders using heuristic search based node stability-indices. Int J Electr Power Energy Syst 46:56–64
11. Injeti SK, Kumar NP (2013) A novel approach to identify optimal access point and capacity of multiple dgs in a small, medium and large scale radial distribution systems. Int J Electr Power Energy Syst 45(1):142–151
12. Kansal S, Kumar V, Tyagi B (2013) Optimal placement of different type of dg sources in distribution networks. Int J Electr Power Energy Syst 53:752–760
13. Katsigiannis Y, Georgilakis P (2008) Optimal sizing of small isolated hybrid power systems using tabu search. J Optoelectron Adv Mater 10(5):1241
14. Keane A, O'Malley M (2005) Optimal allocation of embedded generation on distribution networks. IEEE Trans Power Syst 20(3):1640–1646
15. Khodabakhshian A, Andishgar MH (2016) Simultaneous placement and sizing of dgs and shunt capacitors in distribution systems by using imde algorithm. Int J Electr Power Energy Syst 82:599–607
16. Kim JO, Nam SW, Park S, Singh C (1998) Dispersed generation planning using improved hereford ranch algorithm. Electric Power Syst Res 47(1):47–55
17. Kim KH, Lee YJ, Rhee SB, Lee SK, You SK (2002) Dispersed generator placement using fuzzy-ga in distribution systems. In: IEEE power engineering society summer meeting, vol 3. IEEE, pp 1148–1153
18. Lee SH, Park JW (2009) Selection of optimal location and size of multiple distributed generations by using kalman filter algorithm. IEEE Trans Power Syst 24(3):1393–1400
19. Moradi MH, Abedini M (2012) A combination of genetic algorithm and particle swarm optimization for optimal dg location and sizing in distribution systems. Int J Electr Power Energy Syst 34(1):66–74
20. Nara K, Hayashi Y, Ikeda K, Ashizawa T (2001) Application of tabu search to optimal placement of distributed generators. In: 2001 IEEE power engineering society winter meeting. Conference proceedings (Cat. No. 01CH37194), vol 2. IEEE, pp 918–923
21. Nawaz S, Imran M, Sharma A, Jain A (2016) Optimal feeder reconfiguration and dg placement in distribution network. Int J Appl Eng Res 11(7):4878–4885
22. Poornazaryan B, Karimyan P, Gharehpetian G, Abedi M (2016) Optimal allocation and sizing of dg units considering voltage stability, losses and load variations. Int J Electr Power Energy Syst 79:42–52
23. Prakash K, Sydulu M (2007) Particle swarm optimization based capacitor placement on radial distribution systems. In: 2007 IEEE power engineering society general meeting. IEEE, pp 1–5
24. Raju MR, Murthy KR, Ravindra K (2012) Direct search algorithm for capacitive compensation in radial distribution systems. Int J Electr Power Energy Syst 42(1):24–30
25. Sultana S, Roy PK (2014) Optimal capacitor placement in radial distribution systems using teaching learning based optimization. Int J Electr Power Energy Syst 54:387–398
26. Swarup K (2005) Genetic algorithm for optimal capacitor allocation in radial distribution systems. In: Proceedings of the 6th WSEAS international conference on evolutionary. Lisbon, Portugal, pp 152–159
27. Viral R, Khatod DK (2015) An analytical approach for sizing and siting of dgs in balanced radial distribution networks for loss minimization. Int J Electr Power Energy Syst 67:191–201
28. Zhang X, Karady GG, Ariaratnam ST (2013) Optimal allocation of chp-based distributed generation on urban energy distribution networks. IEEE Trans Sustain Energy 5(1):246–253

Modelling of a Micro-grid for Bali Village Using HOMER Software

Arpita Singh, Omveer Singh, and Ritwik Tripathi

Abstract An economic model which includes diverse sources of electric generation is proposed in this article. This model is designed for an area where adequate supply is not available. This proposed model is developed in the HOMER software scenario. In this system model, renewable energy sources (RESs) replace the conventional energy source due to their benefits. The considered RESs are solar photovoltaic (SPV) and wind turbine generating system. This is also proved by the help of simulation results. RESs being a reasonable investment and a feasible solution would be a good option for generation of electric power at remote locations. Geographically isolated remote rural areas have a large portion of country's population. Village of Maharajganj district, Uttar Pradesh state, India, is considered here for this work. Modelling, simulation, and optimization of the proposed hybrid electric generation system for this particular village is the prime aim of this work.

Keywords Renewable energy sources · Modelling · Optimization · Solar photovoltaic · Wind turbine · Homer software

1 Introduction

The National Renewable Energy Laboratory (NREL) of the USA created the HOMER economic tool for micro-grid optimization model. It helps in designing the structure of micro-grid. The physical behavior of power system, life-cycle cost is modelled, and analyzed by this software. The installing and operating total cost of the system over its life span is the life-cycle cost. This software enables the modeller to look at a wide range of structure choices dependent on their specialized and

A. Singh (✉) · O. Singh · R. Tripathi
Department of Electrical Engineering, Gautam Buddha University, Greater Noida, India
e-mail: arpita.kant@gmail.com

O. Singh
e-mail: omveers@gmail.com

M. N. Favorskaya et al. (eds.), *Innovations in Electrical and Electronic Engineering*, Lecture Notes in Electrical Engineering 661, https://doi.org/10.1007/978-981-15-4692-1_4

financial benefits [1–7]. A nearby load is served by the electricity which is generated by the micro-grid. Such a framework may utilize any blend of electric age and capacity advances and might be network associated or independent, which means separate from any transmission lattice. A few instances of small-scale control frameworks are a sunlight-based battery framework serving a remote burden, a breeze diesel framework serving a disengaged burden, etc. [1]. Out of many possibilities, the optimal system configuration is found and this is done in optimization [5]. Both grid-connected and off-grid micro-power systems are modelled by using this software. Electric loads and thermal loads are served by this system. SPV modules, wind turbines, small hydro, biomass power, micro-turbines, fuel cells, and batteries are the main constituent of the system. An hourly time series simulation (one year) for a particular system configuration is modelled by the usage of this software. Long-term operation simulation of a micro-grid is HOMER's fundamental capability. Two different dispatch strategies can be modelled by this tool. Load following strategy—The battery will be charged by the renewable power sources and the generators will not be charged. Cycle charging strategy—more power than the required one will be produced when the generators operate. Then, the battery bank will be charged by the surplus electricity.

Two purposes are served by the simulation process. Firstly, feasibility of the system is determined. Feasibility of the system is determined by this software tool if the electric and the thermal loads are adequately served. Secondly, life-cycle cost of the system is estimated by this software tool [8, 9].

For comparison of economics of various system configurations, the convenient metric is life-cycle cost. HOMER's optimization process is basically such comparisons. HOMER software tool includes some possible decision variables:

1. Photovoltaic array size.
2. Wind turbines number.
3. Hydro system presence (Only one size of hydro system is considered. The power system should include or not the hydro system is the concern).
4. Each generator size.
5. Batteries number.
6. AC–DC converter size.
7. Hydrogen storage tank size [1].

2 Study Area

In this study area, a remote rural village Bali, a village in Maharajganj district in the Uttar Pradesh is considered for observation. The village is near around 50 km from Indo-Nepal border line. The village is not properly electrified so that replacing the electrification from power grid to a micro-grid-based power system would be a better economic option. Due to isolation, transmission cost increases. The village comprises of flat plains. The village has adequate water and sun light facilities. Basically, researcher belongs from this village. So, the research work is focused for this area upliftment through regular electric supply. Researcher got the load profile for this work from the local substation.

3 System Model

A combination of technologies is being considered [1]. SPV system, wind turbine, batteries, and a diesel generator (back-up) are the technologies which are considered in article [3]. The village demand is alternating current (AC) coupled; at the AC side of the network, the diesel generator is connected and to the DC side, SPV system and conventional batteries are connected. During peak loads and poor resource periods, the hybrid electric generation system is backed up by a conventional diesel generator sets.

3.1 Details About the Identified Village

Village	Bali
District	Maharajganj
State	Uttar Pradesh
Country	India
Latitude	26.7227 N
Longitude	83.36065 E
Total number of houses	205
Total population	1258
Educational facilities (Primary school)	2

3.2 Village Load Assessment

Large quantities of electric energy are not required for this remote area village. Basic electrical appliances are consuming power during midnight hours. Due to this, the power consumption during midnight hours comes down. But during morning hours, the load demand rises up as everyone gets prepared to leave for schools or offices (Fig. 1).

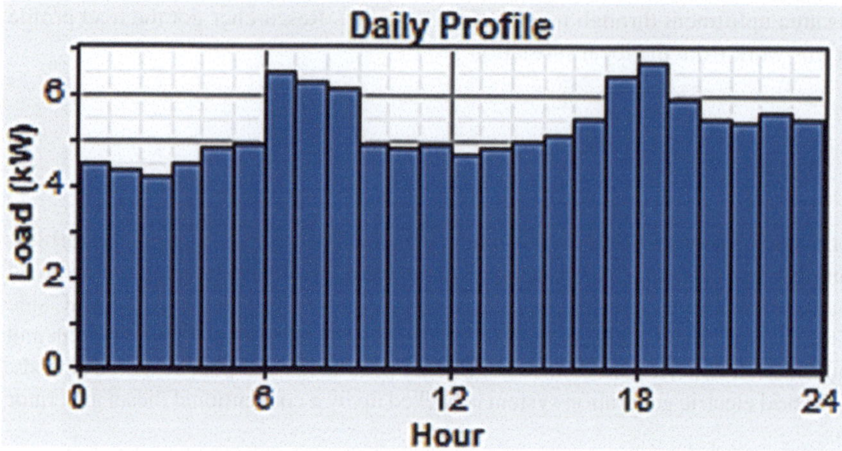

Fig. 1 Proposed residential daily load profile

Table 1 Area average and peak load data

Average (kWh/d)	84.5
Average (kW)	3.52
Peak (kW)	11.5
Load factor	0.307

3.3 Study Area Average and Peak Load

The average and peak load data are presented in Table 1 which is shown below.

3.4 Diesel Generator Modelling

The diesel generator sets which are used are in the range of 16–20 kW. Low cost, ease to operate, and install are the reason behind the usage of diesel generator.

3.5 Battery Modelling

The considered battery bank in this simulation was Trojan L16P model 6 V, 360 Ah. 12–36 units of battery was considered in this simulation.

3.6 *Photovoltaic Array Modelling*

Rated power 10 kW was the size considered with no tracking system allowed.

3.7 *Wind Turbine Modelling*

The hub height 20 m of model Bergey wind power's BWC E-S model was considered. The output is an AC voltage with a rated capacity of 18 kW.

4 Resources Assessment

A remarkable potential in SPV energy is found in Indian village regions. The location of Bali village for which SPV resource was taken is 26.7227 N latitude and 83.36065 E longitudes. 5.01 kWh/m^2/day was the scaled annual average solar radiation. 6.7 kW/m^2 was the highest solar radiation in the month of May. The solar radiation was 3.6 kW/m^2 which lowest in the month of December. The annual average wind speed was 2.21 m/s for the selected location. The selection of SPV and wind turbine is just the case which researchers have taken for this work. Any other renewable source can also be considered and different cases can be made (Figs. 2 and 3).

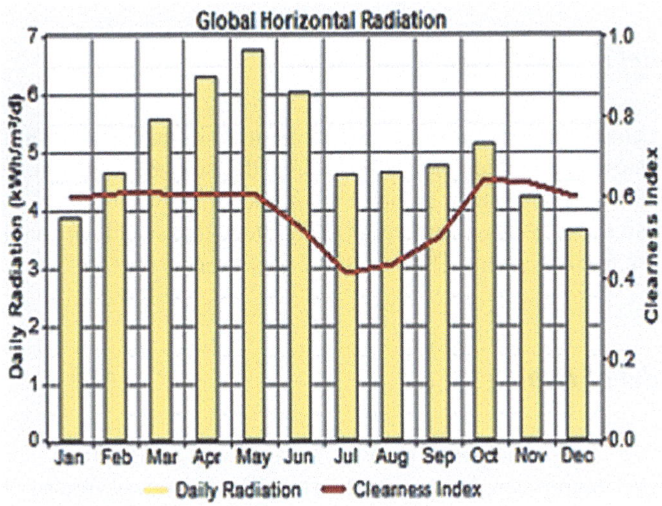

Fig. 2 Selected village solar energy profile

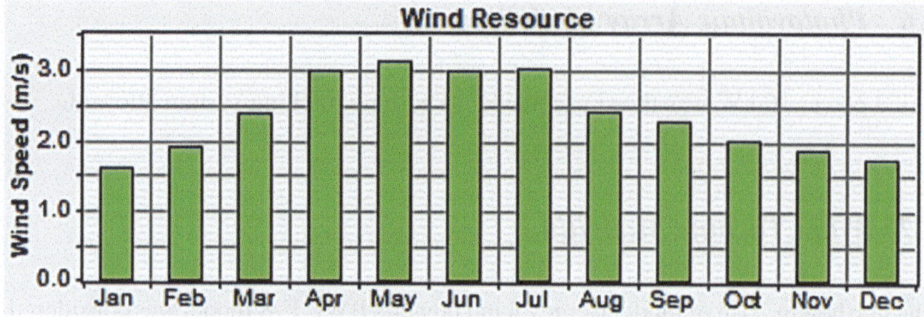

Fig. 3 Selected village wind energy profile

5 Assessment of System Components

Diesel generator sets, SPV module, wind turbine, battery, and power converter are the energy system components which are used for my model [1, 10]. For each of this equipment, number of units to be used, cost, operating hours, etc., needs to be stated clearly in HOMER software platform. In the following sections, description of these components is given. The SPV panel and wind turbine are the two major renewable energy components used in the present simulation. The SPV panel has the other component involved such as a battery Trojan L16P model and inverter. The direct current (DC) is produced by SPV panel and the AC is produced by diesel generator sets and wind turbine [4]. The conversion of DC to AC current is done by converter and vice versa. The proposed system model diagram is shown below in Fig. 4.

5.1 Model of Diesel Generator Sets

For remote electrification, diesel generators are used in this model. The low cost, ease to operate, and install are the reasons behind their usage [1]. In remote villages, the diesel generator sets which are used are in the range of 16–20 kW. $2000/kW is the generator cost in this investigation. The fifteen thousand hours are the rating lifetime taken for this system. The cost of the diesel fuel is 0.93 $/l.

5.2 Model of Battery

In small-scale stand-alone power systems, batteries are an important factor due to their cost [1]. The considered battery bank in this simulation was Trojan L16P model 6 V, 360 Ah. The five years are the estimated lifetime. The one hundred fifty dollars

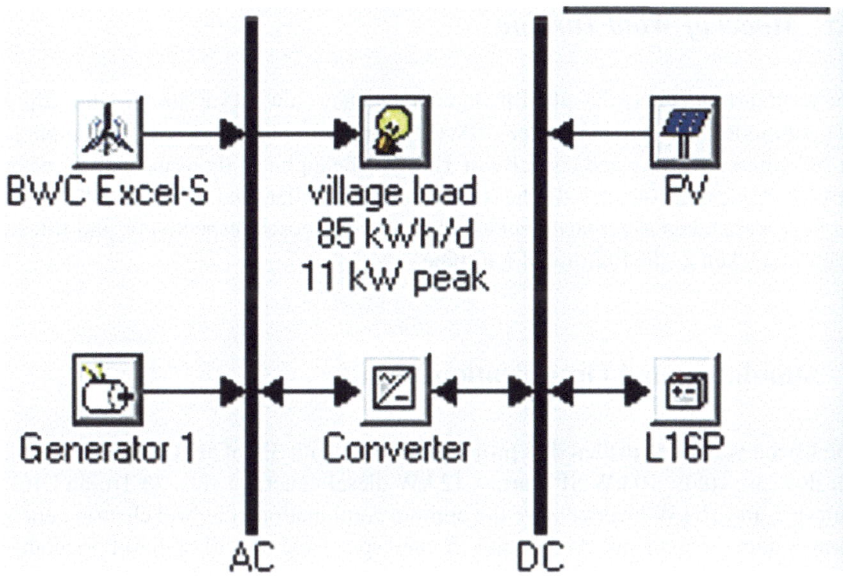

Fig. 4 Proposed economic system model for Bali village

and two dollar per year are the replacement cost and expected maintenance cost, respectively [3]. The twelve to thirty six units of battery range were considered in the simulation time.

5.3 Model of Power Converter

For maintaining energy flow between the DC and AC components, a power electronics converter is needed. The five hundred dollars is taken as installation and replacement cost [6]. An efficiency of 90% with fifteen years is considered as lifetime of a unit.

5.4 Model of SPV Array

The three thousand dollars were taken as one kilowatt solar energy system installation and replacement costs. The ten kilowatt was the size to be considered [1]. The ten years was taken as the lifetime of SPV arrays. In this investigation, no tracking system was utilized.

5.5 *Model of Wind Turbine*

The wind turbines energy availability has dependency on wind variations. Hub height 20 m of model Bergey wind power's BWC E-S model was considered. The output is an AC voltage with a rated capacity of 18 kW. The eighteen thousand dollars were the cost of one unit considered. The twelve thousand dollars and four hundred dollars per year were taken as replacement and maintenance costs, respectively, and fifteen years was taken as the lifetime of a turbine.

6 Simulation and Optimization Results

The fifteen years is considered as project's lifetime with 3% of annual discount rate. For this case study, 10 kW SPV array, 12 kW diesel generator sets, 36 Trojan L16P batteries, and 16 kW inverter were the optimal combination of hybrid electric generation system components. At 2.21 m/s of wind speed and \$0.93/l of diesel price, this system is considered. The \$510329, \$5950, and \$1.294/kWh are the total net present cost, capital cost, and the cost of electricity for this hybrid system, respectively. For generating electricity in Bali village, the wind energy is not a feasible option as per study. The simulation-based optimization result of best electric generating micro-grid system configuration is presented in Table 2. Also, individual electric power production by the proposed electric generation system is in Table 3. Additionally, monthly average electricity production by optimal electric generation system is reflected in Fig. 5.

Table 2 Simulation result of best electric generation system configuration

PV (kW)	XLS	Label (kW)	L16P	Conv. (kW)	Initial capital	Operating cost ($/yr)	Total NPC	COE ($/kWh)	Ren. frac.	Diesel (L)	Label (h)
10		12	36	16	$ 5950	39,456	$ 510,329	1.294	0.38	9371	3011
10	15	12	36	16	$ 23,950	39,143	$ 524,324	1.330	0.54	8276	3010
	15	12	36	16	$ 20,950	45,034	$ 596,633	1.513	0.22	10,924	3364
		12	36	16	$ 2950	49,850	$ 640,199	1.624	0.00	12,767	3740

Table 3 Individual electric power production by the proposed electric generation system

Production	kWh/year	%
PV array	16,104	38
Generator 1	25,923	62
Total	42,026	100

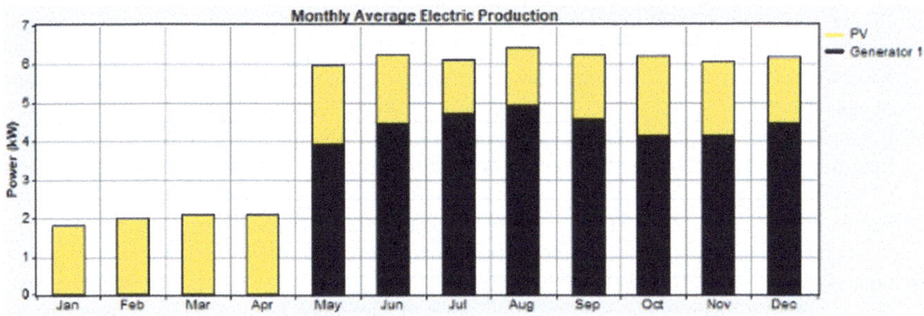

Fig. 5 Monthly average electricity production by optimal electric generation system

7 Conversion of Optimal Result in the Indian Currency

The most optimal configuration net present cost is $510,329 as per United States dollar but this proposed optimal system is configured for Indian village. So, this amount is converted into Indian rupees. The changed currency as per India is Rupees 35900624.49 (Followed by the conversion rate of November 2019).

8 Conclusions

The analysis of the simulation of SPV/wind/diesel hybrid electric generation system has been investigated in this article. From the results, this framework can give great benefit with a sensible venture because of the measure of sustainable power source produced. It uses practically all sustainable power source with a 0.38 inexhaustible portion. Nonetheless, the framework reliability cannot be guaranteed because of variable nature of sun powered radiation accessibility and absence of sufficient wind speed in winter except if other innovation choices are considered. The sun-oriented SPV and diesel generator sets contribute 38 and 62% individually to power age. In the event that the sunlight-based SPV panels are not accessible (or no sun based assets are accessible), the power request can be met with a hybrid electricity generating system involving diesel generators. Be that as it may, the expense of power supply will increment, accordingly making the framework less alluring to clients. At last, on account of the breeze assets in the investigation Bali village are at the lower stage, it is not possible for a breeze turbine to supply the power.

References

1. Hassan Q, Jaszczur M, Abdullateef J (2016) Optimization of PV/WIND/DIESEL hybrid power system in HOMER for rural electrification. J Phys Conf Series

2. Bhattacharyya SC (2012) Energy access programmes and sustainable development: a critical review and analysis. Energy Sustain Dev 16:260–271
3. Khan M et al (2005) Pre-feasibility study of stand-alone hybrid energy systems for applications in newfoundland renewable energy. Renew Energy 30:835–854
4. Barsoum N et al (2007) Balancing cost, operating and performance in integrated hydrogen hybrid energy system. In: IEEE First Asia international conference on modelling and simulation (AMS'07), Thailand
5. Karakoulidis K et al (2011) Techno-economic analysis of a stand-alone hybrid photovoltaic-diesel-battery-fuel cell power system. Renew Energy 36:2238–2244
6. Glatrakos A et al (2009) Sustainable energy planning based on a stand-alone hybrid renewable energy/hydrogen power system: application in Karpathos Island Greece. Renew Energy 34:2562–2570
7. Cristian H et al (2017) Design of hybrid power systems using Homer simulator for different renewable energy sources. In: Proceedings of 9th international conference on electronics, computer and artificial intelligence, Europe
8. Uddin N et al (2019) Optimization and energy management of hybrid renewable power generation using HOMER and FLC. In: 1st International conference on advances in science, engineering and robotics technology (ICASERT 2019), Dhaka
9. Nurunnabi M et al (2015) Grid connected hybrid power system design using HOMER. In: Proceedings of 3rd international conference on advances in electrical engineering, Dhaka, Bangladesh
10. Chandrasekar P et al (2019) Design and performance analysis of hybrid Micro-grid power supply system using Homer pro software for rural village near to Kombolcha town, Ethiopia. Int J Innovative Technol Explor Eng (IJITEE), vol 8, 2278–3075

Transient Stability-Based Security State Classification of Power System Networks Using Kohonen's Neural Network

Rituparna Mukherjee and Abhinandan De

Abstract The paper presents a novel approach for transient stability-based security state classification for modern power system networks using Kohonen's neural network (KNN)-based pattern classifier. Pre-contingency operating conditions of a power system network were used as the input for the KNN. Critical clearing time (CCT) was used as the index for assessment of transient stability condition of the post fault system and classifies the pre-contingency operating states into secure and insecure categories accordingly. The proposed approach has been implemented on the IEEE-39 bus system, and the results demonstrate that the KNN classifier is capable of accurately classifying the power system operating states based on transient stability.

Keywords Power system transient stability · Dynamic security · Classification · Kohonen's neural network

1 Introduction

The operation of a power system requires synchronization in rotational speeds of thousands of large interconnected generating units. The operation also demands that all equipment operates within their physical capability limits regardless of demand fluctuations or eventual outage of a few generators. A typical power system often experiences outage of generators or contingencies, which have the potential to alter the state of operation of the power system. Power system security refers to the 'degree of risk' that a power system poses in surviving against likely occurrence of disturbances such as faults and contingencies, without major interruption in power supply to consumers, at any point of time.

Security analysis is therefore essentiality to assess the robustness of a power system in relation to large variety of severe disturbances which are probable for any network in day-to-day operation. Power system security studies can be broadly categorized as follows: static security analysis (SSA) [1] and dynamic security analysis

R. Mukherjee · A. De (✉)
Indian Institute of Engineering Science and Technology, Shibpur, West Bengal, India
e-mail: abhinandan.de@gmail.com

M. N. Favorskaya et al. (eds.), *Innovations in Electrical and Electronic Engineering*, Lecture Notes in Electrical Engineering 661, https://doi.org/10.1007/978-981-15-4692-1_5

(DSA) [2]. Since it is imminent that the power system will experience a state change during any contingency, it is very important that the system reinstates itself to its normal state in post-contingency condition. In case of SSA, it is assumed that stable operating is achieved and the assessment is mainly focused on whether any physical or security constraints are violated in achieving the post-contingency steady state [3]. However, if there is an occurrence of any severe disturbing event, fast transition to steady-state operating point is highly infeasible. This infeasibility may result in the system achieving a quasi-stable state for a substantially long duration, which may further lead to unintended triggering of system protection and isolation mechanisms, such as relays and circuit breakers. This phenomenon ultimately results in cascaded tripping of major equipment and system outages eventually risking the system security. In the modern competitive power market, the risk of insecurity has even worsened as most power generating units are made to operate close to their maximum capacities and increased complexity of modern large power grids have further worsened the risk of insecurity. Dynamic security analysis [4, 5] which interests itself in the transitional operating state and the quasi-stable state has therefore gained more importance in the modern scenario.

Nonlinear time domain simulation (TDS) is considered one of the most accurate methods for dynamic security analysis under large disturbances (transient stability) [1]. Mathematical modeling of a power system by sets of nonlinear time domain algebraic equations to model generators and other power system equipment to mimic their behaviors is done in this method. The equations are solved by numerical integration methods in order to assess dynamic behavior of a system under credible sets of disturbances and to analyze whether the disturbance will result in loss of transient stability. The major shortcoming of TDS is that it employs resource consuming numerical integrations and it is time intensive, hence rendering it unfit for online applications [6]. Also, TDS is unable to assess relative stability of a power system, which is considered more important than assessing absolute stability. For example, degree of severity of a given contingency can only be judged by its impact on system security relative to others [7].

An alternative approach to the transient stability analysis to determine post-contingency operating point is the direct method using transient energy function [8]. The direct method, which uses reduced order modeling of post-contingency system, fails to achieve the same levels of accuracy. Also, it is not a practical approach for large-scale power systems, which require detailed modeling.

Review of the existing methods for the assessment of dynamic security of electric power systems reveals that none of these traditional approaches can be effectively applied in the online and real-time conditions because of their computation complexity. The complexity of the problem is further aggravated by the fact that most of the critical preemergency states of electric power system, which can potentially lead to large-scale blackouts, are unique and there is no single algorithm to effectively reveal such conditions, fast enough, to be useful in real-time security assessment. Therefore, there is an acute need for developing a fast and reliable method for real-time security monitoring and assessment of the current security level of large power systems. Some research works in the year 1988–89 [9, 10] revealed that effective solution to this problem can be found by application of artificial intelligence (AI) and machine learning (ML) methods. Machine learning and data mining approaches have been tried to develop 'very fast' and 'intelligent' power system security assessment systems [11, 12] and encouraging results were obtained. This was attributed to their general capability of fast recognition of 'patterns' and ability to 'learn through examples,' the unique qualities, which can be effectively utilized to recognize current operating state of a power system. These AI techniques strive to map the inherent relationship between system operating conditions (input) and the dynamic security state of the system (output). By doing so, they can identify and trace back the pre-contingency operating conditions, which led to potential system insecurity and can even predict emergency situations on the basis of certain system security indices.

This paper proposes an AI approach to transient stability-based security analysis of large and complex power systems using KNN-based classification which can alert the system operator about a plausible insecure system operation following severe disturbances and contingencies. The objective behind the development of a security classifier is to predict and classify a power system's present (pre-contingency) operating conditions by assessing the future possibility of the system to become transiently unstable in the events of credible contingencies and outages. For this purpose, Kohonen's neural network-based security classifier (KNNSC) has been trained to act as a 'dynamic security classifier.' By presenting a set of pre-contingency operating variables to the proposed KNNSC, it can classify the power system's dynamic operating states into secure or insecure classes on the basis plausible future transient instability. Synchronized phasor measurement unit (PMU)-based measurements of typical power system operating variables like bus voltage magnitudes, voltage angles and power flow data were used as inputs to the KNNSC. The output produced by the KNNSC is '1' if the system's post-contingency operation is predicted to be insecure or '0' if the system is assessed to be secure under a wide variety of credible contingencies.

2 The Concept of Kohonen's Neural Network-Based Dynamic Security State Classification

The security classifier is intended for prediction and classification of a power system's pre-contingency operating condition by assessing its potential to make the system transiently unstable in the events of future contingencies. To achieve this goal, Kohonen's neural network (KNN) has been trained to act as security classifier (KNNSC). The inputs for the proposed KNNSC are the system variables or attributes measured by PMUs during pre-contingency operating condition to the power system, while the output of the KNNSC is the prediction of the system's dynamic state as secure or insecure [13]. A secure class refers to a pre-contingency operating state in which no future critical line fault or contingency can lead to transient instability. An insecure class, on the other hand, represents a pre-contingency operating state in which there exist at least one line fault or contingency (or more) which may be critical for system's transient stability. A line fault can be considered critical if the same cannot be cleared within the stipulated critical fault clearing time (CCT) of the line [14], as delayed fault clearance can cause potential transient instability of the generators present in the system. In this work, the CCT of different line faults were determined and used as the criterion for determining the criticality of the line faults. The degree of criticality of a pre-contingency operating condition is determined by the number of critical line faults existing for that condition. The proposed KNNSC was trained off-line with the objective to make it recognize the 'degree of criticality' associated with different pre-contingency operating conditions. Once trained, the KNNSC can be implemented online to predict and classify the power system future dynamic security condition into secure and insecure categories solely on the basis of inputs received from the PMU measurements in the pre-contingency operating condition. Figure 1 depicts the conceptual diagram of the proposed KNNSC.

3 Kohonen's Neural Network

Kohenen's neural network (KNN) is an unsupervised neural network which attempts to map the inputs with similar attributes on to an output feature space to form clusters. The algorithm determines 'similarity' between the inputs by measuring the 'Euclidean distance' between different n-dimensional vectors. KNN belongs to the category of competitive learning neural networks (CLNs) having an input layer of linear units and an output layer of units with nonlinear output function as shown in Fig. 2. The neurons in the output layer are extensively interconnected in an 'ON-center' and 'OFF-surround' fashion and produce at equilibrium, larger activation on a single output unit and smaller activation on all other units. This leads to a 'winner-take-all' situation, where only one unit with the largest activation in the output layer becomes sole representative of the presented input pattern. This particular unit is designated as the 'winner' for the presented pattern. The neurons in the output layer of the network thus become spatially organized according to spatial relationship between different input signal patterns or classes of patterns through an unsupervised mapping [13].

4 Off-Line Simulation of the Test System and Training the KNNSC

The IEEE 39-bus system was chosen as an appropriate medium-sized system for all off-line studies, considering its topological spread and intricacy. The IEEE 39-bus is used to test the proposed online fault severity ranking scheme module. This system is commonly known as the 10-machine New England power system. It consists of

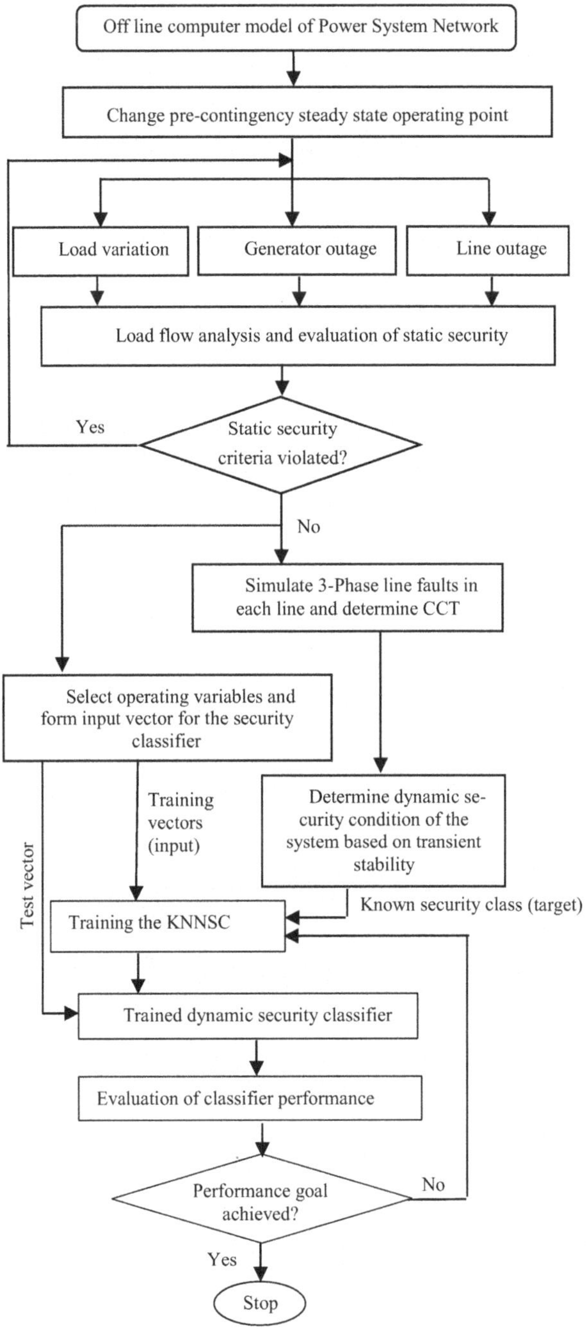

Fig. 1 Conceptual diagram of the proposed KNNSC

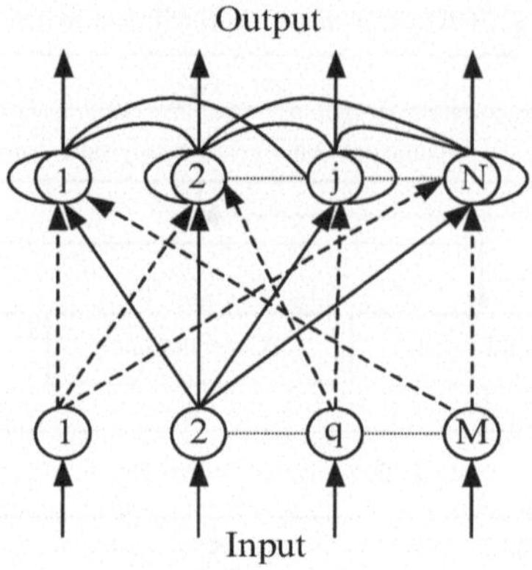

Fig. 2 Typical KNN architecture

ten generators, 34 transmission lines, 12 transformers, 29 load buses and two shunt capacitors. A schematic single line diagram of the system is presented in Fig. 3.

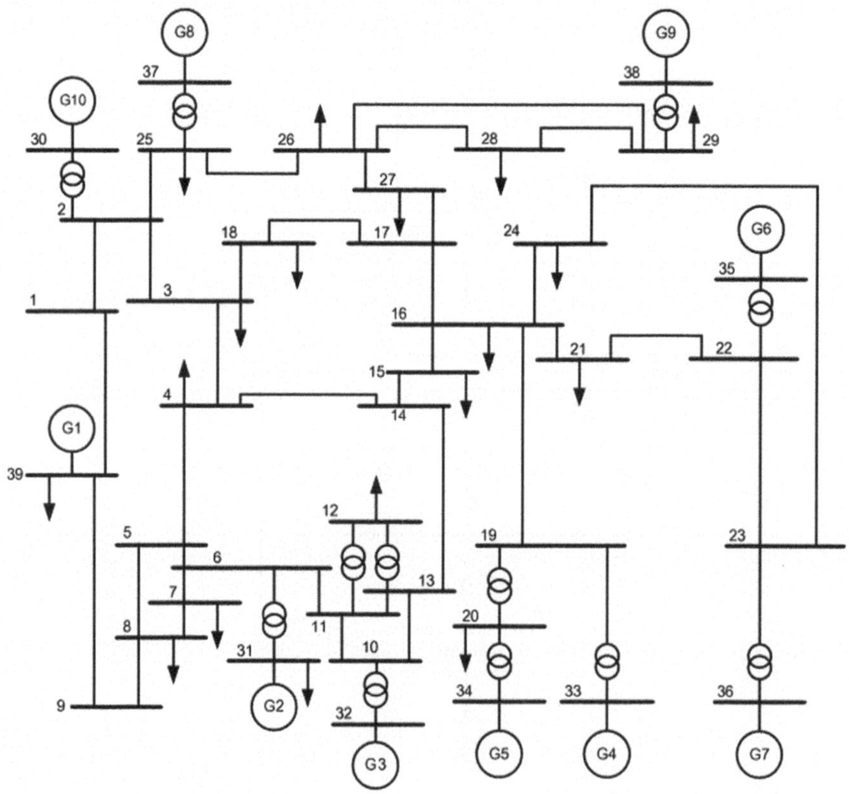

Fig. 3 IEEE 39-bus test system

Creation of large number of diverse and well dispersed pre-contingency steady-state operating conditions (OCs) was achieved by multiple load flow simulations in the IEEE 39-bus test system with variable initial starting conditions such as generation and load variations and changed operating conditions like generator and line outages and their combinations. With the objective to simulate the most practical operating scenarios, simultaneous load variations were considered in about 1/3rd of the total number of system load buses at a time. Out of the total 29 load buses, any arbitrary ten buses were randomly selected at a time as candidates for load variation in five steps (0.5, 0.75, 1.0, 1.25 and 1.5 p.u. of the base load of the buses). The process was then repeated for the next ten arbitrary buses and so on. Repeating the process for 20 times covered all the system load buses. This generated $5 \times 20 = 100$ distinct, well dispersed and yet highly probable loading (and corresponding generation) scenarios. Each of the above 20 loading conditions was them combined with single-generator outage conditions. Considering ten number generators in the IEEE 39 bus system, and 'no generator outage' as yet another scenario, a total of $100 \times 11 = 1100$ unique load generation generator outage scenarios were simulated. Multiple generator outages were not considered, as this would massively proliferate those operating conditions, which are far less credible in real-time operation and hence having large representation of such operating conditions, as training exemplars would be hardly beneficial. Following the same principle, single line outage cases were considered as another form of variation in OC. Considering 34 number of lines in the test system, and taking 'no line outage' as one more condition, combining 34 single line outage cases with 100 load generation patterns, $35 \times 100 = 3500$ mutually exclusive load generation line outage combinations were simulated. Thus, all together $1100 + 3500 = 4600$ unique yet credible initial steady-state OCs were simulated for a broad representation of the complete operational space of the IEEE 39-bus test system.

4.1 Selection of Input Variables and Formation of Input Pattern Vector

Choice of input information plays a crucial role in training the KNNSC. The selection of input information for training the KNNSC began with selection of PMU measurable operating variables, termed as 'primary variables,' which were assessed to be representative of the dynamic characteristic of the system and strongly correlated with the post-contingency system security. Table 1 shows the 146 primary variables selected for the IEEE 39-bus test system.

These variables then constitute the 'input vector' for the KNNSC. The developed input vector has a dimension of 146 and can be expressed as follows: $X = [V_{\mathrm{Bk}}, \delta_{\mathrm{Bk}}, P_{\mathrm{Lin}\,x-y}, Q_{\mathrm{Lin}\,x-y}]$ where the suffix k stands for the kth system bus and $x-y$ refers to the transmission line connecting bus x and bus y. X was determined for all the 4600 representative operating conditions (OCs) mentioned earlier to generate an

Table 1 Initial preselected primary system variables

Primary system variables	Symbol used	Number of variables
Bus voltage magnitude	V_B	39
Bus voltage angle	∂_B	39
Line active power flow	P_{Lin}	34
Line reactive power flow	Q_{Lin}	34
Total		146

exhaustive set of input pattern vectors, which are then passed on to the KNNSC for classification of the OCs based on transient stability.

4.2 *Classification of Training Samples by KNNSC*

The KNNSC was trained off-line using cross-validation with CCT of the line faults referred earlier. Out of the 4600 pre-contingency operating conditions created in Sect. 4, randomly chosen 3000 ones (approximately 2/3rd of total data) were used to form the training set. Remaining 1600 cases (approximately 1/3rd of total data) were preserved as the test samples. Out of the total number of pre-contingency operating conditions (OCs) generated from the simulation, around 26% of the OCs are insecure and the rest 74% are secure OCs. Table 2 summarizes the distribution of the secure and insecure OCs in the training and testing dataset.

The training performance of the KNNSC is demonstrated in the confusion matrix in Table 3. The training accuracy is found to be high. In the training set, 15 OCs out of 720 insecure OCs were misclassified as secure OCs, and the rest 705 OCs were correctly identified as insecure OCs. On the other hand, 45 OCs out of 2280 secure OCs were misclassified as insecure OCs and rest were correctly classified as secure OCs. In Table 3, *TP* represents True Positive: an operating condition is classified as secure when it is actually secure; *FP* represents False Positive: an operating condition is classified as secure when it is actually insecure; *FN* represents False Negative:

Table 2 Training and test data for the KNNSC

Dataset	Count	%
Total no. of OCs	4600	100
No. of insecure OCs	1196	26
No. of secure OCs	3404	74
No. of training OCs	3000	65
No. of test OCs	1600	35

Table 3 Confusion matrix of the KNNSC for 3000 training cases

		Predicted		
		Secure	Insecure	
Actual	Secure	TP = 2235	FN = 45	TP + FN = 2280
	Insecure	FP = 15	TN = 705	FP + TN = 720
		TP + FP = 2250	FN + TN = 750	N = 3000

an operating condition is classified as insecure when it is actually secure; and *TN* represents True Negative: an operating condition is classified as insecure when it is actually insecure.

5 Performance Evaluation of the KNNSC in the Unseen Test Cases

Once the KNNSC was successfully trained, its classification accuracy was evaluated using 1600 unseen test operating cases. Table 4 shows the security state classification results for the 1600 unseen OCs with the help of confusion matrix. The KNNSC could efficiently classify the unseen cases with only 16 misclassifications. A number of metrics were developed and used to assess the performance of the proposed KNNSC. The accuracy '*a*' of the classifier was defined as the probability of performing a correct classification, which is the ratio of the number of correct classifications to the total number of exemplars: $a = \frac{\sum G_{i,i}}{N}$ where $G_{i,i}$ is the ith diagonal element of the confusion matrix, and N is the total number of exemplars. The misclassification rate was defined as $e = 1 - a$ which is the probability of making an incorrect classification. A number of metrics have been used to evaluate the performance of the classifier such as

Classification accuracy = $(\mathrm{TP} + \mathrm{TN})/N$

Positive misclassification rate (PMR) = FP/FP + TP and

Negative misclassification rate (NMR) = FN/FN + TN.

To compare the prediction accuracy of the KNN approach, other popular classification algorithms such as random forest method (RF) [15], support vector machine

Table 4 Confusion matrix of the KNNSC for 1600 random test cases

		Predicted		
		Secure	Insecure	
Actual	Secure	TP = 1184	FN = 48	TP + FN = 1232
	Insecure	FP = 16	TN = 352	FP + TN = 368
		TP + FP = 1200	FN + TN = 400	$N = 1600$

Table 5 Comparison of classification accuracy of different classifiers

Dataset	Metrics	KNN	RF	SGB	SVM	MLS
Training set	Classification accuracy (%)	98	94	89	87	86
	Composite misclassification rate	0.02	0.06	0.11	0.13	0.14
	Positive misclassification rate	0.01	0.09	0.11	0.19	0.2
	Negative misclassification rate	0.06	0.1	0.21	0.2	0.21
Test set	Classification accuracy (%)	96	92	82	89	86
	Composite misclassification rate	0.04	0.08	0.18	0.11	0.14
	Positive misclassification rate	0.01	0.13	0.22	0.19	0.18
	Negative misclassification rate	0.12	0.1	0.29	0.2	0.24

(SVM) [16] and method of least squares (MLS) [17] were also tested. The relative performance of the different classification methods is presented in Table 5. The results indicate that the KNN-based classifier has superior classification performance compared to other methods.

6 Conclusions

The paper presented a pattern recognition approach to transient stability-based security evaluation of power system networks in real-time environment. An efficient Kohonen's neural network-based security classifier (KNNSC) was developed and trained to predict and classify power system's pre-contingency operating states into secure and insecure classes using PMU-based measured typical system variables like voltage, voltage angle and power flow. The KNNSC was trained off-line with the dynamic security calculation results, using critical clearing time (CCT) of line faults as the index of transient stability. The proposed approach was demonstrated on IEEE 39 bus test system with encouraging results. Efficacy of the developed KNNSC was compared with other equivalent classifiers like MLS, RF, SGB and SVM, and KNNSC was found to be superior when tested on IEEE 39-bus system.

References

1. Kundur P (1994) Power system stability and control. McGraw-Hill Education
2. Kundur P, Paserba J, Ajjarapu V, Andersson G, Bose A, Canizares C, Hatziargyriou N, Hill D, Stankovic A, Taylor C, Van Cutsem T, Vittal V (2004) Definition and classification of power system stability IEEE/CIGRE joint task force on stability terms and defintions. IEEE Trans Power Syst 19(3):1387–1401
3. Sauer PW, Pai MA (1998) Power system dynamics and stability. Prentice-Hall, New Jersey
4. Laufenberg MJ, Pai MA (1998) A new approach to dynamic security assessment using trajectory sensitivities. IEEE Trans Power Syst 13(3)

5. Chiang HD, Wang CS, Li H (1999) Development of BCU classifiers for on-line dynamic contingency screening of electric power systems. IEEE Trans Power Syst 14(2):660–666
6. Zhang R, Xu Y, Dong ZY, Wong KP (2015) Post-disturbance transient stability assessment of power systems by a self-adaptive intelligent system. IET Gener Trans Distrib 9(3):296–305
7. Morteza S, Wu NE, Bay JS (2018) Transient stability assessment of large lossy power systems. IET Gener Trans Distrib 12(8):1822–1830
8. James JQ, Hill DJ, Lam AYS, Gu J, Li VOK (2017) Intelligent time-adaptive transient stability assessment system. IEEE Trans Power Syst 33(1):1049–1058
9. Fouad A, Vittal V (1988) Power system transient stability analysis using the transient energy function method. Int J Electr Power Energy Syst 10(4):146–233
10. Hiskens I, Hill D (1989) Energy functions, transient stability and voltage behaviour in power systems with nonlinear loads. IEEE Trans Power Syst 4(4):1525–1533
11. Vu TL, Turitsyn K (2015) Lyapunov functions family approach to transient stability assessment. IEEE Trans Power Syst 31(2):1269–1277
12. Chiang HD, Li H, Tong J (2013) On-line transient stability screening of a practical 14,500-bus power system: methodology and evaluations. High performance computing in power and energy systems, pp 335–358
13. Aghamohammadi MR, Mahdavizadeh F, Bagheri R (2009) Power system dynamic security classification using Kohenen neural networks. In: IEEE/PES power systems conference and exposition
14. Ren LY, Tian F, Yan JF, Yu ZH, Su F, Wu T (2013) Online application and fast solving method for critical clearing time of three-phase short circuit in power system. Int J Smart grid Clean Energy 2(1)
15. Ho TK (1998) The random subspace method for constructing decision forests. IEEE Trans Pattern Anal Mach Intell 20(8):832–844
16. Gomez F, Rajapakse A, Annakkage U, Fernando I (2011) Support vector machine-based algorithm for post-fault transient stability status prediction using synchronized measurements. IEEE Trans Power Syst 26(3):1474–1483
17. Lilley RW (1990) Demonstration of MLS advanced approach technique. IEEE Aerosp Electron Syst Mag 5(5):41–46

A Clustering and Weighted Euclidean Norm-Based Short-Term Load Forecasting of Normal and Anomalous Days

M. Babita Jain and Venu Gopala Rao

Abstract The paper presents a clustering and the weighted Euclidean distance measure-based combined approach for short-term load forecasting (STLF) of a day ahead hourly load of normal and anomalous days. Hourly load, temperature, humidity and day type are selected as the key variables of the data-set used. The input data-set for the clustering algorithm follows distinct day-type similarity criterion for normal and anomalous days. The model uses a two-tier architecture. In tier one, the input data-set is grouped into a pre-defined number of cluster subsets using weather parameter similarity. In tier two, the forecast day cluster subset is narrowed down to obtain five similar days of the forecast day using weighted Euclidean distance norm. The hourly loads of the similar days are then averaged to obtain the actual hourly load forecasts. The technique is tested for the eleven electrical load zones of the New York Independent System Operator (NYISO) over a period of three years. The performance evaluation of the proposed approach is done in comparison with the conventional neural network-based model implemented by the NYISO. Mean absolute percentage error (MAPE) is taken as the performance measure. The proposed technique gives lower MAPE values for all day types, seasons and terrains.

Keywords Clustering · Mean absolute percentage error short-term load forecasting · Weighted euclidean distance measure

1 Introduction

The advent of new technology-driven energy-efficient alternatives has led to the smart grid evolution of the current electrical grids. Smart grids allow for the easy penetration of renewable energy units into the power system, hence reducing the peak

---

M. Babita Jain (✉)
Poornima College of Engineering, Jaipur, Rajasthan, India
e-mail: babita.jain@poornima.org

V. G. Rao
Mody University, Laxmangarh, Rajasthan, India

M. N. Favorskaya et al. (eds.), *Innovations in Electrical and Electronic Engineering*, Lecture Notes in Electrical Engineering 661, https://doi.org/10.1007/978-981-15-4692-1_6

demands. The growth of such smart grids, increasing percentages of renewable energy consumption shares, dynamic tariffing system and the rise of open and competitive energy markets have stiffened the reliability and energy efficiency margins of the utilities. Load evaluation at various stages of power systems is very essential for efficient planning and operation of such complex energy markets. An accurate, efficient and robust load forecasting solution could be of use to the utilities in serving the smart grids effectively whilst maintaining the system reliability and efficiency margins intact.

2 Literature Review

One important aspect of an accurate STLF model is the identification of appropriate variables, and the other aspect is the selection of suitable technique to be used. Considering the first aspect, most commonly used variables in STLF are the weather variables along with the calendar specifics. The effects of temperature and humidity are taken into account in [1], and the impacts of wind speed, humidity and temperature with linear transformation are considered in paper [2]. Different resolutions of calendar data are presented in papers [3, 4]. Availability of the correct forecast data forms a major bottleneck for variable selection.

Considering the second aspect, it has been observed that the early research in the field of short-term load forecasting was completely driven by the standard statistical techniques like the linear and nonlinear regression [5], curve fitting, least square approximation and time series analysis [6]. With the advancement in the field of artificial intelligence, the short-term load forecasting techniques have seen a major breakthrough in terms of dealing with complex load scenarios efficiently and accurately. Various artificial intelligence techniques popularly used for short-term load forecasting include artificial neural networks [7], genetic algorithms [8], fuzzy logic systems [9], support vector machines [10], swarm intelligence [11], data mining [12] and many others. The ever-increasing complexities of the emerging smart grid technologies are demanding very efficient load forecasting tools. Addressing these needs of the power systems, many hybrid techniques have been developed by utilizing the proven strengths of statistical models and the widely emerging generic and efficient artificial intelligence techniques [13].

The research work presented in this paper proposes one such hybrid technique of combining the weighted Euclidean distance measure with the clustered data subsets to obtain hourly load forecasts. Few significant papers implementing hybrid techniques to STLF are discussed in this section with an emphasis on clustering technique. In paper [12], an STLF technique for forecasting the forty-eight half-hourly loads of the forecast day using the clustering-based support vector machines is proposed. A combinatorial model using clustering and fuzzy wavelet-based neural network is proposed in paper [14], and it is validated for a real-time power system of Greece. Paper [15] proposes a robust hybrid model with clustering-based self-organized map and support vector machine (SVM) for STLF of normal and special days. This model is tested on a real-time NYISO data-set for two seasons. Paper [16] identifies the unique daily load profiles for substations using clustering and segregates the customer profiles into specified classes. Paper [17] uses clustering as one of the comparative techniques to validate the performance of the proposed wavelet decomposition model. Clustering is used to extract special days from normal training data-sets in paper [18]. The special days are handled separately in the paper. Paper [19] implements a hybrid model using data mining and function approximation for the STLF of a real power system data-set. Paper [20] presents a hybrid STLF model using clustering and artificial neural networks.

3 Model Synthesis

Apart from addressing the issues of variable selection and technique identification, the model presented in this paper also takes care of the following aspects of an efficient, reliable and robust STLF paradigm that can accommodate the smart grid implementations of the power systems:

1. *Non-stationary nature of load profiles*: Human activity considered in the sense of time resolution is an important factor affecting the electrical load consumption. It depends on the weekday, month specifics of the year and on the holiday information as presented in papers [3, 4]. The aspect of such non-stationary nature of the load data is addressed in the paper by using the day-type similarity criterion for building the input data-set.
2. *Adaptiveness of the STLF model*: The STLF model developed for the normal days may not always be suitable for the load forecasting of anomalous days—including holidays and special days. Most of the existing methods for anomalous load forecasting rely on classifying different anomalous days as being the same while few others employ different models for different day types [21]. The distinct selection criterion applied in the modelling of the input data-sets for normal and anomalous days takes care of the adaptiveness of the proposed model to all day types.
3. *Robustness of the STLF technique*: Atmospheric conditions greatly affect the load forecast technique modelling [22]. Exogenous weather and calendar variables like the temperature [23], humidity, wind speed, day type and few others need to be considered in STLF modelling. Quite frequently such inclusions in modelling make the STLF tools sensitive and parameter dependent. The technique developed in this paper uses only two weather variables for performing the STLF. This makes it a robust model for implementation by the utilities.

The data-set used in the proposed model consists of the hourly loads of the New York power system obtained from the website of the New York Independent System Operator (https://www.nyiso.com) and hourly weather parameters of the NYISO weather zones taken from 'weather underground' which is a commercial weather service, providing real-time weather information (https://www.wunderground.com). The data-set is analysed using various graphical and statistical metrics to understand the hidden patterns of the non-stationary load and for identifying the appropriate parameters impacting the load profiles.

(A) Data Analysis and Parameter Identification

The correlation between the average daily temperature with the load for the New York (NYC) load zone of the NYISO for the year 2014 is shown in Fig. 1. It clearly indicates that the variation in load consumption is very sensitive to changes in temperature. Hence, the inclusion of temperature as one of the forecasting parameters would reduce forecast errors.

Figure 2 depicts the positive impact of increased average humidity on the load consumption behaviour of New York City. So, humidity has also been considered as an important factor influencing the load demand.

Another prominent aspect that affects the daily load curve is the calendar data. To understand this aspect, the load curves of all the Mondays and Sundays of two consecutive months are presented in Figs. 3 and 4. The figures indicate that the load curve patterns of a particular day and its similar days are quite similar. This similarity

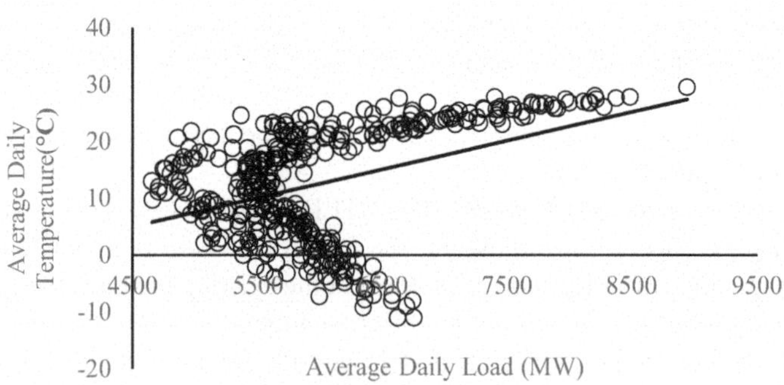

Fig. 1 NYC 2014 annual daily avg. load versus temperature curve

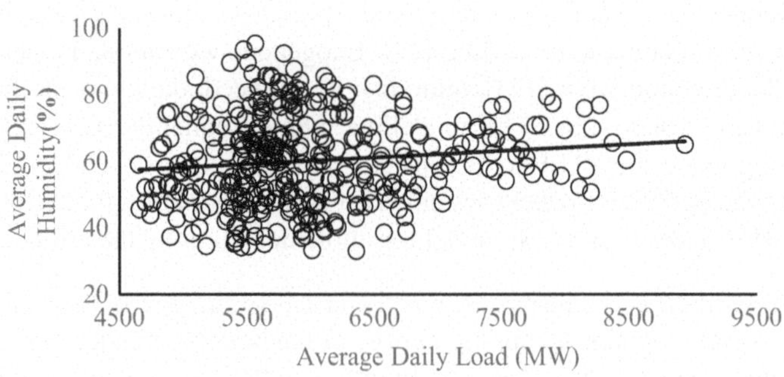

Fig. 2 NYC 2014 annual daily avg. load versus humidity curve

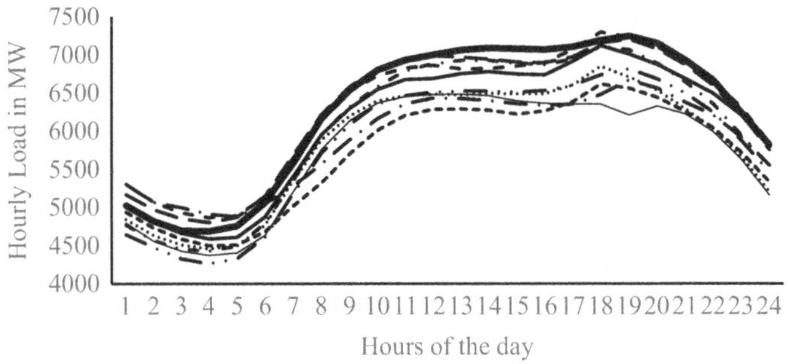

Fig. 3 Daily load curves of NYC 2014 January and February Mondays

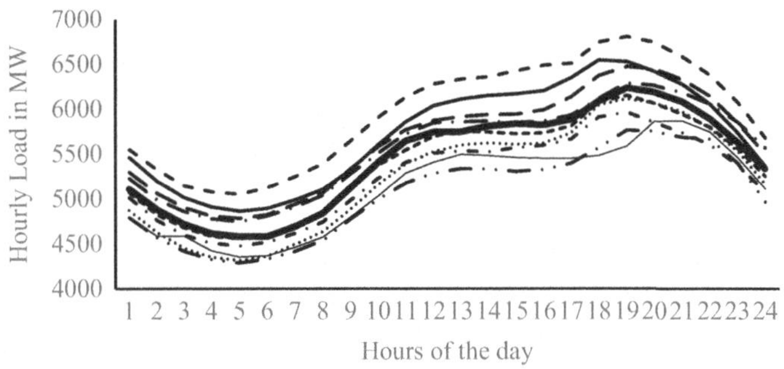

Fig. 4 Daily load curves of NYC 2014 January and February Sundays

is also understood using statistical metric. The statistical index of the percentage of the standard deviation of a particular day's hourly load with respect to its 24-h average load is quite close for days with similar day types and distinct for the continuous days of the week as given in Table 1. The time zone for the similar day's selection is dictated by the seasonal changes.

Table 1 Statistical metrics of full week and similar days

Week days	% SD	Saturdays	% SD	Sundays	% SD	Mondays	% SD
January 1	8.53	January 4	8.50	January 5	9.87	January 6	14.10
January 2	7.47	January 11	9.30	January 12	10.75	January 13	14.69
January 3	7.38	January 18	9.48	January 19	10.47	January 20	15.06
January 4	8.50	January 25	8.36	January 26	9.47	January 27	13.68
January 5	9.87	February 1	9.27	February 2	10.76	February 3	14.18
January 6	14.10	February 8	8.88	February 9	10.01	February 10	13.69
January 7	14.12	February 15	9.09	February 16	10.14	February 17	14.57

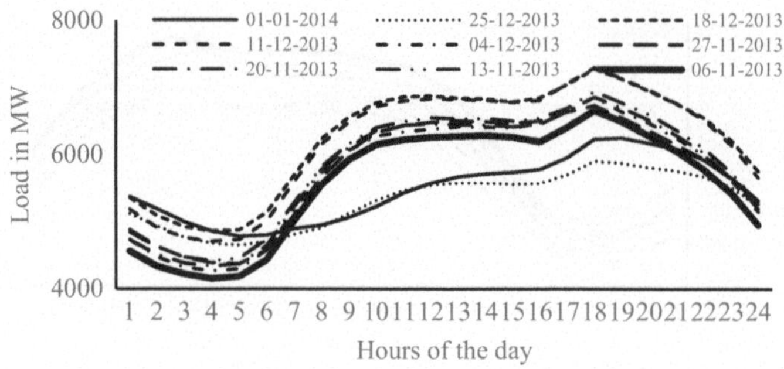

Fig. 5 Daily load curves of 1 January 2014 and its similar days of NYC

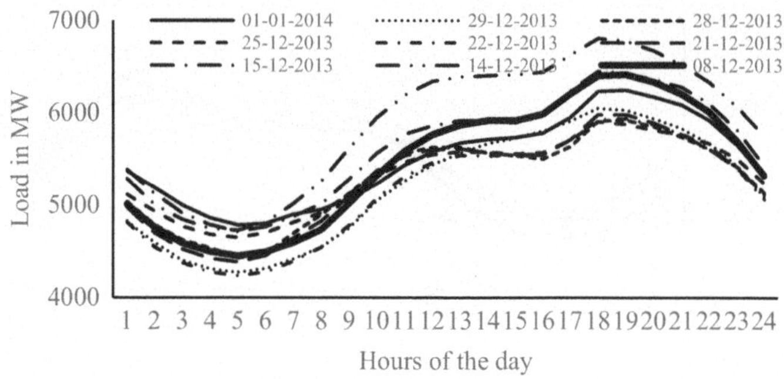

Fig. 6 Daily load curves of 1 January 2014 and weekends and anomalous days in the vicinity of NYC

In the case of anomalous days, the data analysis needs a different approach. From Fig. 5, it can be identified that the anomalous day's (January 1—New Year Day, Wednesday) load curve is different from the load curves of its previous two months same day-type (Wednesdays) load curves. Figure 6 demonstrates close resemblance between the load profiles of January 1 (anomalous day) and other anomalous days and weekends in the similar time zone. This leads to the interpretation that any anomalous day load curve is similar to the load curves of other anomalous days and weekends during the similar time zone. This similarity is also observed in the statistical index of the percentage of the standard deviation of the hourly load of a day with respect to the 24-h average load for that given day. This index value of the anomalous day is very close to the values of other anomalous days and weekends and is distinct for its similar days (same day type) as given in Table 2. Therefore, a common similar day selection criterion for the both normal and anomalous days would not generate good forecast results.

The extensive data analysis has led to the appropriate parameter selection. Average daily temperature and average daily humidity are identified as the weather parameters influencing the electrical load consumption and hence have been used in the modelling of the proposed technique. The selection criterion for the normal days

Table 2 Statistical metrics of anomalous days w.r.t similar days and weekends, anomalous days

Date	Weekend and anomalous days	% SD	Date	Similar day	% SD
01-01-2014	Wednesday	8.531	01-01-2014	Wednesday	8.531
29-12-2013	Sunday	8.990	25-12-2013	Wednesday	8.789
28-12-2013	Saturday	8.871	18-12-2013	Wednesday	7.475
25-12-2013	Wednesday	8.789	11-12-2013	Wednesday	7.553
22-12-2013	Sunday	9.087	04-12-2013	Wednesday	8.139
21-12-2013	Saturday	8.895	27-11-2013	Wednesday	8.100
15-12-2013	Sunday	8.327	20-11-2013	Wednesday	8.079

is based on the similar day-type criterion and for anomalous days is based on the weekends and other anomalous day's similarity criterion as discussed earlier.

(B) **Data-set Dynamics**

For testing and validation purpose, the proposed model considers a five years real-time hourly load data of the NYISO. It is one of the seven ISOs serving two-thirds of the US electrical load. The NYISO control area is divided into eleven load zones known as follows: Capital, Central, Dunwoody, Genesee, Hudson Valley, Long Island, Mohawk Valley, Millwood, New York City (NYC), North and West as shown in Fig. 7.

The entire NYISO load control area has been divided into seventeen different weather stations, namely Albany (ALB), Watertown (ART), Binghamton (BGM), Buffalo (BUF), Elmira (ELM), White Plains (HPN), Islip (ISP), JFK Airport (JFK), LaGuardia Airport (LGA), Massena (MSS), Monticello (MSV), Plattsburgh (PBG),

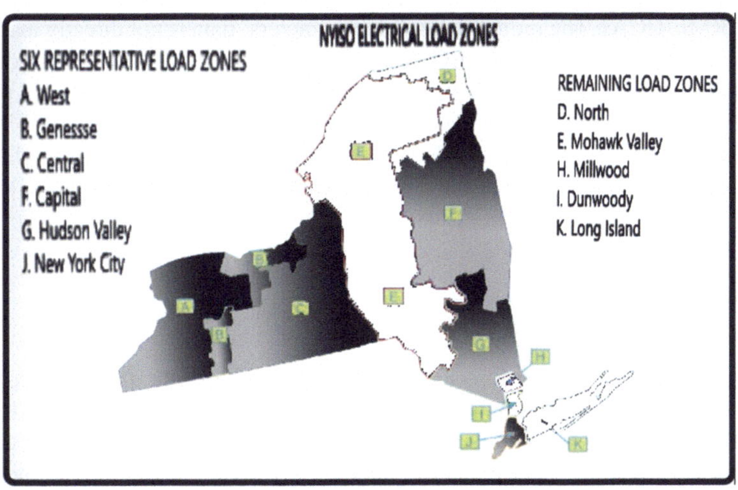

Fig. 7 New York control area electrical load zones

Table 3 Mapping of the NYISO load zones weather data with the weather stations weather data

Load zones	Weather stations	Station weights
West	Buffalo, Elmira and Syracuse	0.91 * BUF + 0.05 * ELM + 0.04 * SYR
Genesee	Elmira, Rochester and Syracuse	0.05 * ELM + 0.85 * ROC + 0.1 * SYR
Central	Binghamton, Elmira, Syracuse and Watertown	0.23 * BGM + 0.14 * ELM + 0.55 * SYR + 0.09 * ART
North	Plattsburgh	1 * PBG
Mohawk Valley	Binghamton, Massena, Monticello, Utica and Watertown	0.2 * BGM + 0.17 * MSS + 0.13 * MSV + 0.35 * UCA
Capital	Albany, Binghamton, Plattsburgh, Poughkeepsie and Utica	0.76 * ALB + 0.03 * BGM + 0.05 * PBG + 0.06 * POU + 0.1 * UCA
Hudson Valley	Newburgh, Poughkeepsie, White Plains and Albany	0.68 * SWF + 0.27 * POU + 0.04 * HPN + 0.02 * ALB
Millwood	White Plains	1 * HPN
Dunwoody	White Plains	1 * HPN
New York City	JFK Airport and LaGuardia Airport	0.21 * JKF + 0.79 * LGA
Long Island	Islip	1 * ISP

Table 4 Data-set division

S. No.	Period	Data-set
1.	01-01-2011 to 31-12-2012	History data-set
2.	01-01-2013 to 31-12-2015	Test data-set

Poughkeepsie (POU), Rochester (ROC), Newburgh (SWF), Syracuse (SYR) and Rome (RME/UCA). The required weather parameters' data is not directly available for eleven load zones of the NYISO. It is obtained by mapping the weather data of seventeen weather stations of NYISO as given in Table 3 (dayahd_schd_mnl.pdf, at https://www.nyiso.com).

The data-set of five years is divided into two parts: part one is the history data-set and part two forms the test data-set. The spread of each data-set is given in Table 4.

4 Model Architecture

The hybrid short-term load forecasting model proposed in the paper has a novel two-tier architecture as shown in Fig. 8. In the first tier, the input data-set is extracted from the history data-set and the K-means clustering algorithm is applied to cluster the input data-set (which includes the forecast day vector) into a pre-defined 'K' number of cluster subsets. The clustering is based on the similarity of the weather parameter metrics. In the second tier, the weights of the Euclidean distance norm

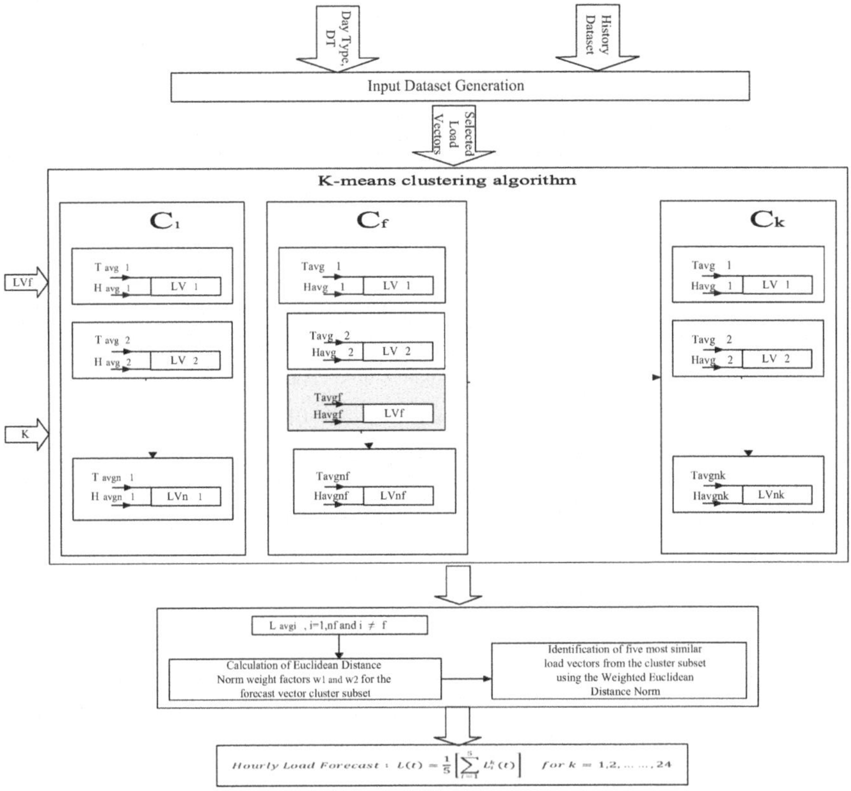

Fig. 8 Detailed architecture of the load forecasting model

are calculated for the forecast cluster subset (cluster subset consisting the forecast day)—'C_f'. The weighted Euclidean distance norm is then applied to 'C_f', and five very similar day profiles of the forecast day are obtained. Finally, the hourly load forecast is obtained by averaging the hourly load values of the five similar days. In entirety, the model clusters the daily load profiles of the input data-set into 'K' cluster subsets and then the identified 'C_f' is narrowed down using the weighted Euclidean distance measure to forecast the hourly loads. The detailed architecture of the model is explained below.

(A) **History Data-set**:

Based on the detailed data analysis, the hourly load, average daily temperature and humidity and day-type values have been taken as the dimensions of the history data-set as given in Table 5.

Table 5 History data-set dimensions

S. No.	Parameter	Detailed description
1	L	Hourly load
2	T_{avg}	Average daily temperature
3	H_{avg}	Average daily humidity
4	D_t	Day type

Table 6 Day-type codes

S. No.	Day	Forecast day-type code	Input data-set day-type code
1	Sunday	000	000
2	Monday	001	001
3	Tuesday	010	010
4	Wednesday	011	011
5	Thursday	100	100
6	Friday	101	101
7	Saturday	110	110
8	Anomalous Day	111	111, 110 and 000

(B) **Tier-One Details**:

1. Input Data-set Generation

The initial stage of the tier one is the building of the input data-set for the clustering algorithm. The input data-set consists of the average daily temperature and humidity values of forecast day and its preceding 110 similar days extracted from the history data-set (Table 3). The similar day selection criterion is different for normal and anomalous days and is given in Table 6. The general form of the input data-set vector is given by (1).

$$\mathrm{LV}_j = \{T_{\mathrm{avg}j}, H_{\mathrm{avg}j}\} \text{ for } j = 1, 2, \ldots, 111 \tag{1}$$

2. *K*-means Clustering

Once the input data-set is generated, the splitting of the input data-set into subsequent cluster subsets is done by the popular *K*-means clustering algorithm. The '*K*' initial cluster subset centroids are randomly generated from the input data-set. The algorithm implements an iterative grouping of data that meets a specified criterion into fixed number of cluster subsets, followed by:

(a) Assignment of data vectors:

In this step, each data vector 'LV_j' is linked to its closest centroid c_i, based on the squared Euclidean distance measure as per (2).

$$\text{argmin}_{c_i \in C} \text{dist}(c_i, \text{LV}_j)^2 \tag{2}$$

where dist $(c_i, \text{LV}_j)^2$ is the standard Euclidean distance measure.

(b) Updating of centroids:

In this step, the cluster centroids are re-calculated by taking the mean of all data vectors linked to that centroid's cluster subset given by (3).

$$c_i = \frac{1}{|S_i|} \sum_{\text{LV}_j \in S_i} \text{LV}_j \tag{3}$$

S_i is the set of data vectors assigned to ith cluster centroid. The algorithm iterates between steps '*a*' and '*b*' until no data vector further changes the previously assigned clusters.

(C) **Tier-Two details:**

After splitting the input data-set into cluster subsets in the tier one, the forecast cluster 'C_f' is identified and its weights for the weighted Euclidean distance norm are determined by least square method based on the regression model given by (4).

$$L_{\text{avg}}^{t} = L_{\text{avg}}^{t-1} + w_1 T_{\text{avg}}^{t} + w_2 H_{\text{avg}}^{t} \tag{4}$$

where L_{avg}^{t} is the average daily load of the current day L_{avg}^{t-1} is the average daily load of the previous day, T_{avg}^{t} is the average temperature of the current day, and H_{avg}^{t} is the average humidity of the current day of forecast cluster load vectors. The forecast day vector is not used for the weights calculation. The selection of the five similar load vectors to the forecast day from 'C_f' is done on the basis of the weighted Euclidean distance norm given by (5).

$$\text{ED} = \sqrt{W_1(\Delta T_{\text{avg}})^2 + W_2(\Delta H_{\text{avg}})^2} \tag{5}$$

where $\Delta T_{\text{avg}} = T_{\text{avg}f} - T_{\text{avg}}^{c}$ and $\Delta H_{\text{avg}} = H_{\text{avg}f} - H_{\text{avg}}^{c}$ and $T_{\text{avg}f}$ and $H_{\text{avg}f}$ are, respectively, the average temperature and humidity of the forecast day, T_{avg}^{c} and H_{avg}^{c} are the average temperature and humidity of the other five similar days of the forecast cluster subset, and w_1,w_2 are the calculated weights of C_f.

The 24-h ahead hourly load forecast $L(t)$ is given by averaging the hourly loads of the selected five similar days as in (6).

$$L(t) = \frac{1}{5}\left[\sum_{i=1}^{5} L_i^k(t)\right] \text{for } k = 1, 2, \ldots, 24 \tag{6}$$

5 Simulation Results

The short-term load forecasting using the proposed technique has been done for all the eleven load zones of NYISO. The STLF results of all day types of the four seasons (winter, spring, summer and fall), spanning over a period of three consecutive years (2013–2015) are used to validate the performance and robustness of the proposed architecture. Mean absolute percentage error (MAPE) has been taken as the performance measure. MAPE is the deviation of the load forecasts from the actual load values as defined in (7).

$$\text{MAPE} = \frac{1}{N}\sum_{i=1}^{N}\frac{\left|P_A^i - P_F^i\right|}{P_A^i} \times 100 \tag{7}$$

'P_A' and 'P_f' are the actual and forecast values of the hourly load, and 'N' is the number of the hours of the day, i.e. 24.

The results obtained have been validated against the load forecasting results obtained by the NYISO load forecasting model used for the real-time load forecasting. The NYISO model is based on the techniques of artificial neural network and statistical regression to forecast the hourly load of the current day and the next six days using the historical load, temperature, dew points, cloud cover and wind speed information. The results corresponding to the MAPE indicate a better forecasting accuracy of proposed model in comparison with the NYISO real-time forecasting model.

As it is difficult to present the forecasting results of all the zones, the results of six representative zones as shown in Fig. 7 are presented in this section. These six representative zones of NYISO selected for result analysis have distinctly varying geographic, demographic and meteorological features. The result analysis of the anomalous days load forecasts is also limited to six selected anomalous days listed in Table 7.

Although Sunday to Saturday, all the seven days as treated as unique days in the proposed model, the results are presented only for four day types ranging from Saturday to Tuesday because of their distinct load profiles. The load patterns of days from Wednesday to Friday are assumed to be similar to Tuesday, since all the four

Table 7 Selected anomalous days and dates of NYISO for 3 years

S. No.	Anomalous day	2013	2014	2015
1	New Year	January-01	January-01	January-01
2	Martin Luther King Jr. Day	January-21	January-20	January-19
3	Lincoln's Birthday	February-12	February-12	February-12
4	Independence Day	July-04	July-04	July-03
5	Columbus Day	October-14	October-13	October-12
6	Veterans Day	November-11	November-11	November-11

days belong to weekdays category. Monday is also a weekday but its results are shown since it follows a slightly different load pattern as it follows a holiday. The actual and forecasted daily average loads, daily average temperature and humidity and the mean absolute percentage error values of the hourly load forecasts of the proposed clustering and weighted Euclidean distance (clustering–WED) norm-based technique and the NYISO real-time forecasting technique for the considered time zone of Genesee load area are given in Table 8. As can be observed from the results in nearly 90% of the cases, the proposed technique's forecasting accuracy is better than the NYISO's real-time load forecasting accuracy. It is significant to note that the MAPE values are less than 3% in most of the cases.

Table 9 presents the comparative result analysis of all seasons yearly average MAPE of six representative load zones of NYISO for the considered 3-year period. The results of the proposed model are more accurate compared to the NYISO model for most of the zones. For the New York City load zone, the NYISO load forecasts have an edge over. Nevertheless, the MAPE values of the proposed model are well within 3%, which happens to be the permissible error limit for good forecasts. The forecast errors for the Capital load zone are shooting above 3% but still are better in comparison with the NYISO actual forecast results.

The anomalous days (Table 7) load forecasting also has been done using the proposed technique. The improvement in the performance accuracy in the anomalous day load forecasts of the proposed technique against the actual load forecasts of the NYISO forecasting technique for the six representative zones can be seen in Fig. 9.

The results show an improvement in the quality of forecasts for most of the zones. The forecast errors for the Capital load zone are slightly above 3% but are still less in comparison with the NYISO actual forecast results.

6 Conclusions

The salient aspects of this research are the hybrid model combining the clustering and weighted Euclidean distance norm; uniqueness followed in the similar day selection criterion; inclusion of the quantitative variables (weather and load) and the qualitative variable (day type) in the model without any complex formulation; model implementation considering the NYISO real-time load data; and its validation for all terrains, seasons and day types, including the anomalous days.

The model synthesis is based on a detailed data analysis performed using graphical and statistical correlation measures for understanding the load dependency on the calendar and weather parameters. The data analysis suggests a distinct similar day-type selection criterion for the anomalous day load forecasting, which is the most difficult part of the STLF.

For the real-time load forecasting application, the proposed model first generates the input data-set from the history data-set using the day-type similarity criterion. Similar days of the forecast day are then grouped into a forecast cluster by clustering the input data-set using the weather parameter similarity. The hourly loads of the

Table 8 Genesee load zone STLF results of four seasons, four representative day types of year 2014

S. No.	Season	Dates	Day type	Actual daily average load (MW)	Forecasted daily average load (MW)	Daily average temperature (°C)	Daily average humidity (%)	Clustering–WED MAPE (%)	NYISO MAPE (%)
1	Winter	February 01–04	Saturday	1121	1095	0	74.65	2.328	6.206
2			Sunday	1087	1068	1.1	81.05	1.784	7.612
3			Monday	1216	1253	−5.7	69.7	3.021	5.486
4			Tuesday	1252	1197	−6.7	70.8	4.330	6.106
5	Fall	March 22–25	Saturday	1062	1049	2.85	65.55	1.209	4.753
6			Sunday	1078	1107	−3.95	62.55	2.715	5.556
7			Monday	1235	1250	−6.15	64.95	1.165	4.268
8			Tuesday	1219	1197	−2.2	53.4	1.806	3.927
9	Summer	August 09–11	Saturday	1050	1019	19.2	66.85	2.978	1.421
10			Sunday	1069	1070	21	68.65	0.060	1.396
11			Monday	1274	1212	22	67.35	1.868	1.883
12			Tuesday	1215	1254	21.8	71.5	3.213	0.895
13	Spring	November 15–18	Saturday	1062	1065	0.7	64.55	0.317	3.881
14			Sunday	1037	1028	1.95	55.95	0.825	4.651
15			Monday	1161	1177	−0.55	87.75	1.405	5.117
16			Tuesday	1243	1214	−5.5	68.45	2.345	6.017

Table 9 Clustering–WED and NYISO comparative yearly all seasons average MAPE of six representative load zones of NYISO for 3 years (2013–2015)

S. No.	Year	Load zone	Clustering–WED yearly all seasons average MAPE (%)	NYISO yearly all seasons average MAPE (%)
1	Y2013	Capital	4.111	4.597
2		Central	3.152	5.618
3		Genesee	2.970	4.225
4		Hud VI	3.531	4.689
5		NYC	2.641	1.736
6		West	2.887	2.898
7	Y2014	Capital	3.402	4.185
8		Central	2.931	5.733
9		Genesee	2.970	4.021
10		Hud VI	3.556	4.561
11		NYC	2.865	1.763
12		West	2.381	3.072
13	Y2015	Capital	4.259	4.766
14		Central	2.399	5.758
15		Genesee	2.945	3.979
16		Hud VI	3.625	5.274
17		NYC	2.701	2.000
18		West	2.417	2.968

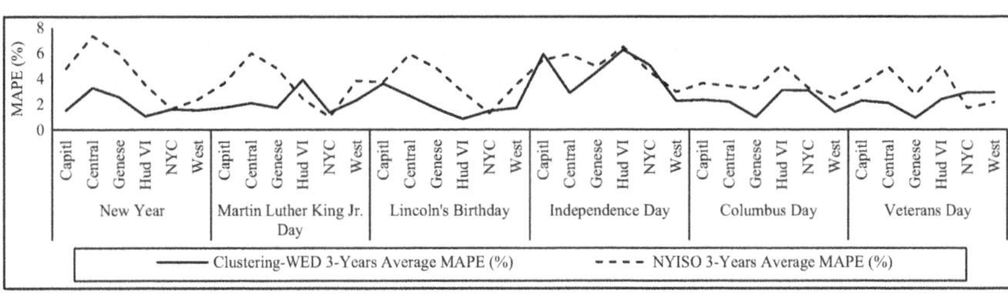

Fig. 9 Clustering–WED and NYISO comparative 3 years (2013–2015) average MAPE of six representative holidays of six representative load zones of NYISO

forecast day are obtained by applying the weighted Euclidean distance measure to the forecast cluster.

The results of the proposed clustering and weighted Euclidean distance norm-based model are compared with the traditional neural network-based model implemented by the NYISO for real-time forecasts. The results obtained by the proposed model in the form of mean absolute percentage error averaged through a year are less than 3.0% for most of the zones. The NYISO real-time forecasting technique results are not close to 3.0% in many cases. Since the clustering–WED technique generates accurate forecasts in all terrains, seasons and day types (also for anomalous days), it can be said that this paper contributes positively to the solution of the STLF problem.

References

1. Khotanzad A, AfkhamiRohani R, Tsun Liang L, Abaye A, Davis M, Maratukulam DJ (1997) NNSTLF-a neural-network-based electric load forecasting system. IEEE Trans Neural Netw 8:835–846
2. Khotanzad A, AfkhamiRohani R, Maratukulam D (1998) ANNSTLF-artificial neural network short-term load forecaster generation three. IEEE Trans Power Syst 13:1413–1422
3. Moghram I, Rahman S (1989) Analysis and evaluation of five short-term load forecasting techniques. IEEE Trans Power Syst 4:1484–1491
4. Amjady N (2001) Short-term hourly load forecasting using time-series modeling with peak load estimation capability. IEEE Trans Power Syst 16:798–805
5. Papalexopoulos AD, Hesterberg TC (1990) A regression-based approach to short-term system load forecasting. IEEE Trans Power Syst 5:1535–1547
6. Hagan MT, Behr SM (1987) The time series approach to short term load forecasting. IEEE Trans Power Syst 2:785–791
7. Hippert HS, Pedreira CE, Souza RC (2001) Neural networks for short-term load forecasting: a review and evaluation. IEEE Trans Power Syst 16:44–55
8. Liao G-C, Tsao T-P (2006) Application of a fuzzy neural network combined with a chaos genetic algorithm and simulated annealing to short-term load forecasting. IEEE Trans Evol Comput 10(3)
9. Ranaweera DK, Hubele NF, Karady GG (1996) Fuzzy logic for short term load forecasting. Int J Electr Power Energy Syst 18:215–222
10. Chen BJ, Chang MW, Lin CJ (2004) Load forecasting using support vector machines: a study on EUNITE competition 2001. IEEE Trans Power Syst 19:1821–1830
11. Jain A, Babita Jain M (2012) Fuzzy modelling and similarity based short term load forecasting using evolutionary particle swarm optimization. IEEE Power and Energy Society General Meeting
12. Jain A, Satish B (2009) Lustering based short term load forecasting using support vector machines. In: Proceedings of IEEE Bucharest Power Tech
13. Tzeng G-H, Wu C-H, Lin R-H (2009) A Novel hybrid genetic algorithm for kernel function and parameter optimization in support vector regression. Expert Syst Appl 36(3):4725–4735
14. Kodogiannis VS, Amina M (2013) Clustering based fuzzy wavelet neural network model for short term load forecasting. Int J Neural Syst 23(5)
15. Fan S, Chen L (2006) Short-term load forecasting based on an adaptive hybrid method. IEEE Trans Power Syst 21(1)
16. Espinoza M, Joye C, Belmans R, De Moor B (2005) Short-term load forecasting, profile identification, and customer segmentation: a methodology based on periodic time series. IEEE Trans Power Syst 20(3)
17. Pandey AS, Singh D, Sinha SK (2010) Intelligent hybrid wavelet models for short-term load forecasting. IEEE Trans Power Syst 25(3)
18. Topalli AK, Erkmen I, Topalli I (2006) Intelligent short-term load forecasting in Turkey. Electr Power Energy Syst 28:437–447
19. Sfetsos A (2003) Short-term load forecasting with a hybrid clustering algorithm. IEE Proceedings of Generation, Transmission and Distribution, vol 150(3)
20. Chicco G, Napoli R, Piglione F (2001) Load pattern clustering for short-term load forecasting of anomalous days. In: IEEE Porto Power Tech conference

21. Kim K-H, Youn H-S, Kang (2000) Short-term load forecasting for special days in anomalous load conditions using neural networks and fuzzy inference method. IEEE Trans Power Syst 15:559–565
22. Rahman S (1990) Formulation and analysis of a rule-based short-term load forecasting algorithm. Proc IEEE 78:805–816
23. Song KB, Ha SK, Park JW, Kweon DJ, Kim KH (2006) Hybrid load forecasting method with analysis of temperature sensitivities. IEEE Trans Power Syst 21:869–876

Wavelet-Based Algorithm for Fault Detection and Discrimination in UPFC-Compensated Multiterminal Transmission Network

J. Pardha Saradhi, R. Srinivasarao, and V. Ganesh

Abstract As interconnected power system transmission line systems are more complex in evaluating system performance, and adding FACTS devices in the power system requires more attention in the analysis. Nowadays FACTS devices play a major role in improving the performance of transmission lines. The evaluation of dynamic behavior during transient conditions is becoming a great difficulty, particularly, for larger networks. Wavelet analysis will give the entire system performance at any part of the system. This paper mainly concentrates on fault detection and its classification of multiterminal networks. It also gives a critical evaluation of nine zones of performance under different fault conditions. Wavelet-based analysis in the presence of UPFC (unified power flow controller) by using Bior 1.5 has been performed. The performance of multiterminal transmission networks with and without controllers under different fault conditions has been estimated.

Keywords FACTS · Wavelet · UPFC · Bior 1.5

J. Pardha Saradhi (✉)
Department of Electrical & Electronics Engineering, Bapatla Engineering College, Bapatla, India
e-mail: pardhasaradhi.jpalli@gmail.com

R. Srinivasarao
Department of Electrical & Electronics Engineering, JNTU, Kakinada, India
e-mail: srinivas.jntueee@gmail.com

V. Ganesh
Department of Electrical & Electronics Engineering, JNTUACEP, Pulivendula, India
e-mail: ganivg@gmail.com

M. N. Favorskaya et al. (eds.), *Innovations in Electrical and Electronic Engineering*, Lecture Notes in Electrical Engineering 661, https://doi.org/10.1007/978-981-15-4692-1_7

1 Introduction

Because of environmental and energy concerns it is very difficult to construct new transmission lines and generation. So instead of constructing new systems, it is essential to increase power transfer capability using existing systems. In order to meet the needs of power transfer it is more important to control the power flow in transmission lines. In addition to this, FACTS devices play a major role in the transmission system. As they are utilized to control the power flow and to change power system parameters. The parameters like line impedances, bus voltages, and phase angles of the power system can be regulated by means of using FACTS devices such as STATCOM, SVC, SSSC, and UPFC. FACTS devices also have the ability to decrease the generation cost, increasing transmission capacities, and improve the stability and security of power systems. The transient and steady-state components of voltage and current signals are affected by compensating devices during fault conditions. These signals will create problems with relay functioning.

The identification and classification of transmission line faults with FACTS devices is a very difficult task. In [1], current and voltage signals are used to find the fault location. But the fault type and the phase in which fault occurs are not reported. In [2], an adaptive Kalman filtering approach is proposed for protecting uncompensated power distribution networks. In [3], an advanced series compensators for compensated transmission systems is employed. But the limitation Kalman filtering is the requirement of a number of different filters to complete the task and also the fault resistance cannot be modeled. Neural networks are applied in [3–5] for pattern recognition but they need large training time and large data and design of a new neural network are needed for each transmission line. In [1–5] different methods based on support vector machines fuzzy logic systems, TT transform, S transform, and wavelet transform are proposed. In these attempts the classification and identification of faulted section is done in a transmission line compensated by TCSC protected by metal-oxide varistor (MOV) or transmission line compensated by series capacitors protected by metal-oxide varistor (MOV) or compensated by both the above-mentioned approaches. The advantage of post fault current and voltage samples are taken from both ends of the line and build a recursive optimization algorithm to identify the fault distance in a transmission line with a series FACTS device. But there is no need of the FACTS device model in this algorithm and it can able to locate the fault without mentioning the type of fault.

According to [6] Power quality conditions and the impact of FACTS devices can be analyzed by using wavelet analysis more effectively. In [7] fault identification in the presence of FACTS devices is obtained by using fuzzy wavelet approach. In [8], wavelet-based entropy algorithm method is applied to find the fault in the presence of FACTS devices has been discussed, the effectiveness of the wavelet entropy algorithm has been checked. In [9], protective gear response is analyzed by using wavelets, which have been discussed.

The UPFC consists of both STATCOM and SSSC which are connected with a common DC link, which allows the bidirectional flow of real power between series output terminals of SSSC and the shunt terminals of the STATCOM. This work performed with UPFC. Protective schemes design in presence of multiterminal network with UPFC is more difficult nowadays. This paper uses Bior 1.5 as a mother wavelet to perform both fault identification and compensation evaluation in the presence of UPFC.

2 Wavelet Transform

A wavelet analysis is nothing but the expansion of functions by means of wavelets, which are created in the form of dilations and translations of a fixed function known as mother wavelet. A mother wavelet is an oscillatory function which has some finite energy and zero average value. It is possible to obtain time and frequency information of a signal using wavelet transform when compared with fourier transformation, which can give only frequency information.

Wavelet transform provides an effective time-frequency representation of signals. All basis functions are formed by shifting and scaling of "mother" wavelet function $\psi(t) \in L^2(R)$

$$\psi_{m,n}(t) = 2^{-\frac{m}{2}}\psi\left(2^{-m}t - n\right)m, n \in Z \tag{1}$$

Signal $f(t) \in L^2(R)$ can be then represented as

$$f(t) = \sum_m \sum_n d_{m,n}\psi_{m,n}(t) \tag{2}$$

where $d_{m,n}$ are spectral wavelet coefficients

$$d_{m,n} = \left\langle f(t), \psi_{m,n}(t)\right\rangle \tag{3}$$

For discrete signals $f(k) \in L^2(Z)$ gives similar results and its equivalent transform is called Discrete Wavelet Transform (DWT).

3 Flowchart for Fault Identification

3.1 Algorithm

Step 1: Initiate I_{a1}, I_{b1}, I_{c1} … I_{a9}, I_{b9}, I_{c9} at all zones.
Step 2: Obtain detailed coefficients at each bus number 1, 2, 3, 4, 5, 6, and 7.

Step 3: Obtain fault index value at each Bus.
Step 4: Highest values fault index in the zone indicate the fault in that Zone.
Step 5: Highest Value in the particular phase will give the faulty phase where Fault occurs.

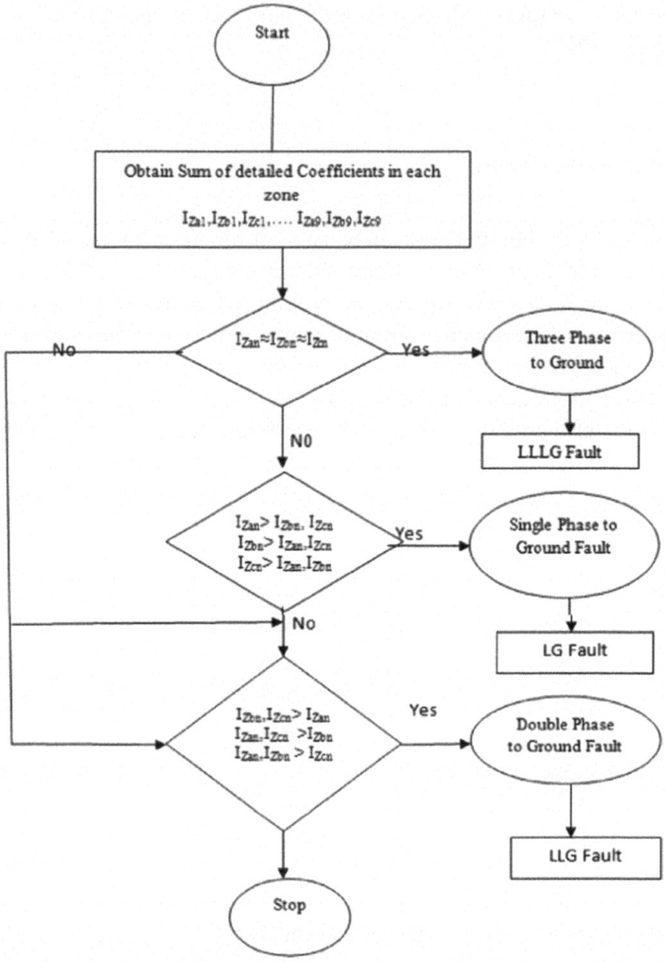

Fig. 1 Flow chart for fault identification

3.2 Flow Chart

4 Test System

4.1 Test System Data and Its Associated Parameters

5 Simulation Results and Analysis

The test system consists of nine zones and seven buses. The length of each zone is shown in the figure. It has five number of DG s and two number of utility grid sources connected. The proposed multiterminal system is operated with 220 kV, 50 Hz. The behaviour of the system is analyzed by using Bior 1.5 mother wavelet detailed coefficients has been calculated and then some of the detailed coefficients have been obtained. Fault analysis carried with the help of wavelet multiresolution analysis for the proposed system with and without UPFC. The performance of the system is studied during line to Ground (LG), Line-Line to Ground (LLG) and Line-Line-Line to Ground Fault (LLLG) at each zone. Coefficients are drawn and tabulated at each bus. Variation of fault index in each bus has analyzed. The analysis also made for different fault inception angles (FIA) and at different distances.

Network during LG Fault in Zone-2, as it has 25 km length, fault analysis has done at different distances and for different fault inception angles. From Table 3 it

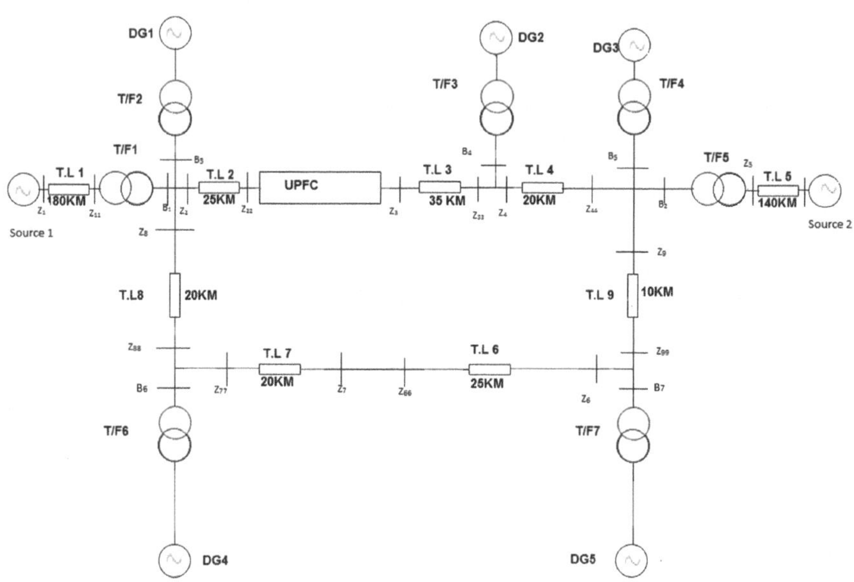

Fig. 2 One line diagram of a test system

Table 1 Test system data

Description	System parameters
Source of supply	**Source 1**: Rated voltage (Vrms) = 400 kV; phase angle of phase A (degrees): 20.2; 3-phase short-circuit level at base voltage(VA): 900 MVA, X/R ratio: 10
	Source 2: Rated voltage (Vrms) = 400 kV; phase angle of phase A (degrees): 20.2; 3-phase short-circuit level at base voltage (VA): 900 MVA, X/R ratio: 10
Distributed generators (DG)	**Generator (DG1)**: Rated power = 189 MVA Line-to-line voltage = 33 kV
	Generator (DG2): Rated power = 50 MVA Rated voltage (Vrms) = 33 kV
	Generator (DG3): Rated power = 200 MVA Rated voltage (Vrms) = 110 kV
	Generator (DG4): Rated power = 200 MVA Rated voltage (Vrms) = 110 kV
	Generator (DG5): Rated power = 150 MVA Rated voltage (Vrms) = 33 kV
Transformer	**Transformer 1**: Nominal power = 900 MVA 400/220 kV
	Transformer 2: Rated power = 200 MVA 33/220 kV
	Transformer 3: Rated power = 100 MVA 33/220 kV
	Transformer 4: Rated power = 100 MVA 33/220 kV
	Transformer 5: Rated power = 200 MVA 400/220 kV
	Transformer 6: Rated power = 200 MVA 110/220 kV
	Transformer 7: Rated power = 150 MVA 33/220 kV
Loads	Load 1: 100 MW Load 2: 25 MW Load 3: 125 MW Load 4: 200 MW Load 5: 100 MW

Table 2 Wavelet information and its associated parameters

Mother wavelet	Bior 1.5
Analyzed information	Detailed coefficients of currents at each bus and zones
Sampling frequency	180 kHz
Actual frequency	50 Hz
Number of samples for cycle	3600

Table 3 LG Fault on Transmission line in Zone-2 without UPFC

FIA		20°	40°	60°	80°
Zones/Phases	Distance (km)				
Zone-1: Phase a	36	1338.427	1342.89	1269.112	1218.884
	72	904.6976	891.4416	935.5863	942.6535
	108	1011.985	943.1618	956.172	940.1689
	144	982.827	838.4148	836.1602	764.5347
	180	921.3867	707.8778	711.8673	644.8466
Zone-1: Phase b	36	1289.998	1295.932	1250.634	1236.385
	72	899.4311	895.4739	804.5511	923.5474
	108	967.2835	947.2077	825.1693	921.0753
	144	938.1209	842.4705	705.167	745.4413
	180	876.7148	711.9523	580.876	553.6225
Zone-1: Phase c	36	181.0136	140.0609	138.8364	148.4208
	72	180.9751	140.0552	138.8456	148.3838
	108	180.944	139.9693	138.7754	148.3702
	144	180.9424	139.9916	138.8042	148.3896
	180	180.8912	139.9425	138.7891	148.3927
Zone-2: Phase a	5	5325.777	5942.902	6243.144	5881.985
	10	5286.246	5137.983	5927.662	6138.758
	15	4267.901	4297.411	5157.559	5000.209
	20	4161.902	4134.042	4384.349	4342.504
	25	3714.04	4040.369	4301.715	4074.849
Zone-2: Phase b	5	602.9014	532.8588	545.8582	520.0408
	10	403.5135	301.5525	280.0188	237.335
	15	226.4366	237.8998	189.8426	162.5452
	20	324.162	227.6833	103.7549	103.3488
	25	313.8367	362.1852	292.0248	357.6616
Zone-2: Phase c	5	609.0574	532.6295	554.8663	533.1901
	10	410.0847	306.9752	265.8487	243.2796
	15	241.3625	244.8704	192.8621	145.0435
	20	323.5173	223.5398	102.1582	102.1004
	25	313.1021	359.939	284.5708	354.1224
Zone-3: Phase a	4	232.0681	156.3246	190.3034	141.1947
	8	360.2451	296.6683	273.1152	199.9896
	12	302.7981	237.6175	167.804	148.7093
	16	690.8967	523.5822	389.8835	319.3736
	20	502.1092	769.6736	657.1128	706.9336

(continued)

Table 3 (continued)

FIA		20°	40°	60°	80°
Zones/Phases	Distance (km)				
Zone-3: Phase b	4	88.58852	104.1699	64.67646	52.09059
	8	109.1919	110.9402	115.9707	120.0759
	12	115.2178	136.103	146.1664	160.5939
	16	238.9	179.2664	175.1665	194.2011
	20	321.4526	342.6262	281.8157	314.3333
Zone-3: Phase c	4	83.82115	104.6752	60.75328	47.86299
	8	110.8297	110.1213	120.3456	122.1523
	12	120.4266	137.1244	143.6077	157.0114
	16	238.5705	180.0519	173.4791	192.4714
	20	329.9034	345.9209	290.1367	317.3911
Zone-4: Phase a	7	532.6241	348.9311	306.3755	231.5944
	14	453.5099	436.2159	401.0555	414.4105
	21	504.0367	374.652	220.8545	263.5287
	28	504.1185	460.9027	390.3962	270.782
	35	470.6284	457.345	489.7008	459.9936
Zone-4: Phase b	7	116.3134	95.57828	84.63895	95.03919
	14	157.872	99.26329	140.6984	161.0165
	21	193.4907	214.9099	198.6255	113.5626
	28	181.0133	177.6535	193.5296	246.8691
	35	469.7538	417.0691	341.1754	321.499
Zone-4: Phase c	7	112.0267	95.78919	88.52096	99.66043
	14	150.6757	96.09216	129.4061	153.7648
	21	176.6552	199.805	184.1064	92.92766
	28	167.3412	173.1193	188.3263	247.5583
	35	473.4147	411.2601	356.2149	326.3267
Zone-5: Phase a	28	385.0973	431.6954	424.9349	421.1367
	56	352.2491	397.267	379.3608	376.8893
	84	323.4669	372.754	362.7732	371.6509
	112	288.2593	333.1941	357.6824	377.2478
	140	289.3428	308.4371	348.8621	371.407
Zone-5: Phase b	28	452.928	514.9861	403.4833	465.1621
	56	423.5847	480.5515	383.8425	446.8402
	84	391.2431	456.0335	374.2237	448.576
	112	357.3886	420.3918	359.4948	440.6106
	140	338.5571	391.7091	330.9635	418.9813

(continued)

Table 3 (continued)

FIA		20°	40°	60°	80°
Zones/Phases	Distance (km)				
Zone-5: Phase c	28	151.6729	102.0236	56.68258	116.5944
	56	151.6744	102.0308	56.68669	116.589
	84	151.6738	102.0231	56.68036	116.5966
	112	151.6677	102.0247	56.68228	116.5964
	140	151.6595	102.0238	56.68666	116.6045
Zone-6: Phase a	5	355.8378	284.2098	308.5839	302.9133
	10	509.6473	387.7858	376.2189	351.3724
	15	481.2672	446.2837	371.9514	271.2653
	20	310.0239	374.683	324.5667	250.4457
	25	569.2888	452.3076	469.2662	497.2072
Zone-6: Phase b	5	104.7985	128.8168	164.0042	120.6892
	10	190.4917	149.4466	134.8459	137.1325
	15	171.6198	158.6567	190.097	180.7123
	20	270.3092	244.7032	195.9436	186.5358
	25	308.7939	336.5116	247.2021	238.5889
Zone-6: Phase c	5	119.6787	137.403	165.1748	129.3649
	10	185.7285	147.3814	140.9089	134.4484
	15	168.5095	158.5218	185.2016	172.2422
	20	254.5742	231.4679	196.2573	189.7845
	25	312.7276	348.5701	248.339	242.7853
Zone-7: Phase a	4	278.0128	194.5997	161.7302	136.9156
	8	242.1587	155.9738	99.49957	120.2239
	12	244.2018	216.4986	137.322	125.3612
	16	184.5326	191.2894	201.3837	195.461
	20	192.6291	256.2483	249.8618	166.0997
Zone-7: Phase a	4	141.53	125.3897	85.27911	70.59581
	8	99.79442	113.4803	81.65146	86.89787
	12	167.7856	128.9359	119.7982	86.10992
	16	188.6617	162.9766	145.5651	127.8336
	20	116.9862	163.2054	117.2235	130.4735
Zone-7: Phase a	4	142.0019	121.6928	85.24108	80.73674
	8	93.55404	112.1002	84.97488	85.07737
	12	170.7214	129.9813	119.1516	90.27449
	16	185.6097	155.0026	145.6415	135.3789

(continued)

Table 3 (continued)

FIA		20°	40°	60°	80°
Zones/Phases	Distance (km)				
	20	123.6192	166.6737	119.1306	142.1886
Zone-8: Phase a	4	1262.262	914.7941	1040.59	876.7483
	8	633.5293	633.4484	484.4508	478.497
	12	270.7869	274.7676	287.1616	269.3898
	16	372.8416	299.9607	187.8446	148.8582
	20	253.2181	258.3597	161.5231	250.8802
Zone-8: Phase b	4	738.9016	673.0562	616.0363	555.1654
	8	392.3671	290.8112	294.6176	266.7699
	12	128.0366	146.1465	145.7985	137.6743
	16	215.2371	177.3551	121.2848	111.7129
	20	148.8235	146.2723	135.7855	112.5016
Zone-8: Phase c	4	735.027	674.8348	610.1455	546.1279
	8	396.8299	295.9845	306.9234	264.4856
	12	132.1732	153.2031	152.1422	140.4799
	16	219.4116	181.2726	128.3871	117.422
	20	128.1035	130.2623	131.6589	105.929
Zone-9: Phase a	2	132.6797	103.8587	70.43439	82.68666
	4	126.2682	146.5265	116.1063	105.7688
	6	137.5003	128.1309	137.8318	94.23266
	8	117.4956	131.7681	121.1288	96.63817
	10	205.9835	103.1875	156.1563	119.3873
Zone-9: Phase b	2	50.86305	48.82548	42.18174	32.93884
	4	65.52933	63.07304	89.25785	64.62524
	6	107.8587	83.47089	79.89563	73.69413
	8	86.14352	84.50055	81.96124	71.09258
	10	92.11515	84.19993	81.10193	61.95255
Zone-9: Phase c	2	47.86158	48.32959	42.6038	30.40071
	4	67.15565	65.79728	89.63213	61.88395
	6	105.6843	82.48872	80.09079	70.90748
	8	83.99967	82.97947	82.16567	72.39437
	10	88.53601	81.7753	76.01193	62.87167

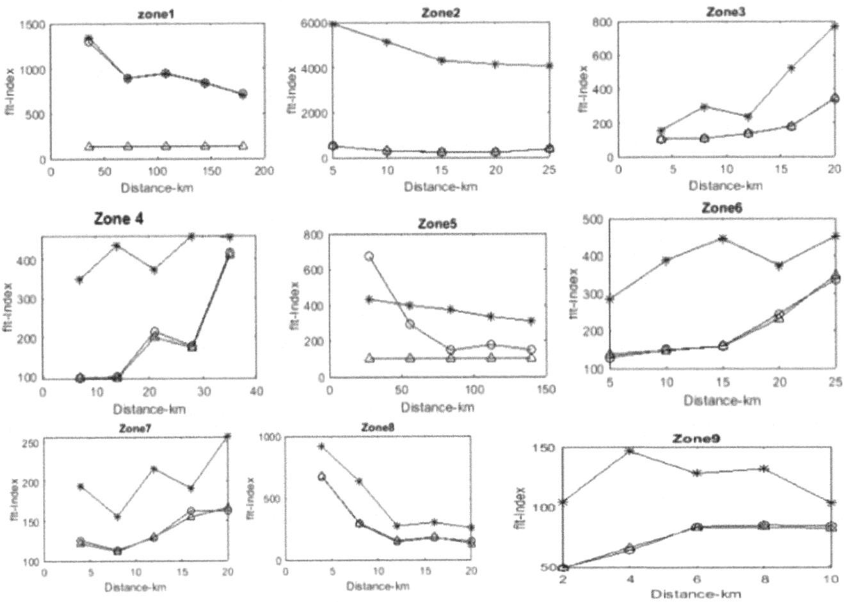

Fig. 3 Variation of fault index in all the zones without UPFC and AG fault in zone-2

is evident that coefficients are high in Zone-2 in phase A. The impact of Fault in Zone-2 is high, coefficients are high for phase A, Hence the fault is of LG type.

From Fig. 3 it clearly represents that the Fault Index value is high for Zone-2 and for Phase A. The compensation of fault can be partly achieved by connecting UPFC in between Zone-2 and Zone-3. Wavelet multiresolution analysis is performed by connecting UPFC between zone-2 and zone-3. The coefficients are taken at all the zones and buses. Loads 1, 2, 3, 4 & 5 are connected at Primaries side of DG's 1, 2, 3, and 4, which are not actually shown in one line diagram.

The compensation of fault current at Zone-2 has been shown clearly. For example, the detailed coefficient vale for AG fault at Zone-2 is at 800 is 6243.144, whereas its value is compensated to 4539.152. Thus UPFC has a considerable effect on Zone-2. Similarly, the coefficients during LG fault at FIA of 20° and at a distance of 5 km are at phase A, phase B, Phase C is 5325.777, 602.9014, 609.057, which shows this value is highest in Phase A. Therefore, it is an LG Fault (Fig. 4).

Figure 5 represents three variations of fault index due to LG Fault in Zone-2 without connecting UPFC. For understanding purpose variation of index values at Bus-1, 2, 3, and 7 have been shown. One phase current got increased much, which can be observed from the diagram.

For understanding purpose variation of index values at Bus-1, 2, 3, and 7 have been shown. Two-phase currents got increased much, which can be observed from Fig. 6 (Fig. 7).

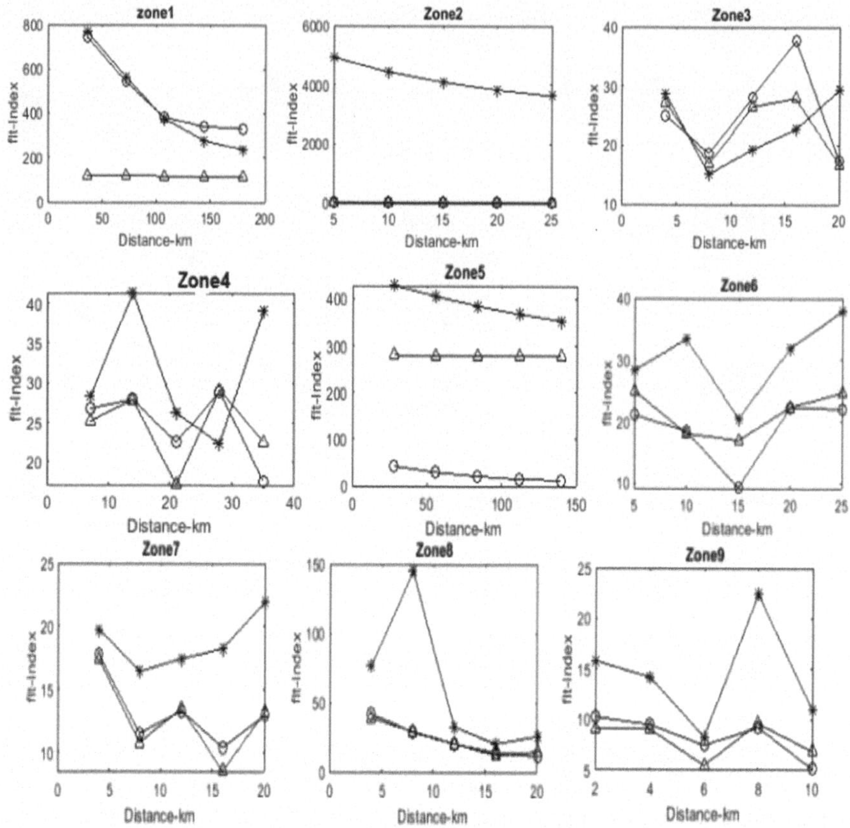

Fig. 4 Variation of fault index in different zones with UPFC and AG fault in zone-2

Figure 8 represents three the phase variations of fault index due to LLG Fault in Zone-2 with connecting UPFC. For understanding purpose variation of index values at Bus-1, 2, 3, and 7 have been shown. Two-phase currents index values variation at different buses has been analyzed, which can be observed from the diagram (Table 4).

Table 5 represents with the sum of detailed coefficients with LLLG in Zone-6, which clearly shows the impact of three-phase fault in Zone-6, the coefficients got increased. As UPFC is connected between Zone-2 and Zone-3, it has an impact on all the zones. The impact of UPFC on Zone-6 due to LLLG fault on Zone-6 can be tabulated in Table 4 (Table 6).

The impact also can be seen for different angles i.e., Fault inception angles. For understanding purposes, only one zone has been shown. Figure 10 represents the impact of UPFC on Zone-6 and Zone-7. From the above figures and tables, It is evident that interconnected networks there is an impact of fault in any Zone reflects fault current on other zones also. At the initial stage by applying fault at each zone, analyzed the detailed coefficients. The fault impact is high in the zone where the fault occurs, whereas there is an impact on other zones. This paper uses UPFC as a

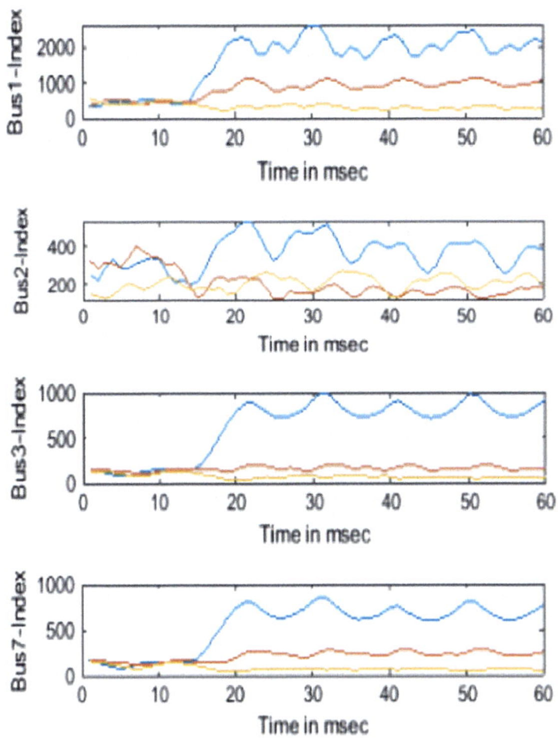

Fig. 5 Variation of fault index at buses 1, 2, 3, 7 due to LG fault in zone-2 without UPFC

Fig. 6 Represents three variation of fault index due to LLG fault in zone-2 without connecting UPFC

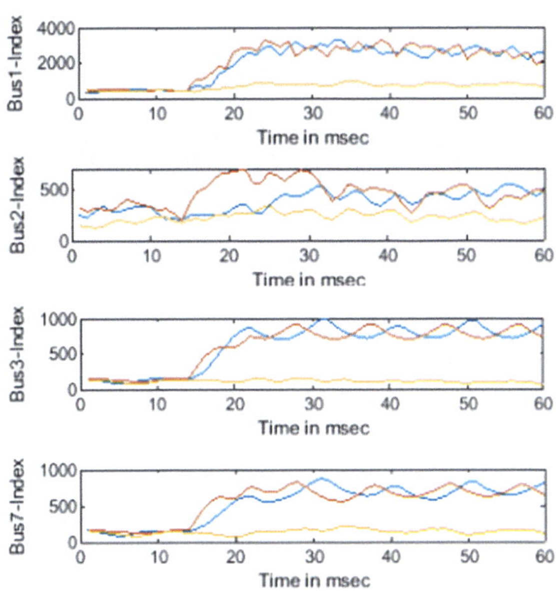

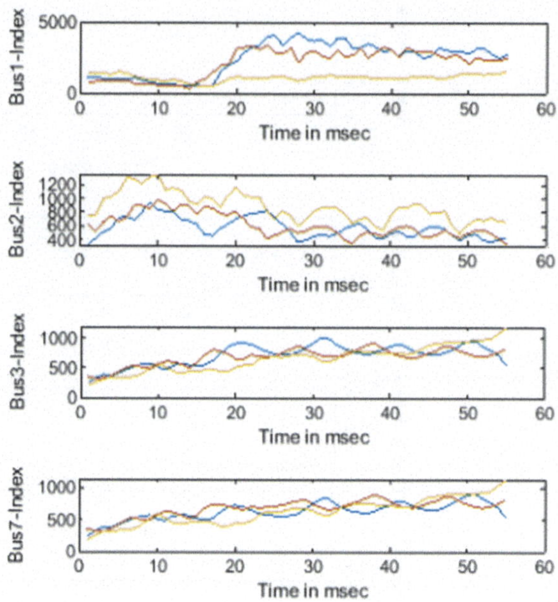

Fig. 7 Variation of fault index at bus 3, 7 due to LLG fault in zone-2 with UPFC

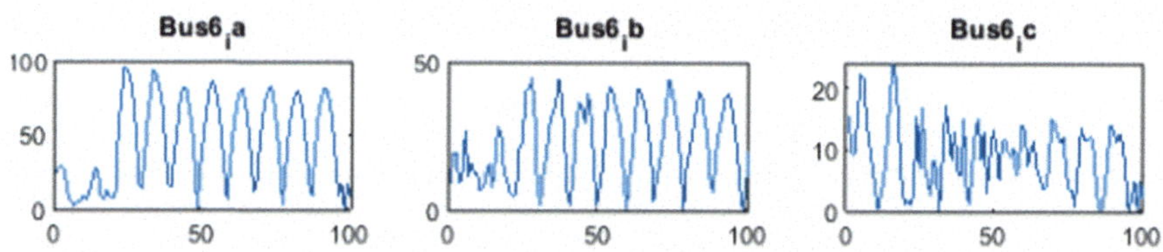

Fig. 8 Variation of effective coefficients of LG fault in zone-2 from terminal-6

compensating device. Even though UPFC is connected between Zone-2 and Zone-3. The fault currents are compensated up to a certain limit, therefore fault current has an impact on other zones too in presence of UPFC (Figs. 9 and 10).

6 Conclusions

The evaluation of dynamic behavior during transient conditions has been studied for multiterminal network. Wavelet analysis will give the entire system performance at any part of the system. This paper mainly concentrates on fault detection and its classification of multiterminal networks. It also gives a critical evaluation of nine zones of performance under different fault conditions. The performance of multiterminal transmission networks with and without UPFC under different fault conditions has been estimated. This algorithm successfully analyzed the different

Table 4 LG Fault on Transmission line in Zone-2 with UPFC

FIA		20°	40°	60°	80°
Zone-s/Phases	Distance (km)				
Zone-2: Phase a	5	4091.516	4920.359	5305.573	4539.152
	10	3914.136	4437.367	4778.623	4298.473
	15	3655.694	4091.654	4365.893	3974.26
	20	3404.021	3831.482	4033.917	3669.852
	25	3210.693	3636.321	3784.54	3429.58
Zone-2: Phase b	5	49.24836	35.88597	23.73107	25.81357
	10	30.64287	29.26032	22.79283	20.67209
	15	24.35378	17.32852	12.20013	18.5182
	20	62.90602	43.49543	42.31739	33.72605
	25	21.96338	19.1193	13.27504	15.52662
Zone-2: Phase c	5	47.61342	30.67786	30.75633	31.71313
	10	36.30296	28.69022	27.67097	23.68694
	15	21.05736	16.59344	19.5321	19.12578
	20	58.38437	42.06938	44.91864	34.02227
	25	21.49128	11.32597	15.00123	15.51111

Table 5 LLLG Fault on Transmission line in Zone-6 without UPFC

FIA		20°	40°	60°	80°
Zone-s/Phases	Distance (km)				
Zone-6: Phase a	5	3796.282	4681.449	5474.372	4144.664
	10	4824.178	4738.143	4644.604	4811.901
	15	4561.506	4749.508	4678.542	5166.384
	20	4307.706	5037.11	4611.306	4993.97
	25	4203.645	4586.955	4725.963	4972.756
Zone-6: Phase b	5	5134.595	4438.365	3978.692	3829.604
	10	5167.076	4423.533	3856.865	3841.574
	15	5185.629	4470.881	3921.195	3926.038
	20	5243.934	4500.099	3981.756	3971.895
	25	5291.249	4555.34	4030.181	3991.938
Zone-6: Phase c	5	3696.025	3339.338	3293.723	4253.44
	10	3879.838	2596.72	4352.824	4199.134
	15	4545.749	3553.078	5023.077	4540.078
	20	4572.546	4135.908	5219.165	4567.857
	25	4640.83	3900.611	4771.929	4486.894

Table 6 LLLG Fault on Transmission line in Zone-6 with UPFC

FIA		20°	40°	60°	80°
Zone-s/Phases	Distance (km)				
Zone-6: Phase a	5	3531.738	4833.433	4906.147	4210.159
	10	3522.45	4771.164	4840.717	4206.845
	15	2920.681	4001.9	4301.455	3937.241
	20	2776.039	3269.824	3662.967	3705.497
	25	2839.302	3291.158	3679.335	3702.119
Zone-6: Phase b	5	4699.737	4061.111	4130.061	3760.77
	10	4633.304	3947.565	4114.084	3749.02
	15	4725.281	3990.008	4126.915	3714.14
	20	5060.88	4217.889	4246.952	3660.499
	25	5447.277	4487.549	4430.945	3714.135
Zone-6: Phase c	5	2631.14	2913.362	4101.797	4427.076
	10	2933.219	2947.628	4327.078	4594.922
	15	3330.649	3040.203	4026.577	3976.125
	20	4252.276	3195.382	3713.94	3083.504
	25	5095.979	3896.149	4361.107	3567.025

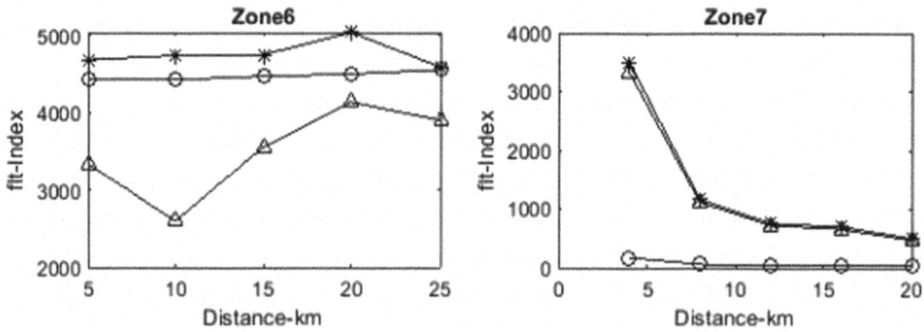

Fig. 9 Variation of fault index LLLG fault in zone-6 without UPFC

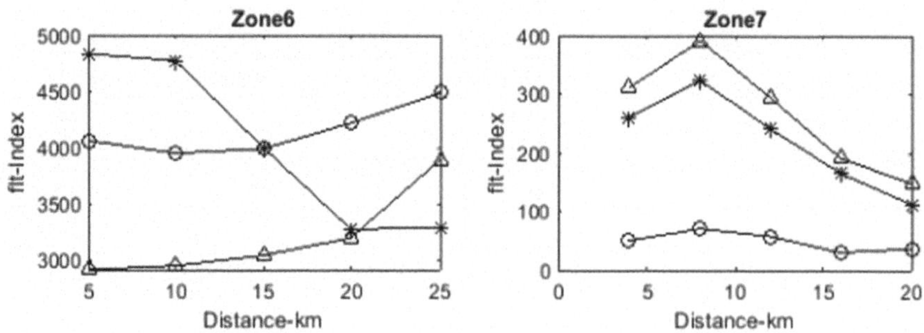

Fig. 10 Variation of fault index LLLG fault in zone-6 with UPFC

faults in all the zones. The proposed scheme is fast and accurate. Even it performs well at different fault inception angles. The effectiveness of the system is obtained by connecting UPFC between Zone-2 and Zone-3 has been evaluated more effectively. Wavelet-based multiresolution analysis is applied to multiterminal interconnected networks. Bior 1.5 is chosen as mother wavelet.

References

1. Sadeh J, Adinehzadeh A (2010) Accurate fault location algorithm for transmission line in the presence of series connected FACTS devices. Int J Elect Power Energy Syst 32:323–328
2. Samantaray SR, Dash PK, Upadhyay SK (2009) Adaptive Kalman filter and neural network based high impedance fault detection in power distribution networks. Int J Elect Power Energy Syst 31:167–172
3. Samantray SR, Dash PK (2008) Pattern recognition based digital relaying for advanced series compensated line. Int J Elect Power Energy Syst 30:102–212
4. Suja S, Jerome J (2010) Pattern recognition of power signal disturbances using S Transform and TT Transform. Int J Elect Power Energy Syst 32:37–53
5. He Z, Gao S, Chen X, Zhang J, Bo Z, Qian Q (2011) Study of a new method for power system transients classification based on wavelet entropy and neural network. Int J Elect Power Energy Syst 33:402–410
6. Pardha Saradhi J, Srinivasa Rao R, Ganesh V (2019) Wavelet based power quality assessment of wind energy source integrated 5-bus system under sudden load conditions in presence of FACTS devices. Int J Eng Adv Technol (IJEAT) 8(6S3). ISSN: 2249-8958
7. Goli RK, Shaik AG, Tulasi Ram SS (2015) A transient current based double line transmission system protection using fuzzy-wavelet approach in the presence of UPFC. Electr Power Energy Syst 70:91–98
8. El-Zonkoly AM (2011) Wavelet entropy based algorithm for fault detection and classification in FACTS compensated transmission line. Electr Power Energy Syst 33:1368–1374
9. Biswajit Sahoo SR, Samantaray (2018) Wavelet-based auto-reclosing technique for TSC compensated lines connecting wind farm, IEEE

Study of Fault Current Limiter—A Survey

K. B. Yadav, Alok Priyadarshi, Shiv Shankar, and Vishal Rathore

Abstract The requirement of electric energy is gaining pace in day-to-day life. Due to which, distributed energy source (DES) has been introduced in power sector. But while connecting DES with the distribution network this will increase the fault current level of the system [1]. Many protective devices like high impedance transformer, series reactor, switchgear, etc., are being used to limit the fault current which make the system costlier, cause low dependability and reduced operational flexibility [2]. As these devices are expensive, it is necessary to protect them from fault. A fault current limiter (FCL) is an alternative and flexible means of standard protective device, and it overcomes the problems occur due to increased fault current levels. This paper focuses on different types of FCL and its application in power system which are still in research.

Keywords High temperature superconductor · Superconducting FCL · Fault ride through · Total harmonic distortion

1 Introduction

Day by day, the requirement of electricity is increasing at a high rate, due to demand in surplus to the supply. Higher the demand higher is the power transfer which creates high fault current and increases the line losses which make the system unstable.

K. B. Yadav · A. Priyadarshi · S. Shankar (✉) · V. Rathore
Department of Electrical Engineering, NIT Jamshedpur, Jamshedpur, Jharkhand, India
e-mail: shiv.3122@gmail.com

K. B. Yadav
e-mail: kbyadav.ee@nitjsr.ac.in

A. Priyadarshi
e-mail: alokpriya4@gmail.com

V. Rathore
e-mail: vishalrathore01@gmail.com

M. N. Favorskaya et al. (eds.), *Innovations in Electrical and Electronic Engineering*, Lecture Notes in Electrical Engineering 661, https://doi.org/10.1007/978-981-15-4692-1_8

Many conventional current limiting equipments like circuit breaker, fuse, transformer, series reactor, etc., are employed to limit the levels of fault current. A fuse is a simple, small in size, and it can handle up to 200 KA, but it is undesirable as the entire power system is shut down on the occurrence of fault, and we have to replace the fuse manually every time. The issue while using the circuit breaker is that it does not operate instantly. The future concern is that the existing circuit breakers cannot interrupt the expected fault current levels. But these conventional devices will cause some problems such as loss of power system, increase in cost and power loss. These devices have large value of fault current, and these devices will activate after passing initial 2 or 3 fault current cycles [3]. There are some other methods for balancing the value of fault current such as by dividing the grid after the occurrence of fault, multibus running or sequential network tripping. But these methods will also create high power loss and lead to instability of the system. The stability and security issues are resolved by the application of fault current limiter. Different types of FCL are discussed in this paper which helps in limiting fault current levels and helps in improving the dynamic behaviour of the system. This study presents a detailed overview of superconducting and non-superconducting FCL applications in different networks of the power system. Current limiters have been applied in both distribution and transmission networks. Therefore, this paper reviews the background of FCL theoretically and its role in power system.

2 FCL and Its Benefits

The Kamerlingh Onnes a nobel laureate Dutch scientist discovered that the mercury has superconductivity property [4]. The resistance of mercury disappeared as the temperature decreases below certain level. Fault current limiter has the ability to limit the fault current and enhance the system stability. The impedance of FCL is high for fault condition whereas zero for unfaulted condition. FCL has the capability to reduce level of fault current within the first cycle of fault current as shown in Fig. 1. FCL is self-activating and limits fault current before first peak. FCL tolerant to a 5 cycle of fault current (100 ms).

2.1 Ideal Superconducting Fault Current Limiter [6]

An ideal FCL shows following properties:

- Under normal operation, it shows zero impedance in the system.
- When fault occurs, it inserts large impedance in the system.
- It operates within the first few cycles of the fault current.
- It returns to its normal state after limiting the fault current within the short interval of time.

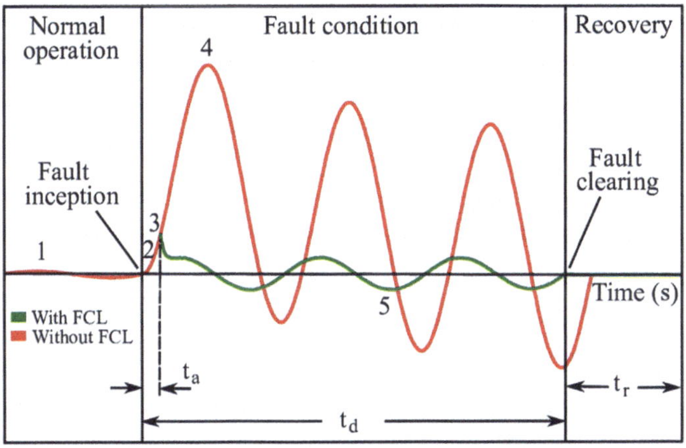

Fig. 1 Current waveform during fault condition with FCL and without SFCL [5]

- It does not affect any relay coordination.

3 Types of FCL

FCL is mainly divided in two categories which are used in power system [5, 7–15].

3.1 Superconducting

SFCL is used to limit the fault current level in place of conventional circuit breakers and is based on high temperature superconductor (HTS). Different types of SFCL are surveyed in this section with their merits and demerits are tabulated in Table 1.

Resistive type FCL

The resistive SFCL element is connected in series with the line, and it works on the principle of zero resistance as shown in Fig. 2. During normal operating condition, the value of current and temperature is less than critical values, and it shows small impedance to the current direction. While faulty condition, the values are higher than the critical value, and it shows large impedance to the current flow (while during faulty condition, the temperature rises due to large current density), this causes transition to occur which results in increase of current with the fault current, and hence, the current is limited. The main advantage of RSFCL is that its design is simple, compact and lightness when compared with other SFCL, and it has quenching problem when several units exist. In INDIA, ministry of power had initiated a national perspective plan for R&D developing 400 V, 800 A RSFCL in phase 1 and 11 kV, 1250 A in phase2 [16].

Table 1 Merits and demerits of different types of SFCL

Types of SFCL	Merits	Demerits	References
Resistive type SFCL	It has faster fault current limiting capability with automatic recovery Due to its simple structure, its cost is low, and size is small	For high voltage, it requires large superconductor Dissipated power and recovery time both are high It cannot be possible to achieve short recovery time by using neutral lines Simultaneously, quenching is not achieved as there is difference in critical current of different units.	[17–30]
Inductive type SFCL	Due to coreless structure, its weight is less, and size is compact	In standby mode, it has high loss due to leakage reactance It requires conventional circuit breaker in order to avoid HTS winding temperature	[31–34]
Hybrid type SFCL	Disadvantage of RSFCL that is quenching problem is removed here We need less number of superconductors for high voltage and current applications	If outage problem is too long, then we need liquid nitrogen	[30, 36–39]
Non-Inductive type SFCL	Its cost is low It takes less recovery time It has less AC losses It can hold high voltage	Its leakage inductance and circulating current are high Quantity of cryogenic is high	[40–42]
Transformer type SFCL	According to impedance ratio of the transformer, it can vary the fault current limiting range With the neutral lines, it can achieve the shortest recovery time	Load of power on SFCL is higher It has high current limiting time	[43]
Flux lock type SFCL	We can vary the operational current There is low load on superconductors	Due to its big size, its cost and weight are high	[44–47]

(continued)

Table 1 (continued)

Types of SFCL	Merits	Demerits	References
Magnetic shield ype SFCL	It does not require any fault detection circuit as magnetic body automatic heated when fault occurs Due to its turn ratio, it has greater flexibility Between SFCL and power network, it provides isolation	During normal operation, it experiences voltage drop It affects the operation of the devices due to its magnetic interference	[48–51]

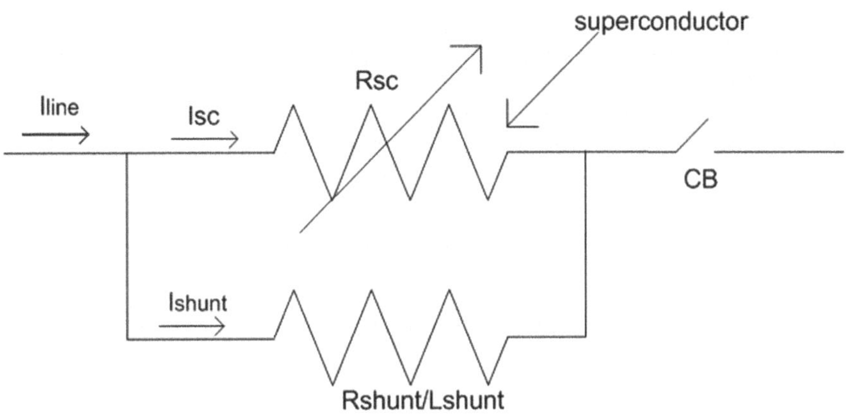

Fig. 2 Simple structure of resistive type SFCL

Inductive type SFCL

It is similar to the transformer where primary is copper winding which is magnetic coupled to superconducting secondary winding which has high temperature superconductor HTS) rings as shown in Fig. 3. Under normal condition, magnetic flux is prevented by HTS ring from entering the iron core due to which the impedance of SFCL is low. During fault condition, the ring becomes quenches, due to that, secondary acts as resistive and the inductance rise immediately, and in that case, the impedance of SFCL is high.

H-type SFCL

HSFCL has primary and some secondary winding as shown in Fig. 4. Superconducting resistive unit is connected in series with the secondary winding. The problem of quenching face by RSFCL is resolved by HSFCL. It improves the dynamic performance of the network and controls the fault current level. Current flows through the power system are without any loss during normal condition as resistance connected in series is zero. While during fault condition, superconducting elements quenched and make the current level low, so it has no effect under normal operating condition.

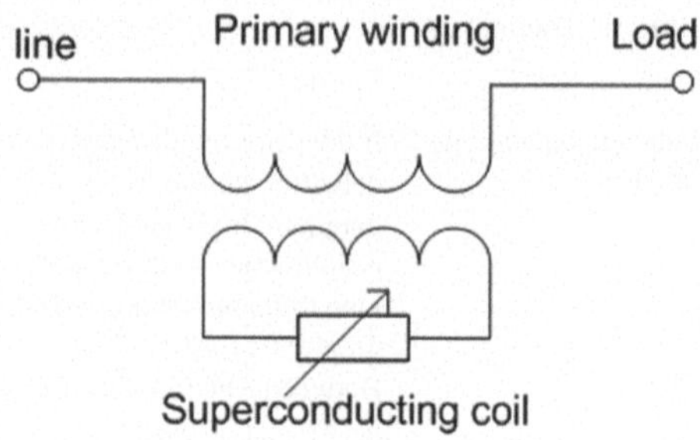

Fig. 3 Simple structure of inductive type SFCL

Fuji and Kansai electric power company jointly developed a 400 V hybrid current limiting device.

Non-Inductive type SFCL
The trigger coil and current limiting coils are the two coils of non-ISFCL which are connected in antiparallel manner and are magnetic coupled as shown in Fig. 5. Coaxial and bipolar are the different configurations of non-ISFCL in which bipolar is superior one with high impedance ratio.

Transformer type SFCL
Power system stability and increase in power transfer capacity have been seen by TSFCL. Load is connected with the primary side, whereas superconductor is connected with secondary side as shown in Fig. 6. Superconductors of secondary side are quenched in faulty condition which decreases the fault current to a lower value. And due to this, fault current of primary is also limited.

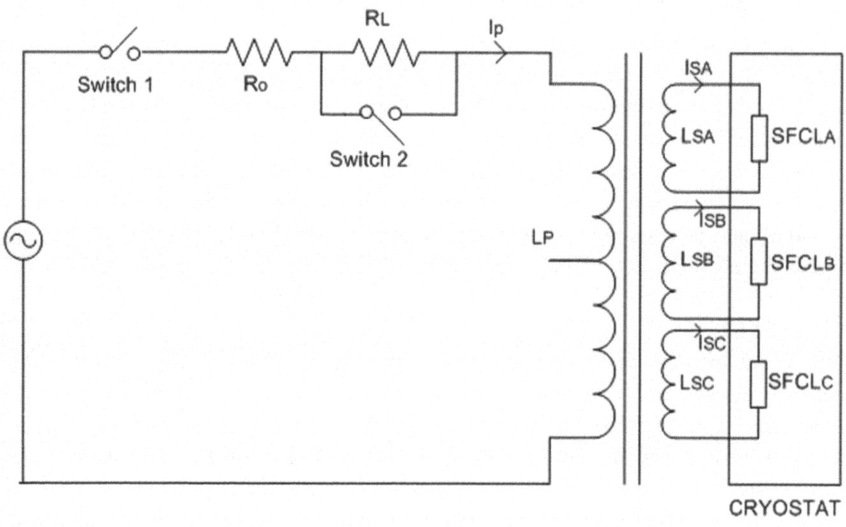

Fig. 4 Simple structure of hybrid type SFCL [35]

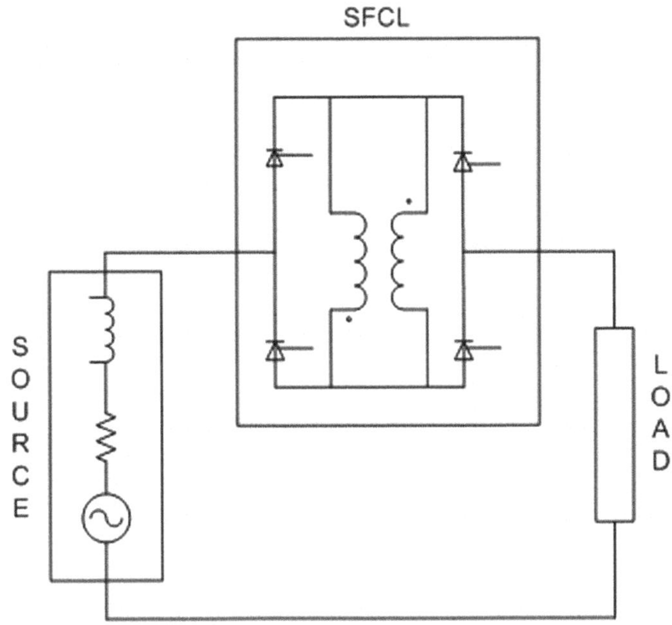

Fig. 5 Simple structure of non-inductive type SFCL

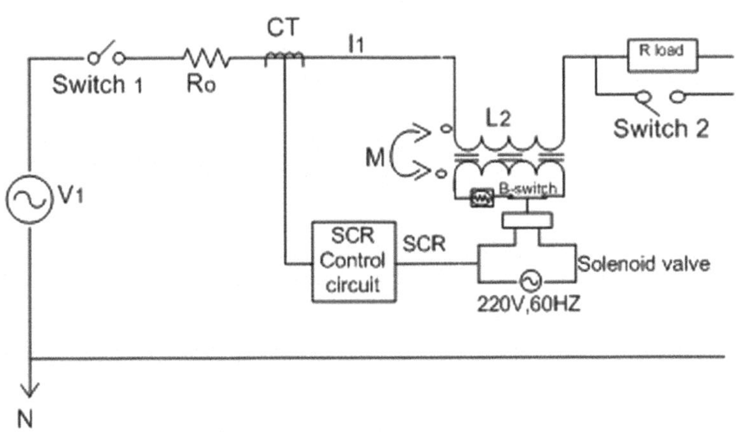

Fig. 6 Simple structure of transformer type SFCL [35]

Flux lock type SFCL

Current limiting and current interrupting are the two main parts of FSFCL as shown in Fig. 7. High temperature semiconductor and two parallel connected paths are in current limiting side, whereas interrupting path consists of over current relay, parallel path and circuit breaker. During normal condition, zero voltage developed across the coil as magnetic fluxes generated by both the coils are cancelled out. But during faulty condition, fault current is limited by voltage developed across the coil.

Magnetic shield type SFCL

This type of SFCL consists of primary copper coil and HTS tube as secondary which is wound around the magnetic coil as shown in Fig. 8. During a fault condition, the normal transition value of superconducting is increased as the current crosses the critical value of HTS element.

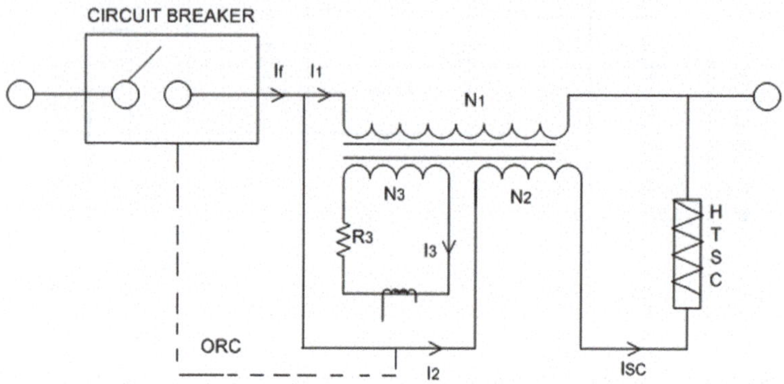

Fig. 7 Simple structure of flux lock type SFCL

Fig. 8 Simple structure of magnetic shield type SFCL

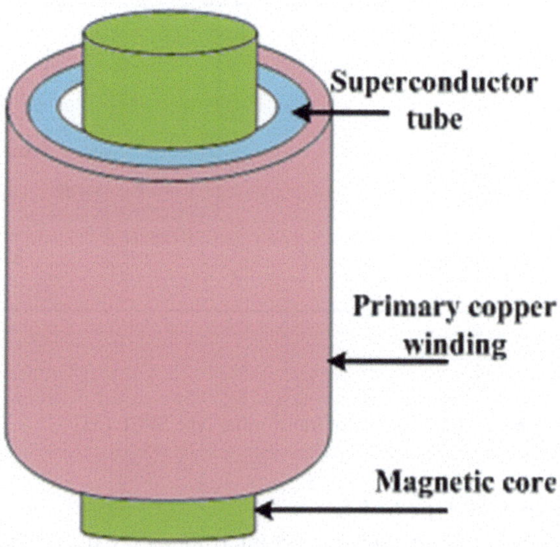

3.2 Non-Superconducting FCL

These types of FCL improve the stability of the power system with lesser cost in comparison with SFCL due to presence of non-superconducting elements. Various types of NSFCL are focused in this section, and their merits and demerits are shown in Table 2.

Series dynamic braking resistor

These types of limiters are mainly used for fault analysis in wind farms. It consists of switch in parallel with the resistor as shown in Fig. 9. On the occurrence of fault, the switch is turn ON or turn OFF. IGBT has fast response, and therefore, it is used as a switch. During normal operation, the braking resistor is bypassed as IGBT switch is turn ON. So, during normal condition, SDBR has no effect on the system. During faulty condition, the common coupling voltage decreases, and it is less than that of reference voltage. The IGBT is turn OFF, and resistor comes in series with the line

Table 2 Merits and demerits of different types of NSFCL

Types of NSFCL	Merits	Demerits	References
SERIES dynamic braking resistor	Negligible effect on the system under normal condition Used for fault detection in wind farms It can vary the resistance to limit the fault level	There is need of control strategy to improve the voltage which is applied to series resistance	[52, 53]
BRIDGE type fault current limiter	This type of SFCL utilize no quenching of superconductor It immediately recovers after the fault	It creates short circuit as one of the semiconductors fails It has high total losses	[54]
Modified bridge type fault current limiter (MBFCL)	It has improved behaviour during transient condition It prevents from sudden voltage drop at the machine	FRT may not be ensured	[57]
DC link fault current limiter (DLFCL)	It improves the FRT capability of the inverter-based distributed generation system The cost is lesser by using non-superconducting inductor	In faulty condition, it limits the inverter current	[59, 60]
Transformer coupled BFCL	It limits the short circuit current Using DC reactor instead SC reduces the cost	Short circuit current increases gradually	[62]

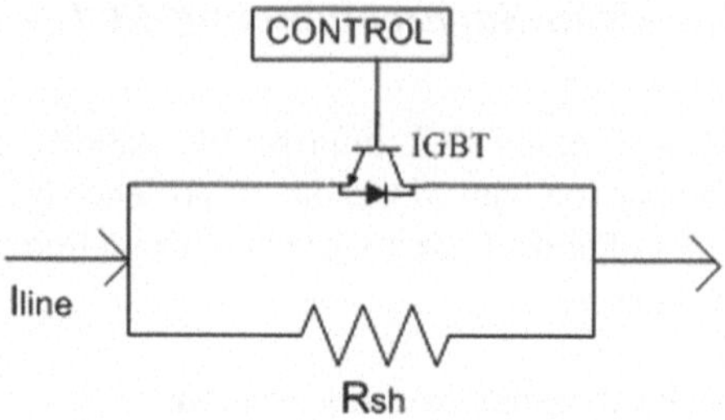

Fig. 9 Simple structure of series dynamic type SFCL

to limit the fault current. This process continues until common coupling voltage is greater than reference voltage and system returns to its normal operation.

Bridge type fault current limiter

BFCL mainly consists of two parts bridge and the shunt part as shown in Fig. 10. Insertion of resistance and inductance during the fault is the main function of BFCL. Under the faulty condition, it inserts resistance and inductance. It does not require any superconducting type of nature, and therefore, its implementation cost is less when compared with other superconducting limiter. The bridge part consists of diode rectifier, a limiting reactor (Ldc) superconductor switch (IGBT), a resistance (Rdc) and a freewheel diode. The bridge part is connected in parallel with the shunt part, and the shunt part consists of series connected resistance and inductance. During normal operation, current flows through the limiting reactor for positive and negative cycles. As the current flows through Ldc is in uniform manner and charge it to peak value of line current and hence it behaves like short circuit with zero voltage drop. Hence, it has no impact in the power system during normal operation. During faulty condition, the IGBT switch is off, and it behaves as open circuit. Therefore, shunt part of bridge comes into operation to limit the fault current.

Modified Bridge Type Fault Current Limiter (MBFCL)

The structure of bridge fault limiter is modified to enhance the voltage drop [55, 56]. This new model is known as modified bridge fault current limiter as shown in Fig. 11. In bridge limiter, resistor and inductor are series connected in shunt part. The inductor is removed in modified bridge as it discharges when the shunt part is disconnected. During normal operation, the resistance of shunt part is bypassed as the bridge of MBFCL is short circuited and the signal of IGBT is high. During faulty condition, shunt resistance is put into the line current to limit the fault current as the bridge of MBFCL is open circuited and a low signal is passed through IGBT.

DC Link Fault Current Limiter (DLFCL)

The DC link is developed for the fault ride through an inverter-based distributed generation system [58]. The non-superconducting DTFCL consists of diode bridge, a small resistance and an inductor as shown in Fig. 12. The impact of DC reactor is negligible during normal condition, whereas during faulty condition, reactor limits the high level fault current.

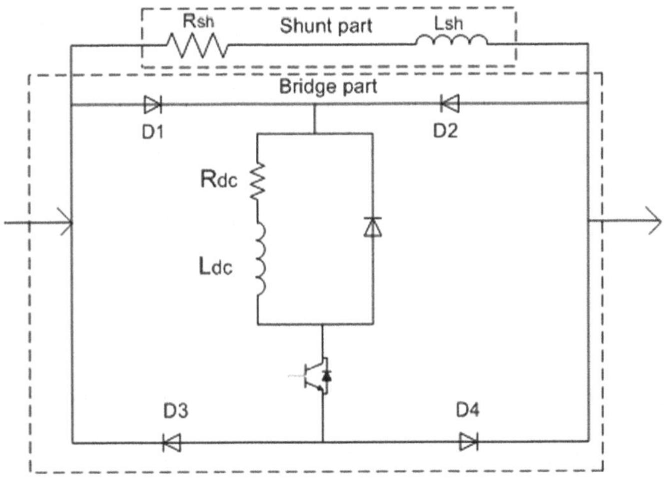

Fig. 10 Simple structure of bridge type limiter

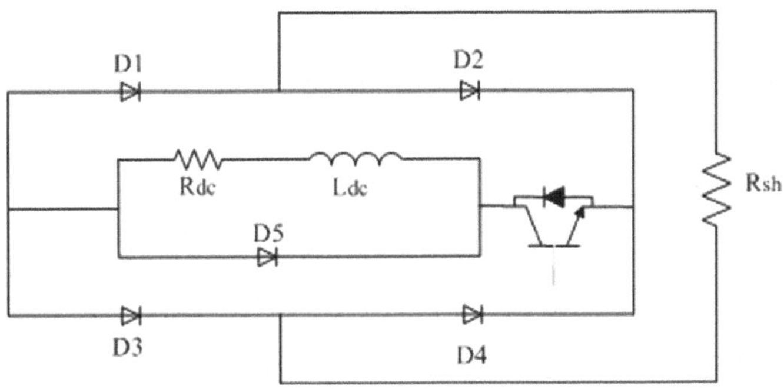

Fig. 11 Simple structure of modified BFCL

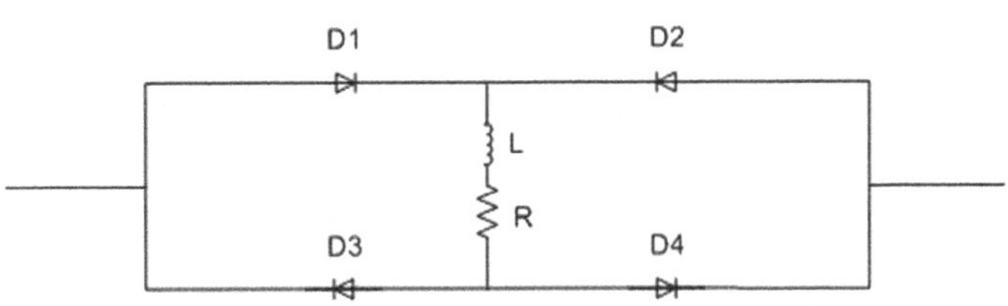

Fig. 12 Simple structure of direct current (DC) link FCL [35]

Transformer coupled BFCL

For the enhancement of low voltage ride through (LVRT) capability, TCBFCL is presented, and it is used for doubly feed induction generator [61]. It consists of diode bridge, a freewheeling diode connected in parallel with DC reactor, and this is

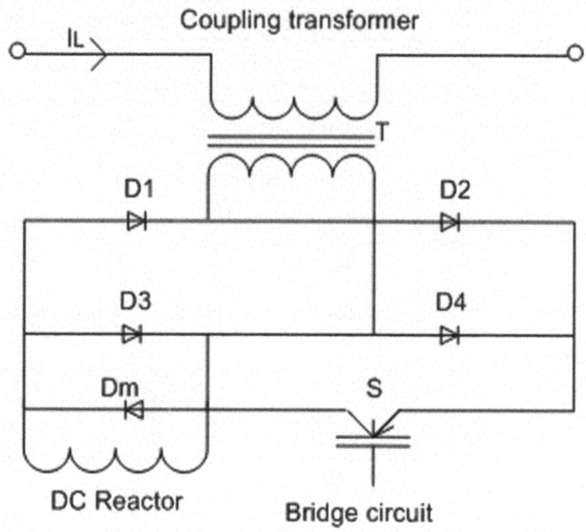

Fig. 13 Simple structure of transformer coupled bridge FCL

connected in series with IGBT switch as shown in Fig. 13. During normal, IGBT is in ON state, and current flows through the DC reactor, whereas during faulty condition, the IGBT is switched off by control circuit, and the bridge circuit is open circuited. Hence, the rise of fault current level is limited by the DC reactor.

Moreover, superconducting and non-superconducting type FCL are compared on the basis of different types of quality parameter and are shown in Table 3.

4 Conclusions

In recent years, FCL technology by superconducting and non-superconducting elements is increasing continuously. FCL provides protection of transmission and distribution elements as it limits the fault current. Lots of research are needed not only to build different FCLs but also how to use them in different operating conditions. FCL technologies continue to make progress in worldwide in limiting the fault current which is increasing due to interconnection of different power system networks. Nowadays, every grid is converting into smart grid, and for that, we need to interconnect different grids. For this, the study of various types of FCL with their advantages and disadvantages is presented in this paper. Finally, a comparative analysis between SFCL and NSFCL is performed which shows that NFSCL is more effective as compared to SFCL on the basis of quality parameters, but economically, SFCL is better. Further, this paper reviews that FCLs play a key role in power system protection and are an anticipated solution for high fault current problems.

Table 3 Comparison between superconducting and non-superconducting FCL

S. No.	Comparison		
	Item	Superconducting	Non-superconducting
1	Voltage distortion [63]	Less	More
2	Current distortion [63]	Less	More
3	THD of load voltage	More due to DC reactor	Less
4	THD of line current [64]	More due to DC reactor	Lesser
5	Feeder power less	More than NSFCL	Negligible
6	Size [15]	Big in size	Small in size
7	Weight [45]	Heavy weight	Less weight
8	Implementation feasibility [64–69]	Implemented in some of the countries	Some of are not implemented in real power network
9	Transient stability performance [64]	Better for other SFCL	Much better than BFCL
10	Structure complexity [61, 64, 70–77]	Highly complex structure	Its structure is simple to design
11	Critical clearing time [64]	The CCT value is less than NFCL	CCT value is high
12	Control circuit [78–81]	Most of them does not require any control circuit	Most of them require control circuit
13	Cost [63, 64, 66, 82–84]	Due to requirement of superconducting nature elements, its cost is more	Does not require non-superconducting elements, so its cost is lesser
14	Efficiency	Lesser than NSFCL	Higher than SFCL

References

1. Yadav S, Choudhary GK, Mandal RK (2014) Review on fault current limiters. Int J Eng 3:4 (2014); Clerk Maxwell J (1892) A treatise on electricity and magnetism, 3rd ed., vol. 2. Oxford, Clarendon, pp 68–73
2. Yadav S, Bharati K, Tewari V Superconducting fault current limiter-a review
3. Khan UA et al (2010) Feasibility analysis of the positioning of superconducting fault current limiters for the smart grid application using simulink and simpower system. IEEE Trans Appl Superconduct 21(3):2165–2169
4. Okakwu IK, Orukpe PE, Ogujor EA (2018) Application of superconducting fault current limiter (SFCL) in power systems: a review. Eur J Eng Res Sci 3(7):28–32
5. Roy F (2010) Modeling and characterization of coated conductors applied to the design of superconducting fault current limiters. No. THESIS. EPFL
6. Vaishnavi BV et al (2016) Superconducting fault current limiter and its application. Int J Sci Eng Res 7(5):126–134
7. Jo HC, Joo SK (2015) Superconducting fault current limiter placement for power system protection using the minimax regret criterion. IEEE Trans Appl Supercond 25
8. Blair SM, Elders IM, Booth CD, Burt GM, McCarthy J, Singh NK (2011) Superconducting fault current limiter application in a power-dense marine electrical system. IET Electr. Syst. Transp. 1:93–102

9. Kim MH, Kim J, You IK, Lim SH, Kim JC (2011) A study on practical impedance of superconducting fault curren tlimiter on bus tie in a power distribution system. J Int Counc Electr Eng 1:54–59
10. Lee J-G, Khan UA, Hwang J-S, Seong J-K, Shin W-J, Park B-B, Lee B-W (2014) Assessment on the influence of resistive superconducting fault current limiter in VSC-HVDC system. Phys. C Supercond. Appl. 504:163–166
11. Jo HC, Joo SK, Lee K (2013) Optimalplacementofsuperconductingfaultcurrentlimiters (SFCLs) for protection of an electric power system with distributed generations (DGs). IEEE Trans Appl Supercond 23:3–6
12. Ye LYL, Lin LLL, Juengst K-P (2002) Applicationstudiesofsuperconductingfaultcurrentlimitersinelectric power systems. IEEE Trans Appl Supercond 12:900–903
13. Hasan M, Rashid G (2016) Fault ride through capability improvement of DFIG based winds farm by fuzzy logic controlled parallel resonance fault current limiter. Electr Power Syst Res 146:1–8
14. Tarafdar MT, Jafari M, Naderi SB (2010) Transient stability improvement using non-superconducting fault current limiter. In: Proceedings of the 1st power electronic and drive systems and technologies conference (PEDSTC), Tehran, Iran, 17–18 February 2010, pp 367–370
15. Hossain ME (2017) Performance analysis of diode-bridge-type non-superconducting fault current limiter in improving transient stability of DFIG based variable speed wind generator. Electr Power Syst Res 143:782–793
16. Hagh MT, Abapour M (2009) Non superconducting fault current limiter with controlling the magnitudes of fault currents. IEEE Trans Power Electron 24:613–619
17. Dixit M et al (2013) Development of 440 V/800 a high temperature superconducting fault current limiter (SFCL). Indian J Cryogen 38(1–4):1–7
18. Sung BC, Park DK, Park JW, Ko TK (2009) StudyonaseriesresistivesFCLtoimprovepowersystemtransient stability: modeling, simulation, and experimental verification. IEEE Trans Ind Electron 56:2412–2419
19. Moghadasi A, Sarwat A, Guerrero JM (2016) Multiobjective optimization in combinatorial wind farms system integration and resistive SFCL using analytical hierarchy process. Renew. Energy 94:366–382
20. Sung BC, Park J (2010) Optimal parameter selection of resistive SFCL applied to a power system using eigenvalue analysis. IEEE Trans Appl Supercond 20:1147–1150
21. Ahn MC, Park DK, Yang SE, Kim MJ, Chang HM, Yoon YS, Seok BY, Park JW, Ko TK (2007) Recovery characteristics of resistive SFCL wound with YBCO coated conductor in a power system. IEEE Trans Appl Supercond 17:1859–1862
22. Zou ZC, Xiao XY, Liu YF, Zhang Y, Wang YH (2016) Integrated protection of DFIG-based wind turbine with a resistive-type SFCL under symmetrical and asymmetrical faults. IEEE Trans Appl Supercond 26:1
23. Zou ZC, Xiao XY, Ou R, Li CS (2015) Low-voltage ride-through capability enhancement of DFIG-based wind turbine with a resistive-type SFCL connected in series with rotor winding. Proc IEEE Int Conf Appl Superconduct Electromagn Dev Shanghai, China 20–23:42–43
24. Morandi A, Imparato S, Grasso G, Berta S, Martini L, Bocchi M, Fabbri M, Negrini F, Ribani PL (2010) Design of a DC resistive SFCL for application to the 20 kV distribution system. IEEE Trans Appl Supercond 20:1122–1126
25. Kim H, Lee JY, Kim HR, Yang SE, Yu SD, Kim WS, Hyun OB, Ko J, Yeom H (2013) An effect of HTS wire configuration on quench recovery time in a resistive SFCL. IEEE Trans Appl Supercond 23:7–10
26. Zhu J, Zheng X, Qiu M, Zhang Z, Li J, Yuan W (2015) Application simulation of a resistive type superconducting fault current limiter (SFCL) in a transmission and wind power system. In: Proceedings of the 7th international conference on applied energy, Abu Dhabi, UAE, 28–31 April 2015, pp 716–721
27. Didier G, Bonnard CH, Lubin T, Leveque J (2015) Comparison between inductive and resistive SFCL in terms of current limitation and power system transient stability. Electr Power Syst Res 125:150–158

28. Mafra GRFQ, Sotelo GG, Fortes MZ, Sousa WTBD (2017) Application of resistive superconducting fault current limiters in offshore oil production platforms. Electr Power Syst Res 144:107–114
29. Behzad S, Negnevitsky M, Jalilian A, Tarafdar M, Muttaqi KM (2017) Low voltage ride-through enhancement of DFIG-based wind turbine using DC link switchable resistive type fault current limiter. Electr Power Energy Syst 86:104–119
30. Lee S, Yoon J, Lee B (2010) Analysis model development and specification proposal of hybrid superconducting fault current limiter (SFCL). Phys C Supercond Appl 470:1615–1620
31. Firouzi M, Gharehpetian GB, Mozafari B (2016) Improvement of powersystem stability by using new switching technique in bridge-type fault current limiter. Electr Power Comp Syst 43:234–244
32. Majka M, Kozak J, Kozak S, Wojtasiewicz G, Janowski T (2015) Design and numerical analysis of the 15 kV class coreless inductive type SFCL. IEEE Trans Appl Supercond 25:1
33. Kozak J, Majka M, Kozak S, Janowski T (2013) Comparison of inductive and resistive SFCL. IEEE Trans Appl Supercond 23:6–9
34. Chen L, Chen H, Shu Z, Zhang G, Xia T, Ren L (2016) Comparison of inductive and resistive SFCL to robustness improvement of a VSC-HVDC system with wind plants against DC fault. IEEE Trans Appl Supercond 26:1
35. Naderi SB, Jafari M, Tarafdar Hagh M (2014) Controllable resistive type fault current limiter (CR-FCL) with frequency and pulse duty-cycle. Int J Electr Power Energy Syst 61:11–19
36. Alam MS, Abido MA, El-Amin I (2018) Fault current limiters in powersystems: a Comprehensive Review. Energies 11(5):1025
37. Choi HS, Cho YS, Lim SH (2006) Operationalcharacteristicsofhybrid-typeSFCLbythenumberofsecondary windings with YBCO films. IEEE Trans Appl Supercond 16:719–722
38. De SK, Raja P (2013) A study on relay coordination in a distribution system with distributed generation and hybrid SFCL. Proc. IEEE AFRICON Conf. Pointe-Aux-Piments, Maurit. 9–12:1–6
39. Kim WS, Hyun OB, Park CR, Yim SW, Yu SD, Yang SE, Kim HS, Kim HR (2012) Dynamic characteristics of a 22.9 kV hybrid SFCL for short-circuit test considering a simple coordination of protection system in distribution networks. IEEE Trans Appl Supercond 22:3–6
40. Kim MJ, Chang HM, Sim J, Yim SW, Hyun OB (2011) Emergency blackout operation of cryogenic system for hybrid SFCL. IEEE Trans Appl Supercond 21:1284–1287
41. Hoshino T, Muta I, Nakamura T, Salim KM, Yamada M (2005) Non-inductive variable reactor design and computer simulation of rectifier type superconducting fault current limiter. IEEE Trans Appl Supercond 15:2063–2066
42. Liang F, Yuan W, Zhu J, Zhang M, Venuturumilli S, Li J, Patel J, Zhang G (2017) Experimental test of two types of non-inductive solenoidal coils for superconducting fault current cimiters use. IEEE Trans Appl Supercond 27:1505–1509
43. Furuse M, Yamasaki H, Manabe T, Sohma M, Kondo W, Yamaguchi I, Kumagai T, Kaiho K, Arai K, Nakagawa M (2007) Current limiting properties of MOD-YBCO thin films stabilized with high-resistivity alloy shunt layer. IEEE Trans Appl Supercond 17:3479–3482
44. Choi HS, Lee JH, Cho YS, Park HM (2009) Recovery behaviors of the transformer-type SFCL with or without neutral lines. IEEE Trans Appl Supercond 19:1793–1796
45. Ko S, Lim S (2016) Analysis on magnetizing characteristics due to peak fault current limiting operation of a modified flux-lock-type SFCL with two magnetic paths. IEEE Trans Appl Supercond 26:4–8
46. Ko SC, Han TH, Lim SH (2013) Study on peak current limiting characteristics of a flux-lock type SFCL with two magnetically coupled circuits. Phys. Procedia 45:305–308
47. Han TH, Ko SC, Lim SH (2013) Current limiting characteristics of a flux-lock type SFCL using two triggered HTSC elements. Phys. Procedia 45:297–300
48. Zhao Y, Saha TK, Krause O, Li Y (2015) Performance analysis of resistive and flux-lock type SFCL in electricity networks with DGs. In: Proceedings of the IEEE power and energy society general meeting, Denver, CO, USA, 22–26 July 2015, pp 1–5

49. Onishi T, Kawasumi M, Sasaki KI, Akimoto R (2002) An experimental study on a fast self-acting magnetic shield type superconducting fault current limiter. IEEE Trans Appl Supercond 12:868–871
50. Heydari H, Abrishami AA, Bidgoli MM (2013) Comprehensive analysis for magnetic shield superconducting fault current limiters. IEEE Trans Appl Supercond 23:5604610
51. Kado H, Ickikawa M (1997) Performance of a high-Tc superconducting fault current limiter-design of a 6.6 kV magnetic shielding type superconducting fault current limiter. IEEE Trans Appl Supercond 7:993–996
52. Janowski T, Kozak S, Malinowski H, Wojtasiewicz G, Kondratowicz-Kucewicz B, Kozak J (2003) Properties comparison of superconducting fault current limiters with closed and open core. IEEE Trans Appl Supercond 13:2072–2075
53. Shafiul AM et al (2018) Protection of inverter-based distributed generation with series dynamic braking resistor: a variable duty control approach. In: 2018 10th international conference on electrical and computer engineering (ICECE). IEEE
54. Okedu Kenneth E (2016) Enhancing DFIG wind turbine during three-phase fault using parallel interleaved converters and dynamic resistor. IET Renew Power Gener 10(8):1211–1219
55. Noe Mathias, Steurer Michael (2007) High-temperature superconductor fault current limiters: concepts, applications, and development status. Supercond Sci Technol 20(3):R15
56. Rashid G, Ali MH (2017) Nonlinear control-based modified BFCL for LVRT capacity enhancement of DFIG based wind farm. IEEE Trans Energy Convers 32:284–295
57. Rashid G, Ali MH (2014) Bridge-type fault current limiter for asymmetric fault ride-through capacity enhancement of doubly fed induction machine based wind generator. In: Proceedings of the 2014 IEEE energy conversion congress and exposition, Pittsburgh, PA, USA, 14–18 September 2014; pp 1903–1910
58. Rashid G, Ali MH (2014) A modified bridge-type fault current limiter for fault ride-through capacity enhancement of fixed speed wind generator. IEEE Trans Energy Convers 29(2):527–534
59. Abapour M, Jalilian A, Hagh MT, Muttaqi KM (2015) DC-link fault current limiter-based fault ride-through scheme for inverter-based distributed generation. IET Renew Power Gen 9:690–699
60. Tarafdar HM et al (2015) Improving fault ride-through of three phase voltage source inverter during symmetrical fault using DC link fault current limiter. In: 2015 Australasian Universities power engineering conference (AUPEC). IEEE
61. Jalilian A, Hagh MT, Abapour M, Muttaqi KM (2015) DC-link fault current limiter-based fault ride-through scheme for inverter-based distributed generation. IET Renew Power Gen 9(6):690–699
62. Guo W, Xiao L, Dai S, Xu X, Li Y, Wang Y (2015) Evaluation of the performance of BTFCLs for enhancing LVRT capability of DFIG. IEEE Trans Power Electron 30:3623–3637
63. Firouzi M, Gharehpetian GB (2017) A modified transformer-type fault current limiter for enhancement fault ride-through capability of fixed speed-based wind power plants. In: 2017 conference on electrical power distribution networks conference (EPDC). IEEE
64. Tarafdar HM, Abapour M (2009) Non-superconducting fault current limiters. Eur Trans Electr Power 19(5):669–682
65. Radmanesh H, Fathi SH, Gharehpetian GB (2015) Novel high performance DC reactor type fault current limiter. Electr Power Syst Res 122:198–207
66. Xue S, Gao F, Sun W, Li B (2015) Protection principle for a DC distribution system with a resistive superconductive fault current limiter. Energies 8:4839–4852
67. Xin Y, Gong WZ, Sun YW, Cui JB, Hong H, Niu XY, Wang HZ, Wang LZ, Li Q, Zhang JY, Wei ZQ (2012) Factory and field tests of a 220 kV/300 MVA statured iron-core superconducting fault current limiter. IEEE Trans Appl Superconduct 23(3):5602305
68. Hyun OB, Yim SW, Yu SD, Yang SE, Kim WS, Kim HR, Lee GH, Sim J, Park KB (2011) Long-term operation and fault tests of a 22.9 kV hybrid SFCL in the KEPCO test grid. IEEE Trans Appl Superconduct 21(3):2131–2134

69. Kim HR, Yang SE, Yu SD, Kim H, Kim WS, Park K, Hyun OB, Yang BM, Sim J, Kim YG (2012) Installation and testing of SFCLs. IEEE Trans Appl Supercond 22:704–707
70. Rashid G, Ali MH (2016) Nonlinear control-based modified BFCL for LVRT capacity enhancement of DFIG-based wind farm. IEEE Trans Energy Convers 32(1):284–295
71. Jung BI, Choi HW, Choi HS (2015) Reduction of the power burden of a transformer-type SFCL using a vacuum interrupter. IEEE Trans Appl Supercond 25:4–7
72. Kim JS, Lim SH, Kim JC (2010) Study on protection coordination of a flux-lock type SFCL with over-current relay. IEEE Trans Appl Supercond 20:1159–1163
73. Ji T, He X, Li X, Liu K, Zhang M (2014) Performance analysis and research on LVRT of PMSG wind power systems with SDBR. In: Proceedings of the 33rd Chinese control conference 2014 Jul 28, IEEE, pp 6953–6958
74. Okedu KE, Muyeen SM, Takahashi R, Tamura J (2012) Wind farms fault ride through using DFIG with new protection scheme. IEEE Trans Sustain Energy 3:242–254
75. Shawon MH, Al Durra A, Caruana C, Muyeen SM (2012) Small signal stability analysis of doubly fed induction generator including SDBR. In: 2012 15th international conference on electrical machines and systems (ICEMS) 2012 Oct 21. IEEE, pp 1–6
76. Okedu KE (2016) Enhancing DFIG wind turbine during three-phase fault using parallel interleaved converters and dynamic resistor. IET Renew Power Gener 10:1211–1219
77. Ali MH, Hossain MM (2015) Transient stability improvement of doubly fed induction generator based variable speed wind generator using DC resistive fault current limiter. IET Renew Power Gener 18:803–809
78. Hussein AA, Hasan Ali M (2016) Comparison among series compensators for transient stability enhancement of doubly fed induction generator based variable speed wind turbines. IET Renew Power Gener 10:116–126
79. Mardani M, Fathi SH (2015) Fault current limiting in a wind power plant equipped with a DFIG using the interface converter and an optimized located FCL. In: The 6th power electronics, drive systems and technologies conference (PEDSTC2015) 2015 Feb 3. IEEE, pp 328–333
80. Zhao Y, Krause O, Saha TK, Li Y (2013) Stability enhancement in distribution systems with DFIG-based wind turbine by use of SFCL. In: 2013 Australasian Universities power engineering conference (AUPEC) 2013 Sep 29, IEEE, pp 1–6
81. Chen L, Zheng F, Deng C, Li Z, Guo F (2015) Fault ride-through capability improvement of DFIG-based wind turbine by employing a voltage-compensation-type active SFCL. Can J Electr Comput Engs 38:132–142
82. Alam MS, Abido MAY (1898) Fault ride-through capability enhancement of voltage source converter-high voltage direct current systems with bridge type fault current limiters. Energies 2017:10
83. Alam MS, Hussein A, Abido MA, Al-Hamouz ZM (2017) VSC-HVDC system stability augmentation with bridge type fault current limiter. In: 2017 6th international conference on clean electrical power (ICCEP), IEEE, pp 531–535
84. Nourmohamadi H, Nazari-Heris M, Sabahi M, Abapour M (2017) A novel structure for bridge-type fault currentlimiter: capacitor based non superconducting FCL. IEEE Trans Power Electron 33:3044–3051

PLL Less Control Algorithm for Power Quality Improvement in Distribution System Using D-STATCOM

Atma Ram, P. R. Sharma, Rajesh Kr. Ahuja, and Nitin Goel

Abstract In this paper, the phase-locked loop less (PLL) control technique has been investigated using four legs VSC-based D-STATCOM for unity power factor (UPF) mode. It is implemented using the MATLAB/Simulink environment. Further without PLL, the performance of D-STATCOM is measured graphically. The proposed technique is suitable for reactive power compensation, source current harmonic elimination and load balancing with linear and nonlinear load. The observation of simulated results of D-STATCOM provides the effective solution for power quality enhancements at steady-state and dynamic load.

Keywords Power quality · Voltage source converter (VSC) · Neutral current compensation · UPF controller

1 Introduction

Day by day, the demand of nonlinear load increases in the distribution system [1]. Various types of nonlinear loads are power electronics-based domestic equipments like computer, LED/LCD, TVs, lighting system and domestic load with SMPS, battery charging circuits, etc., which injects harmonics in the system. This type

A. Ram (✉) · P. R. Sharma · R. Kr. Ahuja · N. Goel
J. C. Bose University of Science and Technology, YMCA, Faridabad, India
e-mail: atma.ram12@gmail.com

P. R. Sharma
e-mail: prsharma1966@gmail.com

R. Kr. Ahuja
e-mail: rajeshkrahuja@gmail.com

N. Goel
e-mail: goel.nitin81@gmail.com

M. N. Favorskaya et al. (eds.), *Innovations in Electrical and Electronic Engineering*, Lecture Notes in Electrical Engineering 661, https://doi.org/10.1007/978-981-15-4692-1_9

of load creates disturbance in distribution system like harmonics and DC offset in system [2, 3]. These harmonics create problems like distortion of bus voltage of the system, takes more amount of reactive power, increases in neutral current, etc. Due to abovementioned reasons, some standards are made on total harmonics of supply current [4]. D-STATCOM is used for minimizing such type of problems. Various types of technique have been developed for control of D-STATCOM [5–8]. The problems explained earlier for which the performance of D-STATCOM will depend upon the control techniques like SRF, IRPT, etc. This paper addresses without PLL control mechanism for four legs VSC-based D-STATCOM with self-regulated DC link voltage. This technique is used for source current harmonics elimination, load balancing and reactive power compensation in the distribution system.

2 Proposed Model

Three-phase four-wire distribution system with four legs VSC-based D-STATCOM is shown in Fig. 1, with linear/nonlinear load. Z_s $(R_s + L_s)$ is the source impedance. L_s and L_f are source inductance and coupling inductance, respectively, connected to AC side of D-STATCOM for harmonic reduction from its output currents. Output currents (I_{Ca}, I_{Cb}, I_{Cc} and I_{Cn}) are injected for reduction harmonics from load currents.

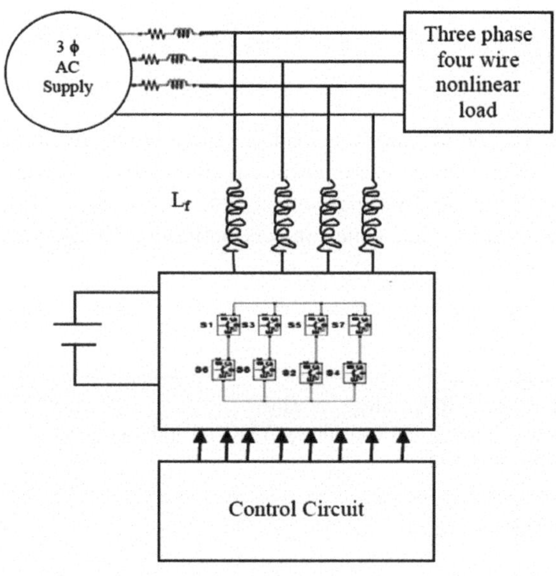

Fig. 1 Proposed D-STATCOM with nonlinear load

3 Proposed Controller

Without PLL, control mechanism used in this paper for generation of reference source currents is shown in Fig. 2. In this control technique, grid voltages (V_{sa}, V_{sb}, V_{sc}), grid currents (I_{sa}, I_{sb}, I_{sc}) and DC link voltage (V_{dc}) of D-STATCOM are taken as feedback signals. Same phase unit templates of grid voltages (V_{sa}, V_{sb} and V_{sc}) are taken as U_{pa}, U_{pb} and U_{pc} and are computed as follows [9].

Magnitude of grid voltage is as

$$V_{PCC} = \left(\frac{2}{3}\left(V_{sa}^2 + V_{sb}^2 + V_{sc}^2\right)\right)^{1/2} \tag{1}$$

where V_{sa}, V_{sb} and V_{sc} are grid phase voltages. The unit template of grid voltages (V_{sa}, V_{sb} and V_{sc}) is taken as

$$U_{pa} = \frac{V_{sa}}{V_{pcc}}, U_{pb} = \frac{V_{sb}}{V_{pcc}}, U_{pc} = \frac{V_{sa}}{V_{pcc}} \tag{2}$$

V_{dc} output DC link voltage of D-STATCOM is taken and passes through LPF to eliminate DC component present in it, and this filter output is subtracted from reference V_{dc}^* to generate an error signal. It is given to discrete proportional–integral (PI) controller to keep DC link voltage constant at changing load conditions. DC link

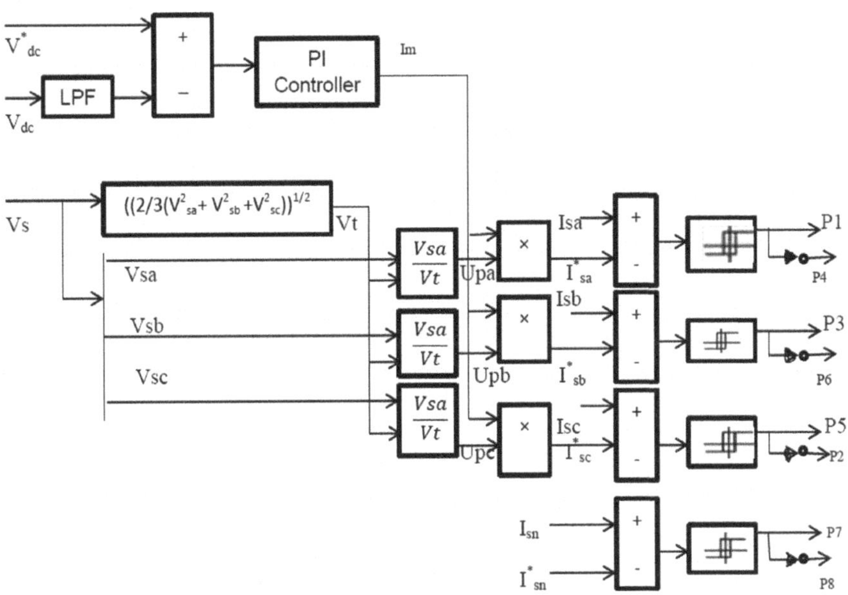

Fig. 2 Reference supply current extraction

error voltage $V_{dc(n)}$ at nth instant is given by equation as below.

$$V_{dc(n)} = V_{dc}^{*} - V_{dc} \quad (3)$$

Output of this discrete PI controller at nth instant is given by

$$I_{mn} = I_{m(m-n)} + K_{pdc}\left(V_{dc\,errm(n)} - V_{dc\,err(n-1)}\right) + K_{Idc} V_{dc\,errm} \quad (4)$$

where
K_{pdc} = discrete PI controller proportional gain
K_{Idc} = discrete PI controller integral gain
Reference currents are taken at any instant.

$$I_{sa}^{*} = I_m \times U_{pa} \quad (5)$$

$$I_{sb}^{*} = I_m \times U_{pb} \quad (6)$$

$$I_{sc}^{*} = I_m \times U_{pc} \quad (7)$$

For the fourth leg, reference current I_{sn}^{*} which is neutral current is extracted by adding grid currents and given by

$$I_{sn}^{*} = -(I_{sa} + I_{sb} + I_{sc}) \quad (8)$$

To calculate the current error signal, reference currents $\left(I_{sn}^{*}, I_{sn}^{*}, I_{sn}^{*} \text{ and } I_{sn}^{*}\right)$ are subtracted from actual grid currents (I_{sa}, I_{sb}, I_{sc} and I_{na}) given below.

$$I_{aerror} = I_{sa} - I_{sa}^{*} \quad (9)$$

$$I_{berror} = I_{sb} - I_{sb}^{*} \quad (10)$$

$$I_{cerror} = I_{sc} - I_{sc}^{*} \quad (11)$$

$$I_{nerror} = I_{sn} - I_{sn}^{*} \quad (12)$$

These error signals are supplied hysteresis current controller. This hysteresis controller generates firing pluses for the IGBT switches of D-STATCOM. I_{sn} is taken as zero.

4 Simulation and Results

Simulink model for four legs VSC-based D-STATCOM is developed with its control technique for three-phase four-wire distribution system in MATLAB using Simulink environment. Response of PLL less control technique for D-STATCOM in time domain is simulated for UPF mode at steady-state and dynamic condition, respectively, with linear and nonlinear load.

4.1 Response of D-STATCOM at Steady State Using Linear Load

Figure 3 shows the waveforms of grid voltage (V_{pcc}), grid current(I_{sabc}), load current (I_{la}, I_{lb}, I_{lc}), inverter currents (I_{inva}, I_{invb}, I_{invc}) and DC link voltage (V_{dc}) at steady-state condition with linear load in UPF mode of D-STATCOM operation. It is found that D-STATCOM gives satisfactory performance in UPF mode in terms of load balancing and constant DC bus voltage by using without PLL control mechanism at steady-state condition with linear load.

4.2 Response of D-STATCOM at Steady State Using Nonlinear Load

Figure 4 shows the waveforms of grid voltage (V_{pcc}), grid currents (I_{sabc}), balanced load currents (I_{la}, I_{lb}, I_{lc}), inverter currents (I_{inva}, I_{invb}, I_{invc}) and DC link voltage (V_{dc}) at steady-state condition with nonlinear load in UPF mode of D-STATCOM operation. Total harmonic of phase "*a*" current I_{sa} of grid which is reduced to 4.22% is shown in Fig. 5. It is observed that the D-STATCOM performance is found satisfactory in UPF mode in terms of load balancing, harmonics elimination and constant DC bus voltage by using without PLL control mechanism at steady-state condition with nonlinear load.

4.3 Response of D-STATCOM at Dynamic Using Linear Load

Figure 6 shows the waveforms of grid voltage (V_{pcc}), grid currents (I_{sabc}), load currents (I_{la}, I_{lb}, I_{lc}), inverter currents (I_{inva}, I_{invb}, I_{invc}) and DC link voltage (V_{dc}) at dynamic condition with linear load by changing the load from three phase to two

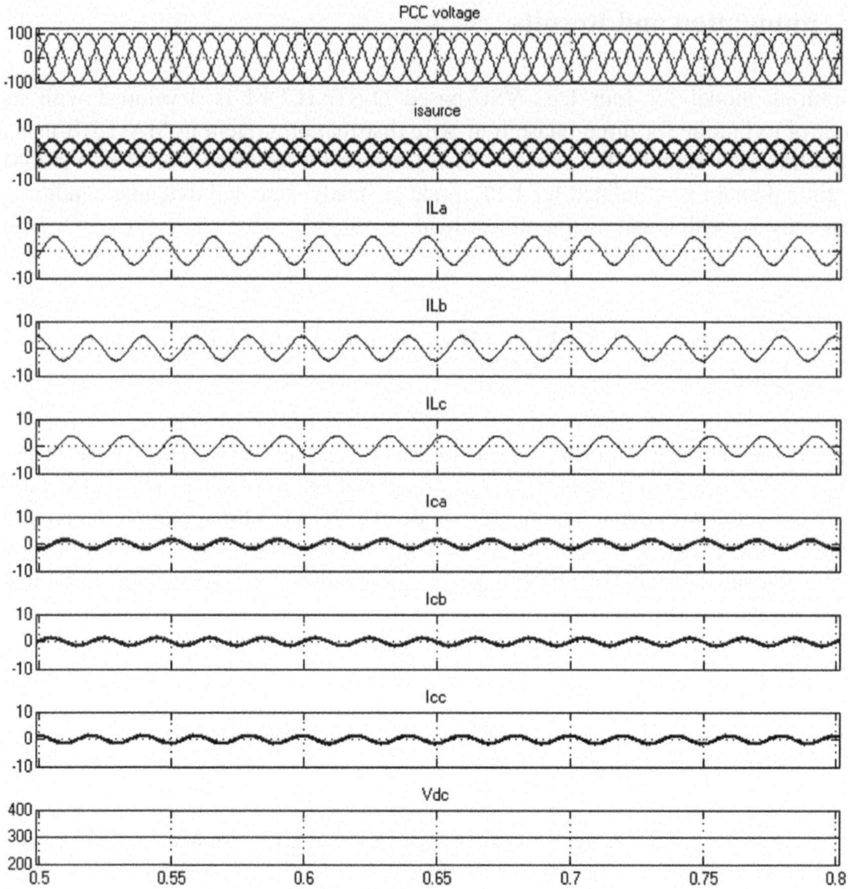

Fig. 3 Response of D-STATCOM at steady-state condition using linear load

phase for time duration 0.6–0.7 s in UPF mode of D-STATCOM operation. The D-STATCOM performs satisfactorily in UPF mode in terms of load balancing and constant DC bus voltage by using without PLL control mechanism dynamic condition with nonlinear load.

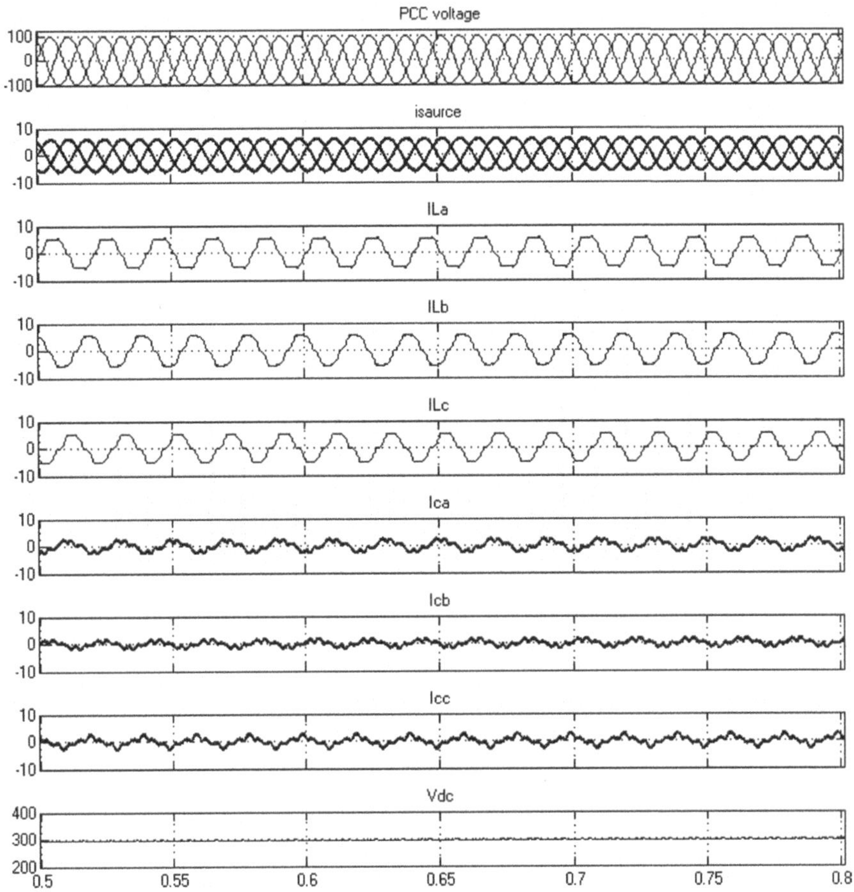

Fig. 4 Response of D-STATCOM at steady-state condition using nonlinear load

4.4 Response of D-STATCOM at Dynamic Using Nonlinear Load

Figure 7 shows the waveforms of grid voltage (V_{pcc}), grid currents (I_{sabc}), load currents (I_{la}, I_{lb}, I_{lc}), inverter currents (I_{inva}, I_{invb}, I_{invc}) and DC link voltage (V_{dc}) at dynamic condition with nonlinear load by changing the load from three phase to two phase for time duration 0.6–0.7 s in UPF mode of D-STATCOM operation.

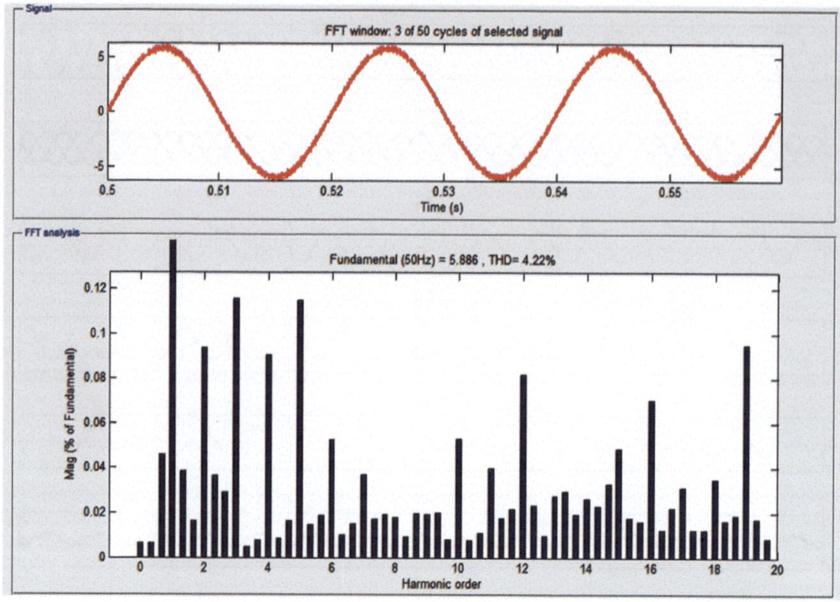

Fig. 5 Total harmonic of grid current I_{sa} for steady-state response under nonlinear load

Total harmonic of phase "*a*" current I_{sa} of grid which is reduced to 4.31% is shown in Fig. 8. The performance of D-STATCOM is found satisfactory in UPF mode in terms of load balancing, harmonics elimination and constant DC bus voltage by using without PLL control mechanism at dynamic condition with nonlinear load.

5 Conclusions

PLL less control technique for four legs VSC-based D-STATCOM is implemented in MATLAB/Simulink environment. The performance of four legs VSC-based D-STATCOM by using without PLL control mechanism is found satisfactory in terms of grid current harmonic elimination, load balancing and constant DC bus voltage at steady-state and dynamic condition with linear and nonlinear load, respectively.

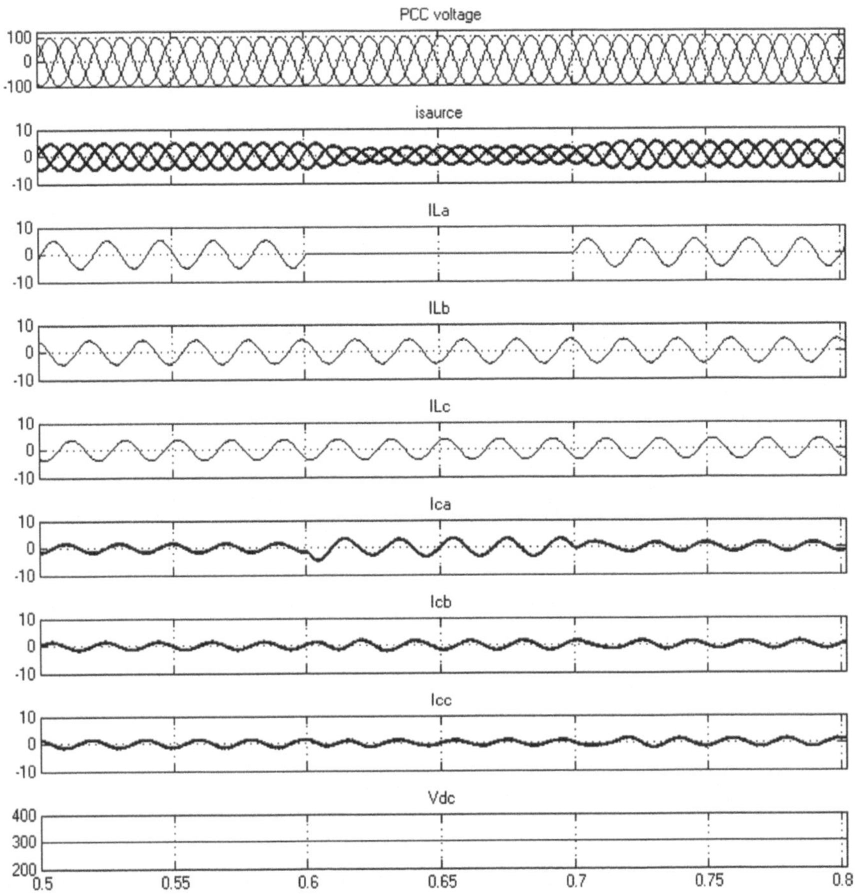

Fig. 6 Response of D-STATCOM at dynamic condition using linear load

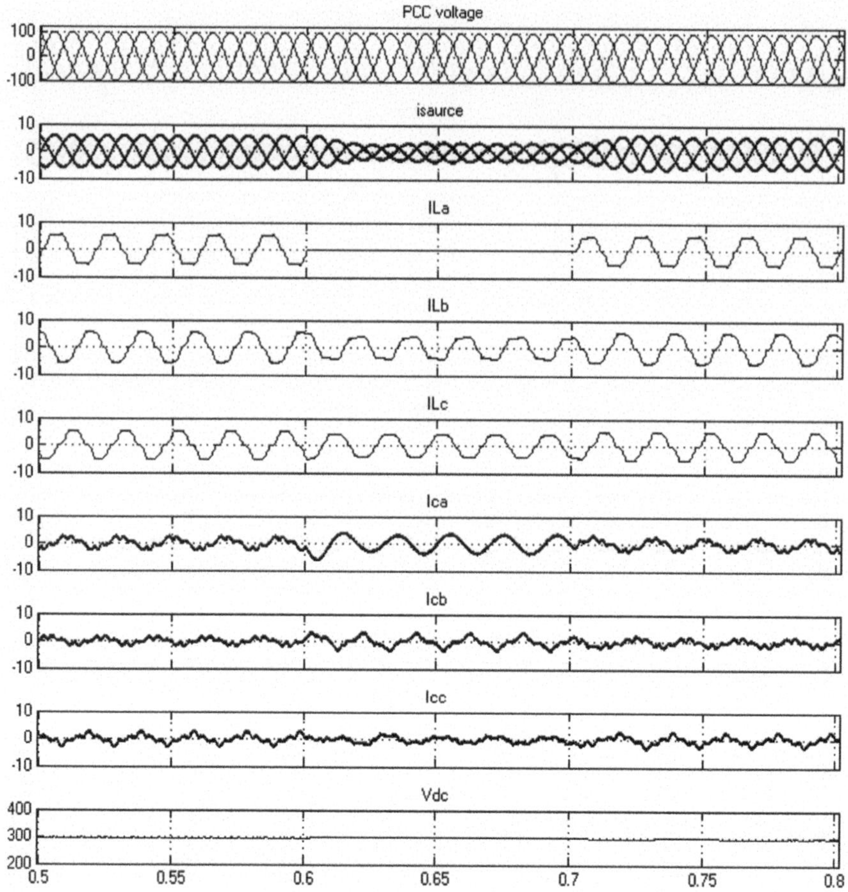

Fig. 7 Response of D-STATCOM at dynamic condition using nonlinear load

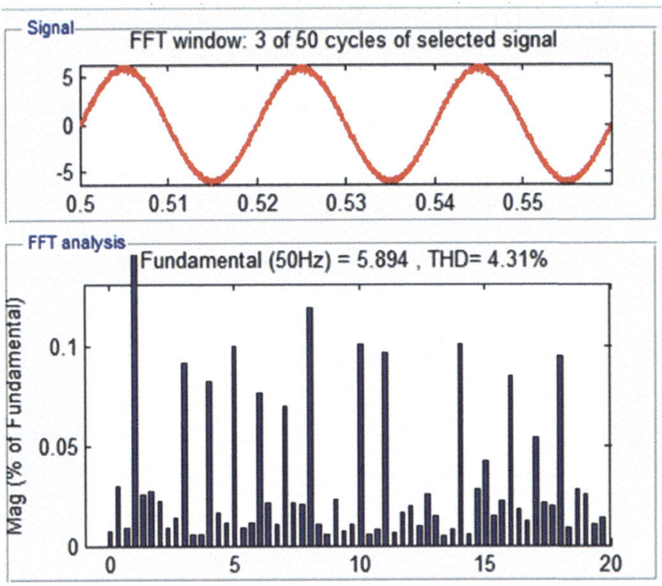

Fig. 8 Total harmonic of grid current I_{sa} for dynamic response at nonlinear load

Appendix

Three-phase AC supply 100 V (L-L), 50 Hz, source impedance Z_s ($R = 0.05\ \Omega$, $L = 0.2$ mH.
(1) Three-phase nonlinear load, $R = 13.66\ \Omega$, $L = 10$ mH,
(2) Nonlinear load (A-N) R $= 15\ \Omega$, $L = 100$ mH.
DC link capacitance and voltage $C_{dc} = 3000\ \mu$F and $V_{dc} = 300$ V, interfacing inductor = 2 mH.

References

1. Dugan RC, Mc Granaghan MF, Beaty HW (2006) Electric power systems quality, 2nd edn. MC Graw Hill, New York
2. Kusko A, Thompson MT (2007) Power quality in electrical system. McGraw Hill Professional, Maidenherd
3. Arrillaga J, Watson NR (2004) Power system harmonics. Wiley, Hoboken
4. IEEE recommended practices and requirement for harmonic control on electric power system, IEEE Std. 519 (1992)
5. Wu JC, Jou HL, Feng YT, Hsu WP, Huang MS, Hou WJ (2007) Novel circuit topology for three-phase active power filter. IEEE Trans Power Del 22(1):444–449
6. Jou HL, Wu KD, Wu JC, Li CH, Huang MS (2008) Novel power converter topology for three phase four-wire hybrid power filter. IET Power Electr 1(1):164–173

7. Singh B, Solanki J (2011) Load compensation for diesel generator-based isolated generation system employing DSTATCOM. IEEE Trans Ind Appl 47(1):238–244
8. Jayaprakash P, Singh B, Kothari DP (2008) Three-phase 4-wire DSTATCOM based on H-bridge VSC with a star/hexagon transformer for power quality improvement. In: Proceedings of IEEE region 10 and the third international conference on industrial and information systems (ICIIS). IEEE, Kharagpur, India, pp 1–6
9. Singh B, Arya SR (2014) Adaptive control of four-leg VSC based DSTATCOM in distribution system. Int J Emerg Electr Power Syst 15(1):93–99

Dynamic Stability Enhancement of a Variable Solar Penetrated Power System by Fractional UPFC-Based Controller

Narayan Nahak, Kritibash Praharaj, and Alok kumar Mishra

Abstract Pertaining to randomness in solar energy, its integration with synchronous generation needs robust controller design for improving small signal stability. Industrial proportional integral and lead-lag controllers are very much popular. But, due to complexity in power system the robustness of controller can be much enhanced with fractional-order controllers. Here, UPFC-PI with fractional lead-lag (PI-FLL) controller is proposed for enhancing small signal stability of variable solar-integrated power system. The controller parameters are tuned by whale optimization technique. It is observed that the stability is hampered with variable solar penetration without effective controller. Time and frequency domain simulations are being carried out to justify the effectiveness of PI-FLL controller. The response predicts that UPFC PI-FLL damps oscillations heavily in contrast to PI and lead-lag controllers optimized by particle swarm, grey wolf, and whale optimizations.

Keywords FACTS · UPFC · PI · Fractional lead-lag · Small signal stability · Damping controller

1 Introduction

Dynamic stability has been a challenging issue for power system operators due to power system oscillation. If the oscillations not damped adequately, then it leads to collapsing of entire system [1]. Again renewable integrations like solar photovoltaic

N. Nahak (✉)
Department of Electrical Engineering, Siksha 'O'Anusandhan Deemed to Be University, Bhubaneswar, Odisha, India
e-mail: narayannahak@soa.ac.in

K. Praharaj · A. Mishra
Department of Electrical and Electronics Engineering, Siksha 'O'Anusandhan Deemed to Be University, Bhubaneswar, Odisha, India
e-mail: kritibashpraharajkp1@gmail.com

A. Mishra
e-mail: alokmishra@soa.ac.in

M. N. Favorskaya et al. (eds.), *Innovations in Electrical and Electronic Engineering*, Lecture Notes in Electrical Engineering 661, https://doi.org/10.1007/978-981-15-4692-1_10

(SPV) with synchronous generations are becoming more critical due to variable solar irradiance. In [2, 3], the effect of heavy solar integration has been presented regarding stabilities issue. In [4], it has been presented that with increase in solar generations in an integrated power system, the oscillation being increased with solar generations, so there should be a limit on SPV generations. In [5], small signal stability studies have been performed with different renewable sources. Synchronous power controller is presented in [6] for improving damping. In [7], PID-supplementary controller with supercapacitor has been proposed for efficient damping. A detailed investigation and damping of electromechanical oscillations have been presented in [8] with stochastically varying solar power generation in a solar-integrated power system. So an effective controller and control algorithms are desirable to create adequate damping.

For long years, PSS has been adapted for damping oscillations, but it has so many demerits including leading power factor operations [9]. FACTS devices put so many advantages over conventional PSS as reported in [10, 11]. UPFC is a challenging FACT device with 3 degrees of freedom and can efficiently damp system oscillations [12, 13]. In [13–15], different control algorithms like DEPSO, GWO, DE-GWO, PSO-IGWO, and GA-GSA have been implemented for tuning UPFC-based controller parameters.

Fractional-order lead-lag controllers are more robust as compared to conventional lead-lag controller, and it can provide flat-phase compensation in a wider frequency range [16, 17]. PI controllers provide rapid dynamic characteristics and rapidly track the systems to reduced dynamic errors. Hence, UPFC-PI with fractional-order lead-lag controller (PIFLL) has been proposed here for damping oscillations of a varying solar penetrated power system. The parameters of the controllers are tuned by an efficient control technique which is whale optimization algorithm (WOA). WOA is a social nature-based technique inspired by a group of humpback whales [18]. It is reported recently which has three basic rules like encirclement, the searching of the prey, and spiraling updation of the position, respectively. It has been implemented for optimal placement of TCSC and SVC in [19]. In WOA, randomly generated solution sets are used with the upgradation of individual solution. In this work, the performance of WOA is compared with PSO and GWO algorithms.

2 Power System with UPFC

The SMIB system along with solar source connected to UPFC is given in Fig. 1, where IEEE-ST1A excitation system is being implemented. The UPFC variables are m_B, δ_B, m_E, and δ_E. By controlling these variables, the power system operation and control can be achieved.

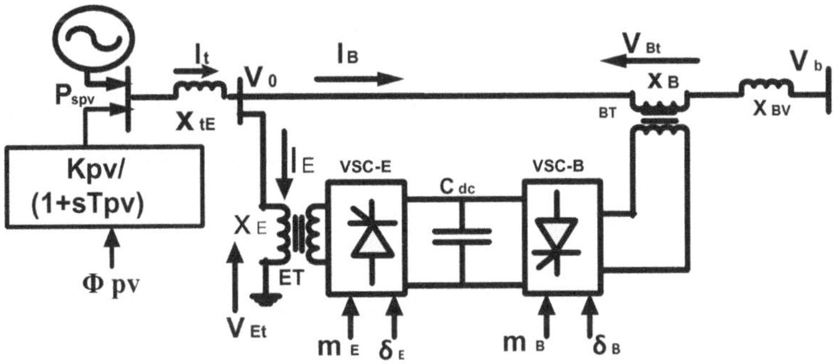

Fig. 1 SMIB system integrated with solar power generation

3 Dynamic Modeling Including UPFC

3.1 Nonlinear Modeling

The following equations [8] represent this modeling.

$$\dot{\omega} = \left(\frac{P_i - P_e - D\Delta\omega}{M}\right) \tag{1}$$

$$\dot{\delta} = \omega_0(\omega - 1) \tag{2}$$

$$\dot{E}_q^{'} = (-E_q + E_{fd})/T_{d0}^{'} \tag{3}$$

$$\dot{E}_{fd} = \left[-E_{fd} + K_a(V_{\text{ref}} - V_t)\right]/T_a \tag{4}$$

$$\dot{V}_{\text{dc}} = \frac{3m_E}{4C_{dc}}\left(I_{\text{Eq}} \sin\delta_E + I_{\text{Ed}}\cos\delta_E\right) + \frac{3m_B}{4C_{\text{dc}}}\left(I_{\text{Bq}} \sin\delta_B + I_{\text{Bd}}\cos\delta_B\right) \tag{5}$$

Active power balance between the VSCs of UPFC is represented in Eq. (6) as

$$\text{Re}(V_B I_B^* - V_E I_E^*) = 0 \tag{6}$$

3.2 Linear Model

The linear modeling of power system around an operating point can be presented as:

$$\dot{\Delta\delta} = \omega_0 \Delta\omega \tag{7}$$

$$\dot{\Delta\omega} = \left(\frac{-\Delta P_e - D\Delta\omega}{M}\right) \tag{8}$$

$$\Delta\dot{E}_q = (-\Delta E_q + \Delta E_{\text{fd}})/T_{d0} \tag{9}$$

$$\Delta\dot{E}_{fd} = \left[-\Delta E_{fd} + K_a(\Delta V_{\text{ref}} - \Delta V_t)\right]/T_a \tag{10}$$

$$\begin{aligned}\Delta V_{\text{dc}} &= K_7\Delta\delta + K_8\Delta E_q^{'} - K_q\Delta V_{\text{dc}} + K_{\text{ce}}\Delta m_E \\ &+ K_{\text{c}\delta\text{e}}\Delta\delta_E + K_{\text{cb}}\Delta m_B + K_{\text{c}\delta\text{b}}\Delta\delta_B\end{aligned} \tag{11}$$

where,

$$\begin{aligned}\Delta P_e &= K_1\Delta\delta + K_3\Delta E_q^{'} + K_{\text{pd}}\Delta V_{\text{dc}} + K_{\text{pe}}\Delta m_E \\ &+ K_{\text{p}\delta\text{E}}\Delta\delta_E + K_{\text{pb}}\Delta m_B + K_{\text{p}\delta\text{b}}\Delta\delta_B\end{aligned}$$

$$\begin{aligned}\Delta E_d &= K_4\Delta\delta + K_3\Delta E_q^{'} + K_{\text{qd}}\Delta V_{\text{dc}} + K_{\text{qe}}\Delta m_E \\ &+ K_{\text{q}\delta_\text{E}}\Delta\delta_E + K_{\text{qb}}\Delta m_B + K_{\text{q}\delta_\text{B}}\Delta\delta_B\end{aligned}$$

4 Solar Power Generation (SPV) Modeling

The SPV generation P_{spv} in watt can be presented as [5].

$$P_{\text{spv}} = \eta_c S_a \varphi_v [1 - 0.005(T_a + 25)] \tag{12}$$

Here, η_c and S_a being efficiency and area are constant, so P_{spv} varies with T_a and ϕ_v, which are temperature and radiation. Temperature T_a is taken as 25 °C. The modeling of SPV is represented by Fig. 2.

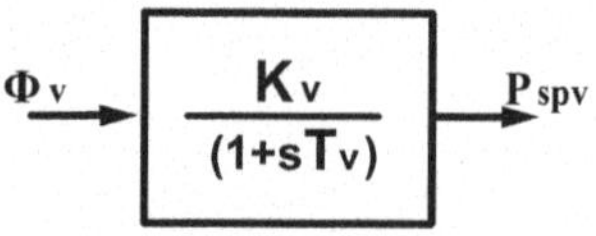

Fig. 2 Modeling of SPV generation

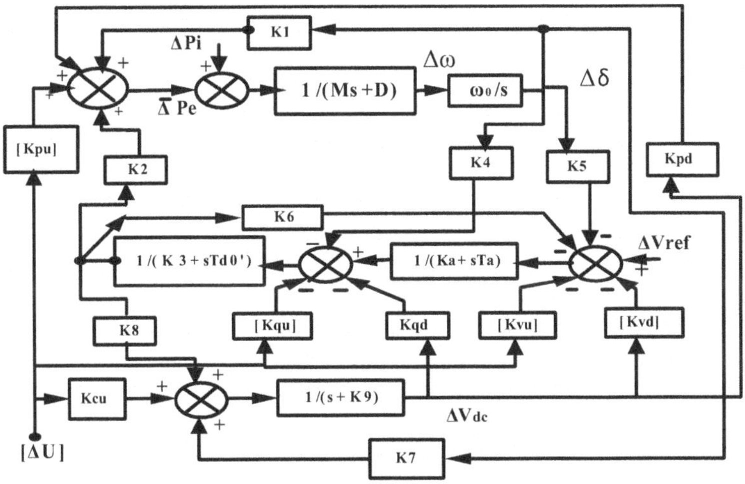

Fig. 3 Modification of Heffron–Phillips model including UPFC

5 Heffron–Phillips Modeling Including UPFC

This modeling as in Fig. 3 consists of 28 constants being calculated at normal operating conditions [14]. The single-machine system's nominal data is being given in the appendix. Input [ΔU] being fed from external control action and [K_{pu}], [K_{vu}], [K_{qu}] represent UPFC.

6 PI with Fractional Lead-Lag (PI-FLL) Controller

We can generalize fractional calculus as the ordinary calculus along with the differentiation and integration of the functional operator D associated with an order p including the integer numbers. Reimann–Liouville (RL) version is the common fractional-order integro-differential operator.

PI controllers provide rapid dynamic characteristics and rapidly track the systems to reduced dynamic errors. So on lead-lag controllers being very efficient to damp oscillations, the closed-loop performance can be further enhanced with fractional-order lead-lag controller. Hence, due to complexity of power system with variable solar energy integration, a new UPFC cascaded PI with fractional lead-lag compensator (PI-FLL) being implemented in this work is presented in Fig. 4, with time delay D of 50 ms so as to access remote signal speed deviation for controller design.

The fractional-order counterpart of a lead-lag compensator $C(s)$ is having the generalized form [16, 17]

Fig. 4 PI with fractional lead-lag (PI-FLL) controller

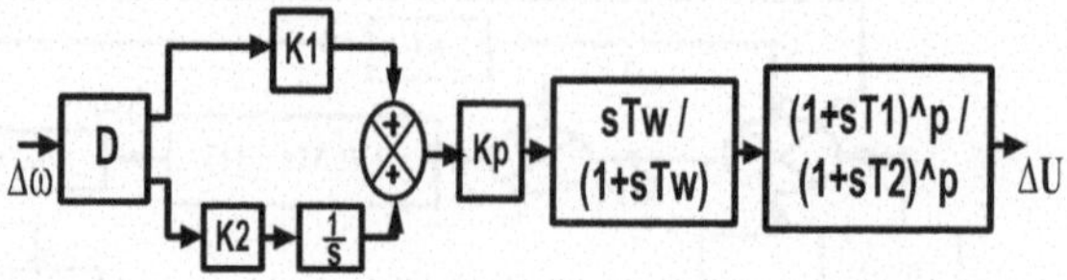

$$C(s) = k\left(\frac{s + 1/\mu}{s + 1/x\mu}\right)^{p} = kx^{q}\left(\frac{1 + s\mu}{1 + sx\mu}\right)^{p}$$
$$0 < x < 1, p \in R \quad (13)$$

where p is the fractional commensurate order of the compensator, and $1/\mu = \omega_z$ and $1/x\mu = \omega_p$ are the zero frequency and pole frequency, respectively.

7 Objective Function

In this work, ITAE objective function has been taken as given by Eq. (14) for which change in speed of generator being taken as input signal.

$$J = \int_{0}^{t\text{sim}} t|\Delta\omega|dt \quad (14)$$

J is to be minimized considering the constraints in parameters. The gains' range is taken from 1 to 10 for K_1, 1–5 for K_2, 1–100 for K_p, and time constants from 0.01 to 1. The range of p is taken from 0 to 1.

8 Whale Optimization Technique

This is a social nature-based technique inspired by a group of humpback whales (Mirjalili and Lewis 2016). Randomly generated solution set is used with the upgradation of individual solution in the group following three basic rules like encirclement, the searching of the prey, and spiraling updation of the position, respectively.

8.1 Encirclement of the Prey

If (p < 0.5 and Abs(A) < 1), then as per the following equations the position of the solution obtained so far is updated.

$$\vec{D} = \left|\vec{C} \cdot \vec{X}^{*} - \vec{X}(t)\right| \tag{15}$$

$$\vec{X}(t+1) = \left|\vec{X} - \vec{A} \cdot \vec{D}\right| \tag{16}$$

where $\vec{X}^{*}$ is the global best solution. A and C are evaluated as per the relations specified in the following equations.

$$\vec{A} = 2 \cdot \vec{a} \cdot \vec{r} - \vec{a} \tag{17}$$

$$\vec{C} = 2 \cdot \vec{r} \tag{18}$$

where a is being reduced to 0 from a starting value of 2 and r is a randomly generated value which is obviously lies within 0 and 1 and including the end points.

8.2 *Prey Searching*

If ($p < 0.5$ and Abs(A) < 1), then like the encirclement phase but instead of the candidate solution a random solution is considered as per the following equations.

$$\vec{D} = \left|\vec{C} \cdot \vec{X}_{\text{rand}} - \vec{X}(t)\right| \tag{19}$$

$$\vec{X}(t+1) = \left|\vec{X}_{\text{rand}} - \vec{A} \cdot \vec{D}\right| \tag{20}$$

Prey searching is implemented in the exploitation phase and executes the global searching work for WOA.

8.3 *Position Updation Using Spiral Method*

If ($p > 0.5$ or $p = 0.5$), then the exploitation work of WOA is utilized in both of the encirclement and spiral updation phases, where the position updating is carried out as per the following equations.

$$\vec{X}(t+1) = \vec{D}'.e^{bl}.\cos(2\pi l) + X^{*}(t) \tag{21}$$

where b bears a constant value and $\vec{D}'$, the ith solution and the best gap distance. The pseudocode of WOA is given in Table 1.

Table 1 Pseudocode of WOA

```
Starting the population of whales X_i (i = 1, 2, ..., n)
The fitness of each searching agent is
found out to be X*=the best searching agent
while (t < maximum number of itera-
tion)
for each searching agent Update a, A,
C, l, p
if1 (p<0.5) if2 (|A| < 1)
        Updation of the place of the present searching agent
        else if2 (|A| ≥ 1)
        A random searching agent (X_rand) is chosen
  Updation of the place of the present searching agent
        end if2
     else if1 (p≥0.5)
          Updation of the place of the present searching agent
     end if1
   end for
  checking if any search agent exceed the findings space and it has
to be followed
  Each searching agent fitness has to be
  found out If there is a better solution, X*
  has to be up to date
  t=t+1
  end while
  return X*
```

9 Result and Discussion

For damping controller design, the speed deviation of generator is taken as input signal to controller. For ITAE criterion, a step change in 10% of input power to generator is considered. To justify the efficacy of proposed UPFC PI-FLL controller variable, SPV penetration has been considered with three cases. The optimized gains are given in Table 2 with PI-FLL and in Table 3, with only PI, lead-lag controller.

Table 2 Optimized parameter with proposed controller

Controller parameters	WOA	GWO	PSO
K_1	8.3160	7.5418	3.5391
K_2	1.640	1.1501	1.5101
K_P	84.3683	83.6862	69.470
T_1	0.6131	0.8114	0.7236
T_2	0.2946	0.6860	0.2266
p	0.4364	0.2219	0.2979

Table 3 Optimized parameter with PI and lead-lag control action

PI controller only		Lead-lag control only		
K_1	K_2	K_p	K_1	T_2
5.6407	0.4226	76.1050	0.5834	0.7652

In case-I, the SPV output power Pspv has been raised to 0.4 pu and the WOA tuned PI-FLL controller being employed for damping resulting oscillation. The PI-FLL controller has also been tuned with PSO and GWO algorithms for comparison purpose. Figures 5 and 6 present speed deviation and bode plot, and Table 4 presents system eigenvalues. The response shows that the PI-FLL-based UPFC damps oscillations much efficiently in comparison to others. In case-II, Pspv is again raised to

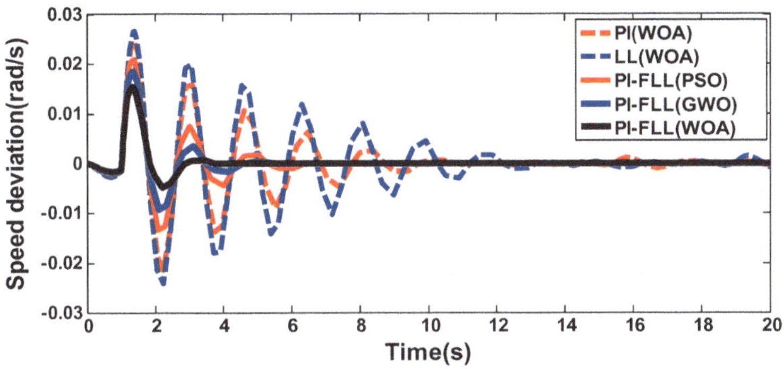

Fig. 5 Speed deviation, Pspv = 0.4 pu

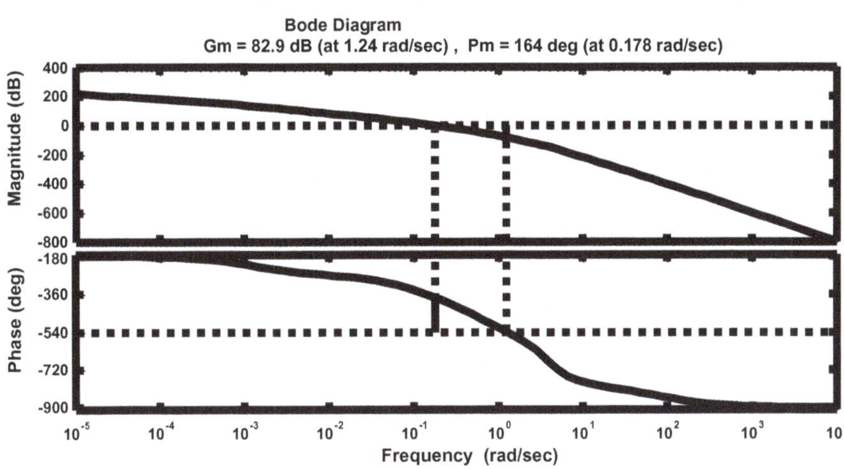

Fig. 6 Bode plot, Pspv = 0.4 pu

Table 4 System eigenvalues with proposed controller

Case-I, $P_{spv} = 0.4$ pu			Case-II, $P_{spv} = 0.8$ pu		
WOA	GWO	PSO	WOA	GWO	PSO
−9.98e+001	−9.98e+001	−9.98e+001	−9.98e+001	−9.98e+001	−9.98e+001
−2.61e+000+1.59i	−1.15e+000+3.95i	−5.96e−001+4.04i	−5.64e+000	−9.14e−001+3.94i	−8.98e−003+4.10e+000i
−2.61e+000−1.59i	−1.15e+000−3.95i	−5.96e−001−4.04i	−2.28e+000+1.62e+000i	−9.14e−001−3.94i	−8.98e−003−4.10e+000i
−1.89e+000	−1.09e+000	−1.59e+000	−2.28e+000−1.62e+000i	−1.54e+000	−4.82e−001
−4.79e−001	−4.79e−001	−4.82e−001	−4.80e−001	−4.79e−001	−1.80e−003
−9.85e−002	−9.92e−002	−9.82e−002	−9.98e−002	−1.01e−001	−2.65e−017
−1.79e−003	−1.79e−003	−1.79e−003	−1.79e−003	−1.80e−003	−5.56e−001
−2.30e−017	−2.38e−017	−3.66e−017	−2.57e−017	−1.14e−017	−2.466e−017
−5.56e−001	−5.56e−001	−5.56e−001	−5.56e−001	−5.56e−001	−5.456e−001

0.8 pu. Figure 7 shows speed deviation, and Table 4 presents system eigenvalues. Figure 8 presents the frequency domain analysis with bode diagram, where the gain and phase margin are 32.1 dB and 58.6°, respectively. The response predicts that oscillations are more aggravated with increased SPV penetration suddenly by large amount, and in this case the PI-FLL controller has been compared with PI and lead-lag controller. The results justified the supremacy of proposed controller. In case-III, the SPV power is randomly changed as given in Fig. 9, and a time domain simulation for 100 s has been performed. The simulation results are given for speed and real power variations in Figs. 10 and 11, respectively, with PI-FLL and lead-lag actions. The results predict that random SPV penetration excites system oscillations, and these oscillations can be much efficiently damped by the proposed controller.

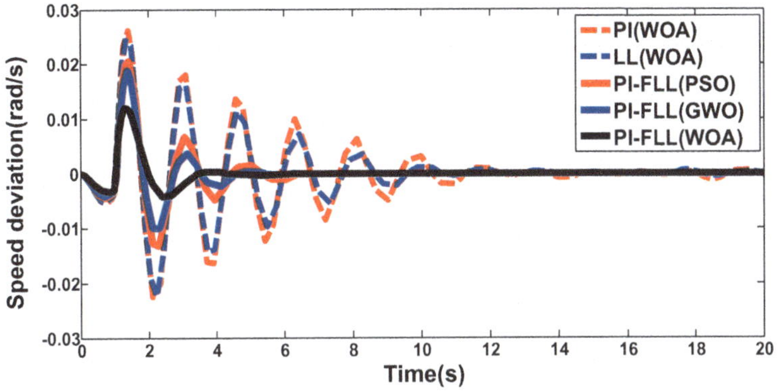

Fig. 7 Speed deviation, Pspv = 0.8 pu

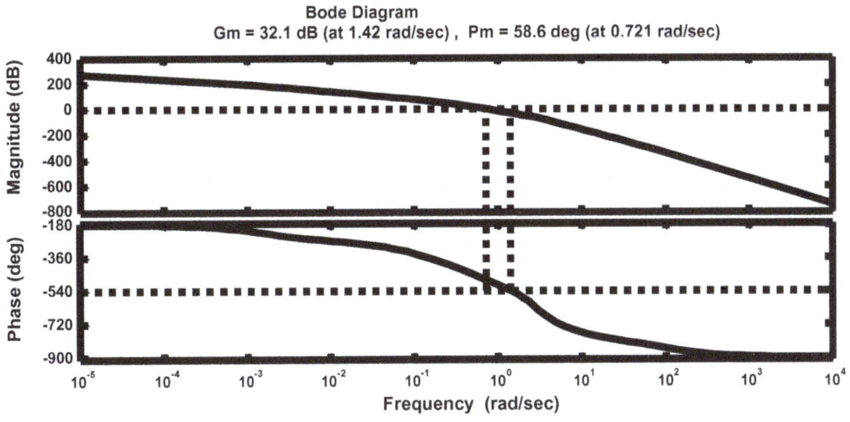

Fig. 8 Bode plot, Pspv = 0.8 pu

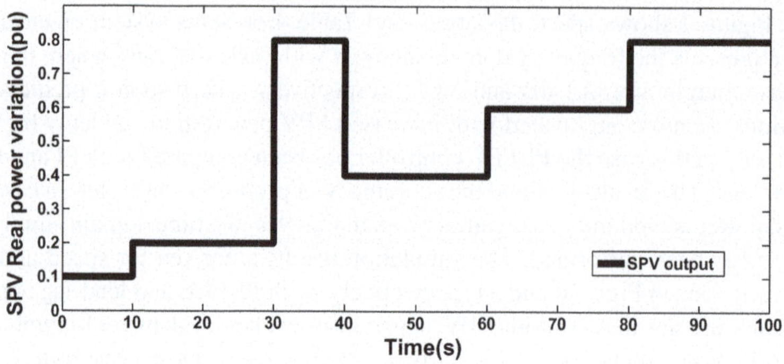

Fig. 9 Random Pspv variations

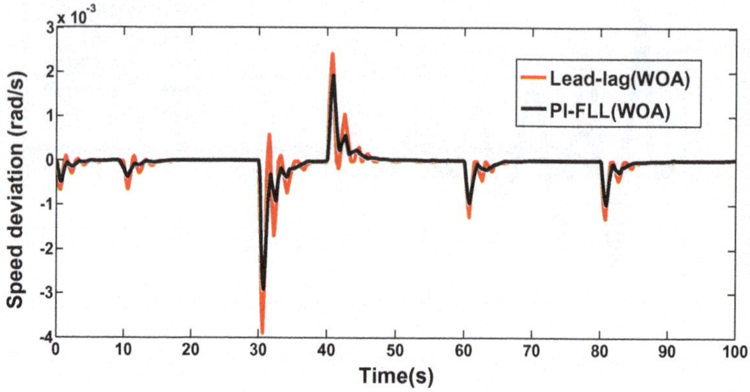

Fig. 10 Speed deviation for random Pspv variation

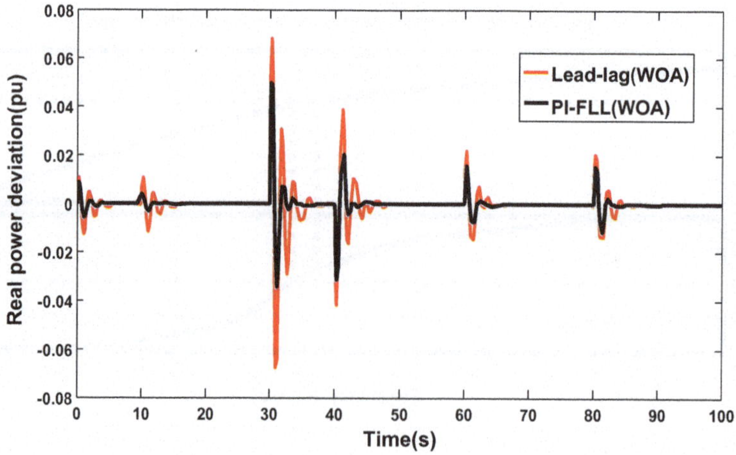

Fig. 11 Real power deviation for random Pspv variation

10 Conclusions

Different case studies were conducted in this work to justify that PI-FLL with unified power flow controller is able to damp oscillations of a variable solar penetrated power system much effectively in comparison to PI and lead-lag controllers. The system response is analyzed in time and frequency domains with step change and random change in SPV generation. Whale optimization algorithm is proposed for tuning controller parameters, and it has been compared with PSO and grey wolf optimization. It is observed that varying SPV integration with conventional power system excites system oscillations, and proposed optimal damping controller is much robust for improving stability of such an integrated power system.

Appendix: SMIB Data

$C_{dc} = 1, H = 4$ MJ/MVA, $K_a = 100, T_a = 0.01, T_{d0} = 5.044$ s,
$D = 0, \delta_0 = 47.13^\circ, V_b = 1, V_{dc} = 2, V_t = 1, X_B = X_E = 0.1,$
$X_{BV} = 0.3, X_d = 1, X_d^{'} = 0.3, X_q = 0.6, X_e = 0.5$

References

1. Padiyar KR (2007) FACTS controllers in power transmission and distribution. New age International (P) Limited
2. Wang L, Lin T (2008) Dynamic stability and transient responses of multiple grid-connected PV systems. In: Proceedings of IEEE PES T&D conference, pp 1–8
3. Tan YT, Krischen DS (2007) Impact on the power system of a large penetration of photovoltaic generation. In: Proceedings of IEEE PES General Meeting, Tampa FL, USA, June, pp 1–8
4. Du W, Wang HF, Dunn R (2009) Power system small-signal oscillation stability as affected by large-scale PV penetration. In: International conference on Sustainable Power Generation and Supply, Nanjing, China, April 2009, pp 1–6
5. Lee DJ, Wang L (2008) Small-signal stability analysis of an autonomous hybrid renewable energy power generation/energy storage system Part I: time-domain simulations. IEEE Trans Energy Convers 23(1):311–320
6. Remon D, Cantarellas AM, Mauricio JM et al (2017) Power system stability analysis under increasing penetration of photovoltaic power plants with synchronous power controllers. IET Renew Power Gener 11(6):733–741
7. Wang L, Vo QS, Prokhorov AV (2018) Stability improvement of a multimachine power system connected with a large-scale hybrid wind-photovoltaic farm using a supercapacitor. IEEE Trans Ind Appl 54(1):50–60
8. Nahak N, Mallick RK (2019) Investigation and damping of low frequency oscillations of stochastic solar penetrated power system by optimal dual UPFC. IET Renew Power Gen 13(3):376–388

9. Noroozian M, Anderson G (1994) Damping of power system oscillations by use of controllable components. IEEE Trans PWRD 9:2046–2054
10. Gyugyi L (1994) Dynamic compensation of ac transmission line by solid-state synchronous voltage sources. IEEE Trans Power Deliv. 9(2):904–911
11. Pandey RK, Singh NK (2009) UPFC control parameter identification for effective power oscillation damping. Electr Power Energy Syst 31:269–276
12. Mallick RK, Nahak N (2016) Hybrid differential evolution particle swarm optimization (DE-PSO) algorithm for optimization of unified power flow controller. In: IEEE UPCON CONFERENCE, IIT BHU Varanasi, India, Dec 2016
13. Khadanga RK, Satapathy JK (2015) A new hybrid GA–GSA algorithm for tuning damping controller parameters for a unified power flow controller. Electr Power Energy Syst 73:1060–1069
14. Nahak N, Mallick RK (2017) Damping of power system oscillations by a novel DE-GWO optimized dual UPFC controller. Eng Sci Tech Int J 20(4):1275–1284
15. Nahak N, Mallick RK (2018) Enhancement of small signal stability of power system using UPFC based damping controller with novel optimized fuzzy PID controller. J Intell Fuzzy Syst 35(1):501–512
16. Abdulkhader HK, Jacob J, Mathew AT (2018) Fractional-order lead-lag compensator-based multi-band power system stabiliser design using a hybrid dynamic GA-PSO algorithm. IET Gener Trans Distrib 12(13):3248–3260
17. Tavazoei MS, Tavakoli-Kakhki M (2014) Compensation by fractional-order phase-lead/lag compensators. IET Contr Theory Appl 8(5):319–329
18. Mirjalili Seyedali, Lewis Andrew (2016) The whale optimization algorithm. Adv Eng Softw 95:51–67
19. Raj Saurav, Bhattacharyya Biplab (2018) Optimal placement of TCSC and SVC for reactive power planning using Whale optimization algorithm. Swarm Evol Comput 40:131–143

Voltage Stability Enhancement Using SVC in PSCAD Software

Mohammad Shabir, Sarfaraz Nawaz, and Ankit Vijayvargiya

Abstract In the modern power system, the need for flexibility, accuracy and fast response is growing every day. Voltage instability affects the system's reliability and security. FACTS devices are used to restore voltage and to control the weak bus. SVC provides the fast acting dynamic compensation in case of severe fault. This paper is focused on voltage stability improvement of IEEE-14 bus test system. Simplified voltage stability index (SVSI) is calculated to identify the weakest bus of the system. All the analysis is being performed using PSCAD simulation software.

Keywords Voltage stability · Voltage deviation · Static VAR compensator · Simplified voltage stability index · Relative electric distance

1 Introduction

The demand for electrical energy is very much intense in the present scenario. Approximately 27% of energy is lost in distribution and transmission in India. India is the top most country in the list of T&D losses [1], which has led the energy system to the transmission and distribution limitation crisis. Such constraints have an even greater impact on stable and safe power supply in energy transmission, typically in

M. Shabir (✉) · S. Nawaz · A. Vijayvargiya
Department of Electrical Engineering, Swami Keshvanand Institute of Technology Management & Gramothan, Jaipur, India
e-mail: mohammadshabir442@gmail.com

S. Nawaz
e-mail: sarfaraz@skit.ac.in

A. Vijayvargiya
e-mail: ankitvijayvargiya29@gmail.com

M. N. Favorskaya et al. (eds.), *Innovations in Electrical and Electronic Engineering*, Lecture Notes in Electrical Engineering 661, https://doi.org/10.1007/978-981-15-4692-1_11

the electric power transmission and distribution network, acceptable voltage variability is much smaller, but due to heavy charging, the voltage rating is not held to the mark [2]. Each time the load or fault increases, and the voltage rate of the system changes. With the voltage fall, the need for reactive power increases. If the demand for reactive energy is not satisfied, the voltage of the bus will decrease further, resulting in a cascading effect on the native regions [3]. Because of any unusual state of fault, the disruption arises in the process, and it goes to transient oscillations. Such unwanted oscillations can alter application performance characteristics. This is therefore essential for monitoring and is accomplished by using the static VAR compensator (SVC) shunt FACTS device designed with auxiliary controllers. SVC will dampen the swings and improve the overall stability of the system [4]. This paper mainly deals with identifying the critical bus by calculating the simplified voltage stability index (SVSI), and voltage enhancement is justified on the critical bus after installation of SVC.

Power system stabilizers (PSS), static VAR compensator (SVC) and shunt static synchronous comparator (STATCOM) addressed various control strategies for damping unwanted oscillations in the power system [5]. Overloaded circuits and buses limit the total transmission power. FACTS technology can control bus voltage, while improving total transmission power (TTC) is a favorable process [6]. A genetic algorithm is used to optimize the FACTS device location. Optimization depends mostly on three parameters: device position, form (shunt or series) and size/value of devices [7]. In [8], many reactive power compensation technologies are summarised and compare the characteristics, operation and applications. The static VAR generators are used in the transmission and distribution network to improve voltage control, stability and power factor [8]. Two forms of FACTS, shunt and series compensation, are efficiently installed using a basic algorithm dependent on heuristic and practical rule that allows us to select the type of FACTS, to determine where FACTS is positioned and managed [9]. The use of PSO to investigate the optimum position and setting parameters of SVC and TCSC controllers eliminates small signal oscillations [10]. This paper explains how SVC's function and structure contain TCR and how it affects the power system's voltage stability. The model has impedance variable that changes with the TCR's firing angle [11].

In this paper, SVC device is used to improve voltage stability of the system. The optimal location of SVC is determined by SVSI. The IEEE 14 bus system is used as test system to check the efficiency of proposed SVSI.

2 Problem Formulation

The voltage instability has a very untoward effect on the reliability of the power system. The aim of this paper study is to eliminate the system's voltage deviation (VD) problem. Voltage deviation is made as small as possible to improve voltage at load bus. Objective function of VD minimization at load bus is defined as

$$VD = \min\left(\sum_{1}^{n} \left| V_{n-V_{\text{refn}}} \right| \right) \tag{1}$$

where

n no. of load buses
V_n voltage magnitude of nth bus
V_{refn} reference voltage of nth bus, usually set to 1 p.u

This improvement of voltage at load buses of the power system is the primary aim, which may be achieved by the connection of SVC at the optimal location. SVC control system is modeled, tuned and optimally placed to analyze the behavior in both steady and dynamic conditions of the system.

3 Mathematical Modeling of SVC

The word SVC was used for shunt linked compensator based on a thyristor without gate turn-off [12]. According to the IEEE standard, SVC is defined as a shunt connected static VAR generator or an absorber whose output is modified to exchange capacitive or inductive current in order to maintain or regulate different electrical power system parameters (Fig. 1).

To explore the effect of SVC on the power system, suitable SVC model is necessary. Here SVC is observed as shunt connected variable susceptance (B_{svc}) that transforms automatically to achieve voltage power. As regards sinusoidal voltage, algebraic equations are defined as

$$I_{\text{SVC}} = jB_{\text{SVC}}V \tag{2}$$

The fundamental frequency TCR equivalent reactance X_{TCR} is

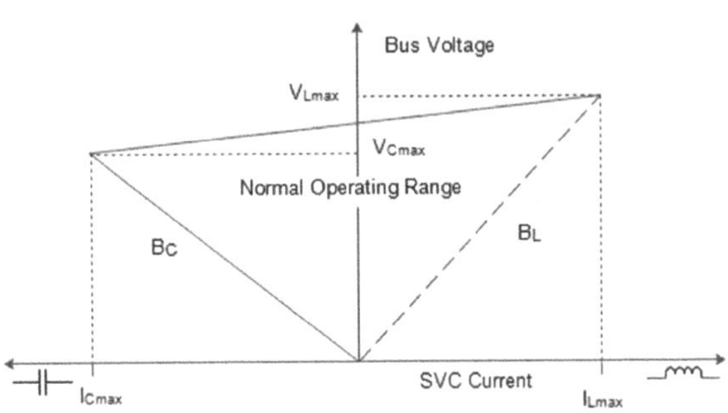

Fig. 1 V–I characteristics

$$X_{\mathrm{TCR}} = \pi X_L/(\delta - \mathrm{Sin}\delta) \tag{3}$$

where $\delta = 2(\pi - \alpha)$, δ is conduction angle and α is firing angle. TCR equivalent reactance X_{TCR} is firing angle (α) terms

$$X_{\mathrm{TCR}} = \frac{\pi X_L}{2(\pi - \alpha) - \sin\alpha} \tag{4}$$

$$X_C = \frac{1}{W_c} \tag{5}$$

at $\alpha = 90°$, $X_{\mathrm{TCR}} = X_L$ means TCR is in fully conducting mode, while at $\alpha = 180°$, $X_{\mathrm{TCR}} = \infty$ means TCR is in blocking mode. Functional reactance of SVC is the parallel combination of X_{TCR} and X_C as.

$$X_{\mathrm{SVC}} = \frac{\pi X_c X_L}{X_C[2(\pi - \alpha) + \sin 2\delta] - \pi X_L} \tag{6}$$

$$Q_{\mathrm{SVC}} = -V_K^2 \left\{ \frac{X_C[2(\pi - \alpha) + \sin 2\alpha] - \pi X_L}{\pi X_C X_L} \right\} \tag{7}$$

As the demand for reactive energy in the bus differs, the susception (B_{SVC}) varies between the boundaries. The reactive energy is, however, proportional to the bus voltage square (V_K). As voltage changes, therefore, the reactive power generated changes.

4 Simplified Voltage Stability Index

In order to detect the loadability of the system, stability indices are introduced. In a power system, indices provide data on the location of voltage instability. Such indices can either show a power system's critical bus or the stability of each line linked in an interconnected network between two buses or measure a system's voltage stability margins. This method is based on the RED theory, which informs us about the nearest generator to load bus.

4.1 *Relative Electrical Distance*

Assume a system of n bus, where 1, 2,……, g are generator buses and $(g + 1)$,……, n are load buses. The admittance matrix of for a given system is

$$\begin{bmatrix} I_S \\ I_L \end{bmatrix} = \begin{bmatrix} Y_{SS} & Y_{SL} \\ Y_{LS} & Y_{LL} \end{bmatrix} \begin{bmatrix} Y_S \\ Y_L \end{bmatrix} \tag{8}$$

here

$[I_S] = [I_1, \ldots, I_g]t$ = injected currents of generator buses, $[I_L] = [I_{g+1}, \ldots, I_n]$ = injected load bus currents,

$[V_S] = [V_{11}, \ldots, V_g]$ = complex generator bus voltage, $[V_L] = [V_{g+1}, \ldots, V_n]$ = complex load bus voltage and $[Y_{SS}]$, $[Y_{SL}]$, $[Y_{LS}]$, $[Y_{LL}]$ = the relevant parts of the Y-Bus matrix network.

The relationship between generator bus and load bus voltages is interpreted mathematically, and we have to drive the matrix $F_{LS} = -[\mathrm{Y}_{LL}]^{-1}\,[Y_{LS}]$ as shown in below equation.

$$[D_{LS}] = \mathrm{abs}[F_{LS}] \tag{9}$$

$[D_{LS}]$ tells about the location of load buses with respect to generators that is called the relative electrical distance [RED]. The [RED] is obtained from $[D_{LS}]$ matrix as

$$[\mathrm{RED}] = \mathrm{I} - [D_{LS}] \tag{10}$$

where I is the unity matrix of size $L \times S$. As electrical distance is found with the [RED], the voltage drop (ΔV_n) is

$$\Delta V_n = \sum_{b-1}^{n_j-1} \left|\vec{V}_b - \vec{V}_{b+1}\right| \cong \left|\vec{V}_g - \vec{V}_l\right| \tag{11}$$

where

V_g Nearest generator voltage
V_1 Analyzed load voltage

In Fig. 2, it is clear that the proposed SVSI index is a measurement-based VSI that needs information from monitoring buses. Clearly, the proposed technique expects of voltage phasors at all the buses of the examined power system to find an area inclined to voltage unsteadiness.

ΔV_n is a simplification of the actual approach defined in [13]. Mostly, voltage stability indices have some degree of inaccuracy due to the different approximation [14]. But SVSI relies on other physical parameters (amplitude and phase of voltage or power) to upgrade its accuracy. To improve the condition of voltage drop in any bus, correction factor (β) for the SVSI is introduced as

$$\beta = 1 - \max(|V_m| - |V_l|)^2 \tag{12}$$

The correction factor on voltage magnitude is correlated with the largest gap. Now the SVSI is given as

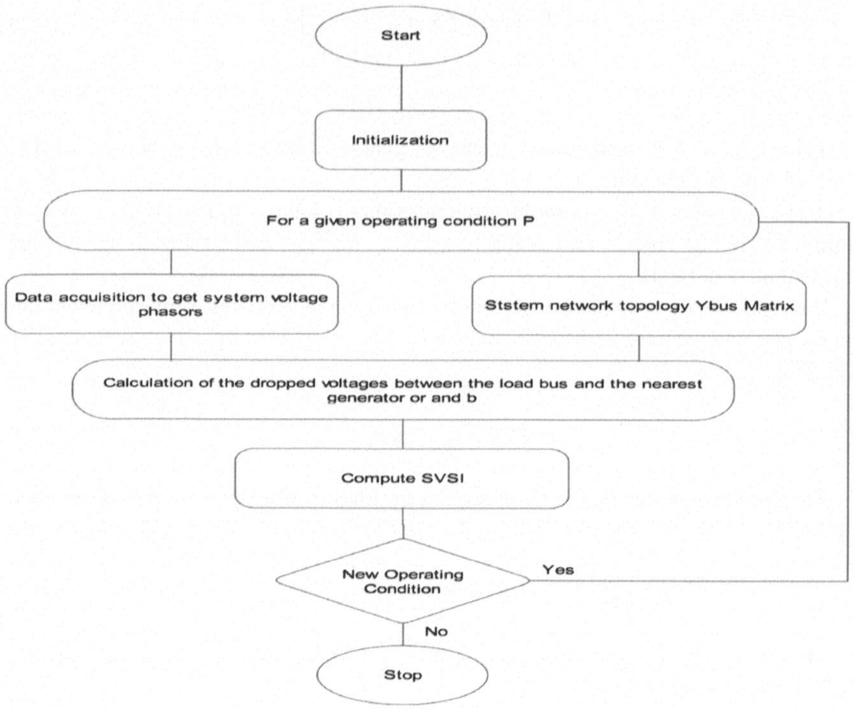

Fig. 2 Flowchart of SVSI

$$\text{SVSI}_i = \frac{\Delta V_i}{\beta * V_i} \tag{13}$$

Due to the depletion in the computational effort, the whole process became simple, so it is called 'simplified' voltage stability index.

5 Simulation Results and Discussions

The IEEE 14 bus system (Fig. 3) was used as a test system to validate SVSI's proposed effectiveness. The PSCAD software simulated the IEEE 14 bus model (Fig. 4). A brief knowledge about the characteristics of each source is given with a base of 100 [MVA] for per unitizing.

Base Case: In base case simulation, we obtain the power flow solution of system as shown in Table 1 and obtained results are tabulated to calculate SVSI of each load bus under steady-state condition.

Step-1: Determine the nearest generator of the load bus by using RED concept. R_{LG} matrix helps in calculating the related electrical distance.

Table 1 Voltage profile under normal condition (in pu)

Bus No.	1	2	3	4	5	6	7	8	9	10	11	12	13	14
Voltage (base case)	1	1.001	0.999	0.971	0.9702	1.008	0.993	1.005	0.988	0.9835	0.991	0.991	0.9862	0.968

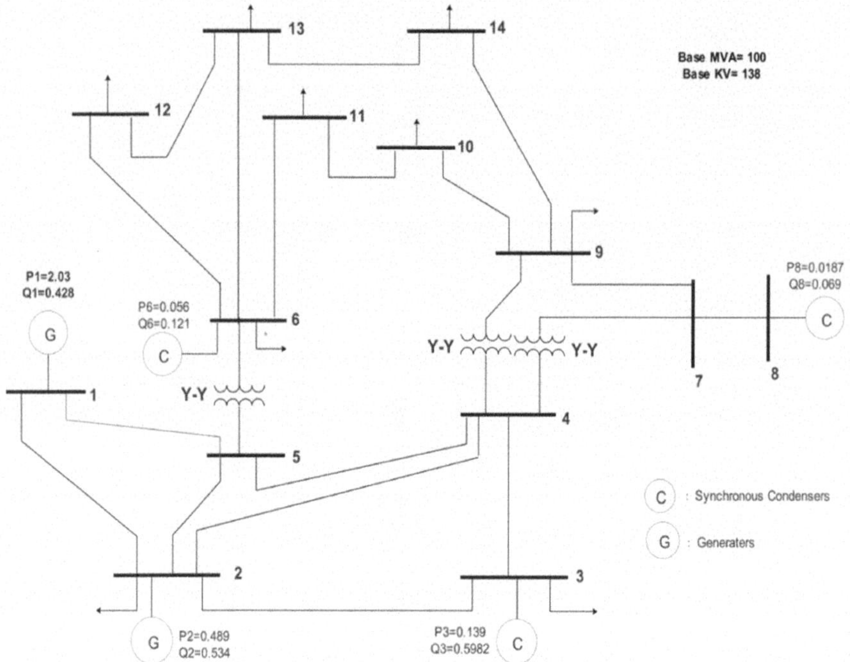

Fig. 3 Single line diagram of 14 BUS system

Step-2: Estimating the SVSI. According to the RLGMatrix, Bus 4 is nearest to the GEN 2. So the voltage drop from GEN 2 to Bus x is calculated as:

$$\Delta V_x^2 = \left|\vec{V}_2 - \vec{V}_x\right|$$

$$\Delta V_x^2 = 1.021 - 0.9969$$

Correction factor is calculated by using the highest difference of voltages as

$$\beta = 1 - (|V_m| - |V_t|)^2$$

$$\beta = 0.9914$$

Here $x =$ Bus 4. Similarly, SVSI is calculated for other load buses as shown in Table 2.

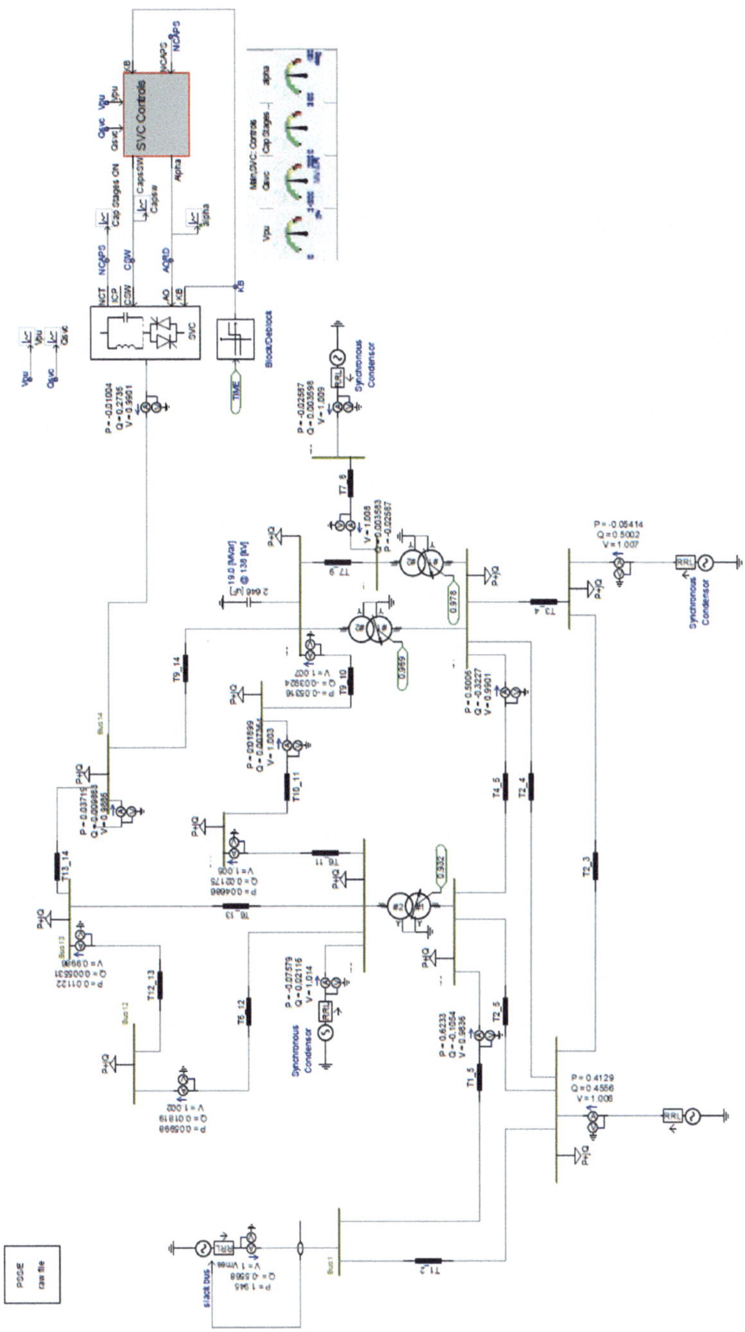

Fig. 4 IEEE 14 bus simulation model in PSCAD software

Table 2 SVSI of load buses

Load bus (x)	4	5	7	9	10	11	12	13	14
SVSI	0.0309	0.0317	0.0121	0.0172	0.0249	0.0163	0.0163	0.0221	0.043

In Table 2, SVSI for all the load buses calculated is shown. The minimum SVSI value is found at bus no 14 (0.043). Hence, the bus no 14 will be the optimal location for SVC installation. After connecting, the SVC voltage profile is improved as we can see in Table 3. From Table 3, it is found that the voltage profile of each bus is enhanced after SVC installation. Voltage deviation without SVC is 0.1442, and voltage deviation after installation of SVC up to 0.0697 is also reduced.

In Table 4, the results of proposed method are compared with the results of latest techniques. It has been observed that SVSI gives better results than any other mentioned technique. Table 4 shows that the voltage deviation of the proposed technique was found to be superior to other techniques.

Voltage deviation of each bus is also decreased after SVC allocation. Now the results of proposed SVSI technique are compared with others [16–18]. Table 4 exhibits the comparison of proposed techniques with other ones.

6 Conclusions and Future Scopes

In this paper, a new index (SVSI) has been presented to determine the location of SVC. The objective of the paper is to improve voltage stability by reducing the voltage deviation. To check the efficacy of the proposed technique, the IEEE 14 bus system has been used as a test system. The voltage deviation of the IEEE 14 bus system is also calculated after SVC installation and compared with other techniques. It has been observed that SVSI gives the optimal location of SVC (i.e., bus no. 14) for IEEE 14 bus system than any other technique. The IEEE 14 bus system has been simulated in PSCAD software.

Table 3 Voltage enhancement after connecting SVC in normal condition

Bus No.	1	2	3	4	5	6	7	8	9	10	11	12	13	14
Voltage (without SVC)	1	1.001	0.999	0.971	0.9702	1.008	0.993	1.005	0.988	0.9835	0.991	0.991	0.9862	0.968
Voltage (with SVC)	1	1.002	1.001	0.9734	0.9722	1.012	0.9993	1.006	0.9996	0.9838	0.9988	0.9988	0.9854	0.998

Table 4 Comparison of proposed SVSI with other techniques

S. No.	Name of technique	Location of SVC	Voltage deviation
1.	Proposed SVSI	14	0.0697
2.	FVSI [2017] [15]	10	0.3667
3.	DA [2018] [16]	9	0.3529
4.	TLBO [2015] [17]	5	0.4993
5.	PSO [2013] [18]	12	0.8952

References

1. Arora MJPS Power transmission and distribution losses in india-a study report. J Curr Sci 20(1)
2. Lakkireddy R, Rastgoufard I, Rastgoufard LP (2015) Steady state voltage stability enhancement using shunt and series facts devices. In: 2015 Clemson University power systems conference (PSC), IEEE, pp 1–5
3. Tomar P, Singhal AK (2019) Power system stability enhancement using UPFC. J Instrum Tech Innov 8(3):1–7
4. Choudekar P, Asija D (2017) Prediction of voltage collapse in power system using voltage stability indices. In: Proceeding of international conference on intelligent communication, control and devices. Springer, Singapore
5. Somasundaram P, Jayakumar V, Sundararaju K (2018) Congestion management using svc under deregulated power system. Int J Pure Appl Mathem 118(20):2307–2317
6. Mithulananthan N, Canizares CA, Reeve J, Rogers GJ (2003) Comparison of PSS, SVC, and STATCOM controllers for damping power system oscillations. IEEE Trans Power Syst 18(2):786–792
7. Ou Y, Singh C (2002) Assessment of available transfer capability and margins. IEEE Trans Power Syst 17(2):463–468
8. Gerbex S, Cherkaoui R, Germond AJ (2001) Optimal location of multi-type facts devices in a power system by means of genetic algorithms. IEEE Trans Power Syst 16(3):537–544
9. Choudekar P, Sinha SK, Siddiqui A (2017) Optimal location of SVC for improvement in voltage stability of a power system under normal and contingency condition. Int J Syst Assur Eng Manag 8(2):1312–1318
10. Dixon J, Moran L, Rodriguez J, Domke R (2005) Reactive power compensation technologies: State-of-the-art review. Proc IEEE 93(12):2144–2164
11. Mahdad B, Bouktir T, Srairi K (2006) Strategy of location and control of facts devices for enhancing power quality. In: MELECON 2006–2006 IEEE mediterranean electrotechnical conference, IEEE, pp 1068–1072
12. Mondal D, Chakrabarti A, Sengupta A (2012) Optimal placement and parameter setting of svc and TCSC using PSO to mitigate small signal stability problem. Int J Electr Power Energy Syst 42(1):334–340
13. Sahu P, Pachori A, Simulink M (2013) Power factor correction using svc with fuzzy logic controller. Int J Enhanced Res Sci Tech Eng 2(4):52–57
14. Kessel P, Glavitsch H (1986) Estimating the voltage stability of a power system. IEEE Trans Power Del 1(3):346–354
15. Dester M, Castro CA (2009) Multi-criteria contingency ranking method for voltage stability. Electr Power Syst Res 79(1):220–225

16. Vanishree J, Ramesh V (2018) Optimization of size and cost of static var compensator using dragon fly algorithm for voltage profile improvement in power transmission systems. Int J Renew Energy Res (IJRER) 8(1):56–66
17. Agrawal R, Bharadwaj S, Kothari D (2015) Optimal location and sizing of svc considering transmission loss and installation cost using TLBO, In: 2015 annual IEEE India conference (INDICON), IEEE, pp 1–6
18. Shende VK, Jagtap P (2013) Optimal location and sizing of static var compensator (svc) by particle swarm optimization (PSO) technique for voltage stability enhancement and power loss minimization. Proc Int J Eng Trends Tech (IJETT) 4(6)

Renewable Energy Source Based Microgrid: Control and Dynamics

Sangeeta Modi and P. Usha

Abstract The major challenge for world's power sector is varying load, low energy efficiency, environmental issues, and low power quality. Microgrid, a smart grid key technology, offers a promising solution to these issues to some extent. By using the renewable energy sources, the generation efficiency can be increased up to 90% without any loss for transmission. This paper presents the performance analysis of a hybrid microgrid under islanded mode. Microgrid under consideration consists of an array of 33 parallel strings of 5 modules connected in series, and maximum power delivered with this combination is 50 KW at 1000 W/m^2 irradiance, and wind turbine of nominal output power of 50 KW at wind speed 12 m/s, a battery of 500 V, 40 Ah rating, an AC load of rating 150 KW, 50 KVAR and a DC load of 200 kW have been considered for the microgrid. Utility grid of three-phase, 25 kV, 50 Hz is connected via three-phase two-winding Star—Delta transformer of 100 kVA, 25 kV/260 V, 50 Hz. In this paper, decoupling PQ control method turns to voltage/frequency (V/f) control for islanded mode of operation. With distributed generation sources supplying loads, a proper load sharing is the objective without overloading the system. Maximum power point tracking (MPPT) controller based on incremental conductance algorithm has been employed.

Keywords Hybrid microgrid · Renewable sources · Green energy · Control of microgrid · Islanded mode

S. Modi (✉)
PES University, Bangalore, India
e-mail: smodi@pes.edu

P. Usha
Dayanand Sagar College of Engineering, Bangalore, India

M. N. Favorskaya et al. (eds.), *Innovations in Electrical and Electronic Engineering*, Lecture Notes in Electrical Engineering 661, https://doi.org/10.1007/978-981-15-4692-1_12

1 Introduction

Generation efficiency of a large power plant generally ranges from 34 to 54%, while the efficiency at customer end is only 30–46% after inclusion of losses. With the use of distributed generation (DG), efficiency can be increased up to 90% without any loss for transmission. Adding a DG source or microsource to the existing grid enhances the reliability and maintaining continuous supply to critical load. In this paper, photovoltaic (PV) and wind energy source-based microgrid are considered along with battery for storage purpose. Interlinking converter, i.e., power electronic-based converter is employed to interface these DGs and utility grid. This converter controls of real and reactive power becomes flexible. Since energy storage system is combined with the DGs, so voltage fluctuation issues can be addressed easily to ensure voltage stability. Local generation of power with the help of distributed generation such as solar and wind is the latest trend. It is more economical, efficient and reliable. Dynamics of control of microgrid are presented in ref [1]. In case of any fault or grid failure, the microgrid can be switched to islanded mode of operation. Integration of various DGs to the utility grid can improve the grid performance, reliability, and flexibility [2].

There are two types of microgrids with respect to supply. One is DC microgrid, and other is AC microgrid. Combination of these AC and DC grid is known as hybrid microgrid. Hybrid microgrid is complex in structure, but it integrates the advantages of its AC and DC microgrids [3]. CERT test bed is presented in reference [4]; in this, effect of integrating DGs has been discussed.

In hybrid microgrid, power can be exchanged via a power electronic interface called interlinking converter. The hybrid microgrid improves the reliability and power quality of the system. To allow bidirectional flow of power, bidirectional converter has been designed which controls the active power transferred to the AC side of the grid from DC side of the grid [6]. It finds it application where quality is of high preference. It also improves the voltage profile.

Droop control method for stabilizing the output power is discussed for the microgrid [7]. Circulating current and its cause have been discussed in [8]. Classification of control strategies is discussed in [9]. Power control and management in microgrid have been discussed in reference [9, 10]. Selection of converter topology depends upon the control objectives [10].

Performance analysis of hybrid microgrid under islanded mode is presented in this work. Energy storage has been clubbed into the DC part of microgrid. Simulation results presented in this work show that continuity of supply to the DC and AC loads was maintained throughout the operation.

Environmental friendliness is another plus point of using clean energy sources as DGs. Any small electric power system which is not dependent on utility grid and is located near customer side to meet the demand of a local customer is referred as distributed generation (DG), whereas a combination of distributed generation technologies with energy storage system is referred as distributed resources (DR).

The major challenge for world's power sector is varying load, low energy efficiency, environmental issues, and low power quality. Microgrid, a smart grid key technology, offers a promising solution to these issues to some extent. Microgrid which can be operated in either mode is a cluster of systematic integration of distributed generation (DG) sources. It is assumed in this work that when any of these sources is in operation, it is giving its maximum possible power. Depending upon the requirement of the load flow, the interlinking converter can be operated as rectifier and inverter. Control functions of a microgrid can be defined at three levels, depending upon the contents of information shared decentralized, distributed, and centralized architecture are there. In this work, centralized architecture is considered. This paper is divided into four sections. Section 1 discusses basics of microgrid and the system architecture. In Sect. 2, control strategy has been discussed. In Sect. 3, simulation results and analysis are presented, and conclusion is given in Sect. 4 (Table 1).

1.1 System Architecture

System under consideration is shown in Fig. 1.

The main feature of the microgrid is to reduce the burden on utility grid. The microgrid under consideration as shown in Fig. 1 includes a PV array, wind turbine, DC-DC converter, and other power electronic interfaces, battery, utility grid, transformer, AC loads and DC loads. Boost converter is considered here for DC-DC converter.

Maximum power point tracking mechanism is added to the boost converter for extracting maximum power from PV array. Bidirectional converter is connected between AC and DC buses to supply the power in either direction. In this work, islanded mode of operation is considered.

Solar arrays consist of various series–parallel combination of solar cells [2]. Solar cells in general have efficiency range from 11% to over 20%. Since power output for PV array and wind turbine is intermittent, DC-to-DC converters are connected to provide constant DC to the inverter. The function of the solar cell is to convert the solar radiation into electricity.

Ideal model, single-diode model, two-diode model, and multiple-diode model are the various mathematical models of solar cell. For most of the applications, single-diode model is used [3, 4].

The modeling equations for ideal solar cell are given as:

$$I_d = I_0 * \left(e^{qV/nkTn} - 1\right) \tag{1}$$

$$I_{\text{ph}} = [I_{sc} + Ki * (T - 298)] * Ir/1000 \tag{2}$$

$$I = I_{\text{ph}} - I_0 * \left(e^{qV/nkT} - 1\right) \tag{3}$$

Table 1 Symbol used

Symbol used	Parameters
I_d	Diode current
I_{ph}	Photo cell current
I_0	Saturation current
Q	Electron charge
V	Voltage
n	Quality factor
k	Boltzmann constant
T	Temperature
I_{SC}	Short-circuit current
K_i	Reference photocurrent
I_r	Irradiation
C_p	Performance coefficient
λ	Blade tip speed ratio
β	Pitch angle
V_{battery}	Battery voltage
V_0	Battery open-circuit voltage
K	Polarization resistance coefficient
i_b	Battery current
R_0	Internal resistance
Q	Battery capacity
k_p and k_q	Droop coefficients
P_0 and Q_0	Active and reactive power
V_0	Inverter voltage at no load
$\sum_{i=0}^{n} P_{gi}$	Power generated by all renewable energy sources
P_{di}	Demanded power/load power
P loss	Power loss

The solar cell is modeled as a current source, and the current I_{ph} is proportional to the irradiance of the sun, as is shown in Eq. (2). Total output current is shown in Eq. (3).

Wind is one of the most wildly adopted renewable energy sources. Wind plant has multiple components. The defining equations for the wind turbine are given below:

$$Cp = 0.22 * (116/\lambda i - 0.48\beta - 5)e - 12.5/\lambda i \tag{4}$$

$$1/\lambda i = 1/(\lambda + 0.008\beta) - 0.0035(\beta 3 + 1) \tag{5}$$

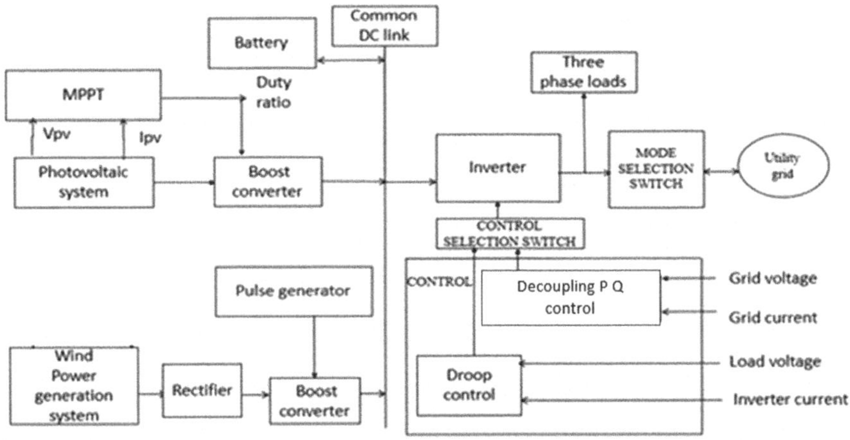

Fig. 1 Block diagram of the system under consideration

The wind plant must have an intelligent control strategy to make sure that the generator output is matched with the input. Pitch control technique is the most widely used and easy-to-implement control for this purpose.

Battery is one of the most common ways of storage used in renewable energy systems. Battery capacity is usually rated to supply the most critical loads in a system [5]; this is done to ensure continuity of supply to these loads. The most widely accepted voltage–current model for constant current discharge is the Shepherd model.

The describing equation for the Shepherd model is given below:

$$V_{\text{battery}} = V_0 - (K * Q/(Q - i_b dt)) * i_b - R_0 * i \tag{6}$$

While designing the battery, it is required to make sure that the battery does not discharge into the system. There should be a charge controller for the battery to always monitor all its parameters like temperature, SOC, voltage, etc. [9].

1.2 *State of Operation and Combination*

Following are the states of the microgrid system under consideration. In this work, islanded states from sr. number 1–6 re-discussed and analyzed. From sr. no 1 to 6, states are considered in this paper with respect to Table 2.

Table 2 States of the microgrid

S. No.	Solar source	Wind source	Battery	Utility grid
Islanded states (1–6)				
1	1	0	0	0
2	0	1	0	0
3	1	1	0	0
4	1	0	1	0
5	0	1	1	0
6	1	1	1	0

2 Control Strategy of Microgrid Under Consideration

Local and centralized control methods are employed for managing the power in a microgrid. Primary objective is to balance the power or one can say power sharing among various distributed generation units connected in the microgrid.

In islanded mode of operation, a microgrid is responsible for both voltage and power control. Control methods of a microgrid consist of DG control and the grid control. These DGs are connected with a bidirectional converter which acts like an inverter under islanded mode of operation. In voltage and frequency control based on droop value, the inverter will regulate the voltage and frequency. This control strategy involves measurement of inverter output voltage and current. The real and reactive power from the DGs would be calculated by the power calculation unit (Fig. 2).

The control system in grid connected mode consists of two control loops—an outer loop for the DC link voltage regulation and inner loop for regulation of active and reactive currents (i_d and i_q). The control consists of (phase-locked loop) PLL synchronization unit which measures the grid voltage and synchronizes the MG voltage with the grid voltage. The DC voltage regulator gives the i_{dref} and i_{qref} which

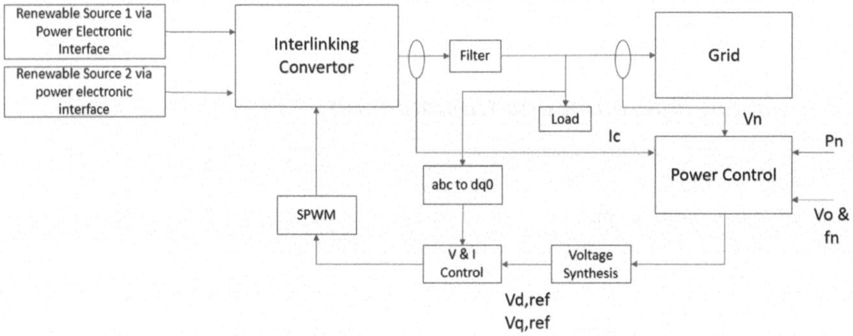

Fig. 2 Control strategy under islanded mode

is set to zero for unity power factor operation. The current controller output gives the V_d and V_q (active and reactive voltages) which are transformed by the inverse Park's transformation to the three-phase reference voltage, V_{abc}. The three-phase reference voltage is then converted to three modulating signals using pulse width modulation (PWM) generator.

The voltage and frequency will be calculated using the reference active and reactive powers entered according to Eqs. (7) and (8). The phase (δ) is obtained by integrating and multiplying the frequency by 2π rad. As shown in Eq. (9)

$$\omega = \omega_0 - kp(P - P_0) \tag{7}$$

$$V = V_0 - kq(Q - Q_0) \tag{8}$$

$$2 * pi * f = d\delta/dt \tag{9}$$

$$kp = -\omega_0 - \omega P_{max} \tag{10}$$

$$kq = -Vo - V_{Qmax} \tag{11}$$

Power balance equations are shown in Eqs. (12)–(14)

$$\sum_{i=0}^{n} P_{gi} = P_{di}; \quad \text{Losses are neglected} \tag{12}$$

$$\sum_{i=0}^{n} P_{gi} = P_{di} + P_{\text{loss}}; \quad \text{With Losses} \tag{13}$$

$$P_{min} \le Pi \le P_{max} \tag{14}$$

Operational constraints are minimum and maximum power supplied by the microgrid sources. Equation 12 shows the limits on power shared by the individual source under consideration. In this work, various DGs are committed for supplying the load as per cost criteria.

In this work, load sharing is based on the availability of DGs, and commitment of the DGs is based on cost criteria. It is assumed here that cost of generating power from solar is least as compared to wind and battery. Then comes wind, if wind and battery both are available to meet the demand, then wind would be preferred.

3 Simulation and Analysis

Simulation studies are performed on the system under consideration.

A microgrid system with 50 kW array, boost converter with 500 V output, bidirectional converter of 260 V, 50 Hz output rating on AC side, and an AC load of 150 KW and 50 KVAR and DC load of 200 kW has been considered.

3.1 The Sequence of Switching Operations of Various DGs Are Simulated as Follows

State 1: (Solar power available).

At $t = 0$ s, the load power required by critical load is 50 kW and is supplied by solar system. Voltage, current, and power associated under this state are shown in Fig. 3.

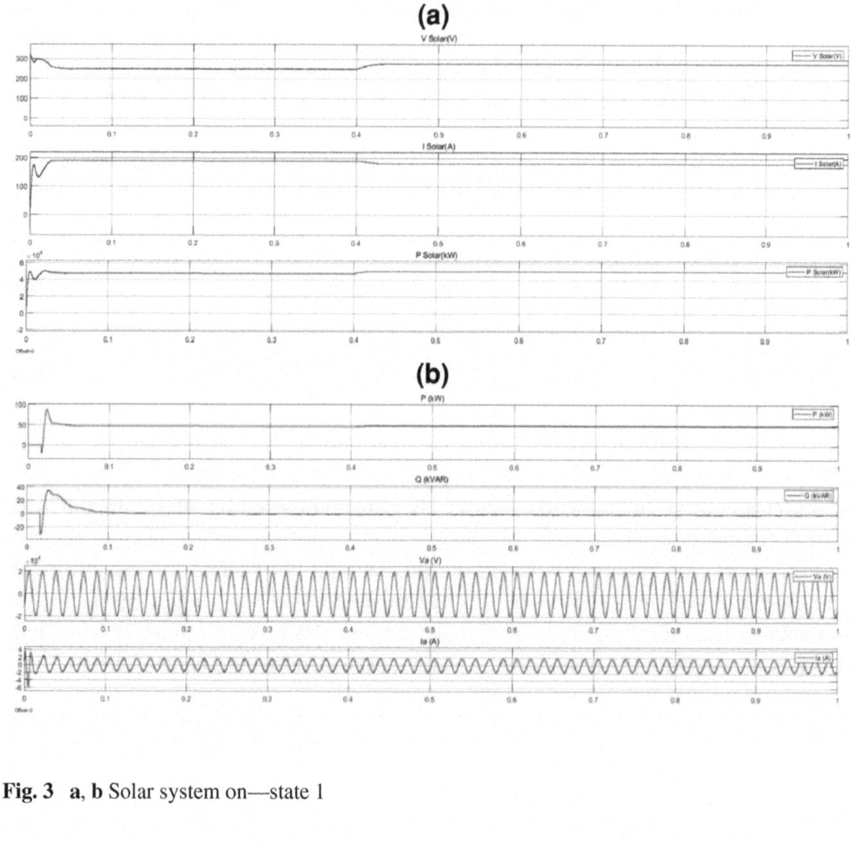

Fig. 3 **a**, **b** Solar system on—state 1

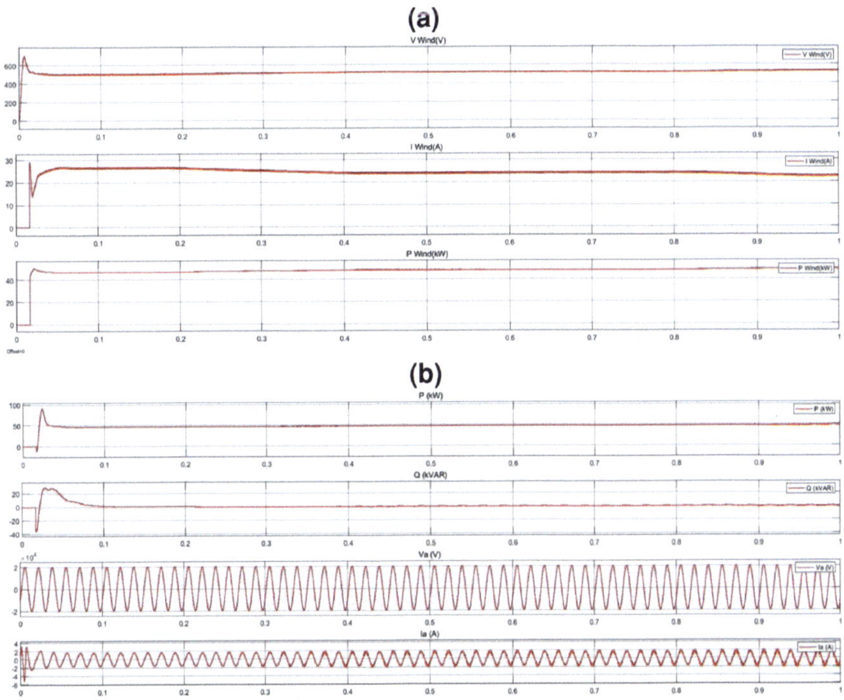

Fig. 4 **a**, **b** Wind system on—state 2

State 2: (Wind power is available).

At $t = 0$ s, the load power required by critical load is 50 kW and is supplied by wind system alone. Voltage, current, and power associated under this state are shown in Fig. 4.

State 3: (Solar and wind available, utility grid-off).

In this, a load of 100 kW is required to supply critical load and load on AC side. To meet this demand, solar and wind are supplying the power required by load as shown in Fig. 5.

State 4: (Solar and battery available, utility grid-off).

At $t = 0.5$ s, the load power requirement is 80 MW. Solar power source is considered as primary source of power in this case. Up to 0.5 s power required by the load is 50 kW which is being supplied by solar source as shown in Fig. 6.

State 5: (Wind and battery available) (Fig. 7).

At $t = 0$ s, power required by the load is 50 kW, and wind source is supplying this power. At $t = 0.5$ s, load requirement is 80 kW as is shown in Fig. 5, and remaining 30 kW load is supplied by battery.

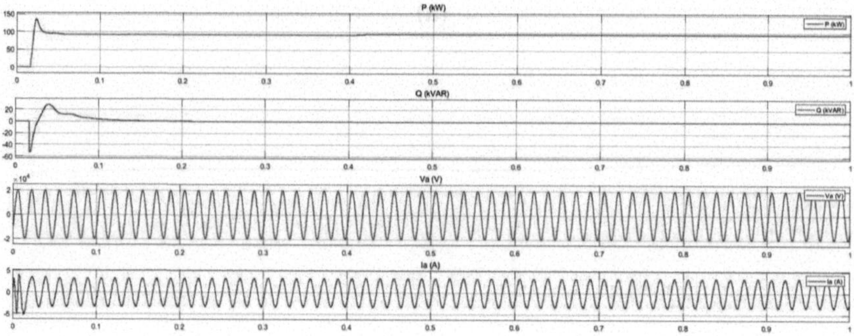

Fig. 5 Solar and wind on—state 3

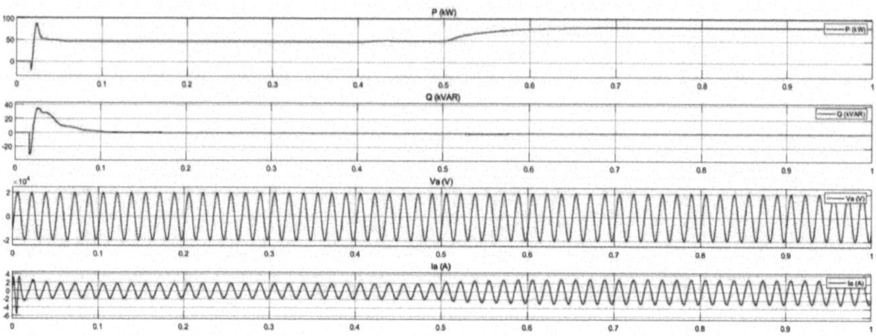

Fig. 6 Solar and battery on—state 4

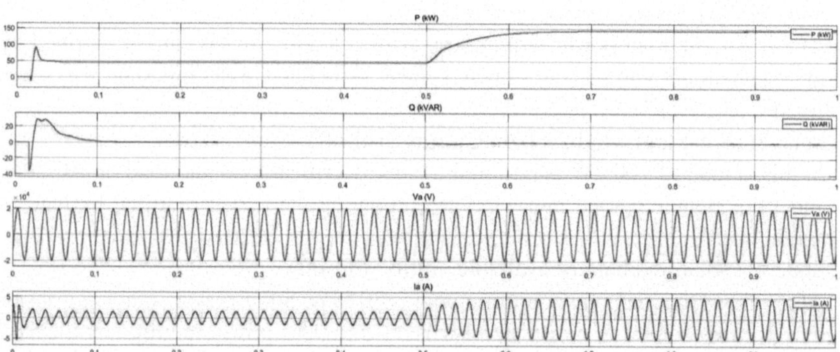

Fig. 7 Wind and battery on: state 5

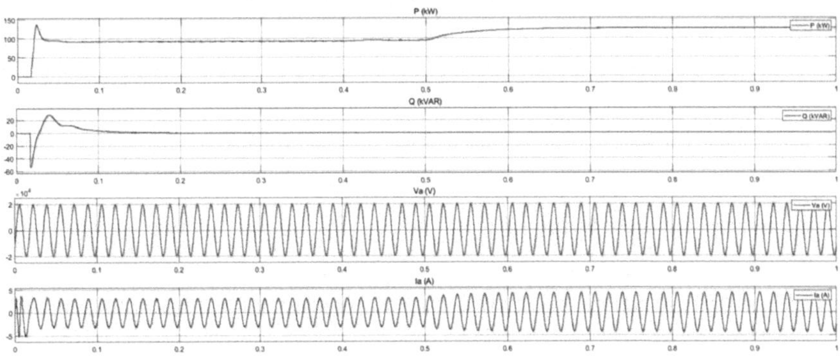

Fig. 8 Solar, wind and battery on: state 6

State 6: (Solar, wind, battery-on, utility grid-off).

At $t = 0$ s, Solar and wind are meeting the demand of 100 kW, and 50 kW is being supplied by solar and wind each. At $t = 0.5$ s, load is increased to 130 kW. Additional 30 kW now is being supplied by the battery as is clear from the graphs shown in Fig. 8.

Voltage, current, active power, reactive power of DG1 and DG2 are shown during islanded mode in Figs. 2, 3, 4, 5, 6, and 7. It can be seen in the waveforms that responses of DGs are settled within 60 ms. However, the voltage and current changes according to the load shared by each source.

4 Conclusions

Simulation studies of microgrid with solar system, wind power system, and battery connected in islanded mode are presented here. In islanded mode, the voltage and frequency regulation is the main criteria. Decoupling PQ control switches to V/f control under islanded mode of operation and presented in this work. Switching sequences under islanded mode is presented. Voltage, current, active power, reactive power of DG1 and DG2 are shown during islanded mode in Figs. 1, 2, 3, 4, 5 and 6. It can be seen in the waveforms that responses of DGs are settled within 60 ms. However, the voltage and current changes according to the load shared by each source. Further work can be carried out on grid connected mode with another distributed generator with the system under consideration. In this work, DG voltage and current with battery characteristics under islanded mode are presented. The performance of the system is analyzed using step responses. The voltages are maintained within the operating limits. The effectiveness of the decoupling PQ control with V/f is presented and analyzed through simulation. This analysis can be applied further to select the protection scheme for the microgrid.

References

1. Lasseter RH, Paigi P (2004) Microgrid: a conceptual solution. In: 30th annual IEEE power electronics specialists conference, Germany
2. Gram Oorja Solutions, A paper titled "Mini-grids in rural India—Reflections of an entrepreneur" and other details available at http://gramoorja.in/microgrid-projects.php
3. Mahmoud MS, Rahman MS, Fouad MS (2015) Review of microgrid architectures-a system of systems perspective. Inst Eng Tech (IET) J
4. Lasseter RH, Eto JH, Schenkman B, Stevens J, Vollkommer H, Klapp D, Linton E, Hurtado H, Roy J (2010) CERTS microgrid laboratory test bed. IEEE Trans Power Del J
5. Freitas DCC, Lima AMN, Morais MRA (2016) Determining lead-acid battery DC resistance by tremblay battery model. 7th Int Renew Energy Congr (IREC) Hammamet 2016:1–4
6. Mohamed A, Elshaer M, Mohammed O (2011) Bi-directional AC-DC/DC-AC converter for power sharing of hybrid AC/DC systems, IEEE
7. Qin LJ, Yang WT Micro-grid converter droop control strategy and simulation. Adv Electr Comput Eng
8. Modi S, Anand A, Usha P (2015) Microgrid control: a comparative study on control strategies for controlling the circulating current. In: 2015 communication, control and intelligent systems (CCIS), IEEE, INSPEC Accession Number: 15872342
9. Unamuno E, Barrena JA (2015) Hybrid AC/DC microgrids—Part II: review and classification of control strategies. Renew Sustain Energy Rev 52(3):1123–1134
10. Eghtedarpour N, Farjah E (2014) Power control and management in a hybrid AC/DC microgrid. IEEE Trans Smart Grid 5(3):1494–1505

Improved Adaptive Control Algorithm of a Grid-Connected PMSG-Based Wind Energy Conversion System

Masood Nazir, Ikhlaq Hussain, and Aijaz Ahmad

Abstract This paper investigates a permanent magnet synchronous generator (PMSG)-based wind energy conversion system (WECS) evaluating various conditions of power flow. The variable parameter reweighted zero-attracting least mean square (VP-RZA-LMS) algorithm is applied to the voltage source inverter (VSI) to account for a variety of power quality problems like compensation of reactive power, power balancing and active power transfer. A Simulink model is developed to validate the same. The proposed control algorithm provides faster convergence speed and is also dependent on a relatively lesser number of parameters.

Keywords Wind energy conversion system (WECS) · Permanent magnet synchronous generator (PMSG) · Power quality · Least mean square (LMS) · Power electronics

1 Introduction

Over the years, the topic of energy crisis has hogged the limelight partly due to the ever-increasing demand of electrical energy and partly due to the rapid extinction of non-renewable energy sources. The swell in pollution levels due to the exploitation of non-renewable energy sources coupled with the surge in the prices of traditional energy sources has compelled researchers to look for viable alternatives. Sustainable

M. Nazir (✉) · A. Ahmad
Department of Electrical Engineering, National Institute of Technology Srinagar, Srinagar, India
e-mail: masoodnazir111@gmail.com

A. Ahmad
e-mail: aijaz54@nitsri.net

I. Hussain
Department of Electrical Engineering, University of Kashmir, Zakura Campus, Srinagar, India
e-mail: ikhlaqiitd2015@gmail.com

M. N. Favorskaya et al. (eds.), *Innovations in Electrical and Electronic Engineering*, Lecture Notes in Electrical Engineering 661, https://doi.org/10.1007/978-981-15-4692-1_13

energy sources, viz solar and wind, are an option in remote areas [1–4] where the cost of electricity by traditional means is a costly affair.

Although different types of generators with multiple converter topologies have been mentioned in the literature [5, 6], it has not been possible to completely penetrate the renewable energy sector yet. However, with proper government support in the form of national schemes and subsidies along with the advancement in power electronics [7], an effort is being made to take the renewable energy generation to the next level. The constant speed wind turbines use permanent magnet synchronous generators (PMSGs) to deliver power at a constant frequency. The advantage that a PMSG [8–11] generator provides over a regular induction generator is that there is no need of external excitation, and it is easier to control. Besides, it provides higher efficiency and reliability as well as its size is considerably small. Also, speed control is not required that is quite a task for variable speed wind turbines [12–14].

A lot of control algorithms have been devised for robust and reliable switching of the voltage source inverter (VSI) [15]. In this paper, the least mean square (LMS) algorithm is improved upon by including powerful adaptive filtering methods for system identification. Till date, several algorithms based on LMS have been proposed like proportionate normalized LMS (PNLMS), zero-attracting LMS (ZA-LMS) and resized zero-attracting LMS (RZA-LMS) [16]. Whereas the former repeatedly adjusts the step size that varies in accordance with the filter coefficient by updating it independently, the latter introduces variable step size strategy as the name suggests. Though ZA-LMS and RZA-LMS perform marginally better than LMS in varied situations, they are relatively tricky to implement because adjusting the step size is not easy.

In this paper, a permanent magnet brushless direct current (PMBLDC)-based wind energy conversion system (WECS) is suggested with a seamless variable parameter ZA-LMS (VP-ZA-LMS) control of VSI. The PMBLDC generator transforms the mechanical energy of wind into electrical energy which is fed to a diode bridge rectifier (DBR) for conversion from AC to DC. This power is fed at the DC bus of the converter by incorporating a boost converter that carries out the tracing of supremum power (MPPT). The advantages of the proposed control algorithm are:

- It provides faster convergence speed.
- This algorithm depends on lesser number of parameters that do not affect the system performance.

The proposed grid-tied WECS is simulated in MATLAB environment.

2 System Topology

A WECS is pictorially represented in Fig. 1. It consists of a wind turbine, PMSG machine and a DBR that is coupled to a converter. The wind energy source is connected at the DC bus of the converter to maintain a constant DC bus voltage. The wind turbine is coupled to a PMSG generator whose output is fed to a DBR. The DC

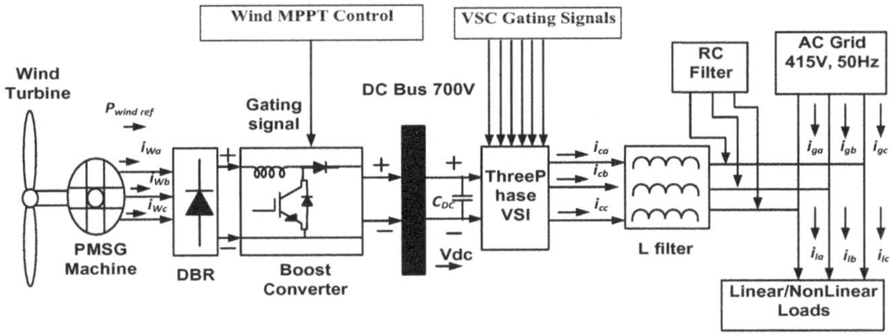

Fig. 1 System topology

power thus obtained is input to a boost converter that is responsible for MPPT. The system is operated in such a way so as to trace continuously its supremum point of power for maximum efficiency. For improvement of power quality and to account for harmonic suppressions, interfacing inductances and RC filter are connected at the point of common coupling (PCC).

3 Modeling and Design of WECS

Different components that comprise a WECS are a wind turbine, PMSG machine and a DBR that is coupled to a converter. The modeling of these components is described in this section.

3.1 Mathematical Model of Wind Turbine

The mechanical energy produced from wind is a function of wind speed [2]. It is given as

$$P_w = \frac{1}{2}\rho A C_P(\lambda, \beta)v^3 \tag{1}$$

$$\lambda = w_m \frac{R}{v} \tag{2}$$

$$C_p = C_1(\frac{C_2}{\gamma} - C_3\beta - C_4)e^{\frac{C_5}{\gamma}} + C_6\gamma \tag{3}$$

where ρ = air density, v = wind speed (m/s), A = area swept by the rotor, C_p = power coefficient, β = pitch angle and R = rotor radius. The wind turbine torque is given as

$$T_m = \frac{p_m}{w_m} \tag{4}$$

The largest value of C_p is $C_{p\max} = 0.48$ and is obtained for pitch angle = 0 and λ = 8.1 so that the power extracted is maximum and is given by

$$P_{\max} = \frac{1}{2}\rho A C_P \frac{R^3}{\lambda_{\text{opt}}^3} w_m^3 \tag{5}$$

3.2 Modeling of PMSG

A PMSG machine of power rating 3.7 kW with the winding resistance of stator as R_{st}, stator flux (ψ_{fsd}, ψ_{fsq}), direct and quadrature axis stator inductances as (L_{sd}, L_{sq}), PM flux linkage as ψ_{pm}, n number of pole pairs and dq axis voltages and currents of stator as V_{sdq}, i_{sdq} coupled with a 10 kW wind turbine comprises a WECS. The mathematical equations are expressed as [14]

$$V_{sd} = -R_{st} i_{sd} - p\psi_{fsd} + n w_m \psi_{fsq} \tag{6}$$

$$V_{sq} = -R_{st} i_{sq} - p\psi_{fsq} - n w_m \psi_{fsd} \tag{7}$$

$$\psi_{fsd} = L_{sd} i_{sd} + \psi_{pm} \tag{8}$$

$$\psi_{fsq} = L_{sq} i_{sq} \tag{9}$$

3.3 Boost Converter Calculations

The duty cycle (D) and the values of capacitor and inductor of the boost converter are calculated as [15]:

$$V_{\text{out}} = \frac{V_{\text{in}}}{1 - D} \tag{10}$$

$$L = V_{in} \frac{V_{\text{out}} - V_{\text{in}}}{i_{\text{ripple}}} f_s V_{\text{out}} \tag{11}$$

$$C = I_{\text{out}} \frac{D}{\delta V f_s} \tag{12}$$

where f_s is 10 kHz, V_{in} is the voltage input to boost converter, V_{out} is the equivalent DC bus voltage and δV is the acceptable voltage distortion content chosen as 1–2% of V_{out}.

3.4 Design of VSC

A three-legged VSC that is meant for compensating reactive power in the system comprises six IGBTs, three AC inductors and a DC capacitor. It is designed for a 415 V, 50 Hz system [15].

3.5 Design of Interfacing Inductance and RC Filter

An RC filter is used to refine the voltage at the point of common coupling (PCC) by draining out noise from it. Considering $R_f C_f = T_s/10$ and switching frequency equal to 2 kHz, the specifications of ripple filter are calculated as $R_f = 10\,\Omega$ and $C_f = 10\,\mu\text{F}$, where R_f, C_f are the resistance and capacitance of the ripple filter and T_s is the switching time [15]. The value of interfacing inductance is taken as 5.5 mH [15].

3.6 Scheme of DC Link Capacitor and DC Bus Voltage

DC link capacitance is calculated as [15]

$$\frac{1}{2}C_{DC}(V_{DC}^2 - V_{DC1}^2) = k3VmIt \tag{13}$$

where V_{DC} is the reference DC voltage and V_{DC1} is the minimum voltage level of the DC bus, m is the overloading factor, V is the phase voltage, I is the phase current and t is the time by which the DC bus voltage is to be recovered. The value of DC link capacitance thus calculated is 10,000 μF.

The least value of DC bus voltage needs to be higher than two times the phase voltage peak. It is calculated as

$$V_{DC} = \frac{2\sqrt{2}V_{LL}}{\sqrt{3}p} \tag{14}$$

where p is the modulation index=1 and VLL is the line to line output voltage. V_{DC} thus obtained is 677.7 V for a line voltage of 415 V and is rounded off as 700V.

4 Control Algorithm

The system is handled by regulating the DC current of the circuit in such a way such that the point corresponding to supremum power (MPP) is traced continuously. The control algorithm consists of wind control and VSC control which are mentioned in the following section.

4.1 Wind Generator Boost Converter Control

The control, shown in Fig. 2, uses a PI controller which reduces the error between the desired current and the actual DC current to zero and in the process draws out maximum generator power. The PI controller input is an error signal that is generated as

$$I_{\text{wind, err}} = \frac{P_{\text{wind, ref}}}{v_{\text{windDC}}} \tag{15}$$

The boost converter utilizes this error signal input to a PI controller to control the output DC current and power.

4.2 Control Algorithm for VSC

In this paper, a VP-RZA-LMS algorithm [16] as shown in Fig. 3 is proposed where the parameters like step size are repeatedly adjusted. This method keeps on minimizing the mean square error (MSE) and converges relatively faster. The simulation results further validate the performance of the algorithm. Initially, the unit template components are calculated as [15]:

$$\mu_{pa} = \frac{v_{sa}}{v_t}; \ \mu_{pb} = \frac{v_{sb}}{v_t}; \ \mu_{pc} = \frac{v_{sc}}{v_t} \tag{16}$$

Voltage at PCC is given by

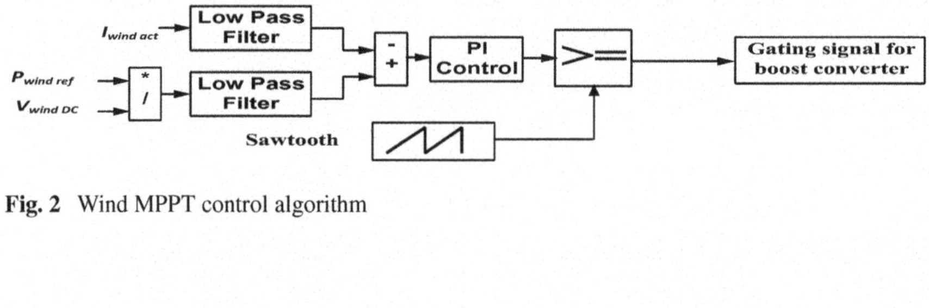

Fig. 2 Wind MPPT control algorithm

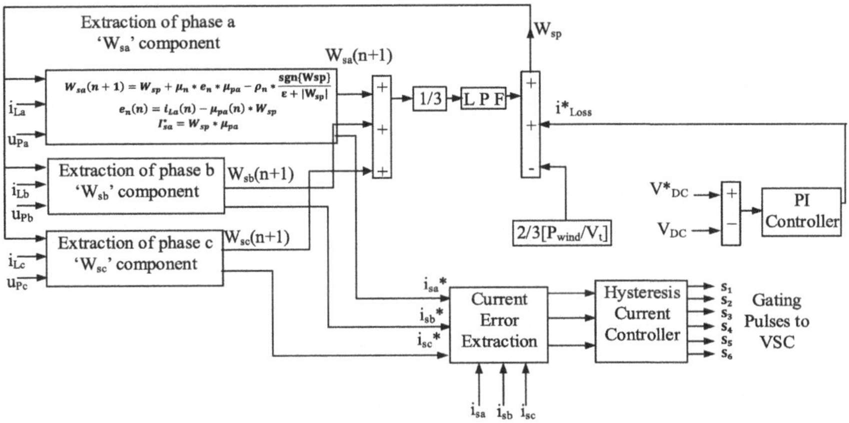

Fig. 3 Proposed control algorithm for VSC

$$V_t = \sqrt{2/3(v_{sa}^2 + v_{sb}^2 + v_{sc}^2)} \tag{17}$$

Next, the corresponding error is obtained by utilizing the unit templates as shown in

$$e_n = i_{La} - \mu_{pa} W_{sp} \tag{18}$$

The RZA-LMS update is obtained by using stochastic subgradient iterations and is given by

$$W(n+1) = W_n + \mu e_n w_n - \rho \frac{\text{sgn}[W_n]}{\epsilon + |W_n|} \tag{19}$$

where the absolute value operator is applied in an element-wise manner and ϵ is a small positive parameter. The parameter w_n is initialized as $w_n = 0.01$.

The active weight components for the three phases are estimated by subtracting w_n from the above equation and subsequently substituting (18) in it. The active weight components are given by equations (20), (21) and (22)

$$W_{sa}(n+1) = W_{sp} + \mu_n e_n \mu_{pa} - \rho_n \frac{\text{sgn}[W_{sp}]}{\epsilon + |W_{sp}|} \tag{20}$$

$$W_{sb}(n+1) = W_{sp} + \mu_n e_n \mu_{pb} - \rho_n \frac{\text{sgn}[W_{sp}]}{\epsilon + |W_{sp}|} \tag{21}$$

$$W_{sc}(n+1) = W_{sp} + \mu_n e_n \mu_{pc} - \rho_n \frac{\text{sgn}[W_{sp}]}{\epsilon + |W_{sp}|} \tag{22}$$

Adding Eqs. (20), (21) and (22), the mean value of the fundamental active component is given as

$$W_p = \frac{1}{3}(W_{sa} + W_{sb} + W_{sc}) \tag{23}$$

The reference supply currents for the three phases are calculated as

$$i_{sa}^* = W_{sp} * \mu_{pa};\, i_{sb}^* = W_{sp} * \mu_{pb};\, i_{sc}^* = W_{sp} * \mu_{pc} \tag{24}$$

5 System Performance

The proposed system consists of a 10 kW wind turbine that is coupled to a 3.7 kW PMSG machine. The output is fed to a bridge rectifier via a boost converter. The VSC feeds power to linear/nonlinear loads. The source voltage is 415 V at 50 Hz. The system is operated under steady-state and dynamic conditions by varying wind speed and incorporating load unbalancing conditions. A Simulink model is designed to validate the results in detail.

5.1 Steady-State Performance Under Nonlinear Load

The conduct of a grid-tied WECS evaluated on the application of a nonlinear load is shown in Fig. 4. It shows the waveforms of source voltage v_{abc}, source current i_{abc}, load current i_{Labc}, converter current i_{vsc}, DC bus voltage v_{DC}, wind velocity v_{wind}, wind current i_{wind} and wind power p_{wind}. It is clear that when the load current (2 A) is less than the converter current (5 A), the wind source supplies power to both load as well as grid. However, when the load current exceeds the converter current, the grid supplies power to both load as well as the wind source.

5.2 Steady-State Performance Under Linear Load

Figure. 5 shows the behavior of a grid-tied WECS evaluated on the application of a linear load. It depicts the waveforms of source voltage v_{abc}, source current i_{abc}, load current i_{Labc}, converter current i_{vsc}, DC bus voltage v_{DC}, wind velocity v_{wind}, wind current i_{wind} and wind power p_{wind}.

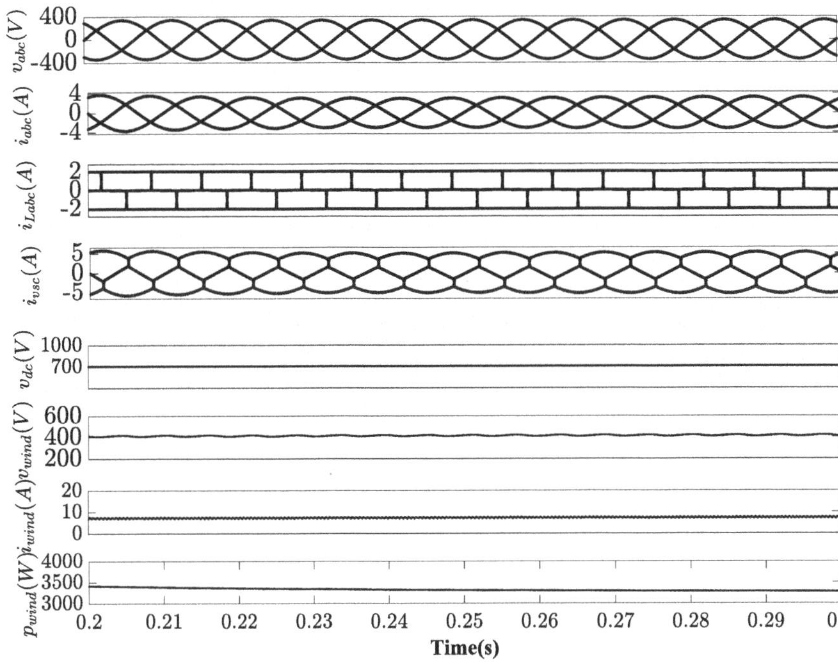

Fig. 4 Steady-state performance under nonlinear load when i_{load} is less than i_{vsc}

Figure 5a depicts that when the load current is less than the converter current, wind supplies power to both load and grid. However, when the load current is greater than converter current as shown in Fig. 5b, grid supplies power to both load and wind energy source.

5.3 Dynamic Performance Under Nonlinear Load

Figure 6 shows the behavior of a grid-tied WECS evaluated on the application of a nonlinear load. It depicts the waveforms of source voltage v_{abc}, source current i_{abc}, load current i_{Labc}, converter current i_{vsc}, DC bus voltage v_{DC}, wind velocity v_{wind}, wind current i_{wind} and wind power p_{wind}. It is observed that at time $t = 0.5$ s, phase B of the load is reconnected to the system, but grid current remains sinusoidal and balanced. It is also observed that DC link voltage is maintained close to reference value under all disturbances.

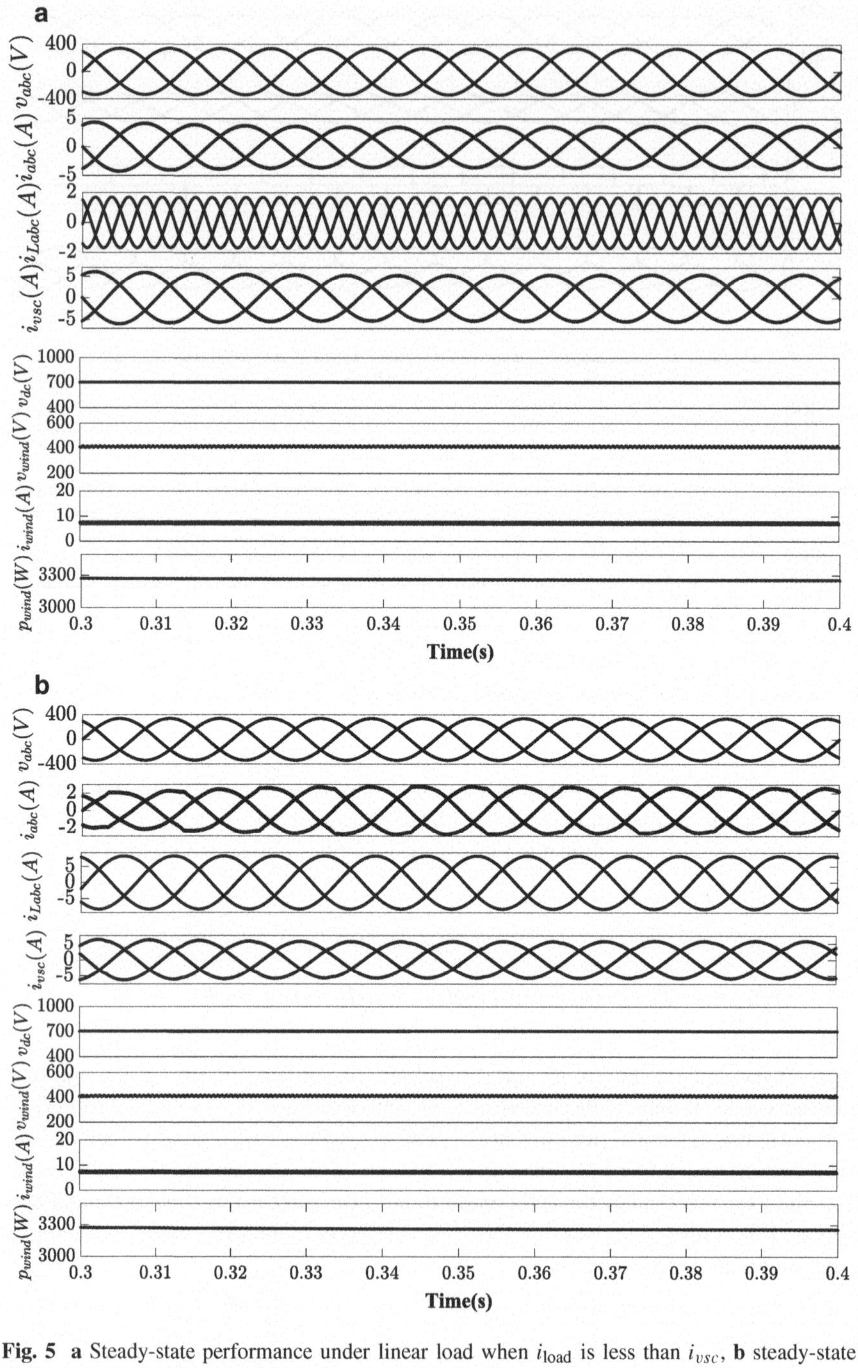

Fig. 5 **a** Steady-state performance under linear load when i_{load} is less than i_{vsc}, **b** steady-state performance under linear load when i_{load} is greater than i_{vsc}

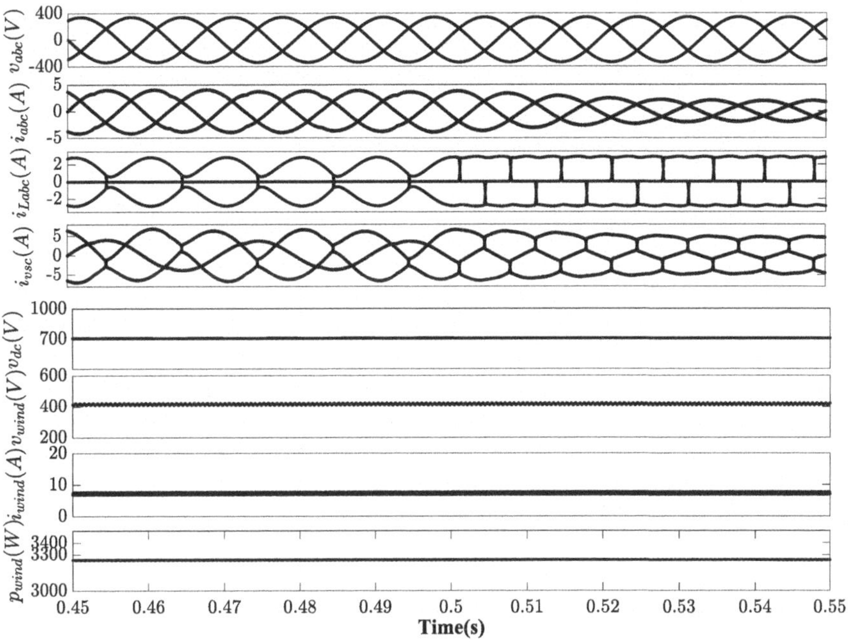

Fig. 6 Dynamic performance under a nonlinear load

5.4 Dynamic Performance Under Nonlinear Load Due to Change in Wind speed

The dynamic performance under a nonlinear load due to change in wind speed is shown in Fig. 7. It shows the waveforms of source voltage v_{abc}, source current i_{abc}, load current i_{Labc}, converter current i_{vsc}, DC bus voltage v_{DC}, wind velocity v_{wind}, wind current i_{wind}, wind power p_{wind}, wind speed v (m/s), rotor speed v_r and torque T_m. It is observed from the figure that wind power is close to 3400 W at a wind speed of 12 m/s, and wind source supplies power to the load. However, at $t = 0.5$ s to $t = 1$ s, when the wind speed is reduced from 12 to 7 m/s, it is observed that the wind power reduces from 3400 to 1500 W. Under such conditions, the wind source is unable to supply power to the load which is evident from the waveforms of Fig. 7. Under such conditions, grid supplies power to the load.

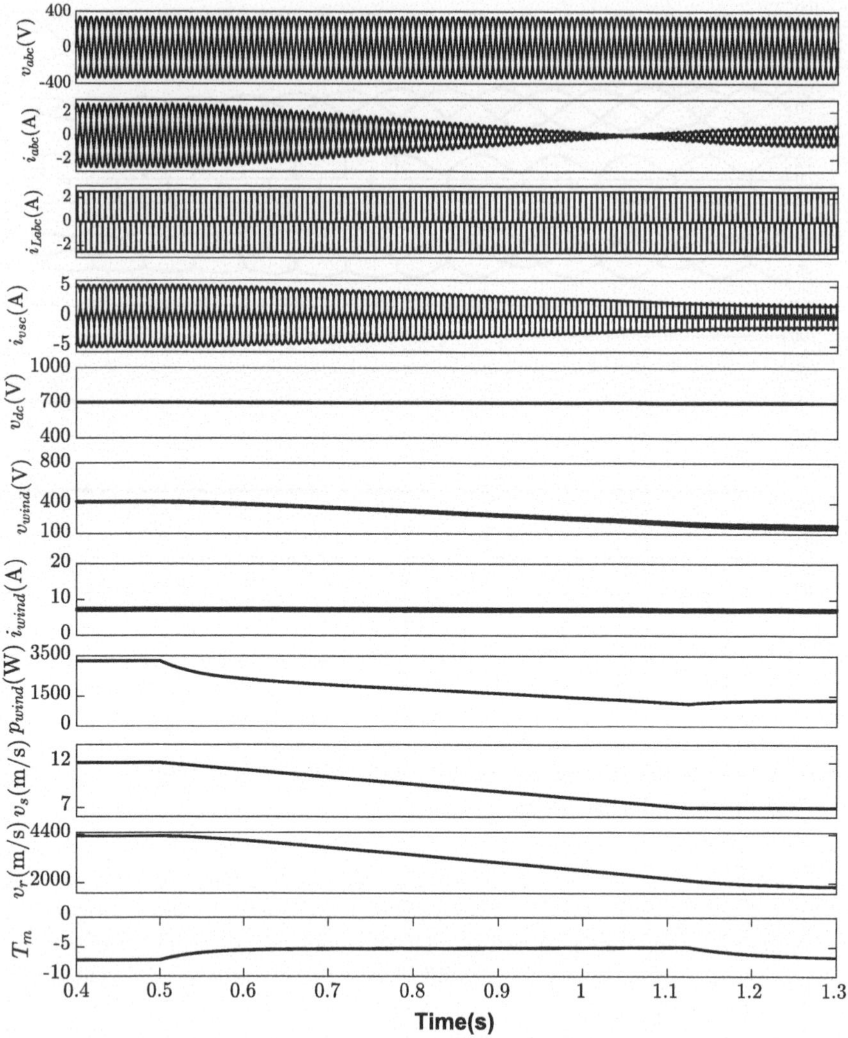

Fig. 7 Dynamic performance under nonlinear load due to wind speed variation

6 Conclusions

The performance of a grid-integrated WECS has been carried out in SIMULINK. The PMSG transforms the mechanical energy of the wind into electrical energy which has been rectified and conditioned and finally fed to the grid. Adaptive control of WECS is incorporated by implementing VP-RZA-LMS algorithm to provide power quality solutions like power balancing, active power transfer and voltage regulation.

Also, favorable results have been obtained under changing wind speeds. Such a setup is very much capable of providing an able backup in remote locations.

Appendix

Wind Turbine: Cut in wind speed = 6 m/s, rated wind speed = 12 m/s, $C_{p\max} = 0.48$, $\lambda_{TSR} = 8.1$, $C_1 = 0.5175$, $C_2 = 116$, $C_3 = 0.4$, $C_4 = 5$, $C_5 = 21$, $C_6 = 0.0069$,
PMSG Rating: 5 hp, 230 V, n = 4, R_s = 1.785 Ω, Stator Phase Inductance = 9.065 mH
System Parameters: Boost converter inductance, $L_b = 10\,\text{mH}$, DC link capacitance, $C_{dc} = 10,000\,\mu\text{F}$, $K_p = 1$, $K_i = 0$, $V_{dc} = 700\,\text{V}$, Adaptive filter constant, $\mu_n = 0.01$, $\rho_n = 0.006$, $\epsilon = 0.002$.
Interfacing inductance, $L_{int} = 5\,\text{mH}$, 3 phase grid voltage, $V_{abc} = 415\,\text{V}$ (rms), $R_f = 10\,\Omega$, $C_f = 10\,\mu\text{F}$, Sampling time, $T_s = 10\,\mu\text{s}$.

References

1. Pimentel D (2008) Biofuels, solar and wind as renewable energy system-benefits and risks. Springer, Colorado
2. Wu B, Lang Y, Zargari N, Kouro S (2011) Power conversion and control of wind energy systems, Wiley, Inc., Hoboken, New Jersey
3. Rashid A, Hasan N, Parvez KT, Maruf (2015) MNI Study and analysis of a small-scale microgrid using renewable energy resources. In: Proceedings of international conference on electrical engineering and information communication technology (ICEEICT). Dhaka, pp 1–4
4. Lasseter RH (2011) Smart distribution: coupled microgrids. Proc IEEE 99(6):10/4-1082
5. Yahia H, Liouane N, Dhifaoui R (2014) Differential evolution method based output power optimisation of switched reluctance generator for wind turbine applications. IET Renew Power Gener 8(7):795–806 September
6. Rezaei E, Ebrahimi M, Tabesh A Control of DFIG wind power generators in unbalanced microgrids based on instantaneous power theory. IEEE Trans Smart Grid PP(99):1–8
7. Mulholland R, McBride V, Vial C, O'Malley A, Bennett D (2015) 2015 top markets report renewable energy. http://www.trade.gov/industry
8. Sun L, Gong C, Han F (2013) Design and optimization of control parameters based on direct-drive permanent magnet synchronous generator for wind power system. In: Proceedings in 8th IEEE conference on industrial electronics and applications (ICIEA). VIC, Melbourne, pp 1238–1243
9. Haque ME, Muttaqi KM, Negnevitsky M (2008) Control of a stand alone variable speed wind turbine with a permanent magnet synchronous generator. In: Proceedings in IEEE power and energy society general meeting—conversion and delivery of electrical energy in the 21st century. Pittsburgh, PA, pp 1–9
10. Pradhan S, Murshid S, Singh B, Panigrahi BK (2018) A robust SMOC for vector controlled PMSG based isolated wind energy generating system. In: IEEE industry applications society annual meeting (IAS). Portland, OR, pp 1–8
11. Pradhan S, Singh B, Panigrahi BK, Murshid S (2019) A composite sliding mode controller for wind power extraction in remotely located solar PV-wind hybrid system. IEEE Trans Ind Electron 66(7):5321–5331 July

12. Singh B, Panigrahi BK, Pathak G (2016) Control of wind-solar microgrid for rural electrification. In: IEEE 7th Power India international conference (PIICON). Bikaner, pp 1–5
13. Mishra S, Hussain I, Pathak G, Singh B (2018) dPLL-based control of a hybrid wind–solar grid connected inverter in the distribution system. IET Power Electron 11(5):952–960
14. Pathak G, Singh B, Panigrahi BK, Chandra A, Al-Haddad K (2016) Wind-PV based microgrid and its synchronization with utility grid. In: 2016 IEEE international conference on power electronics, drives and energy systems (PEDES). Trivandrum, pp 1–6
15. Singh B, Chandra A, Al-Haddad K (2014) Power quality: problems and mitigation techniques. Wiley
16. Jin D, Chen J, Richard C, Chen Jingdong (2018) Model driven online parameter adjustment for zero attracting LMS. Signal Process 152:373–383

Life Cycle Reassessment Strategy for Existing Bulk and Thin-Film Photovoltaic Materials in Indian Context

Manisha Sheoran, Susheela Sharma, and Pancham Kumar

Abstract With the growing population, the energy demands are heightened around the globe. Displacement of conventional energy sources to solar power photovoltaic installations has burst forth solar photovoltaic waste volume to around 1.8 million tons by 2050. Hence a crucial need is to enhance the photovoltaic recycling method to vast industry level. Economically established countries like China, UK, and Japan now have formulated their solar photovoltaic recycling policies while India still needs to shape its solar photovoltaic recycling policy. Inspite of the increasing PV installations, recycling of solar photovoltaic waste presents a supreme and uncommonly platform of recovering valuable materials from the PV waste, and harboring the recycling industry to large scale across the world. This paper also targets the apparently popping up environmental, social, and economic influence of the life span accomplishment of C-Si, GaAs, GaInP, CIGS. To potentially tackle the above-mentioned complications a novel policy framework is proposed for the Indian government to administer.

Keywords Life span accomplishment · Policy · Socio-economic impact

1 Introduction

The solar PV installations will prosper to 637–653 GW by the extremity of 2019. By the 2050 major set up is anticipated to reach 4.7 TW in China and India. Extensive growth of PV materials has occurred across the countries. These increasing set up are contributing to PV waste proliferation. In 2017 China, United States, India, and Japan had the maximum solar PV installations of 53 GW, 10.6 GW, 9 GW, and 7 GW, respectively [1]. India as an economically emerging nation holds large population density in less area of land so it becomes difficult to recycle the total PV installation of 28.18 GW as estimated till March 2019 [2]. Overall 1.8 million tons of solar PV waste will be accumulated till 2050 in India. Accumulation of this huge PV

M. Sheoran (✉) · S. Sharma · P. Kumar
Bhartiya Skill Development University Jaipur, Jaipur, Rajasthan, India
e-mail: manisha.sheoran@ruj-bsdu.in

M. N. Favorskaya et al. (eds.), *Innovations in Electrical and Electronic Engineering*, Lecture Notes in Electrical Engineering 661, https://doi.org/10.1007/978-981-15-4692-1_14

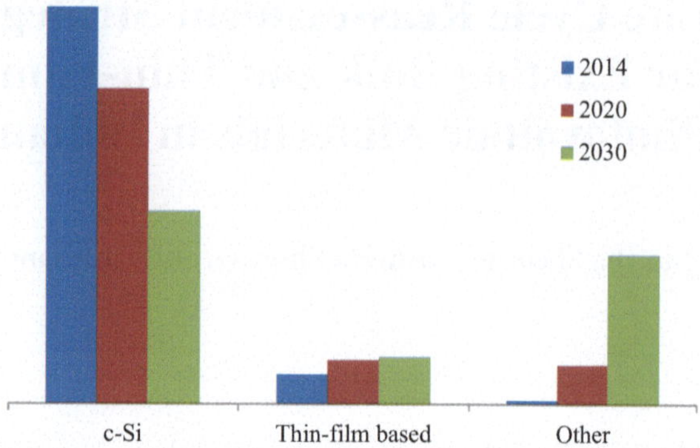

Fig. 1 Market share of PV panels from 2014 to 2030

waste demands a mandatory policy framework design for PV waste recycling [3]. Environmental, social and economic repercussions due to existing solar PV recycling methods like mechanical, thermal, and chemical are investigated thoroughly [4, 5]. Various prior literature studies have been done on the life span reassessment of existing solar PV materials but still their recycling part is lagging behind in terms of policy framework design. Until now landfill dumping is utilized for the namesake of recycling of dismantled and discarded PV materials [6]. Recycling processes which are existing still lack incentive allocation. Lucrative turnout is not obtained from the solar PV recycling process because of very less PV waste generation as the solar PV panels have a life span ranging from 25 to 30 years [7]. Market share of C-Si has been reducing from 2014 to 2030 as depicted in Fig. 1.

First solar had developed methods of recycling thin-film PV materials, however, a great significant contribution has been already made for C-Si [8]. The recycling process is not economical to recyclers as the PV waste reaching to them is less in quantity and they are not provided any incentives for losses in terms of cost after recycling process [9]. Recycling process requires economic aid for various stages involved [10]. C-Si, GaAs, CIGS, GaInP during their manufacturing involves the use of hazardous substances which influence the biotic components [11]. Disposal of the above materials is also a cumbersome task due to further slow leaching of chemicals from discarded solar panels [12, [13].

2 Methodology

Reducing the PV waste material from the faulty solar panels is the foremost step of recycling [14, 15]. Intrinsic composition of solar photovoltaic materials is depicted in Fig. 2 and Table 1.

Till the present time thermal, chemical, and mechanical methods of recycling were used [16]. Si can be regenerated from processed and dismantled solar panels through the laser method to some extent [17]. Further purification of Si is a high

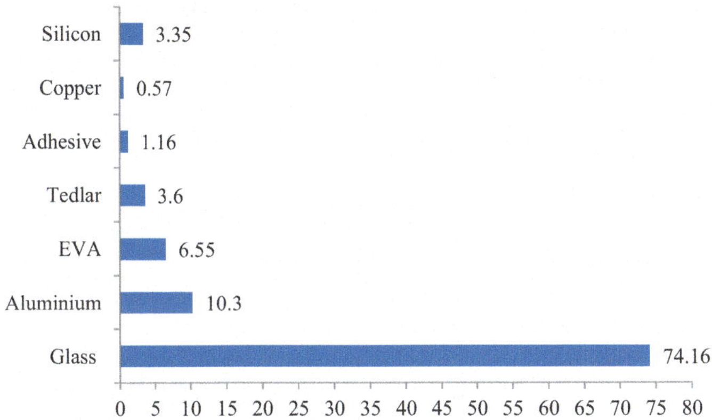

Fig. 2 Intrinsic % C-Si component proportion

Table 1 Intrinsic CIGS and CdTe component proportion

CIGS components	%	CdTe components	%
Glass	84	Glass	95
Aluminium	12	Aluminium	0.35
EVA	3	EVA	3.5
Copper	0.8	Copper	1
Zinc	0.12	Zinc	0.01
Lead	0.0005	Cadmium	0.07
Indium	0.02	Tellurium	0.07

payout step as it requires transportation to China for further processing. Mechanical, thermal, chemical methods are laid out shown in Figs. 3, 4, 5.

2.1 Environmental Aftermath

Various thin-film and bulk materials have been analyzed for their environmental repercussions and are summarized below in Fig. 6.

2.2 Social and Economic Impact

Recycling processes indulge different age group people so it can be taken as a new emerging job sector [18]. Moreover, the recycling companies show a great interest in the extraction of valuable materials like Cu, Ag from the dismantled solar panels [19]. By 2030 estimated recovery of Si is 30.000 tons from the collapsed solar panels

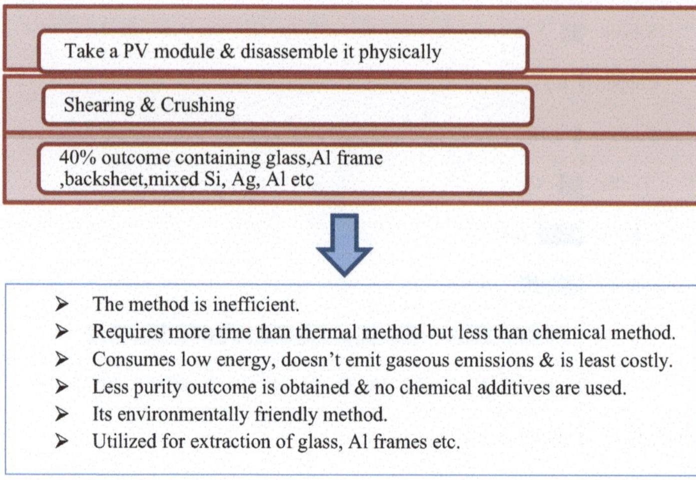

Fig. 3 Mechanical method

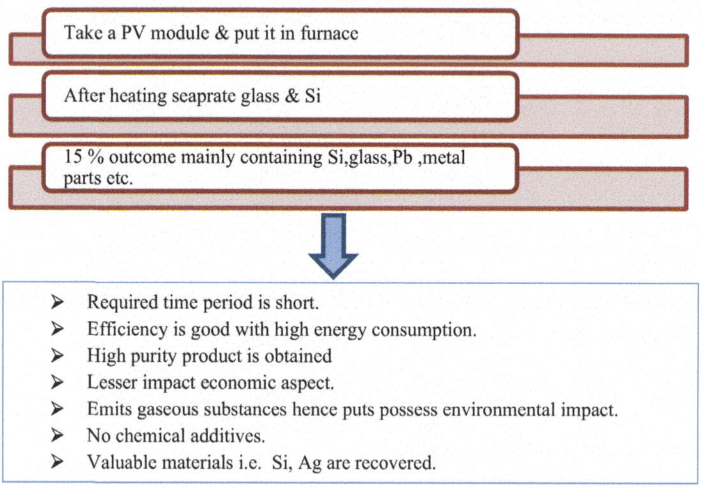

Fig. 4 Thermal method

[2, 20]. Amount of total waste generation by the end of 2050, its recovery %, and total waste recovered in tons is depicted as shown in Figs. 7, 8, and 9.

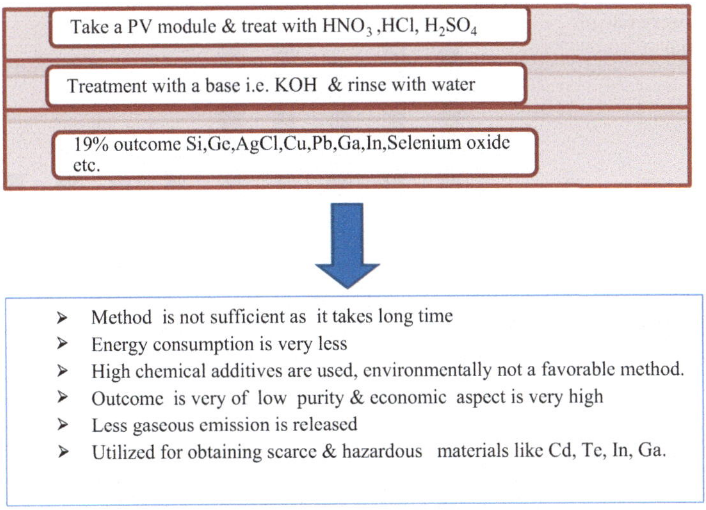

Fig. 5 Chemical method

Fig. 6 Effects on health due to solar photovoltaic materials

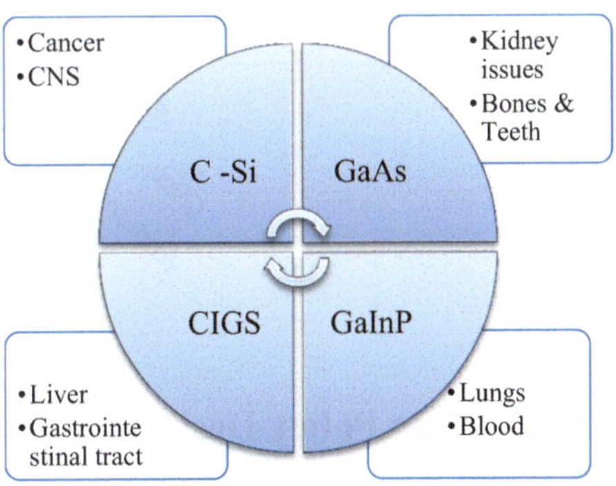

Fig. 7 Waste generated in tons by 2050

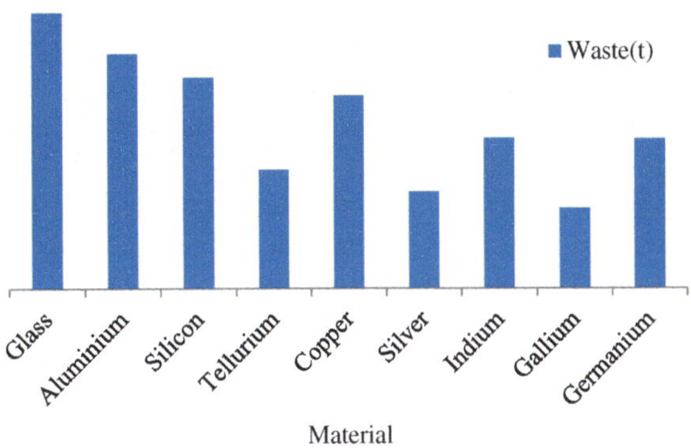

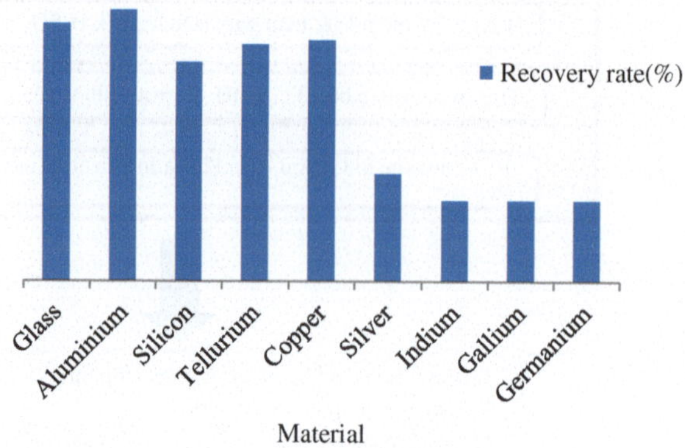

Fig. 8 Recovery rate of waste material

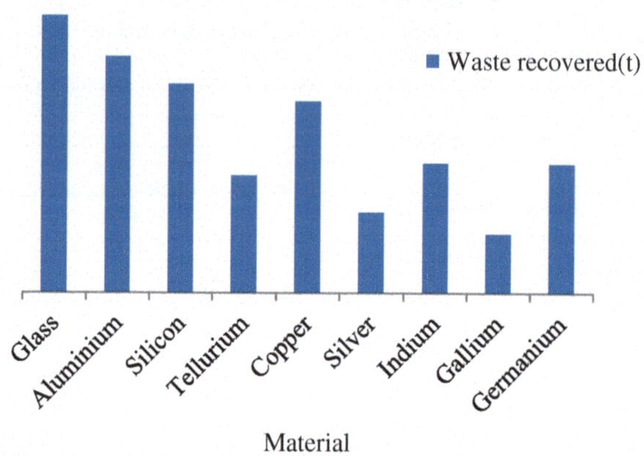

Fig. 9 Waste (in tons) recovered after recycling

3 Recycling Policy Design for Indian Government

With the large accumulation of solar photovoltaic scrap it becomes stringent to framework recycling policy for the Indian government. Bridge to India managing committee aforementioned that solar sector scrap accumulation has heightened to 28 GW in 2019 from a 3 GW in 2014. Presently India lacks proper recycling strategy formulation. Endowment of effective PV scrap recycling policy is to be carried out as yet PV recycling is behind peculiar laws. Solar PV material processing utilizes hazardous substances during various manufacturing steps. If such materials are left without treatment than environmental repercussions are huge [21, 22]. Solar PV recycling legislative recommendations can be urged from members across the globe, as the PV waste is piling across India. The inevitable PV scrap pile up can be handled by administering a strategy. After reaching the last stage of life cycle reassessment accumulated bulk and thin-film waste can be handled by the proposed policy framework for the Indian scenario in Fig. 10.

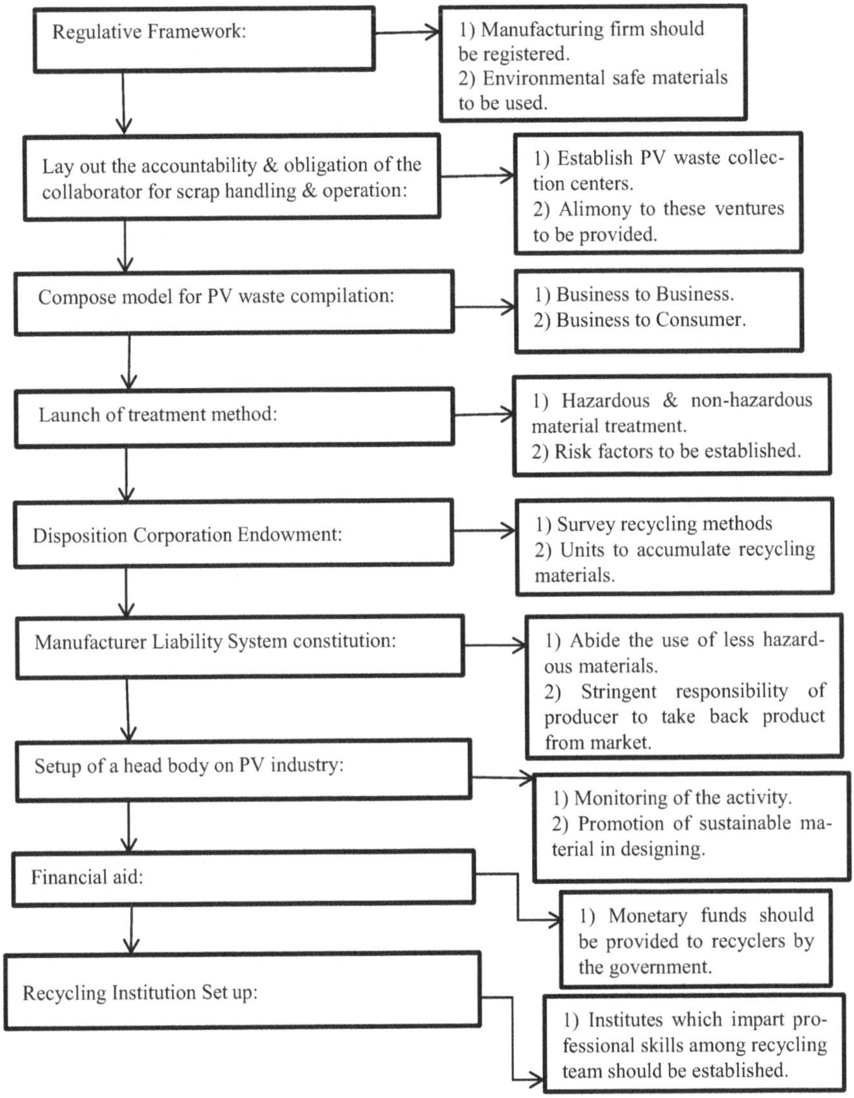

Fig. 10 Proposed policy framework for Indian scenario to handle accumulated bulk and thin-film waste

4 Conclusions

Recycling sector in India is still surviving in the infancy. It's stuck with low outcomes in terms of obtained recycled products. The Indian government needs an effective legislative policy framework for implementation so that the treacherous PV waste which has spawned can be treated without throwing in the garbage. Comparative study of the existing recycling method of PV waste is done by analyzing the mechanical, thermal, and chemical methods. The best method for recycling is a thermal method with some demerits. Everyday new solar plant installation targets are achieved which further increases the waste pile up. By the end of 2050 the global hike in PV waste generation due to C-Si and thin-film PV materials is estimated to reach 4,843,891 and 1,252,617 tons, respectively. Environmental, economic, and social impacts are also studied in a detailed manner in the article. Job industry can also take an impetus due to the recycling process development on a large scale. The novel policy designed hereby can be successfully implemented in both the rural and urban areas of India.

References

1. Suresh S, Singhvi S, Rustagi V (2019) Managing India's PV module waste. Bridge To India, India, Analytical
2. Weckend S, Wade A, Heath GA (2016) End of life management: solar photovoltaic panels. NREL/TP-6A20-73852, 1561525, Aug 2016
3. Strachala D, Hylský J (2017) Methods for recycling photovoltaic modules and their impact on environment and raw material extraction 22(3):13
4. Komoto K, Oyama S, Sato T, Uchida H (2018) Recycling of PV modules and its environmental impacts. In: 2018 IEEE 7th world conference on photovoltaic energy conversion (WCPEC) (a joint conference of 45th IEEE PVSC, 28th PVSEC and 34th EU PVSEC), Waikoloa Village, HI, 2018, pp 2590–2593. https://doi.org/10.1109/pvsc.2018.8547691
5. Latunussa CEL, Ardente F, Blengini GA, Mancini L (2016) Life cycle assessment of an innovative recycling process for crystalline silicon photovoltaic panels. Sol Energy Mater Sol Cells 156:101–111. https://doi.org/10.1016/j.solmat.2016.03.020
6. Lunardi M, Alvarez-Gaitan J, Bilbao J, Corkish R (2018) Comparative life cycle assessment of end-of-life silicon solar photovoltaic modules. Appl Sci 8(8):1396. https://doi.org/10.3390/app8081396
7. Granata G, Pagnanelli F, Moscardini E, Havlik T, Toro L (2014) Recycling of photovoltaic panels by physical operations. Sol Energy Mater Sol Cells 123:239–248. https://doi.org/10.1016/j.solmat.2014.01.012
8. Sinha P, Wade A (2015) Assessment of leaching tests for evaluating potential environmental impacts of PV module field breakage. IEEE J Photovoltaics 5(6):1710–1714. https://doi.org/10.1109/JPHOTOV.2015.2479459
9. Faircloth CC, Wagner KH, Woodward KE, Rakkwamsuk P, Gheewala SH (2019) The environmental and economic impacts of photovoltaic waste management in Thailand. Resour Conserv Recycl 143:260–272. https://doi.org/10.1016/j.resconrec.2019.01.008
10. Goe M, Gaustad G (2016) Estimating direct human health impacts of end-of-life solar recovery. Golisano Institute for Sustainability
11. Fthenakis VM, Kim HC (2011) Photovoltaics: life-cycle analyses. Sol Energy 85(8):1609–1628. https://doi.org/10.1016/j.solener.2009.10.002
12. Sen S, Ganguly S, Das A, Sen J, Dey S (2016) Renewable energy scenario in India: opportunities and challenges. J Afr Earth Sc 122:25–31. https://doi.org/10.1016/j.jafrearsci.2015.06.002
13. Cyrs WD, Avens HJ, Capshaw ZA, Kingsbury RA, Sahmel J, Tvermoes BE (2014) Landfill waste and recycling: use of a screening-level risk assessment tool for end-of-life cadmium telluride (CdTe) thin-film photovoltaic (PV) panels. Energy Policy 68:524–533. https://doi.org/10.1016/j.enpol.2014.01.025

14. Zhang J, Lv F, Ma LY, Yang LJ (2013) The status and trends of crystalline silicon PV module recycling treatment methods in Europe and China'. Adv Mater Res 724–725:200–204. 10.4028/www.scientific.net/AMR.724-725.200
15. Huang B, Zhao J, Chai J, Xue B, Zhao F, Wang X (2017) Environmental influence assessment of China's multi-crystalline silicon (multi-Si) photovoltaic modules considering recycling process. Sol Energy 143:132–141. https://doi.org/10.1016/j.solener.2016.12.038
16. Klugmann-Radziemska E (2012) Current trends in recycling of photovoltaic solar cells and modules waste/Recykling Zużytych Ogniw I Modułów Fotowoltaicznych—Stan Obecny. Chem Didactics Ecol Metrol 17(1–2):89–95. https://doi.org/10.2478/cdem-2013-0008
17. Padoan FCSM, Altimari P, Pagnanelli F (2019) Recycling of end of life photovoltaic panels: a chemical prospective on process development. Sol Energy 177:746–761. https://doi.org/10.1016/j.solener.2018.12.003
18. Quek TYA, Alvin Ee WL, Chen W, Ng TSA (2019) Environmental impacts of transitioning to renewable electricity for Singapore and the surrounding region: a life cycle assessment. J Clean Prod 214:1–11. https://doi.org/10.1016/j.jclepro.2018.12.263
19. Masoumian M, Kopacek P (2015) End of life of management of photovoltaic modules. IFAC 48–24:162–167
20. Karthikeyan L, Suresh V, Krishnan V, Tudor T, Varshini V (2018) The management of hazardous solid waste in India: an overview. Environments 5(9):103. https://doi.org/10.3390/environments5090103
21. Tasnia K, Begum S, Tasnim Z, Khan MZR (2018) End-of-life management of photovoltaic modules in Bangladesh. In 2018 10th international conference on electrical and computer engineering (ICECE), Dhaka, Bangladesh, pp 445–448. https://doi.org/10.1109/icece.2018.8636782
22. Xu Y, Li J, Tan Q, Peters AL, Yang C (2018) Global status of recycling waste solar panels: a review. Waste Manag 75:450–458. https://doi.org/10.1016/j.wasman.2018.01.036

Operation of Droop-Controlled Hybrid Islanded Multiple Subgrids

Harikrishna Muda, T. Mahesh, K. Naga Sujatha, and C. Bhargava

Abstract This paper investigates the issue of power sharing in AC/DC Subgrids (ADSs) formed by interconnection of Interlinking Converters (ICs). Distributed Generations (DGs) are controlled using a droop-based method. This control mechanism of ICs indirectly adjusts the load power of ACS and DCS using droop gains. Further, power sharing among the ACS and DCS is achieved accordingly. A linearized system model and analysis of autonomous operation of hybrid islanded multiple subgrids is developed. The power management schemes presented in the literature are mainly focused on either AC Subgrid (ACS) or DC Subgrid (DCS) conditions. This work proposed use both ac and dc droop gain values to regulate frequency and voltage at the point of IC. Time domain simulations are carried out in MATLAB/SIMULINK software environment and used to validate the proposed control algorithm. A reliable power sharing technique for ADS system is developed to operate at all the operating scenarios of loading and environmental conditions.

Keywords Distributed generations (DGs) · AC/DC sub grids · Bidirectional power sharing analysis · Decentralized control method · Time domain results

---

H. Muda (✉)
National Institute of Advanced Studies, IISc Campus, Bangalore, India
e-mail: harikrishnam2@gmail.com

T. Mahesh
Institute of Aeronautical Engineering, Hyderabad, India

K. Naga Sujatha
Jawaharlal Nehru Technological University, Hyderabad, India

C. Bhargava
Srindhi Institute of Science and Technology, Hyderabad, India

M. N. Favorskaya et al. (eds.), *Innovations in Electrical and Electronic Engineering*, Lecture Notes in Electrical Engineering 661, https://doi.org/10.1007/978-981-15-4692-1_15

Abbreviations

AC	Alternating current
ACSi	*I*th AC Subgrid
ADS	AC and DC Subgrid
DC	Direct current
DCSi	*I*th DC Subgrid
DG	Distributed Generation
IC_i	*I*th Interlinking Converter
P_i	Active power flow through *i*th Interlinking Converter
PV	Photovoltaic
Q_i	Reactive power flow through *i*th Interlinking Converter
n_{ac}	Total number of ac subgrids
n_{dc}	Total number of dc subgrids
n_{ic}	Total number of interlinking converters

1 Introduction

AC/DC Subgrids (ADSs) are evolving towards the coordinate operation of multiple Distributed Generations (DGs) [1]. ADSs are minimizing the power conversion stages and reducing losses by locating generation near demand [2]. An operation of bi-directional Interlinking Converters (ICs) is employed to share power in between AC Subgrids (ACSs) and DC Subgrids (DCSs) [3]. This environment allows the users to effectively operate AC/DC based DGs at distribution network voltage levels. The ADS can be operated in grid-connected mode or islanded mode. The islanded mode provides away to the power in the absence of the grid [4]. In the islanded mode of subgrid operation, the voltage and frequency are negotiated if the loads are not enough to absorb the generated power. Thus, three major parameters such as voltage, angular frequency, voltage and current harmonics in subgrid must be controlled to acceptable standards during islanded mode of operation [5].

The control scheme plays a major role to maintain the acceptable standard values and the islanded mode operation depends on the accurate controls scheme of the DGs in ADSs. Mainly, their control scheme can either be centralized or decentralized. The decentralized control methods which are based on the conventional droop characteristics are widely utilized for power sharing in a predetermined manner [6]. Power sharing among the subgrids will help the customers to rationalize their power consumption and it will lead to a reduction in the peak energy demand. Depending on the measurement of dc voltage and ac frequency at the IC location, the real powers are adjusted in the DCS and ACSs, respectively for the droop control scheme [7].

The concept of droop gain method for autonomous ACS to preserve power sharing in ACSs is introduced in [8]. In [9], an virtual voltage and frequency droop gain frame is presented to improve the stability performance. However, power oscillations between DGs are observed due to the circulating currents in a low inertia of DGs and resistive networks [10]. An increasing in amount of the main DC power sources of electricity such as photovoltaic panels, super capacitors, fuel cell stack and battery energy storage system [2, 3] and loads, such as electric car chargers, water heaters, pool pumps lead to the formation of DCSs. In droop-control based DCSs, the coupling between dc terminal voltage and available power or current output is used. Some of the control issues in the DCS arises from the power sharing controller and their associated control gains [11].

2 Topology of Hybrid Islanded Subgrids

General structure of interconnected DCS and ACSs clusters is shown in Fig. 1. It is to be noted that the two ACSs (ACS_1 and ACS_2) are interconnected through a DCS_1 using ICs (IC_1 and IC_2) to provide bidirectional power flow on present generating and loading conditions of each subgrids. ADS offers several advantages such as integration of ac and dc based DGs, increased reliability of energy utilization and reduced CO_2 gas emissions [6]. Integrated energy storage devices can decrease losses and increase reliability. Whereas ac sources such as wind turbines, diesel generators, synchronous generators would encourage the development of ADSs. The two ACS and DCSs can be treated as ADS system in which the lumped load power (combined loads of ACS and DCSs) can be proportionally shared between ACS and DCSs. Means, the installed power in each subgrid can be utilized more efficiently. For instance, refer Fig. 1a where the load power in the ACS_2 and DCS_1 is unavailable due to its insufficient maximum capacity or other operation constraints. When the generation capacity of ACS_1 is greater than its load power, therefore the DCS_1 and ACS_2 are supported by ACS_1. Because, the ICs are expected to control the power. In ADSs, power sharing is bidirectional between the ACS and DCSs. Figure 1b depicts the power flow scenario when high capacity of generating power available in DCS_1.

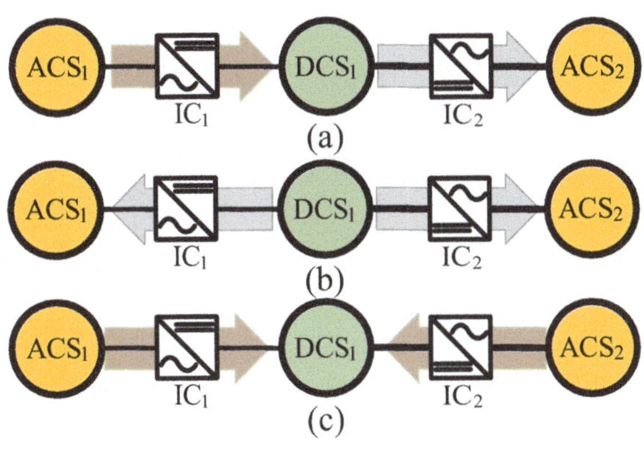

Fig. 1 Possible layout of two AC Subgrids (ACSs) and one DC Subgrid (DCS) interconnection with ICs

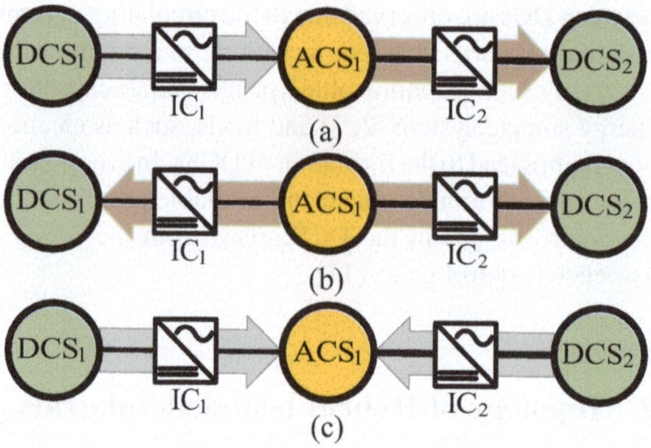

Fig. 2 Possible layout of two DC Subgrids (DCSs) and one AC Subgrid (ACS) interconnection with ICs

Similarly, Fig. 1c shows generation-load balancing within the ADS when there is generation shortfall in DCS_1 with assistance of ACS1 and ACS_2.

As shown in Fig. 2, both DCSs (DCS_1 and DCS_2) are connected to the ACS1 through ICs (IC1 and IC2). The DCSs are interfaced with its neighboring ACS to share the active power among DCS and ACSs. In ADS system, power sharing is directional between the ACS and DCSs as shown in Fig. 2. The necessity of IC is to ensure minimal generation and load curtailment in ADS and equal load power sharing in the ADS based on IC ratings. This analysis has resulted in a novel research significance on decentralized control operation for bidirectional power sharing in ADS system.

A general structure of a cluster of interconnected ac and dc microgrids is shown in Fig. 3. It is to be noted that the two AC (or dc) microgrids are interconnected through AC/AC (or DC/DC) converters to provide bidirectional power flow on present generating and loading conditions of each microgrids. The active power and reactive

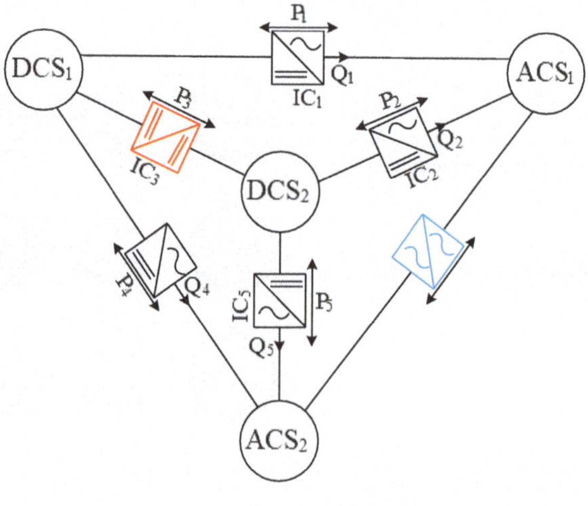

Fig. 3 Possible layout of different ACS and DCSs interconnection with bidirectional power flow

power flowing through ith IC is represented by P_i and Q_i respectively. In this paper we only considered the AC/DC subgrid interconnection scenario. Power sharing in ACS and DCS is controlled autonomously based on their traditional AC/DC droop control characteristics. All ICs can operate by means of a droop control system implemented as discussed in the next section.

Each subgrid consist of converter based DGs empowered by micro-sources, such as PV systems, microturbines, energy storage system, etc. A DC/AC or AC/DC converters are commonly used as an interfacing module to provide bi-directional power flow depending on present generating and loading conditions of each subgrid. The subscript n_{ac}, n_{dc} and n_{ic} indicate the number of ac, dc subgrids and ICs respectively. The active power flowing through n_{ic}th IC is represented by Pn_{ic} as shown in Fig. 5.

Similar to standalone ACS or DCSs, sharing the power between ACS an DCS can be achieved by the conventional droop methods [12–14]. This proposed work uses both ac and dc droop gain values to regulate frequency and voltage at the point of IC. In this paper, the impact of droop gain on the power flow direction of ADSs formed by ACS and DCS clusters interconnected through a group of ICs is considered. The changes in droop gain of ACS and DCSs is studied to assess the power flow direction of ADSs. Further, operating cases such as equal power sharing among ACS and DCSs are considered. The results for power flow direction from ACS to DCS and DCS to ACS are presented.

3 System Under Study

Please Schematic diagram of a ADS system which consisting of a grid, ICs (3 in number), ACSs (AC Subgrids 2 in number) and DCSs (DC Subgrids 2 in number) is shown in Fig. 4. Each subgrid having sources and distributed loads with the

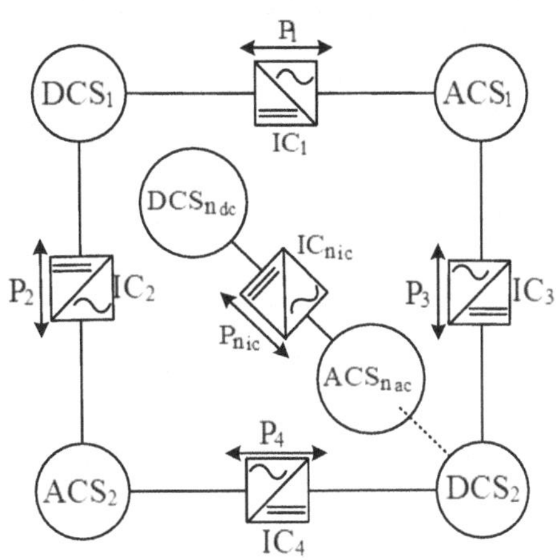

Fig. 4 Simplified diagram of different ac and dc subgrids (ACS and DCSs) interconnection with bidirectional power flow

connecting lines. The ICs are considered to provide bi-directional power among different type of subgrids. The subgrids can be operated in the grid-connected mode as well as islanded mode by closing or opening switch, S as shown in Fig. 3. A dc/dc converter or dc/ac converter is used in order to connect micro-sources to a DCS or ACS respectively. The ACS and DCSs comprise the operating voltage of 220 V, 50 Hz and 400 V respectively. The aggregate maximum power capacity of dc system is 30 KW. Whereas, each ac system has a maximum power capacity of 40 KVA with each ACS capacity of 20 KVA. The sum of distributed loads in each subgrid is taken as 100% of its generation capacity. "The dashed lines are used to mark the border of individual subgrids. The subgrid loads have been increased or decreased such that power is transferred by ICs" (Fig. 5).

In order to analyze behavior of different scenarios, switches (S1 to S6) are deployed among subgrids. This model is adapted from [15, 16]. The electrical connection following subgrids (ACS1, DCS1, and ACS2) and (DCS1, ACS2, and DCS2) are considered as a possible ADSs configuration with the bidirectional power flow among ACS and DCSs. The ACS and DCS system data are referred from [16, 17] respectively.

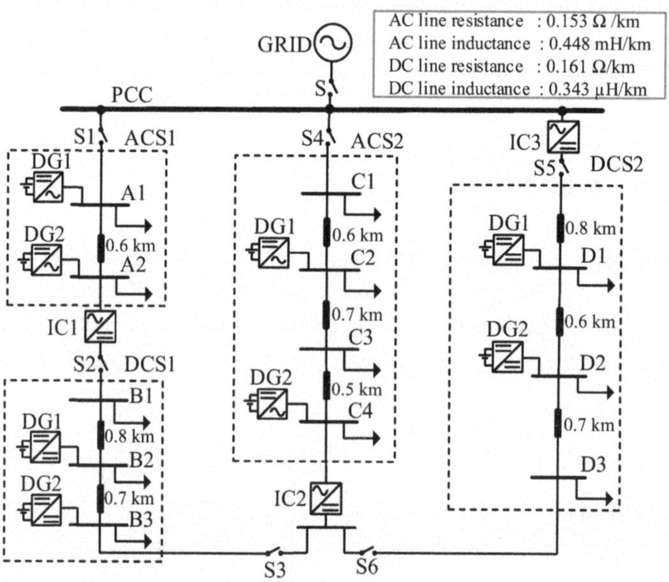

Fig. 5 Hybrid multiple subgrids under study

3.1 Droop Control Method for ACS and DCS

Usually, based on the local measurement of active and reactive power output at the DG terminal, the frequency and AC terminal voltage of the DG are regulated autonomously. Mathematically, the active power/frequency (*P*/*f*)-reactive power/voltage (*Q*/*V*) based droop controllers for the *i*th DG can be expressed as

$$w_i = w_i^* - m_{\text{aci}} P_{\text{aci}} \tag{1}$$

$$E_i = E_i^* - n_{\text{aci}} Q_{\text{aci}} \tag{2}$$

where w_i^* and E_i^* are rated angular frequency and voltage amplitude respectively. P_i and Q_i are active and reactive power outputs respectively. m_i and n_i are active and reactive power droop gains, respectively. w_i and E_i are reference angular frequency and voltage amplitude of the DG respectively.

For the case of DGs in DCS, the coupling between DC terminal voltage and available power output is generally used and mathematically for *i*th DG can be written as

$$V_i = V_i^* - m_{\text{dci}} P_{\text{dci}} \tag{3}$$

where V_i^* and V_i are the rated and reference magnitude of the DC terminal voltage respectively. M_{dci} and P_{dci} are droop gain and active power output of the *i*th DG respectively.

The power sharing between ACS and DCS is achieved base d on the measurement of ac frequency and DC voltages of ACS and DCS. Further, using the droop characteristics shown in Fig. 4 the load capacity in each subgrid can be obtained indirectly. Which is the summation of the individual loads. The equivalent maximum capacity of ACS ($P_{\text{acs}}^{\max}$) can be obtained by summing all of the DG maximum capacity in ACS. While, the maximum capacity of DCS ($P_{\text{dcs}}^{\max}$) is derived from the sum of all of the DG maximum capacity in the DCS.

Based on the ranges specified for the frequency of the *i*th ACS, their active power droop gain (m_{acsi}) can be obtained written as

$$m_{\text{acsi}} = \frac{w_{\max} - w_{\min}}{P_{\text{acsi}}^{\max}} \tag{4}$$

The slopes of the aggregated droop gain (m_{dcsi}) of *i*th DCS can be acquired as

$$m_{\text{dcsi}} = \frac{V_{\max} - V_{\min}}{P_{\text{dcsi}}^{\max}} \tag{5}$$

where $V_{\max}$ and $V_{\min}$ are the maximum and minimum allowed dc voltages of DCS respectively.

3.2 Droop Control for ICs

Droop controllers for IC based on normalized values of ac frequency and the dc voltage were considered and used to achieve flexible power flow among ACS and DCS. Since the ac frequency and dc voltage are independent physical quantities, their common range of pu values are calculated.

The droop control scheme implemented in the IC is shown in in Fig. 6. The r_f l_f and c_f are the resistance, inductance and capacitance of filter, respectively. The DER and ICs are modelled as controllable voltage sources and, hence, fast switching dynamics of the converters are not considered in this paper. The IC voltage is $v_{\rm ic}$, current is $i_{\rm ic}$ and the voltage at the terminal of DCS is $v_{\rm dc}$. For the IC, the output voltage at the ACS terminal is $v_{\rm o}$ and the output current is $i_{\rm o}$. A three-phase phase-locked loop (PLL) method is used to measure the operating frequency of IC and synchronization angle θ is used transform abc stationary frame to dq0 rotating frame. Details of the IC parameters can be found in the Appendix 1.

The control of each individual IC consists of two parts, i.e., the normalization and current controller. The purpose of normalization is to bring the different droop variables used by two consolidated sources (ac frequency and dc terminal voltage) to a common per unit range. Further details can be found in written by authors. Through normalization process, respective ac frequency and dc voltage equations for the ith IC are frames as

$$\widehat{w}_i = \frac{w_i - 0.5 \times (w_{\max} + w_{\min})}{0.5 \times (w_{\max} - w_{\min})} \tag{6}$$

$$\widehat{V}_i = \frac{V_i - 0.5 \times (V_{\max} + V_{\min})}{0.5 \times (V_{\max} - V_{\min})} \tag{7}$$

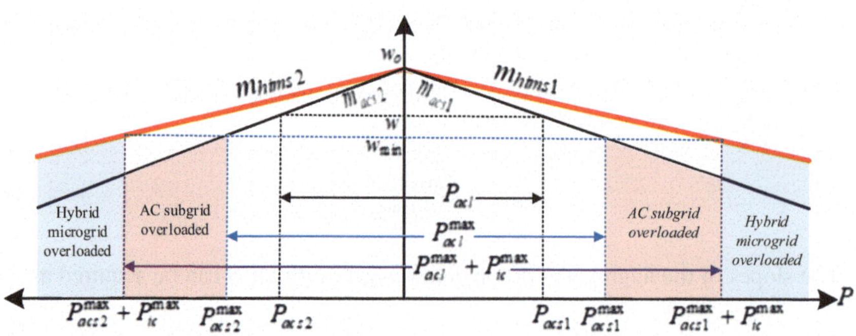

Fig. 6 The representation of the droop gain characteristics of the two ACSs and hybrid islanded multiple subgrids (HIMS) with the loading conditions

where $\widehat{w}_i$ and $\widehat{V}_i$ are the normalized frequency and normalized DC voltage respectively. The current controller is based on traditional proportional integral (PI) controllers.

Similar to (4), the active power droop gain (m_{himsi}) for the ith HIMS, can be rewritten as

$$m_{\text{himsi}} = \frac{w_{\max} - w_{\min}}{P_{\text{acsi}}^{\max} + P_{\text{ic}}^{\max}} \tag{8}$$

4 Simulation Results

In this section time domain simulations are performed to evaluate the performance of the proposed power sharing method. The ADS system under study, shown in Fig. 3 is modeled in the MATLAB/Simulink environment. The electrical connection between subgrids ACS1 and DCS1is considered as a possible AC/DCS configuration with the bidirectional power flow between ACS and DCSs."

To verify the performance of bidirectional power sharing of ICs, MATLAB/simulation is executed with the droop control scheme. The HIMS is simulated by closing the switch, S2 as shown in Fig. 6. With this arrangement, the voltage, frequency, power flows in HIMS are observed and shown in Fig. 6. The first case (I) is shown in Fig. 7, where the plots confirms that the two subgrids are initially experiencing a load demand of 10 kW (1 pu). The droop gain of interlinking converter is chosen to be 0. In third plot (shown in Fig. 7c), the negative value indicates active power flow from DCS1 to ACS1 via IC1. Reactive power demanded by the load in the ac subgrid is found to be 0.26 pu, as shown in Fig. 7f. Since the subgrids are equally loaded, no active power is transferred by the IC, which produce no reactive power too.

The second case (II) shows that ACS1 is overload by 0.4 pu at $t = 0.5$ s as depicted in Fig. 7. The droop gain of IC1 is increased to 0.025. This causes the IC1 to transfer 0.4 pu of active power from the ACS1 to DCS1. Because of that, active power loads in ACS1 and DCS1 are noted to be 1.4 pu and 0.6 pu respectively. The dynamic performance of active power loads is shown in Fig. 7d. In third case (III), the ACS1 is overloaded by 0.8 pu and droop gain of IC1 is considered as 0.05. Therefore, the IC1 starts to operate and transmits 0.8 pu from the DCS1 and ACS1 is maintained healthy at 1.8 pu. For the above two cases, it is to be noted that the DCS1 load are reduced at $t = 0.5$ s and $t = 1.0$ s, respectively.

Before overloading the DCS1, both ACS1 and DCS1 loads are set to 1.0 pu at time $t = 1.5$ s and are found to be static in case IV. The total generation of each subgrid is found to be 1 pu as shown in Fig. 7c. It is to be noted that IC is not transferring any active and reactive powers as shown in Fig. 7d, f respectively. Case V shows load in DCS1 is increased to 1.4 pu at time $t = 2$ s. It can be seen that

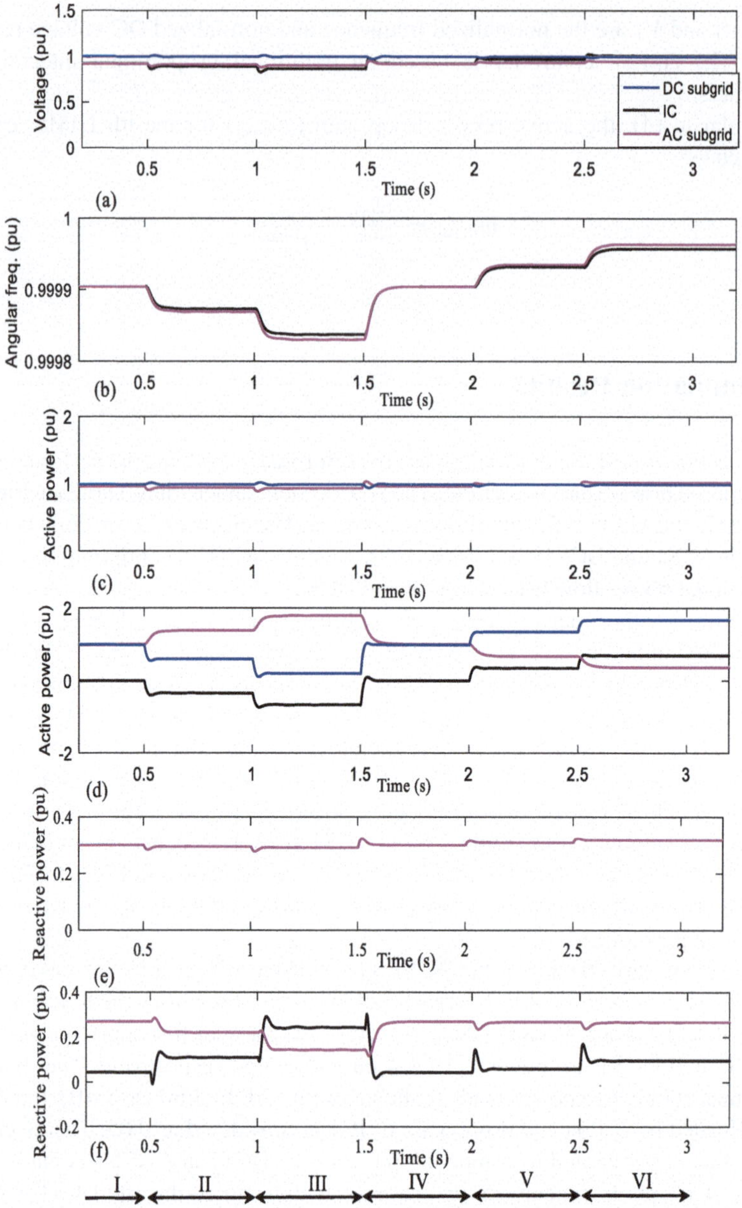

Fig. 7 Simulated waveforms of hybrid microgrids with overload and underload conditions and IC1 droop gain varied from 0 to 0.05 **a** voltage at IC1, DCS1 and ACS1 terminal, **b** angular frequency at IC1 and ACS1 terminal, **c** active power generated by ACS1 and DCS1, **d** active power load consumed by ACS1 and DCS1, **e** reactive power generated by ACS1, **f** reactive power generated by DCS1

IC1 transferred 0.4 pu from ACS1 for DCS1. For final stage, at $t = 2.5$ s, because the loads of DCS1 increased to 1.8 pu which is less than rating of ACS1 and ACS1 should share some active power for the DCS1. The ICs regulated the amount of their transferred active power about 0.8 pu, resulting in proportional active power sharing between the interconnected subgrids. During the time of $t = 2$–2.5 s, IC transferred a very small amount of reactive power. The results are indicated in Fig. 7f.

5 Conclusions

In this paper, the issue of the power sharing in ac/dCSs formed by ACS and DCS clusters interconnected through a group of ICs is considered. The impact of droop gain on the power flow direction formed by ACS and DCS clusters interconnected through a group of ICs is considered. With the addition of ACS, DCS and IC power capacities, the modified droop gain characteristics for the ADS are presented. The power flow direction from ACS to DCS and DCS to ACS are presented. Results demonstrate that as the load power in ACS is increased, power flow from the DCS to the ACS is increased accordingly. Whereas, power flow from ACS to DCS is increased when the load power in the DCS is increased. It is found that an appropriate coordinated power sharing among the ACS and DCSs will maximize the operation of ADS systems.

Acknowledgements The Authors would like to acknowledge the support provided by the collaborative research project under TEQIP-III, JNTU Hyderabad. Project allotment order number: JNTUH/TEQIP-III/CRS/2019/EEE/03.

Appendix 1

Table 1 Parameters of interlinking converter

$w_c = 31.41$ rad s^{-1}, $m_{\text{ic}} = 0.0167$ pu, $L_f = 0.029$ mH, $K_{\text{pc}} = 4.5$, $K_{\text{ic}} = 1600$, $r_{\text{ic}} = 0.23\ \Omega$, $L_{\text{ic}} = 0.35$ mH
$w_{\max} = 319.41$ rad s^{-1}, $w_{\min} = 308.91$ rad s^{-1}
$w'^{\max} = 324.52$ rad s^{-1}
$w'^{\min} = 303.79$ rad s^{-1}
$V_{\text{dc}}^{\max} = 410$ V, $V_{\text{dc}}^{\min} = 390$ V
$V_{\text{dc}}'^{\min} = 420$ V, $V_{\text{dc}}'^{\max} = 380$ V

References

1. Radwan, AAA, Mohamed, YARI (2014) Networked control and power management of AC/DC hybrid microgrids. IEEE Syst J:1–12
2. Kurohane K, Senjyu T, Yona A, Urasaki N, Goya T, Funabashi T (2010) A hybrid smart AC/DC power system. IEEE Trans Smart Grid 1(2):199–204
3. Guerrero JM, Vasquez JC, Matas J, de Vicuna LG, Castilla M (2011) Hierarchical control of droop-controlled AC and DC microgrids—a general approach toward standardization. IEEE Trans Ind Electron 58(1):158–172
4. Harikrishna M, Jena P (2016) Real-time simulation of hybrid microgrid for islanding detection analysis. In: IEEE 6th international conference on power systems (ICPS), pp 1–6
5. Brahma Chary AGR, Naga Sujatha K (2013) Power quality enhancement for flexible operation of grid integrated wind farm. Int J Electr Electron Eng Telecommun 2(4):108–117
6. Kundur P (2006) Power system stability and control, 1st edn. McGraw Hill Education Publishers
7. Loh PC, Member S, Li D, Chai YK, Blaabjerg F (2013) Hybrid AC–DC microgrids with energy storages and progressive energy flow tuning. IEEE Trans Ind Electron 28(4):1533–1543
8. Chandorkar MC, Divan DM, Adapa R (1993) Control of parallel connected inverters in standalone AC supply systems. IEEE Trans Ind Appl 29(1):136–143
9. Li Y, Li YW (2011) Power management of inverter interfaced autonomous microgrid based on virtual frequency-voltage frame. IEEE Trans Smart Grid 2(1):30–40
10. Nikolakakos IP, Zeineldin HH, El Moursi MS, Kirtley JL (2017) Reduced-order model for inter-inverter oscillations in islanded droop-controlled microgrids. IEEE Trans Smart Grid, 1–1
11. Peyghami S, Davari P, Mokhtari H, Loh PC, Blaabjerg F (2017) Synchronverter-enabled DC power sharing approach for LVDC microgrids. IEEE Trans Power Electron 32(10):8089–8099
12. Zhou J, Zhang H, Sun Q, Ma D, Huang B (2017) Event-based distributed active power sharing control for interconnected AC and DC microgrids. IEEE Trans Smart Grid, 1–1
13. Li et al X (2017) A unified control for the DC-AC ICs in hybrid AC/DC microgrids. IEEE Trans Smart Grid 1–1
14. Eghtedarpour N, Farjah E (2014) Power control and management in a hybrid AC/DC microgrid. IEEE Trans Smart Grid 5(3):1494–1505
15. Nikolakakos IP, Zeineldin HH, El-Moursi MS, Hatziargyriou ND (2016) Stability evaluation of interconnected multi-inverter microgrids through critical clusters. IEEE Trans Power Syst 31(4):3060–3072
16. Huang PH, Liu PC, Xiao W, El Moursi MS (2015) A novel droop-based average voltage sharing control strategy for dc microgrids. IEEE Trans. Smart Grid 6(3):1096–1106
17. Pogaku N, Prodanović M, Green TC (2007) Modeling, analysis and testing of autonomous operation of an inverter-based microgrid. IEEE Trans Power Electron 22(2):613–625

Comprehension of Different Techniques Used in Increasing Output of Photovoltaic System

Snigdha Sharma, Manasi Pattnaik, Meenakshi Sarswat, and Lokesh Varshney

Abstract This paper deals with the comparative study of various MPPT techniques and various configurations used for increasing photovoltaic (PV) system output. The problems associated with shading and environmental conditions affecting solar irradiance causes a decrease in efficiency and an increase in cost. Various methods have been discussed in this study to improve the efficiency of solar arrays. The findings of this paper highlight that among maximum power point tracking (MPPT) techniques, intelligent techniques find better scope in the future as compared to offline and online techniques. As far as configurations are considered, Su Do Ku configuration provides better results than other configurations.

Keywords Maximum power point tracking · Series parallel · Total cross-tied

1 Introduction

The total energy consumption on the earth is likely to be tripled as compared to previous decades. The large rate of consumption of fossil fuels is inversely proportional to its life expectancy. Also it has become a terror to our environment. The only possible solution to this present problem is to invest in renewable energy sources, out of which solar/wind energy is most widely used due to its environment-friendly nature, no noise, easy maintenance, and less maintenance cost [1]. The solar panel energy per day is given by the product of solar irradiance and panel wattage. With

S. Sharma
Department of Electrical Engineering, I.T.S. Engineering College, Greater Noida, India

M. Pattnaik
Department of Electrical Engineering, JIMSEMTC, Greater Noida, India

M. Sarswat
Department of Electrical Engineering, IPEC, Greater Noida, India

L. Varshney (✉)
SEECE, Galgotias University, Greater Noida, India
e-mail: lokesh.varshney@gmail.com

M. N. Favorskaya et al. (eds.), *Innovations in Electrical and Electronic Engineering*, Lecture Notes in Electrical Engineering 661, https://doi.org/10.1007/978-981-15-4692-1_16

the advancement in solar technologies, solar power is associated with many real-time objects. However, extracting solar energy from the sun is a very difficult task due to the limited efficiency and increasing cost of solar cells. The efficiency is affected by factors, which include solar radiance, temperature of module, and load impedance. Different techniques, configurations, and their hybrid are being developed to overcome the above two problems. Generally, at the time of installation four different configurations are considered: off-grid configuration, grid-tie configuration, grid-interactive configuration, and grid fallback configuration. Under the situations of shading and discharged batteries, MPPT can be efficiently utilized to extract maximum power. The main components in MPPT configuration include solar panel, controller, battery, and inverter. With the change in sunlight intensity, load characteristics of PV system under consideration changes to keep the output power maximum by various algorithms. Here, consider different MPPT techniques in this paper, which comprises of offline techniques, online techniques, and intelligent techniques.

Rezk et al. [2] presented fuzzy logic controller (FLC) algorithm that has been developed to extract maximum power point (MPP) without requiring accurate system data which proved to be quite good for nonlinear solar array system. This new technique is feasible and gives a fast convergence rate, which is verified under different solar irradiation. Alsumiri [3] introduced a residue theorem based incremental conductance (INC) algorithm to extract MPP and removes fluctuations. Thus, it is superior than perturb and observe algorithms. The modified sine-cosine method was developed in order to track MPP, which is used with grid integration [4]. Danoune et al. [5] configured that short circuit current (SCC or I_{sc}) output voltage is zero which results in losses in the form of heat so it is not considered an economical algorithm. Fractional short circuit current (FSCC) has been developed to overcome this problem but it creates another problem of ripples. Here, Particle swarm optimization (PSO) algorithm is also taken into consideration [6].

To increase the output power numbers of solar cells are connected in various topologies. Partial shading causes mismatch losses in the PV panel, which ultimately results in low efficiency. Due to shading in highly series topology, the whole string can act as a load instead of source; therefore, different configurations are originated from series, parallel, and series-parallel topologies to avoid problem of shading. The main configurations are Series-Parallel (SP) which is derived from series and parallel configurations and Total Cross-Tied (TCT) which overcomes the issues of SP [7–11].

2 MPPT Techniques

MPPT techniques can be classified as offline techniques, online techniques, and intelligent techniques, which are further subdivided into various categories [12–18]. The complete classification is shown by the following Fig. 1.

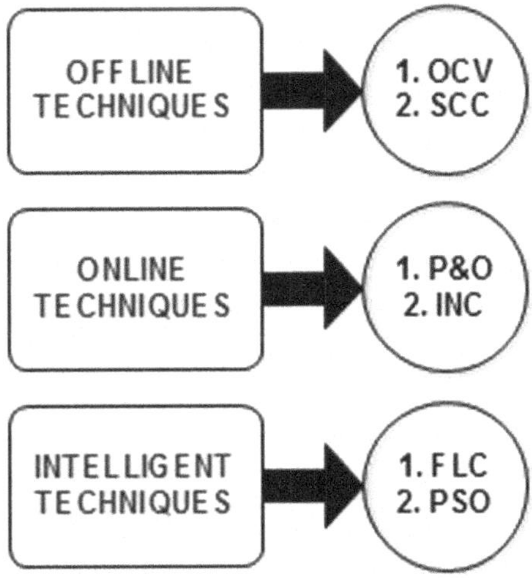

Fig. 1 Classification of MPPT techniques

2.1 Offline Techniques

These techniques are often called indirect techniques and worked in offline mode, which requires prerequisite data of solar panels.

Open circuit voltage (OCV): To determine OCV, the load should be disconnected from PV system, which leads to losses at the time of interruption. To avoid frequent interruptions, reduce the sampling rate but lowering the sampling rate arise a new problem of inaccuracy in determining MPP, thus decreases the efficiency. The drawbacks of V_{OC} can be overcome by determining MPP without separating load from PV system. The use of a photometric sensor to find MPP accurately and gives better efficiency than open circuit voltage method [19]. The flowchart of OCV is shown in Fig. 2 (Fig. 3).

Short circuit current (SCC): This technique is similar to the OCV technique but it is more complex to implement. Both offline techniques suffer from the same problem of power loss while measuring $V_{OC}, I_{SC,}$ and inaccurate MPP tracking. The assumption in the short circuit current algorithm is that maximum current and short circuit current are linear, that is,

$$\text{Im} = \text{KIsc} \tag{1}$$

where $K = 0.78 - 0.92$ [4].

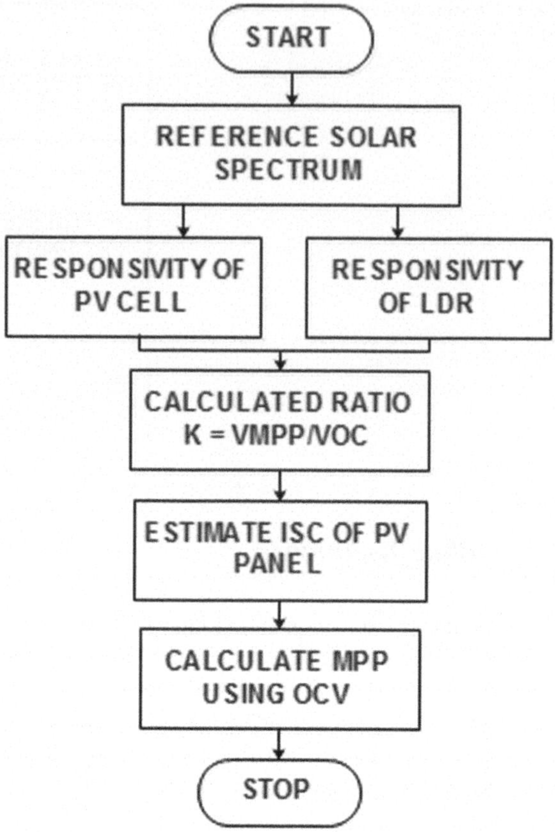

Fig. 2 Flowchart of OCV

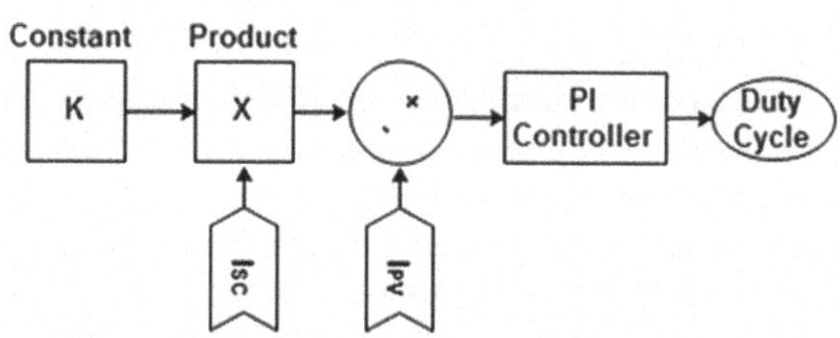

Fig. 3 Block diagram of SCC

2.2 *Online Techniques*

These techniques worked in online mode which measures various related to solar panel data such as in P&O and INC algorithm.

Perturb & Observe (P&O): Most commonly used method adopting hill-climbing strategy having higher efficiency. It gives oscillations in output power. It does not go well with a rapidly changing environment such as temperature and irradiance. The duty cycle is decided by perturb and observe method on the basis of power, voltage,

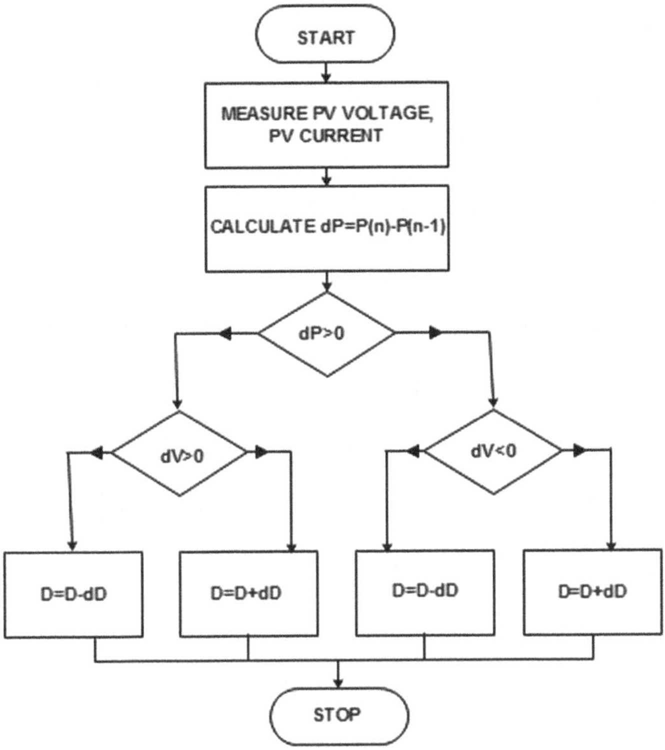

Fig. 4 Flowchart of P&O

and current [20]. The concept of changing the size of step of perturb and observe to overcome its limitations, i.e., it increases tracking speed and reduce oscillations [21]. In Fig. 4, P&O flowchart is shown and D represents duty cycle (Fig. 5).

Incremental conductance (INC): The power change by the change in voltage is determined with an incremental change of conductance. INC depends on the following equation:

$\frac{\mathrm{d}I}{\mathrm{d}V} + \frac{I}{V} = 0$ at MPP
$\frac{\mathrm{d}I}{\mathrm{d}V} + \frac{I}{V} > 0$ that is the left side of maximum power point.
$\frac{\mathrm{d}I}{\mathrm{d}V} + \frac{I}{V} < 0$ that is the right side of maximum power point.

It also gives oscillations in output power but less as compared to P&O method. The only problem in this is complexity and cost.

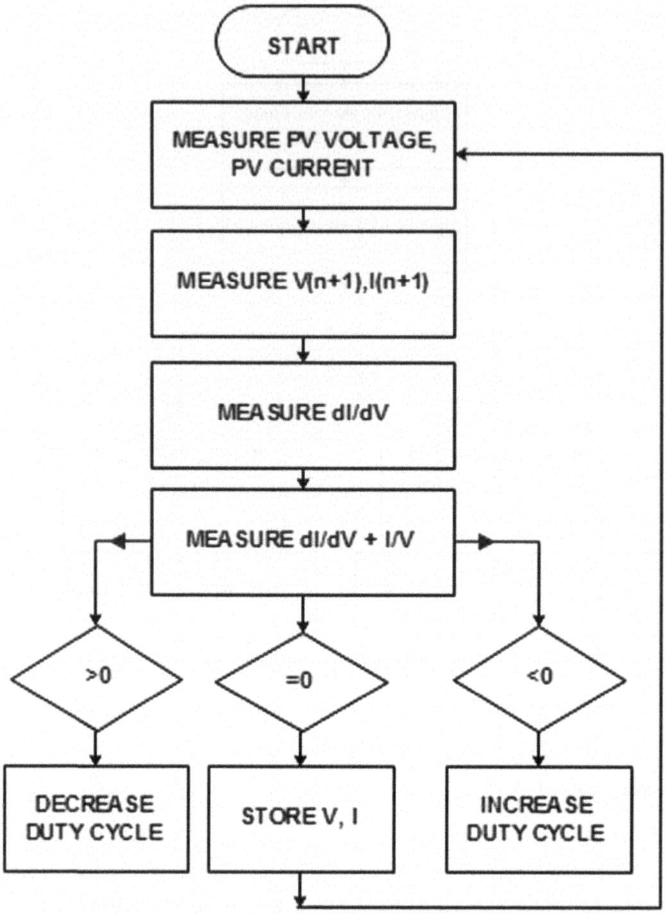

Fig. 5 Flowchart of INC

2.3 Intelligent Techniques

These techniques consist of FLC and PSO, which has high efficiency than offline and online techniques.

Fuzzy Logic Controller (FLC): Fuzzy logic defines the range or degrees of truth or partial truth rather than 0 or 1. FLC is widely used for nonlinear systems which can provide faster response and enhance the stability of the system. It comprises of three stages:

Fuzzification: In this, input values which are numerical in nature are converted into linguistic variables, i.e., +ve big, −ve big, +ve small, −ve small, +ve medium, −ve medium, and zero.

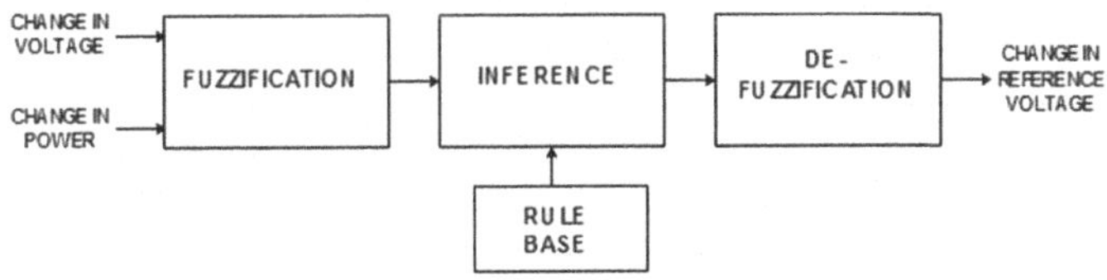

Fig. 6 Block diagram of FLC

Fuzzy inference: This engine works with the help of rules to produce linguistic output. The rules-based table for linguistic variables is described in [22].

Defuzzification: This is used to reconvert linguistic output into a numerical value with the help of the centroid method.

The change in environmental conditions is observed by using proportional integral (PI) controller in addition to the algorithm of the fuzzy system, i.e., adaptive neuro fuzzy inference system (ANFIS) MPPT that leads to increase in performance of PV system. With the use of ANFIS MPPT, there is an improvement in the performance and operation of PV panel system, high rate of convergence, and quick achievement of MPP [23].

Particle swarm optimization (PSO): To search a solution for a particular problem, particles move in search space directed by their own best position or by swarm's best position. The current positions of particles have been replaced by improved positions and this updating continues till the best solution is discovered but this does not always give accurate results. PSO has an advantage of particles following simple method and high converging rate. To vary the velocity and particle's position in PSO algorithm, authors of uses parameters lowest and highest values for every iteration (Figs. 6 and 7).

3 Comparison Between Different MPPT Techniques

All the techniques offline, online, and intelligent are being compared on different parameters, which allow us to know which technique is suitable for which applications. Table 1 gives information regarding the comparison.

The solar array consists of four modules in series forming a string and four of these strings are connected in parallel. The total power capacity of array is 3.2 kW. The specifications of a PV module at standard test conditions are shown in Table 2.

The maximum power at different irradiation levels in MPPT techniques is shown in Table 3. By the following table, it is seen that PSO has maximum power as compared to other techniques (Fig. 8).

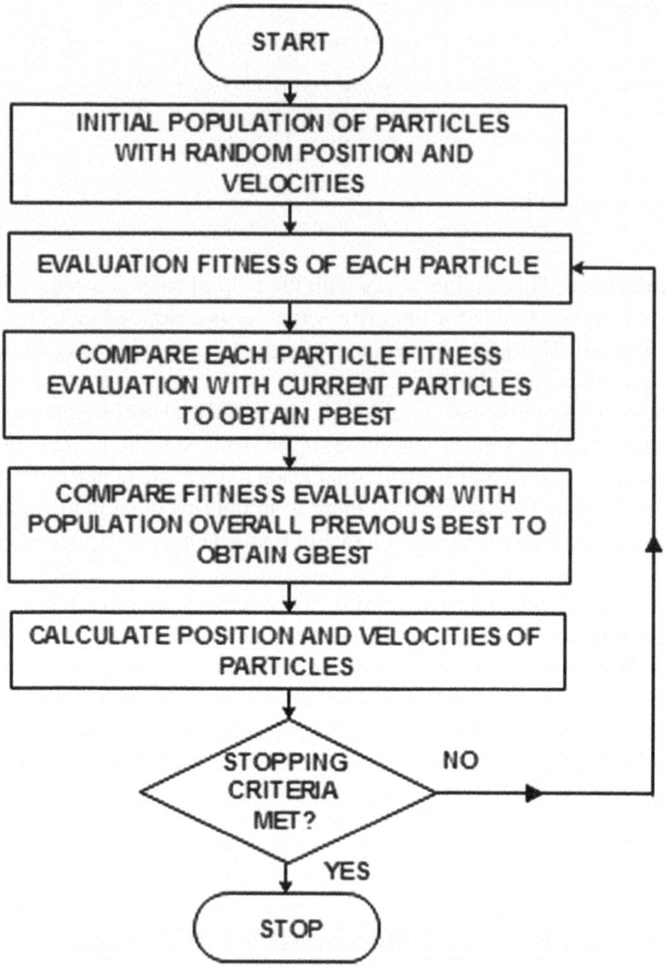

Fig. 7 Flowchart of PSO

4 Configurations of PV System Under Different Shading Patterns

4.1 Shading Pattern

The whole PV system is affected badly in terms of efficiency by a small amount of shading on the panel. Many tools such as obstacle analysis tools can be used for shading analysis. Four types of shading are there: Short wide (SW), Long wide (LW), Short narrow (SN), Long narrow (LN) [24–29] (Fig. 9).

Table 1 Comparison between different MPPT techniques

Parameters	OCV	P&O	INC	FLC	PSO
Complex system	L	L	M	H	H
Convergence speed	L	L	H	H	H
Cost	L	L	L	H	H
Hardware implementation	E	E	M	E	M
Rapid change of irradiance	L	L	H	H	H
Efficiency	L	M	H	VH	VH
Sensed parameters	V	V, I	V, I	V, I	P, V
Algorithm implemented	A, D	A, D	D	D	D
Oscillations at MPP	H	H	M	L	L
Depends on PV specifications	Y	N	N	Y	N

L low, *M* medium, *H* high, *VH* very high, *V* voltage, *I* current, *P* position, *Y* yes, *N* no, *E*-easy, *A* analog, *D* digital

Table 2 PV module specifications

PV model	TP 200
Module dimension	1587 mm × 790 mm × 50 mm
Module weight	16 kg
Pmax	200 W
VMPP	36.6 V
IMPP	5.46 A
Voc	44.4 V
ISC	5.79 A

Table 3 Comparison based on different irradiation level

Irradiation level (W/m2)	OCV	P&O	INC	FLC	PSO
	P (W)	P (W)	P (W)	P (W)	P (W)
400	2952.43	2953.22	2953.73	2954.34	2954.66
600	2995.40	2996.17	2996.71	2997.31	2997.64
800	3039.52	3040.30	3040.83	3041.42	3041.78
1000	3084.35	3085.10	3085.64	3086.24	3086.59

4.2 Series-Parallel (SP)

This configuration is a combination of series and parallel connections. The arrangement of various modules in series and parallel will decide which combination gives the highest output power4 [30, 31] (Fig. 10).

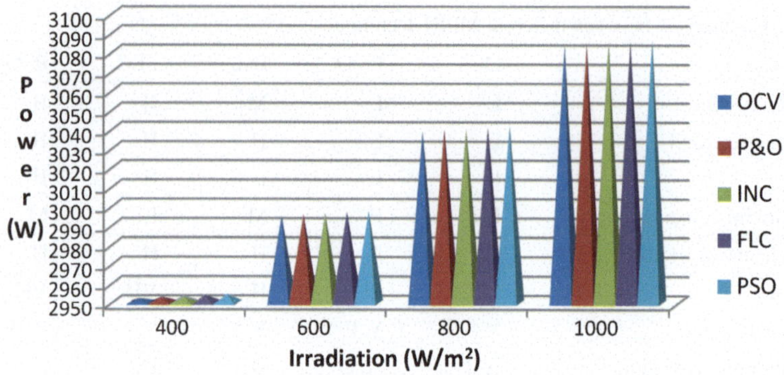

Fig. 8 Power (W) generated in different MPPT techniques at different irradiation level

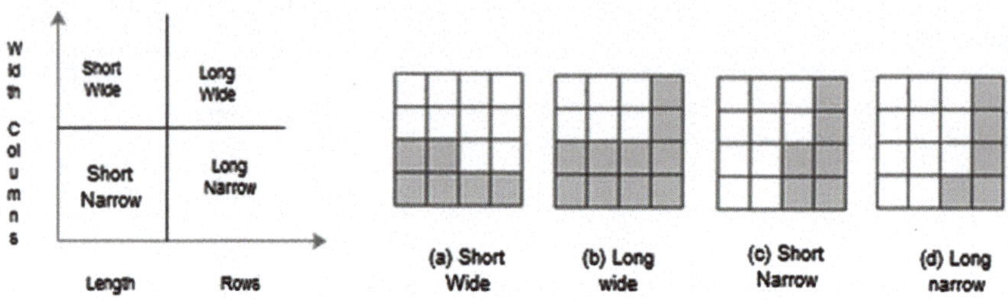

Fig. 9 Types of shading patterns, **a** SW pattern, **b** LW pattern, **c** SN pattern, **d** LN pattern

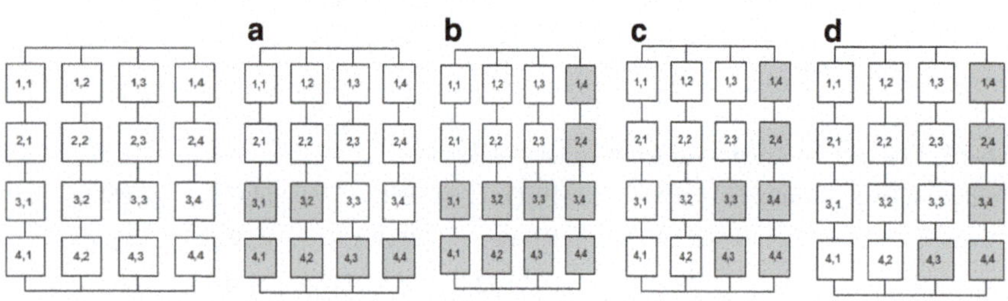

Fig. 10 SP connection, **a** SW pattern, **b** LW pattern, **c** SN pattern, **d** LN pattern

4.3 Total Cross-Tied (TCT)

TCT is an extended version of SP configuration. In this configuration, rows and columns are crossly connected in such a way that across all rows total voltage is equal, and across all columns total current is equal. This scheme works better than SP in shading condition but it has a problem that the numbers of ties are more which increases cable losses.

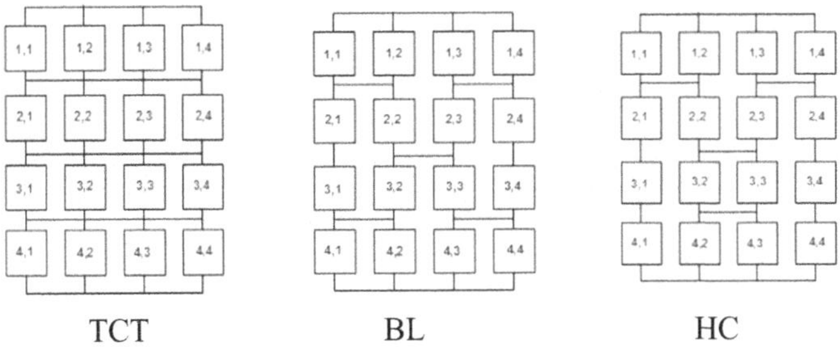

Fig. 11 TCT configurations

Bridge Linked (BL): As its name implies, this configuration is based on the bridge structure. This is derived from TCT with an advantage of a lesser number of ties and cable losses but during shading conditions, it affects overall voltage and current.

Honey Comb (HC): This configuration is also derived from the idea of TCT based on the honeycomb structure. By this configuration, losses in output power can be reduced but it has a limitation that it cannot reduce power losses under all shading conditions (Fig. 11).

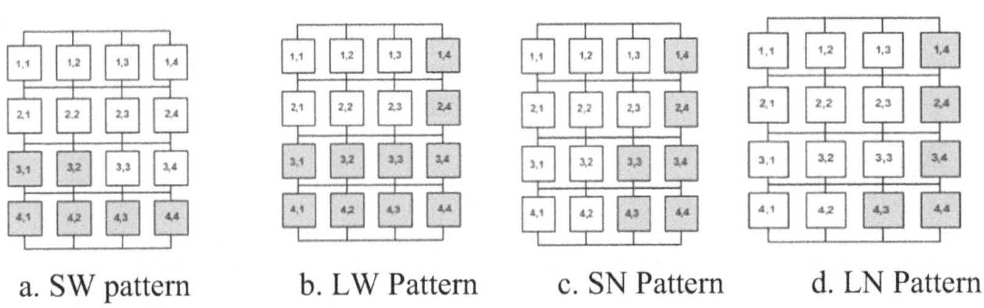

4.4 *Physical Relocation of Module with Fixed Electrical Connections (PRM-FEC)*

This configuration is beneficial over TCT by dispersing the shading part to reduce the impact of shading in the same row or column (Fig. 12).

4.5 *Su Do Ku*

Su Do Ku configuration of 4 × 4 TCT matrix is considered here. This performs well under shading conditions for any configuration for example TCT, in which

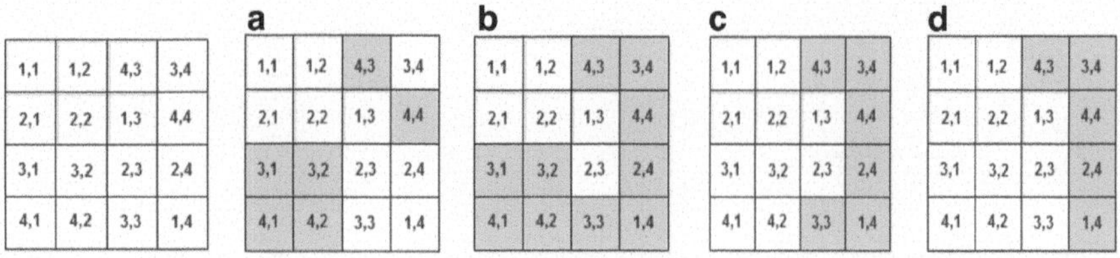

Fig. 12 PRM-FEC pattern, **a** SW pattern, **b** LW pattern, **c** SN pattern, **d** LN pattern

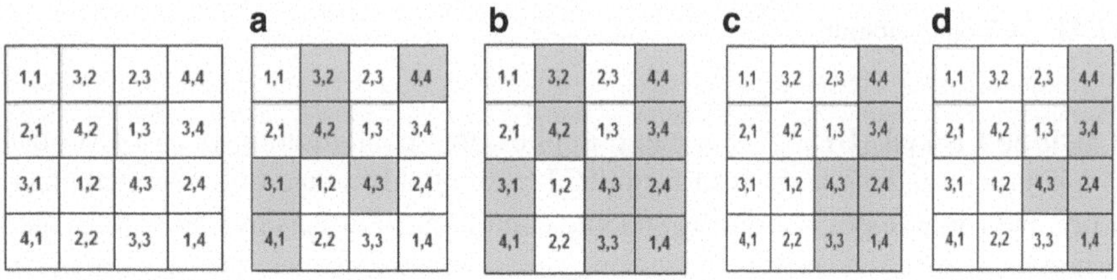

Fig. 13 Su Do Ku, **a** SW pattern, **b** LW pattern, **c** SN pattern, **d** LN pattern

relocation of module occurs irrespective of electrical connections. The disadvantage of this scheme is that it increases the cost of wires used in the physical relocation of modules (Fig. 13).

4.6 Fibonacci Series

In this scheme, solar panels are used on the stem in place of leaves to improve the performance of solar PV panels. The losses occur in this configuration due to the shaded part of higher panels on below panels. The power generated by two identical stages with 1/3 phyllotaxis is 1.5 times of flat surface PV panel power. The Fibonacci number PV module incorporates many features such as number of stages can be increased, reflected light from one cell can be used by other cells, lesser area in shadow and impact of one cell shadow on the other [32].

4.7 Zigzag Pattern

In this technique, reflection of sunlight from one module will fall on another, which enhances the output power. The Zigzag pattern has improved global MPP in comparison with Su Do Ku configuration. Moreover, Zigzag pattern easies the operation of MPPT as it has an advantage of a smooth PV curve [33]. This pattern is advantageous

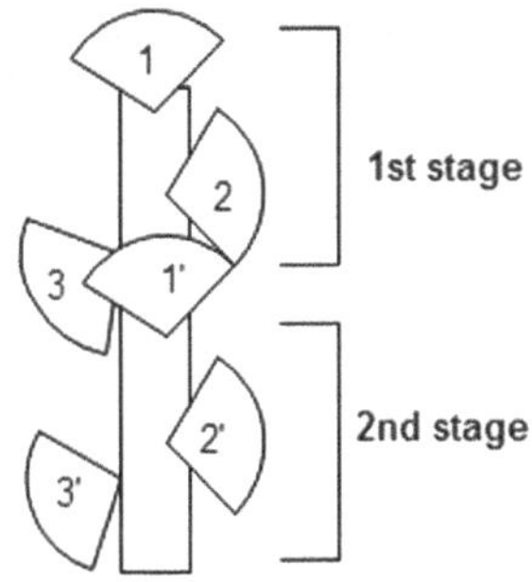

Fig. 14 Arrangement of Fibonacci modules

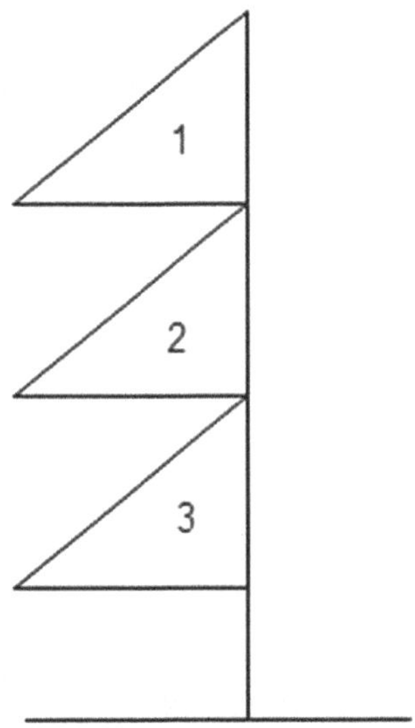

Fig. 15 Zigzag pattern facade of solar panels

for using less area for solar panels but it also has an issue related to short duration usage of these panels for generating electricity (Figs. 14 and 15).

5 Conclusions

The MPPT techniques and configurations described in this paper can raise the reliability, stability, and efficiency of PV system under different conditions for future considerations. It can also combine two or more techniques to produce the desired results. In MPPT, perturb and observe and incremental conductance are the most widely used techniques. P&O is highly efficient but its efficiency is affected by changing the environment. The advantage of residue-based INC method is that the data regarding atmospheric conditions are not needed. Also overall efficiency and tracking MPP have been improved [2]. OCV reduces output power oscillations but its efficiency is less. FSCC has been not found suitable with rapidly changing atmospheric conditions with respect to P&O and INC [4]. However, the hybrid of FSCC and variable step size INC provides better results under rapid irradiance [5]. The fuzzy logic algorithm is simple and adaptive as its input–output membership functions are adjustable to obtain the desired output of the system [1]. Fuzzy logic gives better performance and less oscillation but cannot reaches MPP accurately while PSO provides correct MPP.

In configurations, SP arrangement has an advantage of easy implementation, no redundancy in connections, and reduction in wiring time at installation. The TCT arrangement is better than SP in the following respect: more output power, large fill factor, and low mismatch losses [6]. TCT overcome the drawback of affecting complete string under shading conditions in SP configuration. BL and HC are derived from TCT to reduce the number of ties. Fibonacci structure ensures to gain sunlight on every solar module during shading conditions. Su Do Ku pattern provides better results among all configurations in shading conditions. Zigzag pattern occupies less area is advantageous for a short duration of time till the modules are receiving sunlight.

References

1. Varshney L, Saket RK (2014) Reliability evaluation of SEIG rotor core magnetization with minimum capacitive excitation for unregulated renewable energy applications in remote areas. Ain Shams Eng J 05(03):751–757
2. Rezk H, Aly M, Al-Dhaifallah M, Shoyama M (2019) Design and hardware implementation of new adaptive fuzzy logic-based MPPT control method for photovoltaic applications. Inst Electr Electron Eng J 7:106427–106438
3. Alsumiri M (2019) Residual incremental conductance based nonparametric MPPT control for solar photovoltaic energy conversion system. Inst Electr Electron Eng J 7:87901–87906
4. Padmanaban SK, Priyadarshi N, Bo Holm-Nielsen J, Bhaskar MS, Azam F, Sharma AK, Hossain E (2019) A Novel Modified Sine-Cosine Optimized MPPT Algorithm for Grid Integrated PV System under Real Operating Conditions. IEEE Access J 7:10467–10477
5. Danoune MB, Djafour A, Gougui A, Khelfaoui N, Boutelli H (2018) Study and performance analysis of three conventional MPPT algorithms used in photovoltaic applications. In: International conference on communications and electrical engineering (ICCEE), El Oued, Algeria, 2018
6. Hanan M, Javed MY, Xin AI, Gulzar MM, Ahmad S (2018) A two-stage algorithm to harvest maximum power from photovoltaic system. In: 2nd IEEE conference on energy internet and energy system integration (EI2). Beijing, China, 2018
7. Peter G. Saha AAK (2019) Effects and performance indicators evaluation of PV array topologies on PV systems operation under partial shading conditions. In: Southern African Universities Power Engineering Conference/Robotics and Mechatronics/Pattern Recognition Association of South Africa (SAUPEC/RobMech/PRASA). Bloemfontein, South Africa, pp 322–327, 2019
8. Shivi R, Varshney L, Mishra A, Joshi S (2018) Electric power generation from piezoelectric system under several configurations. In The proceeding of international IEEE conference on computing, power and communication technology. 2018 (GUCON), 28–29 Sep 2018

9. Varshney L (2017) Enhanced power generation from piezoelectric system under partial vibration condition. In: The proceeding of IEEE international women in engineering (WIE) conference on electrical and computer engineering (WIECON-ECE 2017). Women Institute of Technology, Dehradun, Uttarkhand, India, 18–19 Dec 2017
10. Varshney L, Sarswat M, Joshi S, Debnath D (2018) Comparison between several mitigation techniques of inrush current in transformer. In: The proceeding of International IEEE conference on computing, power and communication technology. GUCON, 28–29 Sep 2018
11. Varshney L, Upadhay AP (2017) Comparison of techniques for designing and modeling of high power piezoelectric devices. In: The proceeding of 4th IEEE Uttar Pradesh section international conference on electrical, computer and electronics (UPCON2017). Mathura, India, 2017
12. Sandali A, Oukhoya T, Cheriti A (2014) Modeling and design of PV grid connected system using a modified fractional short-circuit current MPPT. In: International renewable and sustainable energy conference (IRSEC). Ouarzazate, Morocco, 2014
13. Sher HA, Murtaza AF, Noman A, Addoweesh KE, Al-Haddad K, Chiaberge M (2015) A new sensorless hybrid MPPT algorithm based on fractional short-circuit current measurement and P&O MPPT. In: IEEE transactions on sustainable energy 6(4):1426–1434
14. Zhang J, Liu N, Xu J, Zhao S, Xu Y (2019) Novel MPPT method based on large variance GA-RBF. J Eng 2019(16):3365–3370
15. Padmanaban SK, Priyadarshi N, Bhaskar MS, Holm-Nielsen JB, Ramachandaramurthy VK, Hossain E (2019) A hybrid ANFIS-ABC based MPPT controller for PV system with anti-islanding grid protection: experimental realization. IEEE Access J 7:103377–103389
16. Ganthia BP, Pradhan R, Das S, Ganthia S (2017) Analytical study of MPPT based PV system using fuzzy logic controller. In: International conference on energy, communication, data analytics and soft computing (ICECDS). Chennai, India, pp 3266–3269
17. Li X, Wang Q, Wen H, Xiao W (2019) Comprehensive studies on operational principles for maximum power point tracking in photovoltaic systems. IEEE Access J 7:121407–121420
18. Guan T, Zhuo F (2017) An improved SA-PSO global maximum power point tracking method of photovoltaic system under partial shading conditions. In: IEEE international conference on environment and electrical engineering and 2017 IEEE industrial and commercial power systems Europe (EEEIC/I&CPS Europe). Milan, Italy
19. Angelin Grace Stella S, Prasanth BL, MahendraBabu TK, Gnanambal K (2013) Low-cost photometric sensing for maximum PV power generation. In: International conference on power, energy and control (ICPEC). Sri Rangalatchum Dindigul, India, pp 501–504
20. Ramesh T (2018) Solar powered based water pumping system using perturb and observation MPPT technique. In: IEEE international students' conference on electrical, electronics and computer science (SCEECS). Bhopal, India
21. John R, Mohammed SS, Zachariah R (2017) Variable step size perturb and observe MPPT algorithm for standalone solar photovoltaic system. In: IEEE international conference on intelligent techniques in control, optimization and signal processing (INCOS). Srivilliputhur, India
22. Hassan SZ, Li H, Kamal T, Ahmad J, Riaz MH, Khan MA (2018) Performance of different MPPT control techniques for photovoltaic systems. In: International conference on electrical engineering (ICEE). Lahore, Pakistan, pp 1–6
23. Amara K, Fekik A, Hocine D, Hamida ML, Bourennane E-B, Bakir T, Malek A (2018) Improved performance of a PV solar panel with adaptive neuro fuzzy inference system ANFIS based MPPT. In: 7th international conference on renewable energy research and applications (ICRERA). Paris, France, pp 1098–1101
24. Kumar S, Kakkar S, Iqbal A (2017) Effect of topology on the performance of PV module. In: 8th international conference on computing, communication and networking technologies (ICCCNT). Delhi, India
25. Veerasamy B, Takeshita T, Jote A, Mekonnen T (2018) Mismatch loss analysis of PV array configurations under partial shading conditions. In: 7th international conference on renewable energy research and applications (ICRERA). Paris, France, pp 1162–1167
26. Tatabhatla VMR, Agarwal A, Kanumuri T (2018) Improving the output power of solar PV Array under different irradiation conditions. In: IEEE international students' conference on electrical, electronics and computer science (SCEECS). Bhopal, India

27. Mehiri A, Hamid A-K, Almazrouei S (2017) The effect of shading with different PV array configurations on the grid-connected PV system. In: International renewable and sustainable energy conference (IRSEC). Tangier, Morocco. https://doi.org/10.1109/irsec.2017.8477420
28. Iraji F, Farjah E, Ghanbari T (2018) Optimisation method to find the best switch set topology for reconfiguration of photovoltaic panels. IET Renew Power Gener 12(3):374–379
29. Jalil MF, Saxena R, Ansari MS, Ali N (2016) Reconfiguration of photo voltaic arrays under partial shading conditions. In: Second international innovative applications of computational intelligence on power, energy and controls with their impact on humanity (CIPECH). Ghaziabad, India
30. Mochizuki Y, Yachi T (2018) Effective series-parallel cell configuration in solar panels for FPM power generation forest. In: 7th international conference on renewable energy research and applications (ICRERA). Paris, France, pp 294–300
31. Mansur AA, Amin MR, Islam KK (2019) Determination of module rearrangement techniques for non-uniformly aged PV arrays with SP, TCT, BL and HC configurations for maximum power output. In: International conference on electrical, computer and communication engineering (ECCE). Cox'sBazar, Bangladesh
32. Mochizuki Y, Yachi T (2017) Relationship between power generated and series/parallel solar panel configurations for 3D Fibonacci PV modules. In: 6th international conference on renewable energy research and applications (ICRERA). San Diego, CA, USA, pp 126–130
33. Vijayalekshmy S, Bindu GR, Rama Iyer S (2017) Performance comparison of zig-zag and Su Do Ku schemes in a partially shaded photo voltaic array under static shadow conditions. In: International conference on innovations in power and advanced computing technologies (i-PACT). Vellore, India, pp 1–6

Performance Analysis of Self-Excited Induction Generator (SEIG) with ELC for the Wind Energy System

Ravi Chaurasia, Rajkumar Viral, Divya Asija, and Tarannum Bahar

Abstract Self-excited induction generator (SEIG) model is mathematically formulated and examined to see its effect on a small-scale wind plant. This type of small-scale wind generating unit is utilized for remote areas which are far from the grid and provide electricity to the distant people. In today's scenario, small electric generation close to customer's site has expected to demand additional carefulness for usage in remote and rural utilities owing to the financial issues and complexity involved in the grid expansion. Hence, proper standalone schemes using nearby accessible power resources is an optimum possibility for providing adequate generating capacity. It has been done by utilizing green technologies used as alternative energy resources, for instance, wind, small hydro, geothermal, solar, etc. A single-plug setup of such generators is acquainted in such a way that generators speed, line voltage, line current stays in limit with different loading conditions. The Electronic Load Controller (ELC) is mathematically developed to control the performance of SEIG with consideration of different loading conditions. The proposed model consisting of SEIG with ELC is developed in MATLAB environment with the use of Simulink and Sim Power System (SPS) tools. Different parameters such as voltage, current and power characteristics are represented for different loading conditions, and the achieved output clarifies the suitability of SEIG with ELC in small-scale wind energy systems. The achieved outputs also help in determining insulation strength, suitability, shaft strength, selection of rating of capacitors, and designing the protection strategy for the SEIG.

R. Chaurasia
SRM University, Muradnagar, Gahziabad, UP, India
e-mail: ravichaurasia23@gmail.com

R. Viral (✉) · D. Asija
Amity University Uttar Pradesh, Noida, UP, India
e-mail: rviral@amity.edu

D. Asija
e-mail: dasija@amity.edu

T. Bahar
Gautam Budh Univerity, Greater Noida, UP, India
e-mail: tannu007tannu@gmail.com

M. N. Favorskaya et al. (eds.), *Innovations in Electrical and Electronic Engineering*, Lecture Notes in Electrical Engineering 661, https://doi.org/10.1007/978-981-15-4692-1_17

Keywords MATLAB/Simulink · SEIG · Wind Energy System

1 Introduction

Small-scale energy production near customer's locations has expected more curiosity in recent past for the use of remotely located rural populations owing to the budget and intricacy comprised in the grid expansion. Hence, proper standalone schemes using nearby accessible power resources have become a favorite choice. Additionally, wind power schemes may be constructed in remote zones that would be uneconomical to help from a system, or in regions where national power distribution systems is not accessible. Subsequently, minor wind projects typically have negligible electromechanical and civil erection work, and they seem as devising a comparatively small impact on environment in parallel to other renewable power resources such as small hydro, solar, biomass, and ocean. [1].

Expansion of small and moderate speed and advanced synchronous machine, particularly Induction Generators (IG) has realized courtesy to several entities all over the world. IGs propose numerous benefits to small hydro and wind generation systems parallel to other generators existing in the marketplace. It's design, self-running ability, cheaper repair, and less price gains more popularity from the last few spans of wind power makers [2].

2 SEIG

Induction machines are pretty widespread in remote wind energy farms [3, 4]. It has a separately energized AC apparatus. The stator assembly of a 3-Φ Induction machines is coupled to a 3-Φ AC supply, and its rotor terminal obtains power to the stator terminals using the principle of Faraday's laws. Based on the slip characteristics, IMs can operate either in motoring or generating region such as [1, 2, 5–8]:

i. In motor operating region ($0 < s < 1$), rotor runs in the same side as the rotating field generates the current. In this, slip alters '1' to standstill and '0' on the synchronous velocity.
ii. In generator region ($-1 < s < 0$), connections of stator are linked to a voltage generating source with constant output and then rotor runs beyond synchronous velocity using external running source.

SEIG having confined rotor-type design has the capacitors of shunt type, coupled to it for energizing [9, 10]. Shunt capacitors could be of fixed or adjustable values. IG is having quite analogous design as IM with few conceivable changes to improve its' efficiency [11]. In reference to speed all through IG operating process is not somewhat like the synchronous; therefore, it is also known as asynchronous generator. There are many benefits of a SEIG elaborated as: brushless and craggy design, lower cost, of repair and operative easiness, self-defense contrary to any faults, high dynamic response, ability to supply power at variable speed [5, 6, 12–17].

2.1 SEIG with Other Generator

On the basis of necessity traditionally in a wind generating scheme IG, synchronous machine (alternator) or Doubly Fed Induction Generator (DFIG) is most frequently utilized and the prospect researchers aiming on Permanent Magnet Synchronous Generator (PMSG) and SEIG because these machines utilized to exchange the motorized power in blowing wind in electrical power, paralleled to alternator, SEIG having brushless and craggy design and ability to supply energy at variable run and more clearly considerable enhancement in the overall performance [2, 6, 9, 13, 14, 16, 18].

The article is set out as below: The next segment of the study is emphasis on development of mathematical modeling of PMSG. The third section acquaints the readers with the Simulink/MATLAB model of PMSG and its various components. Section four shows the performance of PMSG under various operating conditions by simulation results. The last section summarizes the conclusion of the paper.

3 SEIG Modeling

The deliberated SEIG-ELC model contains of an IG and a capacitors group, customer loads (static along with dynamic), and control system of an ELC is displayed in Fig. 1 [19, 20].

The following assumptions are considered and based on these dynamic models of symmetrical three-phase IM which is derived and developed [16, 21]:

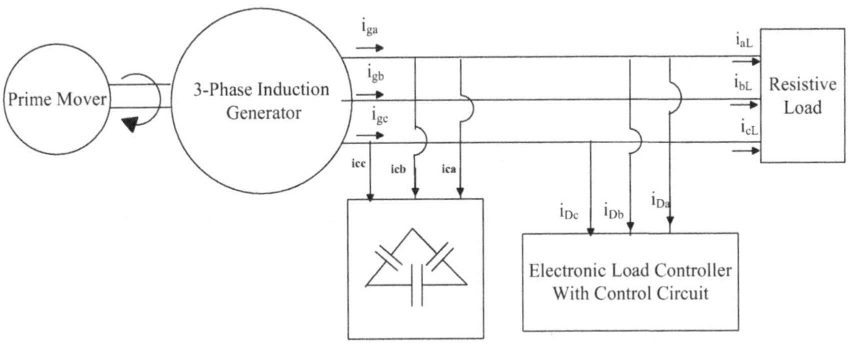

Fig. 1 SEIG with ELC and Load representation illustration for 3-Φ system

i. The alteration in resistance of IM because of the variation in frequency and temperature is ignored.
ii. Time harmonics and space are ignored of generated MMF.
iii. Neglected core losses are considered.

A dynamic model of the 3-Φ squirrel-cage IG is also created considering the static reference frame of *d*-*q* axes [15, 16, 18], and related expressions for mechanically and electrically systems are specified below.

$$T_e = (3P/4)L_m(i_{qs}i_{dr} - i_{ds}i_{qr}) \tag{1}$$

$$T_{shaft} = T_e + j(2/P)pw_g \tag{2}$$

The voltages of *d*- and *q*-axes in static frame are illustrated as:

$$v_{ds} = (2/3)\{(v_q - (v_b/2) - (v_c/2))\} \tag{3}$$

$$v_{qs} = (2/3)\{(\sqrt{3}v_b\Big/2) - \sqrt{3}v_c\Big/2)\} \tag{4}$$

The derivative of the rotor speed from (4) is

$$pw_g = (P/2) + j\{(T_{shaft} - T_e)/J\} \tag{5}$$

where (all the variables are in frame with rotor reference) as: T_e (SEIG electromagnetic torque developed), T_{shaft} (SEIG electromechanical torque), P (number of poles of SEIG), L_m (magnetizing inductance), i_{qs} (stator current of *q*-axis), i_{dr} (rotor current of *d*-axis), i_{ds} (stator current of *d*-axis), i_{qr}, (rotor current of *q*-axis), J (moment of inertia), w_g (of SEIG rotor speed), $p([d/dt])$, v_{ds} (stator voltage: direct axis element), v_{qs} (stator voltage quadrature axis element), v_a v_b v_c (IG terminals voltages).

4 ELC Modeling

The purpose of ELC has to lessen the price of wind power scheme and installed governor with switching of by electrical means. Using an ELC, the electrically connected load at generator side should be fixed despite of the fact that the customer load can be fluctuating in a random style from '0' to rated full load of the system.

The developed SEIG-ELC [13–16] system is revealed in Fig. 2a. It involves of a 3-Φ, Δ coupled SEIG driven by a fixed power prime mover (usually, in an uncontrolled wind generator and turbine) [22].

As the input power is closely fixed, SEIG' output power must be detained constant at all connected loads. The energy in excess of the user load is dumped in a resistance (R_{dL2}) via an ELC linked at the connections of SEIG. The modeled ELC entails of an

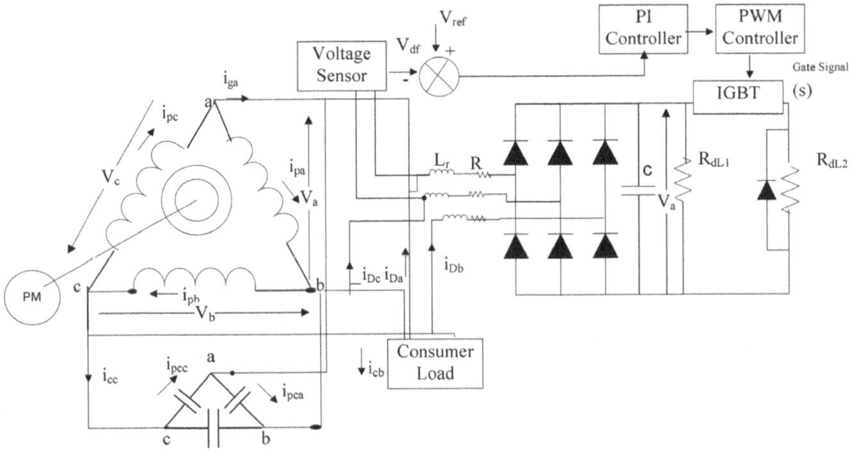

Fig. 2 **a** Three-Phase SEIG with ELC and Load, **b** ELC chopper-based control circuit

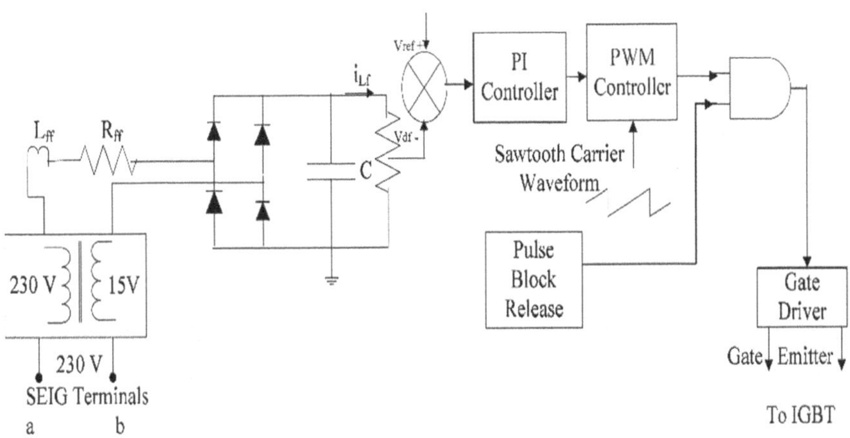

Fig. 2 (continued)

unrestrained rectifier connected in series using a chopper and dumped load (resistors). The SEIG voltage is detected via the sensor and cured through the 1-Φ rectifier for feedback voltage (V_{df}) as presented in Fig. 2b. The output of the PI controller is used to compare using a sawtooth carrier wave in a pulse width modulation (PWM) regulator to regulate the duty cycle [23] of the chopper to produce a gate signal to the Insulated Gate Bipolar Transistor (IGBT), operating as a controlled switch for ELC. The V-I characteristics express the detailed load controller scheme [11, 24] which is illustrated as:

$$v_{\text{max}} = 2R_{\text{f}}i_{\text{d}} + 2L_{\text{f}}Pi_{\text{d}} + v_{\text{d}} \tag{6}$$

Using above equation, the derivative of ELC current (i_d) is expressed as:

$$Pi_{\mathrm{d}} = \frac{(v\mathrm{max} - v_{\mathrm{d}} - 2R_{\mathrm{f}}i_{\mathrm{d}})}{2L_{\mathrm{f}}} \tag{7}$$

where $v_{\max}$ is the highest value of AC voltages of the lines (v_{a}, v_{b}, v_{c}, $-v_{\mathrm{a}}$, $-v_{\mathrm{b}}$, and $-v_{\mathrm{c}}$) reliant on the diode pairs are leading; and v_d is the voltage of DC-link [24]. Charging and discharging of the filter capacitor of ELC (C) are given as [13, 15, 18]:

$$pv_{\mathrm{d}} = \frac{(i_{\mathrm{d}} - i_{\mathrm{L}})}{C} \tag{8}$$

$$i_{\mathrm{L}} = \left\{\left(\frac{v_{\mathrm{d}}}{R_{\mathrm{dL1}}}\right) + S\left(\frac{v_{\mathrm{d}}}{R_{\mathrm{dL2}}}\right)\right\} \tag{9}$$

When, closing 7 opening of switch is representing with $S = 1$ and $S = 0$, respectively. The output voltage SEIG is measured using a stepdown transformer and corrected by a 1-Φ rectifier circuit working as a feedback signal, as given in Fig. 3. The 1-Φ rectifier circuit using for feedback is formulated with the following equations

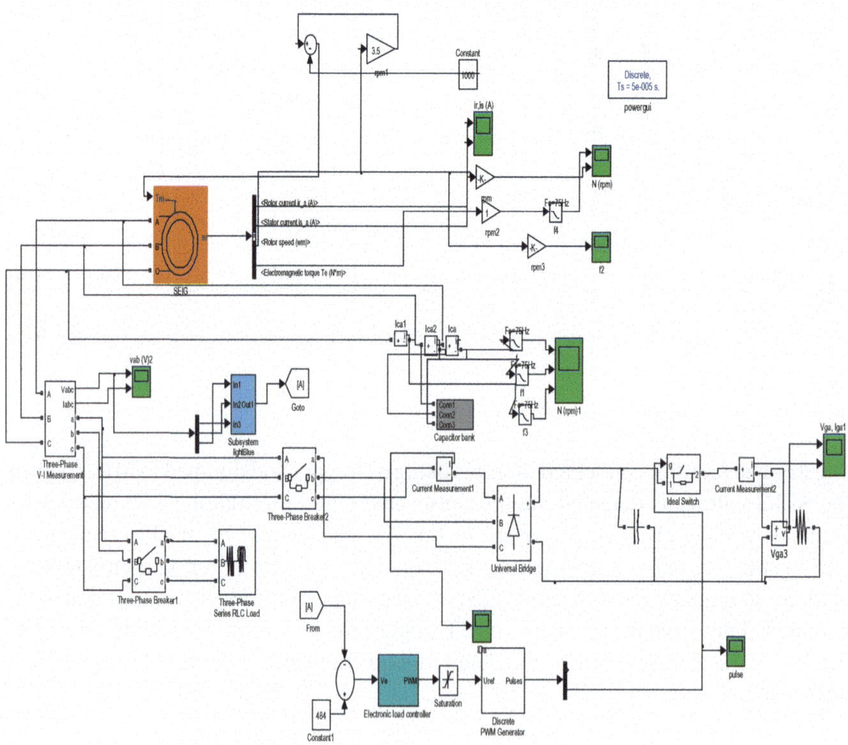

Fig. 3 SEIG with ELC Simulink/MATLAB Model

as:

$$v_{\mathrm{f}} = R_{\mathrm{ff}} i_{\mathrm{df}} + L_{\mathrm{ff}} p i_{\mathrm{df}} + K v_{\mathrm{df}} \tag{10}$$

From (9), current i_{df} derivative is given by

$$p i_{\mathrm{df}} = \frac{(v_{\mathrm{f}} - K v_{\mathrm{df}} - R_{\mathrm{ff}} i_{\mathrm{df}})}{(L_{\mathrm{ff}})} \tag{11}$$

The filtering capacitor C_s) charging and discharging at the output of the 1-Φ uncontrolled rectifier are derived as:

$$p v_{\mathrm{df}} = \frac{(i_{\mathrm{df}} - i)}{(L_{\mathrm{ff}})} \tag{12}$$

Analog voltage controller (PI) is formulated as:

$$p v_{01} = K_1 \frac{(v_{\mathrm{ref}} - v_{\mathrm{df}})}{T_{\mathrm{cc}}} \tag{13}$$

$$v_{02} = K_1 (v_{\mathrm{ref}} - v_{\mathrm{df}}) \tag{14}$$

The output of the PI controller (v_O) is given as:

$$v_0 = v_{01} + v_{02} \tag{15}$$

wherev_{ref} is reference voltage, $K_1 = R_{\mathrm{f}} / R_1$, $T_{\mathrm{CC}} = C_{\mathrm{f}}$ as R_1, $R_{2,}$ and C_{f} are input resistance, feedback resistance, and feedback capacitance, respectively, utilized in PI controller. The output of the PI controller (v_0) is equated with the sawtooth PWM carrier wave and is specified as:

$$v_{\mathrm{st}} = \frac{A_m t}{T_p} \tag{16}$$

whereA_m, *t*, and T_p are an amplitude in volts, time in μs, and of the sawtooth carrier wave, respectively. Thus, mathematical model of the 3-Φ SEIG with ELC is consisting using developed Eqs. (6)–(16). The parameters of ELC are given in Appendix 1 [14–16, 18, 25].

5 Complete Simulink Model of SEIG-ELC

The developed MATLAB/Simulink model is as shown in Fig. 3. The simulation has been performed using MATLAB 2014Ra with ode23t. All electrical scheme is virtual for dissimilar loading conditions and demonstrated by means of resistive and inductive components. The unstable load is shown as described in Fig. 3. Appendix 2 contains entire bounds of the numerous modules of scheme.

6 Results of Simulations

The operating enactment of SEIG has been analyzed under various loading and unloading conditions. The variation of SEIG rotor speed (ω_r), electromagnetic torque (T_e), 3-Φ current (I_{abc}), and 3-Φ stator voltage (V_{abc}) is presented in Figs. 4 and 5 for different loading and unloading conditions. These four cases are considered for study:

6.1 Voltage Build up and Self-excitation Process

Initially, SEIG is connected with an excitation capacitance of 200 μF per phase and started under no load condition. The process of voltage rise-up of all 3-lines at SEIG end, currents of capacitor, torque (electromagnetic), and speed of rotor are as shown in Figs. 4 and 5.

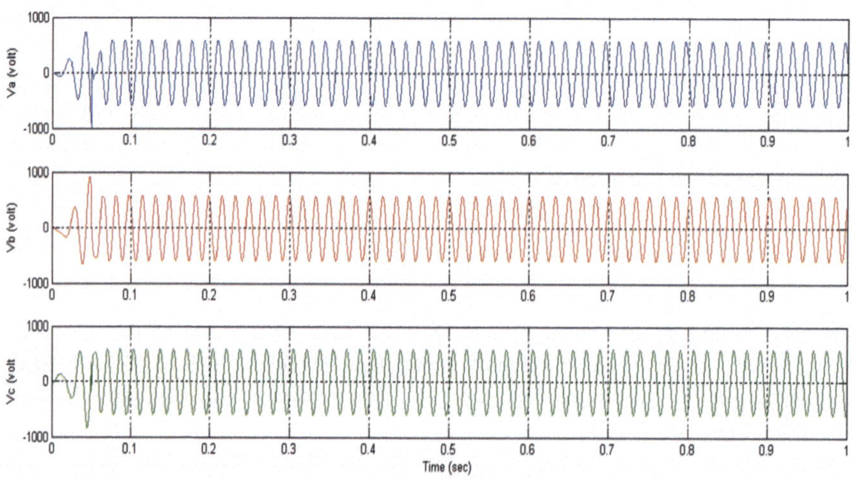

Fig. 4 Line voltages and capacitor currents produced at SEIG

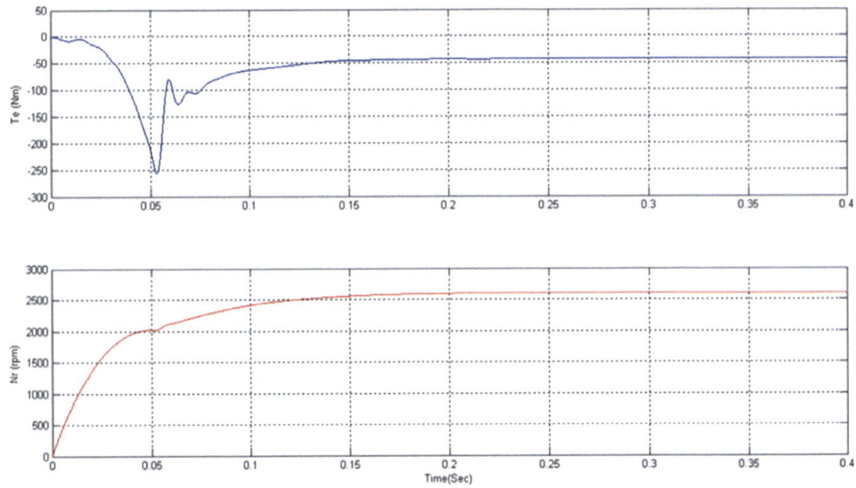

Fig. 5 Torque and rotor speed of SEIG

6.2 SEIG Loading with Resistive Load

Initially starting SEIG under no load condition, a 1.5 kW resistive-type load is applied at $t = 0.05$ s (Appendix 2). At starting, resistive load draws a heavy current but it settles to a steady-state value after 1.2 s. After $t = 1.2$ s, resistive load is again removed from SEIG. The variations of electromagnetic torque, currents, and rotor speed are as shown in Figs. 6, 7, and 8

Since the peak ELC input current of line-A increases due to the increment in the duty cycle of the chopper switch, power is transferred from main load to dump load

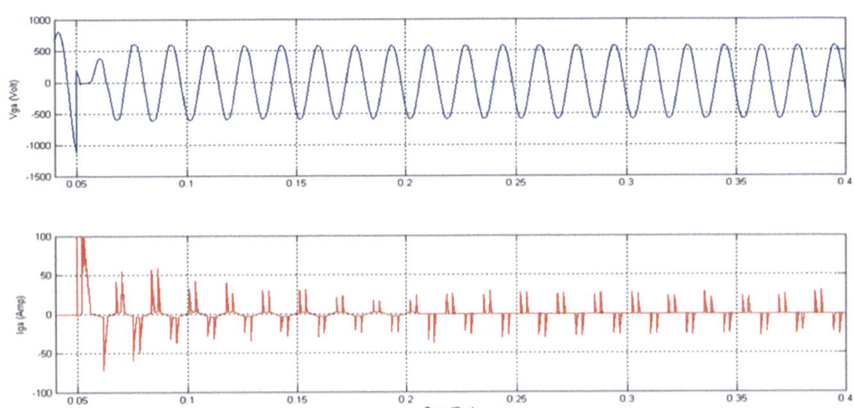

Fig. 6 Voltage and current waveforms of line-A at the SEIG terminal due to sudden application of resistive load

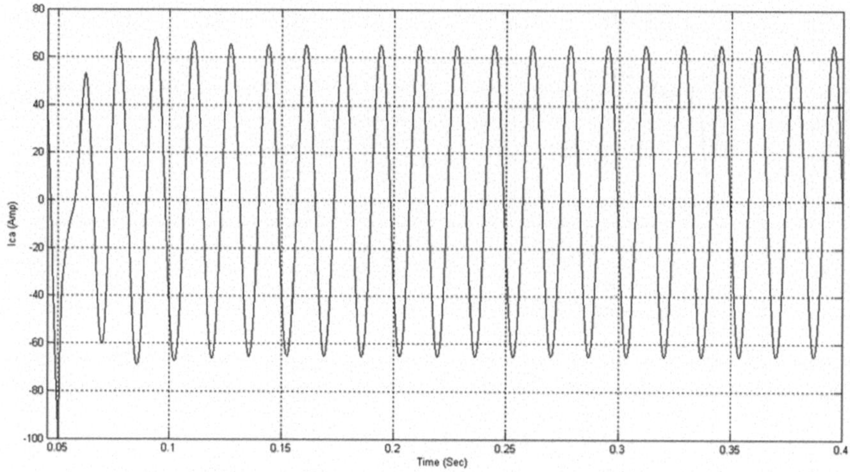

Fig. 7 Capacitor current of line-A due to sudden application of resistive load

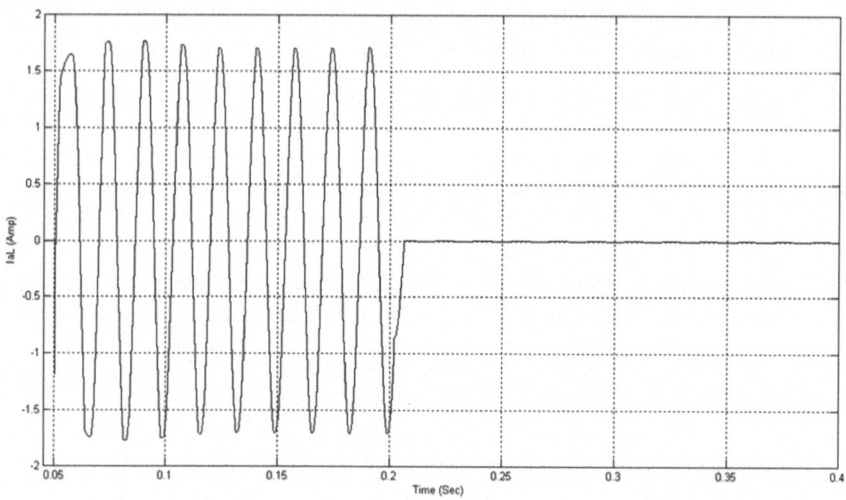

Fig. 8 Waveform of main load current of line-A due to sudden application of resistive load

accordingly, and the total power supplied by the SEIG remains unchanged. Thus, the terminal voltage of the SEIG is maintained constant as shown in Figs. 9 and 10.

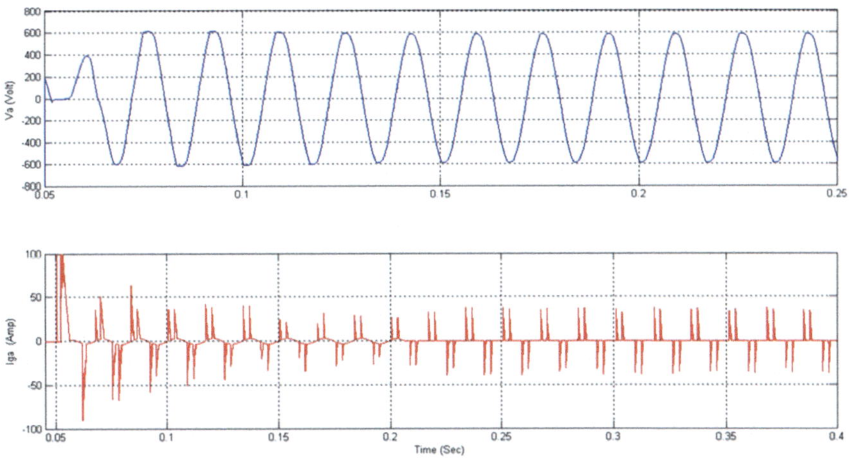

Fig. 9 Voltage and current waveforms of line-A at the SEIG terminals due to sudden application of IM load

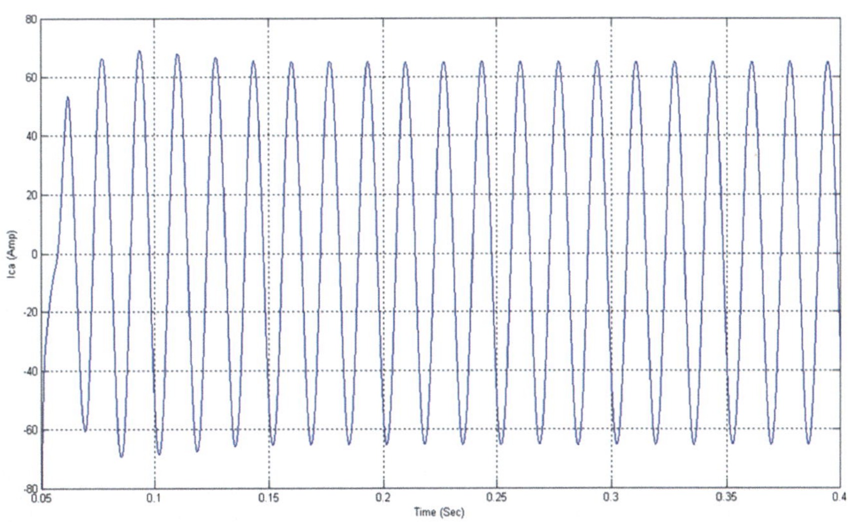

Fig. 10 Capacitor current of line-A due to sudden application of IM load

6.3 Dynamic Loading of PMSG

At starting, SEIG is running on no load condition. When ELC is switched, it draws a heavy current at starting, but it settles to a small steady-state value and takes full load current of SEIG. After that a dynamic IM load of 1HP is applied to the terminals of the SEIG at $t = 0.05$ s. Waveforms corresponding to sudden application of IM load are shown in the Figs. 9, 10, 11, 12, and 13. The magnitude of terminal voltage

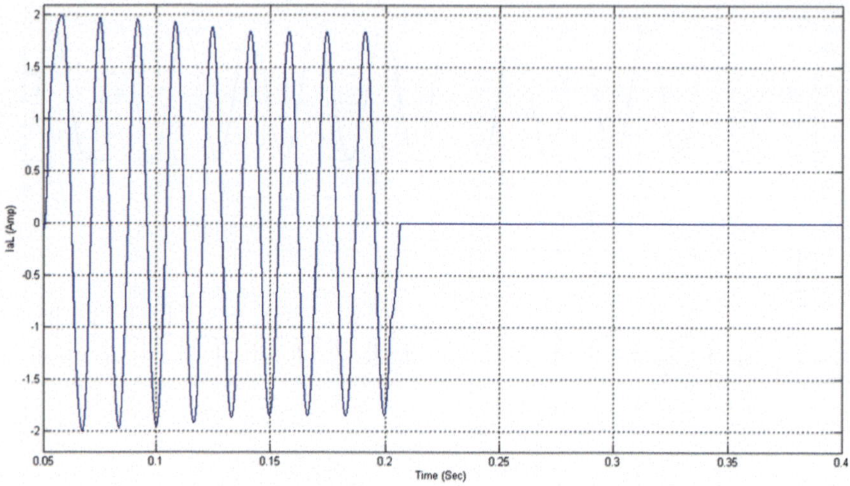

Fig. 11 Waveforms of main load current of line-A due to sudden application of IM load

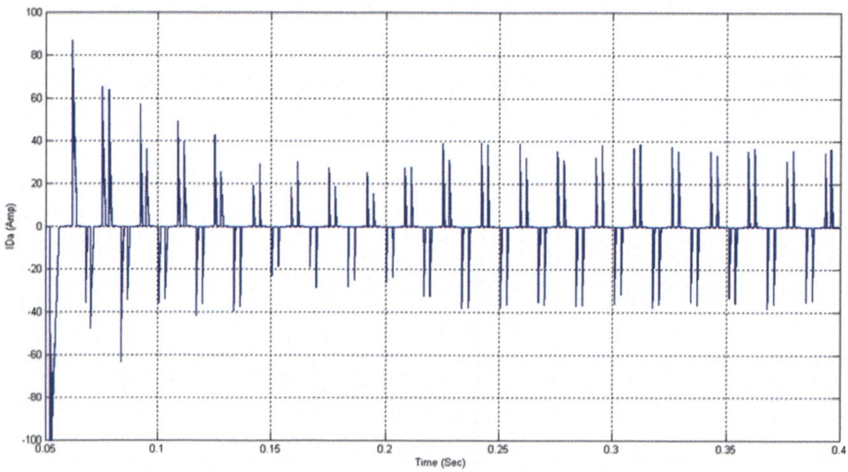

Fig. 12 Waveforms of ELC current of line-A due to sudden application of IM load

and current of SEIG, capacitor current of line-A remain almost constant on sudden

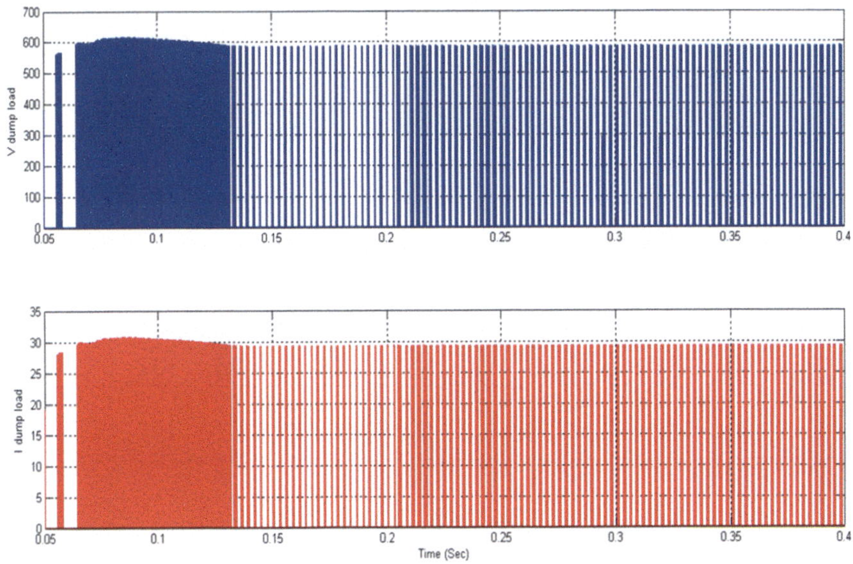

Fig. 13 Voltage and current across the dump load due to sudden application of IM load

application of IM load. Initially, motor load draws high current, but after few cycles, it draws the constant rated current and get settle down to its full load current value. After initially drawing large current, ELC also get settle down to small steady-state value as shown in waveforms of Figs. 11, 12, and 13.

6.4 *Dynamic Unloading of SEIG*

Suddenly, induction motor load is removed from the SEIG at $t = 0.2$ s. Waveforms corresponding to sudden removal of motor load at $t = 0.2$ s. are given in Figs. 14, 15, 16, 17, and 18. The terminal voltage, current of SEIG, and capacitor current of line-A remain same after the removal of dynamic load (IM). The value of load current becomes zero and ELC current increases after sudden removal of IM load on $t = 0.2$ s.

The fluctuating nature of the load active power (P), load reactive power (Q), torque (Ts), and rotor speed (Nr) on application of dynamic loading are as shown in Figs. 14, 15, 16, 17, and 18.

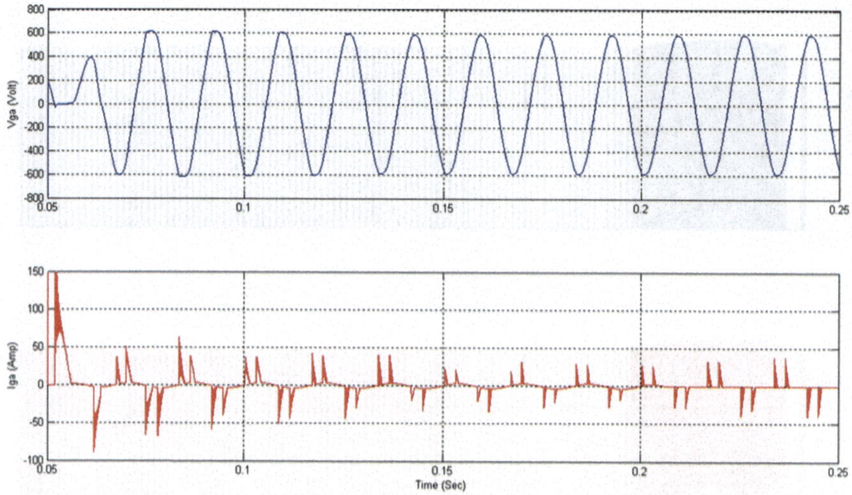

Fig. 14 Voltage and current waveform of line-A at the SEIG terminals due to sudden removal of IM load

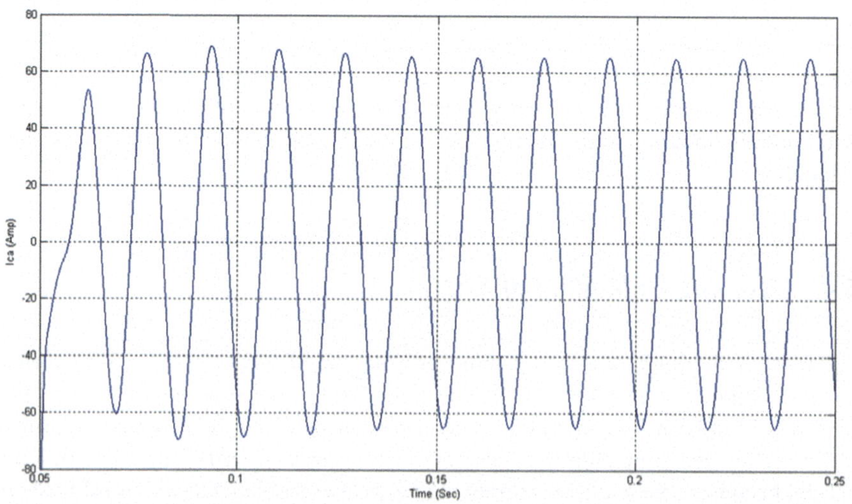

Fig. 15 Capacitor current of line-A due to sudden removal of IM load

7 Conclusions

In this work, the working performance of SEIG has been analyzed in terms of steady-state and transient conditions. The proposed work is most suitable for analyzing several variable parameters for steady-state and transient conditions, such as torque, speed, voltage, and current with different load conditions. The proposed work has

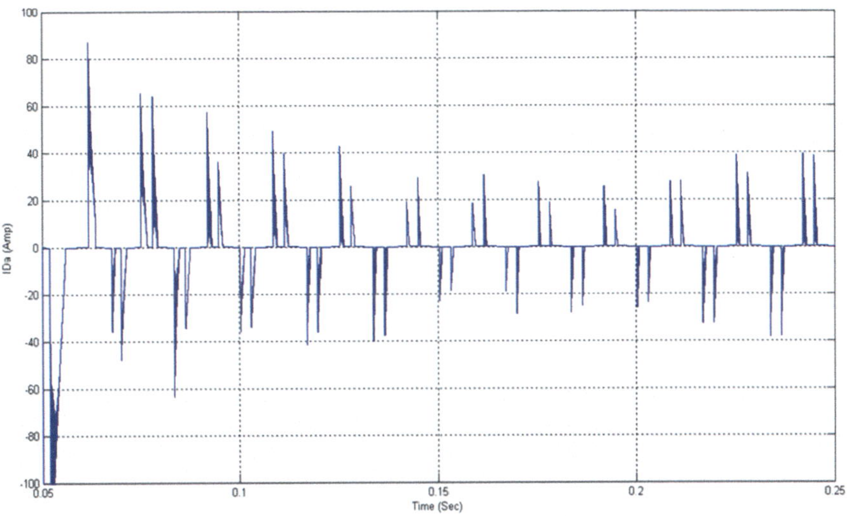

Fig. 16 Waveform of ELC current of line-A due to sudden removal of IM load

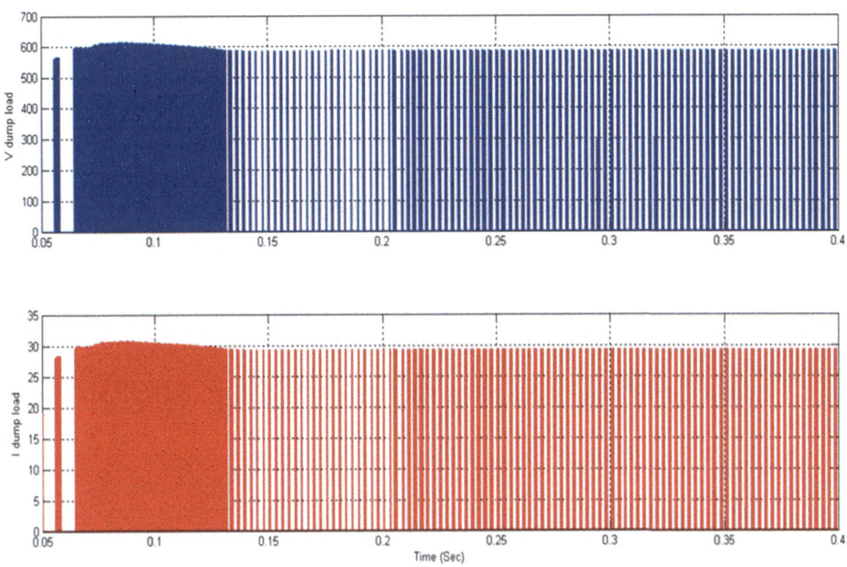

Fig. 17 Voltage and current across the dump load due to sudden removal of IM load

presented that SEIG-ELC system is suitable and adequate option to be utilized along with wind energy systems. It has also indicated the clear increment in the reliability and efficiency of system with proposed model having combination of ELC and SEIG. Thus, this work is of significant practical importance due to usage of wind as a fuel

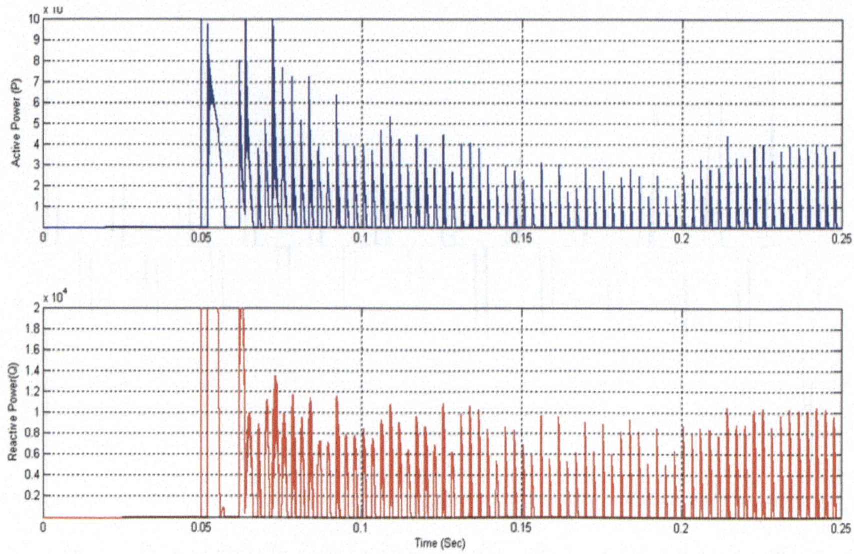

Fig. 18 P and Q generated by IM to the load

available in ample amount unless other fuels which are costly and scarcely available. It has also showed the replacement of traditional generator with SEIG-ELC system.

Appendix 1: Parameters of SEIG [14–16, 18]

$$3.73\,\text{kW}, 415\,\text{V}, 50\,\text{Hz}, Y\text{ - connection}, N = 24,$$
$$R_S = 0.6837\,\Omega, R_r = 0.45, L_{is} = 0.004152\,\text{H},$$
$$L_{ir} = 0.008152\,\text{H}, L_m = 0.011846\,\text{H},$$
$$\text{Inertia(J)} = 0.1\,\text{kg - m}^2, \text{Factor of Friction} = 0.008141, F(\text{N.m.s})$$

Appendix 2: Parameters of ELC [14–16, 18]

Power range = 3.7 kW (selected)
Voltage rage of rectifier and chopper = 900 V,
Current range of rectifier and chopper = 15 A,
Specification of dump load = 15 Ω,
Range of DC filtering capacitor = 380 μF
Resistive Load: 1.5 kW, Dynamic Load: 1 H.P.

References

1. Ministry of New and Renewable. http://mnre.gov.in/related-links/publications/akshay-urja/, Apr 2017
2. https://en.wikipedia.org/wiki/Inductiongenerator#Principle_of_operation. Apr 2017
3. Wang L, Lee CH (2008) Long-shunt and short-shunt connections on a dynamic performance of a seig feeding an induction motor load. IEEE Trans Energy Convers 15(1):1–7
4. Padmanaban S (2005) Triggering market for energy efficiency in India. Renew Energy India 1:17–18
5. http://mnre.gov.in/file-manager/annual-report/20162017/EN/index.Html
6. Jardan RK, Varga PSZ, Nagy I (2016) Novel solutions for high-speed self-excited induction generators. IEEE Trans Ind Electron 63(4):2124–2132
7. L. G. Scherer, R. V. Tambara, and R. F.de Camargo, "Voltage and frequency regulation of standalone self-excited induction generator for micro-hydro power generation using discrete-time adaptive control", IET Renewable Power Generation, Vol. 10, No. 4, pp. 531 – 540
8. Shridhar L, Singh B, Jha CS, Singh BP (1994) Analysis of self-excited induction generator feeding induction motor. IEEE Trans Energy Convers 9(2):390–396
9. Taoufik M, Abdelhamid B, Lassad S (2017) Stand-alone self-excited induction generator driven by a wind turbine. Alexandria Eng J (In Press), Available online 2017
10. Bhattacharya JL, Woodward JL (1998) Excitation balancing of a self-excited induction generator for maximum power output. IEEE Proc Inst Elect Eng Gen Trans Distr 135(2):88–97
11. Krause PC (1986) Analysis of electrical machinery. McGraw-Hill Book Company
12. Bae J-N, Kim Y-E, Yoo C-H Torque control topologies of a self-excited induction generator as an electric retarder. IEEE Trans Energy Convers 31(2):557–565
13. Bae J, Kim Y-E, Son Y-W, Moon H-S, Yoo C-H, Lee J (2015) Self-excited induction generator as an auxiliary brake for heavy vehicles and its analog controller. IEEE Trans Ind Electron 62(5):3091–3100
14. Singh B, Murthy SS, Reddy RS, Arora P (2015) Implementation of modified current synchronous detection method for voltage control of self-excited induction generator. IET Power Electron 8(7):1146–1155
15. Kalla UK, Singh B, Sreenivasa Murthy S (2017) Green controller for efficient diesel engine driven single-phase SEIG using maximum efficiency point operation. IEEE Trans Ind Electron 64(1):264–274
16. Singh GK (2017) Self-excited induction generator for renewable applications. Ref Module Earth Syst Environ Sci
17. Wang L, Su JY (1999) Dynamic performances of an isolated self-excited induction generator under various loading conditions. IEEE Trans Energy Convers 14(1):93–100
18. Ayodele TR, Ogunjuyigbe ASO, Adetokun BB (2017) Optimal capacitance selection for a wind-driven self-excited reluctance generator under varying wind speed and load conditions. Appl Energy 190:339–353
19. Jain SK, Sharma JD, Singh SP (2002) Transient performance of three-phase self-excited induction generator during balanced and unbalanced faults. IEEE Proc Inst Elect Eng Gen Trans Distr 149(1):50–57
20. Taoufik M, Lassad S (2016) Experimental stand-alone self-excited induction generator driven by a diesel motor. J Electr Syst Inf Technol (In Press). Corrected Proof, Available online 2016
21. Bansal RC (2005) Three-phase self-excited induction generators: an overview. IEEE Trans Energy Convers 20(2):292–299
22. Wang L, Deng R-Y (1999) Transient performance of an isolated induction generator under unbalanced excitation capacitors. IEEE Trans Energy Convers 14(1):887–893
23. Rashid MH (1996) Power electronic, circuits, devices, and applications, 2nd edn. Prentice Hall of India Private Limited, New Delhi, India

24. Ahmed T, Noro O, Hiraki E, Nakaoka M (1999) Terminal voltage regulation characteristics by static var compensator for a three-phase self-excited induction generator. IEEE Trans Ind Appl 14(3):298–303
25. Malik NH, Mazi AA (1987) Capacitance required for isolated self excited induction generator. IEEE Trans Energy Convers EC-2(1):62–69

Electrical Energy and Power Generation from Three Different Photovoltaic Module Technology in Composite Climate Condition: A Comparative Study

Rohit Tripathi, G. N. Tiwari, and T. S. Bhatti

Abstract In the present communication, a comparative study has been carried out by considering three different photovoltaic module technologies where silicon monocrystalline based PV cells have been used. In this study, the power and electrical energy have been compared for semitransparent, opaque and bifacial photovoltaic module with a number of cells having 36 cells except bifacial. Bifacial module has 36 cells on front and 36 cells on the back also. Such 36 cells based semitransparent and opaque PV module are eligible to generate 75 Wp. The area of all considered PV modules is 1 m^2. After evaluation the results, it is found that the maximum electrical energy and power has been obtained for bifacial PV module to others due to double area of PV cells on module whereas minimum has been obtained for opaque due to higher solar cell temperature. The electrical energy obtained from bifacial module is found 159% and 107% higher than opaque and semitransparent, respectively.

Keywords Silicon cells · PV technology · Bifacial module · Energy · Power

1 Introduction

Now, day-by-day energy demand increases and it is hard to meet the demand due to limited availability of fossil fuels. To overcome this issue, green energy is the only hope which can be satisfy this demand. Green energy comes from hydro power, wind, bio and solar energy. Solar energy can be obtained and convert into power

R. Tripathi (✉)
School of Electrical Electronics & Communication Engineering, Galgotias University, Greater Noida, India
e-mail: rohittripathi30.iitd@gmail.com

G. N. Tiwari
Bag Energy Research Society (BERS), SODHA BERS COMPLEX, Plot no. 51, Mahamana Nager, Varanasi, UP, India

R. Tripathi · T. S. Bhatti
Centre for Energy Studies, Indian Institute of Technology Delhi, Hauz Khas, New Delhi 110016, India

M. N. Favorskaya et al. (eds.), *Innovations in Electrical and Electronic Engineering*, Lecture Notes in Electrical Engineering 661, https://doi.org/10.1007/978-981-15-4692-1_18

through photovoltaic effect only and used material is called photovoltaic (PV) cell. This PV cell is made of silicon crystalline, mono, ply and amorphous, CIGS and CDTe. But in many researches, it is found that silicon crystalline based PV cells are more dominated than other types of PV cells [1]. Series connection of this PV cells makes PV module. Series and parallel connection of PV module makes PV array [2]. Generally, PV module designed on some different technologies and with different cell materials. Three different technologies are semitransparent, opaque and bifacial PV module. The photovoltaic (PV) has been identified to generate electrical energy. Further to enhance the overall efficiency of semitransparent module based PVT used to generated electrical as well as thermal energy by solar and concentration technology has been developed for further increment. Compound parabolic concentrator has been integrated with PVT collector which is known is PVT-CPC collector [3, 4]. The expression for electrical efficiency of the photovoltaic module (η_c) has been developed for the semitransparent PV module.

Recent research has been presented where extensive work has been studied with concentrated PVT where annual overall energy, exergy and life cycle cost have been presented for New Delhi, India [5, 6]. The photovoltaic modules in the northern hemisphere are placed south-oriented, having an inclination for horizontal surface equal to the latitude of system station [7–9]. So, all the modules are considered to be south-oriented in India at an angle of 30° with the horizontal. The annual solar radiation intensity, annual power output and electrical efficiency have been computed for clear day weather condition for New Delhi climatic condition [10]. The comparative study has been developed for monofacial and bifacial PV modules. The bifacial device found more efficient as a gain of 10% and 15% for 25° and 45° tilt angle, respectively, when compared to the monofacial one [11, 12]. One study has been presented where cost-effectiveness has been made for mono facial and bifacial PV module and it has been found that bifacial is cheap and higher efficiency than monofacial PV module [13].

In this paper, modified analytical expression for electrical efficiency of PV module of all technology have developed and evaluated electrical gain and power. It is also studied that electrical energy enhancement methods through PV technologies.

2 System Description and Principal

In the present study, three different PV module technologies have been considered for evaluation and comparison of electrical energy and power generation. First technology is semitransparent or glass to glass PV module, the second one is opaque PV module and the third one is bifacial PV module. In semitransparent and bifacial PV module, the transparent glass is placed top and bottom both side but the difference is that only top side, Si-based PV cells are placed whereas in bifacial PV module, PV cells are placed at top and bottom both sides. In opaque PV module, PV cells are placed only on top side with glass and Tedlar sheet is placed on the bottom side. The pictures of all considered PV modules are given in Fig. 1.

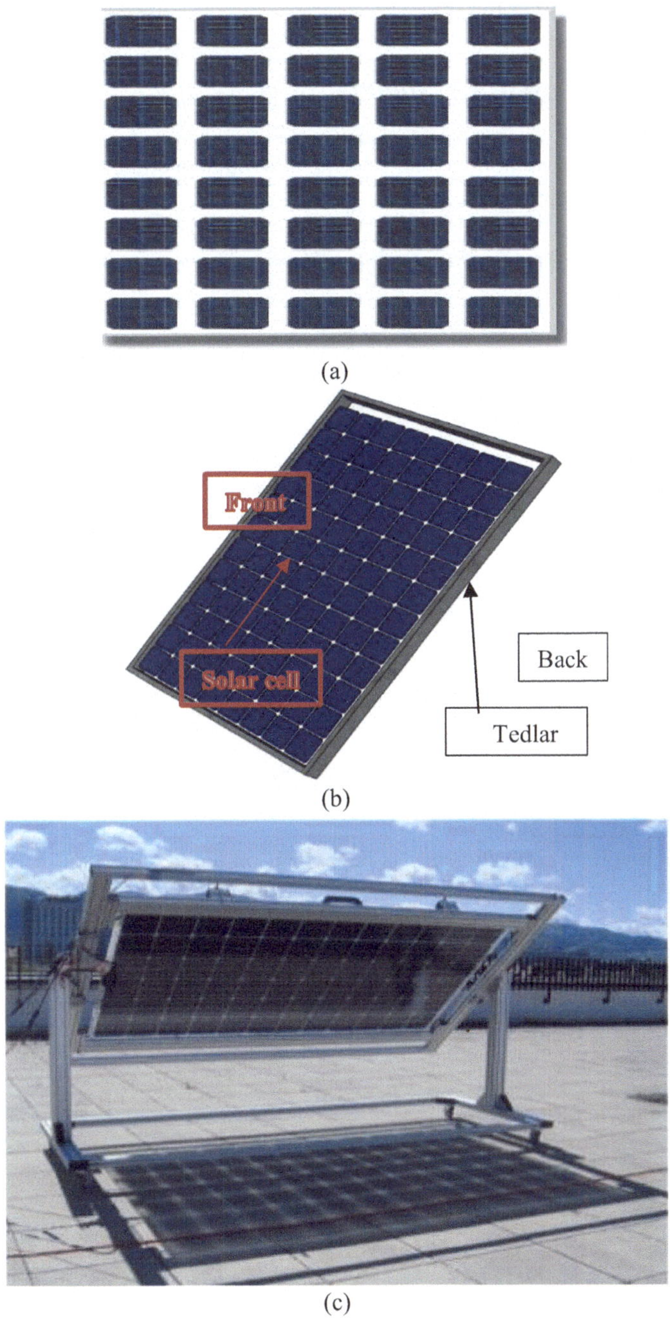

Fig. 1 Pictures of **a** semitransparent PV module, **b** Opaque PV module and **c** Bifacial PV module

3 Modelling and Theory

3.1 Semitransparent PV Module

For the semitransparent monocrystalline-Si photovoltaic module, the energy balance equation can be obtained as [4]

$$\alpha_c \tau_g \beta_c I(t) = U_t (T_c - T_a) + U_b (T_c - T_a) + \eta_c \tau_g \beta_c I(t) \tag{1}$$

where

$$U_t = \left[\frac{L_g}{K_g} + \frac{1}{h_o}\right]^{-1} \text{ and } U_b = \left[\frac{L_g}{K_g} + \frac{1}{h_i}\right]^{-1}.$$

and total overall heat transfer coefficient of PV module as follows:

$$U_L = U_t + U_b \tag{2}$$

From (1), (2), and (3), the expression for cell temperature T_c can be expressed as

$$T_c = \frac{T_a U_L + (\alpha_c \tau_g \beta_c - \tau_g \beta_c \eta_c) I(t)}{U_L} \tag{3}$$

$$T_c - T_a = T_a - T_o + \frac{T_a U_L + (\alpha_c \tau_g \beta_c - \tau_g \beta_c \eta_c) I(t)}{U_L} \tag{4}$$

The photovoltaic cell efficiency is dependent on temperature and it is given as follows [14]:

$$\eta_c = \eta_o [1 - \beta_o (T_c - T_0)] \tag{5}$$

Now putting (3) in (4) we can have the expression of photovoltaic cell efficiency

$$\eta_c = \frac{\eta_o \left[1 - \beta_o \left(T_a - T_o + \frac{T_a U_L + (\alpha_c \tau_g \beta_c) I(t)}{U_L}\right)\right]}{\left[1 - \frac{(\beta_o \beta_c \eta_o \tau_g) I(t)}{U_L}\right]} \tag{6}$$

3.2 Opaque PV Module

The second opaque PV module has been considered for comparison with semitransparent PV module. The same assumptions have been considered for writing energy balance equations.

$$\alpha_c\tau_g\beta_c I(t) + (1-\beta_c)\alpha_T\tau_g I(t) = U_t(T_c - T_a) + U_b(T_c - T_a) + \eta_c\tau_g\beta_c I(t) \quad (7)$$

or,

$$\tau_g I(t)[\alpha_c\beta_c + (1-\beta_c)\alpha_T] = (U_t + U_b)(T_c - T_a) + \eta_c\tau_g\beta_c I(t)$$

or,

$$(T_c - T_a) = \frac{\tau_g I(t)[\alpha_c\beta_c + (1-\beta_c)\alpha_T] - \eta_c\tau_g\beta_c I(t)}{(U_t + U_b)}$$

The solar cell temperature of an opaque PV module can be simplified as follows:

$$T_c = \frac{\tau_g I(t)[\alpha_c\beta_c + (1-\beta_c)\alpha_T] - \eta_c\tau_g\beta_c I(t)}{(U_L)} + T_a \quad (8)$$

With the help of Eq. (7), one can find the electrical efficiency of PV cells in opaque PV module as follows:

$$\eta_c = \frac{\eta_o\left[1 - \beta_o\left(\frac{\tau_g I(t)[\alpha_c\beta_c + (1-\beta_c)\alpha_T]}{(U_L)} + T_a - T_o\right)\right]}{\left[1 - \frac{(\beta_o\beta_c\eta_o\tau_g)I(t)}{U_L}\right]} \quad (9)$$

3.3 *Bifacial PV Module*

$$\begin{aligned}\alpha_c\tau_g\beta_c I(t) + \rho\alpha_c\tau_g\beta_c I(t) &= U_t(T_c - T_a) + U_b(T_c - T_a)\\ &\quad + \eta_c\tau_g\beta_c I(t) + \rho\eta_c\tau_g\beta_c I(t)\end{aligned} \quad (10)$$

where

$$U_t = \left[\frac{L_g}{K_g} + \frac{1}{h_o}\right]^{-1} \text{ and } U_b = \left[\frac{L_g}{K_g} + \frac{1}{h_i}\right]^{-1}$$

And total overall heat transfer coefficient of PV module as follows:

$$U_L = U_t + U_b \quad (11)$$

$$T_c = \frac{\{(1+\rho)(\tau_g\beta_c I(t))(\alpha_c - \eta_c)\} - T_a U_L}{U_L} \quad (12)$$

$$\eta_c = \frac{\eta_o\left[1 - \beta_o\left(\frac{\{(1+\rho)(\alpha_c\tau_g\beta_c I(t))\} - T_a U_L}{U_L}\right)\right]}{\left[1 - \frac{\{(1+\rho)(\beta_o\beta_c\eta_o\tau_g I(t))\}}{U_L}\right]} \tag{13}$$

The module electrical efficiency for all considered technology can thus be written from Eqs. 6, 9 and 13 as:

$$\eta_m = \eta_c\tau_g\beta_c \tag{14}$$

Electrical energy generated form PV module

$$E_{\mathrm{ele}} = \eta_m A_m I(t). \tag{15}$$

4 Results and Discussion

The considered PV modules for the present study have been tilted at latitude angle of New Delhi, 28.7° for maximum absorption of input solar energy. The hourly variation of solar intensity $I(t)$ for four weather conditions in composite climate in Delhi in January month with ambient air temperature has been shown in Fig. 2. It is clearly seen that the maximum irradiance has been found in a clear day than others for the same month.

The silicon crystalline PV cell temperature of all considered PV module technology has been shown in Fig. 3. Here, the PV cell temperature of top and bottom of bifacial module have also shown for better conclusions and effect of cell temperature. It is noted that the maximum temperature has been seen for opaque module whereas

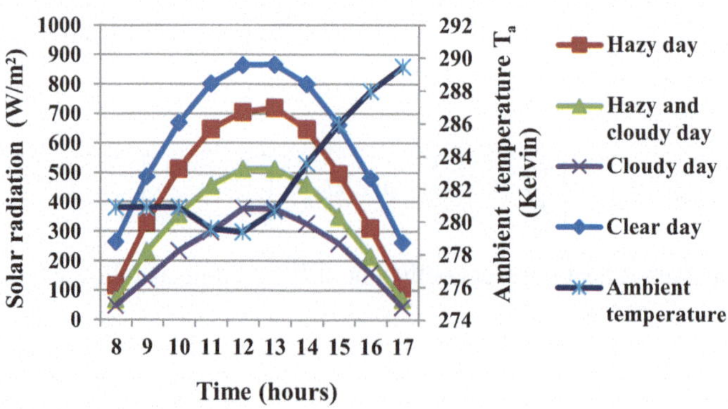

Fig. 2 Hourly variation of solar radiation and ambient air temperature of a day in January month with all-climate condition at New Delhi, India

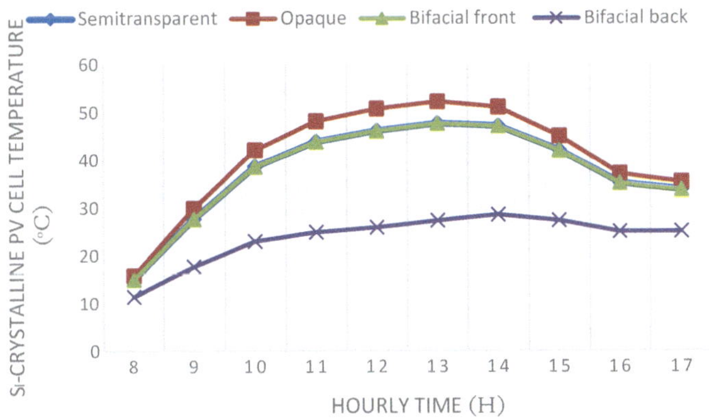

Fig. 3 Hourly variation of silicon crystalline PV cell temperature with all three considered PV module technology for the same climate parameters condition

it is almost same for semitransparent and front of bifacial PV module and minimum for back of bifacial PV module. In this study, one reflector has been placed at facing of backside of bifacial PV module and its reflectivity is 0.7.

The electrical efficiency of Si-crystalline PV cell of all considered module has been shown in Fig. 4. It is clearly seen that maximum efficiency has been found for bifacial module whereas minimum for opaque PV module based cells. Hourly variation of PV cell temperature and electrical efficiency of all considered PV modules have been shown in Fig. 5. Here, it is seen that the electrical efficiency of cell decreases when cell temperature increases. It is mandatory to control cell temperature for gaining high electrical output. The hourly variation of electrical energy and power of all

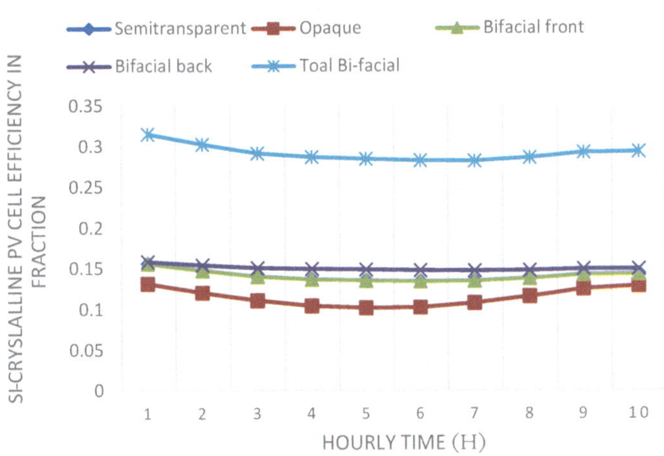

Fig. 4 Hourly variation of silicon crystalline electrical efficiency of PV cell with all three considered PV module technology for the same climate parameters condition

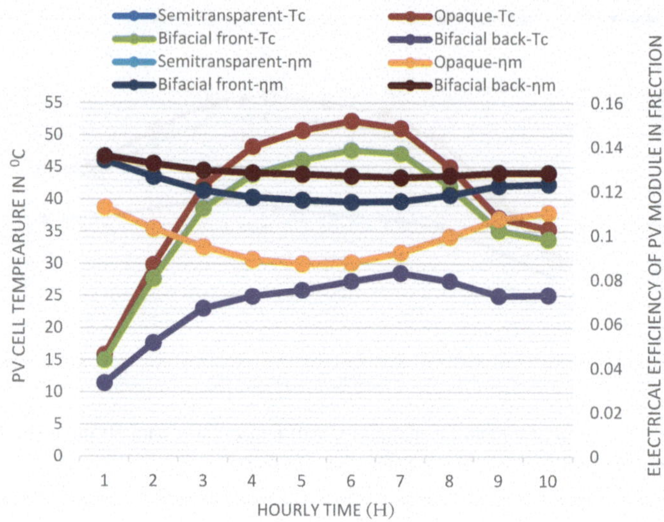

Fig. 5 Hourly variation of silicon crystalline PV cell temperature and electrical efficiency of PV cell with all three considered PV module technology for the same climate parameters condition

considered PV modules have been given in Fig. 6. The maximum electrical energy and power have been found maximum for bifacial PV module and minimum for opaque PV module. Electrical energy and power for bifacial PV module is 61% and 51% higher than opaque and semitransparent, respectively.

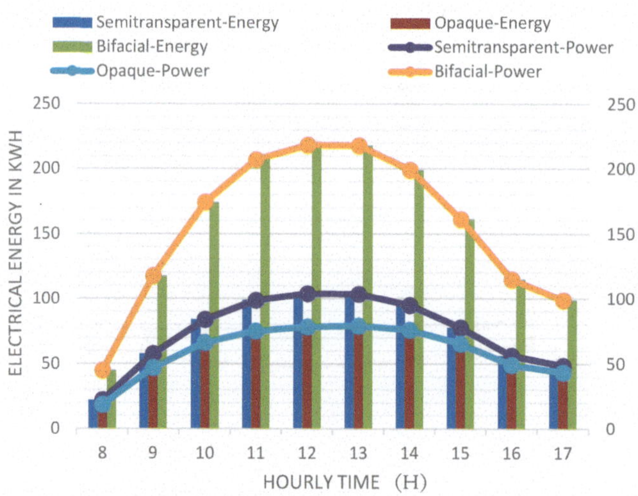

Fig. 6 Hourly variation of silicon crystalline electrical efficiency of PV cell with all three considered PV module technology for the same climate parameters condition

5 Conclusions

Certain conclusions have been made on the basis of the above analysis as follows:

- Electrical efficiency of silicon crystalline PV cells decreases when PV cell temperature increases. It is mandatory to maintain the PV cell temperature with electrical operational range of all considered PV module.
- In the present study, bifacial PV module has been found best among all considered modules and it generates maximum electrical energy and power than others. The daily obtained energy for bifacial is 1557 Wh whereas 749 Wh for opaque.
- Bifacial PV module is more economical than other due to power generation is 2.4 and 1.9 times higher than opaque and semitransparent PV module and it requires the same area with reflector to generate more power.
- Bifacial PV module can be used for snow area or highly cold area where other PV panels are failed due to covering of snow whereas bifacial PV module is best due to back portion is able to generate electrical energy or power.

References

1. Saini V, Tripathi R, Tiwari GN, Al-Helal IM (2018) Electrical and thermal assessment of N partially covered PVT-compound parabolic concentrator collector connected in series, for different solar cell materials. Appl Therm Eng 128:1611–1623
2. Tripathi R, Sharma R, Tiwari GN (2020) Experimental study on PV degradation loss assessment of Stand-Alone Photovoltaic (SAPV) array in field: a new simplified comparative analytical approach. In: Bhoi A, Sherpa K, Kalam A, Chae GS (eds) Advances in greener energy technologies. Green energy and technology. Springer, Singapore
3. Tripathi R, Tiwari S, Tiwari GN (2016) Energy analysis of partially covered number (N) of photovoltaic thermal-compound parabolic concentrator collectors connected in series at constant collection temperature mode. IEEE International Conference on Emerging Trends in Electrical Electronics & Sustainable Energy Systems (ICETEESES) Sultanpur 12–17. https://doi.org/10.1109/ICETEESES.2016.7581344
4. Tripathi R, Bhatti TS, Tiwari GN (2020) Effect of packing factor on electrical and overall energy generation through low concentrated photovoltaic thermal collector in composite climate condition. Mater Today: Proceedings. https://doi.org/10.1016/j.matpr.2020.05.005
5. Tripathi R, Tiwari A, Tiwari GN (2020) Overall performance of N partially covered photovoltaic thermal-compound parabolic concentrator (PVT-CPC) collector with different concentration ratio. In: Singh S, Ramadesigan V (eds) Advances in energy research, vol 2. Springer proceedings in energy, Springer, Singapore
6. Tripathi R, Tiwari GN, Dwivedi VK (2016) Overall energy and exergy performance of partially covered N-photovoltaic thermal (PVT)-compound parabolic concentrator (CPC) collectors connected in series. In: IEEE international conference on power electronics, intelligent control and energy systems, pp 12–17
7. Tripathi R, Tiwari GN (2019) Energy matrices, life cycle cost, carbon mitigation and credits of open-loop N concentrated photovoltaic thermal (CPVT) collector at cold climate in India: a comparative study. Sol Energy 186:347–359
8. Tripathi R, Tiwari GN (2017) Annual performance evaluation (energy and exergy) of fully covered concentrated photovoltaic thermal (PVT) water collector: an experimental validation. Sol Energy 146:180–190

9. Tripathi R, Tiwari GN, Dwivedi VK (2017) Energy matrices evaluation and exergoeconomic analysis of series connected N partially covered (glass to glass PV module) concentrated photovoltaic thermal collector: at constant flow rate mode. Energy Convers Manage 145:357–370
10. Tripathi R, Tiwari GN, Bhatti TS, Dwivedi VK (2017) 2-E (Energy-Exergy) for partially covered concentrated photovoltaic thermal (PVT) collector. Energy Procedia 142:616–623
11. Katsaounis Th, Kotsovos K, Gereige I, Basaheeh A, Abdullah M, Khayat A, Al-Habshi E, Al-Saggaf A, Tzavara AE (2019) Performance assessment of bifacial c-Si PV modules through device simulations and outdoor measurements. Renew Energy 143:1285–1298
12. Rodríguez-Gallegos CD, Bieri M, Gandhi O, Singh JP, Reindl T, Panda SK (2018) Monofacial vs bifacial Si-based PV modules: Which one is more costeffective? Sol Energy 176:412–438
13. Tripathi R, Tiwari GN (2018) Annual energy, exergy, and environmental benefits of *N* half covered concentrated photovoltaic thermal (CPVT) air collectors. In: SenGupta S, Zobaa A, Sherpa K, Bhoi A (eds) Advances in smart grid and renewable energy. Lecture notes in electrical engineering, vol 435. Springer, Singapore
14. Tiwari A, Tripathi R, Tiwari GN (2020) Improved analytical model for electrical efficiency of semitransparent photovoltaic (PV) module. In: Singh S, Ramadesigan V (eds) Advances in energy research, vol 1. Springer proceedings in energy, Springer, Singapore

Piezoelectric Energy Harvesting System Using Railway Tracks

Manvi Mishra, Priya Mahajan, and Rachana Garg

Abstract In this paper, an analysis of two methods of harvesting the electrical power from the two different piezoelectric systems in railway traction system is carried out. The two piezoelectric systems used for analysis are compression type piezoelectric system and cantilever beam type piezoelectric system. These systems are suggested to be embedded on the railway track and attached to the rail tracks to produce the electrical output. The lead zirconate titanate (PZT), which is one of the recent piezoelectric, which accumulates more charge on mechanical disturbance, is used in both the cases. The compression type piezoelectric system is excited by the force applied by the train, and cantilever beam type system is excited by the ground vibrations of the track. The output of two systems is compared for a single unit, and according to this comparison, the array of compression type system has been designed.

Keywords Piezoelectric system (PES) · Lead zirconate titanate (PZT) · Compression type PES · Cantilever beam type PES

1 Introduction

With the advancement in technology and increasing rate of population, the demand of electrical energy is increasing with very high rate. Rising concern about the gap between demand and supply of electricity for masses has highlighted the exploration of the alternate sources of energy and its sustainable use that does not negatively

M. Mishra (✉) · P. Mahajan · R. Garg
Delhi Technological University, New Delhi, India
e-mail: mishmanvi@gmail.com

P. Mahajan
e-mail: priyamahajan.eed@gmail.com

R. Garg
e-mail: rachana16100@yahoo.com

M. N. Favorskaya et al. (eds.), *Innovations in Electrical and Electronic Engineering*, Lecture Notes in Electrical Engineering 661, https://doi.org/10.1007/978-981-15-4692-1_19

impact the environment. This gives the aim to generate electricity from the renewable sources of energy. In India, railway uses around 2–3% of the total electricity produced by the different electrical power generating plants. India has fourth largest railway network in the world of around 63,000 km. In the recent time, the railways are also trying to use more renewable energy for their applications. The different types of renewable energies used by railways could be solar energy, wind energy, supercapacitors, and piezoelectric energy. The energy is collected from the solar panel which could be present outside the train and on the train itself [1]. The wind generator can be present near the railway track or it may be at some distance where the source of high velocity wind is present. The supercapacitors can be present at the railway platforms so that the energy produced due to the acceleration and deceleration of train can be stored in the supercapacitors [2]. One of the novel ways to utilize waste energy to generate electricity is using piezoelectric effect [3]. This technique can be effectively used in railway traction system. As suggested in this research study, the location of piezoelectric unit can be on the railway track and on the side of the railway track. The utilization of waste energy like pressure of the coaches and locomotive, and ground vibrations on the track due to their movement, can produce some amount of electrical energy. The piezoelectric ceramic can convert the mechanical disturbances into the electrical energy. Piezoelectric harvesting system helps to extract such energies which are available, but are not utilized [4, 5]. There are some countries like Japan and Israel researching in this field very effectively to harvest some reasonable amount of electrical energy. East Japan Railway Company (JR-East) is actively involved in the research which aims to have the railway stations and platforms to be environmental friendly and the energy which is being wasted can be utilized to generate electricity. They are working on the technology which aims to generate the electricity by using piezoelectric pads on the ticket gate counters which generate energy when a person walks through it [6, 7]. There is some research using the weight of the human to produce electrical energy by applying the piezoelectric pads on the floor [8]. Besides this, an Israel company known as Innowattech is trying to apply piezoelectric technology to railway tracks and at the sides of the railway tracks for electricity generation [9].

In this paper, two different positions, i.e., on the very side of the railway tracks and on the top of railway tracks are proposed to install the piezoelectric units. The ground vibrations or the force of the train is used to harvest electrical energy. The analysis of two different piezoelectric systems (PES), i.e., compression type piezoelectric system and cantilever beam type piezoelectric system for energy harvesting in railway tracks is carried out. The output of single unit of both the systems has been compared, and based on the results, array of compression type PES has been designed.

2 Piezoelectric Effect

Piezoelectric effect refers to change in electric polarization so that the electric charge can be produced when a mechanical stress is applied to some materials. The word piezoelectric is derived from the Greek piezin, which means to squeeze or press,

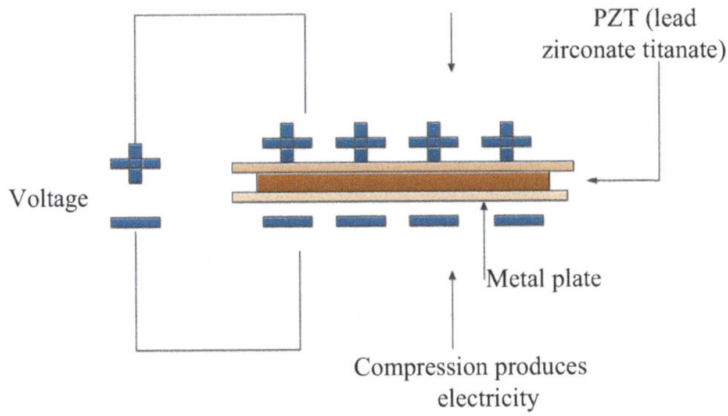

Fig. 1 Basic piezoelectric effect

and piezo, which is Greek for "push." Piezoelectric materials are non-conductive piezoelectric ceramic or crystal is placed between the two metal plates. For piezoelectricity to be generated, material needs to be compressed or squeezed. Mechanical stress applied to the piezoelectric material generates electricity.

PZT can be made from lead zirconate titanate and can produce more voltage than other piezoelectric material with the same amount of mechanical pressure. This effect is the result of change in electric polarization when mechanical effect is applied. The change in electric polarization will result in the change in surface charge density on the crystal faces of piezoelectric unit. Figure 1 shows the basic working of piezoelectric unit on the application of the mechanical disturbance.

Piezoelectric energy harvester is one of the preferable methods for generating electricity from the non-conventional sources of energy. The input of the harvester is mainly the ac current source or the ac voltage source which is analogous to the mechanical force or vibrations. This produces the variations in the piezoelectric layer that lead to change in electric polarization of the material according to the poling directions. There are different types of energy harvesting mode in piezoelectric like d_{31} or d_{33}. The first subscript describes the electric component direction, and second subscript describes the mechanical effect direction. A d_{31} mode of energy harvester is used in case of cantilever beam type piezoelectric system. The d_{33} mode is used in the compression type piezoelectric system [4]. In case of the railway tracks, the train bogie force is used by the piezoelectric unit to produce electrical energy. Similarly, the ground vibrations near the railway tracks should be used by the cantilever units to produce electrical energy.

There are two types of piezoelectric effect discussed:

(i) Compression type
(ii) Cantilever beam type

The compression type piezoelectric system is dependent on the amount of force applied on the piezoelectric unit. The cantilever beam type piezoelectric system is dependent on the acceleration of the train and the frequency of the structure.

2.1 Compression Type PES

The compression type piezoelectric works when the pressure is applied on the piezoelectric unit. For defining piezoelectric property of the material, there are 3 directions, which is to be defined by various constants having subscript with them. 1 indicates x axis, 2 indicates y axis, and 3 indicates z axis. Axis 3 is assigned to the direction of initial polarization of piezo ceramic, and axes 1 and 2 lies in plane perpendicular to the axis 3. The basic electromechanical equations for a piezoelectric material can be written as:

$$\varepsilon_i = S_{ij}^E \sigma_j + d_{mi} E_m \tag{1}$$

$$D_m = d_{mi} \sigma_i + \xi_{ik}^\sigma E_k \tag{2}$$

where the indexes $i, j = 1, 2, \ldots, 6$ and $m, k = 1, 2, 3$ refer to different directions within the material coordinate system, σ is stress vector (N/m^2), ε is strain vector, E is vector of applied electric field (V/m), ξ is permittivity (F/m), d is matrix of piezoelectric strain constants (m/V), S is matrix of compliance coefficient (m^2/N), D is vector of electric displacement (C/m^2).

The piezoelectric constant d is the polarization generated or the electric charge generated per unit of applied mechanical force. Its unit is C/N and it is highly direction dependent. As shown in Fig. 2, the force is applied on the piezoelectric material in the direction 3 which generates the stress

$$\sigma_3 = \frac{F}{lw} \tag{3}$$

which results in electric charge $Q = d_{33} F$, the electric charge flowing through the short circuit. The piezoelectric constant g defines the electric field produced along one direction when the material is stressed in other direction. In case of compression

Fig. 2 Compression type piezoelectric system

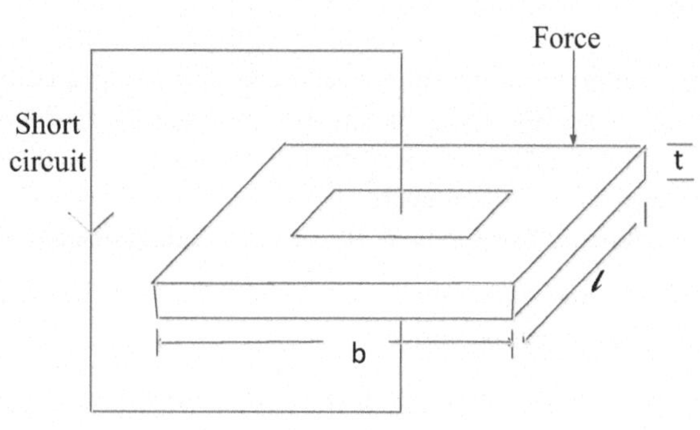

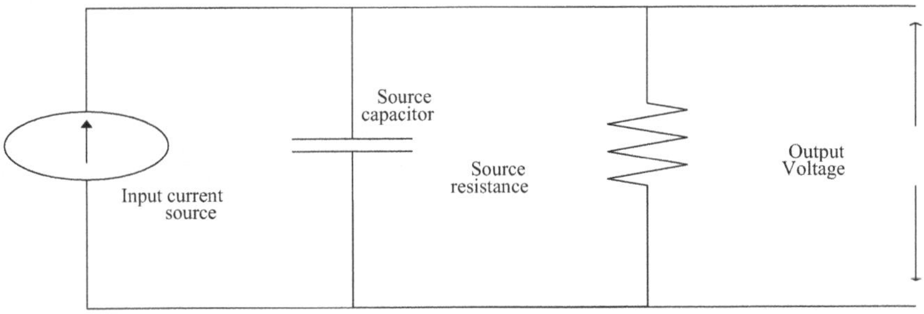

Fig. 3 Equivalent electrical circuit of piezoelectric element

type piezoelectric system, the direction of electric field developed and the stress applied is in the same direction [10]. Therefore, for the force applied in the above figure, the voltage generated is

$$V = \frac{g_{33} F}{t} \tag{4}$$

With this information, the electrical equivalent circuit of the piezoelectric transducer can be derived which is shown in Fig. 3.

The charge accumulated on the plates of electrode and the voltage are proportional to the force applied on the piezoelectric unit. The input current source is in parallel with the source capacitor, which is

$$C = \frac{\varepsilon_{\circ} \varepsilon A}{t} \tag{5}$$

where $\varepsilon_{\circ}$ is the permittivity of free space, ε is the relative permittivity of PZT material, and t is the thickness of electrode.

2.1.1 Design of Compression Type PES for Railway Tracks

The piezoelectric effect which converts the deformation in piezoelectric material due to mechanical effect is known as direct piezoelectric effect. As shown in Eq. (4), the output voltage of the piezoelectric unit in case of compression depends on the applied force on it. Therefore, installing piezoelectric pads under the railway track would help to harvest electrical energy effectively.

If the average weight of each bogie of a train is 40 T_{on} and each bogie has 4 wheels, the force by one wheel is $F = 9.8 \times 10 = 98$ kN. Figure 4 shows the suggested location for piezoelectric pads on the railway track.

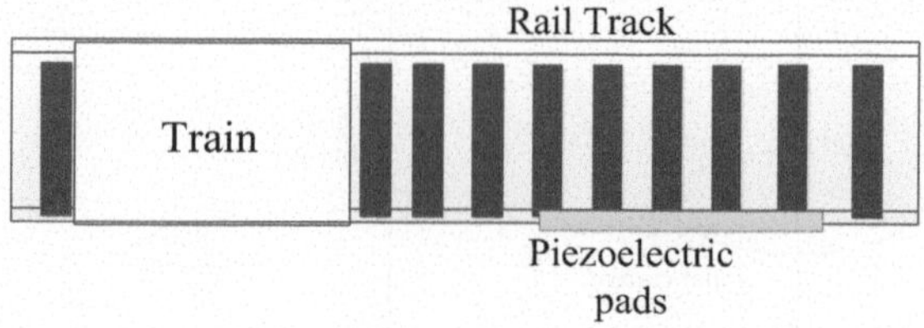

Fig. 4 Location of compression type piezoelectric unit

Table 1 Parameters used for calculation

S. No.	Constants	Value
1	Charge constant (d_{33})	390×10^{-12}
2	Length (mm)	58
3	Width (mm)	15
4	Thickness (mm)	1.82
5	Permittivity of free space ($\varepsilon_\circ$)	8.854×10^{-12}
6	Relative permittivity (ε)	3400
7	Voltage constant (g_{33})	48×10^{-3}

With reference to the values in Table 1, the calculated value of the capacitor is 14.39×10^{-9} F, charge accumulated on one piezoelectric unit is 38.22 μC. The energy generated by the single piezoelectric compression type unit is 0.188 J. The high-speed train will take less than a second to cross over it, so power could be around 0.2 W.

Figure 5 shows the voltage and the current generated in a single compression type piezoelectric unit. Also Fig. 6 shows that as the force increases the energy generated can be increased. Table 2 shows the calculated energy output for different types of trains considering a bogie has 4 wheel system.

2.2 *Cantilever Beam Type PES*

A bimorph cantilever beam type piezoelectric system is one of the structures to generate reasonable amount of output power from the piezoelectric material, because it can handle large amount of strain and variable frequency under vibrating conditions. The harvester system consists of the piezoelectric material, cantilever body having piezoelectric layer and tip mass.

For understanding, the schematic of the system is given in Fig. 7 where m is the mass of the system, k is stiffness of the spring, and c_T is the damping coefficient. The excitation to the system is the external sinusoidal vibrations $x(t) = A \sin(\omega t)$ (where A is the amplitude of the acceleration and ω is the vibration frequency), resulting in the mass moves as $y(t)$ [11–13]. The equation of motion of the system is described in equation:

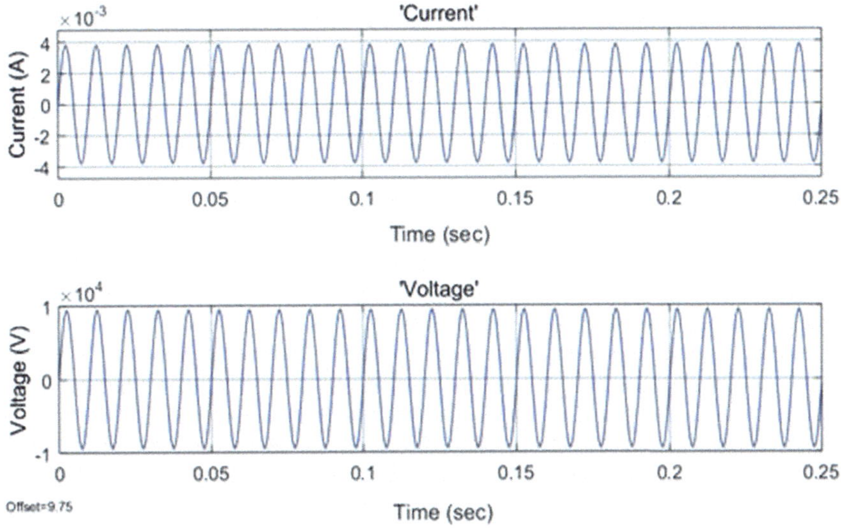

Fig. 5 Output current and output voltage of single PES unit

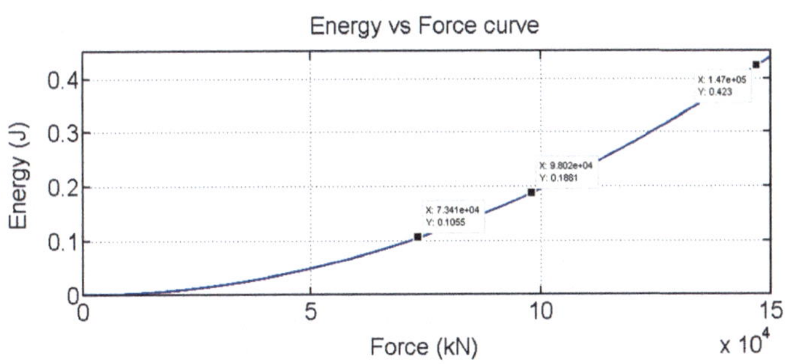

Fig. 6 Energy versus force curve

Table 2 Energy output of different trains

S. No.	Type of train	Weight (Ton)	Energy (J)
1	Suburban	30	0.105
2	Passenger	40	0.188
3	Freight	60	0.42

Fig. 7 Physical model

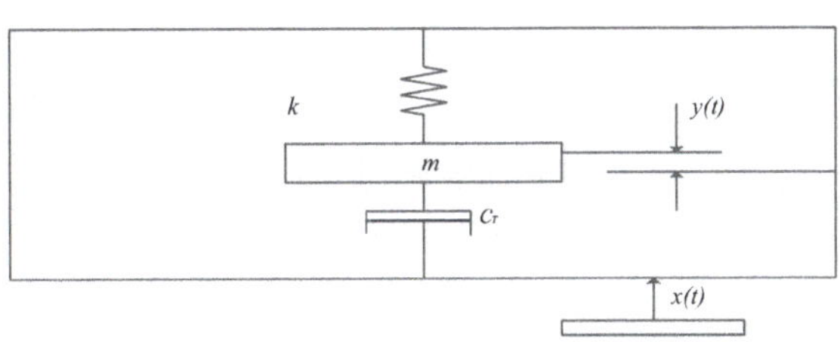

$$m\ddot{y}(t) + c_T\dot{y}(t) + ky(t) = -m\ddot{x}(t) \tag{6}$$

As the vibrations is applied to the system, the energy is extracted from relative moment between the tip mass and the system base, the condition for maximum power is when system frequency matches the natural frequency [14, 15].

$$\omega_n = \sqrt{\frac{K_{\text{eff}}}{m_{\text{eff}}}} \tag{7}$$

where $K_{\text{eff}} = \frac{3EI}{L^3}$ and $m_{\text{eff}} = \frac{33m_b}{140+m_t}$

In the above given expression, I is the moment of inertia of the beam structure, m_b is the mass of the beam, and m_t is the tip mass which has been added to the cantilever structure.

And damping ratio ($\zeta_t = k/2\, m\omega_n$) prevents the power goes to very high value when the harvesting system vibrates at resonant frequency. Figure 7 shows the system of a bimorph piezoelectric energy harvester with tip mass. It can convert the vibrations of the system into the electricity. The open circuit voltage V generated from the harvester system is given by Eq. (8) [13]

$$V = \frac{-d_{31}t_p\sigma_n}{\varepsilon} \tag{8}$$

where t_p is the piezoelectric layer thickness, -d_{31} is the piezoelectric charge constant, ε is the dielectric constant of the piezoelectric material, σ_n is the stress produced in the system structure at resonant frequency and given by:

$$\sigma_n = \frac{3EAh}{4L^2\zeta_t\omega_n^2} \tag{9}$$

where E is the beam effective modulus, h is the piezoelectric layer from the neutral axis, L is the beam length, ζ_t is the system damping ratio, and A is the acceleration of the body.

The power generated from the piezoelectric energy harvester can be calculated by Eq. (10)

$$P = \frac{V_o^2 R_{\text{Lp}}}{(R_s + R_{\text{Lp}})^2} \tag{10}$$

where R_s is the source resistance and R_{Lp} is the load resistance. As we know, the condition of maximum power is when the load resistance equals to the internal source resistance [16–18].

2.2.1 Design of Cantilever Beam Type PES for Railway Track

The cantilever beam type PES has a beam with layer of piezoelectric material present both at the top and bottom surfaces. This system when comes under the vibrating condition produces a voltage. A theoretical model can be designed by calculating the value of structure frequency and voltage harvested by the single unit. Table 3 has been given to specify the values of different constants used in the designing of the system.

The calculated value of frequency of structure is 93.54 Hz and voltage is 22.7 V. Hence, with the source resistance 5 MΩ, the power is 25.76 μW. The location for cantilever type piezoelectric system can be the very side of railway tracks as shown in Fig. 8. These units must be present very near to the railway platforms as the voltage generated in this case is dependent on the acceleration of the train.

The analysis of single unit of cantilever beam type piezoelectric system has been given in Fig. 9:

The power of piezoelectric unit of cantilever type having different acceleration has been given in Fig. 10 with the help of the power vs acceleration curve. The suburban trains have high acceleration than passenger train. With acceleration, the power of the cantilever beam type piezoelectric increases; therefore, it is more suitable for suburban trains.

Table 3 Parameters of cantilever PES

S. No.	Constants	Values
1	E = modulus of beam (GPa)	3.81×10^9
2	I = moment of inertia (mm^4)	0.36
3	L_{eff} = effective length of beam (mm)	36
4	ω_n = frequency of structure	93.54
5	h = distance of piezoelectric layer from neutral axis (mm)	0.3
6	A = acceleration (m/s^2)	2.5
7	ζ_t = damping ratio	0.015
8	d_{31} = piezoelectric strain constant (C/N)	1.75×10^{-10}
9	t_p = thickness of piezoelectric layer (mm)	0.16
10	ε = permittivity of piezoelectric	1.55×10^{-3}
11	m_b = mass of beam (g)	44.3
12	m_t = tip mass (g)	5

Fig. 8 Location of cantilever beam unit

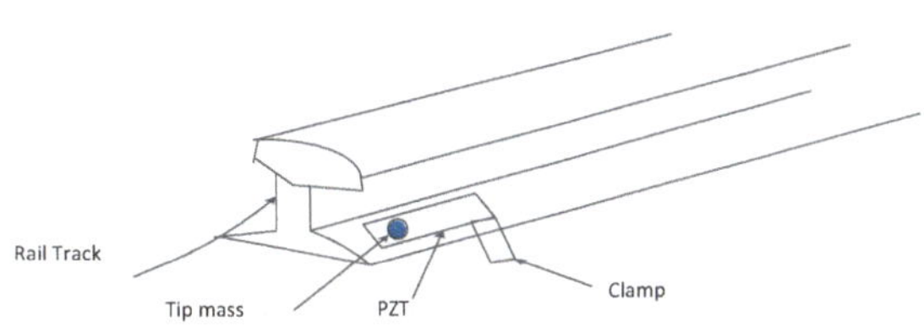

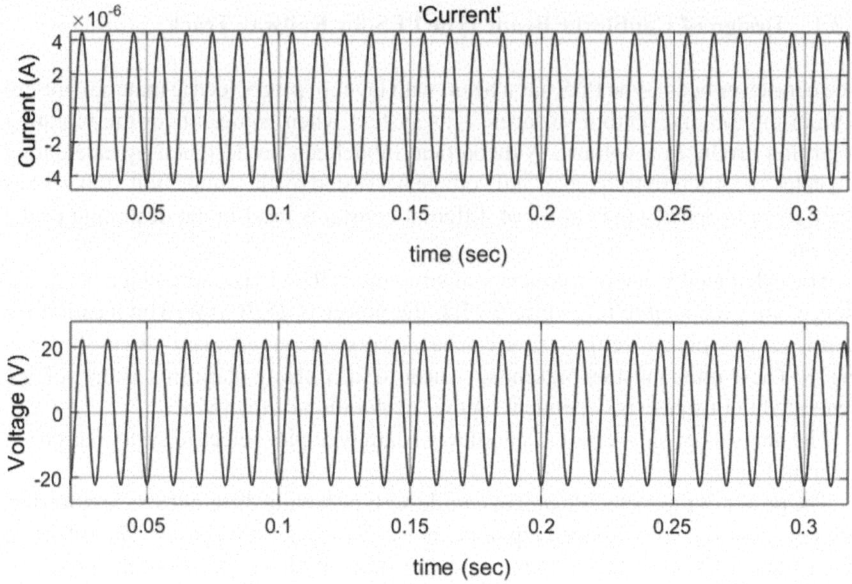

Fig. 9 Output current and voltage of single cantilever PES

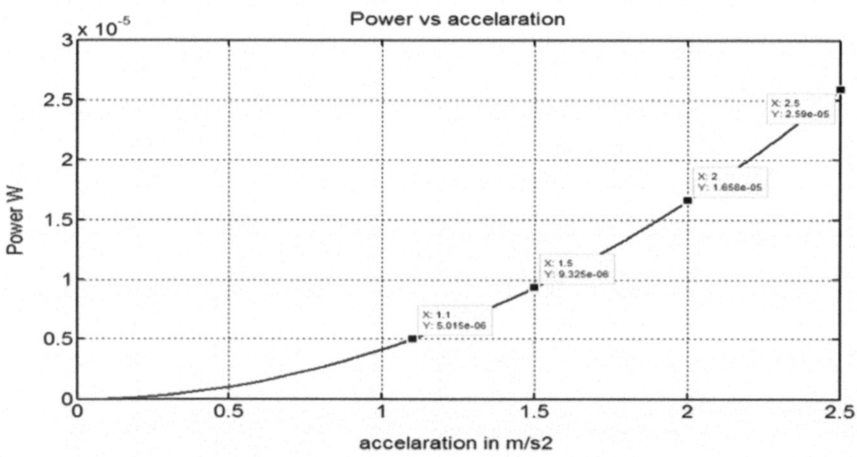

Fig. 10 Power versus acceleration curve

3 Comparison Between Compression Type and Cantilever Type PES

(i) The power produced by the single unit of compression type piezoelectric unit is around 0.2 W, whereas the power produced by the single unit of cantilever beam type piezoelectric system is only 25.76 μW for same system parameters configuration.

(ii) In cantilever type piezoelectric system, the voltage produced is dependent on the force applied to the piezoelectric unit, whereas in cantilever beam type piezoelectric system the voltage is dependent on structure frequency and acceleration of the train.

(iii) The location suggested for the compression type piezoelectric array can be at some distance from the platform. The location suggested for the cantilever beam type piezoelectric system is near the platform to use the acceleration.

Thus, on comparing the power outputs of the individual unit of compression type piezoelectric and cantilever type piezoelectric system, the output of compression type is more as weight of the train is high.

4 Designing of Compression Type Piezoelectric Array

As observed in the above studies, the array of compression type piezoelectric on the railway track could be more helpful. The exact location of compression type piezoelectric array has been shown in Fig. 11.

The main aim of piezoelectric array on the railway track is to generate reasonable amount of kWh when a train crosses over it. The piezoelectric pads are installed at each rail sleeper cross section area. The area of cross section is 34290 mm^2 and the area of one piezoelectric unit is 870 mm^2 [19]. Therefore, at each cross section there will be a piezoelectric pad containing 40 piezoelectric units. Over a stretch of certain kilometer, 500 sleepers have been used for PES. This array is connected in parallel, thus the total charge accumulate wheel a wheel crosses a system is around 0.76C.

The total energy generated by units on both sides = $Q \times V$ = 7.544 kJ. If the train has 4 wheel-axel system and 20 bogies, the total energy generated is 603.57 kJ. The train is traveling with 90 km/h, it takes around 37 s to cross these installed pads, thus total energy generated by one train is 6.2 kWh. If 20 trains cross through that track, then total energy over a day is 84 kWh.

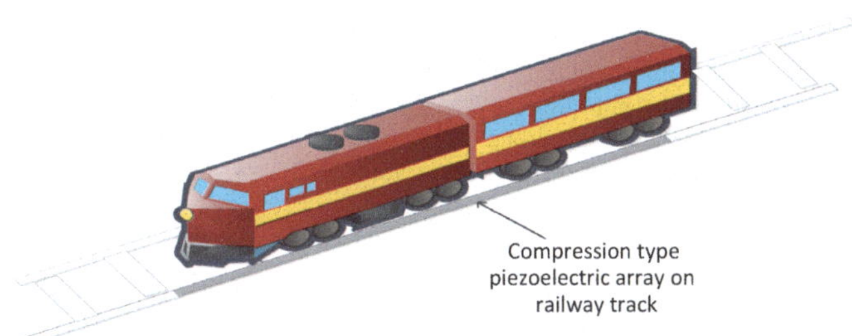

Fig. 11 Implementation of compression type piezoelectric array

5 Conclusions

The aim of the study is to harvest reasonable energy using piezoelectric effect when a train passes through the installed piezoelectric units. The two types of piezoelectric system are studied and compared. It has been observed that the cantilever beam type piezoelectric system is generating very small amount of energy, however, the collective amount of power that can be generated by the cantilever system can be used for signaling purposes near railway platform. Further, power output by single unit of compression type piezoelectric system is found to be more as it is dependent on heavy mass of train. So, an array of compression type PES is designed. The energy generated over a day can be stored and used thereafter, specifically for signal lights and rail road side signs at remote and unpopulated locations and intersections where the cost of providing power using power lines from grid could be high.

References

1. Darshana MK, Karnataki K, Shankar G, Sheela KR (2015) A practical implementation of energy harvesting, monitoring and analysis system for solar photo voltaic terrestrial vehicles in Indian scenarios: a case of pilot implementation in the Indian Railways. In: 2015 IEEE international WIE conference on electrical and computer engineering (WIECON-ECE), Dhaka, pp 542–545
2. Becherif M, Ayad MY, Henni A, Aboubou A (2010) Hybrid sources for train traction: wind generator, solar panel and supercapacitors. IEEE Int Energy Conf 2010:658–663
3. OmPrakash B, Kumar TJ (2017) Hybrid power generation using solar, wind and piezo. Int J Adv Energy Res Tech 5 (11) (IJAERT), November 2017
4. Sodano HA, Inman DJ (2004) Estimation of electric charge output for piezoelectric energy harvesting. Strain J 40(2):49–58
5. Staines G, Hofmann H, Dommer J, Altgilbers LL, Tkach Y (2003) Compact piezo-based high voltage generator - part I: quasistatic measurements. Electromag Phenom 3(3):11
6. Alam KS et al (2012) Modeling and computation of a solar-piezoelectric hybrid power plant for railway stations. In: 2012 international conference on informatics, electronics and vision (ICIEV), Dhaka, pp 155–159
7. Zhang ZW, Xiang HJ, Shi ZF (2015) Modeling on piezoelectric energy harvesting from pavements under traffic loads. J Intell Mater Syst Struct 27
8. Dutta M, Shrimoyee P (2017) Footstep voltage generator using piezo-electric transducers. Int J Sci Eng Res 8(3), March-2017, ISSN 2229-5518
9. Abramovich H, Harash E (2010) Power harvesting from railways: apparatus, system and method. Innowattech Ltd., Patent No.: US 7,812,508 B2, Date of Patent, Oct 12 (2010)
10. Djuguma R, Trivailo P, Graves K (2009) A study of energy harvesting from piezoelectrics using impact forces. Eur Phys J Appl Phys (EPJ AP) 48(1):11101
11. Li Jingcheng, Jang Shinae, Tang Jiong (2012) Design of a bimorph piezoelectric energy harvester for railway monitoring. J Korean Soc Nondestr Test 32:661. https://doi.org/10.7779/JKSNT.2012.32.6.661
12. Karlström A, Boström A (2006) An analytical model for train-induced ground vibrations from railways. J Sound Vibr 292(1–2):221–241
13. Priya S, Chen C (2005) Piezoelectric windmill: a novel solution to remote sensing. Jpn J Appl Phys 44(3):104–107
14. Challa VR, Prasad MG, Fisher FT (2009) A coupled piezoelectric-electromagnetic energy harvesting technique for achieving increased power output through damping matching. Smart Mater Struct 18:095029

15. Roundy S et al (2005) Improving power output for vibration-based energy scavengers. IEEE Pervas Comput 4(1):28–36. https://doi.org/10.1109/mprv.2005.14
16. Li J, Jang S (2012) Design of a bimorph piezoelectric energy harvester for railway monitoring. J Korean Soc Nondestr Test 32(6):661–668, ISSN 1225-7842/eISSN 2287-402X
17. Gegrande G, Chatterjee P, Van de Velde W, Hölscher P, Hopman V, Wang A, Dadkah N, Klein R (2004) Vibration due to a test train at variable speeds in a deep bored tunnel embedded in London clay. In: 11th international congress on sound and vibration. St-Petersburg, Russia
18. Zhu M, Worthington E, Tiwari A (2010) Design study of piezoelectric energy-harvesting devices for generation of higher electrical power using a coupled piezoelectric-circuit finite element method. IEEE Trans Ultrasonic's, Ferroelectr Freq Contr 57:46–51
19. www.railway-technical.com, Last accessed 2/01/2020

Performance Investigation of Different Bypass Diode Topology Based SDK-PV Arrays Under Partial Shading Conditions

Anurag Singh Yadav, Vinod Kumar Yadav, V. Mukherjee, and Santosh Ghosh

Abstract In this paper, a detailed study is carried out on the SuDoKu (SDK) PV array topologies under different shading patterns. The aim of this study is to investigate the impact of different types of bypass diode arrangements connected to the SDK-PV array under non-uniform irradiations, which are without bypass diode (WBD), non-overlapped bypass diode, (NBD) and overlapped bypass diode (OBD). Under partial shaded conditions, the power versus voltage (P–V) and current versus voltage (I–V) characteristics exhibit extreme nonlinearity along with multiple peak points. The modeling of SDK-PV array with various types of bypass diode topologies is carried out in MATLAB/Simulink environment. The result shows that the NBD-based SDK-PV array produced maximum output power and short-circuit current, amongst all the bypass topologies studied.

Keywords Bypass diode · Partial shading · Photovoltaic array

1 Introduction

The output power of photovoltaic (PV) arrays can be reduced by several environmental factors [1, 2], but the mismatch power losses and partial shading conditions are most dominant ones [3, 4]. Under partial shading conditions, the PV modules get reverse biased, which implies that PV modules work as a load and not as a power generators [4, 5]. The shaded PV modules may be irreversibly damaged due to hotspot phenomenon, when reverse bias voltage exceeds its reverse breakdown

A. S. Yadav · V. Mukherjee
Electrical Engineering Department, Indian Institute of Technology (ISM), Dhanbad, Jharkhand 826004, India

V. K. Yadav
Department of Electrical Engineering, Delhi Technological University, New Delhi 110042, India

S. Ghosh (✉)
Corporate R&D Department, Kirloskar Brothers Limited, Pune 411045, India
e-mail: Santosh.ghosh@kbl.co.in

M. N. Favorskaya et al. (eds.), *Innovations in Electrical and Electronic Engineering*, Lecture Notes in Electrical Engineering 661, https://doi.org/10.1007/978-981-15-4692-1_20

voltage of the PV cells [1, 3–8]. To reduce the reverse voltage across the cell or group of cells under partial shading condition, and thereby avoid the hotspot phenomenon traditionally a bypass diode is connected across one-third of the cells in a module. However, when the bypass diodes conduct under partial shading condition, multiple peaks are generated in the *I–V* and *P–V* curves [3–5]. And hence the performance of PV arrays largely depends upon the connection topologies of bypass diodes [3–11]. Two different types of bypass diode connection topology employed, which are overlapped bypass diode (OBD) and non-overlapped bypass diode (NBD) topology. In case of NBD topology, bypass diodes are connected across separate set of PV cells, whereas in case of OBP topology, the bypass diodes are connected across mutual set of PV cells, as illustrated in Fig. 1. The analysis of PV array system with NBD in partial shading conditions has been addressed by most of the researchers [11–22]. The mathematical modeling of PV system with OBD topology is extremely complicated, and very few studies are available in the literature focused on the same [23–29]. Moreover, researchers have also proposed various techniques for reducing the impacts of hotspot under partial shading condition, and maximizing the *I-V* and *P–V* characteristics of PV systems without considering a bypass diode [24–35], i.e., without bypass diode (WBD) topology. In this paper, three types of bypass diode topologies connected to SDK-PV array are analyzed under partial shading conditions.

Following this brief introduction, the rest of the paper is organized as follows. In Sect. 2, the mathematical modeling of PV arrays with various bypass diode connections is described. Various types of shading patterns considered in the present work are also discussed. Subsequently, the results and discussion are delineated in Sect. 3, and finally, Sect. 4 concludes this paper.

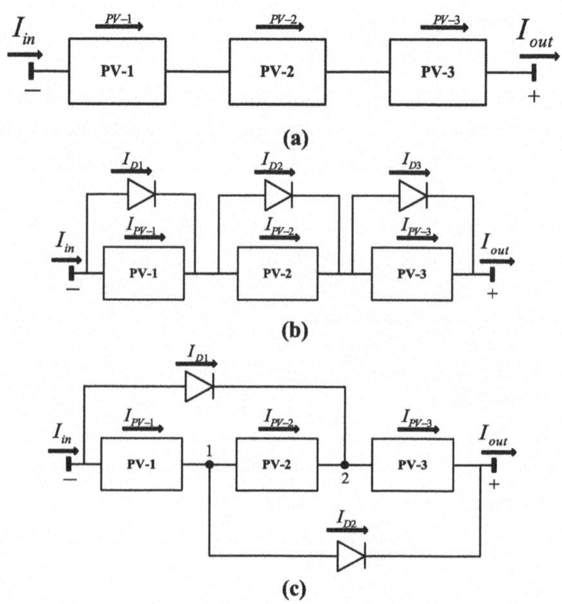

Fig. 1 Bypass diode topologies **a** WBD, **b** NBD, and **c** OBD

2 Modeling of PV Arrays with Various BD Topologies

Different bypass diode topologies, used for mitigating the impacts of partial shading as described in Sect. 1, are illustrated in Fig. 1.

The array output current for WBD topology is mathematically expressed as

$$I_{\text{out(WBD)}} = I_{\text{PV}-1} = I_{\text{PV}-2} = I_{\text{PV}-3} \tag{1}$$

$$I_{\text{out(WBD)}} = S_{\text{fn}} I_{\text{PV}-n} \tag{2}$$

Whereas the output current for NBD-based PV system may be presented as

$$I_{\text{out(NBD)}} = I_{\text{PV}-1} + I_{\text{D1}} = I_{\text{PV}-2} + I_{\text{D2}} = I_{\text{PV}-3} + I_{\text{D3}} \tag{3}$$

The total module current is dependent upon the shading coefficients (S_{c}) and the output current can be presented as

$$I_{\text{out(NBD)}} = S_{\text{c}} \sum_{n} I_{\text{PV}-n} + I_{\text{Dn}} \tag{4}$$

Similarly, for OBD topology, by applying KCL in nodes 1 and 2 of Fig. 1c we get

$$\begin{aligned} I_{\text{PV}-1} &= I_{\text{PV}-2} + I_{\text{D2}} \\ I_{\text{out}} &= I_{\text{D2}} + I_{\text{PV}-3} \end{aligned} \tag{5}$$

From Eqs. (1) and (2), we get

$$I_{\text{out}} = I_{\text{PV}-1} + I_{\text{PV}-3} - I_{\text{PV}-2} \tag{6}$$

The approximated values of currents under short-circuited situations may be obtained as.

$$\left.\begin{aligned} I_{\text{PV}-1(\text{SC})} &\cong S_{f(\text{PV}-1)} I_{\text{m}} \\ I_{\text{PV}-2(\text{SC})} &\cong S_{f(\text{PV}-1)} I_{\text{m}} \\ I_{\text{PV}-2(\text{SC})} &\cong S_{f(\text{PV}-1)} I_{\text{m}} \end{aligned}\right\} \tag{7}$$

Using Eqs. (7), (6) may be rewritten as,

$$\left.\begin{aligned} I_{\text{out}} &= S_{f(\text{PV}-1)} I_{\text{m}} + S_{f(\text{PV}-3)} I_{\text{m}} - S_{f(\text{PV}-2)} I_{\text{m}} \\ I_{\text{out}} &= \left(S_{f(\text{PV}-1)} + S_{f(\text{PV}-3)} - S_{f(\text{PV}-2)}\right) I_{\text{m}} \end{aligned}\right\} \tag{8}$$

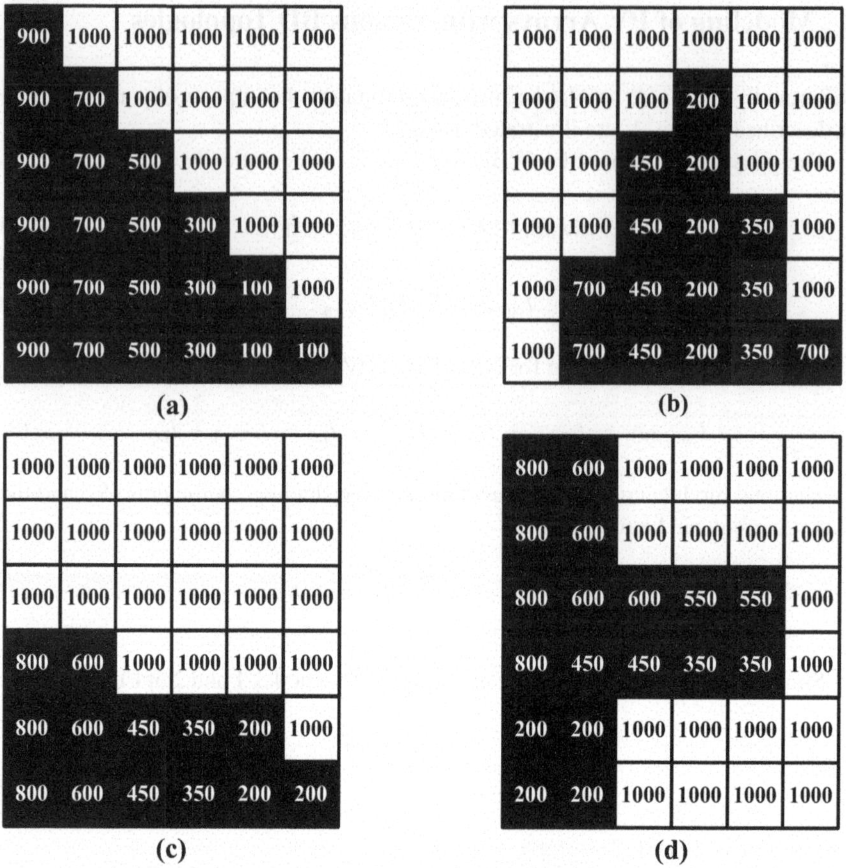

Fig. 2 Various shadings patterns on SDK-PV array **a** Case a, **b** Case b, **c** Case c, and **d** Case d

2.1 Partial Shading Conditions and Its Impact

The various possibilities of shading patterns on PV arrays, considered in the present work, are illustrated in Fig. 2.

3 Results and Discussion

3.1 P-V *and* I-V *Characteristics of SDK-WBD, SDK-OBD, and SDK-NBD PV Arrays Configurations*

Figures 3 and 4 show the *P–V* and *I–V* profiles of SDK-WBD, SDK-OBD, and SDK-NBD configurations of PV arrays, respectively. The SDK-WBD configuration

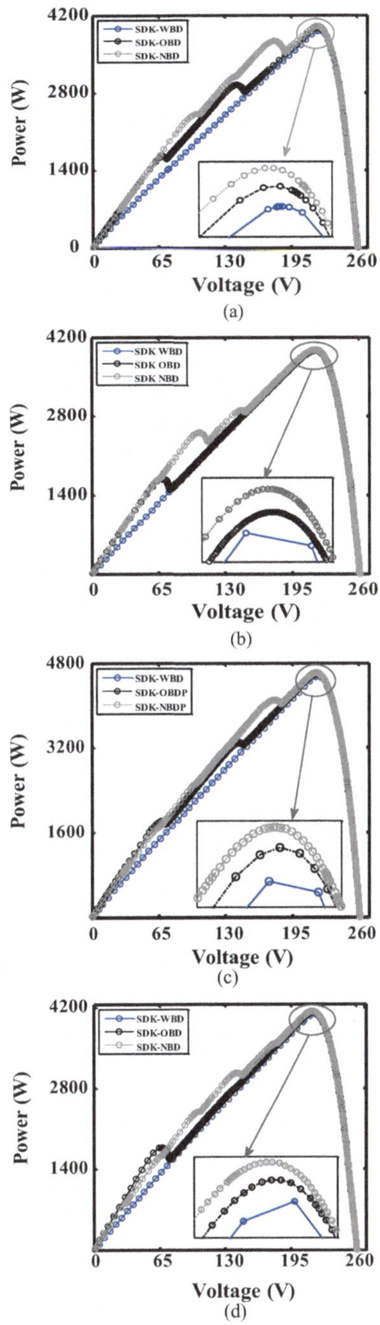

Fig. 3 LMPP and GMPP points of SDK-PV arrays for different shading patterns **a** for Case a, **b** for Case b, **c** for Case c, and **d** for Case d

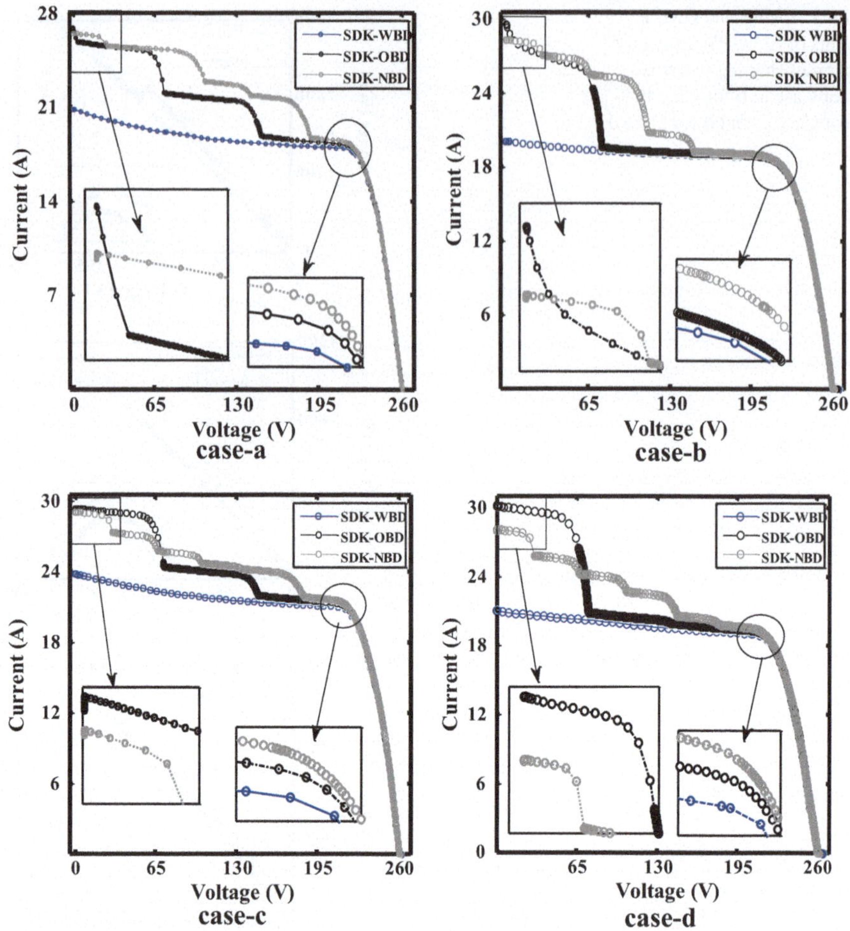

Fig. 4 Comparative I–V characteristics under different shading patterns

shows that only one peak is generated in the power and current profile, under all the shading patterns illustrated in Fig. 2a–d. And these peak power points in terms of absolute values are 3926 W, 3959 W, 4545 W, and 4078 W, respectively.

Under shading pattern—Case a (Fig. 2a), three peak points are generated in SDK-OBD configuration, whereas in case of SDK-NBD, five peak points are generated. And in SDK-OBD configuration, the first peak power point is generated of 1660 W at 68.71 V, second peak power point of 2944 W is generated at 141.2 V, and the third peak power point of 3966 W is generated at 222.2 V (refer Fig. 3a). From these values, it is evident that the third peak power point (i.e., 3966 W) is the global maximum power point (MPP) (GMPP) and other points are local MPP (LMPP). In case of SDK-NBD PV configuration, five peak power points of 561.03 W, 2317.53 W, 3032 W, 3717 W, and 4003 W are found to have occurred at 21.43 V, 93.6 V, 135 V,

175 V, and 221.4 V, respectively. The GMPP of SDK-NBD PV array configuration is found to be higher than that of the SDK-OBD PV array configuration.

For shading pattern–Case b (refer Fig. 2b), the behavior of SDK-OBD and SDK-NBD PV configurations has two and five peak power points, respectively (refer Fig. 3b). In this case, the LMPP of SDK-OBD configuration occurred of 1632.9 W at 63.34 V and GMPP occurred of 3971.24 W at 218.2 V. For SDK-NBD PV configuration, five peak power points of 740.81, 1596.43, 2473.30, 2899, and 3982.40 W are generated at 26.88 V, 60.22 V, 99.73 V, 143.2 V, and 216.2 V, respectively. And the maximum generated power in SDK-OBD PV configuration is lesser than that of the SDK-NBD configuration.

For shading patterns—Case c (see Fig. 2c) and Case d (see Fig. 2d), SDK-OBD and SDK-NBD PV configurations performance is illustrated in Fig. 3c and d, respectively. The peak power points observed in SDK-OBD of 1722.46 W, 3263.28 W, and 4590.14 W at 60.8 V, 139.1 V, and 219.1 V, respectively, for shading pattern—Case c. While for shading patterns—Case d, the peak power points of 1680.82 W, 2772.91 W and 4105.36 W occurred at 58.77 V, 138.3 V and 217.1 V, respectively. In case of SDK-NBD PV array configuration, six peak power points are generated for both shading patterns—Case c and Case d. For shading pattern—Case c, peak power points of 765.75 W, 1637.44 W, 2457.17 W, 3211.56 W, 4052.95 W, and 4605.09 W are generated at 27.03 V, 61.19 V, 97.16 V, 132 V, 172.1 V, and 216.1 V, respectively.

While for shading—Case d (as illustrated in Fig. 2d), peak power points of 734.09 W, 1561.16 W, 2350.40 W, 3042.6 W, 3549.66 W, and 4123.93 W occurred at 26.89 V, 62 V, 99.72 V, 138.3 V, 176.6 V, and 215.18 V, respectively.

Figure 4 illustrates comparative *I–V* profile of all the PV arrays configurations under all four different shading patterns described in Fig. 2. It may be observed from Fig. 4 that the short-circuit current of SDK-WBD configuration is lesser than that of the SDK-OBD and SDK-NBD PV array configurations. The short-circuit current of SDK-NBD configuration is very closer to the SDK-OBD array configuration but lesser in absolute terms. From Fig. 4b and c, it may be observed that the values of short-circuit current of SDK-OBD configuration are higher than the other types. The absolute values of peak power and short-circuit current of all PV array configurations are summarized in Tables 1 and 2, respectively.

Table 1 GMPP positions for various natured PV arrays

Shadow problems	SDK-WBD	SDK-OBD	SDK-NBD
	P_{GMPP} (W)	P_{GMPP} (W)	P_{GMPP} (W)
Case—a	3926	3958	4003
Case—b	3959	3971	3984
Case—c	4545	4589	4615
Case—d	4078	4103	4125

Table 2 Current position for various natured PV arrays

Shadow problems	SDK-OBD	SDK-NBD
	I_{st} (A)	I_{st} (A)
Case—a	26.83	26.42
Case—b	29.55	28.9
Case—c	29.23	29.02
Case—d	29.23	27.98

4 Conclusions

In this paper, the performance of SDK-PV array topology under different shading patterns is investigated for different bypass circuits, which are without bypass diode (EBD), overlapped bypass diode (OBD), and non-overlapped bypass diode (NBD) configurations. From this investigation, it is evident that the performance of WBD-based SDK-PV array is poorer than other considered SDK configurations.

From the *I*–*V* characteristics of SDK arrays, it is observed that the OBD-based SDK configuration generates more short-circuit current compared to NBD-based SDK array for all types of shading patterns considered in the present work. For NBD-based SDK configuration, the short-circuit current values are higher by 1.52%, 2.19%, 0.71%, and 4.27%, respectively, than OBD configuration. Hence, this simulation-based investigation reveals that the NBD-based SDK configuration performs better than OBD and WBD-based SDK configuration under the impact of different shading patterns.

References

1. Ghosh S, Yadav VK, Mukherjee V (2019) Impact of environmental factors on photovoltaic performance and their mitigation strategies—a holistic review. Renew Energy Focus 28:153–172
2. Ghosh S, Yadav VK, Mehta G, Mukherjee V, Birajdar R (2015) Status check: journey of India's energy sustainability through renewable sources. Control of Power and Energy System, IFAC-PapersOnLine 48:456–461
3. Ghosh S, Yadav VK, Mukherjee V (2020). A novel hot spot mitigation circuit for improved reliability of PV module. IEEE Trans Device Mater Reliab https://doi.org/10.1109/tdmr.2020.2970163
4. Ghosh S, Yadav VK, Mukherjee V (2019) Improvement of partial shading resilience of PV array though modified bypass arrangement. Renew Energy 143:1079–1093
5. Ghosh S, Yadav VK, Mukherjee V (2018) Evaluation of cumulative impact of partial shading and aerosols on different PV array topologies through combined Shannon's entropy and DEA. Energy 144:765–775
6. Ghosh S, Yadav VK, Mukherjee V, Yadav P (2017) Evaluation of relative impact of aerosols on photovoltaic cells through combined Shannon's entropy and Data Envelopment Analysis (DEA). Renew Energy 105:344–353
7. Alonso-García MC, Ruíz JM (2006) Analysis and modelling the reverse characteristic of photovoltaic cells. Sol Energy Mater Sol Cells 90:1105–1120

8. Ziar H, Afjei E (2012) Analysis of thermal effect on solar cells with considering bypass diodes at shadow condition through binary coding method. Int Rev Modelling Simul 5(1):412–417
9. Ziar H, Mansourpour S, Afjei E, Kazemi M (2012) Bypass diode characteristic effect on the behavior of solar PV array at shadow condition. In: Proceeding of 3rd International Conference on Power Electron. Drive Syst. Technol., Feb. 2012, pp. 229–233
10. Díaz-Dorado E, Suárez-García A, Carrillo C, Cidrás J (2011) Optimal distribution for photovoltaic solar trackers to minimize power losses caused by shadows. Renew. Energy 36:1826–1835
11. Díaz-Dorado E, Cidrás J, Carrillo C (2014) Discrete I–V model for partially shaded PV-arrays. Sol Energy 103:96–107
12. Zheng H, Shuhui Li S, Challoo R, Proano J (2014) Shading and bypass diode impacts to energy extraction of PV arrays under different converter configurations. Renew Energy 68:58–66
13. Alonso-García MC, Ruiz JM, Chenlo F (2006) Experimental study of mismatch and shading effects in the I–V characteristic of a photovoltaic module. Sol Energy Mater Sol Cells 90:329–340
14. Alonso-García MC, Ruiz JM, Herrmann W (2006) Computer simulation of shading effects in photovoltaic arrays. Renew Energy 31:1986–1993
15. Bai J, Cao Y, Hao Y, Zhang Z, Liu S, Cao F (2015) Characteristic output of PV systems under partial shading or mismatch conditions. Sol Energy 112:41–54
16. Bishop J (1988) Computer simulation of effects of electrical missmatches in PV Cell interconnection circuits. Solar Cells 25:73–89
17. Deline C, Dobos A, Janzou S, Meydbray J, Donovan M (2013) A simplified model of uniform shading in large photovoltaic arrays. Sol Energy 96:274–282
18. Jena D, Ramana VV (2015) Modeling of photovoltaic system for uniform and nonuniform irradiance: a critical review. Renew Sustain Energy Rev 52:400–417
19. Karatepe E, Boztepe M, Çolak M (2007) Development of a suitable model for characterizing photovoltaic arrays with shaded solar cells. Sol Energy 81:977–992
20. Vemuru S, Singh P, Niamat M (2012) Modeling impact of bypass diodes on photovoltaic cell performance under partial shading. In: 2012 IEEE international conference on electro/information technology (EIT), pp 1–5
21. Petrone G, Ramos-Paja CA (2011) Modeling of photovoltaic fields in mismatched conditions for energy yield evaluations. Electr Power Syst Res 81:1003–1013
22. Greacen C, Green D (2001) The role of bypass diodes in the failure of solar battery charging stations in Thailand. Sol Energy Mater Sol Cells 70:141–149
23. Tian H, Mancilla-David F, Ellis K, Muljadi E, Jenkins P (2013) Determination of the optimal configuration for a photovoltaic array depending on the shading condition. Sol Energy 95:1–12
24. Ziar H, Farhangi S, Asaei B (2014) Modification to wiring and protection standards of photovoltaic systems. IEEE J Photovoltaics 4 Nov 2014
25. Ziar H, Nouri M, Asaei B, Farhangi S (2014) Analysis of overcurrent occurrence in photovoltaic modules with overlapped by-pass diodes at partial shading. IEEE J Photovoltaics 4:713–721
26. Martínez-Moreno F, Muñoz J, Lorenzo E (2010) Experimental model to estimate shading losses on PV arrays. Sol Energy Mater Sol Cells 94:2298–2303
27. Quaschning V, Hanitsch R (1996) Numerical simulation of current-voltage characteristics of photovoltaic systems with shaded solar cells. Sol Energy 56:513–520
28. Díaz-Dorado E, Suárez-García A, Carrillo C, Cidrás J (2010) Influence of the shadows in photovoltaic systems with different configurations of bypass diodes. Speedam 2010:134–139
29. Díaz-Dorado E, Suárez-García A, Carrillo C, Cidrás J (2010) Influence of the PV modules layout in the power losses of a PV array with shadows. EPE-PEMC, pp 134–139
30. Silvestre S, Boronat S, Chouder A (2009) Study of bypass diodes configuration on PV modules. Appl Energy 86:1632–1640
31. Gao L, Dougal RA, Liu S, Iotova AP (2009) Parallel-connected solar PV system to address partial and rapidly fluctuating shadow conditions. IEEE Trans Industr Electron 56:1548–1556
32. Ziar H, Afjei E, Siadatan A, Arjhangmehr A (2011) Different diode configurations evaluation in photovoltaic arrays using binary coding method. In: Proceedings of IEEE conference on electrical machines and power electronics and electromotion, pp 682–686

33. Paja CAR, Bastidas JD, Montes AJS, Gispert FG, Goez M (2012) Mathematical model of total cross-tied photovoltaic arrays in mismatching conditions. In: Proceedings of IEEE workshop on circuits and systems, pp 1–6
34. Wang YJ, Shu-Syuan Lin SS (20120 Analysis of a partially shaded PV array considering different module connection schemes and effects of bypass diodes. In: Proceedings of IEEE conference on utility exhibition on power and energy systems, pp 1–7
35. Rodriguez JDB, Paja CAR, Montes AJS (2013) Reconfiguration analysis of photovoltaic arrays based on parameters estimation. Simul Model Pract Theory 35:50–68

Synchronization, Fault Detection of PV Array and Grid with MPPT Techniques Using MATLAB/Simulink

Rohit Tripathi, Sanjay Agrawal, Rashmi Agraawal, Devender Singh, and R. N. Shaw

Abstract In this paper, a solar cell (PV) system with different MPPT techniques connected to grid is introduced. There is a rise in the rapid development of high-penetration solar energy due to an increase in interest as well as investment in renewable energy. There are numerous ways in which a PV array can be interfaced with the grid. The lookup table PV panel with a centralized inverter topology has been developed in this paper. Since solar arrays output varies with change in solar irradiation as well as climate surroundings, the maximum power point tracking (MPPT) algo is adopted in DC-to-DC converter to enable PV arrays system to operate at maximum power point output. In order to control the inverter, the incremental conductance algorithm and fractional open voltage method is employed. Scheme for fault detection and system reconnection after fault is presented and simulated in MATLAB.

Keywords Distributed generation · PV array · Micro-grid · Smart grid · Lookup table

R. Tripathi (✉) · R. N. Shaw
School of Electrical Electronics & Communication Engineering, Galgotias University, G. Noida, India
e-mail: rohittripathi30.iitd@gmail.com

S. Agrawal
Electrical Engineering Department, Indira Gandhi National Open University, IGNOU, New Delhi, India

R. Agraawal · D. Singh
Electrical Engineering Department, JC Bose University of Science and Technology, Faridabad, India

R. Tripathi
Centre for Energy Studies, Indian Institute of Technology Delhi, Hauz Khas, New Delhi 110016, India

M. N. Favorskaya et al. (eds.), *Innovations in Electrical and Electronic Engineering*, Lecture Notes in Electrical Engineering 661, https://doi.org/10.1007/978-981-15-4692-1_21

1 Introduction

Demand for electricity doubles in every decade but in the last decade, it increases 2.5 times. With the energy issues rising rapidly, it becomes important for the world to begin with new energy and related technology [1–3]. These days, the power generation of PV systems and other renewable energy has been on a large scale and has become part of the development strategy which leads to the development of photovoltaic industries. At present, India's installed power generation capacity stands at 228,722 MW, while by the year 2030, and the total demand for electricity is expected to cross 950,000 MW. Out of the electricity consumed in India, a major portion is generated by conventional electricity generation sources such as thermal power plants and hydroelectric power plants [4, 5]. Thermal power plants, hydroelectric power plants, and nuclear power plants contribute about 65%, 22%, and 3%, respectively, of the total power generated while the rest is generated from other alternate sources like solar, wind, biomass, etc. India's National Solar Mission is 100,000 MW by 2022. This much of power needs to be connected with the utility grid. However, due to its different characteristics as compared to conventional power generation technologies, the security, stability, reliable operation, etc., of a grid-connected PV power station (power system energized photovoltaic panels which are connected to utility grid) have becomed a new challenge [6–9]. There are various methods of synchronization described earlier. After synchronization, controlling of the grid is necessary for which we require mathematical modeling of PV cell [10–12]. Use of power generated by different source needs to be utilized in an efficient way. Smart can be used to utilize the generated power efficiently, which in future induces transients due to frequent switching of loads.

This paper is to develop and study the Simulink model of PV array using lookup table power generation connected to grid and islanding mode operation and also the design of MPPT and inverter controller so that output of inverter can connect with the utility grid. Fault analysis and circuit breaker automation have been presented using overvoltage and undervoltage fault detection method.

2 PV Cell

A PV cell converts solar irradiance into electrical energy/power. The material of a solar cell is silicon, generally, due to high energy output. The number of cells can be grouped in order to form a module. Modules are grouped to form PV arrays. Array is a PV panel (with no. of cells connected in series and/or parallel) or a group of panels [3]. Therefore, the term array used henceforth means any solar cell module has several basic cells.

The pictures of all considered PV modules are given in Fig. 1.

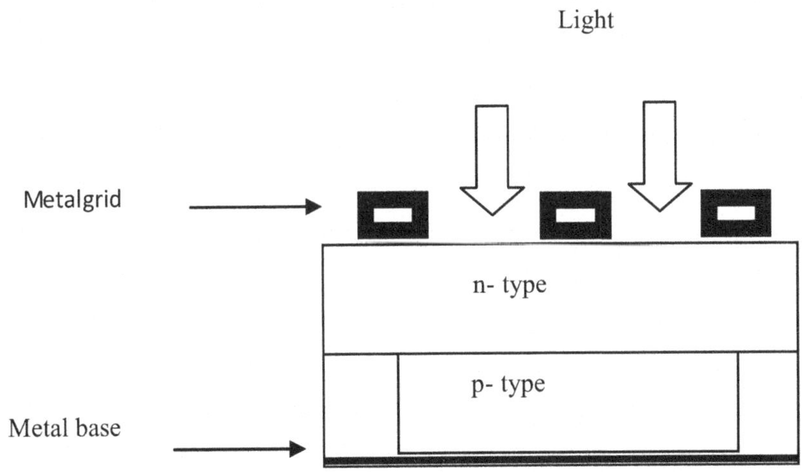

Fig. 1 Single solar cell: internal diagram

3 Equivalent Electric Circuit of PV Cell

A PV module can be usually seen as a flat plate with two connector or terminals, producing a current, I, at a voltage V. The electrical characteristics of a solar cell can be obtained by an electric circuit consisting of a diode and resistors as follows [3] (Fig. 2).

3.1 Current source: The photocurrent source can be described by the relation:

$$I_{\text{ph}} = A_{\text{cell}} H \xi \tag{1}$$

where A is the cross-sectional area of cell, H is the solar radiation, and ξ is the cell response in A/W.

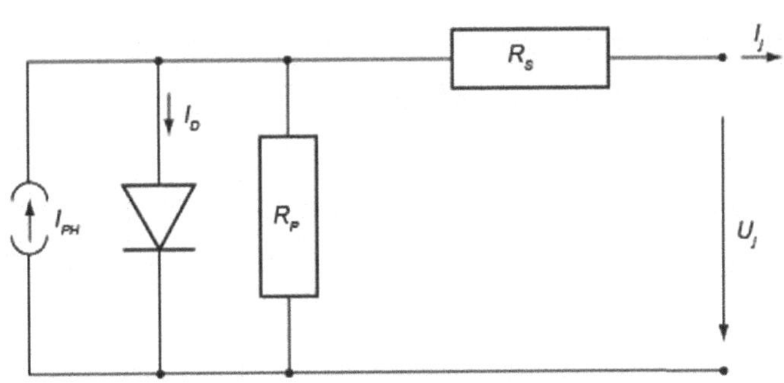

Fig. 2 Equivalent circuit of PV cell

3.2 Diode: Diode is a nonlinear element which accounts for dependency on energy bandgap and the losses which occur due to recombination holes and electrons. It is characterized by the reverse current, I_0.

3.3 Shunt resistor, R_p: Shunt resistor in the equivalent circuit represents the copper loss in conductors due to its resistance.

3.4 Serial resistor, R_s: Non-ideal conductors losses incurred can be represented as.

The relation of I with U of an individual cell is expressed as:

$$I_{\text{cell}} = I_{\text{ph}} - I_0[\exp\frac{(U_{\text{cell}} + I_{\text{cell}}R_s)}{U_T} - 1] - \frac{U_{\text{cell}} + I_{\text{cell}}R_s}{R_p} \quad (2)$$

with thermal voltage:

$$U_r = \frac{q\text{k}T}{e} \quad (3)$$

where T is the temperature (in Kelvin), k is the Boltzmann constant which is equal to 1.38e–23, and e is the elementary charge which is equal to 1.602e–19. By the use of this formula, we can compute maximum power points, as well as behavior under different temperature conditions (Fig. 3).

PV array is developed using the lookup table, and the original data can be used to develop the PV array so that its behavior can match the practical PV module. PV array

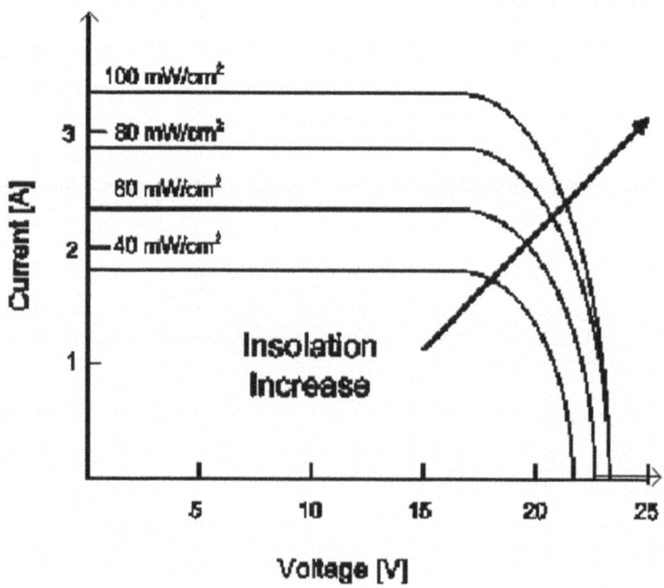

Fig. 3 *V-I* characteristics of PV cell [8]

has nonlinear characteristics and it can be seen that at a particular voltage and current, maximum power from the array can be obtained. In other words, the output from the array is dependent on the load. To make it independent of load, some DC–DC converters are needed which can take the care of load so that the output power from the array become independent of the load. This method of achieving maximum power is called maximum power delivering on point to point. In this study, 360 cells of 0.6 V open-circuit voltage are connected in series. The lookup table is constructed using 1000 w/m^2 irradiation, more via characteristics can be added at different irradiations.

4 Maximum Power Point Tracking and Various Methods

Maximum power point tracking (MPPT) is an electronic tracking—usually digital. The output of the panels is control of the charge controller. In this section, the theory and operation of MPPT as used in solar electric charge controllers have been covered. An MPPT is an electronic DC-to-DC converter which optimizes the match between the utility grid and the solar array. It converts a high DC voltage output from solar panels to the low voltage needed to charge batteries [7].

The main categories of MPPT algorithms are model-based algorithms, training-based algorithms, and searching algorithms [7].

1. Fractional short-circuit method
2. Fractional open-circuit voltage
3. Perturb and observe P&O/hill climbing
4. Incremental conductance INC.

Two methods are used for the MPPT, and Simulink model for the two methods used is shown below (Figs. 4 and 5).

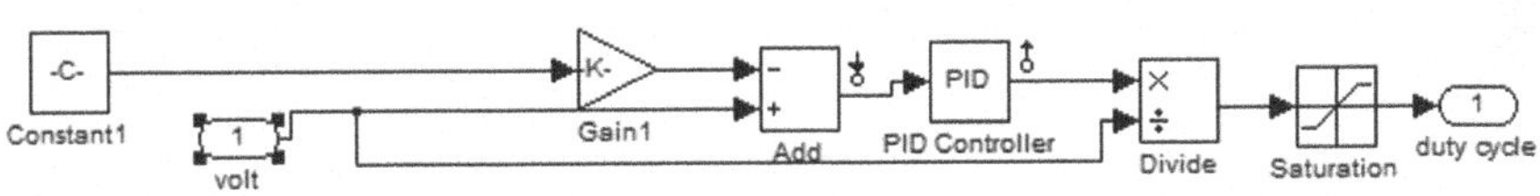

Fig. 4 Fractional open-circuit voltage [3]

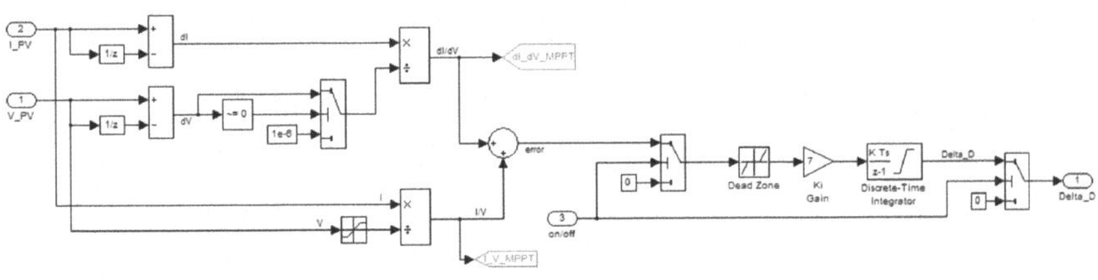

Fig. 5 Incremental conductance Simulink model

5 Synchronization for Three-Phase DPGS

Few methods have been noticed as per earlier research [1–6]:

(a) Zero-crossing method
(b) Phase-locked loop method.

Phase-locked loop method is used for the synchronization of inverter output to the grid.

6 Results and Discussions

The output power is dependent on load, and we need some DC–DC converter which can perform this task so that the power becomes independent of load. DC–DC buck and boost converter can be used for this purpose. It can be noticed that with the change in the duty cycle of the converter, the output power also changes. Also, as the duty cycle increases, the output power first increases and after a particular duty cycle, it starts decreasing, and this behavior is analogous to the behavior of PV array with different load.

Above waveforms in Fig. 6 show that with a change in the load the output power from the PV array also increases; it first increases with an increase in load and the n decreases with increase in load. That is, the output power depends on the load value, for a particular load it gives the maximum power output.

In Fig. 7, it shows that MPPT controller adjusts the duty cycle automatically such that the power output remains the same independent of the load with a change in

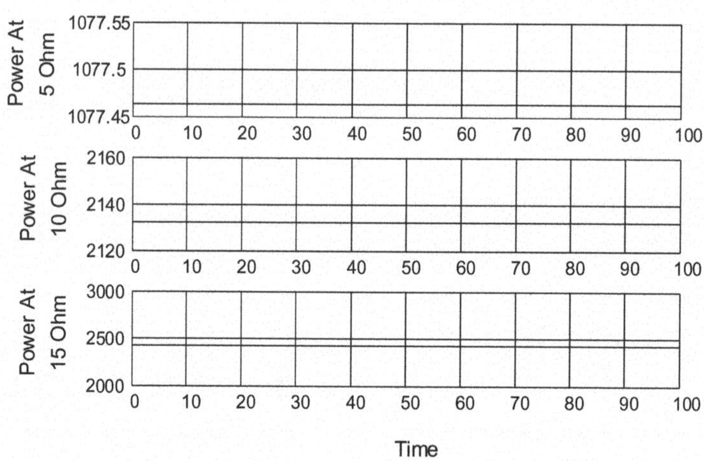

Fig. 6 Power output with different load

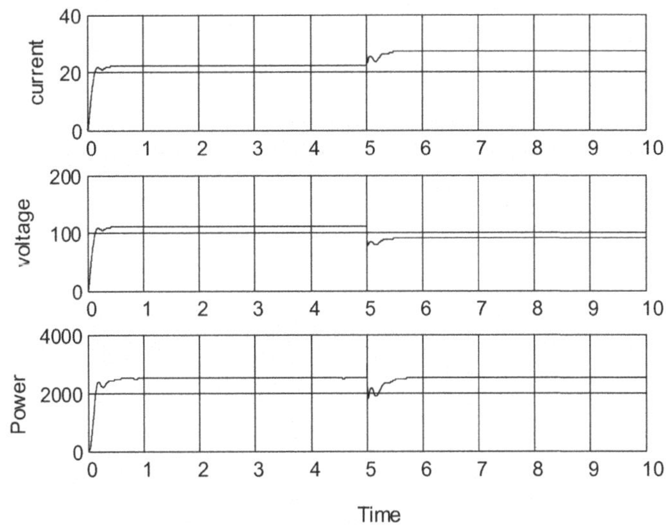

Fig. 7 Power output with Inc MPPT

load. Also, it should be noticed that the power is exactly 2500 W that is equal to the maximum power that array can provide. From this, we can conclude that our MPPT is working and performing their work.

The output from the DC–DC converter is fed to inverter after passing through LC filter. The average voltage source converter is used for the simulation purpose detailed model and can be used for the harmonic analysis purpose. For triggering the gate pulse of inverter, we require a voltage regulator which gives the triggering pulse to inverter so that the output can be maintained at 1 per unit. PI controller is used for the active power control and reactive power control V_d is responsible for the active power flow and V_q is responsible for the reactive power.

With a PI voltage controller using abc-to-dq and dq-to-abc transformations, the load voltage is regulated at 1 pu (380 V rms). The voltage regulator's first output is a vector having three modulating signals utilized by PMW generator to produce 6 IGBT pulses. The modulation index is returned by second output. The discrete 3-phase PWM pulse generator is available in the Extras/Discrete Control Blocks library. The Extras/Measurements and Extras/Discrete Control block libraries are utilized to build a voltage regulator.

Above waveform in Fig. 8, shows that the power is taken by the 1000 W load. This operation is called islanding mode of operation. Inverter works in islanding mode when there is fault in the grid side or the power output from the inverter is such that the power can only be fed to local loads, also system should be such a way that inverter output in exact synchronism with the grid, i.e., the phase, frequency, and magnitude of voltage should be same. Hence, the inverter is working satisfactory and ready to connect with the grid. Phase-locked loop method is used for the synchronism of the inverter and the Grid. Vabc of grid is used for the PLL block. Above waveforms

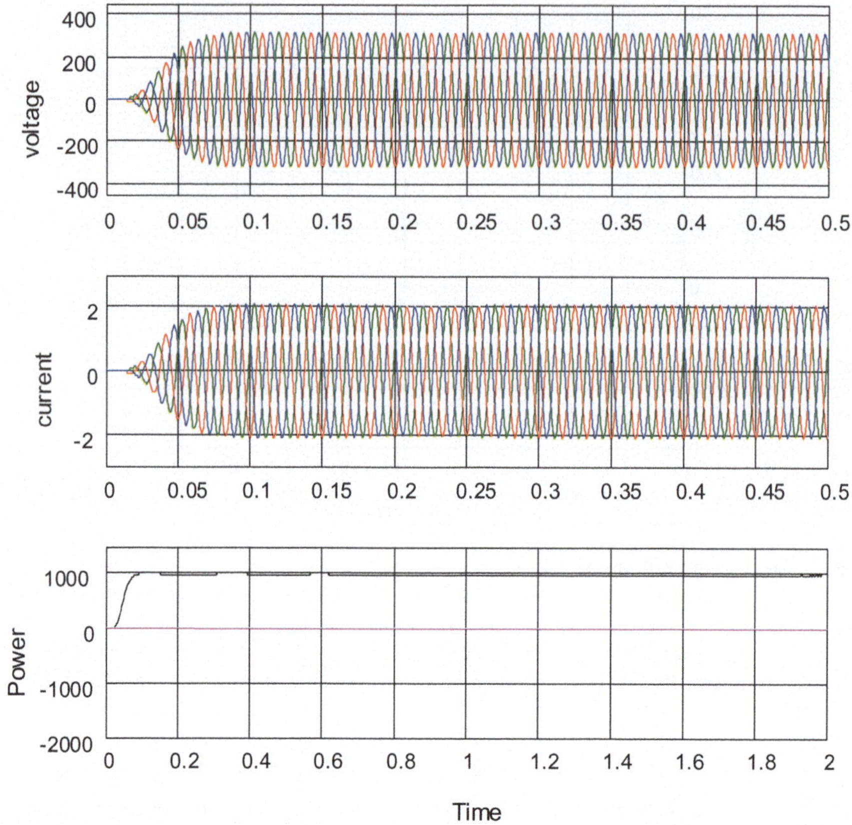

Fig. 8 Waveform of voltage, current, and power across 1000 W load

through Fig. 9 show the voltage, current, and power and phase voltage of grid, due to small differences in phase of voltage at PV side and grid side. Grid connected with PV at 0.5 s current balances after the output hooked up with the grid (Fig. 10).

Distributed system should be such that it should be disconnected from the system as soon as the fault occurs in the grid and should remain disconnected till the fault exists, also system should be able to reconnect with the grid when fault clears. For such type of operation, circuit breaker should be self-operated, and overvoltage and undervoltage method are used to detect the fault condition. For simulation of such system, two ideal three-phase sources of equal rating are taken which are capable of delivering any amount of load and can absorb the power. Both the sources have their local load and critical load. The Critical load is the common load for both the sources which should be supplied all the time.

Due to the slow response of PV and inverter operation, ideal source is taken instead of PV source. AT 0.05 s, the distributed generated source is connected to the main utility grid and provide power to the test load, i.e. power is exported to the grid. At

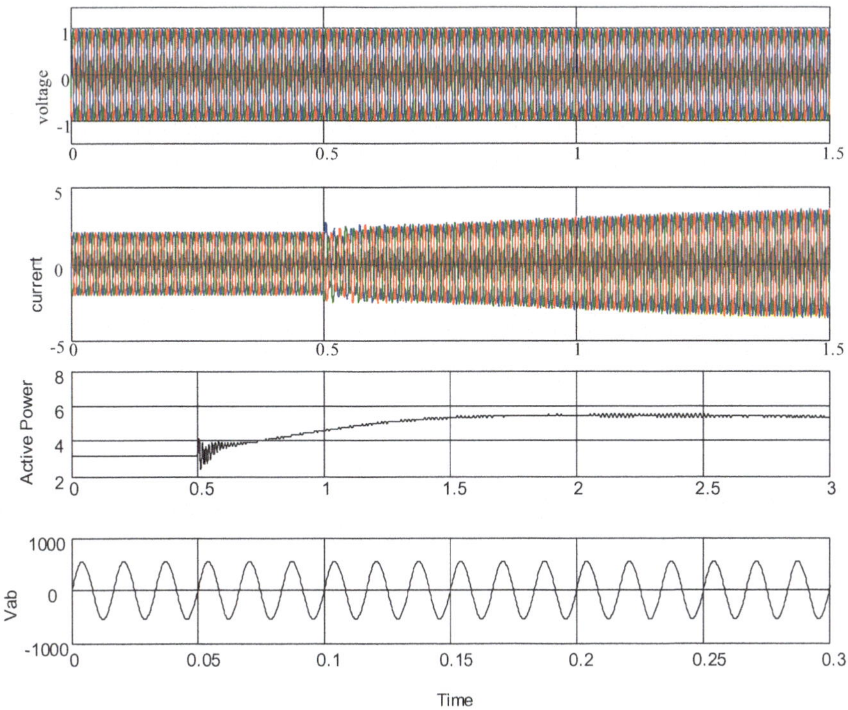

Fig. 9 Voltage, current, and power and phase voltage of grid

0.1 s, three phase fault occur in main grid and it gets disconnected from the load as soon as load get disconnected the power is given by the source to test load or critical load. AT 0.15 s, fault is cleared and critical load again connects with the main utility grid.

It can be noticed that after clearing of fault at 1.5 s main grid circuit breaker is closed and it remains open till the fault is not cleared. In this way, automatic switching of circuit breaker can be achieved. There is sag in power at the time of operation of the circuit breaker. This induces all types of harmonics in the system which need to be removed with filters and other fact devices.

7 Conclusions and Future Scopes

A method of modeling photovoltaic array using lookup table is presented and results of the lookup table validate the behavior of the PV array. Lookup table of any PV module for different irradiation can be formulated and used for simulation in MATLAB.

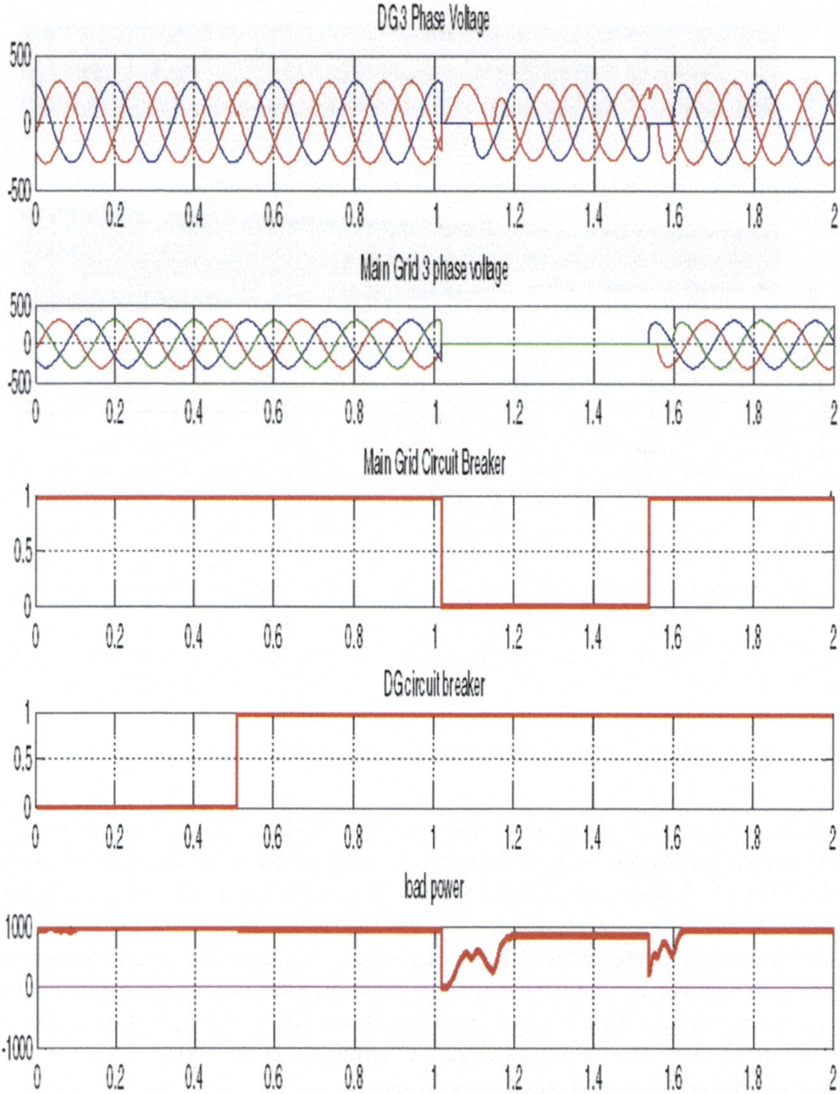

Fig. 10 Indices of voltage, current, and power and phase voltage of grid

Practical data of module can be used to formulate the lookup table. Also, the PV connected to the grid and islanding mode of operation is presented, various waveforms under different condition are quite satisfactory and more quick response can be achieved using a high-efficiency controller like neuro-fuzzy, fuzzy, and ANN instead of PI controller used for inverter control and MPPT controller.

Self-operation of circuit breaker under a fault condition is presented which can be used in any distributed generation, and switching of circuit breaker induces the harmonics in the system which is not desirable and need to be eliminated. For detection of fault on the grid side may be detected by observing the ripples in the voltage of DC-link capacitor and other methods can be determined in future for fault detection.

References

1. Timbus A, Liserre M, Teodorescu R, Blaabjerg F (2005) Synchronization methods for three phase distributed power generation systems—an overview and evaluation. IEEE 36th power electronics specialists conference, Recife, pp 2474–2481. https://doi.org/10.1109/PESC.2005.1581980
2. Blaabjerg F (Fellow, IEEE), Teodorescu R (Senior Member, IEEE), Liserre M (Member, IEEE), Timbus AV (Student Member, IEEE) (2006) Overview of control and grid synchronization for distributed power generation systems. IEEE Trans Ind Electron 53(5) (Oct 2006)
3. *"Mathematical Modeling of Photovoltaic Module with Simulink"* N. Pandiarajan and Ranganath Muthu Department of Electrical & Electronics Engineering SSN College of Engineering Kalavakkam -603110, Tamil Nadu, India, International Conference on Electrical Energy Systems (ICEES 2011), 3–5 Jan 2011
4. Järventausta P*, Repo S*, Rautiainen A*, Partanen J** (2011) Smart grid power system control in distributed generation environment. Tampere University of Technology, P.O. Box 692, 33101 Tampere IEEE 2011
5. Tripathi R, Tiwari GN (2018) Annual energy, exergy, and environmental benefits of N half covered concentrated photovoltaic thermal (CPVT) air collectors. In: SenGupta S, Zobaa A., Sherpa K, Bhoi A (eds) Advances in smart grid and renewable energy. Lecture Notes in Electrical Engineering, vol 435. Springer, Singapore
6. Guo X-Q, Wu W-Y, Gu H-R. Phase locked loop and synchronization methods for grid interfaced converters: a review. Yanshan University PrzeglądElektrotechniczny (Electrical Review), ISSN 0033-2097, R. 87 NR 4/2011
7. Manohar K, Sobha Rani P (2012) Mppt and simulation for a grid-connected photovoltaic system and fault analysis. Int J Eng Sci (IJES) 1:158–166
8. Tripathi R, Tiwari GN (2017) Annual performance evaluation (energy and exergy) of fully covered concentrated photovoltaic thermal (PVT) water collector: an experimental validation. Sol Energy 146:180–190
9. Chowdhury P, Koley I, Sen S, Saha PK, Panda GK (2013) Modelling, simulation and control of a grid connected non-conventional solar power generation system using Matlab. Int J Adv Res Elect Electron Instrum Eng 2(4) (Apr 2013)
10. Xia M, He X, Zhang X. Design and implementation of a control strategy for microgrid containing renewable energy generations and electric vehicles. School Of Electrical Engineering, Beijing Jiaotong University, No. 3 Shang Yuan Cun, Hai Dian District, Beijing 100044, China, Hindawi Publishing Corporation mathematical problems in engineering, vol 2013, Article ID 686508
11. Tripathi R, Tiwari GN, Dwivedi VK (2017) Energy matrices evaluation and exergoeconomic analysis of series connected N partially covered (glass to glass PV module) concentrated photovoltaic thermal collector: at constant flow rate mode. Energy Conv Manage 145:357–370
12. Yuan Z (2013) PV array simulator design based on S3C2440. Int J Autom Power Eng (IJAPE) 2(4). Department of Automation, Donghua University of Shanghai (201620), China, May 2013
13. Tripathi R, Tiwari S, Tiwari GN (2016) Energy analysis of partially covered number (N) of photovoltaic thermal-compound parabolic concentrator collectors connected in series at constant collection temperature mode. IEEE International Conference on Emerging Trends in Electrical Electronics & Sustainable Energy Systems (ICETEESES), Sultanpur, pp 12–17. https://doi.org/10.1109/ICETEESES.2016.7581344

14. Dinakaran C, Eswaramma K (2014) Modeling and simulation of grid connected fuel cell distributed generation using Quasi Z-source inverter. Int J Sci Res 3(7) ISSN No 2277 – 8179 (July 2014)
15. Tripathi R, Tiwari GN, Dwivedi VK (2016) Overall energy and exergy performance of partially covered N-photovoltaic thermal (PVT)-compound parabolic concentrator (CPC) collectors connected in series. IEEE International Conference on Power Electronics, Intelligent Control and Energy Systems (ICPEICES), Delhi, pp 1–6. https://doi.org/10.1109/ICPEICES.2016.7853669
16. Tripathi R, Tiwari GN (2019) Energy matrices, life cycle cost, carbon mitigation and credits of open-loop N concentrated photovoltaic thermal (CPVT) collector at cold climate in India: a comparative study. Sol Energy 186:347–359

MATLAB Based Modeling of Conventional and Fuzzy Logic Controller MPPT Techniques for Solar PV System

Ipuri Anil, Pavan Fuke, and Anil Kumar Yadav

Abstract To mitigate problems like energy shortage and environmental degradation caused by conventional energy sources, it needs to shift from conventional energy to non-conventional energy sources. The solar photovoltaic (PV) technology emerges as one of the leading non-conventional energy sources. To operate a PV system at its maximum efficiency, it requires the operating module at maximum power output point (MPP). There are many conventional as well as advanced maximum power point tracking (MPPT) techniques developed and implemented for a solar PV system. This work provides a comparative analysis of the conventional and fuzzy logic controller (FLC)-based MPPT techniques for a solar PV system. In this study, performance of two conventional techniques is considered, i.e., perturb and observe (P & O) and incremental conductance (IC) MPPT technique. The analysis of the PV system is performed for different solar irradiance levels. The comparative study is evaluated for complexity, tracking time, efficiency, range of power, voltage, and current oscillations. The DC-DC boost converter is used for operation of the proposed PV system at MPP. The implementation of the entire system and control algorithms has been modeled in MATLAB Simulink.

Keywords Solar photovoltaic · Maximum power point tracking · Perturb and observe · Incremental conductance · Fuzzy logic controller

1 Introduction

Now, the population in the world is increasing at a considerable amount, and electricity usage is also increasing. But, the conventional resources are depleting day by day, so it needs to shift on non-conventional resources. As solar energy is available

I. Anil (✉) · P. Fuke
Centre for Energy and Environmental Engineering, NIT Hamirpur, Hamirpur 177005, India
e-mail: ipurianil@gmail.com1

A. K. Yadav
Department of Electrical Engineering, NIT Hamirpur, Hamirpur 77005, India

M. N. Favorskaya et al. (eds.), *Innovations in Electrical and Electronic Engineering*, Lecture Notes in Electrical Engineering 661, https://doi.org/10.1007/978-981-15-4692-1_22

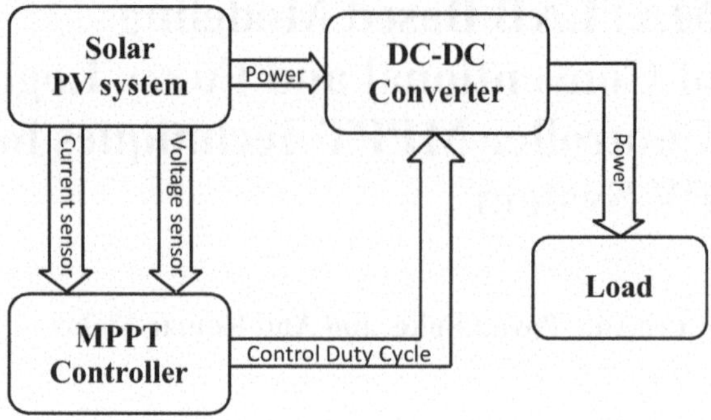

Fig. 1 Block diagram of solar PV system with MPPT controller

in high amounts and its various advantages over other non-conventional sources, it is emerging as a principal source of energy [1]. The unalterable environmental factors affect the power yield generation of the solar PV system. To raise the use of the PV system, it needs to focus on performance and efficiency improvement. The solar PV module characteristics are nonlinear, and at one particular point, it operates at MPP. For getting high efficiency, i.e., the best performance of the module, it requires to operate it on MPP continuously [2]. As power delivered by a module depends on the load connected across it, the load connected across the PV system is continuously changing. So, in favorable environmental conditions also, it is not possible to extract maximum power. For continuously operating PV systems at MPP irrespective of load condition, MPPT techniques are used. In recent days, many conventional MPPT techniques like P&O, IC, and voltage/current reference are available [3]. Due to technological development, many advanced intelligent MPPT techniques are invented. Each of these techniques has its own merits and demerits. The comparative study of different MPPT techniques is requiring for selecting the most superior one for a solar PV system. Such that the technique should give optimum results at any outdoor condition, especially for variable solar irradiance.

The MPP extraction is performed by continuously changing the duty cycle applied to converter. It is extracted from the MPPT algorithm and fed into the system as shown in Fig. 1. In this paper, a comparative analysis is carried out between conventional, i.e., P&O and IC, and FLC based MPPT techniques. In the proposed system, DC-DC boost converter is used for simulation performed in MATLAB simulink, such that it is connected in betwixt of the PV array and resistive load. The system simulation is performed at various solar irradiance levels 0, 550, 650, 750, 850, 1000 W/m^2.

2 PV System Modeling

A PV system consists of different components according to its application. An inverter is used in a grid-connected system to convert the DC power into AC power. In the case of the standalone system, a storage device, i.e., the battery, is used to store the PV power. In this paper, we analyze the standalone system such that system output is connected to fix load similar to a battery. An intermediate DC–DC boost converter is used to enable the MPPT technique in the system.

2.1 Solar PV Module

A solar cell is a semiconductor device which converts incident solar irradiance into electrical output without any intermediate process. In solar modules, numerous solar cells are connected in succession and parallel circuits to get the needed electric rating. To use a PV module for real field applications, it requires interconnecting many modules in succession and parallel, which form a PV array. The modeling of the PV module requires representing it in terms of an electrical circuit as represented in Fig. 2, with mathematical equations.

$$I = I_{\mathrm{L}} - I_{\mathrm{D}} - I_{\mathrm{sh}} \tag{1}$$

$$I = I_{\mathrm{L}} - I_0\left[\exp\left(\frac{V + IR_s}{nV_{\mathrm{T}}}\right) - 1\right] - \frac{V + IR_{\mathrm{s}}}{R_{\mathrm{sh}}} \tag{2}$$

I_{L} = Photocurrent (A); I_0 = Diode reverse saturation current (A)
R_{s} = Series parasitic resistance (Ω); R_{sh} = Shunt parasitic resistance (Ω)
n = Diode ideality factor.

From Fig. 2, it is seen that a solar cell is fundamentally a p–n junction diode which is connected parallelly with a controlled current source. When input solar radiations are received on top of the solar cell, then a photo-generated voltage is induced. When any external load connects across the cell, then the current I starts flowing through the load. From (1), (2), it saw that load current I depends on photocurrent I_{L}, which is chiefly affected by solar irradiance and junction temperature. The series and shunt parasitic resistance is representing ohmic and recombination losses in solar cells, respectively [4].

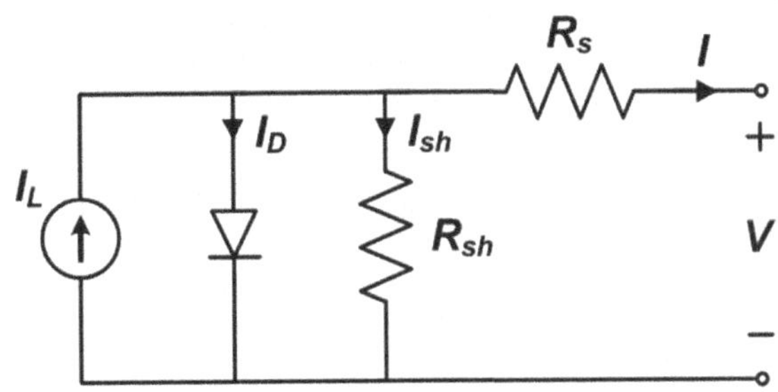

Fig. 2 Equivalent diagram of solar PV cell

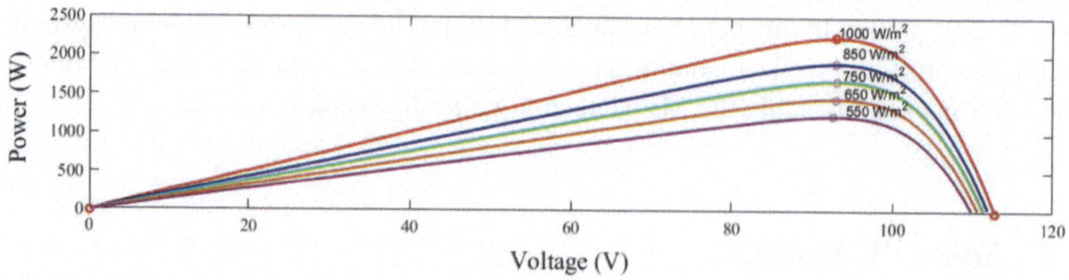

Fig. 3 P–V characteristic of a PV system

The proposed system chosen for modeling consists of a 2250 W_p solar PV array. The PV modules used in this study consists of 60 polycrystalline solar cells electrically connected to generate a peak power of 250 W_p. In this system, totally nine modules are connected in a 3 × 3 series–parallel combination. The *P–V* characteristic of the mentioned system simulation with different irradiance levels is shown in Fig. 3.

2.2 DC–DC Boost Converter

In the proposed system, the DC–DC boost converter is placed in the middle of the load and source. The peak power has been received by a load from the supply when the impedance of source and load has same value [5]. As load resistance is continuously varying, so to extract maximum power irrespective of load condition, impedance matching is required. It can be achieved by regulating the duty cycle of converter in range of 0–1 according to various atmospheric conditions. For reverse current protection, diode is used, and to boost voltage as well as to smooth current and voltage signals, combination of indictor and capacitor is used.

2.3 MPPT Technique

The power yield of the solar PV system is determined by the type of module technology used and the environmental condition. But, the power extracted from the system depends on the load connected across it. For maximum power extraction from the PV system, it always requires to work at optimum load value to satisfy maximum power transfer condition. The load connected across the system is continuously changing, so every time, it is not feasible to run a system at MPP. So to mitigate this problem, various MPPT algorithms are used [6]. It varies the duty cycle, so at any load condition, the resistance across the PV system seems as equals to source resistance.

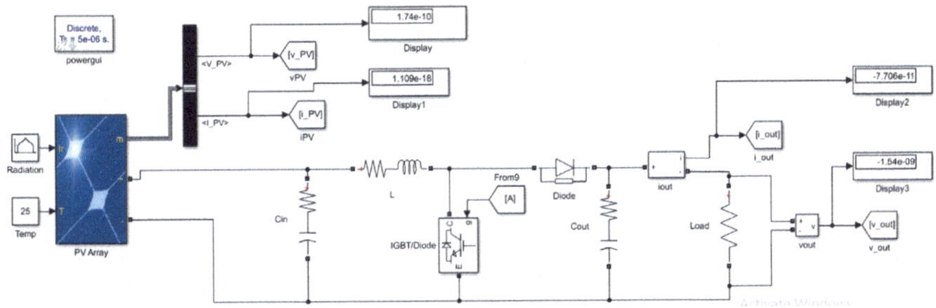

Fig. 4 Simulink model of the Solar PV system

The solar PV system output is mostly affected by input solar irradiations. As it is continuously changing, it results in a rapidly changing system output. It requires that the MPPT technique is such that it gives better performance at changing irradiance conditions with better efficiency. In this study, we analyze the P&O, IC, and FLC based MPPT techniques at changing irradiance conditions. Figure 4 shows the Simulink model of proposed system used for all three mentioned MPPT techniques.

3 Perturb and Observe MPPT Technique

It is the most frequently used MPPT technique. Because of its simplicity, less measured parameters requirement, and simple implementation, currently this technique is frequently used for a solar PV system. It is alike as the hill-climbing/mountain-climbing technique. The only dissimilarity is in hill climbing to its simplicity, less measurable parameters. and less method duty cycle directly changes with the use of slope value +1, 0 or -1 [7]. It is a self-optimizing process in which the operating point is perturbed with fixed step size by continuously rising or reducing the duty cycle. The basic control idea in this technique is such that initially terminal voltage and current are computed to determine the actual power value P_{k}. Then, this actual power value is compared with preceding power value $P_{(k-1)}$. If actual power value is larger than earlier value, then carry on with duty cycle perturbation in the same direction or else move in the opposite direction. This operation is repeated every time by considering the perturbation of duty cycle ΔD. At steady-state conditions, the algorithm oscillates all over the MPP. In the case of rapidly changing environmental conditions, problems like power loss and response delay are observed in this system.

4 Incremental Conductance MPPT Technique

The crucial pitfall of the P&O MPPT technique is that it is unable to track rapidly changing solar irradiance conditions, and it continuously perturbs the system, which leads to creating oscillations in power output. Both of these disadvantages are eliminated in the IC method. When the slope of the P–V curve is zero, then terminal voltage is considered as MPP voltage. At this point, the algorithm stops perturbing the operating point. If this case is not satisfied, then the operating point direction perturbation depends on the instantaneous conductance of the PV array, i.e., the relationship between $\frac{dI}{dV}$ and $-\frac{I}{V}$ [8]. If the terminal voltage is smaller than MPP voltage, then the duty cycle varies such that reference voltage, i.e., terminal voltage, is increased by small dV value. On the other hand, if the terminal voltage is higher than MPP voltage, then the duty cycle varies such that the terminal voltage is decreased by small dV value. The algorithm of P&O and IC MPPT technique is showed by red and blue color in Fig. 5.

$$\frac{dI}{dV} = -\frac{I}{V} = \text{At MPP}; \frac{dI}{dV} > -\frac{I}{V} = \text{Left of MPP}; \frac{dI}{dV} < -\frac{I}{V} = \text{Right of MPP}$$

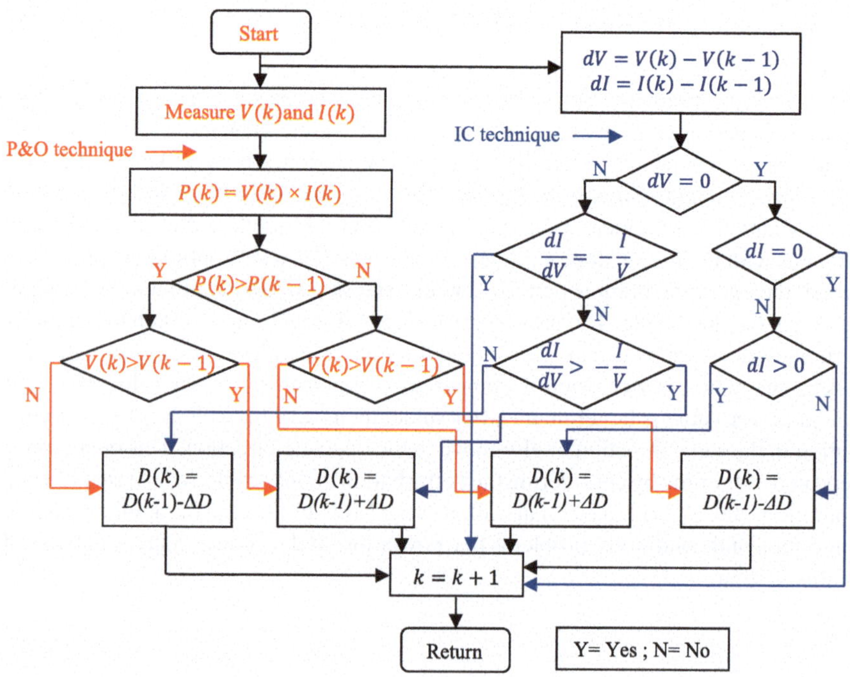

Fig. 5 Flowchart of P&O and IC MPPT technique

5 Fuzzy Logic-Based MPPT Technique

Due to higher oscillation and power loss in the PV system, it operates with P&O MPPT technique and complexity in IC method. There is a need for advancement in the MPPT techniques, which has a small step size, low power loss, and less oscillation. All these properties are found in a FLC based technique. The chief upper hand of this technique is that it does not require understanding of the mathematical model of the system [9]. The fuzzy controller comprises of a fuzzy interface engine, membership functions, rule base, and defuzzification, as shown in Fig. 6.

The FLC uses two linguistic inputs, error E and change in error CE, and it also uses single output $\mathrm{d}D$ with sample space k. The error input is the ratio of deviation in output power to deviation in terminal voltage, and it indicates the operating point location on the P–V curve. The second input CE is error at kth sample space to the previous error, and it indicates the active direction of the operating point.

$$E(k) = \frac{\mathrm{d}P}{\mathrm{d}V}(k) = \frac{P_{\mathrm{PV}}(k) - P_{\mathrm{PV}}(k-1)}{V_{\mathrm{PV}}(k) - V_{\mathrm{PV}}(k-1)} \tag{3}$$

$$\mathrm{CE}(k) = E(k) - E(k-1) \tag{4}$$

where $P_{\mathrm{PV}}(k)$ and $V_{\mathrm{PV}}(k)$ are the immediate output power and the terminal voltage of the PV system. In this fuzzy logic controller, the span of interest of input as well as output linguistic variable is divided into five levels shown as ML: minus large, MS: minus small, ZE: zero, PS: plus small, PL: plus large. The fuzzy rules used to make a relation between input and output signal, such that the output signal is expressed in terms of an input signal. For this, fuzzy rules required to insert in FLC block. The fuzzy control rules for the mentioned PV system use the Mamdani method as given in Table 1. The membership function of inputs as well as output describe each of these acronyms and are presented in Fig. 7.

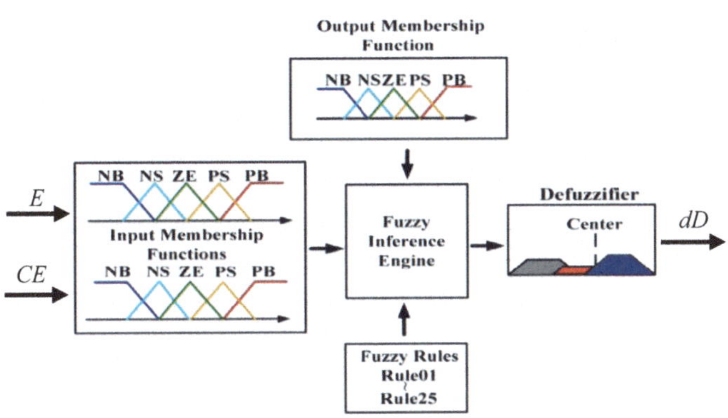

Fig. 6 Block diagram of fuzzy logic control

Table 1 Fuzzy control rules

CE E	ML	MS	ZE	PS	PL
ML	ZE	ZE	PL	PL	PL
MS	ZE	ZE	PS	PS	PS
ZE	PS	ZE	ZE	ZE	MS
PS	MS	MS	MS	ZE	ZE
PL	ML	ML	ML	ZE	ZE

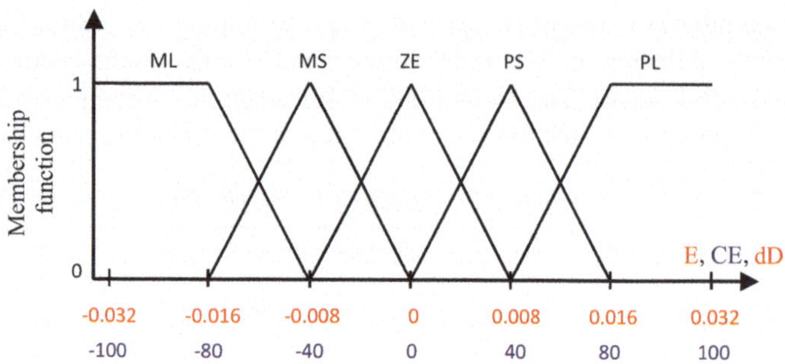

Fig. 7 Membership functions of input and output variables

Defuzzification is the final step of FLC based MPPT technique. In this, the linguistic value of the output signal is converted into a crisp value by using the membership function, as mention in Fig. 7. The defuzzification is carried out by the most famous center of gravity method to use the crisp get the crisp value of duty cycle D, which further given to the gate signal of converter [10].

6 Simulation Results

In this proposed model, we test the dynamic performance of P&O, IC, and FLC based MPPT technique developed in MATLAB Simulink. The simulations were presented for variable irradiance levels of 0, 550, 650, 750, 850, 1000 W/m^2. The perturbation of the duty cycle set at a perturbation time of 0.05 s tracks the MPP when irradiance changes. At a fixed irradiance level, the operating point oscillates about MPP value. The amplitude of oscillation depends on the size of the duty cycle perturbation. The obtained results for P&O, IC, and FLC based MPPT techniques represent in Fig. 8a, b, and c, respectively.

The voltage, current, and power response of the proposed system without MPPT and in addition to this parameter efficiency of the system with P&O, IC, and FLC based MPPT techniques, it is given in Table 2. The parameter variation for different solar irradiance is illustrated.

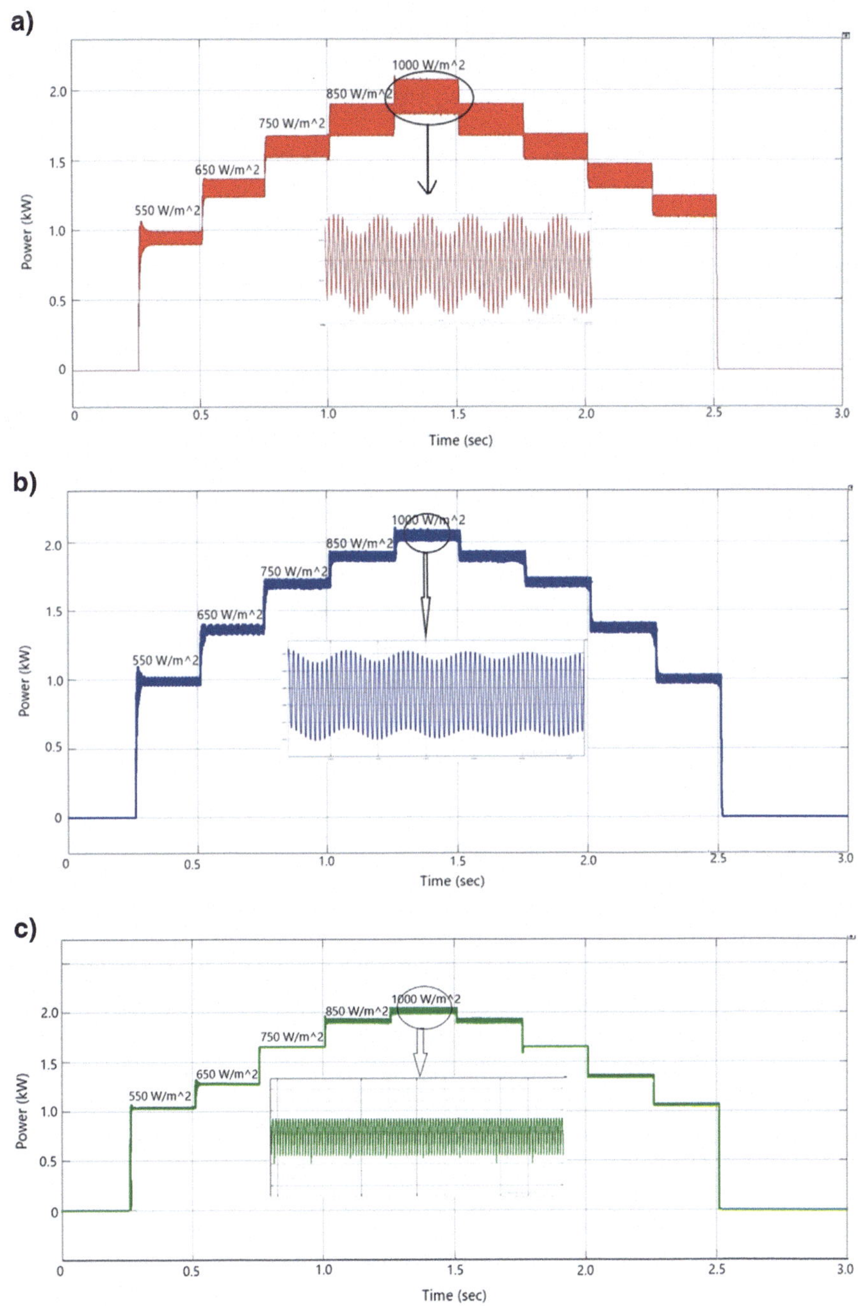

Fig. 8 **a** Output power by using P&O technique, Fig. 8(b). Output power by using IC technique, Fig. 8(c). Output power by using a FLC technique

Table 2 Simulation results of P&O, IC, and FLC based MPPT techniques

Input	PV panel output (without MPPT)			System output with P&O MPPT		
Irradiance (W/m^2)	Voltage (V)	Current (I)	Power(W)	Voltage (V)	Power (W)	Efficiency (%η)
0	0	0	0	0	0	0
1000	99.58	20.54	2045	202.8	1993.0	97.46
850	95.04	20.36	1935	194.42	1876.9	97.00
750	94.07	17.93	1687	183.50	1621.4	96.11
650	90.2	15.96	1440	173.1	1340.9	93.12
550	71.38	15.27	1090	136.23	1000.7	91.81

Input	System output with IC MPPT			System output with FLC based MPPT		
Irradiance (W/m^2)	Voltage (V)	Power (W)	Efficiency(%η)	Voltage (V)	Power (W)	Efficiency (%η)
0	0	0	0	0	0	0
1000	202.4	2005.3	98.06	201.8	2016.5	98.61
850	194.2	1887.7	97.56	194.1	1898.4	98.11
750	184.3	1637.7	97.08	184.2	1649.5	97.78
650	167.4	1354.2	94.04	164.5	1366.3	94.88
550	138.6	1010.1	92.67	140.2	1020.3	93.61

7 Conclusions

In this paper, modeling and simulation of a solar PV system using a DC–DC boost converter have been performed using MATLAB Simulink. The proposed system is simulated for P&O, IC, and FLC based MPPT technique. After evaluating all simulation results, we observed that the step size required in the P&O technique is more than IC and fuzzy logic technique. The fuzzy logic technique seems to be more complicated, but its operating time is very less as compared to P&O and IC technique. The output voltage and power curves with the fuzzy logic-based technique have fewer oscillations. In the P&O technique, the operating point is continuously oscillating about MPP. In the IC technique, output power oscillations are less than P&O but more than fuzzy logic-based technique. At every irradiance level, the dynamic response of the FLC based technique improved with higher percentage efficiency as compared to the P&O and IC technique. From all simulation results, we can conclude that for variable irradiance condition, the FLC based technique has superior performance than P&O and IC method, i.e., conventional MPPT technique.

References

1. Subudhi B, Pradhan R (2012) A comparative study on maximum power point tracking techniques for photovoltaic power systems. IEEE Trans Sustain Energy 4(1):89–98
2. Rezk H, Eltamaly AM (2015) A comprehensive comparison of different MPPT techniques for photovoltaic systems. Sol Energy 112:1–11
3. Pathak PK, Yadav AK, Alvi PA (2020) Advanced solar MPPT techniques under uniform and non-uniform irradiance: a comprehensive review. Solar Energy Eng 142(4):040801–26
4. Blange R, Mahanta C, Gogoi AK (2015) MPPT of solar photovoltaic cell using perturb and observe and fuzzy logic controller algorithm for buck-boost DC–DC converter. In: IEEE international conference on energy, power and environment: towards sustainable growth (ICEPE), pp 1–5
5. Pathak PK, Yadav AK (2019) Design of battery charging circuit through intelligent MPPT using SPV system. Sol Energy 178:79–89
6. Abderezak L, Aissa B, Hamza S (2015) Comparative study of three MPPT algorithms for a photovoltaic system control. In: IEEE world congress on information technology and computer applications, pp. 1–5, (2015)
7. Al-Gizi, A.G. and Al-Chlaihawi, S.J.: Study of FLC based MPPT in comparison with P&O and InC for PV systems. In IEEE International Symposium on Fundamentals of Electrical Engineering, pp. 1–6, (2016)
8. Saxena, A.R. and Gupta, S.M.: Performance Analysis of P&O and Incremental Conductance MPPT Algorithms Under Rapidly Changing Weather Conditions. Journal of electrical Systems, vol. 10, no. 3, (2014)
9. Jena, P. K., Mohapatra, A. and Choudhary, P.: Comparative study of solar pv mppt by perturbation and observation and fuzzy method. In IEEE Uttar Pradesh Section International Conference on Electrical, Computer and Electronics Engineering (UPCON), pp. 515–518, (2016)
10. Bouchafaa F, Hamzaoui I, Hadjammar A (2011) Fuzzy Logic Control for the tracking of maximum power point of a PV system. Energy Procedia 6:633–642

Economical Unit Commitment Solutions Through Renewable Energy Management Centers for Reliable Integration of Large-Scale Renewable Energy Resources

E. Sheeba Percis, A. Nalini, C. Chelladurai, T. Jenish, and J. Jayarajan

Abstract This paper aims at achieving economical and secure commitment of conventional thermal power generating units in a power system integrated with renewable energy sources (RES) by harnessing the renewable energy generation potential to its maximum. The data forecasting tool of renewable energy management center (REMC) is used for closer prediction of RES availability which helps to achieve much improved unit commitment (UC) solution of conventional thermal units apart from intermittent nature of power output obtained from RES. Large-scale penetration of RES is achieved with the advancement in technology supports better optimization of economical dispatch and reduces the thermal plant running hours. This supports considerable reduction in fossil fuel requirements, carbon foot print, and green house emission, and causes lesser impact on global warming, climate change, and improves quality of atmospheric air. This paper analyzes present need of REMCs to integrate large-scale renewable into the grid and to reduce the RES curtailment due to poor prediction of RES availability.

Keywords Renewable energy · Renewable energy management center · Forecasting · Wind power · Solar power · Unit commitment · Grid

1 Introduction

Today's large-scale energy utilization, faster depletion of fossil fuels, energy security needs, and environment safety concerns encourage the world toward renewable sources and smart power systems. India has been blessed with huge potential of renewable energy sources. Projected solar potential in India is about 100 GW at 20–30 MW/km^2, and the wind potential is about 103 GW at 80 m hub height. To

E. Sheeba Percis (✉) · A. Nalini · T. Jenish · J. Jayarajan
EEE Department, Dr. M.G.R. Educational and Research Institute, Maduravoyal, Chennai, India
e-mail: sheebapercis.eee@drmgrdu.ac.in

C. Chelladurai
Tamil Nadu Transmission Corporation Ltd., Chennai, India

M. N. Favorskaya et al. (eds.), *Innovations in Electrical and Electronic Engineering*, Lecture Notes in Electrical Engineering 661, https://doi.org/10.1007/978-981-15-4692-1_23

Table 1 Installed renewable power capacity as on November 2017

Type of RERs	Capacity in MWp
Hydel	4399.35
Bio-power	8181.7
Solar	16,611.73
Wind	32,746.87
Total	61,939.65

enhance the utilization green energy availability, the Government of India (GOI) has taken various initiatives to encourage RES generation programs mainly from wind, solar, bio-gas, and hydel sources.

As per Ministry of New and Renewable Energy (MNRE)—Year End Review 2017 report [1], India attains global fourth and sixth position in global wind and solar power installed capacity by November 2017, a total of 62 GW renewable power installed, of which 27 GW installed since May 2014 and 11.79 GW since January 2017 as seen in Table 1. This report shows the faster RES integration in Indian power network.

The difficulty in managing the variability and uncertainty of RE generation calls for an accurate forecast of wind and solar patterns [2, 3] which can be met through establishment of RE management centers [4–6] to plan the system scheduling. Increase in level of RES penetration may cause substantial variation in the total generation leading to wider load-generation gap. To handle these imbalances, forecasting and dispatching of renewable power [7, 8] will be an important aspect for their reliable and secure integration with the grid.

Establishment of dedicated renewable energy management centers to facilitate large-scale integration of renewable into the grid, equipped with advanced forecasting tools, smart dispatching solutions, and real-time monitoring of RE generation, can closely coordinate with the grid operations team for safe, secure, and optimal operations of the overall grid. Renewable energy management center (REMC) is the "hub" for all information regarding RE power generation in its area of responsibility which could be on SLDC, RLDC, or even on NLDC level.

The aim of this paper is to prioritize and utilize RE sources as base loads and optimize thermal units in unit commitments scheduling and dispatching [9, 10] during normal operation sequences with reasonable margin of spinning reserve to manage RES generation variability. Weather plays major role in estimating wind and solar generation patterns, and hence, it is very challenging to predict [11–13] exact generation of RE sources output.

2 Functionalities of REMC

Dedicated renewable energy management centers are mainly entrusted with forecasting, dispatching, and balancing of RE generation by coordinating with respective LDCs and RE developers, and thus, variability of RE generation can easily be

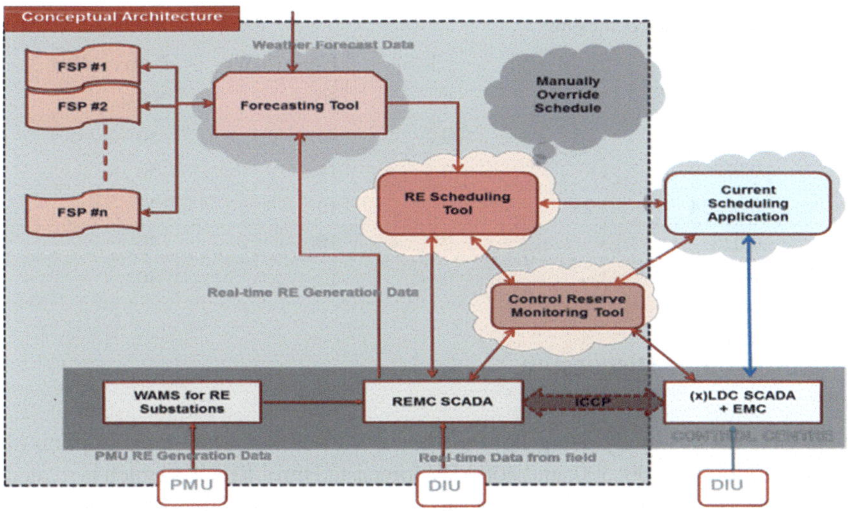

Fig. 1 Scheme of logical interconnectivity between various software modules

handled [14, 15]. Forecasting of RE generation can be done in different time scales ranging from few seconds to hours to days. The data obtained from different time scales are used for various applications, viz dynamic operation of power system, optimal unit commitment solutions, prediction on reserve requirement for balancing the system (Fig. 1).

Further, they serve as a central repository and provide reliable RE generation data for the respective LDCs, SCADA/EMS, and power market mechanism. The hierarchy followed for REMCs are such that the data can be compiled at state, regional, and national level for different applications.

3 Unit Commitment Formulation

Out of several evolution algorithms [16–18] used for solving unit commitment solutions, modified PSO method [19] is preferred in this project. The primary objective of this paper is to prioritize and utilize RE generation as base loads and optimize thermal units in unit commitments scheduling and dispatching during normal operation sequences with reasonable margin of spinning reserve to manage RES generation variability so as to satisfy the required load demand at minimum operating cost while satisfying all units and system equality and inequality constraints.

The single line diagram of an IEEE-30 bus system with generator cost coefficients and generating unit's data are provided in Table 2. The unit commitment solution is arrived for IEEE 30 bus test network with six generating units using particle swarm optimization method.

Table 2 Generator cost coefficients

Unit	a1(MW/$)	b1 (MW/$)	C1 ($)	Pimin (MW)	Pimax (MW)
G1	0.00375	2	0	50	200
G2	0.01750	1.75	0	20	80
G3	0.06250	1	0	15	50
G4	0.00834	3.25	0	10	35
G5	0.025	3	0	10	30
G6	0.025	3	0	12	40

4 Results and Discussion

Simulation for this study using PSO method is carried out with an assumption of RES penetration of 200, 300, and 400 MW forecasted thro' REMCs which is treated as base load in order to commit the thermal units, for total load demand of 600 MW. Minimum fuel cost and power loss calculation for this study for various scenarios are obtained using MATLAB algorithm and tabulated as follows in Table 3 and Fig. 2.

Scenario—I: In this case, optimal allocation of thermal units for power demand of 600 MW with a RES base load of 200 MW is considered.
The cost characteristics for thermal unit's contribution of 400 MW with RES share of 200 MW for a load demand of 600 MW are shown below:
Scenario—II: In this case, optimal allocation of thermal units for power demand of 600 MW with a RES base load of 300 MW is considered as in Table 4.

Table 3 Optimal scheduling of generators for scenario—I

G1 MW	G2 MW	G3 MW	G4 MW	G5 MW	G6 MW	Loss MW	Total MW	Cost $/h
200	75.03	47.85	33.41	21.60	35.53	13.44	413.4	1303.38

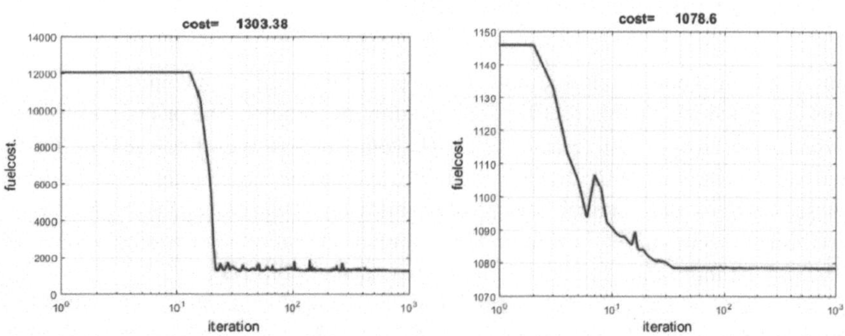

Fig. 2 Cost characteristics for thermal units contribution of 400 MW with RES share of 200 MW and for thermal units and RES share of each 300 MW

Table 4 Optimal scheduling of generators for scenario—II

G1 MW	G2 MW	G3 MW	G4 MW	G5 MW	G6 MW	Loss MW	Total MW	Cost $/h
133.2	75.36	28.88	10.0	30.0	30.39	7.68	307.86	1078.59

Table 5 Optimal scheduling of generators for scenario—III

G1 MW	G2 MW	G3 MW	G4 MW	G5 MW	G6 MW	Loss MW	Total MW	Cost $/h
70.49	60.19	23.76	10.00	19.75	18.79	2.91	203.0	646.65

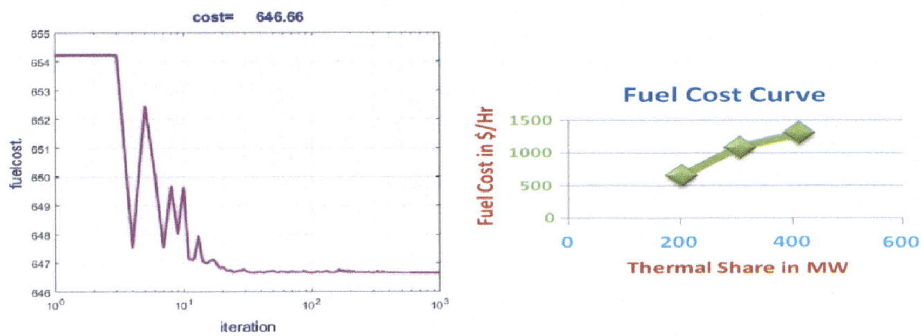

Fig. 3 Cost curve for thermal unit's contribution of 200 MW with RES share of 400 MW combined share of thermal units and RERs to meet the load demand of 600 MW

The cost characteristics for thermal unit's contribution of 300 MW with RES share of 300 MW for a load demand of 600 MW are shown below

Scenario—III: In this case, optimal allocation of thermal units for power demand of 600 MW with a RES base load of 400 MW is considered as seen in Table 5. The cost characteristics for thermal unit's contribution of 200 MW with RES share of 400 MW for a load demand of 600 MW are shown below in Fig. 3.

From the above study results, it is clearly evident that the integration of large-scale RES is much useful for optimum allocation of thermal unit's commitment and to bring down the plant operating cost at the sensible level.

5 Conclusions

India, being the top five countries with renewable installed capacity worldwide, has articulated to boost renewable energy through various policies. With the advancement in renewable energy technologies, declining trend in energy costs and better environment considerations, India has decided to achieve an ambitious target of 175 GW by 2022 through RE. Proposed unit commitment solution is useful for reduction in capital expenditure investment, reduction in need for conventional generation, increased value of the electricity generated by variable RE generation, and reduction in suboptimal operation of conventional generation due to reduced need for ramping. The study results obtained for IEEE 30 bus system using PSO can be verified and compared with other meta-Heuristic approaches, as future scope. Also, the simulation study can be carried out on real-time network with real-time inputs from REMC.

References

1. Ministry of power, India study report on GREENING THE GRID: pathways to integrate 175 Gigawatts of renewable energy into India's electric grid, vol I National Study (2017)
2. Andrade JR, Bessa RJ (2017) Improving renewable energy forecasting with a grid of numerical weather predictions. IEEE Trans Sustain Energy
3. Percis ES et al (2016) Electric vehicle as energy storage for grid connected solar power system. IJPEDS, pp 567–575 (2016)
4. Tummuru NR et al (2015) Dynamic energy management of renewable grid integrated hybrid energy storage system. IEEE Trans Industr Electron, 7728–7737
5. Abbas F et al (2016) Hybrid energy management system with renewable energy integration. Int Conf Comput Intell Model Simul (CIMSim)
6. Simopoulos D, Kavatza S (2005) Consideration of ramp rate constraints in unit commitment using simulated annealing. IEEE Power Tech, Russia 1–7 (27–30 June, 2005)
7. Bu S, Yu FR, Liu PX (2011) Stochastic unit commitment in smart grid communications. IEEE INFOCOM
8. Senjyu T, Shimabukuro K, Uezato K, Funabashi T (2002) A fast technique for unit commitment problem by extended priority list. IEEE/PES, T & D conference and exhibition 2002, vol 1, pp 244–249, 6–10
9. Raglend IJ, Padhy NP (2006) Solutions to practical unit commitment problems with operational, power flow and environmental constraints. IEEE Power Eng Soc General Meeting
10. Gast N, Tomozei D-C, Le Boudec J-Y (2014) Optimal generation and storage scheduling in the presence of renewable forecast uncertainties. IEEE Trans Smart Grid 1328–1339
11. Naoi S et al (2017) Demand and supply simulations considering detailed forecast, scheduling and control functions for Japanese power system with a massive integration of renewable energy sources. IEEE Manchester Powertech
12. Lenzi V, Ulbig A, Andersson G (2016) Impacts of forecast accuracy on grid integration of renewable energy sources. IEEE Greenoble Conf
13. Ariel S, Luis AC, Jorge RV (2019) A hybrid methodology for a contingency constrained economic dispatch under high variability in the renewable generation. IEEE Latin Am Trans 17(10):1715–1723
14. Deutsche Gesellschaft für Internationale Zusammenarbeit (GIZ) GmbH study report on Detailed Project Report for Establishment of Renewable Energy Management Centres (REMC) (2017)
15. Central Electricity Authority Study Report on Large Scale Grid Integration of Renewable Energy Sources—Way Forward (2016)
16. Mantawy AH, Abdel-Magid YL, Selim SZ (1998) A simulated annealing algorithm for unit commitment. IEEE Trans Power Syst 13(1):197–204
17. Sisworahardjo NS, El-Keib AA (2002) Unit commitment using the ant colony search algorithm. Large Eng Syst Conf Power Eng LESCOPE 02, 2–6
18. Sheeba Percis E, Manivannan S (2017) Realistic modelling and control of PV and multiple electric vehicle based microgrid. J Comput Theor Nanosci 2976–2980
19. Optimal solution to economic load dispatch using modified particle swarm optimization method, Danson Kimanthi, pp 1–43 (2014)

Design of Robust PID Controller for Improving Voltage Response of a Cuk Converter

V. Bharath Kumar, Godavarthi Charan, and Y. V. Pavan Kumar

Abstract Power electronics based cuk converter finds many practical and industrial applications such as renewable energy systems, hybrid electric vehicle, electronic voltage regulators, etc., to provide required voltage regulation between input DC stage and output DC stage. For the effective operation of all these kinds of practical applications, it seeks a quick and smooth response from a cuk converter along with the regulated DC output magnitude. The converter control is achieved through a closed-loop PID (proportional cum integral cum derivative) controller. Hence, the efficacy of the overall converter depends on how well the PID gain specifications are designed. There are many methods available for the tuning of PID controller gain specifications, but, the identification of better methods is a tedious task. Further, a method identified as best for a given system can't be generalized as it may not suit for other systems. Hence, a better method for PID tuning has to be identified with respect to each and every system. Keeping this in mind, this paper provides a comprehensive investigation of all the key tuning methods and suggests the best one for cuk converter control with the help of time-domain performance index and frequency-domain stability index. The analysis is done using MATLAB/Simulink software models.

Keywords Cuk converter · Power converter · DC voltage regulation · PID tuning

V. Bharath Kumar · G. Charan · Y. V. Pavan Kumar (✉)
School of Electronics Engineering, Vellore Institute of Technology—Andhra Pradesh (VITAP) University, Amaravati 522237, India
e-mail: pavankumar.yv@vitap.ac.in

V. Bharath Kumar
e-mail: bharath.18bec7093@vitap.ac.in

G. Charan
e-mail: charan.18bec7075@vitap.ac.in

M. N. Favorskaya et al. (eds.), *Innovations in Electrical and Electronic Engineering*, Lecture Notes in Electrical Engineering 661, https://doi.org/10.1007/978-981-15-4692-1_24

Abbreviations

CC Cohen–Coon
CHR Chien–Hrones–Reswick
IAE Integral Absolute Error
ISE Integral Square Error
ISTE Integral Square Time Error
ITAE Integral Time Absolute Error
KCL Kirchhoff Current Law
KVL Kirchhoff Voltage Law
PI Pessen Integral
PID Proportional cum Integral cum Derivative
PO Pessen Overshoot
TL Tyreus Luyben
WJC Wang–Jang–Chan
ZN-1 Ziegler–Nichols Method 1
ZN-2 Ziegler–Nichols Method 2
ZNM Ziegler–Nichols Modified

1 Introduction

Voltage regulation for DC power system is mainly done with the help of DC/DC converter. One such important and useful converter is cuk converter. These types of converters are mainly used in solar [1–4], wind energy systems as a voltage regulator in which magnitude of input DC voltage depends on intensity of sunlight and wind flow, so in order to nullify these voltage variations, cuk converter is used. Having constant input and output current makes this converter more useful to the variety of applications, whereas, it is not possible in buck or boost converters. Ideally, the expected output response of this converter is constant DC, but, practically there will be some ripple content in the response. So, an L-C filter is normally used to compensate for the ripple content. The specialty of cuk converter is to transform the input voltage to the output voltage with opposite polarity. The main difference in this converter is having different types of passive components (capacitor), which facilitates this converter useful for many applications. Besides, a variety of controllers were implemented in literature as mentioned follows to obtain fruitful output from the converter.

In [5], ZN-1 method for tuning PID was implemented for DC motor. In [6], comparative analysis on the novel controller and conventional PID controller was presented for the power converter. In [7], Fuzzy logic controller for the cuk converter was compared with PI and PID controller. Besides, modelling of FO-PID controller via genetic algorithm was implemented for cuk converter in [8], whose effectiveness in controlling the cuk converter response was validated in [9]. In [10], analysis of cuk converter performance based on genetic algorithm was reported, which results in high gain values. However, this method leads to more settling time in the response.

So, for detailed analysis and identification of robust PID controller for improving voltage response of a cuk converter, this paper implements all the key conventional PID tuning methods by using design formulas given in [11–15]. The robust method was identified with the help of transient and steady-state specifications.

2 Modelling and Control of Cuk Converter

Cuk converter is a DC to DC converter, which changes the output voltage to greater or lesser than the magnitude of input voltage. The magnitude of output voltage for the cuk converter can also be a negative value. In cuk converter, the power from input side to output side is transferred through the intermediate capacitor. The circuit of cuk converter is the union of buck converter as well as boost converter as shown in Fig. 1.

Where V_s = Source Voltage, L_s = Inductor on source side, C_l = Interlinking capacitor, L_{os} = Inductor on output side, C_{os} = Capacitor on output side, R_L = Load resistance.

The cuk converter circuit comprises mainly two inductors in which one inductor is connected to the input and the other is connected to the output, two capacitors in which one is connected as an intermediate capacitor between the input and the output and the other is connected across the output, a MOSFET is connected across the input inductor and the voltage source, and a diode is connected across output inductor and output capacitor [16–18]. It operates in two modes, namely switch on

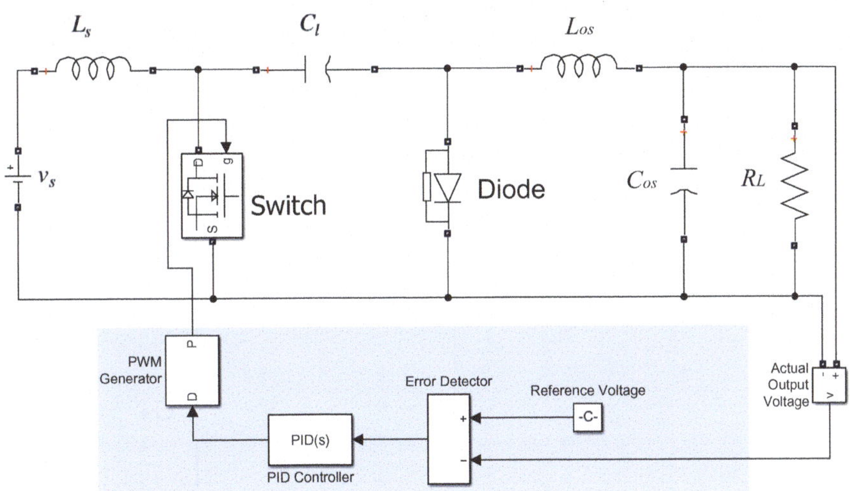

Fig. 1 Circuit diagram of cuk converter

and switch off modes. When the switch is ON, the diode acts as an open circuit. The voltage stored in the interlinking capacitor is shifted to the inductor, capacitor, and load which is present at the output side. When the switch is OFF, the diode acts as closed circuit and the intermediate capacitor is charged from the input supply and the energy stored in the inductor connected on source side.

Modelling of cuk converter is slightly different from other converter circuits due to the reason of changing circuit configuration with respect to change in switch operation [19]. The switch ON and switch OFF equivalent circuits are shown in Figs. 2 and 3, respectively. Equations (1)–(6) represents the mathematical relations during switch ON conditions, where, (1) gives voltage across the inductor at source, (2) gives KVL for loop-2, (3) gives current through interlinking capacitor, (4) gives KCL to loop-3, (5) and (6) gives the cumulative state-space model for switch ON circuit.

$$v_{\mathrm{s}} = L_{\mathrm{s}} \frac{\mathrm{d}i_{\mathrm{Ls}}}{\mathrm{d}t} \tag{1}$$

$$v_{\mathrm{Cl}} - v_{\mathrm{Cos}} = L_{\mathrm{os}} \frac{\mathrm{d}i_{\mathrm{Los}}}{\mathrm{d}t} \tag{2}$$

$$i_{\mathrm{Los}} = -C_{\mathrm{l}} \frac{dv_{\mathrm{Ls}}}{\mathrm{d}t} \tag{3}$$

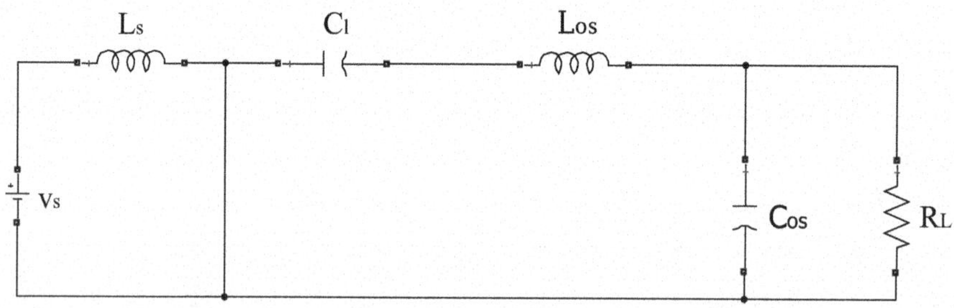

Fig. 2 Circuit diagram of cuk converter during switch ON

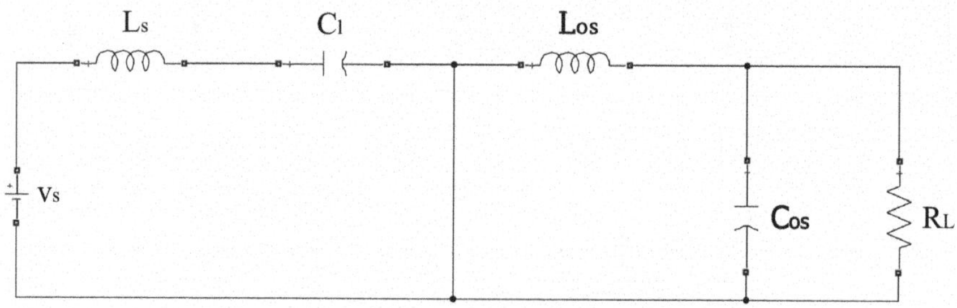

Fig. 3 Circuit diagram of cuk converter during switch OFF

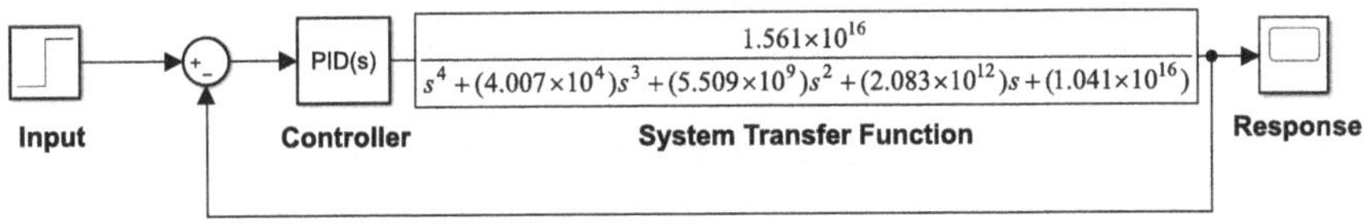

Fig. 4 MATLAB/Simulink model of the cuk converter control system

$$i_{\mathrm{Los}} - \frac{v_{\mathrm{Cos}}}{R_{\mathrm{L}}} = C_{\mathrm{os}} \frac{\mathrm{d}v_{\mathrm{Cos}}}{\mathrm{d}t} \tag{4}$$

$$\dot{X} = A_1 X + B_1 U \tag{5}$$

$$\begin{bmatrix} \dot{i}_{\mathrm{Ls}} \\ \dot{i}_{\mathrm{Los}} \\ \dot{v}_{\mathrm{C1}} \\ \dot{v}_{\mathrm{Cos}} \end{bmatrix} = \begin{bmatrix} 0 & 0 & 0 & 0 \\ 0 & 0 & (1/L_{\mathrm{os}}) & (-1/L_{\mathrm{os}}) \\ 0 & (-1/C_1) & 0 & 0 \\ 0 & (1/C_{\mathrm{os}}) & 0 & (-1/R_{\mathrm{L}} C_{\mathrm{os}}) \end{bmatrix} \begin{bmatrix} i_{\mathrm{Ls}} \\ i_{\mathrm{Los}} \\ v_{\mathrm{C1}} \\ v_{\mathrm{Cos}} \end{bmatrix} + \begin{bmatrix} (1/L_{\mathrm{s}}) \\ 0 \\ 0 \\ 0 \end{bmatrix} [v_{\mathrm{s}}] \tag{6}$$

Similarly, (7)–(12) represents the mathematical relations during switch OFF conditions, where, (7) gives voltage across the output inductor, (8) gives KVL for loop-1, (9) gives current through interlinking capacitor, (10) gives KCL to loop-3, (11) and (12) gives the cumulative state-space model for switch OFF circuit. Further, the output state equation is given by (13) and (14). Finally, by applying state-space averaging method on (6), (12), and (14), the overall system transfer function is derived as (15) and the overall system simulation block diagram is shown in Fig. 4.

As shown in Figs. 1 and 4, a PID based closed-loop control is used to control the response of the cuk converter according to the user requirement. The smooth response of the system can be achieved through proper design of PID controller specifications. Hence, various PID tuning methods given in Fig. 5, which were discussed in [20, 21] are implemented. Finally, a robust method is identified out of these methods with the help of transient and steady-state performance of the cuk converter.

$$v_{\mathrm{Cos}} = -L_{\mathrm{os}} \frac{\mathrm{d}i_{\mathrm{Los}}}{\mathrm{d}t} \tag{7}$$

$$v_{\mathrm{s}} - v_{\mathrm{C1}} = L_{\mathrm{s}} \frac{\mathrm{d}i_{\mathrm{Ls}}}{\mathrm{d}t} \tag{8}$$

$$i_{\mathrm{Ls}} = C_1 \frac{\mathrm{d}v_{\mathrm{Ls}}}{\mathrm{d}t} \tag{9}$$

$$i_{\mathrm{Los}} - \frac{v_{\mathrm{Cos}}}{R_{\mathrm{L}}} = C_{\mathrm{os}} \frac{\mathrm{d}v_{\mathrm{Cos}}}{\mathrm{d}t} \tag{10}$$

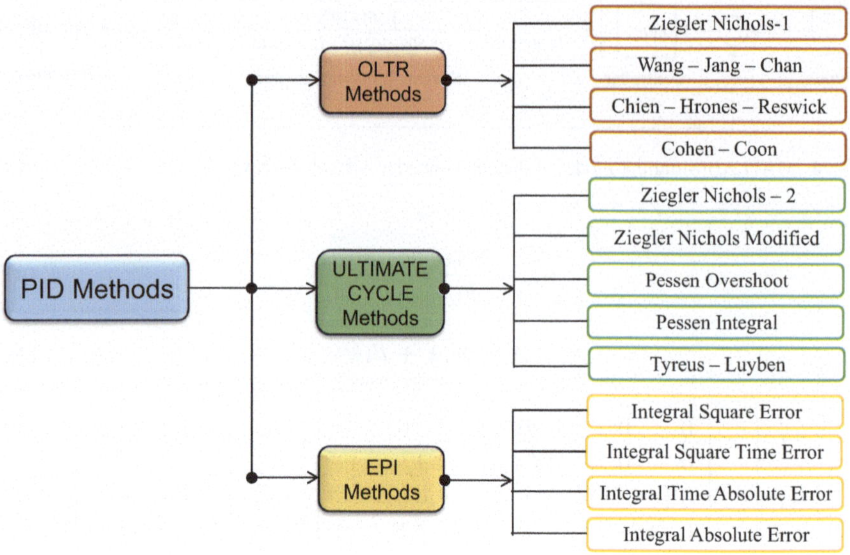

Fig. 5 Various PID tuning methods

$$\dot{X} = A_2X + B_2U \tag{11}$$

$$\begin{bmatrix} \dot{i}_{Ls} \\ \dot{i}_{Los} \\ \dot{v}_{Cl} \\ \dot{v}_{Cos} \end{bmatrix} = \begin{bmatrix} 0 & 0 & (-1/L_s) & 0 \\ 0 & 0 & 0 & (-1/L_{os}) \\ (-1/C_l) & 0 & 0 & 0 \\ 0 & (1/C_{os}) & 0 & (-1/RC_{os}) \end{bmatrix} \begin{bmatrix} i_{Ls} \\ i_{Los} \\ v_{Cl} \\ v_{Cos} \end{bmatrix} + \begin{bmatrix} (1/L_s) \\ 0 \\ 0 \\ 0 \end{bmatrix} [v_s] \tag{12}$$

$$Y = CX + DU \tag{13}$$

$$\begin{bmatrix} v_{os} \\ i_s \end{bmatrix} = \begin{bmatrix} 0\,0\,0\,1 \\ 1\,0\,0\,0 \end{bmatrix} \begin{bmatrix} i_{Ls} \; i_{Los} \; v_{Cl} \; v_{Cos} \end{bmatrix}^T \tag{14}$$

$$\frac{v_{os}}{v_s} = \frac{1.561 \times 10^{16}}{s^4 + (4.007 \times 10^4)s^3 + (5.509 \times 10^9)s^2 + (2.083 \times 10^{12})s + (1.041 \times 10^{16})} \tag{15}$$

3 Simulation Results and Analysis

Each different method will produce different gain values [proportional gain (k_p), integral gain (k_i), derivative gain (k_d)] for PID controller. However, it is always preferred to have lesser gain values to reduce the practical hardware implementation cost. So, using this idea, the gain values computed by various methods are tabulated as shown in Table 1 for their cumulative comparison. From this, it is observed that CHR method possesses lower k_p and k_d values and ZN-1 method possesses lower k_i value when compared to all other methods.

3.1 Time-Domain Transient-State Analysis

Any system can be represented in mathematical form as differential equations with reference to function of time, whose analysis is called its time-domain analysis. The time-domain based transient response plots obtained with different methods are shown from Figs. 6, 7, 8, 9 and 10, whose cumulative quantitative results by using time-domain performance specifications are given by Table 2. In order to consider the best method, rise time, delay time, %peak overshoot, and settling time must be low. From Table 2, it is observed that ZN-2 method possess better response in terms of rise time, TL method possess better response in terms of settling time, and PI method possess better response in terms of delay time.

Table 1 PID gain values calculated with different methods

Category	Method Name	K_p	T_i	T_d	K_i	K_d
OLTR methods	ZN-1	0.0011563	2.000537	0.500207	0.000577995	0.00057839
	WJC	0.2023476	0.509815	0.009457	0.39690398	0.00191374
	CHR	0.0003746	0.009637	0.500267	0.0388675	0.0001874
	CC	0.1058977	0.77666	0.018309	0.136350155	0.00193894
Ultimate cycles methods	ZN-2	10.503552	0.000435	0.000108	24146.09655	0.00114226
	ZNM	3.501184	0.00087	0.00029	0.00029	0.00101534
	PO	5.7769536	0.000435	0.00029	13280.35310	0.0016753
	PI	12.254144	0.000348	0.000130	35213.05747	0.0015991
	TL	7.877664	0.001914	0.000138	4115.81191	0.0010878
Error performance index methods	ISE	0.0065633	0.000260	0.29084	−25.1860	0.00190888
	ISTE	0.0065257	0.000406	0.24894	−16.0505	0.0016245
	ITAE	0.0108895	0.002430	0.445	4.475892	0.00484599
	IAE	0.0020535	0.000885	0.751883	2.319774011	0.00154403
Superior method		CHR			ZN-1	CHR

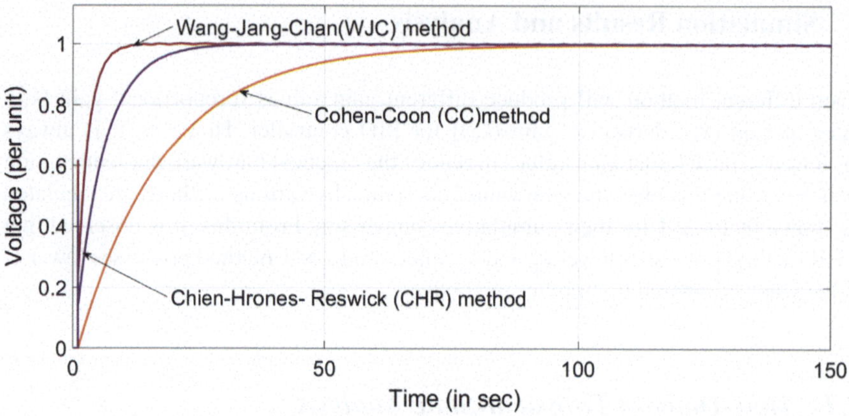

Fig. 6 System response for WJC, CC, CHR methods

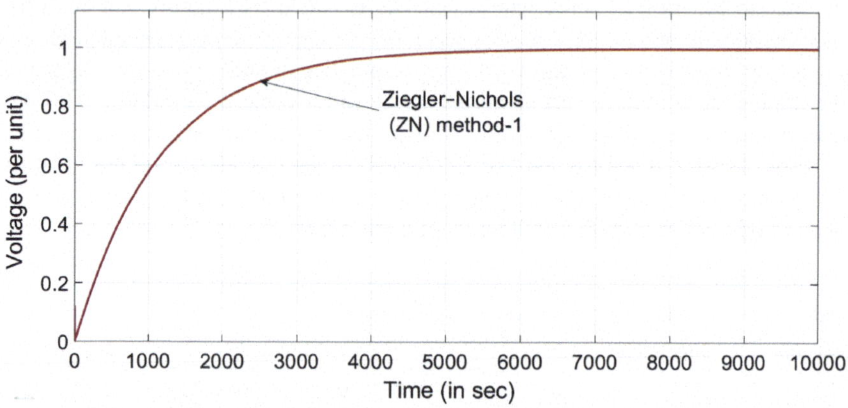

Fig. 7 System response for ZN-1 method

Figures 6 and 7 show the output responses of OLTR methods. In these, WJC method gives a better response by taking less transient time compared to remaining OLTR methods, whereas ZN-1 gives the worst performance by taking the more transient time. Figures 8 and 9 show the output responses of EPI methods. In these responses, ITAE and IAE methods give good responses among the other ITAE methods, whereas ISTE and ISE methods do not reach to steady-state. Figure 10 shows UC methods, in which, ZNM and TL methods give the best responses than the remaining methods.

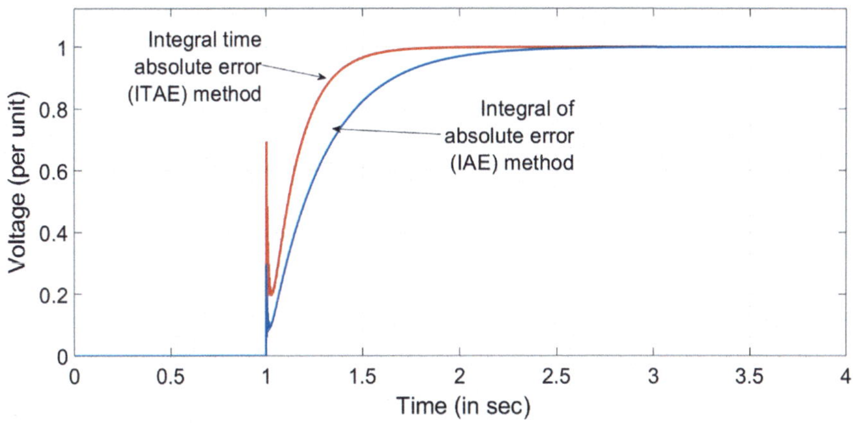

Fig. 8 System response for ITAE, IAE methods

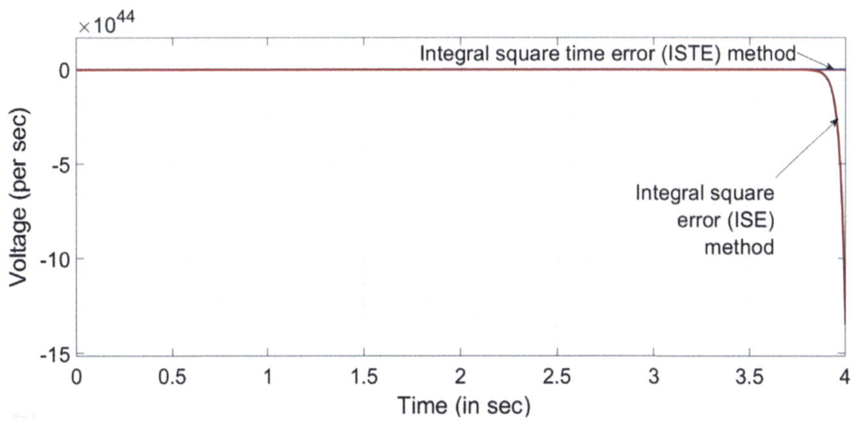

Fig. 9 System response for ISTE, ISE methods

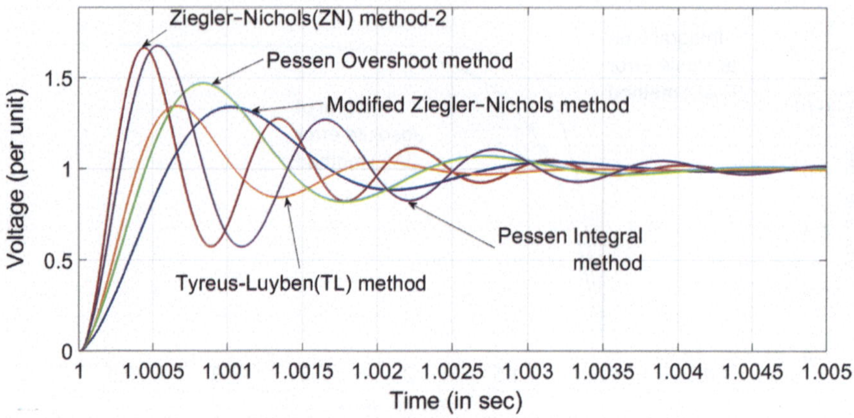

Fig. 10 System response for ultimate cycles (UC) methods

Table 2 Time-domain specifications calculated with different methods

Category	Method name	Rise time	% peak overshoot	Settling time	Delay time	Peak time
OLTR methods	ZN-1	2537.56	0	4519.3	802.66	0
	WJC	4.47305	0	8.968	1.0013	0
	CHR	37.847	0	67.977	12.9085	0
	CC	12.264477	0	22.2628	1.00182	0
Ultimate cycles methods	ZN-2	0.0001566	66.51	1.004048	1.000188	1.000569
	ZNM	0.00043	34.05	1.003556	1.000339	1.0010019
	PO	0.000194	67.63	1.00461	1.000258	1.0008442
	PI	0.000328	47.08	1.00389	1.00018	1.00052
	TL	0.000276	34.68	1.002905	1.0002306	1.0006685
Error performance index methods	ISE, ISTE	Unstable				
	ITAE	0.344067	0	1.5746	1.00108	–
	IAE	0.661435	0	2.12208	1.202385	–
Superior method		ZN-2		TL	PI	

3.2 *Frequency-Domain Steady-State Analysis*

Analyzing the steady-state and transient-state system response in terms of frequency and phase called frequency-domain analysis. The criteria for commenting on whether the system is stable is by using phase margin (PM) and gain margin (GM) values. So, to compute these values, the frequency response plots are given in Figs. 11, 12, 13, 14, 15, 16, 17, 18, 19, 20, 21, 22 and 23 that are obtained using different PID

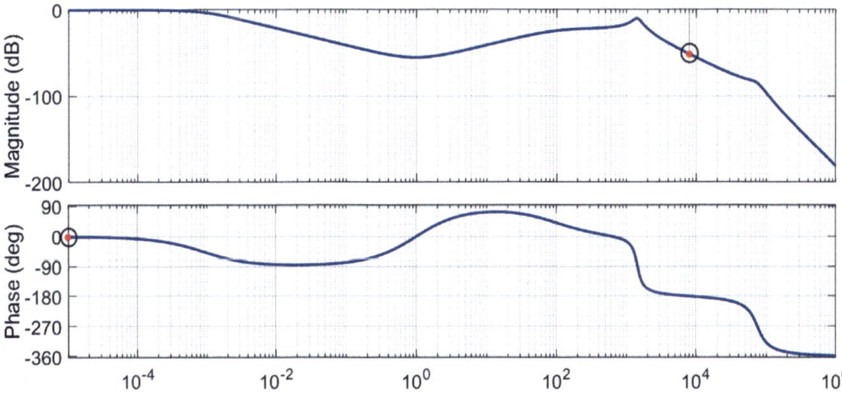

Fig. 11 System stability response obtained with ZN-1 method

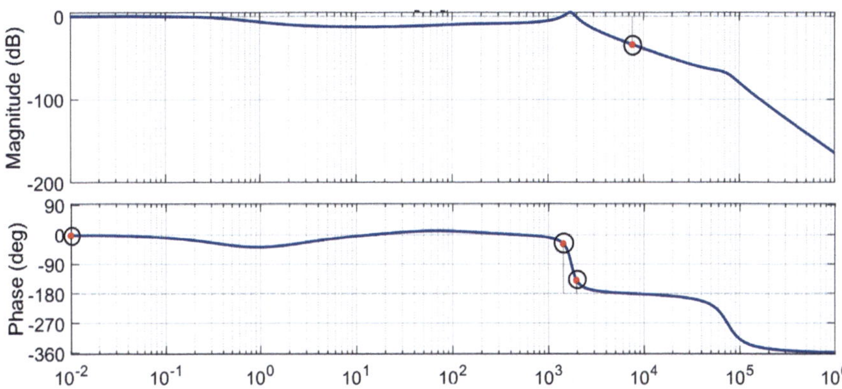

Fig. 12 System stability response obtained with WJC method

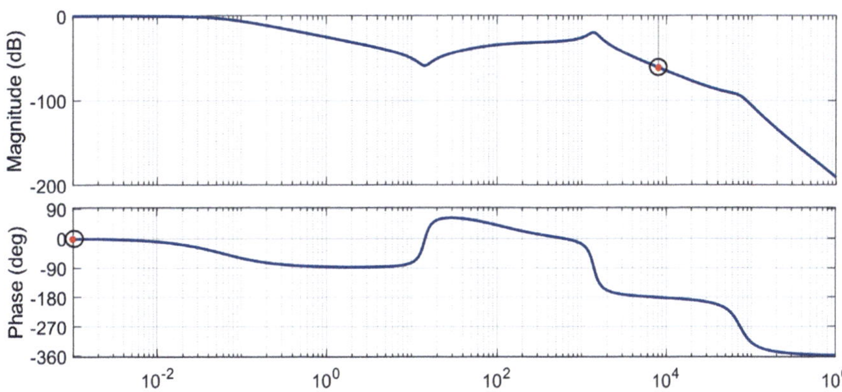

Fig. 13 System stability response obtained with CHR method

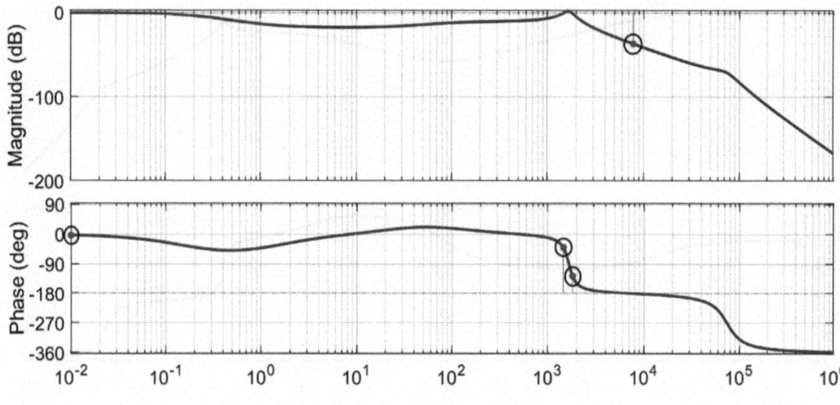

Fig. 14 System stability response obtained with CC method

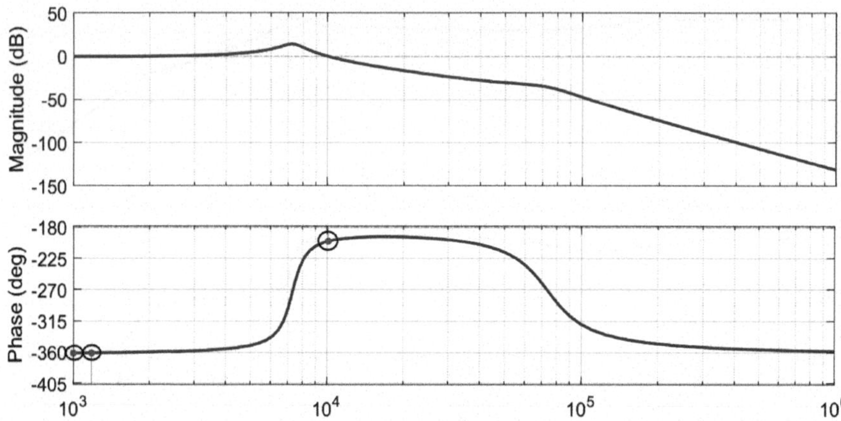

Fig. 15 System stability response obtained with ZN-2 method

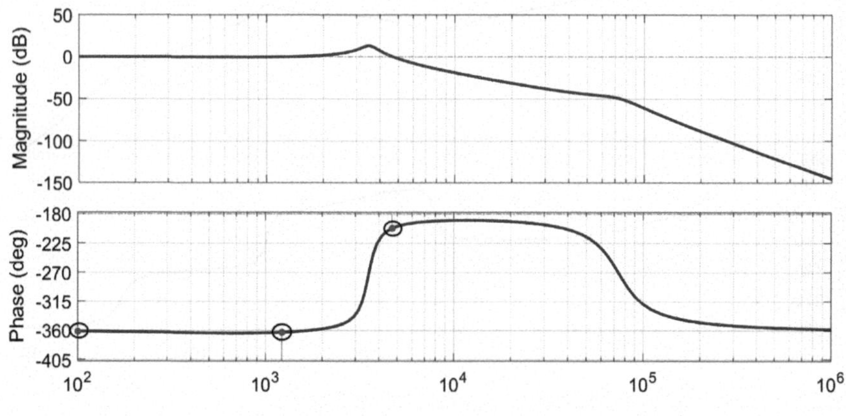

Fig. 16 System stability response obtained with modified Ziegler–Nichols method

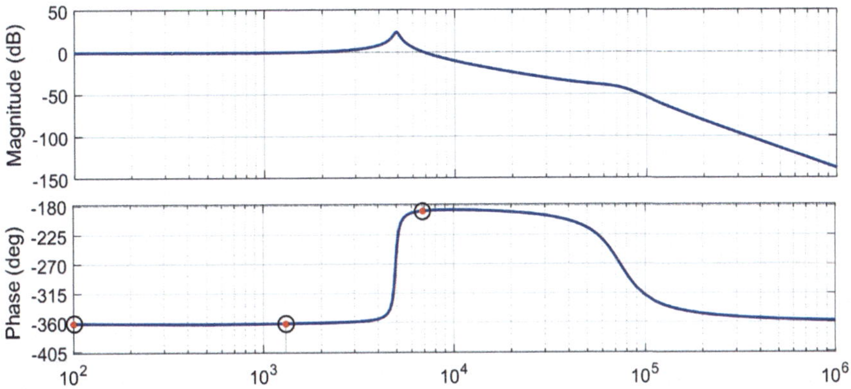

Fig. 17 System stability response obtained with TL method

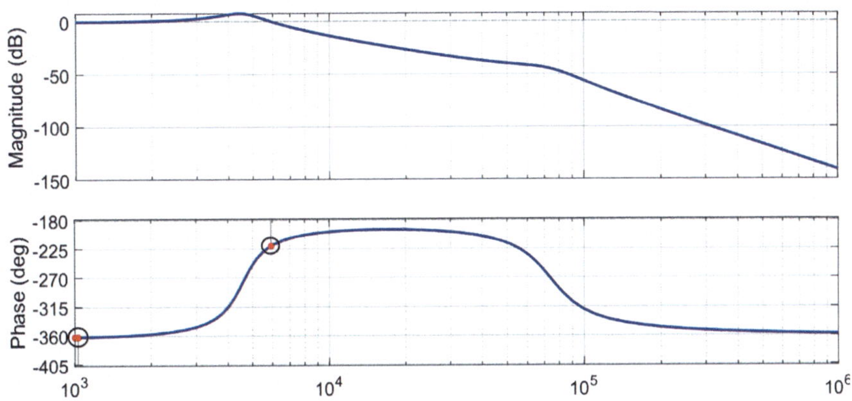

Fig. 18 System stability response obtained with PO method

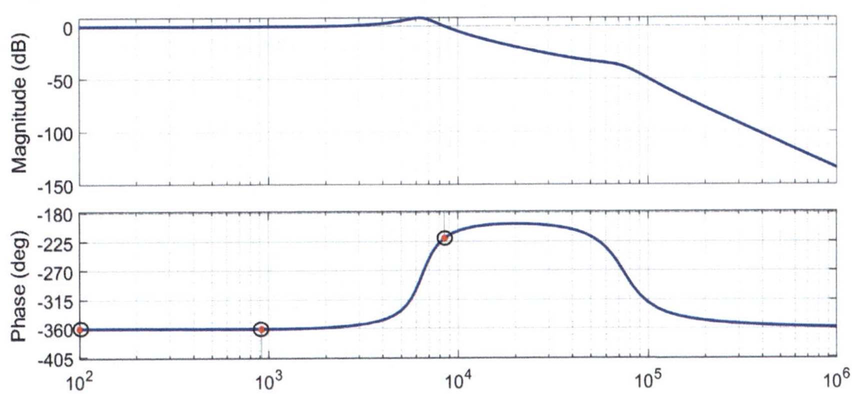

Fig. 19 System stability response obtained with PI method

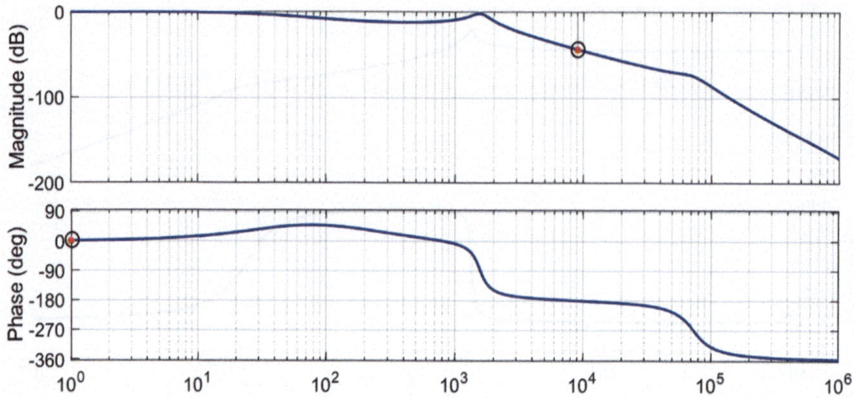

Fig. 20 System stability response obtained with ISE method

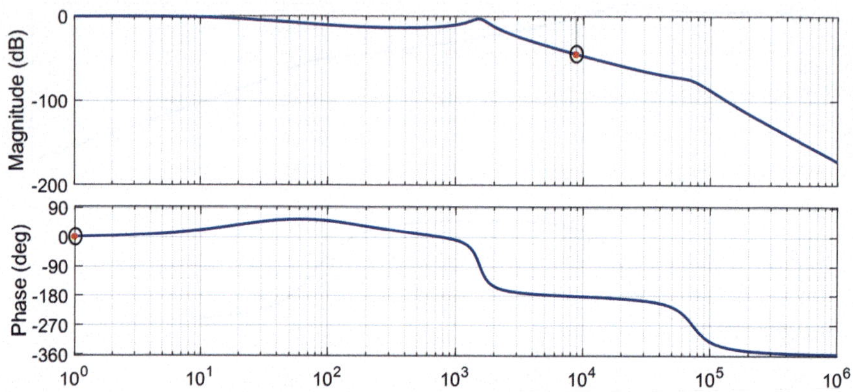

Fig. 21 System stability response obtained with ISTE method

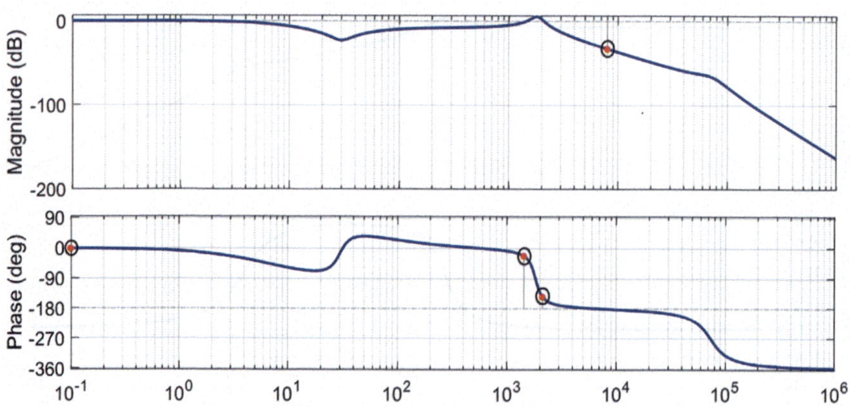

Fig. 22 System stability response obtained with ITAE method

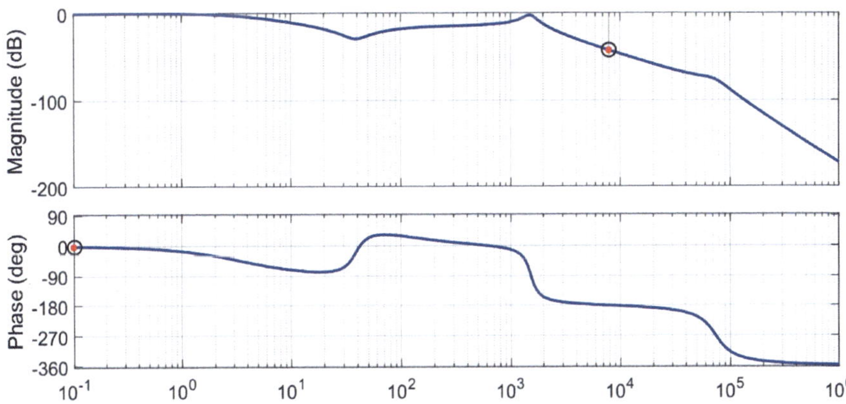

Fig. 23 System stability response obtained with IAE method

tuning methods. All these computed values are tabulated in Table 3 to compare all the methods. From this, it is observed that, in terms of GM values, CHR is considered to give a more stable response and in terms of PM values, ITAE is considered to give a more stable response.

Figures 11 and 12 show the frequency response for ZN-1 and WJC methods. In ZN-1 method PM value is −180 and GM value is 51.4 and in WJC method PM values are −180, 41.3, 152. Generally, the system which is having high PM and GM values are considered to be a stable system. By comparing in terms of GM and PM values for the above two methods, WJC method is the better one for this system. Similarly, Figs. 13 and 14 shows frequency response for CHR and CC methods. In CHR method, PM value is −180 and GM value is 61.2 and in CC method, PM values are −180, 54.2, 141 and GM value is 36.5. So, CC method is getting a better response when compared with CHR method with respect to GM and PM values.

Figures 15, 16 and 17 show frequency response plots for ZN-2, ZNM and TL methods, respectively. In ZN-2, PM values are −180, −19.8, 180 and GM value is infinite. In ZNM method, PM values are −180, −21.7, 177 and GM value is infinite. In TL method, PM values are −180, −8.07, 179 and GM value is infinite. Similarly, Figs. 18 and 19 show the frequency response for PO and PI methods. For PO method, PM values are −180, −39.9, 179 and GM value is infinite. For PI method, PM values are −180, −39.5, 179 and GM values are infinite. Hence, the stability of the system can't be determined as all the above-mentioned methods possess infinite GM values.

Figures 20 and 21 shows the frequency response for ISE and ISTE methods. For ISE method, PM value is −180 and GM value is 42.9 and for ISTE method, PM value is −180 and GM value is 43.8. By considering GM value, ISE is getting good response as it has high GM value. But, when comparing in terms of PM value, none of the methods are said to be stable. Figures 22 and 23 shows the frequency response of ITAE and IAE method. For ITAE method, GM value is 32.6 and PM values are −180, 36.6, 158 and for IAE method, GM value is 42.6 and PM value is 180. By comparing PM and GM values, IAE is producing a better response.

Table 3 Phase margin and gain margin values calculated with different methods

Category	Method name	Gain margin in (dB)	Phase margin in (deg)
OLTR methods	ZN-1	51.4	−180
	WJC	33.8	−180
			41.3
			152
	CHR	61.2	−180
	CC	36.5	−180
			54.2
			141
Ultimate cycles methods	ZN-2	INF	−180
			−19.8
			180
	ZNM	INF	−180
			−21.7
			177
	PO	INF	−180
			−39.9
			179
	PI	INF	−180
			−39.5
			179
	TL	INF	−180
			−8.07
			179
Error performance index methods	ISE	42.9	−180
	ISTE	43.8	−180
	ITAE	32.6	−180
			36.6
			158
	IAE	42.6	−180
Superior method		CHR	ITAE

4 Conclusions

In this paper, different PID tuning methods are compared with respect to time-domain and frequency-domain responses to design the best controller for cuk converter. By comparing all the time-domain responses, WJC, ZNM, and ITAE methods for PID tuning leads to a better response. Besides, WJC method has more transient time than the ZNM and ITAE methods. ZNM has 30.4% peak overshoot whereas WJC and

ITAE methods have no peak overshoot. By comparing frequency-domain responses, WJC and ITAE shows the best responses for this system. So, it is concluded that ITAE method is the best choice for the design of PID controller for cuk converter application with less transient time, zero peak overshoot, less controller gains, more phase margin and more gain margin values.

References

1. Peracaula J, Bordonau J, Canellas JJ (1991) Microcomputer control of DC/DC converter for photovoltaic applications. In: 6th Mediterranean electronical conference, pp. 214–217, Slovenia
2. Chafle SR, Vaidya UB, Khan ZJ (2013) Design of cuk converter with MPPT technique. Int J Innov Res Electr Instrum Control Eng 1(4):161–167
3. Anto EK, Asumadu JA, Okyere PY (2016) PID control for improving P&O-MPPT performance of a grid-connected solar PV system with Ziegler-Nichols tuning method. In: IEEE 11th conference on industrial electronics and applications (ICIEA), pp. 1847–1852, Hefei
4. Gupta BSSY, Lakshmi SRP (2014) PAnalysis and design of cuk converter using PI controller for PV application". IJSRD-Int J Res Dev 2(2):669–672
5. Meshram MP, Rohit KG (2012) Tuning of PID controller using Ziegler-Nichols method for speed control of dc motor. In: IEEE-international conference on advances in engineering, science and management (ICAESM-2012), pp. 117–122, Nagapattinam, Tamil Nadu
6. Algamluoli AF (2019) Novel controller for DC-DC cuk converter. In: 1st global power, energy and communication conference (IEEE GPECOM2019), Cappadocia, Turkey
7. Rakshit S, Maity J (2018) Fuzzy logic controlled cuk converter. In: International conference on communication and signal processing, pp. 0771–0775, Chennai, India
8. Tiwari S, Rayeen Z, Hanif O (2018) Design and analysis of fractional order PID controller tuning via genetic algorithm for cuk converter. In: IEEE 13th international conference on industrial and information systems (ICIIS), Rupnagar, India
9. Hanif O, Rayeen Z, Tiwari S (2018) Design and comparative analysis of fractional order controllers for cuk converters. In: IEEE 8th power india international conference (PIICON), Kurukshetra
10. Poudeh MB, Eshtehardiha S, Ershadi MH (2008) Optimizing the classic controllers to improve the cuk converter performance based on genetic algorithm. In: International conference on smart manufacturing application, KINTEX, Gyeonggi-do, Korea
11. Lin JM, Kao PF, Cho KT (2012) Ziegler-Nichols based intelligent controller design of a SPM system. In: International conference on automatic control and artificial intelligence (ACAI), pp. 2272–2275, Xiamen
12. Tripathi A, Verma RL, Alam MS (2013) Study and analysis of various tuning methods of PID controller for AVR system. Int J Res Electr Eng 1(1):93–98
13. Tasoren AE, Orenbas H, Sahin S (2018) Analyze and comparison of different PID tuning methods on a brushless DC motor using atmega328 based microcontroller unit. In: International conference on control engineering & information technology (CEIT), Istanbul, Turkey
14. Tan W, Liu J, Chen T, Marquez HJ (2006) Comparison of Some well-known PID tuning formulas. Comput Chem Eng 30(9):1416–1423
15. Srinivas P, Lakshmi KV, Kumar VN (2014) A comparison of PID controller tuning methods for three tank level process. Int J Adv Res Electr Electron Instrum Eng 3(1):6810–6820
16. Zhang Z, Cuk S (2000) A high efficiency 500 W step-up cuk converter. In: 3rd international power electronics and motion control conference (IPEMC), Beijing, China
17. Verbytskyi I, Bondarenko O, Vinnikov D (2017) Multicell-type current regulator based on cuk converter for resistance welding. In: 58th international scientific conference on power and electrical engineering of Riga Technical University (RTUCON), Riga

18. Ponmani P, Priya MSD (2015) Hybrid solar-wind energy system with MPPT using cuk-sepic fused converter. Int J Eng Res Technol (IJERT) 4(1):502–507
19. Mokal BP, Vadirajacharya K (2017) Extensive modeling of DC-DC cuk converter operating in continuous conduction mode. In: International conference on circuits power and computing technologies [ICCPCT], Kollam
20. Murthy BV, Pavan Kumar YV, Kumari UVR (2012) Application of neural networks in process control: automatic/online of PID controller gains for ±10% disturbance rejection. In: IEEE international conference on advanced communication control and computing technologies, Ramanathapuram
21. Prusty SB, Padhee S, Pati UC, Mahapatra KK (2015) Comparative performance analysis of various tuning methods in the design of PID controller. In: Michael faraday IET international summit: MFIIS, Kolkata, India

Modified Cascaded Reversing Voltage Multilevel Inverter Using Optimal Switching Angle Technique for Photovoltaic Applications

Manoj Hegde, S. Nagaraja Rao, and M. S. Indira

Abstract This paper presents a single-phase Modified Cascaded Reversing Voltage Multilevel Inverter (MCRVMLI) to generate stepped output voltage using various Switching Angle Techniques (SATs) for Photovoltaic (PV) applications. The MCRVMLI consists of two units namely, Level Generating Unit (LGU) and Polarity Changing Unit (PCU) to generate required steps in the output. The suggested MCRVMLI requires only '$(m-1)$' power switches, whereas the traditional cascaded MLI and cascaded based multilevel DC-link requires '$2(m-1)$' and '$(m+3)$' power switches respectively to generate the required number of steps in the output phase voltage. Therefore, the MCRVMLI is cost-effective and requires less installation area as the number of levels increases. MCRVMLI has improved performance as SATs are implemented in terms of Total Harmonic Distortion (THD) and fundamental output voltage. In this paper, equal-phase, half equal-phase and step-pulse-wave SATs are investigated to determine the optimal SAT. The MCRVMLI has been simulated in MATLAB/SIMULINK for 9-level using equal-phase, half-equal-phase and step-pulse-wave SATs to find an optimal SAT and the results are validated with theoretical values. Further, the suggested MCRVMLI with optimized SAT has been simulated for 11-level, 13-level, 15-level and 17-level to obtain THD as per the IEEE-519 standards without filter circuits.

Keywords MCRVMLI · Equal-phase · Half-equal-phase · Step-pulse-wave SAT · THD

M. Hegde · S. N. Rao (✉) · M. S. Indira
Department of Electrical Engineering, Ramaiah University of Applied Sciences, Bengaluru, Karnataka, India
e-mail: nagarajarao.ee.et@msruas.ac.in

M. Hegde
e-mail: hegdemanoj619@gmail.com

M. S. Indira
e-mail: indira.ee.et@msruas.ac.in

M. N. Favorskaya et al. (eds.), *Innovations in Electrical and Electronic Engineering*, Lecture Notes in Electrical Engineering 661, https://doi.org/10.1007/978-981-15-4692-1_25

1 Introduction

Renewable Energy Sources (RESs) have emerged in recent years as a major alternative to conventional energy sources. Among the RESs, Photovoltaic (PV) cell energy has distinct characteristics of low carbon emissions with minimal maintenance. Power inverters are generally integrated with PV to convey green power. Since the conventional two-level inverter configuration has more number of power switches and filter circuits, it leads to increase in cost, complexity, switching losses and harmonics. Hence, the two-level inverters are being replaced by low switching or fundamental switching frequency Multilevel Inverters (MLIs) [1].

MLI is the integration of several levels of DC step voltages. It reduces the THD, electromagnetic interference, switching frequency and voltage stress on the power switches. As the number of voltage levels increase, the harmonic distortion decreases reducing the need for filters at the output [1, 2]. The conventional MLIs are categorized as Diode Clamped MLI (DCMLI) [3], Flying Capacitor MLI (FCMLI) [4] and Cascaded H-Bridge MLI (CHBMLI) [5]. The CHBMLI has many advantages like fewer power switches, no extra clamping diodes and capacitors when compared to DCMLI and FCMLI. In addition, due to the presence of capacitors in DCMLI and FCMLI, there is voltage unbalance in the output waveform. However, the traditional MLIs require more power switches i.e., '$2(m-1)$' by increasing the number of voltage levels[2]. Cascaded based RVMLI require only '$(m+3)$' power switches [6, 7]. Though, the cascaded based reverse voltage MLI involves more power switches and driver circuits particularly at higher levels i.e., beyond five. Therefore, researchers seek new and promising MLI topology to meet the demands [8]. As the conventional and cascaded based reverse voltage MLIs requires a greater number of power switches beyond five levels.

Though the THD is reduced, the switching frequency stress on the switches is not reduced and this may lead to failure. Thus, there is an urge for new topologies and a driver circuit that reduces the voltage stress and THD on the power switches to considerable state.

The MLIs can have the enormous performance by employing various modulation techniques such as Sinewave Pulse Width Modulation (SPWM) [9], Selective Harmonic Elimination (SHE-PWM) [10], Space Vector Modulation (SVM) [11] and Switching Angle Technique (SAT) and others [12, 13]. The SPWM and SVM control scheme having higher switching frequency is used extensively in industrial applications leading to more switching losses [14]. These losses are high compared to the fundamental switching scheme. In this paper, step-pulse-wave SAT is proposed for the reduction in THD and compared with equal-phase and half-equal-phase SATs [15].

2 Structure of the Proposed MCRVMLI System

The MCRVMLI topology block diagram is shown in Fig. 1.

2.1 *Modified Cascaded Reversing Voltage MLI (MCRVMLI)*

The MCRVMLI consists of eight power switches and four DC sources as shown in Fig. 2. The required levels are generated by level generating switches $L_1 - L_4$ and the polarity by switches $P_1 - P_4$. The switches are triggered according to Table 1 (Figs. 3 and 4).

The maximum output phase voltage of MCRVMLI is the addition of all the DC sources and is given as:

$$V_{0,\max} = \sum_{i=1}^{4} V_{dci} \tag{1}$$

As the DC sources are symmetrical, the output levels are synthesized into the stepped form using the LGU and PGU switches to synthesize the required positive and negative levels [4]. Equation (1) can be rewritten as:

$$V_{0,\max} = \sum_{i=1}^{n} +V_i, \quad \text{If} \quad P_1, P_2 = 1 \tag{2}$$

$$V_{0,\max} = \sum_{i=1}^{n} -V_i, \quad \text{If} \quad P_3, P_4 = 1 \tag{3}$$

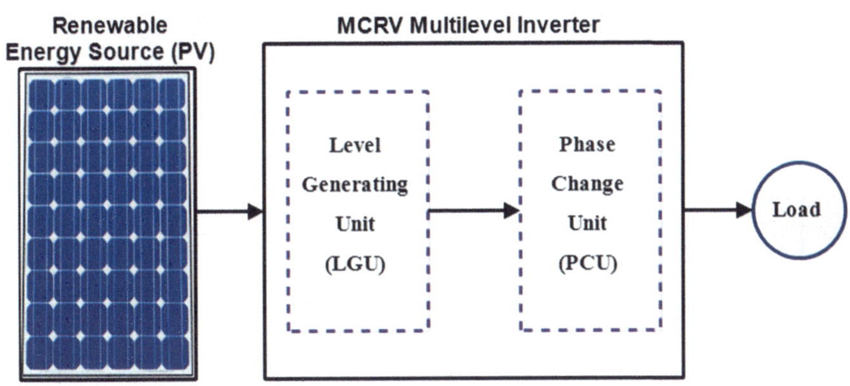

Fig. 1 Block diagram of MCRVMLI topology

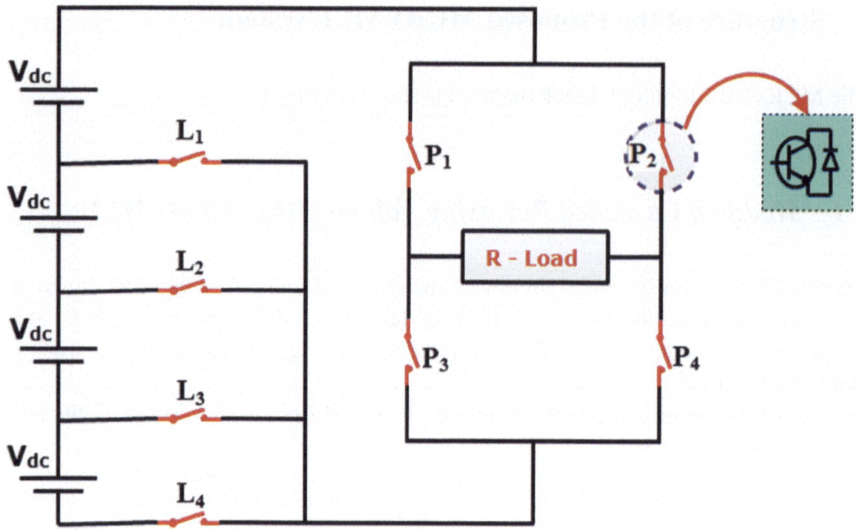

Fig. 2 Proposed MCRVMLI topology

Table 1 9-level MCRVMLI switching table

Output voltage levels	Switch states							
	LGU Switches				PCU switches			
	L1	L2	L3	L4	P1	P2	P3	P4
$4V_{dc}$	✗	✗	✗	✓	✓	✗	✗	✓
$3V_{dc}$	✗	✗	✓	✗	✓	✗	✗	✓
$2V_{dc}$	✗	✓	✗	✗	✓	✗	✗	✓
$1V_{dc}$	✓	✗	✗	✗	✓	✗	✗	✓
0	✗	✗	✗	✗	✓	✗	✓	✗
$-1V_{dc}$	✓	✗	✗	✗	✗	✓	✓	✗
$-2V_{dc}$	✗	✓	✗	✗	✗	✓	✓	✗
$-3V_{dc}$	✗	✗	✓	✗	✗	✓	✓	✗
$-4V_{dc}$	✗	✗	✗	✓	✗	✓	✓	✗

Where '✗' = OFF and '✓' = ON

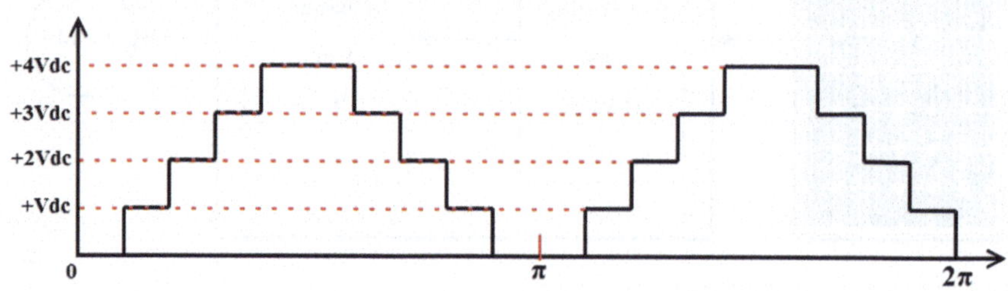

Fig. 3 Output of LGU

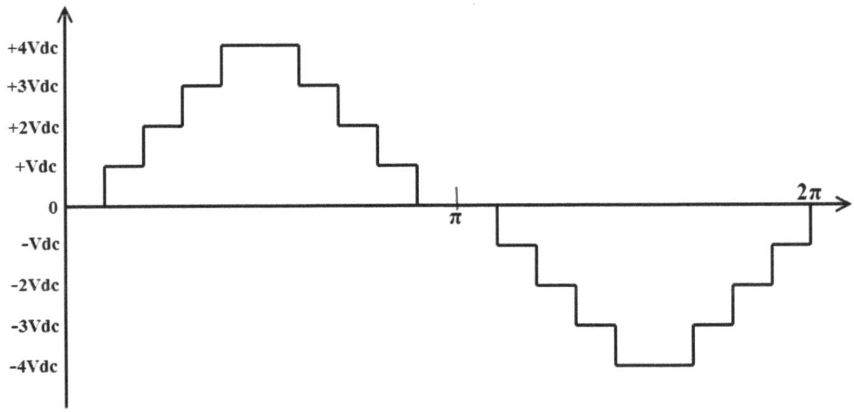

Fig. 4 Output of PCU

The switching voltage stress is more in the existing RVMLI topology as it consists of twelve power switches. A MCRVMLI is proposed with a lower switching voltage stress.

The levels of the output voltage are synthesized according to Eqs. (2) and (3). The generalized voltage levels are be obtained using the following equation:

$$N_L = (2S + 1)^H \tag{4}$$

where 'S' is the number of DC sources.

'H' is the number of full-bridge circuits.

The number of power switches required for the MCRVMLI topology is determined using Eq. (5)

$$Ns = ((m - 1)/2) + 4 \tag{5}$$

3 Switching Angle Techniques for MCRVMLI

The overall efficiency of the MCRVMLI topology is influenced by the modulation technique employed. A good modulation technique gives the best result. In this case, the SAT plays a vital role. The SAT controls the output phase voltage %THD and $V_{\rm rms}$ of the proposed MCRVMLI. In the present work a Step Pulse Wave SAT (SPWSAT) triggers the level generating switches in MCRVMI. The performance is compared with Equal Phase SAT (EPSAT) and Half Equal Phase SAT (HEPSAT) [14, 15]. Figure 5 shows the generation of a 9-level step voltage waveform using SAT.

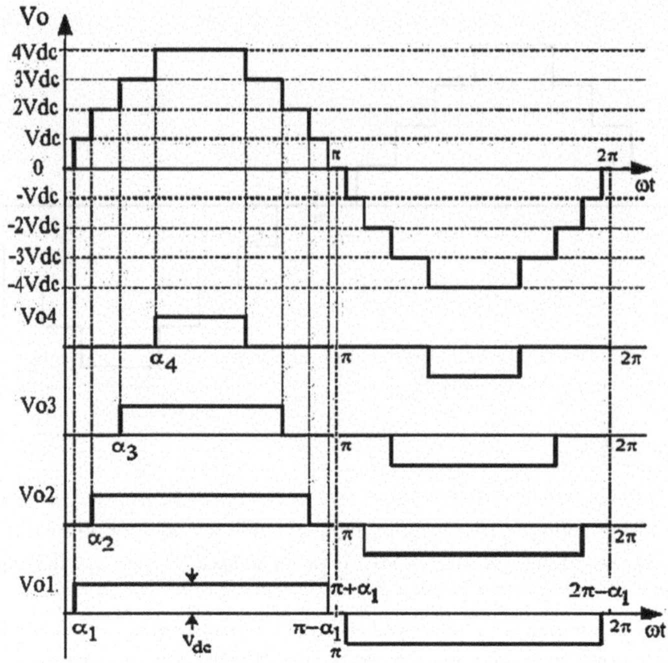

Fig. 5 Generation of 9-level stepped voltage waveform using SAT

The switching angles in a SAT must be determined during the period of the first quadrant for a m-level step waveform ($(m - 1)/2$). These switching angles are the main angles i.e., $\alpha_1, \alpha_2, \alpha_3$ and α_4 for 9-level and these determine the other angles. The following condition must be met by these angles [15]:

$$0 \leq \alpha_1 \leq \alpha_2 \leq \alpha_3 \leq \alpha_4 \leq \frac{\pi}{2} \tag{6}$$

3.1 *Equal Phase SAT*

The main angles in an EPSAT are given by Eq. (7) by averaging the distribution of the switching angles from 0° to 180°:

$$\alpha_i = i\left(\frac{180^{\circ}}{m}\right) \tag{7}$$

where $i = 1, 2, \ldots, (m - 1)/2$ and m = number of levels.

3.2 Half Equal Phase SAT

The switching angles are rearranged to obtain a better and wider output waveform by using:

$$\alpha_i = i\left(\frac{90^\circ}{\left(\frac{m+1}{2}\right)}\right) \tag{8}$$

3.3 Step Pulse Wave SAT

In this technique, the switching angles are obtained by calculating the volt-second areas of the reference voltage waveform to the step output phase voltage waveform of MCRVMLI by the equations [16].

$$V_{\text{ref}} = 4\left(\frac{4V_{\text{dc}}}{\pi}\right)(\text{Mi Sin}\omega t) \tag{9}$$

$$V_{\text{out - phase}} = i.V_{\text{dc}}(1 \le i \le 4) \tag{10}$$

where the modulation index is 'Mi' and the integer number is 'i'.

The conducting angle in a 9-level MCRVMLI using SPWSAT are obtained by the following equations:

$$\alpha_1 = \frac{16\text{Mi}}{\pi}\left\{\cos\left(\sin^{-1}\left(\frac{\pi}{16\text{Mi}}\right)\right) - 1\right\} + \sin^{-1}\left(\frac{\pi}{16\text{Mi}}\right) \tag{11}$$

$$\begin{aligned}\alpha_2 = {} & \frac{16\text{Mi}}{\pi}\left\{\cos\left(\sin^{-1}\left(\frac{2\pi}{16\text{Mi}}\right)\right) - \cos\left(\sin^{-1}\left(\frac{\pi}{16\text{Mi}}\right)\right)\right\} \\ & + 2\sin^{-1}\left(\frac{2\pi}{16\text{Mi}}\right) - \sin^{-1}\left(\frac{\pi}{16\text{Mi}}\right)\end{aligned} \tag{12}$$

$$\begin{aligned}\alpha_3 = {} & \frac{16\text{Mi}}{\pi}\left\{\cos\left(\sin^{-1}\left(\frac{3\pi}{16\text{Mi}}\right)\right) - \cos\left(\sin^{-1}\left(\frac{2\pi}{16\text{Mi}}\right)\right)\right\} \\ & + 3\sin^{-1}\left(\frac{3\pi}{16\text{Mi}}\right) - 2\sin^{-1}\left(\frac{2\pi}{16\text{Mi}}\right)\end{aligned} \tag{13}$$

$$\alpha_4 = \frac{4\pi}{2} - \frac{16\text{Mi}}{\pi}\left\{\cos\left(\sin^{-1}\left(\frac{3\pi}{16\text{Mi}}\right)\right)\right\} - 3\sin^{-1}\left(\frac{3\pi}{16\text{Mi}}\right) \tag{14}$$

3.4 THD Calculation of a 9-Level SAT Enabled MCRVMLI

The expression of THD for the proposed MCRVMLI is given by:

$$\text{THD} = \sqrt{\left(\frac{V_{\text{rms}}}{V_1}\right)^2 - 1} \tag{15}$$

Where 'V_1' is the fundamental component and
'V_{rms}' is the output phase voltage.
The values of V_1 and V_{rms} are obtained using Eqs. (16) and (17).

$$V_{\text{rms}} = V_{\text{dc}}\sqrt{\left[\frac{2}{\pi}\cdot\left(\begin{array}{c}(\alpha_2-\alpha_1)+4(\alpha_3-\alpha_2)+9(\alpha_4-\alpha_3)\\ +16(\frac{\pi}{2}-\alpha_4)\end{array}\right)\right]} \tag{16}$$

$$V_1 = \frac{4V_{\text{dc}}}{\pi\sqrt{2}}[\cos\alpha_1 + \cos\alpha_2 + \cos\alpha_3 + \cos\alpha_4] \tag{17}$$

By substituting Eqs. (16) and (17) into (15), the THD for the proposed MCRVMLI is given by:

$$\text{THD} = \sqrt{\left[\frac{\pi}{4}\cdot\frac{\begin{array}{c}(\alpha_2-\alpha_1)+4(\alpha_3-\alpha_2)\\ +9(\alpha_4-\alpha_3)+16(\frac{\pi}{2}-\alpha_4)\end{array}}{(\cos\alpha_1+\text{Cos}\alpha_2+\text{Cos}\alpha_3+\text{Cos}\alpha_4)^2}\right] - 1} \tag{18}$$

Table 2 represents the theoretical values of V_{rms} and %THD for the conventional and proposed SATs for the 9-level MCRVMLI along with switching angles. From

Table 2 Theoretical THD, V_{rms} and Conducting angles for a 9-level MCRVMLI

Modulation technique	Switching angles (in degrees)				%THD (Theoretical)	V_{rms} (V)
	α_1	α_2	α_3	α_4		
EPSAT	20	40	60	80	25.37	123.12
HEPSAT	18	36	54	72	22.06	137.5
SPWSAT	7	21.66	8.04	9.76	9.19	166.3

the theoretical analysis of %THD and V_{rms}, it is inferred that the lower %THD and maximum output RMS voltage is achieved by using SPWSAT when compared to equal phase and half equal phase SATs.

4 Simulation Results of a MCRVMLI for EPSAT, HEPSAT and SPWSAT Techniques

The simulation results for a 9-level MCRVMLI using EPSAT, HEPSAT and SPWSAT and its THD spectrum are shown in Figs. 6, 7, 8, 9, 10, 11, 12, 13 and 14. The simulation parameters of MCRVMLI are given in Table 3.

The unipolar output across LGU of a HEPSAT–MCRVMLI is shown in Fig. 7. The 9-level output phase voltage and its THD spectrum are shown in Figs. 8 and 9. The magnitude of the output voltage and its RMS value are noted as 194.5 V and 137.5 V respectively. The THD of the proposed HEPSAT–MCRVMLI is 22.04%.

The 9-level stepped phase output voltage and its %THD spectrum is shown in Figs. 13 and 14. The magnitude of the output voltage and its RMS value are 235.8 V and 166.38 V respectively. The SPWS at.%THD of a SPWSAT–MCRVMLI is 9.11%.

The simulation results and theoretical calculations indicate that the SPWSAT technique gives lower THD compared to EPSAT and HEPSAT for the proposed MCRVMLI topology. The drawbacks of EPSAT and HEPSAT are, the output phase voltage of both EPSAT and HEPSAT resemble a triangular waveform with a higher THD and lower fundamental value. As the SPWSAT gives lesser THD and good fundamental value for the proposed MCRVMLI, the paper extends up to 17-level to

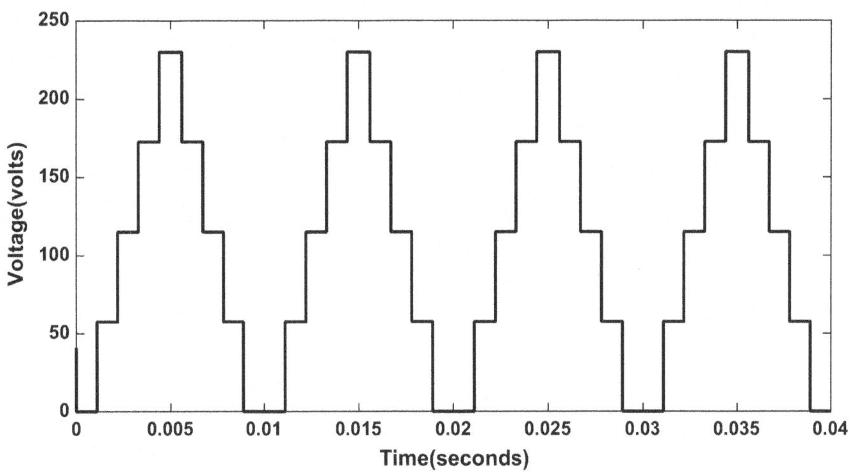

Fig. 6 Output across the LGU of 9-level EPSAT–MCRVMLI

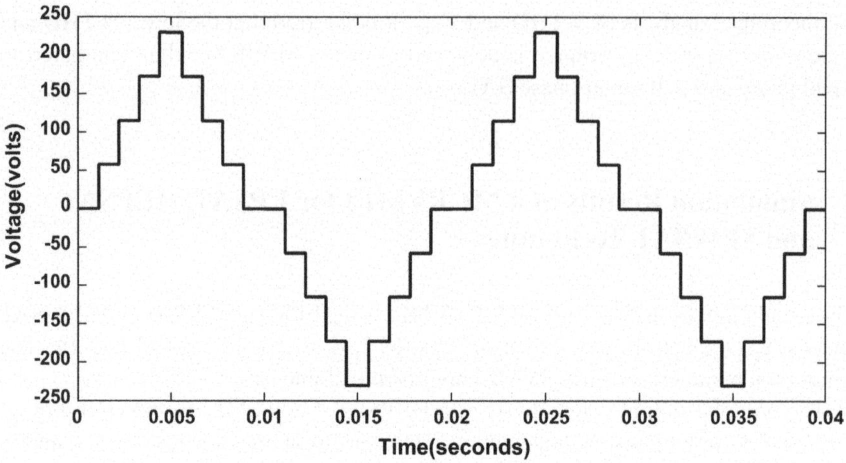

Fig. 7 Output across PCU of a 9-level EPSAT–MCRVMLI

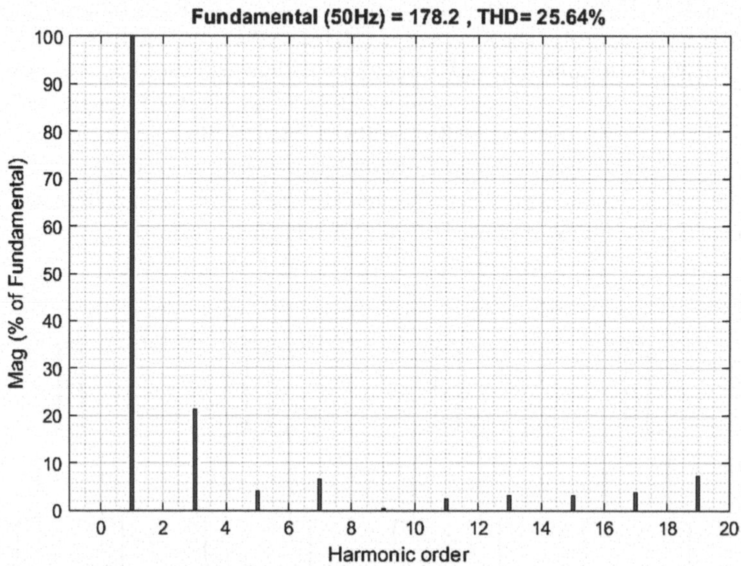

Fig. 8 THD spectrum of a 9-level EPSAT–MCRVMLI

reduce the THD as per the IEEE standard 519. Using Eqs. (4) and (5) the number of levels and its switches required is given in Table 4.

Table 5 gives the angles with theoretical %THD and V_{rms} of the levels.

The simulation results of MCRVMLI topology using SPWSAT for 11-levels and extension to 17-levels with their respective THD spectrum are shown in Figs. 15, 16, 17, 18, 19, 20, 21 and 22.

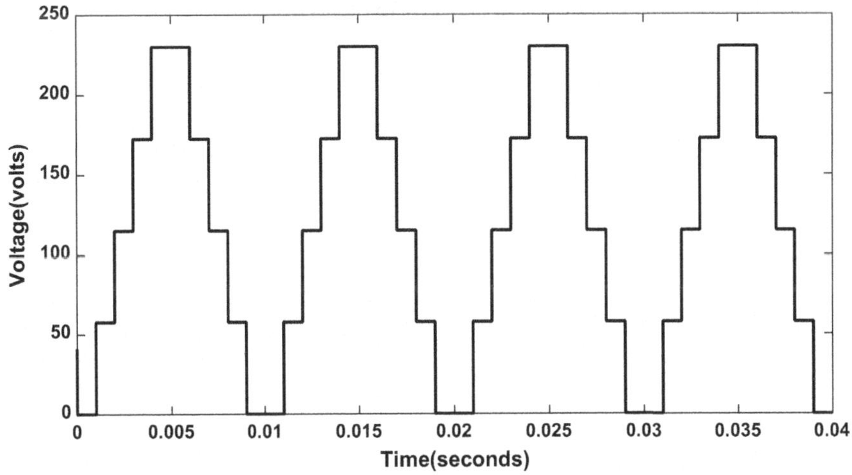

Fig. 9 Output across LGU of a 9-level HEPSAT–MCRVMLI

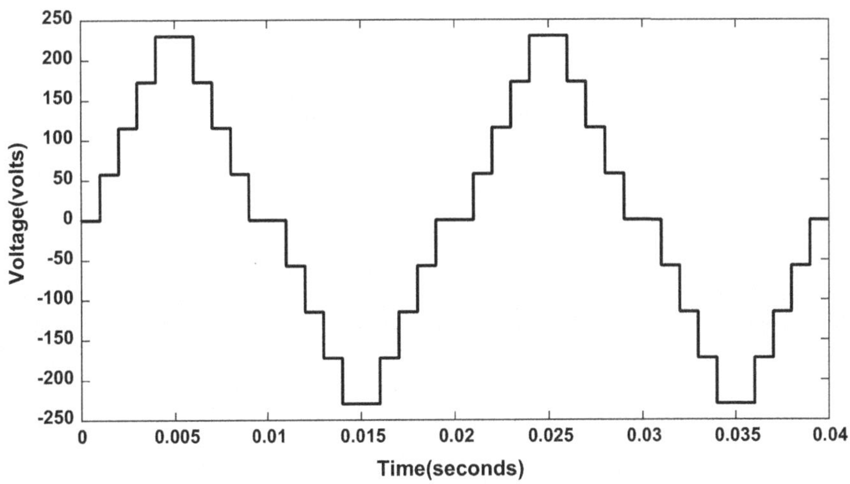

Fig. 10 HEPSAT–MCRVMLI 9-level output

From Figs. 15, 16, 17, 18, 19, 20, 21 and 22, the output phase voltage and its %THD of the MCRVMLI topology was analyzed to generate 11-level, 13-level, 15-level and 17-levels respectively and the respective %THD values are 7.31%, 6.22%, 5.38% and 4.93%. The results have shown that SPWSAT can be of paramount importance in achieving the minimum %THD value by extending the number of levels.

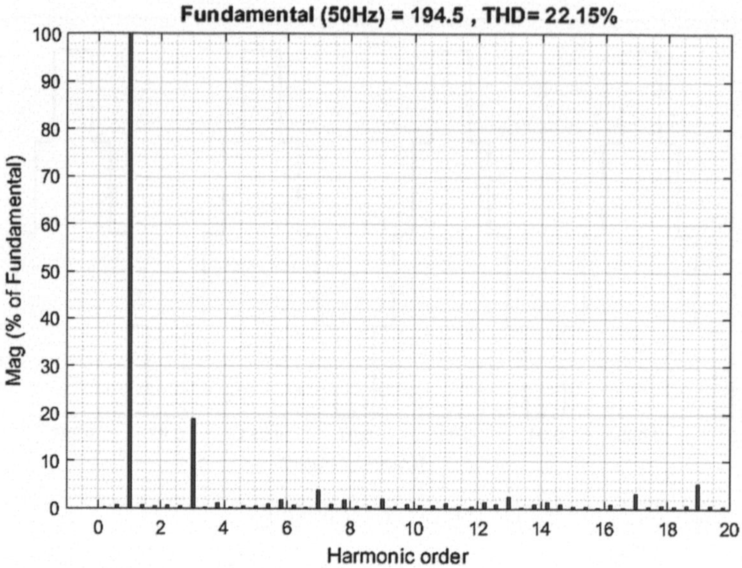

Fig. 11 THD spectrum for a 9-level HEPSAT–MCRVMLI

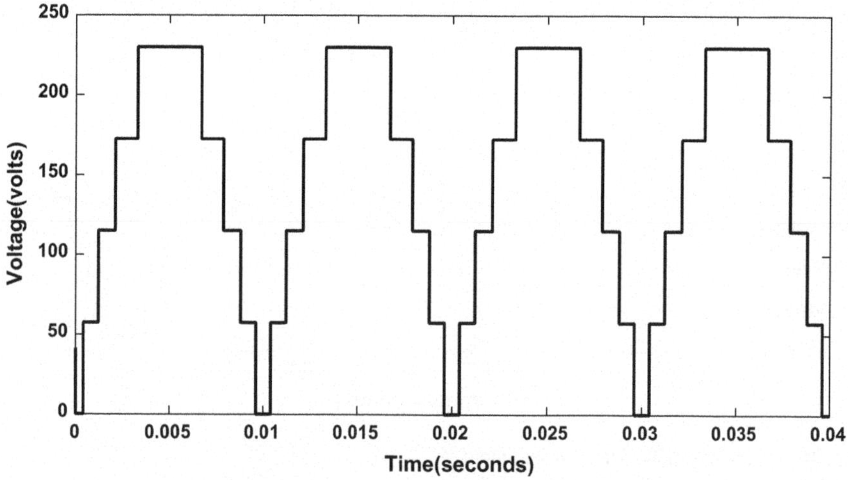

Fig. 12 Output across LGU of a 9-level SPWSAT–MCRVMLI

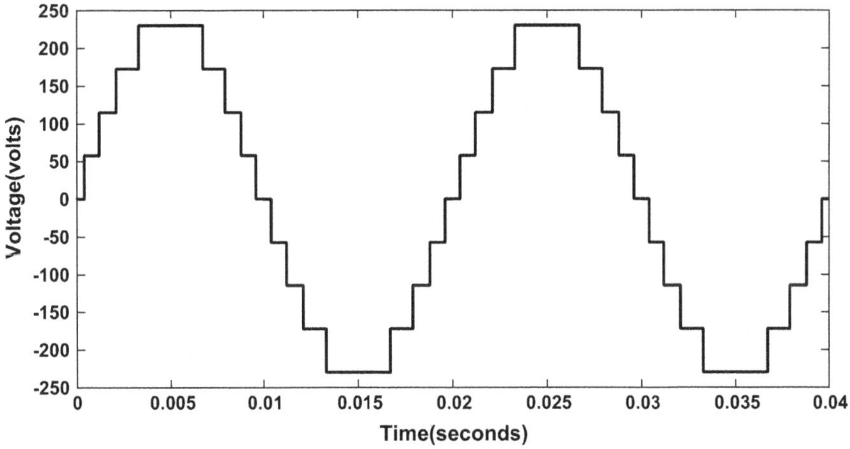

Fig. 13 SPWSAT–MCRVMLI 9-level output

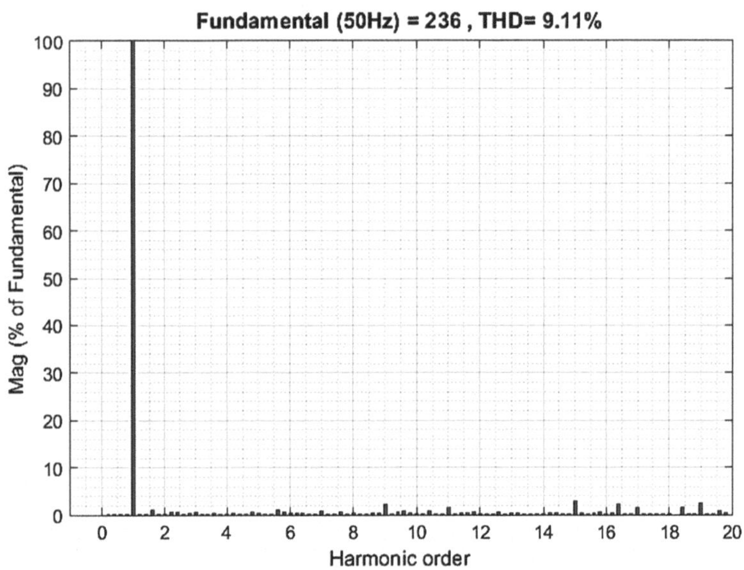

Fig. 14 THD spectrum of a 9-level SPWSAT–MCRVMLI

Table 3 Simulation parameters

Parameter	Value
Resistive load	100 Ω
DC voltage (V_{dc})	57.5 V
Frequency	50 Hz
Mi	0.8

Table 4 Number of switches required

No. of levels	No. of switches
9	8
11	9
13	10
15	11
17	12

Table 5 Switching angles, %THD and V_{rms} for proposed MCRVMLI

No. of levels	Switching angles (in degrees)								%THD (Theoretical)	V_{rms} (V)
	α_1	α_2	α_3	α_4	α_5	α_6	α_7	α_8		
9	7	21.66	38.04	59.76	×	×	×	×	9.19	166.3
11	5.64	17.16	29.47	43.58	62.35	×	×	×	7.63	167.3
13	4.66	14.22	24.18	35.03	47.56	64.15	×	×	6.35	166.6
15	4.02	12.14	20.58	29.42	39.2	50.64	65.5	×	5.6	166.3
17	3.6	10.8	18	25.2	34.2	43.2	52.2	63	5.05	167

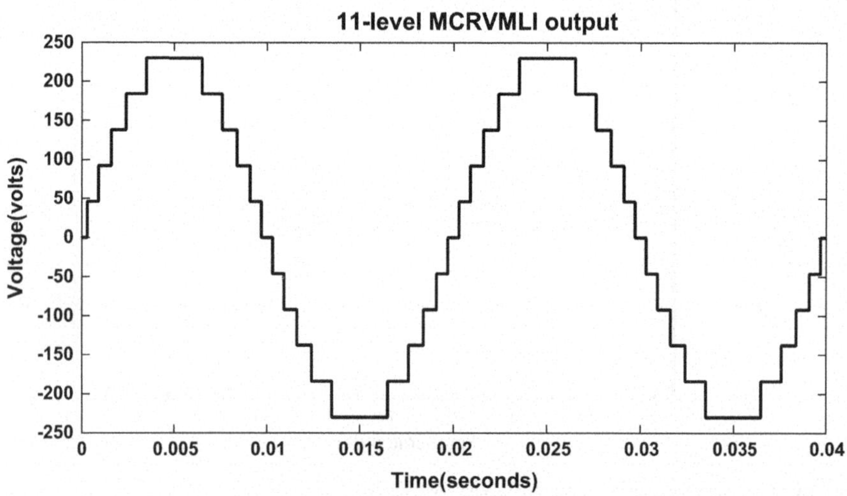

Fig. 15 11-level SPWSAT–MCRVMLI output phase voltage

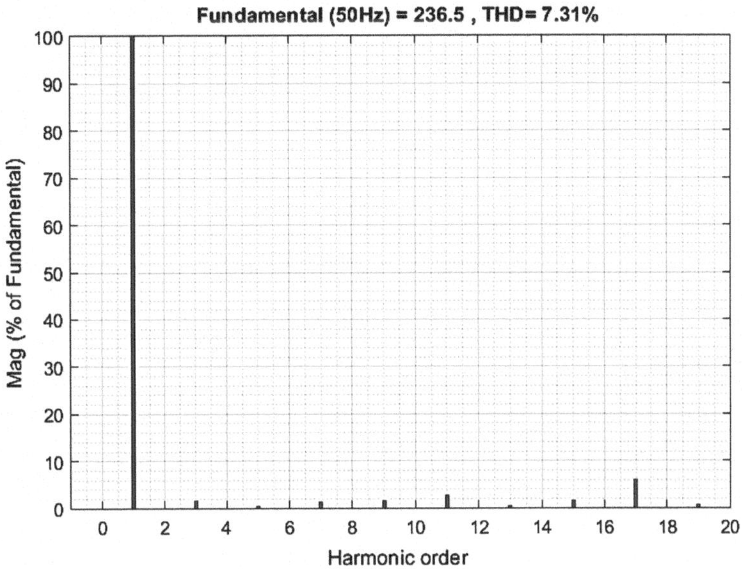

Fig. 16 THD spectrum of 11-level SPWSAT–MCRVMLI

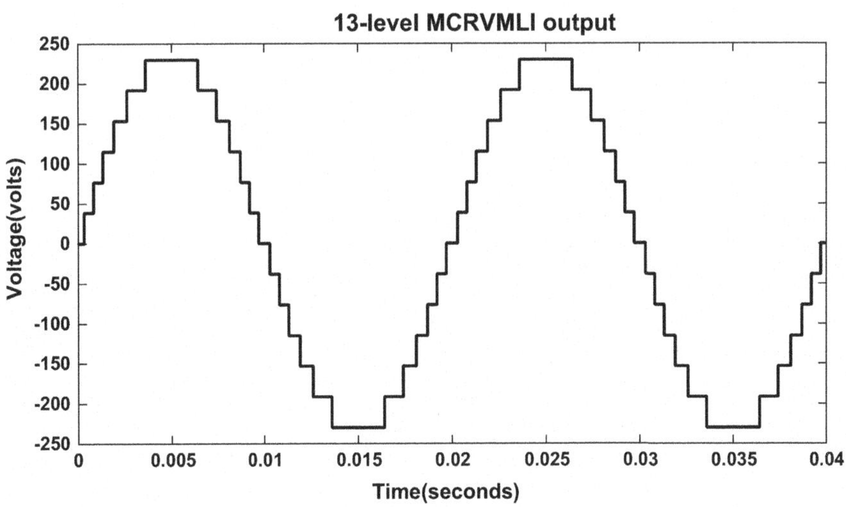

Fig. 17 13-level SPWSAT–MCRVMLI output phase voltage

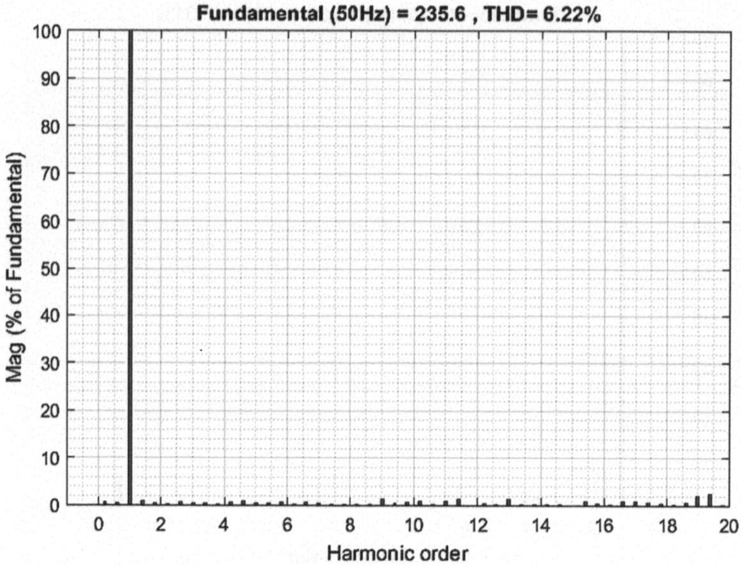

Fig. 18 THD spectrum of 13-level SPWSAT–MCRVMLI

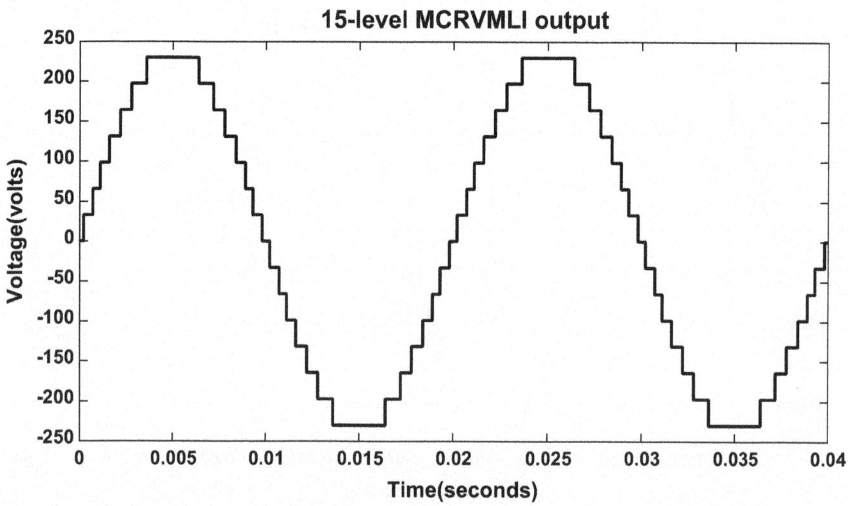

Fig. 19 15-level SPWSAT–MCRVMLI output phase voltage

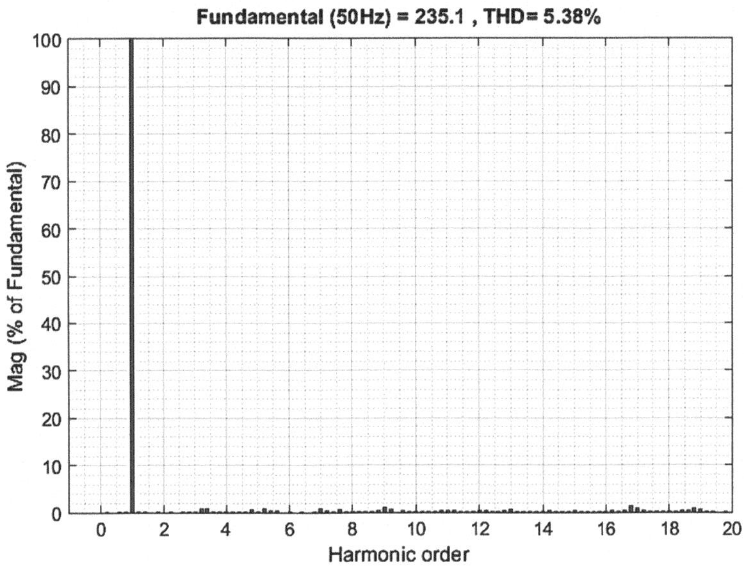

Fig. 20 THD spectrum of 15-level SPWSAT–MCRVMLI

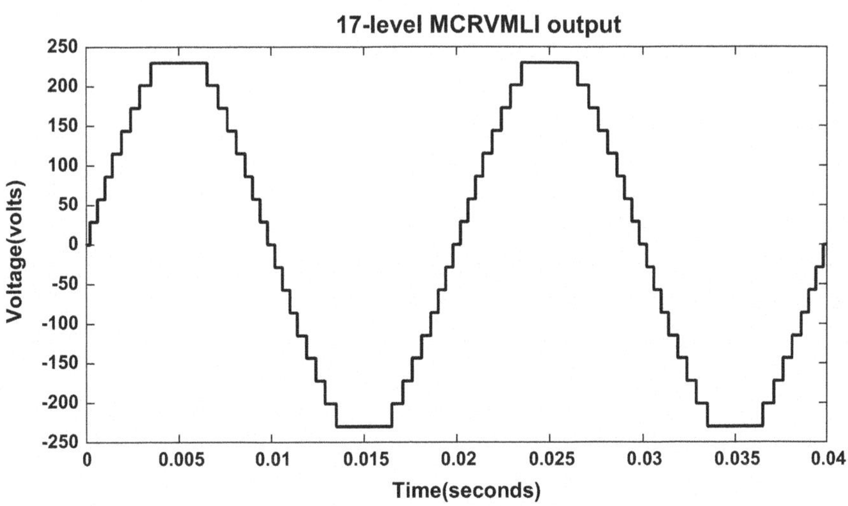

Fig. 21 17-level SPWSAT–MCRVMLI output phase voltage

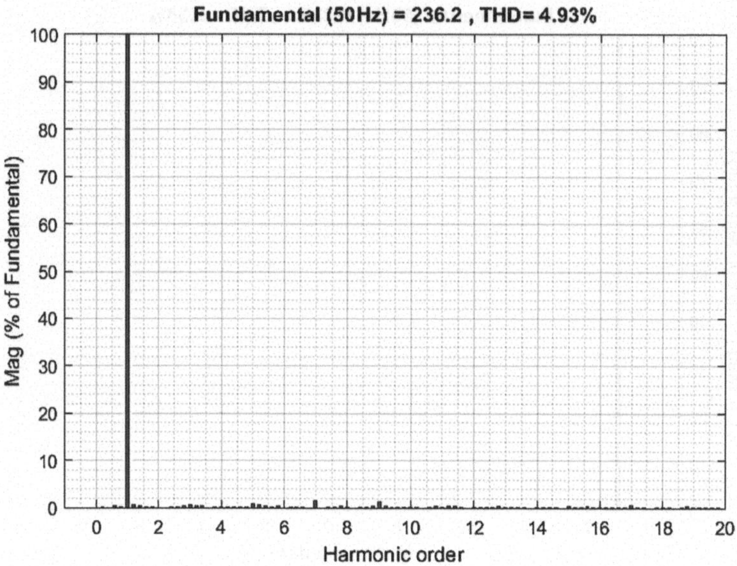

Fig. 22 THD spectrum of 17-level SPWSAT–MCRVMLI

5 Comparison of MLI Topologies with MCRVMLI

The comparison of MCRVMLI topology with the conventional topologies are shown in Table 6. The plot of number of levels against the component count, explains the need for reduced switch count. In case of a 13-level waveform phase voltage, diode clamped, flying capacitor and cascaded topologies require 72, 60 and 54 components respectively, whereas cascaded RVMLI requires 38 components and modified

Table 6 Comparison of components requirement per phase

Topology/ components	Conventional MLIs			RV MLIs	
	DC	FC	Cascaded	Cascaded RVMLI	Modified cascaded RVMLI
Main switch	$2(m-1)$	$2(m-1)$	$2(m-1)$	$(m-1)+4$	$((m-1)/2)+4$
Main diodes	$2(m-1)$	$2(m-1)$	$2(m-1)$	$(m-1)+4$	$((m-1)/2)+4$
Clamping diodes	$(m-2)$	0	0	0	0
Clamping capacitors	0	$(m-2)$	0	0	0
Voltage splitting capacitors	$(m-1)$	0	0	0	0
DC sources	Single	Single	$(m-1)/2$	$(m-1)/2$	$(m-1)/2$
Total components	$6(m-1)$	$5(m-1)$	$9(m-1)/2$	$(5m+11)/2$	$(3m+13)/2$

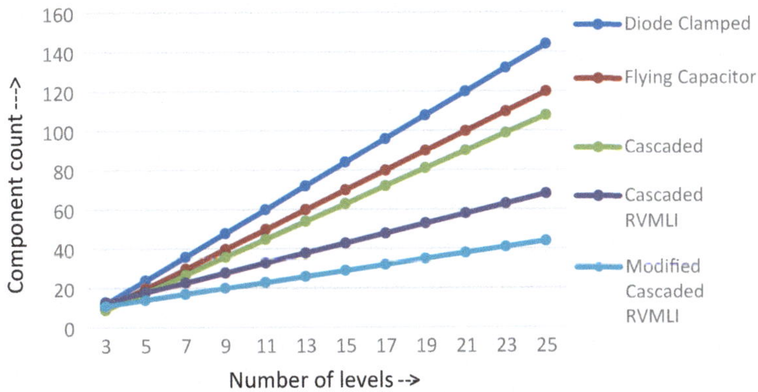

Fig. 23 Number of levels against component count

cascaded RVMLI requires only 26 components. Thus, it is observed that the modified cascaded RVMLI has a smaller number of power switches and component count than the other topologies. Figure 23 shows the number of levels against the component count.

6 Conclusions

In this paper, a single-phase MCRVMLI is proposed to generate 9-level stepped output voltage for photovoltaic applications. The MCRVMLI employs only 8 switches to achieve the 9-level output which results in reduction of size, less installation area and cost-effective. In addition, EPSAT, HEPSAT and SPWSATs are investigated to verify the performance of MCRVMLI in terms of fundamental voltage and THD. The SPWSAT has resulted in a higher output voltage and lower %THD when compared to EPSAT and HEPSAT. Also, the SPWSAT technique involves repetitive trigonometric functions, making it easier to determine the switching angles. Further, the proposed MCRVMLI topology using SPWSAT technique has been analyzed for the higher levels of 11, 13, 15 and 17-levels. The %THD reduces with increase in the number of levels. The proposed SPWSAT-MCRVMLI with reduced switch count and lower %THD may be recommended for industrial and residential applications.

Acknowledgements Authors would like to sincerely thank the Honorable Vice Chancellor and Management of RUAS for providing all facilities required to carry out this research work.

References

1. Rodríguez J, Lai JS, Peng FZ (2002) Multilevel inverters: a survey of topologies, controls, and applications. IEEE Trans Ind Electron 49(4):724–738
2. Colak Ilhami, Kabalci Ersan, Bayindir Ramazan (2011) Review of multilevel voltage source inverter topologies and control schemes. Energy Convers Manag 52(2):1114–1128
3. Thomas RV et al (2015) Identification of optimal SVPWM technique for diode clamped multilevel inverter based induction motor drive. In: 2015 IEEE international conference on electrical, computer and communication technologies (ICECCT). IEEE, New York
4. Shanthi B, Natarajan SP (2009) Comparative study on unipolar multicarrier PWM strategies for five level flying capacitor inverter. In: 2009 International conference on control, automation, communication and energy conservation. IEEE, New York
5. Reddy VNB, Rao SN, Babu CS (2013) Emphasis of modulated techniques for cascaded multilevel inverters fed drive using FPGA. In: 2013 international conference on power, energy and control (ICPEC), pp 686–692. IEEE, New York
6. Rao SN, Kumar DA, Babu CS (2013) New multilevel inverter topology with reduced number of switches using advanced modulation strategies. In: IEEE international conference on power, energy and control (ICPEC). IEEE, New York, pp 693–699
7. Sulake NR, Venkata AKD, Choppavarapu SB (2018) FPGA implementation of a three-level boost converter-fed seven-level DC-link cascaded H-bridge inverter for photovoltaic applications. Electron 7(11)
8. Sulake NR, Pranupa S, Indira MS (2019) Cross connected source based reduced switch count multilevel inverter topology with fault tolerance. In: IEEE global conference for advancement in technology (GCAT). IEEE, New York
9. Sabarad J, Kulkarni GH (2015) Comparative analysis of SVPWM and SPWM techniques for multilevel inverter. In: 2015 international conference on power and advanced control engineering (ICPACE). IEEE, New York
10. Fei Wanmin, Xiaoli Du, Bin Wu (2009) A generalized half-wave symmetry SHE-PWM formulation for multilevel voltage inverters. IEEE Trans Industr Electron 57(9):3030–3038
11. Gupta KK, Jain S (2018) Control of multilevel inverters with reduced device count. In: Modeling and control of power electronics converter system for power quality improvements. Academic Press, Cambridge, pp 121–149
12. Shalini BAP, Sethuraman SS (2012) Cascaded multilevel inverter for industrial applications. In: International conference on advances in information technology and mobile communication. Springer, Berlin, Heidelberg
13. Reddy VNB, Sai Babu C, Nagaraja Rao S (2010) Comparison of modulation techniques for multilevel inverter fed permanent magnet synchronous motor. Int J Eng Sci Technol 2(10):5206–5214
14. Luo FL (2013) Investigation on best switching angles to obtain lowest THD for multilevel DC/AC inverters. In: 2013 IEEE 8th conference on industrial electronics and applications (ICIEA). IEEE, New York
15. Nagaraja Rao S, Ashok Kumar DV, Sai Babu C (2018) Implementation of multilevel boost DC-link cascade based reversing voltage inverter for low THD operation. J Electr Eng Technol 13(4):1528–1538
16. Kang D-W et al (2005) A simple method for acquiring the conducting angle in a multilevel cascaded inverter using step pulse waves. IEE Proc-Electric Power Appl 152(1):103–111

A Novel Discontinuous Phase-Controlled Converter for Suburban/EMU DC Tractions Drives

M. S. Jamil Asghar

Abstract Single-phase AC-to-DC phase-controlled semi-converters are used for DC traction motors. During starting, due to low voltage requirements, the switching angle of converters is high. Thus the line power factor (PF) becomes very low and total-harmonic distortion (THD) increases significantly. This problem is severe for suburban or electric-multiple unit (EMU) trains, which has to stop and accelerate regularly after a short interval. To solve these problems, a novel discontinuous phase-controlled converter circuit is proposed for DC motors of locomotives (motorized coach). This multi-step multi-converter circuit topology is used for a single load (DC motors). At the time of starting, the switching angle is kept at zero degree and the output voltage is controlled in steps to improve PF and to reduce THD. As the voltage requirement increases with the speed pickup, the thyristor pairs are connected at different AC voltage level, switched sequentially with higher voltage and at zero switching angle. Thus, PF is controlled close to unity and THD is maintained to a low level. The experimental results validate the performance of the proposed controller.

Keywords DC drives · Phase controlled converters · Discontinuous phase control (DPC) · Total harmonic distortion (THD) · Power quality (PQ)

1 Introduction

Many traction systems employ AC-DC thyristor based semi converters. Such locomotive with capacities as large as 5 MW and higher are powered from a single-phase supply. At the time of starting, low output DC voltage is required. Thus in an AC-DC semi-converter, total harmic distortion (THD) increases and power factor (PF) becomes low [1]. At lower speed, hence at low speed during the acceleration, the switching angle (α) of the conventional AC-DC semi converter is high (Table 1). Motors draw large line current at poor PF, and THD becomes high. For traction electric locomotives, during starting or acceleration period, a constant torque (or

M. S. Jamil Asghar (✉)
Department of Electrical Engineering, Aligarh Muslim University, Aligarh 202002, India
e-mail: sjasghar@gmail.com

M. N. Favorskaya et al. (eds.), *Innovations in Electrical and Electronic Engineering*, Lecture Notes in Electrical Engineering 661, https://doi.org/10.1007/978-981-15-4692-1_26

Table 1 Performance of a semi-converter for a highly inductive load

α (in deg.)	V_o (in %)	I_1 (in %)	I (in %)	THD (in %)	PF (in %)
0°	100.0	90.03	100	48.34	90.03
22.5°	96.19	88.30	93.54	34.95	92.58
45°	85.35	83.17	86.60	28.98	88.73
67.5°	69.13	74.85	79.05	33.95	78.73
90°	50.00	63.66	70.71	48.34	63.66
112.5°	30.86	50.01	61.23	70.63	45.37
135°	14.64	34.45	50.00	105.16	26.36
157.5°	03.80	17.56	35.35	174.69	09.69

armature current) is maintained [2]. The voltage is gradually increased for a constant acceleration. Thus speed and power input to motor increase linearly. Initially due to low voltage requirement, α is kept high. Moreover, the tendency for wheel slip is lowest when all motors are connected in parallel [2]. However it causes enormously large amount of current drawn by all motors at low voltage, simultaneously at low PF and high THD. If the PF is allowed to fall (during the acceleration period), the voltage drop and transmission line losses will be too high and transmission line and substation equipment may be over loaded. Thus, measures are taken to ensure that the PF does not fall below 0.8 [2]. The circuit, which operates and carries signals is called track circuit and its supply consists of low voltage DC or AC. One of the running rail is used as earth for the main power circuits. Usually, the same running rail is used as the return line for track circuit [2]. As the power and track circuit share the same return conductor (earthed running rail), power circuit harmonics can cause mal-operation of the control signals, which could be dangerous. Further, the high-frequency harmonics in the source or line current during acceleration produce electro-magnetic interference (EMI). These harmonics in source or line current and terminal voltage have a number of undesirable effects such as malfunction of electronic equipment connected to the line, excitation of system resonance, overloading of capacitors, decrease in efficiency owing to increase in losses due to harmonic currents and skin effect, saturation of transformer and EMI with computer and telecommunication lines and equipment [3].

In case of suburban trains this problem is more severe. The distance between consecutive stations is usually very short (few kilometers or less). After each halt, during starting (or acceleration) period, the problems of low PF and high line current harmonics occur. These problems are solved in two steps. First, the static VAR compensation is used to maintain PF above 0.8, and filters are used to reduce the harmonics to the extent that the possibility of mal-operation of signals is completely eliminated and noise in the neighboring telecommunication lines are reduced to a tolerable level. There are several methods available to reduce harmonics using additional filter circuits [4–7]. However, it increases on-board weight and occupy more space. For railway applications, some specific schemes were also proposed [8, 9].

Secondly, two-stage series converters are used [2]. However, the improvement in performance is not significant. For example, a PF close to unity is obtain only near, 50% of the output voltage. While at starting with small DC output voltage, PF remains poor and THD is significantly high. In the proposed paper, these problems are solved by discontinuous phase-controlled switching using novel converter topologies. These problems are solved by using a multi-step converter circuit for single load. A reduced AC input voltage, according to the desire load current, is supplied to converter such that the switching angle is maintained at zero.

2 Conventional Semi-converters

Figure 1 shows the basic speed control scheme for a separately excited DC drive system used in traction. As the power factor of a semi converter is better than a full converter for the same load, therefore single phase transformer based semi-converter configuration is preferred for the traction applications [2]. Figure 2 shows the conventional AC-to-DC semi-converter configurations for an armature circuit of a separately exited DC motor. Figure 3 shows the waveforms of a conventional AC-to-DC semi-converter for a highly inductive ripple free load current.

At the time of starting the current is high and continuous. For continuous conduction of current, the average value of output or armature voltage is given by

$$V_a = \frac{1}{\pi}\int_{\alpha}^{\pi} V_m \sin\omega t(\mathrm{d}\omega t)$$

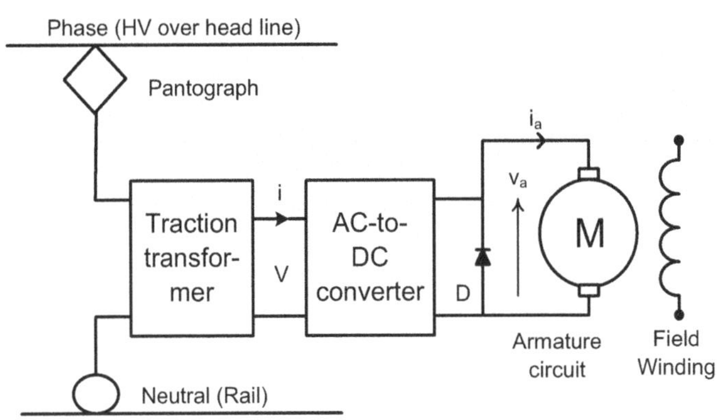

Fig. 1 Basic scheme of a separately exited DC motor based traction system

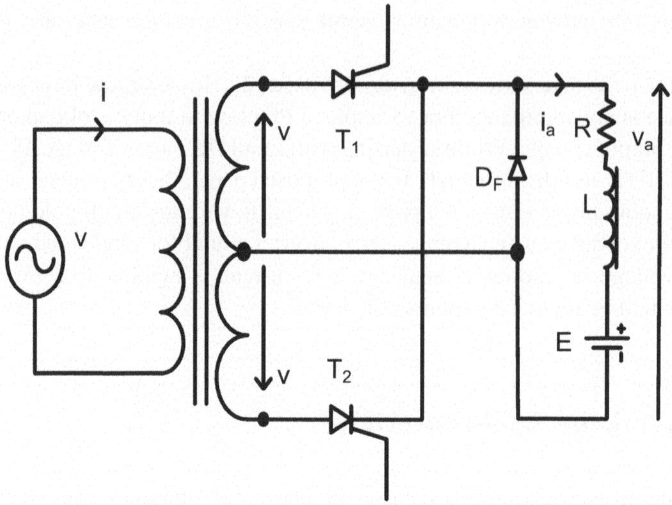

Fig. 2 Basic topology of transformer based semi-converter for separately exited DC motor

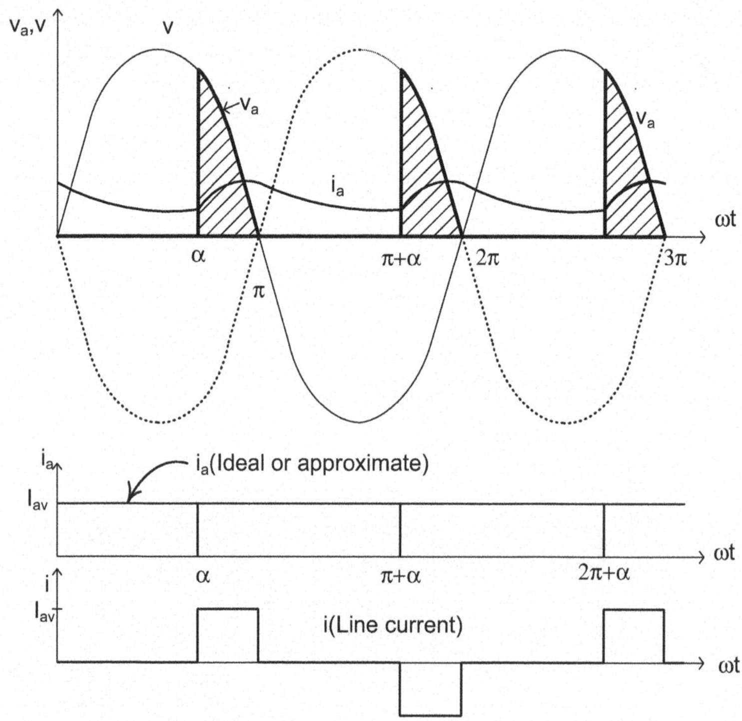

Fig. 3 Waveform of a semi-converter for a highly inductive load current

$$V_o = V_a = \frac{\sqrt{2}V}{\pi}(1 + \cos\alpha) \tag{1}$$

At the time of starting, the armature current is very high. Either thyristors conduct or the free-wheeling diode conducts. The load or armature current due to presence of inductance of armature circuit, becomes almost constant and the semi-converter operates in continuous conduction mode (Fig. 3). For continuous conduction and high magnitude of armature current, it becomes almost ripple free. Thus, for a particular input AC voltage (V), the average values of the load or armature current (I_a), THD the input PF are given by

$$I_a = \frac{V_a - E}{R_a} = \frac{1}{R_a}\left(\frac{\sqrt{2}V}{\pi} + \frac{\sqrt{2}V}{\pi}\cos\alpha - E\right) \tag{2}$$

The input power factor (PF) and the total harmonic distortion (THD) depend upon the input line current (i) at AC side as shown in Fig. 3. During the conduction period of thyristors both i and i_a are same. For a particular input AC voltage (V) the voltage equation is given by,

$$\sqrt{2}V\sin\omega t = L\frac{\mathrm{d}i_a}{\mathrm{d}t} + Ri_a + E \quad \alpha \le \omega t \le \pi \tag{3}$$

where, $\phi = \tan^{-1}(\omega L/R)$, while R, L and E are resistance, inductance and back-emf of the armature circuit, respectively.

It gives the expression of the armature current as [2],

$$i_a = i = \frac{\sqrt{2}V}{Z}\left[\sin(\omega t - \phi) - \sin(\alpha - \phi)\exp\{(\alpha - \omega t)/\tan\phi\}\right] - \frac{E}{R}\left[1 - \exp\{(\alpha - \omega t)/\tan\phi\}\right] \tag{4}$$

where, $E = K\omega_m$, while K and ω_m are torque-constant and angular-speed of the motor, respectively.

The relation of speed and torque (T) is given by

$$\omega_m = \frac{\sqrt{2}V}{\pi K}(1 + \cos\alpha) - \frac{R_a}{K^2}T \tag{5}$$

It is quite difficult to use the expression of i given in (4) to find PF and THD. Thus as shown in Fig. 3, the line current can be considered as constant. Thus, PF and THD can easily be found as [1],

$$\mathrm{PF} = \frac{2(1 + \cos\alpha)}{\sqrt{2\pi(\pi - \alpha)}} \tag{6}$$

$$\mathrm{THD} = \sqrt{\frac{\pi(\pi - a)}{8\cos^2\left(\frac{\alpha}{2}\right)} - 1} = \sqrt{\frac{\pi(\pi - \alpha)}{4(1 + \mathrm{Cos}\alpha)} - 1} \tag{7}$$

It is evident from (1), at very low armature voltage at starting, α is high. The overall performance of converter becomes very poor (Table 1). It is evident from (6) to (7), for high value of α, PF is very low and THD is high. At this condition, for a highly inductive load (ripple-free starting armature current), the line current has almost constant magnitude and therefore, as given in Table 1, THD could be more than 100% and PF less than 50% [1].

3 Discontinuous Phase Control

Here, a multi-step multi-converter circuit is proposed for a single load system (Fig. 4). Different reduced AC voltages are supplied according to the desire output or armature voltage in $\boldsymbol{n}$ steps. The requirement of armature voltage varies which depends upon E or ω_m as given in (2) and (5). As a reduced AC voltage (lower than the rated one) is applied, therefore from (1) for the same voltage, α also reduces. Thus, from (6) and (7), PF becomes high and THD becomes low.

When motor is started from rest, the starting torque is kept in between 150 and 200%, and I_a also varies proportionally between 150 and 200% [2]. According to

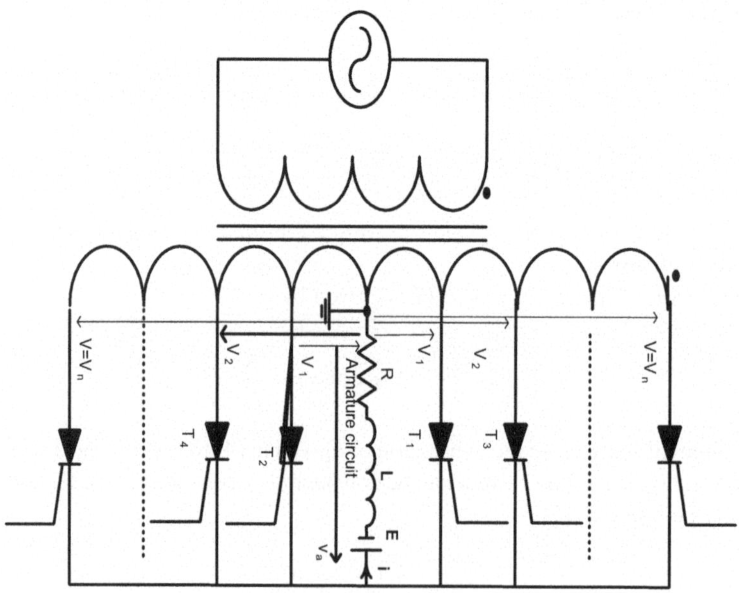

Fig. 4 The proposed discontinuous phase-controlled converter circuit

the armature current requirement at the starting condition with $E = 0$, as given in (2), a reduced voltage V_1 is applied by switching the thyristor pair T_1–T_2 (Fig. 4). Thus at $\alpha = 0$, the performance of the converter improves drastically. Thus, PF becomes high i.e. near 0.9.

Let I_{a1}, is the armature current of the DC motor at the time of starting ($E = 0$) and it is twice of the rated current (I_{ar}). Therefore, starting torque is also twice of the rated value.

For the maximum PF, $\alpha = 0$, then the required AC voltage (V_1) for this armature current is given by

$$I_{a1} = 2I_{\mathrm{ar}} = \frac{1}{R_a}\left(\frac{\sqrt{2}V_1}{\pi}(1 + \cos\alpha) - E\right), \quad I_{a1} = \frac{1}{R_a}\left(\frac{2\sqrt{2}V_1}{\pi}\right)$$

where $V_1 < V$ and $\alpha = 0$.

$$V_1 = \frac{\pi I_{\mathrm{ar}} R_a}{\sqrt{2}} \tag{8}$$

When motor picks up speed, I_a reduces ($I_a < 2I_{\mathrm{ar}}$), which is allowed to reduce up to I_{ar} level. Thus, the starting torque also reduces to the rated value. At this condition, the speed of the motor, $\omega_{m1} = E_1/K$. Thus, for $\alpha = 0$,

$$I_{\mathrm{ar}} = \frac{1}{R_a}\left(\frac{2\sqrt{2}V_1}{\pi} - E_1\right)$$

$$E_1 = \left(\frac{2\sqrt{2}V_1}{\pi} - R_a I_{\mathrm{ar}}\right)$$

And

$$\omega_{m1} = \frac{1}{K}\left(\frac{2\sqrt{2}V_1}{\pi} - R_a I_{\mathrm{ar}}\right) \tag{9}$$

At this instant, the armature current (hence the torque developed) reduces from two times to one (rated value). Therefore, to raise it again such that $I_{a2} = 2I_{\mathrm{ar}}$, input AC voltage is raised to V_2. For this purpose, another pair of thyristors, T_3–T_4 is triggered at $\alpha = 0$. Since T_3 and T_4 are operating at higher voltage ($V_2 > V_1$), they easily commutate $T_1 - T_2$ pair (line commutation). Now the armature current is given by

$$I_{a2} = 2I_{\mathrm{ar}} = \frac{1}{R_a}\left(\frac{2\sqrt{2}V_2}{\pi} - E_1\right) \quad \text{where} \quad V_1 < V_2 < V \tag{10}$$

$$V_2 = \frac{\pi}{2\sqrt{2}}(2I_{\text{ar}}R_a + E_1) = \frac{\pi}{2\sqrt{2}}(2I_{\text{ar}}R_a + K\omega_{m1}) \tag{11}$$

Again the next pair of thyristors (T_5 and T_6) can be triggered at $\alpha = 0$, when motor picks up speed ($\omega_{m2} = E_2/K$) and I_a reduces to I_{ar}. The same procedure can be followed till the motor speed reaches to the desired level or up to the rated voltage, V. Thus PF remains close to 0.9 (approximately) and THD below 48%.

4 Experimental Results

A 230 V, 2A, ½ hp, 1500 rpm DC motor is used for the test. The full-load torque is obtained at the armature current = 2.0 A. At 150% load torque (3A), the motor is started. Figure 5 shows the performance of conventional semi-converter used for starting of the separately exited DC motor with 150% torque. The switching angle and THD of the conventional converter at rated voltage are $\alpha = 126°$ and 77.0% (Fig. 5). The expression of PF in the generalized form, for any converter with any type of load is given by [1],

$$\text{PF} = \text{Distortion Factor} \times \text{Fundamental PF}$$

$$\text{PF}=\frac{1}{\sqrt{1+\text{THD}^2}}\cos\phi_1$$

For a conventional semi-converter, with $\alpha = 126°$ and THD = 77%, PF = $\frac{1}{\sqrt{1+\text{THD}^2}}\cos(-\alpha/2) = \frac{1}{\sqrt{1+0.77^2}}\cos 54° = 0.359$

While for the same operating condition, the switching angle and THD of the proposed discontinuous phase-controlled converter are $\alpha = 0°$and THD = 37.3% (Fig. 6). Thus for this case, PF is given by

$$\text{PF} = \frac{1}{\sqrt{1+\text{THD}^2}}\cos(-\alpha/2) = \frac{1}{\sqrt{1+0.373^2}}\cos 0° = 0.937$$

Thus, in case of the proposed converter, PF improves and THD reduces.

5 Conclusions

In this work a novel AC-to-DC phase-controlled converter topology is proposed to improve power quality of a semi-converter and for better performance of DC drive exclusively used for EMU traction applications. At the time of starting condition,

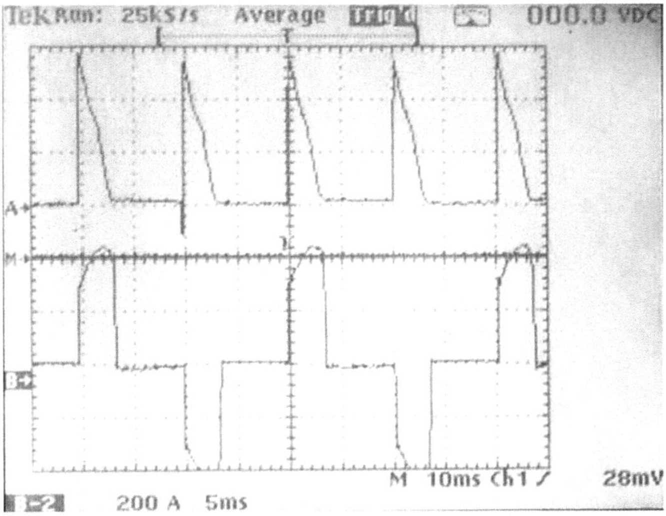

(a)

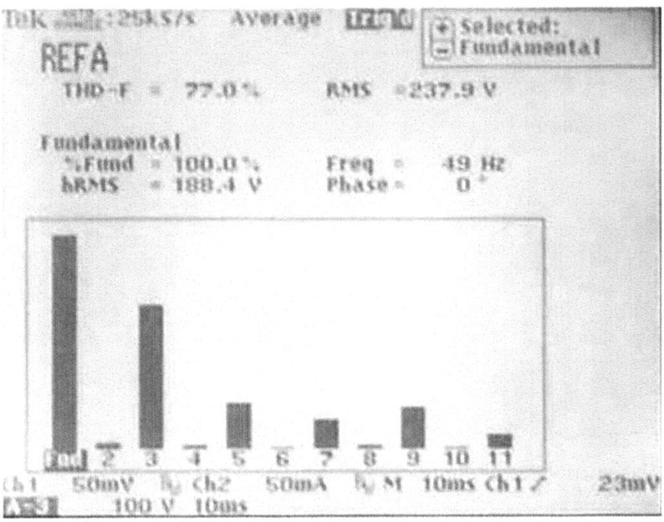

(b)

Fig. 5 Load voltage, line current (**a**), and its harmonic spectrum of conventional semi-converter at starting condition with armature circuit load ($I_a = 3\text{A}$) (**b**)

with the help of the proposed converter, PF is increased and THD is reduced drastically. Here, even the PF becomes 0.937, higher than 0.9 which is the ceiling of a conventional converter. Here THD is also below 48%.

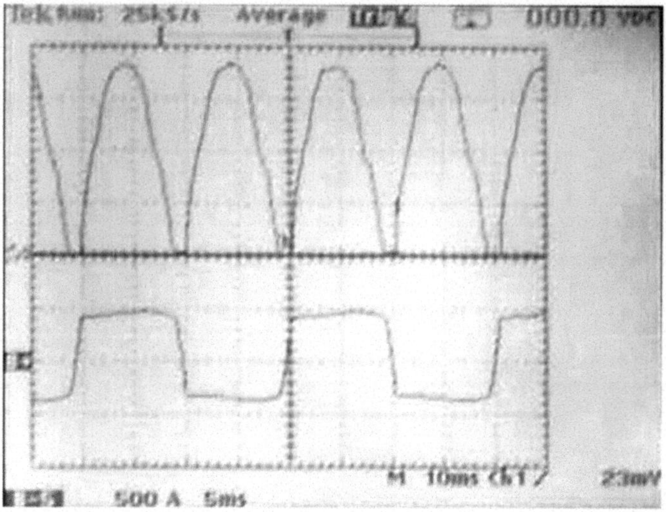

(a)

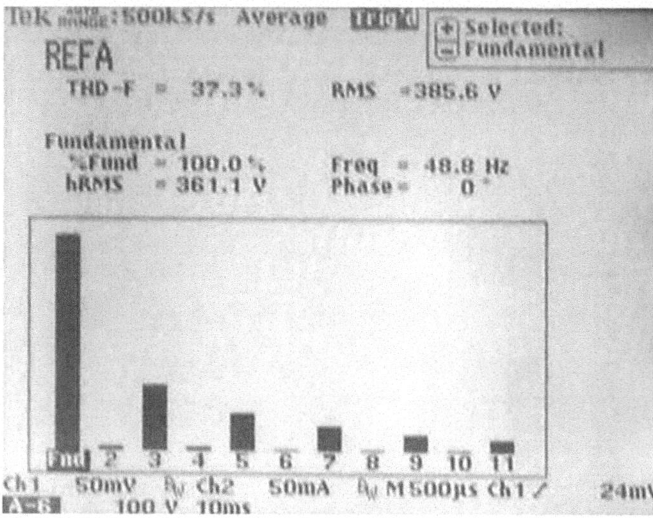

(b)

Fig. 6 Load voltage, line current (**a**), and its harmonic spectrum of the proposed converter at starting condition with armature circuit load ($I_a = 3\text{A}$) (**b**)

Acknowledgements The work has been done with the support of fund under SAP program of the University Grants Commission (UGC), Government of India.

References

1. Jamil Asghar MS (2004) Power electronics. Prentice-Hall of India Pvt. Ltd., New Delhi
2. Dubey GK (1995) Fundamental of electrical drives, 2nd edn. Narosa Publishing House, New Delhi
3. Jamil Asghar MS (1996) Elimination of inrush current of transformers and distribution lines. In: IEEE international conference on power electronics, drives and energy systems (PEDES'96), pp 976–980
4. Dixon J, Morán L, Rodríguez J, Domke R (2005) Reactive power compensation technologies: state-of-the-art review. Proc IEEE 93(12):2144–2164
5. Shawon MH, Hanzelka Z, Dziadecki A (2015) Voltage-current and harmonic characteristic analysis of different FC-TCR Based SVC. In: 2015 IEEE power tech conference, pp 1–6
6. Araya PM, Castro JM, Nolasco JC, Palma-Behenke RE (2011) Lab-scale TCR-based SVC system for educational and DG applications. IEEE Trans Power Syst 26(1):3–10
7. Rahmani HS, Al-Haddad K (2010) A hybrid passive filter configuration for VAR control and harmonic compensation. IEEE Trans Ind Electron 57(7):2419–2434
8. Roudsari HM, Jalilian A, Jamali S (2018) Flexible fractional compensating mode for railway static power conditioner in a V/v transition power supply. IEEE Trans Ind Electron 65(10):2963–2974
9. Chen M, Chen Y, Wei M (2019) Modeling and control of hybrid power quality compensation system for 25 kV electrified railway. Energies 12(17):3033. http://doi.org/10.3390/en121/3033

Modified Isolated Triple-Active Bridge Bidirectional DC–DC Converter for Energy Storage Application

Chandra Sekhar Nalamati and Rajesh Gupta

Abstract The growing applications of environmental friendly green energy sources, hybrid electric vehicles, smart electrical energy grid, and energy storage have motivated the research world toward the bidirectional DC–DC converters. Among them, dual-active bridge (DAB) and triple-active bridge (TAB) are popular for smooth bidirectional high-power transferring capability. Both DAB and TAB operating with single-phase shift control (SPSC) suffer with the large circulating current and reduced soft switching operation. In this paper, a modified isolated triple-active bridge (MITAB) bidirectional DC–DC converter for high-power storage application has been proposed to improve the TAB converter performance by adding two extra switches and corresponding operation. The converter's two output ports are connected in parallel to improve the power handing capability. Proposed converter operation with SPSC has been presented and also compared with the conventional TAB. The simulation results using PSCAD has been presented to verify the proposed converter operation.

Keywords Triple-active bridge (TAB) · Bidirectional converters · Battery energy storage · Phase shift modulation

C. S. Nalamati (✉) · R. Gupta
Electrical Engineering Department, M. N. National Institute of Technology Allahabad, Prayagraj, India
e-mail: chandu3072002@gmail.com

R. Gupta
e-mail: rajeshgupta@mnnit.ac.in

M. N. Favorskaya et al. (eds.), *Innovations in Electrical and Electronic Engineering*, Lecture Notes in Electrical Engineering 661, https://doi.org/10.1007/978-981-15-4692-1_27

1 Introduction

The ever-increasing power generation–demand gap, growing dependency on environmental friendly green energy sources, hybrid electric vehicles, smart electrical energy grid, and energy storage have encouraged the world research community toward the power electronics-based efficient bidirectional DC–DC and DC–AC power flow controllers [1, 2]. The large voltage difference in the electrical energy storage devices and the available DC bus have made isolated bidirectional DC–DC converters popular, and mostly the DAB based DC–DC converters, due to their smooth bidirectional power transfer capability with galvanic isolation. But these converters also suffer with the high peak current stress and large circulating current with the basic SPSC [3, 4]. Various control algorithms with increased degree of freedom has been proposed in the literature to improve its performance. The DAB converters are useful in the single input–single output applications [5–7].

The necessity of integration and power flow control among various available energy resources, loads, and various energy storage devices has made the multi-port converters a viable solution [8, 9]. In the literature, multi-port converters with unidirectional as well as bidirectional capability has been presented. The multi-port converters with bidirectional power flow capability have been very useful in the high-power energy storage applications. A galvanically isolated three-port bidirectional DC–DC converter called as triple-active bridge (TAB) has been presented in the literature that can exchange power among the available three ports with source, loads, and energy storage devices [10–12]. The limitations of basic SPSC with TAB converter such as circulating current, limited soft switching range, and high peak current stress have been minimized with the new control algorithms in the literature [13, 14]. However, the existing modulation method minimizes the circulating power either at the high-voltage input side or low-voltage energy storage side. The TAB converter topology, one-port with full-bridge inverter and other port with diode rectifier, has capability to reduce the circulating power at the low-voltage port. However, this topology works with unidirectional power flow only. The TAB converter with three-level pulse width modulation control methods has the capability to minimize the circulating power but with lesser power transfer capability [15–17].

The performance of the conventional TAB converter can be improved with the proposed modified isolated TAB (MITAB) bidirectional DC–DC converter with minimization of the circulating power at the high-voltage port side and also at the low-voltage ports and corresponding peak current stress. The converter operation with SPSC method waveforms has been presented.

2 Proposed Modified Isolated Triple-Active Bridge Bidirectional DC–DC Converter Topology

2.1 MITAB Converter Topology

Figure 1 depicts the applications of the conventional TAB converter with various power generation sources such as solar PV, wind, and fuel cell, and with the large energy storage bank. The galvanically isolated TAB converters help in efficient usage of the available resources, loads, and the storage bank. The TAB converter acts as an efficient power electronic interface between the DC micro-grid and the low-voltage distributed energy storage DC bus.

The conventional TAB bidirectional DC-DC converter has been shown in Fig. 2. It consists of three ports with three power electronic semiconductor switches based full-bridge inverters having three-winding high-frequency transformer for interfacing and providing isolation among the three different sections of source, load, and energy storage bank, or combination of the either depending on the application. Figure 2 has one high-voltage side port for connecting high-voltage DC bus and two low-voltage ports to connect two low-voltage energy storage sections or loads or one storage section with parallel connection of two low-voltage ports. The three-winding high-frequency transformer provides electrical galvanic isolation and also matches the high voltage–low voltage ports with the help of its turn ratio. The limitations of the TAB converter with the SPSC method have been discussed in the following section in detail.

Figure 2b shows the proposed MITAB bidirectional DC–DC converter topology circuit diagram having two more controlled power electronic semiconductor switches (H_X and B_X) one at the high-voltage port side and second one at the low-voltage port end [6]. In order to improve the power handing capability and reduce the device stress of the low-voltage side inverter section, the low-voltage side two inverter sections are connected in parallel, and large battery storage section has been connected to it.

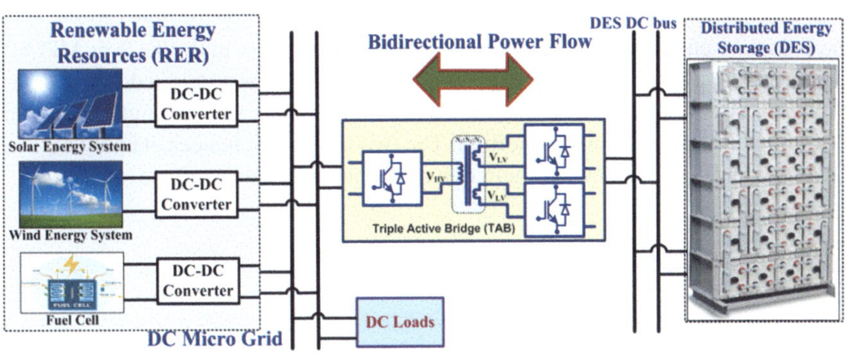

Fig. 1 Generalized diagram of application of triple-active bridge converter for interfacing distributed energy storage system with renewable energy sources

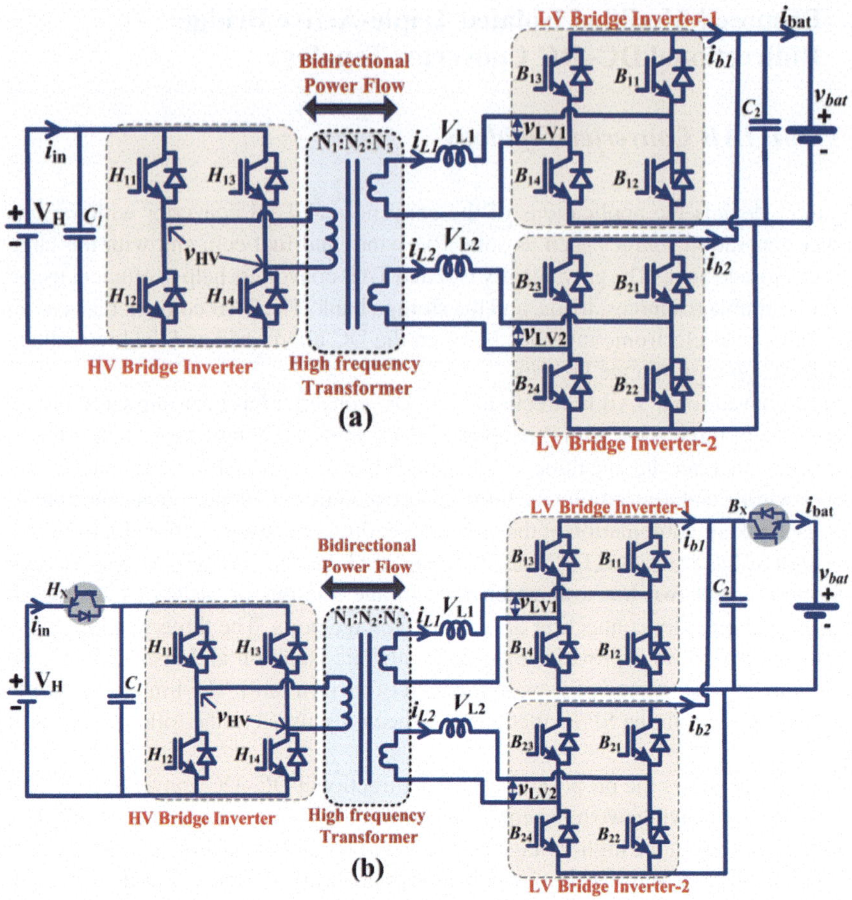

Fig. 2 Converter topologies of conventional triple-active bridge converter (**a**), and proposed isolated triple-active bridge bidirectional converter (**b**)

The inductors on the low-voltage side help to transfer power in the TAB or MITAB converter. The converter having two sources V_H and V_{bat} are supplied at the high-voltage and low-voltage sides respectively. The battery acts as both source and load depending on the operational condition. The two filter capacitances, namely C_1 and C_2, at two ends of the converter help to have ripple-free stable voltage.

2.2 Principle of Operation and Control

The proposed MITAB converter has capability to operate in both the forward and reverse power flow operation. Figure 3 depicts the SPSC waveforms of the conventional TAB converter in Fig. 3a and proposed MITAB bidirectional DC–DC converter in Fig. 3b. In forward power flow operation, full-bridge inverter output voltage v_{HV} in the high-frequency transformer's primary side leads the low-voltage port voltage v_{LV} with desired phase angle shift as shown in Fig. 3. However, in the reverse power flow operation, the voltage v_{LV} leads the voltage v_{HV}, and power flows from low-voltage battery port to the high-voltage port.

In the conventional TAB converter with SPSC operation (Fig. 3a), v_{HV} and v_{LV} are two-level high-frequency square waveforms, and both the output battery current i_{bat} and the input current i_{in} suffer with the large negative current portion called circulating current that degrades the converter performance and also result in high peak current stress. In the proposed MITAB converter, for the forward operation (Fig. 3b), v_{HV} is two-level high-frequency square waveform, whereas v_{LV} is the three-level high-frequency waveform. Both the current waveforms i_{bat} and i_{in} are free from the negative current portion.

The elimination of the circulating current helps in enhanced converter performance with reduced losses. In forward mode, the circulating current entering from the battery has been restricted by the switches of B_{13} and B_{11} in the LV bridge inverter-1 and by the switches of B_{23} and B_{21} in the LV bridge inverter-2, and is turned off. The diode of the switch H_X restricts the current flow into the high-voltage source. In the reverse power flow mode, the upper switches in the HV bridge inverter are disabled, and diode of H_X conducts and helps to achieve elimination of the circulating current in the reverse operation. Hence, the proposed MITAB converter eliminates the circulating current in both forward and reverse power flow operation.

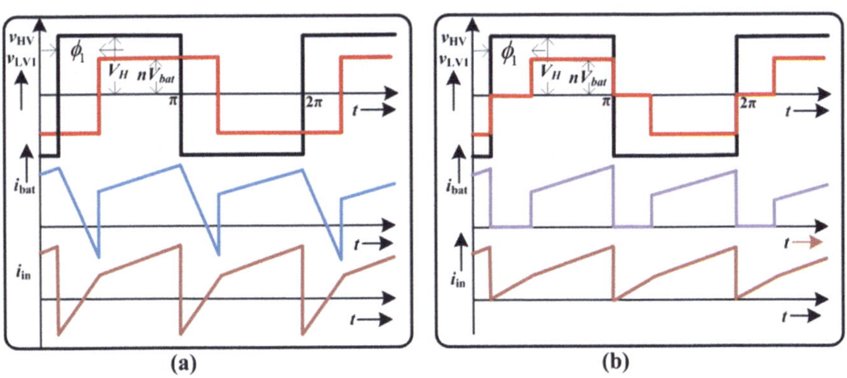

Fig. 3 SPSC method waveforms for conventional triple-active bridge converter (**a**), and proposed MITAB bidirectional converter (**b**)

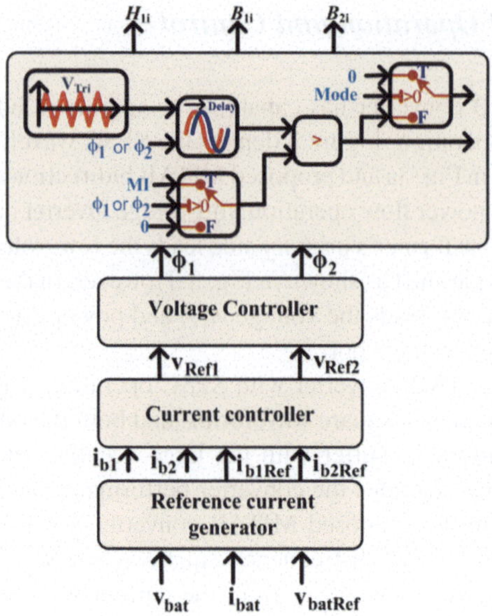

Fig. 4 MITAB converter control strategy

2.3 Converter Control Strategy

In the proposed MITAB converter, the controller design plays a vital role to achieve desired circulating current elimination operation. Figure 4 depicts the control strategy for the proposed MITAB converter with the SPSC method. The controller logic with the measurements of the actual converter parameters and reference parameter values generates the desired phase angle shift and disables the corresponding switches in the converter to achieve the zero circulating current operation. By implementing the operation discussed in the previous section, both the forward mode as well as reverse mode of operation has been achieved.

3 Simulation Results

The proposed converter operation with single-phase shift control in the forward mode of operation has been verified through PSCAD/EMDTC software-based simulations. The conventional TAB converter is also simulated with the SPSC method and compared the performance of the TAB converter with the proposed MITAB converter. The simulation parameters are given in Table 1.

Table 1 MITAB converter simulation parameters

Parameter	Numerical value
MITAB inductance (2No.)	10 μH
Switching frequency (f_{sw})	20 kHZ
Nominal battery voltage and capacity	120 V, 650 Ah
MITAB transformer rating	100 kVA
Transformer turn ratio	25:1:1

Figure 5 shows the simulation results of the TAB bidirectional DC–DC converter with SPSC technique. The two-level voltage square waveforms and inductor current waveform ($v_{HV/n}$, v_{LV}, v_L, i_L) depicted in Fig. 5a, show that the HV bridge leads the LV bridges by a phase angle shift of 90°, and the power is transferred from the HV side to the LV side. Figure 5b depicts the LV bridge inverters output current waveforms and converter input current waveform (i_{b1}, i_{b2}, i_{bat}, i_{in}). It is observed that even though SPSC method is simple, it suffers with large circulating current in the both the output battery current i_{bat} and the input current i_{in}. The circulating current in the output port affects the battery life. The currents i_{b1} and i_{b2} are the LV bridge inverters output currents and i_{bat} is the sum of the inverter output currents.

The performance of the proposed MITAB bidirectional converter with SPSC technique is shown in Fig. 6. As the proposed control logic switch off the upper switches of the LV bridge inverters, the HV inverter output voltage is of two level, whereas LV bridge inverter output becomes three level (Fig. 6a). The shape of the inductor current also changes. Figure 6b depicts the LV bridge inverters output current waveforms and converter input current waveform (i_{b1}, i_{b2}, i_{bat}, i_{in}). It is observed from the magnitude that the peak current is also reduced. Both ends of the proposed converter produces zero circulating in input as well as at output ends.

4 Conclusions

The single-phase shift operation of the dual-active bridge and triple-active bridge results in the high circulating current in both the input side and output side. It also has decreased soft switching range. The proposed MITAB bidirectional DC–DC converter eliminates the circulating current in both the ends and peak current stress also reduced. These benefits of the proposed converter result in better performance compared with the TAB converter with improved power transmission capability.

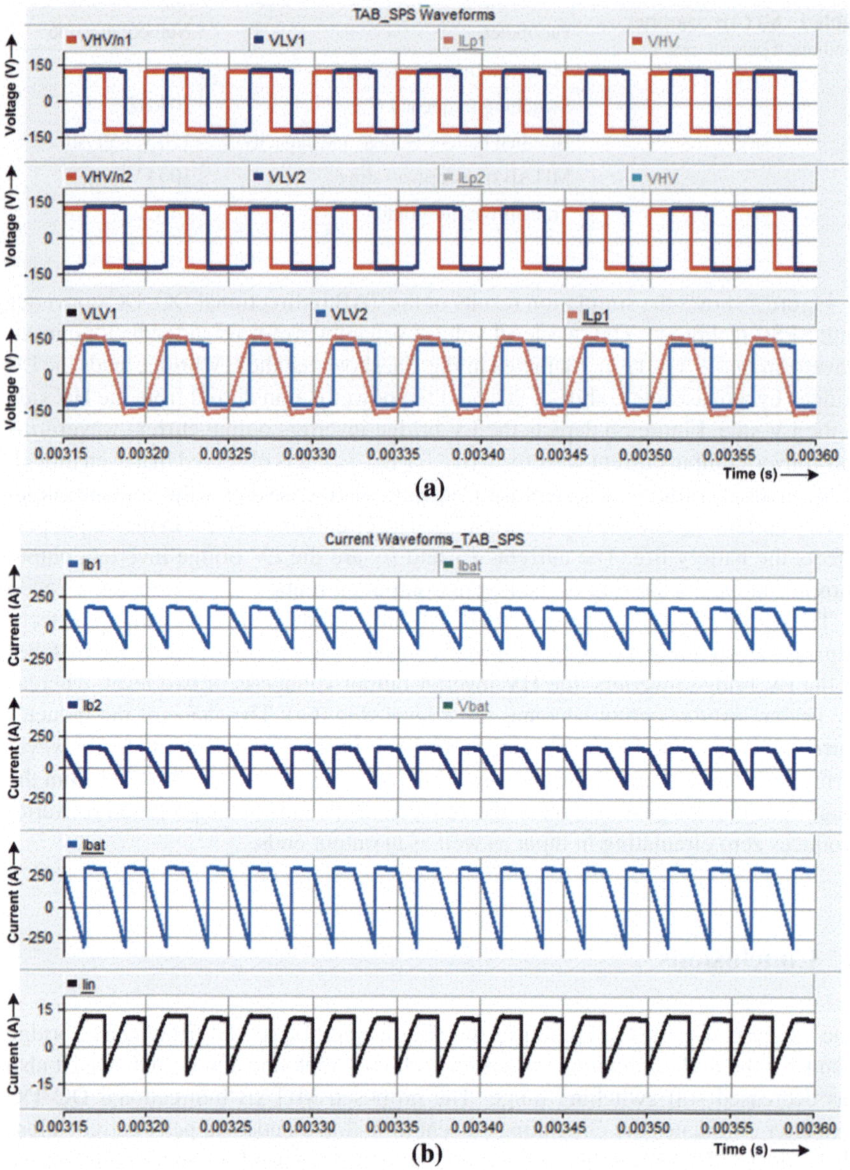

Fig. 5 SPSC method waveforms of conventional TAB converter voltage waveforms ($v_{HV/n}$, v_{LV}, v_{L}, i_{L}) (**a**), and converter current waveforms (i_{b1}, i_{b2}, i_{bat}, i_{in}) (**b**)

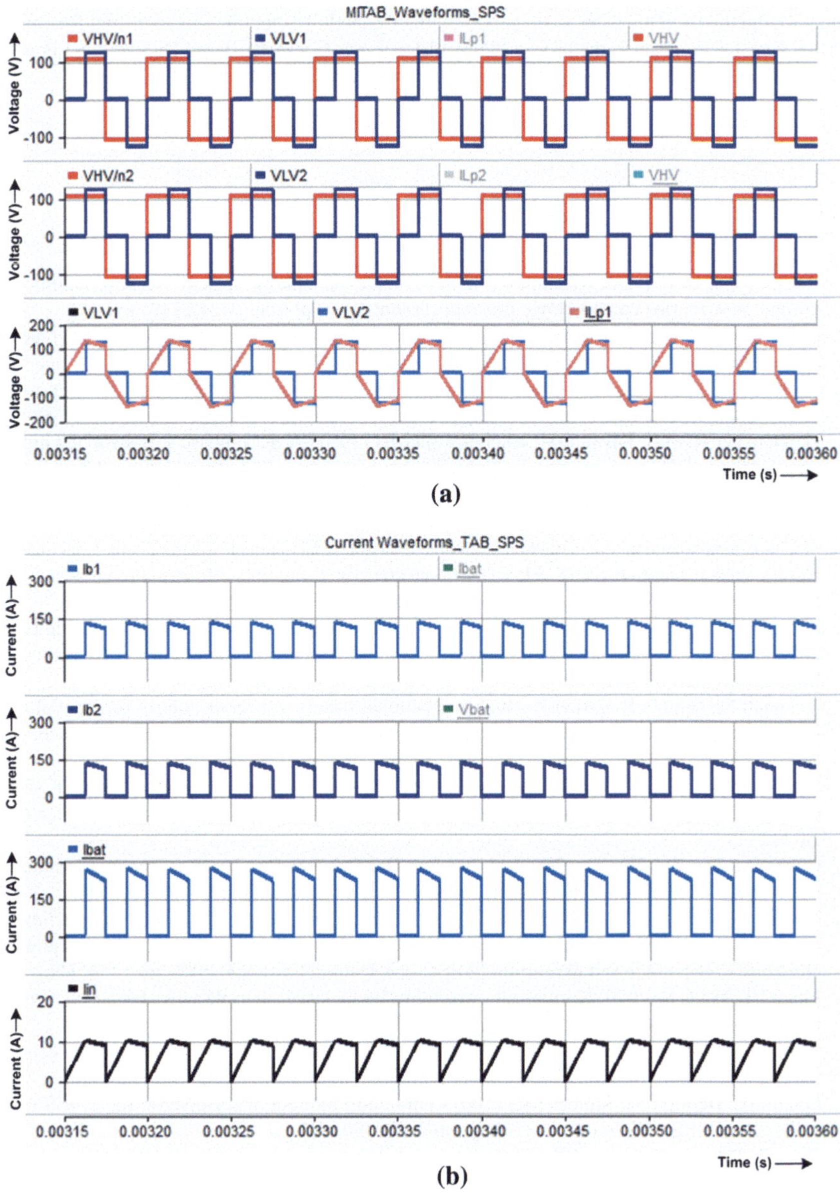

Fig. 6 SPS control of the proposed MITAB bridge bidirectional converter, voltage waveforms ($v_{HV/n}$, v_{LV}, v_L, i_L) (**a**), and converter current waveforms (i_{b1}, i_{b2}, i_{bat}, i_{in}) (**b**)

Acknowledgements This work is supported by the Council of Scientific and Industrial Research (CSIR), New Delhi, through the project "Power Electronics Converter Based Bidirectional Power Flow Controller for Energy Storage Applications," Sanction No. 22(0691)/15/EMR-II.

References

1. Herran DS, Tachiiri K, Matsumoto K (2019) Global energy system transformations in mitigation scenarios considering climate uncertainties. Appl Energy 243:119–131
2. Koohi-Fayegh S, Rosen MA (2020) A review of energy storage types, applications and recent developments. J Energy Storage 27:101047
3. Kroposki B, Pink C, DeBlasio R, Thomas H, Simões M, Sen PK (2010) Benefits of power electronic interfaces for distributed energy sources. IEEE Trans Ind Electron 25(3):901–908
4. De Doncker RWAA, Divan DM, Kheraluwala MH (1991) A three-phase soft-switched high-power-density DC/DC converter for high-power applications. IEEE Trans Ind Appl 27(1):63–73
5. Kheraluwala MH, Gascoigne R, Divan DM, Baumann E (1992) Performance characterization of a high-power dual active bridge DC-to-DC converter. IEEE Trans Ind Appl 28(6):1294–1301
6. Karthikeyan V, Gupta R (2018) FRS-DAB converter for elimination of circulation power flow at input and output ends. IEEE Trans Ind Electron 65(3):2135–2144
7. Zhao B, Song Q, Liu W, Sun Y (2014) Overview of dual-active-bridge isolated bidirectional DC–DC converter for high-frequency-link power-conversion system. IEEE Trans Power Electron 29(8):4091–4106
8. Nalamati CS, Gupta R (2018) Isolated bidirectional battery converter control for standalone solar PV applications. In: 2018 IEEE IEEMA international engineer infinite conference, pp 1–5
9. Karthikeyan V, Gupta R (2018) Multiple-input configuration of isolated bidirectional DC-DC converter for power flow control in combinational battery storage. IEEE Trans Ind Informat 14(1):2–11
10. Chen YM, Liu YC, Wu FY (2002) Multi-input DC/DC converter based on the multiwinding transformer for renewable energy applications. IEEE Trans Ind Appl 38(4):1096–1104
11. Tao H, Kotsopoulos A, Duarte JL, Hendrix MAM (2006) Triple-half-bridge bidirectional converter controlled by phase shift and PWM. In: 2006 IEEE 21st annual IEEE applied power electronics conference and exposition (APEC'06), Dallas, TX, USA, 2006, pp 1256–1262
12. Pham VL, Wada K (2020) Normalization design of inductances in triple active bridge converter for household renewable energy system. IEEJ J Ind Appl 9(3):227–234
13. Yu Y, Masumoto K, Wada K, Kado Y (2018) A DC power distribution system in a data center using a triple active bridge DC-DC converter. IEEJ J Ind Appl 7(3):202–209
14. Duarte JL, Hendrix M, Simoes MG (2007) Three-port bidirectional converter for hybrid fuel cell systems. IEEE Trans on Power Electron 22(2):480–487
15. Zhao C, Round SD, Kolar JW (2008) An isolated three-port bidirectional DC-DC converter with decoupled power flow management. IEEE Trans Power Electron 23(5):2443–2453
16. Nishimoto K, Kado Y, Wada K (2018) Implementation of decoupling power flow control system in triple active bridge converter rate 400 V, 10 kW, and 20 kHz. IEEJ Jour of Ind Appl 7(5):410–415
17. Yu Y, Wada K (2018) Simulation study of power management for a highly reliable distribution system using a triple active bridge converter in a DC microgrid. Energies 11(11):3178

A Novel Approach for Real-Time Drowsiness Detection and Alert to Driver

Shivanand Pandey, Ritika Bharti, Prachi Verma, and Tauseef Ahmad

Abstract Drowsiness is the situation just before sleep. It could be due to lack of sleep, continuous long working hours, time of day and physical and mental state. It limits the ability to concentrate while driving. It leads to additional symptoms, such as forgetfulness or falling asleep at inappropriate times. Drowsiness at the wheel has been a severe problem which can be controlled using drowsiness detection system. This project is based on drowsiness detection using behavioral measures which include physical traits of human body such as facial expressions and head movement. This paper proposes an efficient and accurate system to detect driver's drowsiness by using three effective algorithms simultaneously, i.e., eye blinking, PERCLOS and head tilt techniques. A final triggering variable is obtained as resultant of contributing three algorithms. Each algorithm is set to produce the value of an intermediate variable up to a defined value, i.e., threshold value. These intermediate variables affect the value of final triggering variable to reach its threshold value. If the final variable reaches its threshold, it can be used to generate any type of alert.

Keywords Drowsiness · Eye blinking · EAR · PERCLOS · Haar features · Viola–Jones algorithm · Dlib · OpenCV

S. Pandey · R. Bharti · P. Verma · T. Ahmad (✉)
Department of Information Technology, Rajkiya Engineering College, Azamgarh, India
e-mail: tauseefahmad@zhcet.ac.in

S. Pandey
e-mail: codesp7784@gmail.com

R. Bharti
e-mail: ritikabharti9415@gmail.com

P. Verma
e-mail: prachiverma1144@gmail.com

M. N. Favorskaya et al. (eds.), *Innovations in Electrical and Electronic Engineering*, Lecture Notes in Electrical Engineering 661, https://doi.org/10.1007/978-981-15-4692-1_28

1 Introduction

It is wisely said, "Journey is as important as destination" [1]. One safer drive can make a huge difference. The number of traffic accidents is increasing day by day which has become a serious problem. Car accident is the major cause of death in which around 1.3 million people die every year. Majority of these accidents are caused because of distraction or the drowsiness of driver.

Car driving is a complex and potentially a risk activity. Alertness and awareness are necessary to maintain while driving. Sleep deprivation causes drowsiness, fatigue and slow reaction time which affects driver's performance. It can be harmful for health and can result in road accidents. Drowsiness is the situation just before sleep. It could be due to lack of sleep, continuous long working hours, time of day and physical and mental state. It limits the ability to concentrate while driving. It leads to additional symptoms, such as forgetfulness or falling asleep at inappropriate times. No one can tell when the person will feel sleepy. Drowsiness at the wheel can result in accidents that can cause severe harm to human life, natural resources and physical assets. Since there is no scale of drowsiness, it is a very big challenge to measure and detect it.

To avoid mishappenings due to drowsiness, it is necessary to detect it in time. Drowsiness detection is a technology which helps to prevent accidents caused by driver getting drowsy. It is necessary to come with an efficient technique to detect drowsiness with high accuracy as soon as driver feels sleepy. This can prevent large number of accidents to occur. Drowsiness detection system will alert the driver and prevent road accidents. We can use a number of different techniques for analyzing driver's drowsiness.

Drowsiness can be detected by recognizing the facial expressions and physical movements. Face detection can be done by using different algorithms such as Viola–Jones algorithm. This algorithm is an object detection framework to provide competitive object detection rates in real time. It can be used to detect variety of objects. This algorithm looks for specific Haar feature to detect face in an image.

In the proposed system, three intermediate variables and one final variable are used to determine the state of triggering the alarm. Three intermediate variables correspond to three different techniques, and the final variable is resultant of these entire three variables.

Some of very efficient drowsiness detection techniques are as follows (Fig. 1).

1.1 Eye Blinking-Based Technique

This technique monitors eye blinking rate of the driver as this parameter is different in drowsy state than in normal state. Thus, drowsiness can be detected.

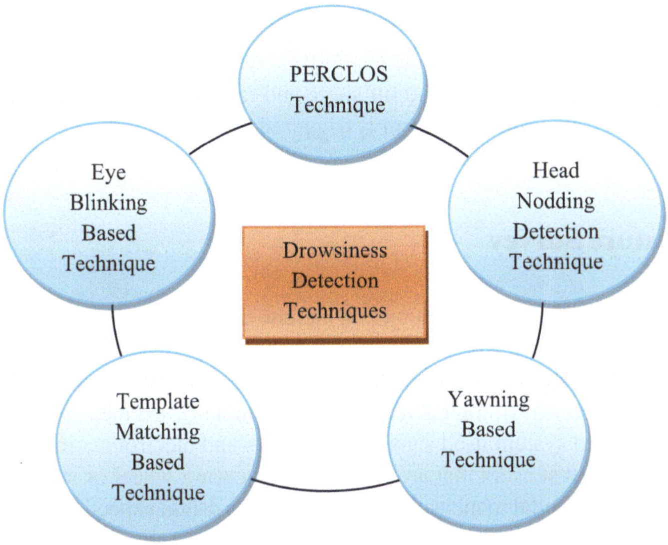

Fig. 1 Techniques of drowsiness detection

1.2 Template Matching Technique

This method uses pre-trained templates of open and closed eyes. If the driver closes his/her eyes for some particular time, then the system will generate the alarm.

1.3 PERCLOS Technique

The PERCLOS technique measures the percentage of time that an eye is closed in a given period. When the PERCLOS score crosses threshold mark, an alarm will be generated.

1.4 Yawning-Based Technique

This technique detects yawning based on opening rate of mouth and amount changes in mouth contour area.

1.5 Head Nodding Detection Technique

This technique determines the head tilt angle and rate. When the head angle and tilt rate go beyond certain limit, an alarm is triggered.

2 Literature Survey

Recent researches on drowsiness detection include different methodologies and algorithms that led to significant results. Researchers have been trying to prepare different detectors [2] and proposed three measures to develop the detection systems, i.e., behavioral measure, physiological measure and vehicle-based measures. Physiological measure-based system used sensing electrodes and electronic circuits. Vehicle-based measure used sensors that are installed on steering wheel or steering column and on acceleration pedal to measure the variability in driving speed. The researchers have considered only small movement of steering wheel that is 0.5°–5° which is needed to adjust the vehicle position within the lane [3]. Behavioral approaches can be used to determine drowsiness level by measuring driver's abnormal behaviors [3]. The authors have suggested developing a system to analyze images of face captured by video camera using image processing technology. According to [4], drowsiness can be detected by checking visual changes in eye locations using the proposed horizontal symmetry of eyes. This method detects eye blinking via standard webcam in real time at 110 fps for a 320 × 240 resolution. It detects the eye blink with 94% accuracy with a 1% false rate. In driver fatigue detection systems, eye region is always processed for symptom extraction because the most important symptoms are related to eye's activity. Therefore, eye detection is required before processing of eye region. Eye detection method can be divided into three categories, i.e., method based on imaging in IR spectrum, feature-based methods and other methods [5]. In the series of lots of work on face and eye detection, the authors in [6] tested eye blinking-based method for detecting driver's drowsiness on 15 adult volunteers' eyes by using LabVIEW^{R} software. In [7], they focus on integral image, AdaBoost algorithm and region of interest [7]. Template matching is a high-level machine vision technique that allows to identify the parts of an image (or multiple images) that match the given image pattern. The various methodologies of template matching are—naive template matching, image correlation matching, pattern correlation images. In [8], the authors compared the accuracy of PERCLOS technique and gaze estimation technique that resulted in 83.64% and 80.5% accuracy, respectively [9]. In [9], two steps are proposed to detect the yawning. The first step involves detection of hole in the mouth. This step is not based on mouth's location. Location of mouth used to verify the hole is included in the second step [10]. According to researchers in [10], tilting head by angle of just 10° can make a person appear drowsy.

3 Drowsiness Detection Techniques and Algorithms

Various techniques which are used in implementing the drowsiness detection system are discussed below:

3.1 Eye Blinking Technique

Eye blinking technique is an effective technique to detect the drowsiness. According to researches, the human eye blinks for 14–17 times per minute. The rate of blinking increases significantly for a drowsy person. This fact can be used easily to detect the drowsiness in 85% cases because when driver feels sleepy his/her eye blinking and closure rates are different from normal situations. In this technique, eye closure duration is used along with blinking rate. When the eye blinking and closure rates of the driver, on the wheel, exceed threshold value, it may result in severe accidents.

In order to localize the eyes, it is necessary to detect face and facial landmarks in captured image. To detect the eye, AdaBoost and Haar feature algorithms need to be implemented to train the Haar cascade classifier. In this process, two images are required—one containing background without object and the other with object.

Blink Detection. To detect the blink, a metric called eye aspect ratio (EAR) was introduced in [11]. Eye blinking computation involves various methods of combination of

- Eye localization
- Thresholding to find the whites of the eyes
- Identifying appearance of white region of eyes.

This method of localizing eyes is fast, efficient and easy to implement. As it is discussed above, that facial landmark detection is used to localize face, including eyes, eyebrows, nose, ears and mouth, so we can obtain facial structures with the help of their indexes [11]. Eyes can be represented with six coordinates, i.e., starting clockwise from the left corner of eye around the remainder section of eye as shown in Fig. 2. In [11], the authors derived the equation that defines the relation called eye aspect ratio:

Fig. 2 Eye landmarks

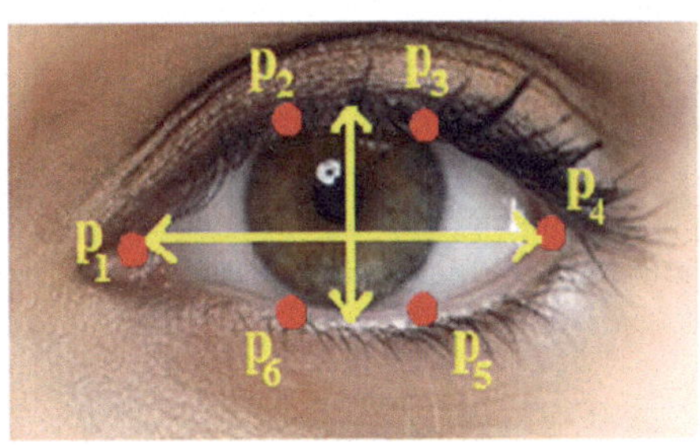

$$\text{EAR} = \frac{(||p2 - p6|| + ||p3 - p5||)}{(2||p1 - p4||)} \tag{1}$$

where $p1, \ldots, p6$ are 2D facial landmark indexes. The numerator and denominator of the equation define the distance between the vertical and horizontal landmarks, respectively. The equation contains one set of horizontal points and two sets of vertical points.

The equation provides the relaxation from traditional methods of image processing. It calculates the aspect ratio to determine whether eyes are closed or not. Generally, the eye's aspect ratio is constant in case of open eyes; if aspect ratio falls to zero, it means the blink is detected.

3.2 Percentage Eye Closure (PERCLOS) Technique

This technique was introduced by researchers in 1994. PERCLOS technique calculates "the percentage of time that an eye is closed in a given period." It can be measured using non-intrusive computer vision methods. The frame rate greater than 4 fps may be used for better accuracy. According to research, human blinks once in every 4–5 s. The technique compares this data with real-time blinking rate to detect the drowsiness.

To train our model for required results, we use three types of images as data set

- Fully open eyes
- Partially open eyes
- Fully closed eyes.

The estimation of PERCLOS technique is done by eye state classification. To detect eyes, we can use any one of the different algorithms of facial landmark detection, i.e., Viola–Jones algorithm, local binary patterns histograms (LBPH), histogram of oriented gradients (HOG).

The histograms of data set images and real-time streaming images can be used for comparison, but according to research due to real-time situations like illumination and noise this method is feasible to detect drowsiness in real time. Viola–Jones algorithm gives higher accuracy in different illumination and physical situations, so it is used to detect drowsiness in real time.

The camera captures the live stream video which is processed to yield PERCLOS value. When the resultant value reaches more than threshold value in one minute window, then an alarm is triggered to alert the driver.

Figure 3 shows the grayscale images of closed and open eye frames, and corresponding histograms are obtained with the help of OpenCV function calcHist which shows the significant change in values of pixel intensity frequency.

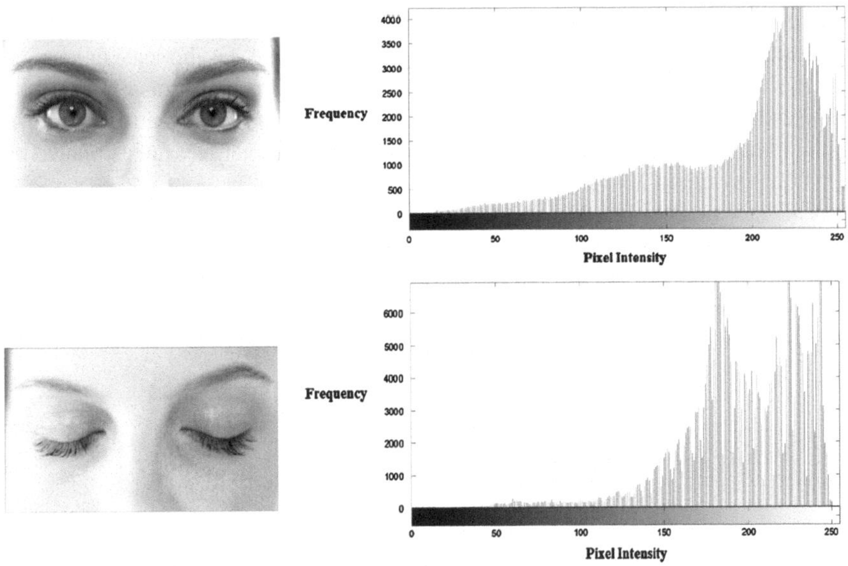

Fig. 3 Comparison of open and closed eye histograms

3.3 *Head Tilt Technique*

The head position of the driver changes very often while driving. Head tilt angle and its frequency can be used to detect when the driver feels drowsy. Head tilt technology measures the rotation angle. When the head goes beyond certain angle, a triggering variable is initiated. The motion of head with respect to camera can be estimated by using the concept of rotation and translation.

POSIT. Head tilt technique can be implemented by using pre-proposed algorithm, i.e., POSIT. Pose with iteration is an abbreviation of POSIT. It estimates the 3D pose of an object including rotation and shift over *X*-axis/*Y*-axis/*Z*-axis [11]. In this algorithm, rotation of head is calculated by using initialized standard 3D coordinate and current 2D shape coordinate. The correlation between 2D coordinates and corresponding point in 3D space are used to train model. Head tilt technique uses six unknown variables, three for translation t_x, t_y, t_z and other three for rotation r_x, r_y, r_z, i.e., to determine the change in position of head.

Proposed formula for the algorithm is given below

$$s\begin{bmatrix} u \\ v \\ 1 \end{bmatrix} = \begin{bmatrix} f_x & 0 & c \\ 0 & fy & cy \\ 0 & 0 & 1 \end{bmatrix} \begin{bmatrix} r_{11} & r_{12} & r_{13} & t_1 \\ r_{21} & r_{22} & r_{23} & t_2 \\ r_{31} & r_{32} & r_{33} & t_3 \end{bmatrix} \begin{bmatrix} X \\ Y \\ Z \\ 1 \end{bmatrix} \tag{2}$$

In the above matrix, (X, Y, Z) shows the coordinates of 3D space and (u, v) shows the point of 2D. Points (c_x, c_y) are the center points of image, s is the scale factor, and (f, f_y) is the focal length present between the pixels. These are intrinsic parameters of the camera matrix. The extrinsic parameters of rotation and translation matrix $[R \mid t]$ are r_{ij} and t_{ij}, respectively. The rotation and translation matrices are estimated by POSIT algorithm in 3D space.

A 3D face model is used to estimate head poses. The 3D face model is of average size. The coordinates of 3D space and facial landmarks are identical. The coordinates of facial landmarks in 3D space are the input to POSIT.

The pre-proposed algorithms used to implement the above techniques are discussed below.

3.4 Viola–Jones Algorithm

The Viola–Jones algorithm is used for human face detection. Viola–Jones algorithm uses the property of human face having lighter and darker regions such as the eye region darker than the bridge of the nose and cheeks. This algorithm uses Haar-like features to detect face, so it does not perform multiplication. Cascade of Haar classifier increases the computational process. This framework is used in real time, and training is slow but detection is fast. The algorithm comprises different steps. The first step in face detection is used to generate integral image. This algorithm uses rectangular features. Next step is to run AdaBoost training and creating classifier cascades.

The algorithm works on very small regions to detect a face with the help of facial features. Since the algorithm is applied on different face sizes, it checks all the specific positions. The values determined by rectangle features specify the presence (or absence) of characteristics of particular region such as location of border between light and dark regions indicated by two-rectangle feature. Haar-like feature for each region is calculated by moving a window of predefined size over the desired image. The difference obtained is compared to target threshold to distinguish among objects and non-objects. Due to its low detection quality and being a weak learner or classifier, it needs large number of features to locate an object with sufficient accuracy. Therefore, to make Haar-like feature a strong learner or classifier it is organized in classifier cascades.

Haar-like Features: Haar features are represented by rectangular boxes as shown in Fig. 4. Rectangular box is further divided into black and white rectangular shapes. In a detection window, Haar feature takes adjacent rectangular regions into account at particular location and sums up the pixel intensities in each region. Haar features can be categorized into major three-rectangle features.

- Two-rectangle feature
- Three-rectangle feature
- Four-rectangle feature.

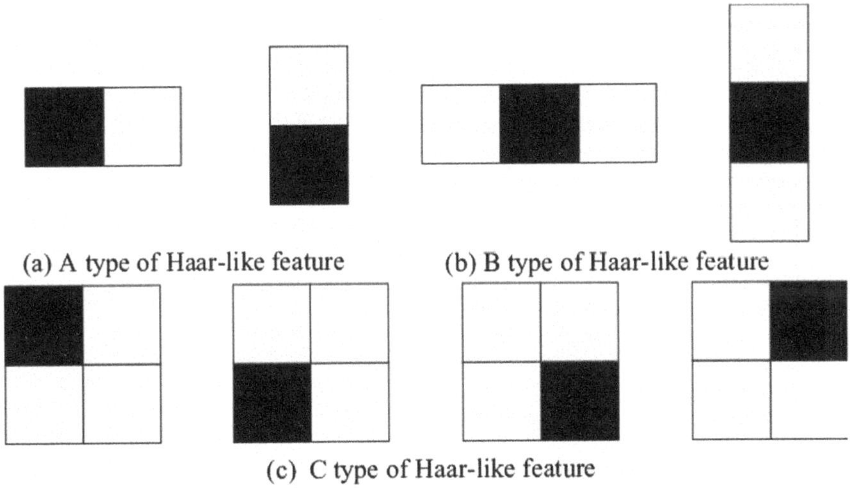

Fig. 4 Haar-like features

Two-rectangle feature is given by difference between the additions of pixels present inside these two rectangles. Subtraction of two external rectangular zones from the summation of pixels inside the center rectangle gives three-rectangle features. Four-rectangle features are calculated by difference between diagonal pairs of rectangles.

Further researches introduced tilted Haar-like feature which helped to improve the dimensionality of rectangle feature. This concept indicates the presence of edge at 45° in two-rectangle tilted feature.

- Detection window is used for object detection.
- Set the minimum and maximum window size, and choose the sliding step size.
- Slide the chosen window vertically and horizontally with same sliding step.
- Apply the face recognition filter. If it gives positive answers, the face is detected in the current window.
- Increase the size of the window, and go to step 2 until the window size reaches maximum.
- If all the classifiers give positive answers, the face is recognized, otherwise not.

3.5 AdaBoost Algorithm

AdaBoost is an acronym for adaptive boosting. It is used with different algorithms to improve their performance. It creates strong classifier by combining weak classifiers. A model is built from training data, and this model is made error free or corrected to make the second model. To predict the training set perfectly, a maximum number of models are added. AdaBoost algorithm is developed for binary classification problems to enhance the performance of decision trees or machine learning algorithms. This algorithm is susceptible to noise and outliers. It selects training facts on the basis of accuracy of previous training and retrains the algorithm iteratively. To select the training facts, AdaBoost assigns weight to each fact. The classifier having more accuracy is assigned higher weight so that it impacts final output with higher probability. Classifier with 50% accuracy is assigned zero weight, and negative weight is given to classifier of accuracy less than 50%.

Functional libraries used to implement the algorithms are discussed below:

3.6 Dlib

Detecting faces and processing images is easy to perform through computer vision libraries, but these are not sufficient to get detailed information while working on human faces. To work on detailed information of human faces, a pre-trained landmark detector is needed which is handled by Dlib. Dlib is an open-source C++/Python library. It is licensed under the boost software license. It is used to implement machine learning such as classification, regression, clustering, data transformation and structured prediction and support functionality like threading, networking, image processing, data compression and integrity algorithms. All classes and functions of Dlib are documented which can be found on its home page. It provides good framework to develop machine learning applications. Dlib supports different operating systems, Linux, Windows, OS X.

Facial landmarks: Facial landmarks help in localizing facial features such as eyes, eyebrows, nose, mouth and jawline. These facial regions are used for face alignment, head tilt estimation, eye blink detection, etc. Here, Dlib, OpenCV and Python are used for extracting facial landmarks for an image using shape prediction methods.

It is a two-step mechanism.

Step 1: Estimating face location
Step 2: Detecting facial features of the face from region of interest.

To locate face, OpenCV's built-in Haar features can be used. A pre-trained HOG and linear SVM object detector can be used for face detection. Whatever method is used, most important thing is to obtain face bounding box. Each facial feature is surrounded by (x, y) coordinates which are used as training set on an image. Detection of facial features depends on the probability of distance between the pairs of pixels of an image. Facial landmark detector is used to detect facial landmarks of 68 (x, y) coordinates that map to facial structure on the face in real time and give high-quality results.

Figure 5 depicts the facial landmarks with the help of pre-trained shape predictor. It maps 68 coordinate points on the driver's face.

- To localize right eyebrow, 18–22 landmark dots are recognized.

Fig. 5 Shape predictor (facial landmarks)

- To localize left eyebrow, 23–27 landmark dots are recognized.
- To localize right eye, 37–40 landmark dots are recognized.
- To localize left eye, 43–46 landmark dots are recognized.
- To localize mouth, 49–68 landmark dots are recognized.
- To localize jawline, 1–17 landmark dots are recognized.
- To localize nose, 28–36 landmark dots are recognized.

4 Concept of Effective Detection and Workflow

Eye blinking detection technique uses 68 facial landmark dots to detect face and features of face such as eyes, nose, jawline, eyebrows and mouth in the proposed system. Twelve facial landmark dots (37, 38, 39, 40, 41, 42 for right eye and 43, 44, 45, 46, 47, 48 for left eye) are used to detect eyes. From Fig. 1, the difference between the middle points of upper eyelid $((P2 + P3)/2)$ and lower eyelid $((P5 + P6)/2)$ is used to estimate whether the eye is open or closed. When the difference is greater than 0, then the eye is open, and when it reaches 0, the eye is closed.

$$x = \frac{P2 + P3}{2} \tag{3}$$

$$y = \frac{P5 + P6}{2} \tag{4}$$

$$z = x - y \tag{5}$$

If the difference becomes 0 more than 14 times within a minute, then intermediate threshold variable for eye detection technique is increased by 1.

For intermediate value of "PERCLOS technique variable," we use pre-trained model of HOG and SVM to detect open, closed, partially closed eyes. The number of frames per minute captured by camera is analyzed to determine the state of eyes. In general, human eye takes 1/3 of a second to blink once. If the time of closeness of eyes exceeds 14* (1/3), i.e., 5 s, then the value of PERCLOS intermediate variable is increased to threshold and supposed to affect the resultant variable.

In head tilt technique, the six unknown variables (translation variables and rotation variables) determines the change in the position. If the angle of movement changes to certain angle with respect to camera, the intermediate variable of Head tilt Technique is increased up to threshold value and contribute in resultant of final variable.

By contribution of three different variables, the final variable is obtained as a resultant and determines the state of drowsiness.

The flowchart in Fig. 6 depicts the whole procedure in the précised manner to find out the resultant value. This chart represents the simultaneous flow of three techniques. First of all face detection is performed to start the process of two techniques, eye blinking and PERCLOS. To start the process of third technique, i.e., head tilt, 3D points are detected.

In order to complete the process of eye blinking technique, eye detection is performed and EAR is calculated. The EAR threshold is compared to current EAR value; if the current EAR reaches the threshold, then resultant value will be increased; otherwise, it will calculate EAR of next frame. In order to complete the procedure of PERCLOS technique, pre-trained templates are matched with detected frame. If frame matches with closed eye template, the intermediate value is increased by 1, and if the frame does not match with the closed eye template, the intermediate value becomes 0. The idea behind the PERCLOS technique is to calculate the duration of time for which the eyes are closed. This duration is determined by intermediate value. For third technique, after detecting 3D points corresponding 2D points are recognized. 2D points are used to detect the positions of different facial landmarks and thus provide the angle with the help of rotational and translation vector. If the angle changes in different frames more than the threshold value, head tilt intermediate value increases.

After calculating all three intermediate values, the resultant value is generated. If the resultant value reaches the final threshold value, alarm will be triggered to alert the driver.

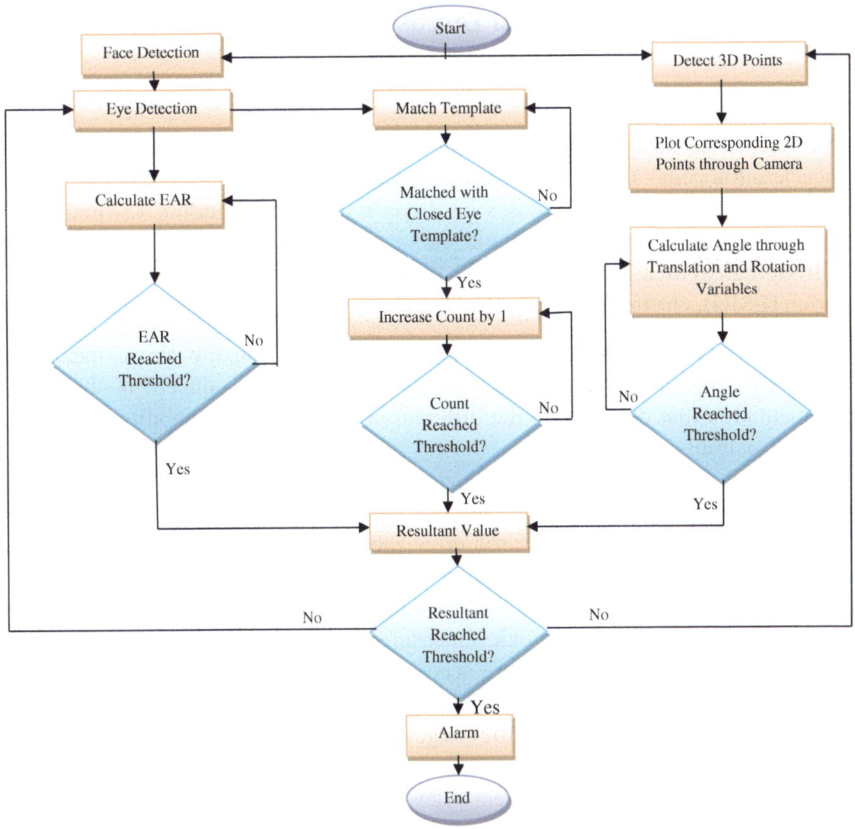

Fig. 6 Workflow diagram

4.1 Hardware Setup

A Web camera will be installed in the vehicle and will focus on facial expression and head movement of the driver. Web camera will be connected to any logical device, i.e., Raspberry Pi, that will receive and process the streaming images to recognize changes in facial expressions and head tilt angle. Images are captured using the camera at fixed rate of 20 fps. The images are sent to the detection application where the images will be analyzed using the algorithm discussed above. The triggering device will depend on resultant variable. The setup may contain triggering device, i.e., LED screen or a buzzer. LED screen will display the message, and audio alarm will be generated by buzzer.

5 Experimental Result and Discussion

Drowsiness detection systems have been developed on the basis of different measures.

- **Vehicle-based measures**. It focuses on steering wheel movement, pressure on the acceleration pedal, deviation from lane positions, etc.
- **Behavioral measures**. It focuses on driver's behavior including yawning, eye closure and blinking, head tilt angle, etc.
- **Physiological measures**. It deals with the electrocardiogram (ECG), electromyogram (EMG), electrooculogram (EoG) and electroencephalogram (EEG).

Existing systems of drowsiness detection developed on the above measures possess their own advantages and disadvantages. The main concern is about the implementation that depends on the different criteria such as reliability, accuracy, cost and feasibility.

The proposed system works on behavioral measures and focuses on eye closure duration, head movement, eye blinking rate; these measures satisfy almost all the criteria of reliability. In vehicle-based drowsiness detection system, steering angle sensor is mounted on the steering column to measure driver's steering behavior that may fail in different situations such as narrow lane or high traffic congestion. Systems developed on the basis of physiological measures need electronic circuits, and in few of the systems electronic helmets are needed to check the consciousness of driver. But the system proposed in this chapter has the capability to overcome the problems related to narrow lane and heavy traffic congestion, and it does not need any electronic circuits besides Raspberry Pi, Web camera, buzzer, LED screen.

The drowsiness detection systems developed till now on behavioral-based measures are using either only head movement or only facial expressions. The proposed system is based on both, head movement and facial expressions. In order to detect closed eyes, two techniques eye blinking and PERCLOS are used and to increase the accuracy of drowsiness detection head tilt technique with POSIT algorithm is used.

After the concepts and theory work, finally this chapter reaches to experiments and results. The following section shows the experimental results of supporting algorithms along with final experimental table of drowsiness detection system.

5.1 Viola–Jones

The algorithm detects the face and draws rectangle over the region of interest (RoI). After detecting the face, we can extract the eyes and can highlight the region of eyes as shown in Fig. 7.

Table 1 shows experimental result of Viola–Jones algorithm. It contains face angle, i.e., frontal view ranging from 0° to 25° clockwise and anticlockwise with respect

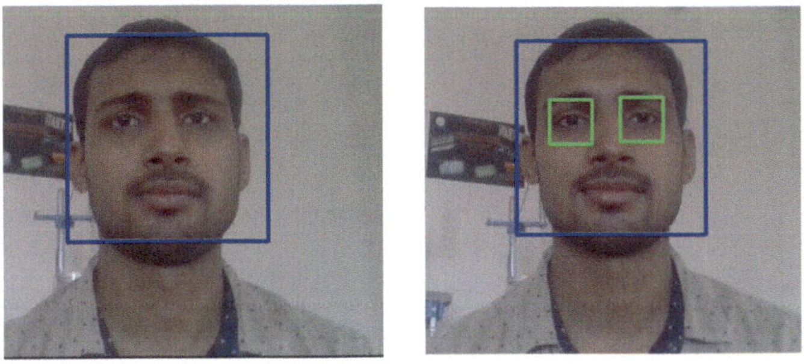

Face Detection Face and Eye Detection

Fig. 7 Output of Viola–Jones algorithm

Table 1 Experimental result of Viola–Jones algorithm

Face angle	Illuminance	Face detected	Eye detected
Frontal view	High	Yes	Yes
Frontal view	Moderate	Yes	Yes
Frontal view	Low	Yes	Yes
Side view	High, low, moderate	No	No

to camera and side view ranging from 25° to 90° clockwise and anticlockwise with respect to camera.

5.2 *Histogram Comparison*

The histograms given below are the compared histograms of two continuous frames. Histogram details of eyes can be used to detect the state of eyes. There is a sudden change in histogram details such as frequency of pixel intensity. These changes can be quantified by using OpenCV function calcHist; for instance, Fig. 8a shows the histogram of open eye frame (eye frame shown in Fig. 3). The intensity of open eyes

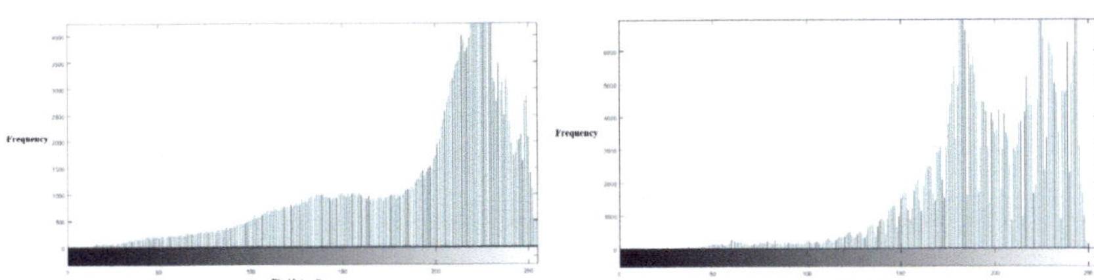

Fig. 8 Comparison of open and closed eye histograms

Table 2 Experimental result of PERCLOS

Status of eyes	State of illuminance	Outcome
Open	High	Detected
Open	Moderate	Detected
Open	Low	Detected
Open	Extremely low and uneven	Not detected
Close	High	Detected
Close	Moderate	Detected
Close	Low	Detected
Close	Extremely low and uneven	Detected

increases gradually. Figure 8b shows histogram of closed eye in which the curve shows the closed eye intensity. Sudden change in frequency values can be seen in closed eye histogram; for instance, it is visible in frequency–pixel intensity graph from 150 to 250 on pixel intensity axis.

$$\text{PERCLOS} = \left(\frac{\text{The sum of frames when eyes are closed}}{\text{The interval of frames}}\right) * 100 \tag{6}$$

Table 2 contains the experimental result of PERCLOS technique. It shows the outcome of PERCLOS algorithm in different illuminance conditions and eye states.

Face and Eye Detection with Dlib: Face detection based on Dlib library and shape detector can detect the face and eyes in low illuminance also. The shape detector helps to locate eyes and face in different illumination conditions and different face angles. Figure 9 shows the real-time detection of face and eyes with Dlib.

Table 3 shows the experimental outcome of face and eye detection with Dlib in different illuminance states, i.e., high, moderate and low, and face angle, i.e., frontal and side view.

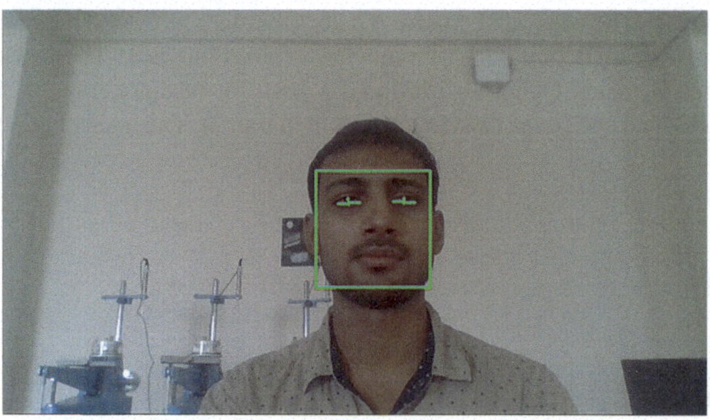

Fig. 9 Face and eye detection with Dlib

Table 3 Experimental result of face and eye detection with Dlib

Face angle	Illuminance	Face detected	Eye detected
Frontal view	High	Yes	Yes
Frontal view	Moderate	Yes	Yes
Frontal view	Low	Yes	Yes
Side view	High	Yes	Yes
Side view	Low	Yes	Yes
Side view	Moderate	Yes	Yes

Table 4 Combined experimental result

Face angle	Illuminance	Face detected	Resultant reached threshold	Alarm status
Frontal view	High	Yes	Yes	Triggered
Frontal view	Low	Yes	Yes	Triggered
Frontal view	Moderate	Yes	Yes	Triggered
Frontal view	Extremely low	Yes	Yes	Triggered
Side view	High	Yes	Yes	Triggered
Side view	Low	Yes	Yes	Triggered
Side view	Moderate	Yes	Yes	Triggered
Side view	Extremely low	Yes	Almost reached	Not triggered

5.3 *Final Result*

Overall contribution of three variables influences the final variable that, on reaching threshold, triggers the alarm. The following table shows when the alarm will be triggered in different situations which give the efficiency of approximately 87.5%.

The combined result of the detection system is shown in Table 4. It shows the different conditions which determine whether the alarm will trigger or not on the basis of comparison between threshold and resultant values.

6 Conclusions and Future Scopes

The proposed system uses three techniques for drowsiness detection of driver which contribute to more accuracy than existing systems. Although it has some limitations, the algorithms used make it efficient and reliable. This system provides non-intrusive and greater precision. It continuously checks the driver's alertness level in real time and triggers a signal in case when the drowsiness is detected by Web camera. It also offers easy setup and better speed. This system is beneficial for accident prevention which can save lives and decrease the number of road accidents in the future.

Although this chapter can help in detecting driver's drowsiness, the system may face some challenges while implementing it. One of the most happening issues could be the reflection of light from driver's glasses. Lightning background and climatic conditions are the serious issues that affect in detecting driver's drowsiness. As the system works on pre-trained data, it may face minor problems due to change in facial and eye characteristics. It may face some problems to handle the time duration when driver is not facing the camera. In such situation, alarm will be triggered unnecessarily.

The proposed system can be extended in the future by implementing different ML algorithms to security. By implementing Web application and different APIs of Python such as REST API, we can send the signal to traffic control rooms and to nearby tolls. By using some map APIs (such as Google Maps), we can get accurate location of vehicles.

The system can be extended by implementing IoT ideas to help "specially abled" people to drive. The system can be implemented in medical field for the routine checkups. The proposed system can be used in switching between automatic driving and manual driving on the basis of drowsiness and fatigue detection. The proposed system implements behavioral approach to detect drowsiness. In addition to this, other physiological and vehicle-based measures can be implemented along with discussed approach to increase the accuracy. Further work on this literature may focus on weather sensing.

References

1. Fahkry T, Pradhan T, Bagaria AN, Routray A (2012) The journey is more important than destination. In: Measurement of PERCLOS using eigen-eyes. IEEE proceedings of 4th international conference on intelligent human computer interaction, Kharagpur, India (December 27–29, 2012)
2. Tarun G, Gupta S (2013) Road accident prevention system using driver's drowsiness detection. Int Adv Res Comput Eng Technol (IJARCET) 2
3. Aidman E, Chadunow C, Johnson K, Reece J (2015) Real-time driver drowsiness feedback improves driver alertness and self-reported driving performance. Accident Anal Prevention 81:8–13
4. Danisman T, Bilasco IM, Djerabe C, Ihaddadene N (2010) International conference on machine and web intelligence, pp 230–233
5. Thummar S, Kalariya V (2015) Int J Eng Res General Sci 3(1)
6. Ma'touq J, Al-Labulsi J, Al-Kazwini A, Baniyassiem A, Al-Haj Issa G, Mohammad H (2014) Eye blinking-based method for detecting driver drowsiness. J Med Eng Technol
7. Viola P, Jones M (2001) Robust real-time object detection. In: Second international workshop on statistical and computational theories of vision-modeling, learning, computing, and sampling. Vancouver, Canada, (July 13, 2001)
8. Darshana S, Fernando D, Jayawardena S, Wickramanayake S, DeSilva C (2014) Efficient PERCLOS and Gaze measurement methodologies to estimate driver attention in real time. In: 2014 fifth international conference on intelligent systems, modelling and simulation
9. Abtahi S, Hariri B, Shirmohahammadi S (n.d.) Driver drowsiness monitoring based on yawning detection. Distributive Collaborative Virtual Environment Research Laboratory. University of Ottawa, Ottawa, Canada

10. Witkower Z (2014) Literature review: registered nursing (RN) leadership in healthcare. David Couper Consulting, Los Angeles, CA
11. Soukupova and Cech (2013) Real-time eye blink detection using facial landmarks. In: 21st computer vision winter workshop, Luka Cehovin, Rok Mandeljc, Vitomir Struc. Rimske Toplice, Slovenia (February 3–5, 2016)

An Attention Arousal Space for Mapping Twitter Data

Divya Rajput and Seema Verma

Abstract Not every trending tweet commands your attention! Overflow of social media data will soon require mechanisms to identify the severity of a situation to evaluate if a tweet must be given ample attention. News about bomb blasts, flood, epidemic outbreak affects several sectors and impacts the local economy. Such news is covered by media and government sources which forms meta-information with tweets. Most of the algorithms, both traditional and deep learning can decipher if the opinion is positive, negative, or neutral, they cannot allocate attention to the opinion. Compared to the binary classification of positive and negative sentiments, this article proposes a glove-text-CNN algorithm and attaches meta-information to predict the attention accuracy of sentence-level data. Correlation is being established among the various parameters of a trending hashtag and the most important parameter is being reported. The primary contribution is a modified neural algorithm to define an Attention Arousal Space that is designed to capture five major insights of sentence-level documents by attaching meta-information, to include media score, time decay, retweet, favorite scores and government source to modify original algorithm. The sentiment classification algorithm used for this research is Text-CNN with glove word embeddings, which was trained on 2,000,000 sentiment marked tweets from SemEval datasets. Sentiment classification training accuracy was 98.6% and validation set accuracy was 87.13%. Our experiments show accuracy over Hierarchical Attention Networks (HAN) which proves the effectiveness of this modified algorithm with meta-information for capturing attention.

Keywords Attention score · Sentiment analysis · Machine learning · Twitter data

---

D. Rajput (✉) · S. Verma
Banasthali Vidyapith, Banasthali, Rajasthan 304022, India
e-mail: rajput.divya@gmail.com

M. N. Favorskaya et al. (eds.), *Innovations in Electrical and Electronic Engineering*, Lecture Notes in Electrical Engineering 661, https://doi.org/10.1007/978-981-15-4692-1_29

1 Introduction

Sentiment Analysis is the categorization of the polarity of the given text in phrase, sentence or in the form of a document. Its basic aim is to decide whether the expressed opinion in the text is positive, negative or neutral. In today's scenario, enormous size of data has been recorded through social media like Twitter, Facebook, blogs, etc. which if analyzed properly can benefit business and society as a whole. In this article, we will be focusing on sentiment classification of tweets (posts in Twitter) and our dataset will consist of tweets trending with media and government sources.

1.1 Twitter

Twitter is an online social networking site or a popular microblogging site that involves people from every sphere of life. It enables registered users to read as well as write short messages known as 'Tweets' but unregistered users can only read them. News and opinions spread faster on Twitter; with a user base of 330 million people, Twitter has at least 10 times more subscribers than any national/international newspaper. With a global user community that can share any current news, their personal opinions in return help in understanding the severity or even normalcy of a situation more effectively. Classification of tweets in positive, negative and neutral can aid businesses, various political groups in gaining a better interpretation of the public's reaction towards them. With more real-time big data being generated, users will find it hard to allocate attention to critical issues vs say celebrity news and severe situations vs say ceremonies that go trending on social media.

In this article, we are exploring ways to capture intelligence from tweets, which are necessarily sentence-level data, by utilizing meta-information to classify and create an attention space of a trending tweet.

1.2 Deep Learning Algorithms

In the last few years, Sentiment Analysis has been performed using deep learning algorithms with improved results, not only on text level data but efforts have been initiated to extract sentiments from image and video (action) data as well [30]. Researchers have also reported hybrid CNN and RNN which have outperformed all other deep neural networks [23]. Till early in this decade, most of the machine learning algorithms operated on 1 or 2 hidden layers for feature representation and/or analysis, majorly due to the complexity of using more than two hidden layers which resulted in sluggish ML systems. With high computation power and emergence of GPUs and TPUs which are used for training 'Deep Neural Networks' for learning multiple layers of data, deep learning has outgrown all other forms of machine learning to produce high-end prediction analysis. Deep learning has found applications in computer vision, speech recognition, NLP, action recognition from videos, opinion mining to name a few [33]. Big data analysis has produced better results with deep learning.

While a conventional CNN considers whole text as input, our solution can input a single hashtag to the web interface and evaluate its attention score. The dimension of attention refers to modeling a single hashtag as input with CNN considering full text with its pre-processor Global Vector (GloVe) being a count-based co-occurrence model; while HAN presumes that not all parts of the document are relevant for determining attention. We conducted experiments on standard data of 2 Million (1.6 Million and 40 k as test data) fetched from SEMEVAL to train our algorithms. Glove-text is the pre-processors to modified CNN for sentence-level classification. Modifications were introduced for the addition of attention-specific parameters. This resulted in feature extraction by encoding word semantics and our web interface resulted in search (hashtag) query retrieval. In the present work, the modifications to activation functions and nodes to each layer contributed to improvements in the accuracy and resulted in attention score.

2 Prior Work

To support the concept of Twitter sentiment analysis, several research works have been conducted. Some of the recent work done on the application of deep learning to NLP problems are given below:

Word2Vec is a feature learning technique that contains Continuous Bag-of-Words model (CBOW) and Skip-Gram model (SG) [14]. Global Vector (GloVe) is another technique trained on non-zero inputs of co-occurrence matrix.

Convolutional Neural Network (CNN) [3, 4, 25, 32] is an orderly procedure for mixing of two sources of information. Pre-processed data is fed to different layers of filters for input scanning, followed by subsampling layer to reduce the number of features. This layer retains the most useful information and thereafter the CNN utilizes a highly connected layer as classifier. CNN has found usage in sentiment keyword engineering, commonly also known as Feature Engineering. Even the n-gram can determine the topic of the document by the use of CNN [33].

Recurrent Neural Network (RNN) [2, 9, 12] utilizes its internal memory to process sequential information. The connection between neurons forms a directed cycle, with every layer in RNN processing one word of a sentence. RNN, however, has limitations of vanishing gradient and that prompted researchers to develop higher versions of RNN such as Bidirectional RNN, Deep Bidirectional RNN and the most commonly used Long Short-Term Memory (LSTM). These advanced RNNs require a lot of training data. To improve long-range dependencies further, the attention mechanism has been proposed in bidirectional neural networks [29, 33].

LSTM Network can manage long term dependencies. It has four layers set up in a special manner and has a hidden state and a cell state. It utilizes a forget gate to operate the cell state and then decides the information to be stored via the input gate. Finally, a vector of new values will be added to the cell state and LSTM will combine those values to update the state and then decides the output, based on the cell state [21].

Memory Networks (MemNN) are specialized neural networks that can answer questions. MemNN encodes the question into a vector form, then finds evidence, reads a sentence at a time, and updates memory on the basis of sentence illustration. One of the four components, the O component uses the vector to scan through the memory and create an output vector which is processed by the R component to give the final output [27]. Advancements such as End-to-End Memory Network (MemN2N) have been also proposed by researchers [20].

2.1 Sentence-Level Sentiment Analysis

Sentence-level classification is primarily used to identify whether the sentence is subjective or objective, and then to determine the positive, negative or neutral sentiments of the sentence. Sentence-level big data is conveniently available through tweets and the ease of its availability and fetching data to conduct feature detection and extraction has led to the application and development of many classifiers. Initially ANN and SVM were applied to conduct sentiment analysis on big data and their performances have been compared [18]. *When the number of terms in the data set is increased above 1000, the training time of ANN and the running time of SVM increases.* However, ANN has an advantage of converging if the number of hidden layers is changed with improvement in its activation function whereas the result of SVM converges to remain the same. SVM however is more stable in an unbalanced dataset whereas ANN outperformed SVM in accuracy of balanced data [15].

Recently, RNN and CNN have become popular as they do not require feature extraction. The input to RNN and CNN is word embeddings with encoded semantic and syntactic information. This enables us to learn the inherent relationships among words in a sentence. Noted research on sentence-level classification is cited as under:

Dynamic CNN [10] exhibited the ability to capture word relations through semantic sentence modeling, utilizing a subsampling K-max pooling operator. Few other variants have also been proposed such as CNN-rand (random initiation of word embeddings), CNN-static (fixed and pretrained word embeddings), CNN-non-static (fine-tuned word embeddings) and CNN-multichannel (word embeddings in several sets) [11]. In 2014, Santos and Gatii have used deep CNN for short text sentiment analysis. Such a model called the CharSCNN can extract relevant features from short text [19].

The initial experiments with Long Short-Term Memory have been reported by Wang and Nyberg by predicting Twitter data opinion polarity. They have simulated the interaction of words during tweets composition itself. By introducing two-way connections in the hidden layers, the LSTM can be expanded to a bidirectional LSTM [24].

Thereafter, emerged an era of hybrid architectures such as the CNN-LSTM model and CNN-RNN combination model for short texts [26], LSTM and CNN-based deep neural network to figure out the emotional quotient in online user comments [7], encoding syntactic tags in LSTM to improve phrase and sentence illustration [8], multi-layered perceptron architecture for effective classification of financial short blogs and news [1].

Modified CNN (such as semi-supervised) has deployed a two-step process [6] by screening the review ratings and then labelled sentences for fine improvements. Yu and Jiang have learned from both labelled and unlabeled data by combining two CNNs to learn hidden features in sentences from cross-domains [31]. Researchers have also experimented to understand the semantic relationship between user posts and their social connections by introducing a recurrent random walk network learning model [34]. Linguistically regularized LSTM which can combine sentiment lexicon, negative words, high-intensity words into the LSTM have developed the ability to accurately measure sentiments in sentences [17].

Much recently in 2018, Stanford researchers have used dank learning to generate meme and image captioning to create sentiments or opinions about images. The process involves feeding an image caption to an encoder-decoder, gate for data recall, capturing semantic relationship by utilizing vector embeddings and picking up encoder embeddings by using attention mechanism without getting into memory issues [16].

An algorithm that implements "*Hierarchical Attention Networks*" achieves an accuracy score of 71% on paragraph text trained data was found most suitable to the current problem to detect the severity of a situation [29]. To achieve a high speed of training and great accuracy, researchers have proposed a combination of convolutional filters and deep learning hierarchical architecture for self-attention models [5].

2.2 *Sentiment Analysis of Twitter Data with Value-Adds*

Literature review was conducted on a few additional sources which utilized special techniques of NLP and additional Twitter features for reference to the current problem at hand.

A supervised, labelled data solution was proposed by Tian et al. to find the best community-based answer after learning from data and feature identification [22].

Researchers in social science utilized the many twitter features for non-text data such as pictures to figure out demographic info about users [13].

Wu et al. conducted research on text to speech synthesis by utilizing Deep Neural Networks [28]. They further experimented on hidden activation layers by stacking many frames and reporting improvements.

3 Methodology

Sentiment Analysis via the attention score of Twitter data (tweets) is about analyzing the severity of a trending situation. Overall procedure for assessing severity or normalcy of a situation was devised. An algorithm was designed to weigh in meta-information related to individual tweets, meta-information includes source information or Twitter handle, media information, retweet count, favorite count and time of post. Experiments were conducted on a variety of hashtags and varying the multipliers for each of the meta-information (Fig. 1).

3.1 Training the Algorithm

The algorithm works by taking input in the form of a hashtag keyword, retrieves tweets on that hashtag, finds the sentiment score of individual tweets, then applies the modification for government source and picture and returns with a final verdict regarding the hashtag. The first algorithm results are based on 100,000 tweet data retrieved from SemEval tasks. A comparison of improvements to accuracy as a result of the processing can be seen below. These results are as per experiments with HAN and glove-text-cnn, which was conducted on subset of SemEval dataset and consists of 20,000 (train on 16,000 and validate on 4,000) tweets (Fig. 2).

3.2 Sentiment Score for Modified Algorithm

The modified algorithm works by taking input in the form of a hashtag keyword, retrieves tweets on that hashtag, finds the sentiment score of individual tweets, then applies the modification for government source and returns with a final verdict regarding the hashtag. For sentiment analysis of tweets, modifications to existing algorithms such as Text-CNN with glove vectors was done.

Overall procedure for assessing severity or normalcy of a situation was devised. An algorithm was designed to weigh in meta-information related to individual tweets, meta-information includes source information or Twitter handle, media information, retweet count, favorite count and time of post.

3.3 Hashtag Analysis Using Modified Algorithm

Starting with a hashtag (extendable to a collection of related hashtags), recent tweets related to that hashtag were downloaded using standard Twitter developer api. This returned with 500–4000 tweets depending on the hashtag which has been posted

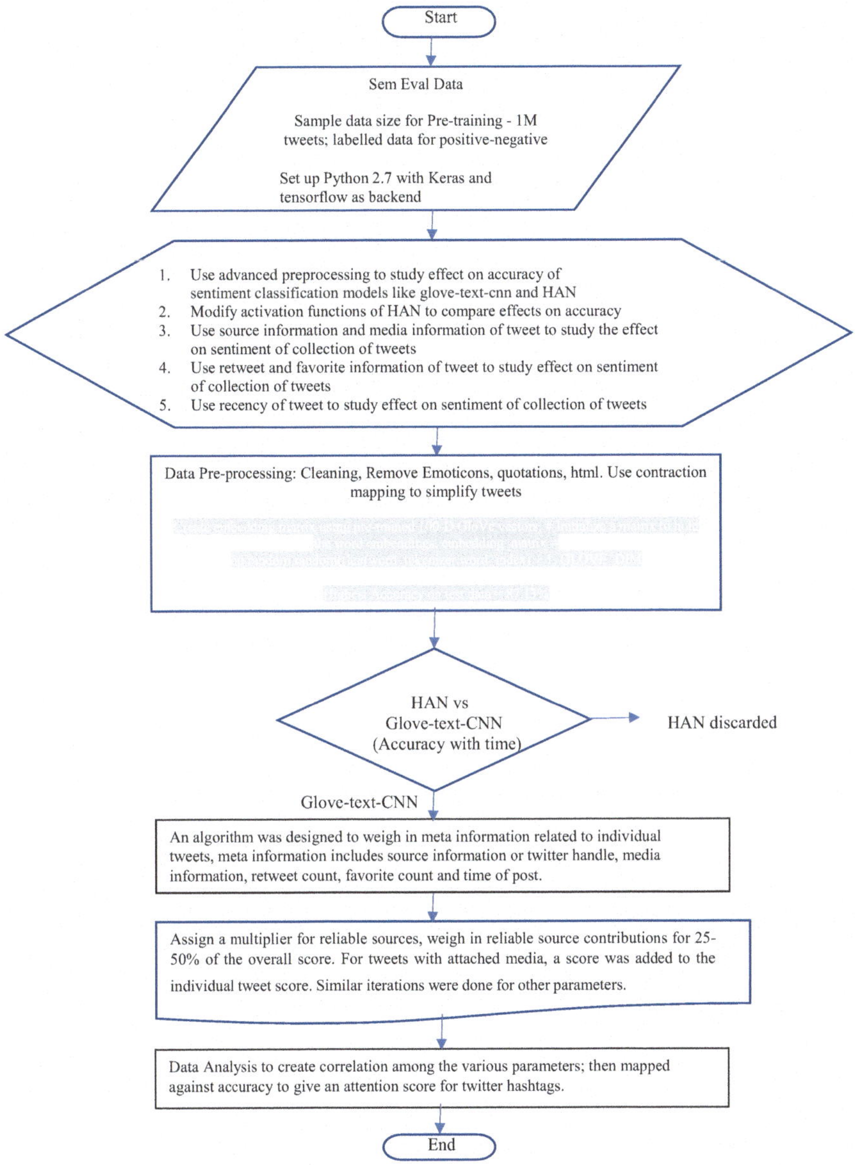

Fig. 1 Methodology flowchart

on Twitter in the past seven days. Meta-information for these tweets was extracted from returning json and a score was assigned to each tweet for each of the meta-information. A sentiment analysis was done on the collection of tweets and a sentiment score of 1 or -1 was assigned to each tweet. A meta-analysis modification was done on each tweet to assign a final score for the tweet, contributed by source

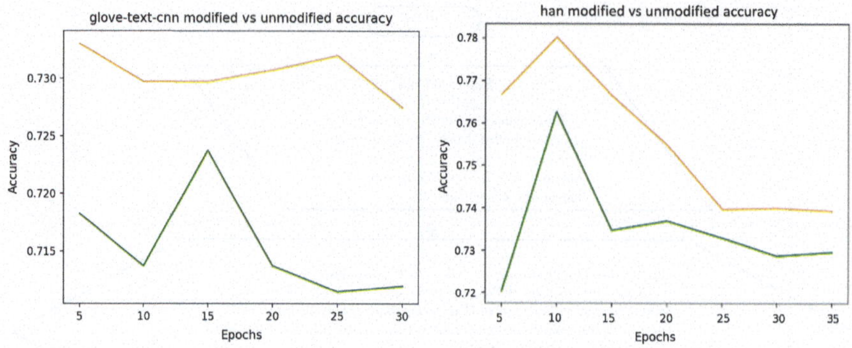

Fig. 2 A comparison of accuracy with glove-text-CNN and HAN

authenticity information, media information, retweet count, favorite count and time of post. Then a cumulative score for the hashtag was calculated, which was the sum of the scores of all tweets in the sample set.

3.4 Establishing Correlation Among Parameters Used for Sentiment Analysis

Twitter data has many parameters that can establish the authenticity of trending data. While most of the prior research has focused on sentiment scores (positive, neutral and negative), very few have combined the effect of additional parameters, such as pictures, source, time of tweet, retweet count, favorite etc.. As this research focuses on objective data, it will not consider the impact of pictures as an additional parameter but will consider only those parameters which can establish objective information.

3.4.1 Preprocessing

Preprocessing was improved in the algorithms used for sentiment analysis to pre-refine data before applying classifiers. Using python library spacy, texts were lemmatized and stemmed, and using a contraction mapping all common apostrophe words were converted to simpler form. This was done with the objective of standardizing text to a standard form. Contraction mapping normalizes the data by reducing punctuation separated words with full words, this improves the training data and model accuracy.

Sentiment Analysis on tweets and results modified by attaching meta-information (attached media, the government handles, retweet, time decay).

The algorithm works by taking input in the form of a hashtag keyword, retrieves tweets on that hashtag, finds the sentiment score of individual tweets, then applies the modification for government source and picture and returns with a final verdict regarding the hashtag. The first algorithm results are based on 100,000 tweet data retrieved from SemEval tasks.

The modified algorithm works by taking input in the form of a hashtag keyword, retrieves tweets on that hashtag, finds the sentiment score of individual tweets, then applies the modification for government source and returns with a final verdict regarding the hashtag.

```
#!/usr/bin/python
import sys
import os
import time
import datetime
import tensorflow as tf
import numpy as np
import data_utils as utils
from tensorflow.contrib import learn
from text_cnn import Text-CNN
from data_utils import IMDBDataset
```

For sentiment analysis of tweets, modifications to existing algorithms such as Text-CNN with glove vectors was done.

3.4.2 Modification Using Data Source

An important source of trust is the Twitter handle from which a tweet is posted. For example, a government source can be assumed to be more responsible and tweets from their official accounts can be considered reliable information. Using this inspiration, a list of government Twitter handle recognised by the Indian government was prepared, for example, the ministry of health, ministry of foreign affairs, prime minister office.

Other noteworthy sources of tweet include accounts that have huge number of followers, which in turn represent views of a large group of people. So, accounts with 500,000+ followers were considered trustworthy and hence their contribution to overall severity or normalcy of a situation weighed more. Experiments were conducted using this hypothesis, by assigning a multiplier of 20 for reliable sources, and then by weighing in reliable source contributions for 25–50% of the overall score. It was observed that tweets accompanied by media like pictures and videos gave significant contributions towards prediction of final verdict. For tweet with an attached media, a score of 10 was added to the individual score of the tweet. Results demonstrated effect of media information on changing the final score.

A comparison of unmodified and modified is given in the following tables to establish correlation among those parameters (Tables 1 and 2).

```
mud = float(1.0) + float(((float(cad[k]) - l0)/(h0 - l0)) * 5) +
float(((float(rcd[k]) - l2) / (h2 - l2)) * 10) + float(((float(fcd[k]) - l3) / (h3 - l3)) * 10)
if ghd[k] == 1:
    mud += 10
if med[k] == 1:
    mud += 10

red[k] = float(red[k]) * mud
```

Fig. 3 Calculation for reweighing of tweet sentiment

Table 1 Government data sources

hashtag	Positive score	Negative score	Modified verdict (%)
WC	1176.5	−394.5	74.89
G20	4170.33	−2129.66	66.20
Budget2019	9034.18	−7316.31	55.25
WorldBank	279	−327.5	46.00
Bitcoin	6954.0	−7317.5	48.73

Table 2 If the tweet had a mention of media information

hashtag	Positive score	Negative score	Modified verdict (%)
WC	2454.0	−412	85.62
G20	13,547	−963	93.36
Budget2019	22,923	−6502	77.90
WorldBank	519	−164	75.99
Bitcoin	21814.0	−5219	80.69

3.4.3 Retweet and Favorite

Retweet count and favorite count provide useful information regarding general approval or disapproval of a tweet. This motivated the research to make use of this information when evaluating results on a set of tweets. Experiments were done with different multipliers for retweet count. For example, a score out of 5 or 10 was experimented with. It was observed that retweet count can go very high in certain tweets and maximum achievable multiplier for retweet count was normalized. This was done by finding the highest and lowest retweet count in the tweets being analyzed, then assigning a score out of 5 based on retweet count. So, score corresponding to retweet count is calculated as multiplier * (retweet count–lowest retweet count)/(highest retweet count–lowest retweet count). So, retweet count was thus calculated and added to final sentiment score of tweet (Table 3).

Similarly, for favorite count, the highest and lowest favorite count was determined, and corresponding score was calculated in a similar way to retweet count, and this contributed to the final score. Technically, final score of a tweet can be $1.0 + 10.0$ for retweet $+10.0$ for favorite count + other modifiers according to meta-information.

Table 3 If tweet had a favorite or was retweeted

hashtag	Positive score	Negative score	Modified verdict (%)
WC	2860.67	−135.09	95.49
G20	15495.69	−1087.59	93.44
Budget2019	17614.57	−8497.19	67.46
WorldBank	1356.51	−1521.4	47.14
Bitcoin	10406.59	−4151.25	71.48

The effect of meta-information like retweet count and favorite count was studied by varying the multipliers and those findings have been reported.

3.4.4 Recency of Tweets

To maintain effect of recency of tweets, the time of tweet is of significance. More recent tweets are given priority over older tweets. This is done by decaying the score of older tweets compared to recent ones. Experiments were conducted to study what multiplier should be used for time decay (Table 4).

Since it is only possible to get tweets for the past seven days through the standard developer API of Twitter, the effect of time decay factor could not contribute much. But it was clear that it will play an important role in determining changing trends when computed for tweets from a longer duration. Calculation for reweighing of tweet sentiments

- More recent tweets are given higher weights
- A tweet with attached media has higher weightage
- A tweet with government handle has higher weightage.

Table 4 If there was a time decay

hashtag	Positive score	Negative score	Modified verdict (%)
WC	9484.00	−1702.68	84.78
G20	28279.48	−7253.71	79.59
Budget2019	65905.92	−27412.37	70.62
WorldBank	3432.98	−1603.26	68.17
Bitcoin	77475.81	−27758.99	73.62

Table 5 Comparison of parameters

hashtag	Source	Media	Time decay	Favorite/Retweet
WC	74.89	85.62	84.78	95.49
G20	66.20	93.36	79.59	93.44
Budget2019	55.25	77.90	70.62	67.46
WorldBank	46.00	75.99	68.17	47.14
Bitcoin	48.73	80.69	73.62	71.48

Table 6 Correlation among data parameters

Parameter	Correlation
Source and media	0.74
Source and time decay	0.94
Source and favorite	0.91
Media and time decay	0.78
Media and Favorite	0.87
Time decay and favorite	0.94

4 Results and Analysis

Data analysis was performed on the scores reported by applying modifications to the given parameters and attention scores have been reported in this chapter. The findings of the data analysis have led to interesting conclusions, both common to all the hashtags as well as few specific to some hashtags. Given the correlation established among the various parameters in the last chapter, hashtags (keywords) were selected which could reflect the impact of most of our parameters (Tables 5).

To find out the relevancy of these parameters with each other and if they would make sufficient contribution to create an attention score, we calculated correlation among their scores.

As can be observed from Table 6, the correlation among any two parameters is between 0.7 and 1.0 suggesting high linear correlation. Therefore, these parameters are being considered for algorithmic and modified algorithmic applications to calculate sentiment analysis on Twitter data.

5 Outcome

The sentiment classification algorithm used for this research is Text-CNN with glove word embeddings, which was trained on 2,000,000 sentiment marked tweets from SemEval datasets. Sentiment classification training accuracy was 98.6% and validation set accuracy was 87.13%. Improvements were made to preprocessing of data

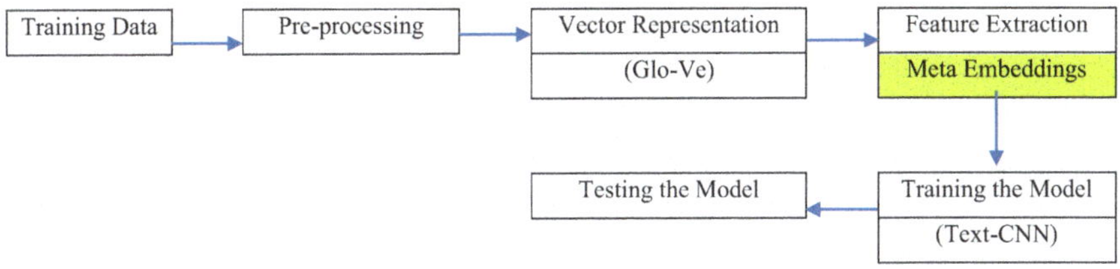

Fig. 4 Modified Glove-Meta-Text-CNN algorithm for creating attention space

by introducing Meta-embeddings for the classifier to handle Twitter data. Original algorithm was built for classifying reviews data.

As described in the work plan, twitter data was extracted for recent trending tweet sets within a period of 7 days on the following keywords:

1. WorldCup
2. G20 [Conference]
3. Budget 2019
4. World Bank
5. Bitcoin

For data from the above keywords, Meta-embedded-glove-text-CNN returned multiple scores; as much score is closer to 1, the hashtag will command attention. The Meta-embedded-CNN algorithm, therefore, can calculate attention scores for a given hashtag to find out whether it is trending and commands attention (Fig. 4).

References

1. Akhtar M, Kumar A, Ghosal D, Ekbal A, Bhattacharyya P (2017) A multilayer perceptron based ensemble technique for fine-grained financial sentiment analysis. In: Proceedings of the 2017 conference on empirical methods in natural language processing, pp 540–546
2. Cho K, Merrienboer B, Gulcehre C, Bahdanau D, Bougares F, Schwenk H, Bengio Y (2014) Learning phrase representations using rnn encoder-decoder for statistical machine translation pages. In: Proceedings of the 2014 conference on empirical methods in natural language processing (EMNLP). Association for Computational Linguistics, Doha, Qatar, pp 1724–1734
3. Conneau A, Schwenk H, Barrault L, Lecun Y (2017) Very deep convolutional networks for text classification. In: ACL-EACL, pp 1107–1116
4. Dos Santos CN, Gatti M (2014) Deep convolutional neural networks for sentiment analysis of short texts. In: COLING, pp 69–78
5. Gao S, Ramanathan A, Tourassi G (2018) Hierarchical convolutional attention networks for text classification. In: Proceedings of the 3rd workshop on representation learning for NLP. Association for Computational Linguistics, Melbourne, Australia, pp 11–23
6. Guan Z, Chen L, Zhao W, Zheng Y, Tan S, Cai D (2016) Weakly-supervised deep learning for customer review sentiment classification. In: Proceedings of the twenty-fifth international joint conference on artificial intelligence
7. Guggilla C, Miller T, Gurevych I (2016) CNN- and LSTM-based claim classification in online user comments. In: Proceedings of COLING 2016, the 26th international conference on computational linguistics: technical papers, pp 2740–2751

8. Huang M, Qian Q, Zhu X (2017) Encoding syntactic knowledge in neural networks for sentiment classification. ACM Trans Inf Syst (TOIS) 35(3)
9. Irsoy O, Cardie C (2014) Opinion mining with deep recurrent neural networks. In: EMNLP, pp 720–728
10. Kalchbrenner N, Grefenstette E, Blunsom P (2014) A convolutional neural network for modelling sentences. In: Annual meeting of the association for computational linguistics. Maryland
11. Kim Y (2014) Convolutional neural networks for sentence classification. Retrieved from https://arxiv.org/abs/1408.5882
12. Liu P, Qui X, Huang X (2016) Recurrent neural network for text classification with multi-task learning. In: IJCAI, pp 2873–2879
13. McCormick TH, Lee H, Cesare N, Shojaie A, Spiro ES (2015) Using twitter for demographic and social science research: tools for data collection and processing. Sage
14. Mikolov T, Sutskever I, Chen K, Corrado GS, Dean J (2013) Distributed representations of words and phrases and their compositionality. In: Advances in neural information processing systems 26 *(NIPS 2013)*
15. Moraes R, Valiati JF, Neto WG (2013) Document-level sentiment classification: An empirical comparison between SVM and ANN. Expert Syst Appl 40(2):621–633
16. Peirson VAL, Tolunay EM (2018) Dank learning: generating memes using deep neural networks. Retrieved from https://arxiv.org/pdf/1806.04510.pdf
17. Qian Q, Huang M, Lei J, Zhu X (2017) Linguistically regularized LSTMs for sentiment classification. Retrieved from https://arxiv.org/abs/1611.03949
18. Rajput D, Madhukar M, Verma S, Manisha (2015) Sentiment analysis on big data using machine learning for holiday destinations 2015. In: IEEE European modelling symposium. Madrid
19. Santos CN, Gatti M (2014) Deep convolutional neural networks for sentiment analysis of short texts. In: Proceedings of COLING 2014, the 25th international conference on computational linguistics: Technical Papers, pp 69–78
20. Sukhbaatar S, Szlam A, Weston J, Fergus R (2015) End-to-end memory networks. In: Advances in neural information processing systems 28 (NIPS 2015)
21. Tai KS, Socher R, Manning CD (2015) Improved semantic representations from tree-structured long short-term memory networks. In: Annual meeting of the association for computational linguistics
22. Tian Q, Zhang P, Li B (2013) Towards predicting the best answers in community-based question-answering services. In: International AAAI conference on web and social media. AAAI
23. Van VD, Tahi T, Nghiem M-Q (2017) Combining convolution and recursive neural networks for sentiment analysis. In: SoICT '17. Vietnam
24. Wang D, Nyberg E (2015) A long short-term memory model for answer sentence selection in question answering. In: Proceedings of the 53rd annual meeting of the association for computational linguistics 7th international joint conference on natural language processing (Short Papers). Beijing, pp 707–712
25. Wang T, Wu DJ, Coates A, Ng AY (2012) End-to-end text recognition with convolutional neural networks. In: ICPR. IEEE, New York, pp 3304–3308
26. Wang X, Jiang W, Luo Z (2016) Combination of convolutional and recurrent neural network for sentiment analysis of short texts. In: Proceedings of COLING 2016, the 26th international conference on computational linguistics: technical papers, pp 2428–2437
27. Weston J, Chopra S, Bordes A (2015) Memory networks. In: ICLR
28. Wu Z, Valentini-Botinhao C, Watts O, King S (2015) Deep neural networks employing multi-task learning and stacked bottleneck features for speech synthesis. In: 2015 IEEE international conference on acoustics, speech and signal processing. IEEE Explore, Brisbane, Australia
29. Yang D, Yang DC, He X, Smola A, Hovy E (2015) Hierarchical attention networks for document classification. Retrieved from https://www.cs.cmu.edu/~hovy/papers/16HLT-hierarchical-attention-networks.pdf

30. You Q, Luo J, Jin H, Yang J (2009) Robust image sentiment analysis using progressively trained and domain transferred deep networks. In: Twenty-Ninth AAAI conference on artificial intelligence
31. Yu J, Jiang J (2016) Learning sentence embeddings with auxiliary tasks for cross-domain sentiment classification. In: Proceedings of the 2016 conference on empirical methods in natural language processing. Austin, pp 236–246
32. Zeng D, Liu K, Lai S, Zhou G, Zhao J (2014) Relation classification via convolutional deep neural network. In: COLING, pp 2335–2344
33. Zhang L, Wang S, Liu B (2018) Deep learning for sentiment analysis: a survey. Wires Data Mining Knowl Discovery 8(4)
34. Zhao Z, Lu H, Cai D, He X, Zhuang Y (2017) Microblog sentiment classification via recurrent random walk network learning. In: Proceedings of the twenty-sixth international joint conference on artificial intelligence

Review of SLAM Algorithms for Indoor Mobile Robot with LIDAR and RGB-D Camera Technology

Chinmay Kolhatkar and Kranti Wagle

Abstract Simultaneous localization and mapping are techniques of mapping the homogeneous environments by localizing the sensors' position in an environment with centimetre accuracy. This paper reviews the different techniques used in mapping and localization of mobile robot and designing of low-cost mobile platform with sensors like RPLIDAR and Microsoft Kinect. This paper also discussed the use of Robot Operating System and different packages and topics used for implementation of SLAM. Along with software implementation of ROS, this paper also covers the comparison of different hardware boards such as NVIDIA Jetson Nano and NVIDIA TK1 GPUs for running heavy and complex CUDA algorithms. Lower slave boards such as Arduino Mega and STM32 are also discussed in this paper. This paper also focuses on use of different localization techniques such as AMCL, ORB-SLAM, Hector SLAM, Gmapping, RTAB-MAP and particle filter SLAM. Robot Operating System played the crucial role in all the complex processing and communication between different running nodes. Autonomous navigation is achieved using ROS navigation stack, and simulations for the same are done in Rviz and gazebo real-time environments. 2D point cloud-based algorithms with laser scanner are compared against the 3D visualization techniques. At the end, clear comparison between all the benchmark algorithms such as Hector SLAM, Gmapping, RTAB-MAP, ORB-SLAM and **ZEDfu** is done for clear understanding while selecting an algorithm for future research.

Keywords SLAM · ROS · LIDAR · Kinect · Navigation stack · AMCL · Gmapping · ORB-SLAM · Mobile platform · Rviz

C. Kolhatkar (✉) · K. Wagle
Fr. Conceicao Rodrigues College of Engineering, Bandstand, Mumbai University, Bandra (W), Mumbai, Maharashtra 400050, India
e-mail: chinmayhere84@gmail.com

M. N. Favorskaya et al. (eds.), *Innovations in Electrical and Electronic Engineering*, Lecture Notes in Electrical Engineering 661, https://doi.org/10.1007/978-981-15-4692-1_30

1 Introduction

With advances in electronics and computer vision technology, service robots are becoming part of our daily life. Service robots are usually deployed as assistants in indoor environments, and few examples of such robots are "Roomba" vacuum cleaning robot, Personal Robot PR2, Pepper service robot, Husky (UGV) and many more [1]. GPS-based navigation system cannot provide centimetre accuracy and is also not suitable for indoor environment; hence, navigating the mobile robot in office or home-style environment becomes a challenge. Problem of navigation becomes more difficult when environment is complex and unknown. When robot does not know neither map nor location, formerly problem is called as SLAM problem. SLAM technique allows the use of sensors like LIDAR, RGB-D cameras, IMU, etc., in order to construct the map of surrounding environment while simultaneously navigating the robot towards the target location. SLAM problem requires accurate sensors and precise real-time algorithms. There is verity of real-time SLAM algorithms like SLAM algorithms based on monocular camera called mono SLAM, parallel tracking and mapping (PTAM), oriented fast and rotated brief (ORB-SLAM), etc., where SLAM algorithms implemented using RGB cameras are known as stereo SLAM or visual SLAM (V-SLAM) algorithms [2]. RGB cameras with depth-sensing capabilities like Microsoft Kinect are mostly used in SLAM-based navigation; because of their depth-sensing capabilities they can be used to map the 3D environment. On the other hand, LIDARs are 360 degree rotating horizontal scanners which can scan the environment in 2D, and LIDARs are expensive but most effective [2] while mapping the environment and constructing map at speed with minimal computational power. Review is targeted to address the problem faced by the newcomers in the field of ROS and SLAM with real-world robotics platforms. Most of the well-known and premade robotics platforms such as TurtleBot, Kobuki and Husky are very costly and can cost thousands of dollars. Rather, making your own platform with the help of low-cost controller such as Arduino and high-end controller such as NVIDIA CUDA GPU can make a good combination which is much more sufficient for entry-level person/research to perform any research on ROS and robotics platform. Platform can be designed with help of DC geared motors, stepper motors along with the encoders for odometry purposes. Additionally, 6 to 10 DOF IMU sensor can be added to improve localization and mapping quality. Sensors such as RPLIDAR A1 can be used as 2D planner mapping sensor, and sensors like Microsoft Kinect and Asus cams can be used as 3D depth-sensing mapping cameras for excellent dense map building and accurate navigation along with features such as loop closure. Additionally, sensors such as stereo cameras and ZED cameras can be used to improve the map quality and density; also, more features can be detected which improves the quality of navigation. 2D techniques such as Gmapping and Hector SLAM can be used for 2D mapping and navigation of robotic vehicle, whereas techniques such as RTAB-MAP, DPPTAM and ORB-SLAM can be used for 3D mapping and navigation of robotic vehicles. This review is targeted to explore all these techniques and their pros and cons. This paper refers to AMCL, Hector SLAM, Gmapping, ORB-SLAM,

RTAB-MAP, DPPTAM, ZEDfu and usage of sensor and quality of each sensor with different algorithms in different scenarios. We have also discussed the hardware utilized by the other researchers in order to implement the SLAM algorithms, and comparative analysis of different SLAM techniques is also done. Finally, conclusion is drawn which can give a clear idea to the newcomer while selecting the particular technique and utilization of hardware in order to reduce the cost of set-up and cost of experimentation.

2 Mobile Robot Platform (System Set-Up)

For autonomous navigation in indoor environment, mobile robotic vehicle is designed with multiple sensory inputs and actuating outputs. Mobile robot also called unmanned ground vehicle is designed to move freely with its own computational power. The UGV or mobile robotic platform is designed with differential drive base which is having two driving wheels and two auxiliary wheels (omnidirectional). For driving wheel, either DC geared motors with encoders or stepper motors are used. According to the author, stepper motors have smaller accumulated error [1]. For driving these motors, a motor driver with suitable ampere ratings is used. In all the designs, the mobile robotic platform is divided into two sections. One is bottom layer called as low-level computing layer and upper layer which is called as high-performance processing layer. The lower layer consists of some slave device like microcontroller/microprocessor, such as Arduino and STM32 which has two main functions:

1. Collect data from on-board sensors like wheel encoders and inertial measurement unit (IMU), and send that data in appropriate format to upper processing layer.
2. Controlling the motor directions (PWM control) as per the commands from upper processing node.

Usually, upper layer consists of a processor and GPU for processing complex data such as data from camera modules, LIDAR and ROS. Upper controller configuration used by Ilmir Z. Ibragimov and Ilya Afanasyev uses Intel Core i3 CPU 3.6 GHz 4 core processor and Geforce GT740M supporting CUDA-based GPU [2]. Kartik Madhira and Dr. Jignesh Patel performed a quantitative study of mapping and localization algorithms on ROS-based differential drive robot. They have designed a robot which utilizes the RPLIDAR A2 as laser scanner and 560 CPR 300RPM encoder motors for odometry calculations. Their design utilizes CUDA-based NVIDIA Jetson TK1 CPU and Arduino Mega 2560 as lower controller [3]. Other robotics platform and computational configurations are presented in Table 1. The Robot Operating System (ROS) is running on the upper layer processor, upper layer basically processes the data from LIDAR and other sensors like RGB-D, and it also collects the data such as encoder data (for odometry) from lower layer and sends this data to the master ROS connected node which may be a laptop or a PC [4]. In ROS language, the on-board upper processing layer is called as slave layer/slave node working on ground vehicle

Table 1 Different robotic platforms and their configurations with utilized sensors and ROS packages

Hardware	Author					
	Ilmir Z. Ibragimov and Ilya M. Afanasyev	Manuel Gonzalez Ocando and Novel Certad	Yong Li and Changxing Shi	Sukkpranhachai Gatesichapakorn and Jun Takamatsu	Kartik Madhira and Dr. Jignesh Patel	Maksim Filipenko and Ilya Afanasyev
Processor utilized	Intel Core i3 processor	PC/laptop for 3D reconstruction of map over Wi-Fi ROS node	Industrial computer	Intel NUC Kit D54250WYK and Raspberry Pi 3 model B+	NVIDIA TK1 and laptop with 1 GB GPU 4 Gb RAM Core i3	Quad ARM A57
GPU utilized	Geforce GT740M CUDA	Not used	–	–	NVIDIA Kepler GK20a GPU with 192 CUDA cores	NVIDIA Maxwell with 4 GB of RAM
Laser scanner	Hokuyo UTM-30LX^2	Hokuyo URG-04LX-UG01	RPLIDAR	SICK LMS100–10000	RPLIDAR A2	Hokuyo UTM-30LX
RGB-D camera technology	(1) Microsoft Kinect (2) Stereolabs ZED camera (3) Basler acA2000-50gc GigE	Not used	Microsoft Kinect	ASUS Xtion PRO Live	Not used	Basler acA1300-200uc Stereo camera ZED camera
Communication protocol used	–	RS232	Serial port	USB to RS232	–	Serial
Motor and motor driver/lower controller board	–	Omron Adept mobile robot	Stepper motor drive with STM32f103 ARM cortex M3 as lower controller	Raspberry Pi as lower controller	560 CPR 300 RPM encoder motors, Arduino Mega 2560	Traxxas 7407 Radio-Controlled Car Model

(continued)

Table 1 (continued)

Hardware	Author					
ROS package, nodes and frameworks	Rosbag, hector_slamORB-SLAM2, ZED, RTAB-MAP, OctoMap	RosAria, Move_base, Rviz, Smach viewer, OctoMap Node	Rviz tool	Tf2_ros, SLAM_gmapping, teleop_twist_keyboard, AMCL	ROS framework based on ClearkBot, /ticks, /cmd_vel, /odom, hector_slam, slam_gmapping, coreslam and amcl	nav msgs/OccupancyGrid, geometry msgs/PoseWithCovarianceStamped, sensor msgs/Image, sensor msgs/LaserScan, base_link.

and the ROS node running on laptop or PC is called as master node. The master node is usually used for monitoring the different sensor data visually in real time; for example, the tools like ROS visualizer (Rviz) can be used to see what exactly LIDAR and RGB-D cameras receive in real time. The master node basically subscribes to the topic published by the slave node (robot) over Wi-Fi in same network configuration. In construction of this mobile robotic platform, sensors such as RGB-D cameras, LIDARs, IMUs and motor encoders are used which we will discuss now.

Yong Li and Changxing Shi used STM32f103 as a lower controller which is loaded with μC/OS-II, it takes the command from upper controller for controlling the speed of the stepper motors, and the speed of the stepper motors is controlled by the variable frequency control [1]. Mobile robot system uses both RGB-D camera and LIDAR at different floor levels; usually, LIDAR is placed on the top of the robot to avoid blocking its range, as LIDAR rotates all 360 degree and its range should not be blocked. As LIDAR is placed high at height of 5 cm, this scheme basically allows the robot to avoid obstacle even below 5 cm in range. 2D LIDAR is used as an "ObstacleCostmapPlugin", and RGB-D camera is used as "voxelCostmapplugin"; further PCL is used for converting depth-sensing data to point cloud data. Openni2_launch package is utilized for handling PCL with Kinect.

3 Navigation and Localization Techniques (Implementation)

Kalman filter, extended Kalman filter SLAM (EKF-SLAM) and particle filter algorithms are known as baseline algorithms amongst many SLAM-based algorithms in mobile robotics applications. ROS supports multiple navigation and localization algorithms, and navigation stack can be utilized in order to navigate the robot towards its destination using multiple techniques. Algorithms such as Hector SLAM, Gmapping, RTAB-MAP, ORB_SLAM and AMCL are used to generate map of the environment and to localize and navigate the robot in environment. Now, we will discuss few algorithms used in localization and autonomous navigation of robot. Hector SLAM and Gmapping algorithms are mostly used with range sensors/LIDARS, while ORB-SLAM and RTAB-MAP algorithms are mostly used with RGB-D camera set-up along with range sensors.

3.1 Extended Kalman Filter SLAM (EKF-SLAM)

The most common and baseline algorithm used in robotics is EKF-SLAM algorithm which can be implemented either for multiple sensor fusion or for state estimation [3]. Kalman filter basically works on the principle of weighted mean of the measurement and predicted Gaussian distribution. Kalman filter has assumption that the state of the robot is linear, i.e. noise free, and distribution is not multimodal but always Gaussian. But the sensor data is always noisy which causes distribution to be nonlinear and non-Gaussian. Hence in EKF-SLAM first nonlinear distribution is converted to Gaussian, linear distribution. But the main problem with EKF-SLAM is it requires simultaneous updating of many matrix elements which increases the computation complexity linearly with complexity of environment, i.e. number of landmarks in the environment.

3.2 Particle Filter SLAM

Although EKF-SLAM is baseline for many SLAM techniques, most practical SLAM techniques such as Hector SLAM and Gmapping use the particle filter as the baseline algorithm. Particle filter such as Rao-Blackwellized particle filter which is baseline for Gmapping algorithm uses large number of particles to represent the pose of the robot and map with the platform [3].

3.3 AMCL

AMCL is a probabilistic localization system for a robot moving in 2D. It implements the adaptive (or KLD-sampling) Monte Carlo localization approach, which uses a particle filter to track the pose of a robot against a known map [3]. AMCL takes in a laser-based map and laser scans and transforms messages, and outputs pose estimates. The particle filter data is visualized in Rviz visualization tool and represented with green colour where trajectory is represented with red colour. The start and end point for navigating the robot towards its destination can be given by clicking on particular point in the map [5].

3.4 Hector SLAM

Hector SLAM is one of the famous algorithms used for 2D map building and localization in mobile robots. Hector SLAM is an algorithm specifically designed for range sensors. Hector SLAM works based on 2D laser scan data obtained from LIDAR sensor. Hector SLAM does not require odometry information; it only requires 2D laser scan data [2]. Algorithms are able to build 2D map and localize the robot at same frequency as scanning frequency of LIDAR. Hector SLAM is used to build the occupancy grid map from the laser scan data, and it publishes this information to the nav_msgs/OccupancyGrid topics [2]. Hector SLAM combined with IMU uses scan matching with Kalman filter to compute the robot pose in full 3D environment. When Hector SLAM is visualized in Rviz, each cell of the map is painted with different colours such as occupation probability when coloured black means cell is busy, light grey means cell is free and dark grey means cell is not scanned yet. Green line in Hector map represents the UGV trajectory. According to the author, Hector SLAM algorithm has very high update rate and low measurement noise. The author also says the estimation and orientation are through relative overlapping of beam endpoints with the current map [3]. Hector SLAM uses a Gaussian–Newton minimization which is an updated version of Newton method; it has the advantages that second derivatives need not be computed. Also, it takes less energy to avoid changes in dynamic environment. However, this method does not solve the loop closure problem.

3.5 Gmapping

Gmapping algorithm is the most widely used algorithm in the field of mobile robotics. Gmapping algorithms are based on Rao-Blackwellized particle filter (RBPF) [6]. Gmapping algorithm creates the 2D occupancy grid-based map. This algorithm uses the particles as representation of map. These particles are regarded as hypotheses that means particle can either be true or be false. Each particle obtains its measurement from laser scanned data. RBPF filters these particles to obtain the most probable particle that is considered as the map and trajectory by the algorithms. RBPF is used to solve the metric (grid-based) SLAM problem where the inputs are sensor reading $z_0, z_1, z_2, \ldots, z_n$, odometry or control signal $u_1, \ldots, u_t$ and output is trajectory estimation of robot $x_1, \ldots, x_t$ and map M. The joint posterior is given by $(x1{:}\, t, M \mid z1{:}\, t, u1{:}\, t-1)$. In order to simplify the posterior calculation, two steps are conducted; firstly, estimated trajectory is calculated from odometry data and after that simply $(M \mid x, z1{:}\, t)$. Then, the joint posterior simply becomes equal to $(x1{:}\, t, M \mid z1{:}\, t, u1{:}\, t-1) = (M \mid x, z1{:}\, t).\ (x1{:}\, t \mid z1{:}\, t, u1{:}\, t-1)$; then, individual map for all particles is built [7]. ROS uses "Gmapping" package and slam_gmapping node to start map building process. This node takes data from sensors/laser scan messages which are basically sent from LIDAR to ROS stack. The information from these messages is used to transform the scan into odometry tf frame.

3.6 ORB_SLAM

Oriented fast and rotated brief (ORB-SLAM) algorithm is a fast feature detection algorithm. ORB-SLAM is a feature-based method which maps the environment in 3D space using RGB-D cameras. ORB-SLAM uses bundle adjustment algorithm to create a 3D environment by extracting features from different images and place them in 3D. ORB-SLAM basically tracks the ORB features of the robot's pose, calculates the camera trajectory and recovers a sparse 3D scene of environment. This detector algorithm works in real time and creates sparse point cloud as the map. The main feature of ORB-SLAM algorithm is loop closure detection, keyframe selection and localization for each frame. ORB_SLAM library is used in ROS for monocular trajectory tracking. Visualization tool such as Rviz can be used to visualize the ORB-SLAM map in 3D; also, ORB_SLAM library has its own GUI for visualization [8]. The red points represent the point cloud of generated map, green line represents the trajectory, and blue rectangles represent the planes in which camera shoots at a particular frame. According to the author, ORB-SLAM is very robust and provides the good approximation of robot trajectory.

3.7 RTAB-MAP

Real-time appearance-based mapping algorithm is visual SLAM algorithm which creates the map of environment in 3D using graph-based SLAM approach. The algorithm relies on graph-based SLAM and global loop closure detection to compute robot pose and map. The map is stored as graph with robot poses, associated images as nodes and odometry/loop closure transformation as edges [8]. It can be used with any sensor providing 3D scanning information; mostly, stereo RGB-D cameras like Microsoft Kinect are used with this algorithm. This algorithm provides excellent loop closure and pose estimation. The loop closure detection algorithm uses "bag-of-words" algorithm to compare the images captured of current location with the previously visited location images [8]. Then, each such closure helps in optimizing the graph. The algorithm can be implemented with or without the odometry information. According to the author, Maksim Filipenko and Ilya Afanasyev algorithm sometimes fails in pose estimation when robot moves closer to monotonous walls. But the failure recovery system is able to detect the failure and determine the robots' pose properly. According to the author, the system is fairly accurate and robust. It solves the problem of localization more accurately than LIDAR without additional manipulation. RTAB-MAP is integrated with ROS via RTAB-MAP ROS package. Algorithm provides the 3D map in the form of dense point cloud and 2D map in the form of occupancy grid which can be visualized using Rviz tool of ROS. According to Ilmir Z. Ibragimov and Ilya Afanasyev during their test, UGV motion trajectory deviates from the marked trajectory but the Kinect RGB-D camera and camera-based RTAB_MAP odometry give results which are close to the ground reality.

3.8 DPPTAM

Dense piecewise planar tracking and mapping from monocular sequence is a new method of real-time direct visual SLAM. DPPTAM uses an assumption that homogeneous and monochrome regions belong to approximately same plane [5]. According to authors Maksim Filipenko and Ilya Afanasyev, DPPTAM is not enough robust method as it lost tracking during turns. Also, author Ilmir Z. Ibragimov

Table 2 Comparison of different SLAM algorithms based on pose estimation trajectory accuracy and map building

SLAM algorithm	2D/3D	LIDAR/RGB-D or both	Odometry/IMU	Map building density	Pose estimation	Loop closure algorithm
AMCL	2D	LIDAR	Odometry	Sparse	Good	None
Hector SLAM	2D	LIDAR	IMU	Sparse	Poor	None
Gmapping	2D mono	LIDAR	Odometry	Sparse	Good	None
ORB-SLAM	3D	RGB-D cameras	Better with data association	Semi-dense	Good	FabMap
RTAB-MAP	3D mono/stereo	RGB-D	Better with data association	Semi-dense	Good	Bag of words
DPPTAM	3D mono	Direct RGB-D	Odometry	Dense	Poor	None
ZEDfu	Stereo 3D	RGB-D	Odometry	Dense	Good	None

concluded from the DPPTAM real-time trajectory and marked white line trajectory are mismatched; hence, it is a failure visual odometry method in indoor navigation environment.

3.9 *ZEDfu*

ZEDfu is a stereo visual SLAM method. Stereolabs developers provided cameras with various tools and interfaces. **ZEDfu** stereo camera can be used for generating 3D map of the environment in the form of point cloud. ZED SDK has the integration with ROS which is performed with zed-ros-wrapper package [2]. Authors Ilmir Z. Ibragimov and Ilya Afanasyev performed the experiment with ZEDfu tool and recorded point cloud is store in SVO format. According to them, the map is dense and even obstacles such as chairs and table are clearly visible which are not visible in case of Hector SLAM case. To build the UGV trajectory, ZED camera visualization data and position tracking data from OpenGL window are used. According to the author, the trajectory calculated using ZEDfu tool matches with the ground trajectory and only has difference of 10 cm (Table 2).

4 Future Scope

SLAM is a baseline algorithm which can be used for autonomous localization and navigation of mobile robot. SLAM algorithms such as Hector SLAM and Gmapping are highly dependent on sensors' accuracy; hence, the work can be done in direction to reduce sensor noise and improve accuracy of these algorithms. Data association is a crucial part of mobile robotics and machine vision. If sensory data from LIDAR and RGB-D cameras is combined together, it will improve the accuracy of the navigation. Algorithms such as RRT, PRM, Dijkstra and A*can be used in association with the existing SLAM algorithms in order to improve navigation. Along with SLAM algorithm, machine learning algorithms especially deep reinforcement learning algorithms can be developed in order to avoid the dependencies for preamp building and accuracy can also be improved to some extent. Deep reinforcement algorithms such as DQN, DDPG, actor critic and SARSA can also be deployed in order to improve SLAM algorithms' accuracy and improve navigation. Problem of loop closure can be easily tackled with improved sensor quality and sensor data association. There are various other possible ways which can also be deployed such as ceiling feature matching and landmark tracking [9] which are used in low-cost dust cleaning robot to improve the overall system performance and to make mobile robot cost-effective. SLAM is a benchmark technique in the field of robotics and computer vision, and it is been used from several years and will be used in future, good example of which is Xiaomi using the SLAM in their smart dust cleaning robots such as Xiaomi RoBoRock S650.

5 Conclusions

1. Low-cost and high-performance hardware can be built by using a low-cost LIDAR and sensor like Microsoft Kinect. Optionally to improve performance of experiment, odometry can be added by using the motor encoders and IMU sensors.
2. A ready-made platform such as Omron Adept mobile robot can be used to design a low-cost mobile SLAM-based robot hardware [6].
3. Even though it is possible to place a system like laptop on mobile robot, it is feasible to use small on-board computer like Jetson TK1 or Intel NUC to reduce the robot's weight and mobility constraints [3].
4. As ROS navigation stack performs heavy computation during navigation and processes so much of data from all nodes, it is feasible to use on-board GPU [3].
5. ROS open-source repositories and navigation stack packages greatly reduce the code length and shorten the development time [1].
6. LIDAR is 2D sensor that comes out to be fast and precise amongst all for 2D map building and navigation experiments.
7. As LIDAR is single-line laser radar, only obstacles of the level of LIDAR are detected and obstacles on the other levels cannot be effectively avoided [4].
8. Use of Kinect sensor with laser range sensor can highly improve the obstacle avoidance capabilities of mobile robot.
9. Odometry sensor seems to be noisy, and the algorithms such as Gmapping are highly affected by the measurement noise.

10. Rough terrain areas cause robot wheels to stuck in place but as motors kept running Rviz showed robot is still moving since the wheels are still rotating. Causing serious errors in scan matching and mapping inaccuracies [4].
11. Gmapping algorithms worked well in all scenarios come out to be best for 2D planner navigation despite the fact that not originally designed for RGB-D cameras [10].
12. Although Gmapping comes out to be accurate, it fails to detect some obstacles like chairs and tables considering them as homogeneous walls [4].
13. Both Gmapping and Hector SLAM realize the map correctly, as Gmapping algorithm needs odometry information which puts certain limit on the robot's speed. Navigation-wise Hector SLAM works much faster and with lesser noise in maps, but Hector SLAM requires higher laser scanning frequency.
14. During semi-outdoor environment, laser sensor is not able to provide accurate scan due to bright light originating from outdoor area. This causes SLAM algorithm to misinterpret the orientation of robot [4].
15. Faster movement of robot while map generation causes poor map building and causes error in scan matching [4].
16. No monocular SLAM algorithm estimates absolute scale of the received map and localization, but stereo and RGB-D sensors give good result as compared to mono sensors in map building and localization.
17. ORB-SLAM comes out to be better amongst the other visual SLAM/odometry algorithms, and it is more robust in homogeneous environments as compared to DPPTAM. DPPTAM provides dense map [5].
18. ORB-SLAM method should be used if high-performance requirement is required and environment does not have any flat monochrome objects.
19. DPPTAM method should be used if a dense area map is required to build an obstacle map or the environment does not contain enough features and also hardware has enough performance power. Lack of feature points affected the DPPTAM method adversely [8].
20. Zed camera odometry also shows good results with errors close to LIDAR data; however, it loses tracking over sharp turns [2].
21. RTAB-MAP can be considered as good solution for 3D mapping scenarios with loop closure detection and good feature of recovery in case of moving out of trajectory [11].

References

1. Li Y, Shi C (2018) Localization and navigation for indoor mobile robot based on ROS. IEEE, New York
2. Ibragimov IZ, Afanasyev IM (2017) Comparison of ROS-based visual SLAM methods in homogeneous indoor environment. In: 14th WPNC. IEEE, New York
3. Madhira K, Patel J (2017) A quantitative study of mapping and localization algorithm on ROS based differential robot. In: NUiCONE. IEEE, New York

4. Syaqur WA, Kamarudin K (2018) Mobile robot based simultaneous localization and mapping in unimap's unknown environment. IEEE, New York
5. da Silva BMF, Xavier RS (2017) Experimental evaluation of ROS compatible SLAM algorithm for RGB-D sensors. IEEE, New York
6. Cheng Y, Wang GY (2018) Mobile robot navigation based on LIDAR. IEEE, New York
7. Eliwa M, Adham A (2017) A critical comparison between Fast and Hector SLAM algorithms. REST J Emerging Trends Modell Manuf
8. Filipenko M, Afanasyev I (2018) Comparison of various SLAM systems for mobile robot in an indoor environment. IEEE, New York
9. Hwang S-Y, Song J-B (2013) Clustering and probabilistic matching of arbitrarily shaped ceiling features for monocular vision-based SLAM. Adv Robot 27(10):739–747
10. Gatesichapakorn S, Takamatsu J (2019) ROS based autonomous mobile robot navigation using 2D LiDAR and RGB-D Camera. IEEE, New York
11. Yagfarov R, Ivanou M (2018) Map comparison of LIDAR based 2D SLAM algorithms using precise ground truth. In: ICARCV. IEEE, New York
12. Ocando MG, Certad N (2017) Autonomous 2D SLAM and 3D Mapping of an environment using a single 2D LIDAR and ROS. IEEE, New York
13. Megalingam RK, Teja CR (2018) ROS based autonomous indoor navigations simulation using SLAM algorithm. Int J Pure Appl Math 118(7):199–205

Detection of Deepfakes Using Visual Artifacts and Neural Network Classifier

M. A. Sahla Habeeba, A. Lijiya, and Anu Mary Chacko

Abstract The field of Artificial Intelligence is so advanced that it made the creation and modification of synthetic images and videos very easy. Tampering of videos attained a new level of refinement due to the deep learning techniques and the availability of high computing power. This contributes to the 'deepfake' era. Deepfake is a term coined for the fake videos created using deep learning techniques. With this method, one can create fake videos of people that they never did by replacing their face in some other real videos. There is a great danger of misusing this technique to disseminate false information or fake news. Thus the detection of deepfakes is critical to protect the people's pride and trust in the digital content. Most of the works in detecting deepfakes are using deep learning methods. In this paper, we are proposing an approach to identify deepfake videos with very less computational power. The proposed method exploits visual artifacts present in the face regions in the generated deepfakes. We use a three-layer neural network to classify the videos as deepfake or real. As a second step of confirmation, the variance of laplacian is calculated for different patches in the face, and based on their comparison, detection of deepfakes is assured. Our approach is tested in two datasets, UADF dataset, and the latest DeepFakeDetection dataset released by the Google AI team. The proposed method achieves better results in terms of computational requirements and accuracy, and are explained in detail in the analysis section.

Keywords Deepfakes · Neural network · Video manipulation

M. A. Sahla Habeeba (✉) · A. Lijiya · A. M. Chacko
Computer Science and Engineering Department, National Institute of Technology Calicut, Kozhikode, Kerala, India
e-mail: sahla_p180044cs@nitc.ac.in

A. Lijiya
e-mail: lijiya@nitc.ac.in

A. M. Chacko
e-mail: anu.chacko@nitc.ac.in

M. N. Favorskaya et al. (eds.), *Innovations in Electrical and Electronic Engineering*, Lecture Notes in Electrical Engineering 661, https://doi.org/10.1007/978-981-15-4692-1_31

1 Introduction

Manipulations of digital images and videos attained a new level of sophistication due to the recent advances in deep learning tools and inexpensive availability of high computing power. By using Convolutional Autoencoders and Generative Adversarial Network (GAN) Models, anyone can produce fake images and videos with significant realistic effects. Users can create fabricated images very quickly by using the available camera Apps. The fantastic effect of this trend is Deepfakes; the fake videos created using deep learning techniques. Using this, we can replace the face of a person in a video with the face of someone else and thereby creating his/her fake videos. Misusing this technique is a great danger because it will disseminate false information or fake news. It will severely affect people by injuring their dignity and self-respect, and it also decreases trust in digital content. Due to these factors, the identification of integrity and trustworthiness of visual content became very important.

Deepfake technology first appeared as a script for the generation of face-swapped adult content in late 2017 by a Reddit user named 'deepfakes' [1]. Other than the malicious effects, Deepfake technology can also be used for numerous benefits for the fashion and entertainment industries. Some of the good applications are enhancing the efficacy of video conferencing, creating video sequences for movies, etc. Since the 1990s, there have been several approaches that target video sequence manipulations. Thies et al. suggested Face2Face [2], a reenactment scheme that can change facial movements in video streams of various kinds. Facial changes in aging can be generated using Generative Adversarial Networks (GANs) [3]. Deep interpolation [4] demonstrates significant outcomes in changes in facial characteristics such as age, facial hair, or mouth. Various public implementations for the deepfake generation are available. FakeApp[1] is an application program for desktops for creating deepfakes. Faceswap github[2] also provides the code for deepfake creation. Stankiewicz [5] suggests the 2020 US presidential election campaign might be greatly affected by deepfakes. The biggest risk is that deepfakes can be used to humiliate people. In the coming days journalists and politicians will be slandered by malicious deepfakes. Because of the high social impacts, deepfake detection is a highly demanded research area. "Media Forensics" is a funded project by US government. Microsoft, Facebook, and Kaggle recently announced deepfake detection challenges. Google AI team released a large DeepfakeDetection dataset to support the fight against deepfakes [6]. The proposed method is based on the finding that many visual [7] and warping [8] artifacts are present in the generated deepfakes. Visual inconsistencies like flickering, boundary effects, insufficient details in the eye and teeth area may present between the swapped faces and the rest of the scene. After extracting features that represent these artifacts, a multi-layer feed-forward neural network is used to classify them as deepfake or real. To reduce the False positives and False negatives, we add a refinement step. The Variance of Laplacian in the forehead area, in eyes and

[1] https://www.malavida.com/en/soft/fakeapp.

[2] https://github.com/deepfakes/faceswap.

teeth, is calculated and compared to ensure the detection of deepfakes. Our approach needs less computational requirements when compared to other methods of deepfake detection using deep learning techniques.

2 Related Works

2.1 Deepfake Video Generation

The core idea in deepfake is to train two autoencoders in parallel. An autoencoder network consists of an encoder followed by a decoder. The encoder performs dimensionality reduction. The decoder uses the variables produced by the encoder and creates an approximation of the original input. So there will be two encoders and two decoders. The same weights will be shared by both of the encoders (in effect, a single encoder), and the two decoders use different weights. Suppose we want to create deepfake of ‘B’ from the videos of ‘A’, many samples of ‘A’ and ‘B’ is used to train the network. The single encoder will produce all the latent faces by identifying common features (e.g. number and position of eyes, nose, etc.) from both face samples. Decoder is only trained with faces from one of the subjects. So two different decoders will be used for ‘A’ and ‘B’. After training, we generate a latent representation of the original subject’s (A) face present in the video. These will be then given to the decoder network trained on faces of the subject we want to insert (Decoder of B) in the video [9], and faces will be swapped. This swapping will be repeated for each frame in the video, thereby creating the deepfake of ‘B’. Figure 1 shows the deepfake generation process.

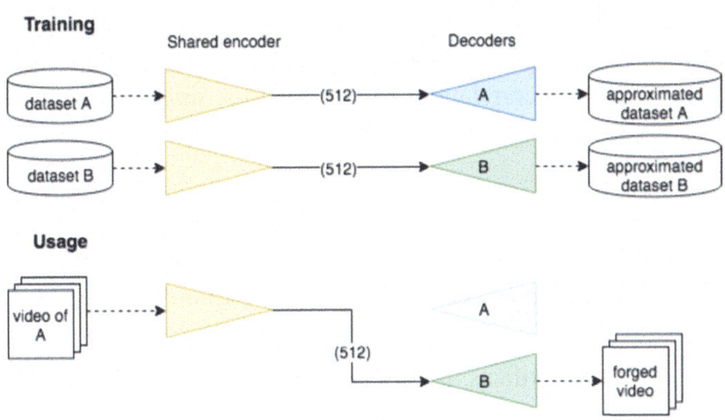

Fig. 1 Principle of deepfake [1]

2.2 Works in Deepfake Detection

Most of the works for deepfake detection use deep learning methods, which need high computational power and a large dataset. Variations in lighting conditions and usage of distinct video codecs make it hard for autoencoders to generate real-looking faces under all circumstances. This generally results in swapped faces which are visually incompatible with the remainder of the scene. Güera et al. [9] used Recurrent Neural Network, which exploit intraframe inconsistencies and temporal inconsistencies present in deepfake videos. Sabir et al. [10] showed that a combination of recurrent convolutional model and face alignment approach improves upon the state-of-the-art. Due to the limitation of computing resources and production time, existing DeepFake techniques can only generate face images of a fixed size and resolution. So an affine warping should be there to match the configuration of the source's face. This warping creates distinct artifacts, and they are used in [8] to detect DeepFake Videos. This work also uses a Convolutional Neural Network (CNN) to detect such artifacts by comparing the generated face areas and their surrounding regions. Matern et al. [7] proposed simple features to characterize visual artifacts in geometry and illumination estimation. Then the videos are detected as fake using logistic regression or neural networks with these features. However, this method is not applicable to videos in low resolution. Li et al. [11] proposed a paper based on the observation that individuals depicted in the first generation of deepfakes either didn't blink or didn't blink at the expected frequency. This method is less efficient as blinking was integrated in the next generation of deepfake synthesizing techniques. The same team used the comparison of head poses in [12]. They estimated head poses from two different portions of a face. One using facial landmarks from the whole face and the other from the central face region only. The difference of these is fed to an SVM classifier to differentiate the original image from the Deepfake. This approach is not effective at detecting lipsync. Agarwal et al. [13] exploit the facial expressions and movements that typify an individual's speaking pattern. As an individual speaks, they have distinct facial expressions and movements, which will not be present in deepfake videos. So these correlations can be used for authentication and thereby detecting deepfakes.

3 The Proposed Method of Detecting Deepfakes Using Visual Artifacts and Neural Network Classifier

The relevance of Visual Artifacts in the detection of Deepfakes is explained in this section. The proposed methodology for deepfake detection is discussed in the following section.

3.1 Relevance of Visual Artifacts in the Detection of Deepfakes and Face Manipulations

The advancements in the deepfake generation technology can create deepfakes which are very much realistic. But some visual artifacts can still be identified in the face region where the swapping had done.

In a standard video due to the lighting in the surrounding there will be a specular reflection in the eye. This catch-light reflections in the eye add life to the video. When producing a fake video, this effect should be there to look like real. For that, the incident illumination that was present in the actual video has to be transferred to the fake. Deepfakes are created using deep learning method, and the network itself will learn all these from the data given to it during training. But if the learned illumination details are wrong; then, it will lead to related artifacts like lack of proper specular reflection in the eyes. The videos generated by Deepfake techniques show unconvincing specular reflection. Most of the time, reflection in the eyes are missing or appear as a white blob or a dull appearance of the eyes [7]. There is a unique geometry for different parts of the face, and when we create fake videos, the estimation of facial geometry in a correct manner is very crucial. Some geometry is missing in the deepfakes that are generated until now. In deepfakes, teeth appear as a single white blob instead of individual teeth [7].

It is tough to get the source and target samples with the same viewing and lighting conditions. This will limit the autoencoder in creating realistic faces, and visual inconsistencies will be present between swapped faces and the rest of the scene. Since the encoder does not know the skin or other scene information, there will be boundary effects in deepfakes. The autoencoder used in one frame is unaware of the face it has created in the previous frame. This will lead to inconsistent illuminations between scenes in different frames, thereby creating flickering phenomenon in the face region [14]. Due to the limitation of computing resources and production time, existing DeepFake techniques can only generate face images of a fixed size and resolution. So an affine warping is done to match with the configuration of the source's face. This warping creates distinct artifacts and resolution inconsistency [8] in face regions.

These artifacts, i.e., missing reflections, missing details, especially in the eye and teeth areas, and the resolution inconsistency in the face areas have been used in this work for the detection of deepfake videos.

3.2 Proposed Detection Methodology

To detect deepfake videos, we propose some features that exploit the visual artifacts present in the generated deepfakes. These features are then fed to a multi-layered feed-forward neural network to identify the missing details in the eye and teeth areas, and it will detect whether the video is deepfake or not. Initially, we converted the videos into frames. Then a median filter is used to remove the speckle noise that may present in them. Face regions in the frames are detected using dlib library. The face detector is based on the classic Histogram of Oriented Gradients (HOG) feature combined with a linear classifier, using a sliding window detection scheme. To extract the landmarks, we used the trained model by Kazemi et al. [15]. Using this trained facial shape predictor with 68 landmarks, we extracted the eye and teeth regions from the faces. To detect deepfakes, we use the blurring effect in the eye and teeth areas present in the deepfake videos. So the features are the landmarks corresponding to the eye and teeth areas. These features are fed to a three-layered feed-forward neural network which updates the weights using the backpropagation algorithm and optimised using sigmoid optimizer with momentum. We trained the neural network with the labeled features. After proper training, the network can detect the lack of details present in the eye and teeth areas. At the testing stage, we extracted the same features from the video and use the trained neural network for classification. We checked the label predicted by the neural network for all of the frames. If the number of frames having blurry eyes and teeth is above a threshold, then that video may be deepfake. To confirm that video as deepfake, one more step is done. We Compared the average sharpness metric of the eye and teeth area with the sharpness metric of another region in the face, which is not affected by the face-swapping of deepfake generation pipeline. Sharpness metric is calculated using the Variance of Laplacian [16]. Laplacian is used to measure the 2nd derivative of an image, and it highlights regions with rapid intensity changes. The Laplacian $L(x, y)$ of an image with pixel intensity values $I(x, y)$ is given by:

$$L(x, y) = \frac{\delta^2 I}{\delta x^2} + \frac{\delta^2 I}{\delta y^2}$$

Variance of the Laplacian gives a focus measure of the image and can be considered as a sharpness metric. High variance in the image indicates the widespread of responses, shows the image is a normal in-focus image. But if there is very low variance, then the responses will be of little spread and shows the absence of edges in the image. As we know, if an image is blurred, it will contain fewer edges. The average value of the Variance of Laplacian is found for eye and teeth regions in a frame. Then we extracted a patch from the forehead area of the face, which is not usually affected by the deepfake generation pipeline. Variance of Laplacian is estimated for this patch also. If the average of the sharpness metric for the eye and teeth area is very less compared to that of the forehead area, we can confirm that the video is deepfake. If both the sharpness metrics are almost similar, or if the eye area contain convincing specular reflection and individual teeth are clearly visible in most of the frames, that video is classified as real one. The overall procedures in the proposed approach is shown in Fig. 2.

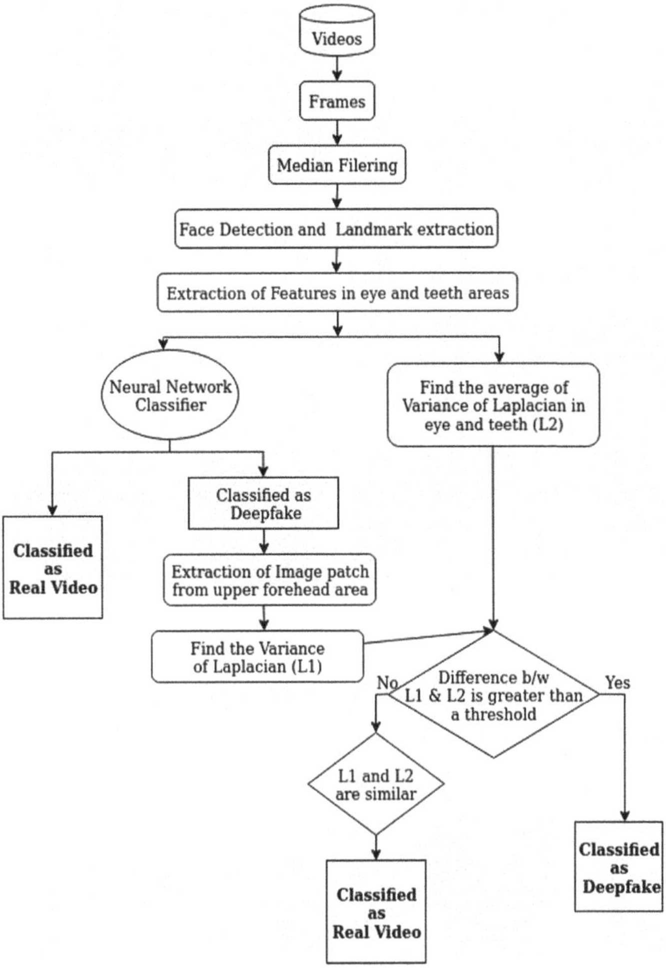

Fig. 2 Block diagram of proposed approach

4 Experiments and Result Analysis

4.1 Datasets Used

4.1.1 UADFV Dataset

This Dataset consists of 49 real and fake videos prepared by Li et al. [11]. Various interviews and presentation episodes were collected from the web and generated fake videos using the DeepFake algorithm. Samples from UADF dataset is given in Fig. 3.

Fig. 3 Samples from UADF dataset, 1st row—deepfake frames and 2nd row—real frames

Fig. 4 Samples from FaceForensics++ dataset, 1st row—deepfake frames and 2nd row—real frames

4.1.2 DeepFakeDetection Dataset in the FaceForensics Benchmark

This is a vast dataset created by the Google AI team in collaboration with Jigsaw, Technical University of Munich, and the University Federico [6]. They released the dataset containing real and fake videos in the last week of September 2019. This dataset is freely available to the research community and can be downloaded from FaceForensics github page. They incorporated paid actors and created thousands of deepfakes by using open source DeepFake generation algorithms. Figure 4 shows sample images from the FaceForensics++ dataset.

4.2 Result Analysis

In this section, we discuss our results and also compare our approach with [7] proposed by Matern et al., eye aspect ratio (EAR) based method [17] and LRCN [11] and CNN based method. The last two methods use deep learning techniques; hence they need very high computing requirements.

We divide the dataset into Training, Validation and Testing sets in the ratio 60:20:20. 'Accuracy' and 'Area Under Curve (AUC)' are considered for the performance analysis. Accuracy is the percentage of videos correctly detected as deepfake. AUC is the area under the curve of the plot, which is drawn using False Positive Rate (*FPR*) in the x-axis and True Positive Rate (*TPR*) in y-axis at different points in [0, 1]. True Positive Rate corresponds to the proportion of deepfakes that are correctly detected as deepfakes, concerning all deepfakes. False Positive Rate is the proportion of real videos that are mistakenly detected as deepfakes, with respect to all real videos.

For the UADF dataset, the proposed neural network based architecture got 92% accuracy in Testing. During the training the accuracy was 98% and accuracy in Validation was 94%. Figure 5 shows the ROC curve plotted for the UADF dataset. The AUC value we got for our detection technique in UADF dataset is 0.96. LRCN and CNN approach got 99% and 98% accuracy respectively on UADF dataset. But they use Deep learning techniques which demand high computing power. The proposed method took 3 h for training in 8GB Core i5 machine.

We also tested our model with 200 compressed videos taken from the DeepFakeDetection dataset released by Google AI team in Faceforensics++ [6]. For that dataset we got an accuracy of 86%. This newly released dataset contain lesser visual artifacts. Some of the videos have shown teeth area clearly. This is the reason for reduction in the accuracy. Figure 6 is the ROC curve plotted for the DeepFakeDetection dataset provided in Faceforensics++. The AUC value we got for our method is 0.91. Table 1 show the comparison of AUC values got for the FaceForensics++ dataset by the proposed method and Matern et al. [7]. Table 2 shows the Accuracy got for UADF dataset by eye aspect ratio (EAR) based approach [17], LRCN [11] and CNN based method. The values shown are taken from their papers. LRCN and

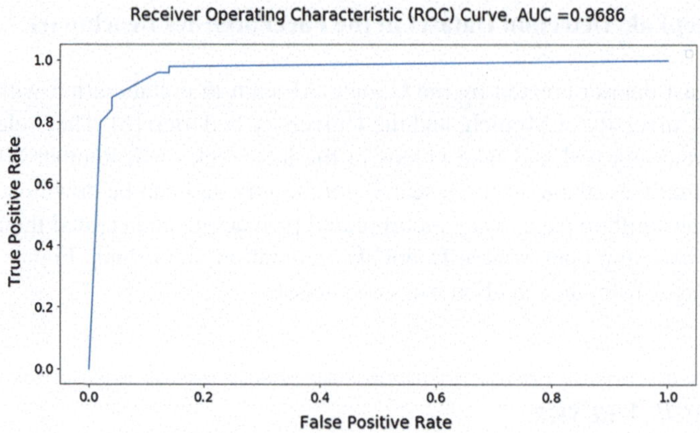

Fig. 5 ROC of proposed approach on UADF dataset

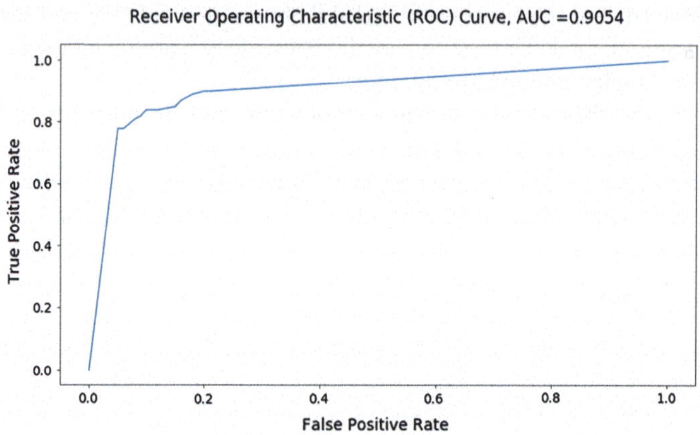

Fig. 6 ROC of proposed approach on the dataset in Faceforensics++

Table 1 Comparison with [7]

Approach	AUC value
Method proposed by Matern et al. [7]	0.82
Proposed approach	**0.90**

CNN methods have higher accuracy, but the advantage of our approach is the lesser computational requirements. Since these methods use deep learning techniques, they need high computational power.

Table 2 Comparison of accuracy

Approach	Accuracy for UADF dataset
EAR [17]	0.79
LRCN [11]	0.99
CNN [11]	0.98
Proposed approach	**0.92**

5 Conclusions

In this paper, an approach for the detection of deepfakes is proposed. The method exploits the visual and warping artifacts present in the face regions of the deepfakes. We used a three-layer neural network to identify the lack of details in the eye and teeth area. The videos with blurred eye and teeth regions are further processed using the concept of Variance of Laplacian. Videos with clear and sufficient details in the eye and teeth area are classified as Real videos. If the average of the sharpness metric for the eye and teeth area is very less compared to that of the forehead area, we can confirm that the video is a deepfake. If both the sharpness metrics are almost similar, then it is identified as a real video. The main advantage of the proposed method is computational requirements are very less. Most of the existing deepfake detection methods are based on deep learning techniques, and they need high computational power for execution. For the UADF dataset, the maximum accuracy we got is 92%. We got the minimum False Positive Rate for the same dataset as 0.02 and the maximum True Positive Rate as 0.98. For the DeepFakeDetection dataset provided in the Faceforensics++, we got maximum accuracy as 87%. We got the minimum False Positive Rate for the same dataset as 0.08 and the maximum True Positive Rate as 0.92. By analyzing the results, we can conclude that the results we got are better with less computational requirements. The limitation is that the proposed method may not be robust for the videos with shallow resolution. For future work, we plan to come up with a robust strategy that can also work in videos with very low resolution.

References

1. Afchar D, Nozick V, Yamagishi J, Echizen I (2018) Mesonet: a compact facial video forgery detection network. In: IEEE international workshop on information forensics and security (WIFS). IEEE, pp 1–7
2. Thies J, Zollhofer M, Stamminger M, Theobalt C, Niessner M (2016) Face2face: real-time face capture and reenactment of rgb videos. In: The IEEE conference on computer vision and pattern recognition (CVPR)
3. Antipov G, Baccouche M, Dugelay J (2017) Face aging with conditional generative adversarial networks. In: 2017 IEEE international conference on image processing (ICIP), pp 2089–2093
4. Upchurch P, Gardner J, Pleiss G, Pless R, Snavely N, Bala K, Weinberger K (2017) Deep feature interpolation for image content changes. In: The IEEE conference on computer vision and pattern recognition (CVPR)

5. Stankiewicz K (2019) Three threats posed by deepfakes that technology won't solve. Technol Rev
6. Rössler A, Cozzolino D, Verdoliva L, Riess C, Thies J, Nießner M (2019) Faceforensics++: learning to detect manipulated facial images. arXiv:1901.08971
7. Matern F, Riess C, Stamminger M (2019) Exploiting visual artifacts to expose deepfakes and face manipulations. In: IEEE winter applications of computer vision workshops (WACVW). IEEE, pp 83–92
8. Li Y, Lyu S (2018) Exposing deepfake videos by detecting face warping artifacts. arXiv:1811.00656
9. Güera D, Delp EJ (2018) Deepfake video detection using recurrent neural networks. In: 2018 15th IEEE international conference on advanced video and signal based surveillance (AVSS), pp 1–6
10. Sabir E, Cheng J, Jaiswal A, AbdAlmageed W, Masi I, Natarajan P (2019) Recurrent convolutional strategies for face manipulation detection in videos. Interfaces (GUI) 3:1
11. Li Y, Chang M-C, Lyu S (2018) In ictu oculi: exposing AI created fake videos by detecting eye blinking. In: IEEE international workshop on information forensics and security (WIFS). IEEE, pp 1–7
12. Yang X, Li Y, Lyu S (2019) Exposing deep fakes using inconsistent head poses. In: ICASSP 2019—2019 IEEE international conference on acoustics, speech and signal processing (ICASSP), pp 8261–8265
13. Agarwal S, Farid H, Gu Y, He M, Nagano K, Li H (2019) Protecting world leaders against deep fakes. In: Proceedings of the IEEE conference on computer vision and pattern recognition workshops, pp 38–45
14. Güera D, Delp EJ (2018) Deepfake video detection using recurrent neural networks. In: 2018 15th IEEE international conference on advanced video and signal based surveillance (AVSS). IEEE, pp 1–6
15. Kazemi V, Sullivan J (2014) One millisecond face alignment with an ensemble of regression trees. In: Proceedings of the IEEE conference on computer vision and pattern recognition, pp 1867–1874
16. Pech-Pacheco JL, Cristóbal G, Chamorro-Martinez J, Fernández-Valdivia J (2000) Diatom autofocusing in brightfield microscopy: a comparative study. In: Proceedings 15th international conference on pattern recognition. ICPR-2000, vol 3. IEEE, pp 314–317
17. Soukupova T, Cech J (2016) Eye blink detection using facial landmarks. In: 21st computer vision winter workshop. Rimske Toplice, Slovenia

On-Road Crime Detection Using Artificial Intelligence

Gopichand G, Vijayakumar, and Naga Swetha Pasupuleti

Abstract The law enforcement of AI/ML used for crime detection, predication and forensic analysis is revolutionizing fast to protect the supply chain and possibly save lives. In recent years, a lot of on-road crimes have been increased. In this paper, we propose a new approach which can easily trace the details of the people on-road immediately to reduce the rate of crimes. We discuss various techniques to get the details about the driver by detecting the number plate of the vehicle. Once the number plate is detected, it then automatically searches the RTO database and provides the details of the vehicle owner. Face detection and mapping through the existing face data and detects the person's identity by using the traffic cameras, and by this, we can save the victims from kidnapping and several other road-related crimes.

Keywords Bagging · Bootstrap aggregation · Principal component analysis (PCA) · Pattern recognition

1 Introduction

Vehicle license plate numbers are utilized for unmistakable verification of vehicles wherever all through the globe. Despite the way that the way it is shaped wherever all through the globe changes, it basically comes down to mixes of letters all together

The original version of this chapter was revised: The Author (Gopichand G) name and his affiliation corrections have been incorporated. The correction to this chapter is available at https://doi.org/10.1007/978-981-15-4692-1_60

Gopichand G
School of Computer Science and Engineering (SCOPE), Vellore Institute of Technology, Vellore 632014, Tamilnadu, India

Vijayakumar
School of Computer Science and Engineering, Galgotias University, Uttar Pradesh, India

N. S. Pasupuleti (✉)
School of Civil Engineering, Galgotias University, Uttar Pradesh, India
e-mail: swetha.pasupulethi@galgotiasuniversity.edu.in

M. N. Favorskaya et al. (eds.), *Innovations in Electrical and Electronic Engineering*, Lecture Notes in Electrical Engineering 661, https://doi.org/10.1007/978-981-15-4692-1_32

and numbers [1]. Label number detection is a senseless picture preparing system to print out/show the characters in the label subsequent to deal with a given image [2]. In this paper, unprecedented approach is utilized, as in, picture is gained, locale of intrigue is singled out, and characters are divided. There are in like way estimations which depend upon a mix of morphological frameworks, division and edge unmistakable verification. Different advances solidify expansion, separating, smoothing, and division of characters [3].

Face area procedure modified system for unmistakable evidence of a person in perspective of the information contained in a dim scale picture. Various strategies have been made to handle this issue, for instance, principal component analysis (PCA) [14], dynamic link matching, and face recognition using neural networks [4]. An extensive bit of the practical face affirmation structures requires a face acknowledgment stage to recognize the territory of the face inside a source picture [5]. Face affirmation structures in like manner institutionalize the size and acquaintance of the face with achieves more power. The institutionalization procedures use the zone of the immense facial component, for instance, eyes, nose, or mouth. The essentialness of effective facial component disclosure for both distinguishing proof and affirmation has achieved the headway of an extensive variety of facial segment area counts [6, 7]. The geometric component-based philosophies are the soonest approaches to manage go up against affirmation and acknowledgment. These systems were based on distinguishing solitary features, for instance, eyes, ears, head diagram and mouth, and evaluating different properties, for instance, eyebrow thickness and their vertical position or nose position and width, in a component vector that is used to address a face [8].

From the first component, vectors of test images and the photos/pictures in the database are gained. Second, an equivalence measure between these vectors, much of the time a base partition control, is used to choose the character of the face. Key section examination (i.e., PCA) is an essential authentic dimensionality diminishing methodology that has possibly transformed into the most understood and by and large used strategy for depiction and affirmation of human appearances. PCA, through the KL change, can remove most quantifiably imperative information for a game plan of pictures as a course of action of Eigenfaces when which can be used both to see and duplicate go up against pictures. Starting late, some gathering-based face affirmation structures are exhibited. In this paper, we proposed a novel social affair-based face affirmation system which relies upon K-nearest neighbor classifier and sacking. The execution of 97.5% exactness is proficient over the ORL stand up to database [9].

The paper organized

2 Related Work

In [10], Chhoriya et al. have proposed and altered license plate recognition which utilizes infrared cameras to find pictures under moved lighting and climate conditions. The goal of this paper is to finished K-means clustering algorithm for license plate extraction and maximally over the best domain for name division, template arranging strategy for name insistence, and moreover divide in toll court and ceasing ranges regularly by seeing the number plates of vehicles which thusly decrease the advancement and utilization of time in toll stations.

In [3], Chow and Li have proposed the setting module which is a low-determination module which characterizes a face format as far as power valley districts. The valley areas are recognized utilizing morphological separating and 8-associated blob shading. The goal is to create a rundown of estimated confront areas positioned by confront probability. The nitty gritty investigation is left for the high-determination eye and mouth modules [15]. The go for both is to affirm and also refine the areas and states of their individual highlights of intrigue.

In [11], Sarnay and Ancy have proposed a modified license plate recognition which uses infrared cameras to find pictures under moved lighting and atmosphere conditions. The objective of this paper is to realize K-means clustering algorithm for license plate extraction and maximally ludicrous region for label division, template organizing system for label affirmation, and moreover portion in toll court and stopping zones normally by perceiving the number plates of vehicles which consequently diminish the development and usage of time in toll stations.

In [12], Chang et al. have proposed LPR technique containing two rule modules: a label discovering module and allow number ID module. The past portrayed by soft controls tries to think labels from a data picture, while the last conceptualized in regards to neural subjects intends to recognize the number present in a tag.

In [13], Rahman et al. have proposed a sharp and clear figuring is shown in this paper for vehicle's label affirmation structure. In light of illustration organizing, this estimation can be associated for steady area of labels for social event data for investigating or for some application specific purposes. The proposed structure has been prototyped using C ++ and the exploratory results have been showed up for affirmation of Alberta labels.

3 Proposed Methodology

This section, the proposed method of architecture is shown in Fig. 1.

The input from camera image captured is of minimum 13 megapixel clarity by the traffic cameras and special security cameras to train the dataset.

In the preprocessing of image evolution, from the auto picture, the number plate is removed and it is honed utilizing any accessible honing channels, with the goal that we can unmistakably perceive every one of the characters.

In the extraction of number plate progression, morphological image handling is finished. At that point, it can separate the clamor and subtract it from the past image, with the goal that every one of the gaps that were available in the picture gets filled and the commotions are evacuated (Fig. 2).

The segmentation and recognition of characters procured image got from past advance at that point experience the jumping box technique. Here, the image is

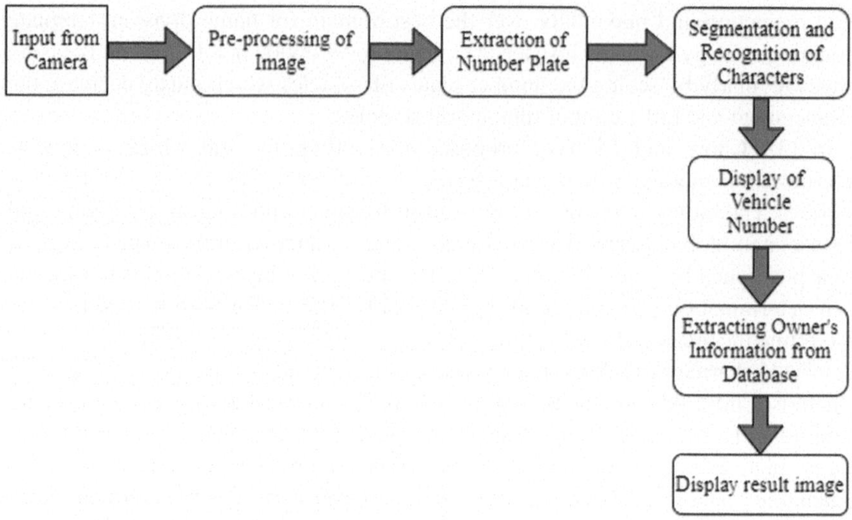

Fig. 1 Proposed method for data processing

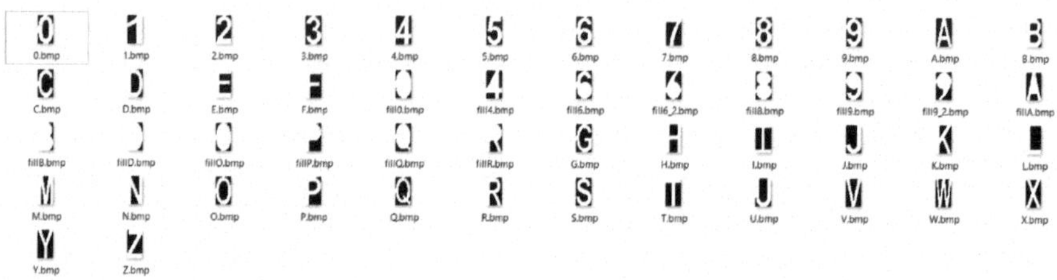

Fig. 2 Display of number plate representation

separated with the assistance of boxes. This is done to distinguish singular characters. The individual characters are separated and the zone with significant letters will be chosen as appeared.

Then, the display of number plate template matching is used to convert the recognized characters to a string of characters.

The extracting owner information message as content report is then placed in as a contribution to the current database application to separate the data about the vehicle and it is proprietor. Like the proprietor name, proprietor send, producer show, shading, fuel compose, motor limit, enrollment date, registration specialist, vehicle age, chassis number, and vehicle class.

4 Algorithm

Algorithm 1 : Vehicle details generation process

Input: Car image (.jpg) :Label

Output: Text number : Text Document

Step 1: Initialize select an image for morphological processing.

Step 2: Fill the object in selected image using bwmorph() function.

Step 3: Next, storage the data in region wise using props() function.

Step 4: Then select the area from region using cat() function.

Step 5: Select the car number text using connec() function.

Step 6: save data successfully

The K-nearest neighbor (KNN) classifier is an expansion of the basic closest neighbor (NN) classifier framework. The KNN classifier works in light of a basic nonparametric choice. Each question image is inspected in light of the separation of its highlights from the highlights of different pictures in the preparation database. The closest neighbor is the picture which has the base separation from the question picture in the component space. The separation between two highlights can be estimated in light of one of the separation capacities, for example, city piece removes d_1, Euclidean separation d_2 or cosine separate $d_{\cos}$:

$$d_1(x, y) = \sum_{i=1}^{N} |x_i - y_i|$$

$$d_2(x, y) = \sqrt{\sum_{i=1}^{N} |x_i - y_i|}$$

$$d_{cos}(x, y) = 1 - \frac{\hat{x} \cdot \hat{y}}{|x| \cdot |y|}$$

The KNN algorithm calculates and uses the tests of K nearest to detect the image. The KNN classifier is implemented by taking a unique value of K, the quality and quantity of the performed tests, and their dispersion topologically over the space of the component.

An ensemble classifier is a method of wrapping the data for developing the accurate data from a group of classifiers by taking a mixture of their output results by the voting methods. Generally, a normal type of learning system consists of feature detection and decision making units separately for a classifier. The classifier finally examines all the trained data against the decision function. A normal learning type of algorithm enables the parameters or boundaries to the classifier depending on the trained data which results certain rate of accuracy. Then, the results of the data tested are predicted by the system. In most cases, it is shown as 20, a committee of classifiers that can develop a better and proper predictions than the normal individual classifiers.

Bagging which is also called as or "bootstrap aggregation," it is an effective algorithm of machine learning. In this process, we use a different case of averaging the data, which is also being used for decision tree type of classification, but it can also be used with various other types of classification models. This method uses various other types of training data to work with different data models and with various other models of classifier. Every other dataset can be trained with the sampling technique and replacement of the data.

5 Algorithm

Algorithm : recording faces

Input: Training data set of faces that has to be recorded as sample database.

Output: final image Classified-features.

Step 1: Initialize cam and record the sample data faces data = [] ,ix = 0.

Step 2: while True: ret, frame = cam.read()

Step 3: if ret == True:

gray = cv2.cvtColor (frame, cv2.COLOR_BGR2GRAY)

faces = face_cas.detectMultiScale(gray, 1.3, 5)

Step 4: for (x, y, w, h) in faces:

face_component = frame[y:y+h, x:x+w, :]

fc = cv2.resize(face_component, (50, 50))

if (ix%10) == 0 and len(data) < 20: data.append(fc)

cv2.rectangle(frame, (x, y), (x+w, y+h), (0, 255, 0), 2)

ix += 1 cv2.imshow('frame', frame)

Step 5: if cv2.waitKey(1) == 27 or len(data) >= 20:else: print "error

Step 6: save data successfully

After saving the different samples of data, now, we perform KNN algorithm to recognize the face and map to recorded face and give details.

Algorithm: face recognition and mapping

Input: Training data set of images or faces from the cam recorder.

Output: final identified-Classified-features.

```
Step 1: Initialize and load the data set into linear vectors and create a mapping detail to every vector.
Step 2: compute distance
distance(x1, x2):
np.sqrt(((x1-x2)**2).sum())
Step 3: knn(x, train, targets, k=5):
  m = train.shape[0]
dist = []
step 4:   for ix in range(m):
dist.append(distance(x, train[ix]))
dist = np.asarray(dist)
indx = np.argsort(dist)
sorted_labels = labels[indx][:k]
counts = np.unique(sorted_labels, return_counts=True)
return counts[0][np.argmax(counts[1])]
Step 5: if ret == True:
gray = cv2.cvtColor(frame, cv2.COLOR_BGR2GRAY)
faces = face_cas.detectMultiScale(gray, 1.3, 5)
Step 6: after this the image is to be processed back and map to the detected data details.
```

---

6 Results

The special technique works very efficiently and can be used in many parts of areas where license plates are to be mapped with the details of the vehicle owner and also recognizes the face of the driver and maps to the database, and the system can be used in many developing countries with more crime rate instead of regular traffic cameras. The image capturing process can be done as represented in Fig. 3.

In this process of results and analysis, totally, 50 images were used to test the proposed methodology for number plate recognition using artificial intelligence technique. Moreover, the proposed method can be treated for only producing result of displayed image to recognize the car number plate with help of automation method.

Figure 4 shows the need of make a database of numbers were taken with make

Fig. 3 Image capturing process using number plate recognition

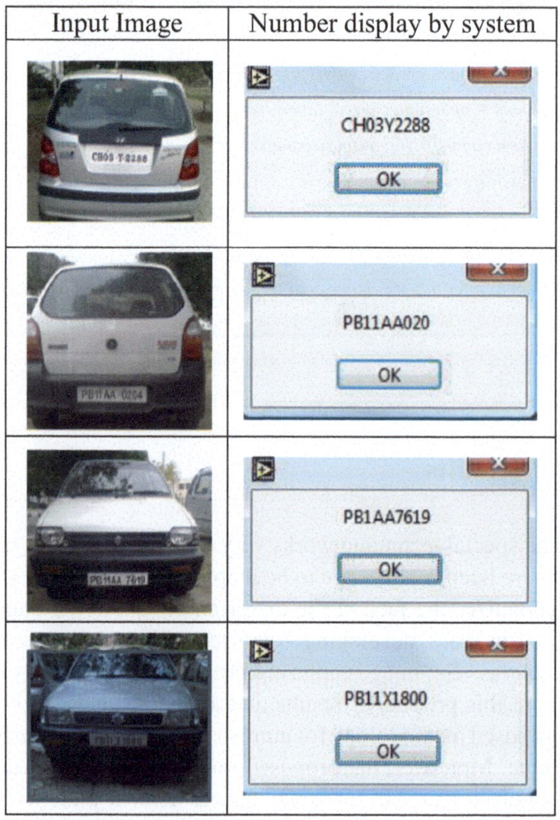

Fig. 4 Results of four out of 50 sample images tested by proposed method

automatic software app to check against the database each time it recognize any number plate from the cars. With help of this, we can reduce the time complexity in toll plazas to pay the toll fee in highways. This can be done for also home land security control process environment.

This device can not only be used for decreasing crime but also for various other fields like speeding tickets, parking tickets, and toll tickets.

7 Conclusions

Although the algorithm looks complicated but the fact is it detects the vehicle details easily with few limitations in general, and if the number plate is with cracked and other combination of colors, the system might show the error, or if there are similar types of alphabets and the clarity of the image is not that good with differentiating lines like, capital O and capital D, and capital B, capital O and number 0, these may not perform well without the clarity of the image.

Also, we have assumed that the number plates are of standard form; if the plates or of new design or stylish, the system cannot process and detect it. We cannot detect the faces through mask and sometimes, it may not work efficiently with twins and people who look alike, but we can still overcome it by training the variant data using machine learning techniques.

In the future, this type can be considered for stains, smudges, blurred images, and different styles of fonts can be taken from cars. Also, it can be extended to minimize the errors due to them.

References

1. Saranya K, Ancygloria C (2014) License plate recognition for toll payment application. 713–717
2. Rahman CA, Badawy W, Tn C, Radmanesh A, Tp C (2003) A real time vehicle's license plate recognition system. 4–7
3. Chow G, Li X (1993) Towards a system for automatic facial feature detection. Pattern Recogn 26(12):1739–1755
4. Pathak VM, Satonkar SS, Khanale PB (2015) An efficient approach for 3D face recognition using ANN based classifiers
5. Chhoriya P, Paliwal G, Badhan P (2013) Image processing based automatic toll booth in indian conditions. 3(4):410–414
6. Chang S, Chen L, Chung Y, Chen S, Member S (2004) Automatic license plate recognition. 5(1):42–53
7. Anagnostopoulos CE, Anagnostopoulos IE, Psoroulas ID, Loumos V, Kayafas E (2008) License plate recognition from still images and video sequences : a survey. 9(3):377–391
8. Turk MA, Pentland AP (1991) Face recognition using eigenfaces. In: Computer vision and pattern recognition, 1991. Proceedings CVPR'91., IEEE computer society conference on, IEEE, pp 586–591

9. Ebrahimpour H, Kouzani A (2007) Face recognition using bagging KNN. In: International conference on signal processing and communication systems (ICSPCS'2007) australia, gold coast, pp 17–19
10. Chhoriya P, Paliwal G, Badhan P (2013) Image processing based automatic toll booth in indian conditions. In: International Journal of Emerging Technology and Advanced Engineering, vol 3, no 4
11. Saranya. K, Ancy Gloria C License plate recognition for toll payment application
12. Chang S-L et al. (2004) Automatic license plate recognition.In: IEEE transactions on intelligent transportation systems, vol 5, no 1, pp 42–53
13. Rahman CA, Badawy W, Radmanesh S (2003) A real time vehicle's license plate recognition system.In: Advanced Video and Signal Based Surveillance, 2003. Proceedings. IEEE Conference on. IEEE
14. Yong-dong W, Dong-wei X, Peng P, Yi L, Gui-jun Z, Xiao X (2019) Kernel PCA for road traffic data non-linear feature extraction. IET Intel Transport Syst
15. Suvorov R (2015) The use of eye tracking in research on video-based second language (L2) listening assessment: a comparison of context videos and content videos. Language Testing. 32(4):463–483

Digital Transformation Using Immersive Technologies in Manufacturing and Utilities

Anurag Choudhry and Anshu Premchand

Abstract Immersive technologies have a tremendous ability to be a growth driver in digital transformation journey for manufacturing and utilities industries. This is an emerging technology area which has been in research and development for many years. The paper looks at the relationship between immersive technologies (augmented reality, virtual reality, and mixed reality), drivers and trends, some use cases applicable particularly in manufacturing and utilities industries followed by the approach for creating immersive applications and representative architecture. This paper also highlights the leading development tools available in the market and factors that may be considered before selecting the right development kit.

Keywords Augmented reality · Virtual reality · Mixed reality · Immersive technologies · Immersive experience · Augmented experience

1 Introduction

In the current global business environment, manufacturing and utilities industries are facing tough challenge to create new and innovative products and/or services at lower cost at an improved time to market. In order to be efficient and competitive, manufacturers must re-imagine their processes. Using immersive technologies, manufacturing and utilities processes can be simulated and improved before executing. This would enable various activities such as planning, designing, and so on right at the first instance and would eliminate the need to rework.

A. Choudhry (✉)
Solution Architect Research & Innovation, Manufacturing & Utilities Business Group, Tata Consultancy Services, Delhi, India
e-mail: anurag.choudhry@tcs.com

A. Premchand
Agile & DevOps Practice Lead, Manufacturing & Utilities Business Group, Tata Consultancy Services, Chennai, India
e-mail: anshu.premchand@tcs.com

M. N. Favorskaya et al. (eds.), *Innovations in Electrical and Electronic Engineering*, Lecture Notes in Electrical Engineering 661, https://doi.org/10.1007/978-981-15-4692-1_33

As virtual reality, augmented reality, and mixed reality are becoming mature, manufacturers are adopting these technologies for implementing several use cases such as training, product design and development, field services, worker safety, and so on. With the advancement in technology, new generation reality devices and SDKs are available for creating innovative applications. As industry thought leaders and numerous analysts are predicting that these technologies will take off, we will get into new realities.

In manufacturing and utilities industries, every moment is of importance, production floors never stop running and a small mistake in product design can cost the manufacturer millions. AR is a next-generation technology that augments not just things around a person but in manufacturing and utilities context, it provides greater insights into products and machines. These technologies play a critical role in the blending of virtual and physical worlds. Immersive technologies have the potential to be game changers in the area of ensuring physical safety and monitoring of physical assets such as power plants, power distribution lines, water pipes, gas distribution networks, and so on.

2 Immersive Technologies

An immersive technology is the one that integrates virtual information into the real-world environment so that a user can engage naturally with fused reality. Immersive technologies include virtual reality, augmented reality, and mixed reality.

Virtual reality is a computer-generated digital environment which replaces the user's real-world environment. This is totally immersive environment where everything is computer generated. Virtual reality is not new technology and has been around for many decades. However, with advancement in technology, experience has improved significantly in recent years.

Augmented reality is the integration of digital content with the physical environment. In this technology, user sees the real-world objects with additional information that are augmented by creating a virtual object. Virtual objects are created by computers to enhance user's sensory perception. Users can not differentiate between real world and virtual world which is augmented by computers.

Mixed reality is fusion of real world and digital world. Mixed reality creates new environments and visualizations and enables interaction amongst the elements of both real and digital worlds in real time.

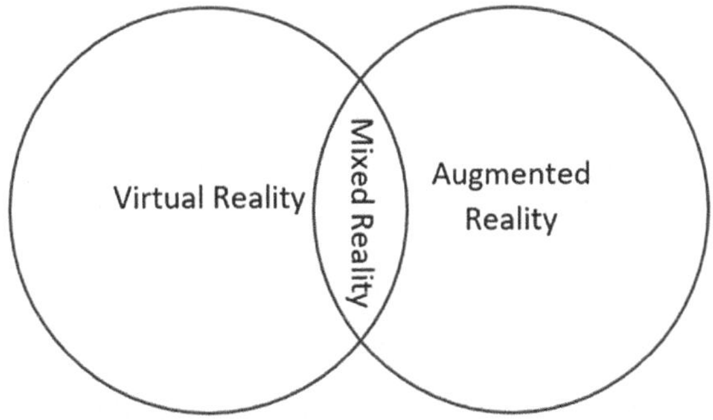

Fig. 1 Relationship between AR, VR, and MR

3 Relationship Between Augmented Reality (AR), Virtual Reality (VR), and Mixed Reality (MR)

Virtual reality (VR), augmented reality (AR), and mixed reality (MR) are not the same; Virtual reality is the umbrella term used to describe the other two technologies.

VR brings you into the digital world, AR superimposes digital content over physical world around us, and MR combines virtual world with the physical world.

VR limits the view of the world whereas AR opens the field of vision with many more possibilities. MR is the abstract concept of VR and AR. MR is not a technology rather a combination of both VR and AR.

For better explanation, VR, AR, and MR are depicted in Venn diagram as in Fig. 1. Mixed reality is the area that is common ground to augmented reality and virtual reality.

4 Differences Amongst VR, AR, and MR

Table 1 given below summarizes the difference between VR, AR, and MR.

To summarize the relationship between VR, AR, and MR—if one needs a special device to experience the reality, then it is VR and if mobile/tablet/computer vision is used, then it is AR. Alternatively, if a device such as HoloLens is required to scan the physical objects around one's surroundings and interaction with both the physical and virtual objects in real time is possible, then it is MR.

Table 1 Characteristics of AR, VR, and MR

Characteristics	VR	AR	MR
User aware of the physical world		✓	✓
Augments the physical world with digital content		✓	✓
User can interact with the physical world and virtual world in real time		✓	✓
Fused holograms with the physical world			✓
Physical objects and virtual objects can interact with each other in real-time			✓
Brings you into a virtual world	✓		✓
Replaces the physical world around you	✓		
User fully immersed with virtual environment	✓		
Fully artificial environment	✓		
Special hardware devices are required to experience	✓		✓
Aims at enhancing human experience by providing deeper interaction	✓	✓	✓

5 Drivers and Trends

In past few years, immersive technologies have been refined and transformed into powerful tools that are driving innovation and transforming businesses across industry verticals such as manufacturing, utilities, education, health care, entertainment, and so on.

Following are key drivers for the growth of immersive technologies:

5.1 Cost-Effective Solutions

Using immersive technologies, cost effective solutions are being developed in various industry verticals. For instance, in aviation industry, training solutions are being created to train maintenance engineers and pilots where virtual environments are created which are like real-world environments using AR and VR. Trainings using AR/VR solutions are very cost effective as compare to the training using actual aircrafts. Immersive technology-based solutions reduce cost significantly in numerous manufacturing and utilities processes. Many large corporations such as Boeing, Caterpillar, BMW, Ford, AGL Energy, National Grid and so on have been using AR/VR for years and have created solutions that are saving cost and time in many ways.

5.2 *Short Learning Curve*

Immersive technologies-based solutions are easy to use and intuitive in nature. The researchers opine that the crunch of resources that manufacturing and utilities industries are witnessing due to large number of skilled resources retiring is leading to creation of innovative solutions for hastening the learning curves for new recruits. The new generation of workers has grown up with the digital devices and is fond of digital interfaces. Organizations are making their jobs intuitive by providing tools that they are comfortable with. An added advantage is the fact that these tools make learning curve very short. Therefore, rate of adoption of immersive technologies will be higher for the new generation of workforce in manufacturing and utilities industries

5.3 *Reduction in Maintenance Time*

Timely maintenance is important for any machine to work without interruption. Immersive technology-based solutions enable technicians and engineers to find out errors and fix them quickly. Timely maintenance reduces machine downtime and thus increases the productivity of the production line it is a part of. BMW [1] has developed AR–VR solutions to help technicians which enable them to gain faster access to technical information. BMW's Technical Information System (TIS) allows technicians to access all technical data required to fix any issues with a BMW car without having to the shop floor.

5.4 *Better Remote Assistance*

For the longest time, no one could imagine how immersive technologies would improve field services and remote assistance. Sometimes, for complex issues service, a technician may need help from an expert or engineer remotely. BMW [1] has provided TSARAVision Smart Glasses to their dealers and workshops; using TSARAVision, technicians can connect with experts who are at remote locations via a hands-free video link and can fix the fix issues faster and in a more efficient manner. Experts can project step-by-step technical information inside the technician's glasses and can also take screen shots for better visibility and understanding.

According PwC report [2], VR and AR combined sales forecast to $150 billion by 2020 and AR alone can hit to $120 billion. Further, this report suggests that AR and VR applications will mostly be used in product design and development. Other VR and AR applications will be developed for repair, maintenance, skills development, worker safety, and field services. According to another report by Allied Market Research [3], market size of AR and VR was $11.35 billion in 2017 which they forecasted to become \$571.42 billion by 2025, growing at a CAGR of 63.3% from 2018 to 2025.

All the efforts that have been put into research and thought leadership in immersive technologies have started paying off in past few years. Innovation in computer vision and hardware devices enabled immersive technologies are adding immense value to business and taking user experience at new levels.

6 Some Use Cases of Interest

6.1 Design and Development

There are several stages in product development life cycle. Immersive technologies have potential to enable almost every stage of product creation. For instance, in automotive industry, converting a concept to final product is a rigorous process which includes the involvement of different stakeholders and several revisions in the initial concept before freezing the design of the car and takes several months and years in some cases. Using immersive technology-based solutions, manufacturers can re-imagine this entire process. AR/VR/MR solutions can increase the collaboration between stakeholders, and they can visualize the final product easily in a more intuitive way. This will also enable stakeholders to share their feedback and advice quickly which will reduce the rework and increase productivity.

Automotive companies are leveraging immersive technologies to make design processes more efficient. For instance, Ford [4] has developed an immersive system Ford immersive vehicle environment (FIVE) which converts designs (CAD—computer-aided design) into a virtual vehicle. This enables engineers to view the interior and exterior down to half a millimetre. This solution allows collaboration with other stakeholders across geographies; designers can discuss and take their requirements and feedback within the tool which saves costs significantly, including for essentials like travel and co-location and above all, saves time.

6.2 Quality Assurance

Quality assurance (QA) is one the key use cases of immersive technologies. Some manufacturers around the world have already started exploring these technologies for quality inspections. Quality engineers in an automotive manufacturer plant can take pictures of parts or assemblies to compare with the pictures provided by an AR overlay. AR solution can enable QA engineers to easily identify parts which are out of specification(s). This can improve the inspection process significantly as well as quality of the end-product.

6.3 Automation

Many strategists, technologists, and industry thought leaders believe that immersive technologies and automation are competing, but in a positive way. Augmented reality solutions provide information to workers based on the data collected from (Internet of things (IoT) sensors which gathers information about the environment of the device. With evolution of emerging technology areas such as artificial intelligence, augmented reality, and Internet of things, one can imagine and create solutions which could not be envisioned earlier. One can ask an employee with a certain skill to work on an activity and by augmenting her/him with AI using AR and IoT, the employee's skill levels can be elevated. The researchers envisage that with more automation in the industry, workers will move up in skill matrix and AR will be the key enabler in automation.

6.4 Talent Development

Creation of training material in traditional way is very time consuming. It is also pertinent to mention that conventional learning methods are also unexciting for the new generation of workers in most industries. Immersive technology-based solutions are an excellent choice for training. AR solutions enable trainers to create AR experiences rather than creating lengthy documents. Using sensors and computer vision, audio and visuals are captured including manufacturing and utilities processes. AR solutions also connect workers in the plant with an expert who can be at remote location. These solutions even give flexibility to employees who are going to retire and want to continue to work.

6.5 Safety

In manufacturing and utilities industries, safety is very critical and of utmost importance. Everybody who works in plants, power generation units, distribution centers, and so on must be ready for emergency situations such as runaway, chemical leaks, explosions and so on. Several other scenarios such as plant shutdown or machinery upgrade have heightened security and safety requirements. Immersive technology-based solutions can be very effective for these scenarios by providing virtual environments for practice. Every worker can be trained for emergency situations and behavior appropriate to respond to such situations can be taught easily and intuitively. Immersive digital twins could be a milestone for training where physical environments cannot be made available.

7 Approach for Creating an Immersive Application

This section outlines an approach that can be followed for creating an immersive application. It is a high-level approach but provides a clear path to create an immersive application.

7.1 Discovery

This is the first and foundational step which is started with the definition of business goals and use cases. Next steps of this phase are selection of devices and software development kits (SDKs) based on the business needs and use case followed by the technical requirements analysis.

7.2 Design

This is related to the design of immersive experience which includes user flows, storyboarding, user experience (UX), and creation of prototype. This provides a feel to the user on how product/application will look like.

7.3 Asset Creation

This includes the creation of 3D objects, videos, audios, graphics, and animations according to the stories finalized in previous step.

7.4 Build and Test

In this phase, developers create custom application using assets created in previous step and write code to implement the use case/story. Simulators can be used to test the developed use case.

8 Representative Architecture

Figure 2 shows a representative architecture to create an augmented reality application [5]. Here, the researchers have considered an AR mobile application. This architecture will help AR architects and developers to understand different components/layers/services of the said AR mobile application.

8.1 Application

This layer consists of the entire logic related to application. For instance, a remote assist application, this includes presentation components (3D objects, scenes, and so on), business rules, and cross-cutting concerns.

8.2 AR Services/Components

This layer consists of AR service components. Registration and interaction are related to the detection of position of physical world elements. Tracking service provides geographical (longitude and latitude) information to registration service. Registration services align physical and virtual content/objects in the same space. Without knowing where physical objects are in space and where they are moving (tracking), it is not possible to overlay digital content on physical objects. Display service accesses

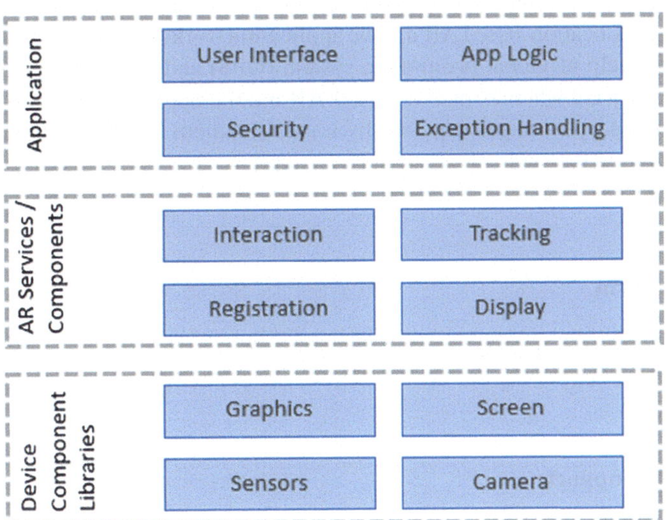

Fig. 2 AR mobile application—Representative architecture

the camera and shows captured images. Also, display services match camera parameters of physical objects with virtual ones and render digital content on physical objects. Once the content is rendered properly, a user can start interacting with it.

8.3 Device Component Libraries

Although these libraries are not part of AR application, they play a critical role in the working of the AR application. This includes OS libraries which enable developers to access the functionality of different device components such as camera, sensors, and so on. For instance, Google Android Application Programming Interface (APIs) is used to access the camera on android-based devices.

8.4 Development Tools

Selection of right SDKs and tools sets are essential for any software application and same applies to immersive applications as well. Following are some key considerations for selecting the right SDKs:

8.4.1 Application Type

One should decide the type of application that needs to be created. For instance, in augmented reality space, there are two types of applications; location-based and marker-based. Location-based AR mobile application works by detecting user's position with the help of global positioning system (GPS) and overlays digital content on physical objects whereas marker-based AR mobile application works on image recognition concept. Images captured by a user's camera are analyzed against the markers stored in the application and such an application overlays digital content on physical object.

8.4.2 Platform

In case of a native application, platform selection (Android, iOS, Windows) decision is critical.

8.4.3 GPS Support

Without geolocation support, a location-based application cannot be created. So, one needs to consider this parameter while selecting the SDKs.

8.4.4 Cost

If application is for a simple demonstration or R&D (research and development) purpose, open-source SDKs may be considered. Some available SDKs have different licensing models such as free, commercial, enterprise, and so on which can be selected based on the use case.

8.4.5 Image Recognition & Tracking

This is a key capability. SDKs must have these features which allow applications to recognize images, objects, and places.

Some of the leading SDKs available in the market are Vuforia, Google ARCore, Apple ARKit, ARToolKit, Wikitude, Maxst, Microsoft MixedReality ToolKit.

9 Recommendations and Conclusions

Immersive technologies (AR, VR & MR) have moved beyond R&D phase to the center stage in manufacturing and utilities industries. It is quite evident from drivers and trends that industries across the board have started building strategies for creating immersive experiences. The researchers recommend a deeper inspection of the use cases mentioned in this paper, given the benefits of using immersive experiences. The researchers believe that immersive technologies are going to be key enablers in digital transformation journeys for industries across the spectrum. The researchers have also discussed an approach and representative architecture as well as some development tools selection considerations for creating immersive applications.

Immersive technologies are going to become much bigger than what is currently being visualized. These technologies are going to revolutionize the way we interact with computers (phone, computer, television, and so on) in a big way.

References

1. https://www.bmwblog.com/2019/06/11/bmw-technicians-now-have-augmented-reality-smart-glasses-to-rely-on/
2. https://www.pwc.com/us/en/industrial-products/publications/assets/augmented-virtual-reality-next-manufacturing-pwc.pdf
3. https://www.alliedmarketresearch.com/augmented-and-virtual-reality-market
4. Kim K (2016) Is Virtual reality (VR) becoming an effective application for the market opportunity in health care, manufacturing, and entertainment industry?. Euro Sci J 12(9) ISSN: 1857 – 7881 (Print) e - ISSN 1857– 7431
5. Grasset R, Grubert J (2013) Augmented reality for android application development. Packt Publishing

Non-parametric Frontier Analysis Models for Relative Performance Evaluation

Shubham Gupta, Vinod Kumar Yadav, Santosh Ghosh, and D. Saravanan

Abstract Data envelopment analysis (DEA) model was introduced nearly half a century ago for evaluation of relative efficiency of entities having homogeneous set of input and outputs. Classical DEA has been applied to wide and diverse fields of research since its inception, due to its applicability to universal problems without any prior knowledge of functions of entities understudy or the interrelation between the inputs and outputs. However, it has certain shortcomings and several extensions of the basic DEA model have been introduced in the literature. This paper discusses the basic concept of the classical DEA models, its advantage and disadvantages vis-à-vis more recent extensions of this model.

Keywords Data envelopment analysis · Classic DEA model · Return to scale

1 Introduction

Since its introduction by Charnes et al. [1] in 1978, data envelopment analysis (DEA) has gained popularity as an effective mathematical tool for computing relative efficiency of homogenous decision-making units (DMUs). It is a simple, but extremely important development in the field of relative performance evaluation, as it enables the generation of performance indices, slack (deficit) analysis of DMUs, having multiple inputs and outputs, without having any information regarding the underlying functional relation between the inputs and outputs. It has been applied to wide and diverse fields of study by research from around the globe, viz. power generation [2], power transmission and distribution utilities [3], banking [4], education [5],

S. Gupta (✉) · V. K. Yadav
Department of Electrical Engineering, Delhi Technological University, New Delhi 110042, India
e-mail: shubham.gupta6691@gmail.com

S. Ghosh
Corporate R&D Department, Kirloskar Brothers Limited, Pune 411045, India

D. Saravanan
School of EE&C Engineering, Galgotias University, Greater Noida 203201, India

M. N. Favorskaya et al. (eds.), *Innovations in Electrical and Electronic Engineering*,
Lecture Notes in Electrical Engineering 661, https://doi.org/10.1007/978-981-15-4692-1_34

hospitals [6], sports [7], supply chain [8], agro-industry [9], and many new fields are adopting it day by day. Recently, DEA method is being applied in some very novel and innovative ways as well. For example, it has been applied for analyzing, indexing, and evaluating experimental results pertaining to performance of photovoltaic cells [10, 11] under different environmental conditions, very recently.

Several DEA models have been proposed by researchers so far. However, the application of traditional or classical DEA method in real-world problems still dominates the literature, though lack of discrimination in efficiency ranking is a concern pertaining to these models [10–12]. Thus, to overcome this issue, many authors suggested various extensions in basic DEA models, which are additive model [13], cross-efficiency model [14], multi-criteria DEA [15], slack-based model [16], free disposal hull model [17], and super efficiency model [18] improved DEA model by Shannon's entropy [10, 11].

In the present paper, aspects of fundamental DEA models like constant return to scale (CRS), variable return to scale (VRS), their limitations and weight assignment to parameters. However, the present work is not intended to provide complete coverage of literature as it is voluminous [19–21]. The remaining paper is organized as follows: Sect. 2 explains the mathematical modeling of classical DEA models (radial and non-radial). The result and discussion of a case study are presented in Sect. 3 followed by the conclusion in Sect. 4.

2 Mathematical Model of DEA

Earlier, DEA did not get much attention due to the limitation of its assumption of constant return to scale (CRS). Many economists preferred to resort to statistical methods over the DEA for performance assessment. Later on, Banker et al. [22] included convexity constraint in the basic CCR model to assert the variable return to scale (VRS) property. The concept of VRS can be understood through the production function graph (Fig. 1) at ease.

From Fig. 1, it may be noted that for a DMU the ratio of *Y2/Y1* is greater than *X2/X1,* it is said to be the increasing return of scale (IRS), while on the other hand *Y3/Y2* is less than *X3/X2,* is said to be the decreasing return of scale (DRS). Mathematically, the same can be expressed as:

If $\sum_{j=1}^{n} \lambda_j < 1$, implies increasing return of scale.

If $\sum_{j=1}^{n} \lambda_j > 1$, implies decreasing return of scale.

Where $\sum_{j=1}^{n} \lambda_j = 1$ is a convexity constraint in BCC model and implies that DMU is efficient.

Fig. 1 Production function

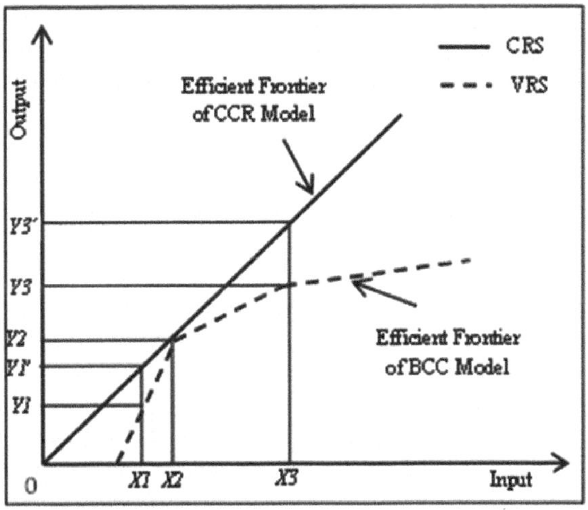

3 DEA Methodology

3.1 CCR DEA Model

To measure the relative efficiency of n DMUs: $DMU_j(j = 1, 2, \ldots, n)$ using similar m inputs $x_{ij}(i = 1, 2, \ldots, m)$ to produce s outputs $y_{rj}(r = 1, 2, \ldots, s)$, the mathematical formulation is given as

$$\text{Max}\theta_0 = \frac{\sum_{r=1}^{s} v_r y_{ro}}{\sum_{i=1}^{m} u_i x_{io}} \tag{1}$$

s.t.

$$\frac{\sum_{r=1}^{s} v_r y_{rj}}{\sum_{i=1}^{m} u_i x_{ij}} \leq 1, \ \ j = 1, 2, \ldots, n \tag{2}$$

$$u_i, v_r \geq 0 \quad \forall i \text{ and } r \tag{3}$$

In (1) to (3), v_r and u_i are the multiplier weights to outputs and inputs, respectively. The above model can be written in its dual form, which is as follows:

Envelopment version of input-oriented CCR model:

$$\text{Min}\theta_i \tag{4}$$

s.t.

$$\sum_{j=1}^{n} \lambda_j x_{ij} - x_{io}\theta_i \leq 0 \forall i \tag{5}$$

$$-\sum_{j=1}^{n} \lambda_j y_{rj} + y_{ro} \leq 0 \forall r \tag{6}$$

$$\lambda_j \geq 0 \forall j \tag{7}$$

Envelopment version of output-oriented CCR model:

$$\text{Max}\theta_o \tag{8}$$

s.t.

$$\sum_{j=1}^{n} \lambda_j x_{ij} - x_{io} \leq 0 \forall i \tag{9}$$

$$-\sum_{j=1}^{n} \lambda_j y_{rj} + \theta_o y_{ro} \leq 0 \forall r \tag{10}$$

$$\lambda_j \geq 0 \forall j \tag{11}$$

where θ_i and θ_o are the efficiency score in case of input-oriented and output-oriented models. λ_j, is the dual variable for j_{th} DMU. The DMU is said to be efficient when score obtained from the above model is equal to 1.

3.2 BCC DEA Model

BCC model is a variable return to scale model proposed by Banker et al. [22]. The addition of convexity constraint (15) allows DMUs to become efficient despite operating at different scales. The BCC efficiency (technical) can be evaluated by solving the following linear program (LP):

Envelopment version of input-oriented BCC model:

$$\text{Min}\theta_i \tag{12}$$

s.t.

$$\sum_{j=1}^{n} \lambda_j x_{ij} - x_{io}\theta_i \leq 0 \forall i \tag{13}$$

$$-\sum_{j=1}^{n} \lambda_j y_{rj} + y_{ro} \leq 0 \forall r \tag{14}$$

$$\sum_{j=1}^{n} \lambda_j = 1 \tag{15}$$

$$\lambda_j \geq 0 \forall j \tag{16}$$

Envelopment version of output-oriented BCC model:

$$\text{Max}\theta_i \tag{17}$$

s.t.

$$\sum_{j=1}^{n} \lambda_j x_{ij} - x_{io} \leq 0 \forall i \tag{18}$$

$$-\sum_{j=1}^{n} \lambda_j y_{rj} + \theta_o y_{ro} \leq 0 \forall r \tag{19}$$

$$\sum_{j=1}^{n} \lambda_j = 1 \tag{20}$$

$$\lambda_j \geq 0 \forall j \tag{21}$$

The efficiency evaluated through above model is also known as technical efficiency. Here, if (15) is substituted by $\sum_{j=1}^{n} \lambda_j \leq 1$ than model assumes non-increasing return to scale (NIRS) or $\sum_{j=1}^{n} \lambda_j \geq 1$ assumes non-decreasing return to scale (NDRS).

3.3 Additive DEA Model

This non-radial DEA model is introduced by Charnes et al. in [13]. Radial and non-radial are two ways to evaluate in DEA. BCC and CCR models are either input-oriented (the inputs are reduced by keeping the outputs constant) or output-oriented (the outputs are maximized keeping the inputs constant), called as radial models. The additive model combines orientations, viz. input and output in a single DEA model. This model suggests reducing input and increasing output simultaneously for inefficient DMUs to move to the efficient frontier.

A CCR or BCC model can be modified into an additive model. Here, a BCC model altered in an additive model presented as:

$$\text{Max } P_0 = \sum_{i=1}^{m} S_i^- + \sum_{r=1}^{s} S_r^+ \tag{22}$$

s.t.

$$\sum_{j=1}^{n} \lambda_j x_{ij} - x_{io} + S_i^- = 0, \forall i \tag{23}$$

$$\sum_{j=1}^{n} \lambda_j y_{rj} - y_{ro} - S_r^+ = 0, \forall r \tag{24}$$

$$\sum_{j=1}^{n} \lambda_j = 1 \tag{25}$$

$$\lambda_j \geq 0; S_i^- \geq 0; S_r^+ \geq 0 \tag{26}$$

where S_i^- and S_r^+ are the input (excess) and output (deficit) slacks, respectively, and for zero slacks, the DMU_o is said to be additive-efficient.

4 Results

To illustrate the methodology presented in paper, a data set for 27 DMUs each having two inputs (*x1 & x2*) and two outputs (*y1 & y2*) is taken from [23], which is summarized in Table 1. It is noted that, the data is used for example purpose only; no comparison has been done in this paper with the originally published results by the authors. The efficiency results are obtained through application of each DEA model is presented in Table 2. Columns 2 and 3 present CCR efficiency scores for input-oriented (IO_CCR) and output-oriented (OO_CCR) models, respectively. Results computed from the input (IO_BCC) and output-oriented (OO_BCC) BCC models

Table 1 Data set having 27 DMUs [23]

DMUs	*x1*	*x2*	*y1*	*y2*
1	7.2	0.15	60	1.35
2	4.8	0.05	6	1.1
3	5	1.27	45	1.27
4	7.2	0.025	1.5	0.66
5	9.6	0.25	50	0.05
6	1.07	0.1	1	0.3
7	1.76	0.1	5	1
8	3.2	0.1	15	1
9	6.72	0.2	10	1.11
10	2.4	0.05	6	1
11	2.88	0.5	30	0.9
12	6.9	1	13.6	0.15
13	3.2	0.05	10	1.2
14	4	0.05	30	1.2
15	3.68	1	47	1
16	6.88	1	80	1
17	8	2	15	2
18	6.3	0.2	10	1
19	0.94	0.05	10	0.3
20	0.16	2	1.5	0.8
21	2.81	2	27	1.7
22	3.8	0.05	0.9	1
23	1.25	0.1	2.5	0.5
24	1.37	0.1	2.5	0.5
25	3.63	0.2	10	1
26	5.3	1.27	70	1.25
27	4	2.03	205	0.75

are presented in the 4th and 5th columns, respectively. Finally, the additive efficiency score (ADD) is appended. In this paper, the evaluated results are not intended to make any conclusions regarding DMUs like which is most efficient or least efficient ones. The purpose is to discuss the different aspects related to the classical DEA model.

From Table 2, it may be noted that the efficiency measures assessed by OO_CCR are greater than or equal to one, which is equal to the ratio of 1 to efficiency computed by IO_CCR. It may be justified by (1) in Sect. 2, where θ_i and θ_o denote denominator and numerator (in multiplier form). IO_CCR and OO_CCR dictate the overall efficiency while IO_BCC and OO_BCC portray the technical efficiency. The weak discrimination among DMUs can be thoroughly investigated from the results as more

Table 2 Efficiency scores obtained from different DEA models

DMUs	IO_CCR	OO_CCR	IO_BCC	OO_BCC	ADD
1	1	1	1	1	1
2	0.9097	1.0993	0.9074	0.9167	0.2219
3	0.5288	1.8909	0.6667	0.843	0.843
4	1	1	1	1	1
5	0.5924	1.6882	0.5938	0.7384	0.699
6	0.4824	2.0731	0.8649	0.6888	0.0834
7	1	1	1	1	1
8	0.7825	1.2779	0.7829	0.8571	0.7849
9	0.3814	2.622	0.3912	0.8204	0.1776
10	1	1	1	1	1
11	0.6713	1.4896	0.6766	0.8024	0.6731
12	0.1024	9.769	0.1419	0.1257	0.1076
13	1	1	1	1	1
14	1	1	1	1	1
15	0.6125	1.6326	0.6237	0.7823	0.7809
16	0.6035	1.657	0.6039	0.801	0.634
17	0.4045	2.4719	1	1	1
18	0.3652	2.7381	0.3668	0.7458	0.1841
19	1	1	1	1	1
20	1	1	1	1	1
21	0.8515	1.1743	1	1	1
22	0.8333	1.2	0.9132	0.8333	0.0637
23	0.6943	1.4403	0.9234	0.858	0.6418
24	0.6361	1.572	0.8465	0.7343	0.6016
25	0.5533	1.8072	0.5562	0.7932	0.2626
26	0.581	1.7211	0.7707	0.9035	0.9035
27	1	1	1	1	1

than one DMU is found efficient in all models of classic DEA; a major setback of classical DEA models.

"A CCR efficient DMU is also a BCC efficient but a BCC efficient DMU is not necessary to be a CCR efficient" can be verified from the results given in Table 2. Due to having same efficiency frontier, a BCC efficient DMU is also an additive-efficient. Table 3 shows the RTS properties for all 27 DMUs, which are evaluated using concept mention in Sect. 2.

The correlation between input and output parameters have been drawn in Fig. 2. It is said that correlation coefficient should be positive to maintain isotonic relationship [24].

Table 3 Return to scale

DMUs	RTS	DMUs	RTS	DMUs	RTS
1	CRS	10	CRS	19	CRS
2	IRS	11	IRS	20	CRS
3	DRS	12	IRS	21	DRS
4	CRS	13	CRS	22	IRS
5	IRS	14	CRS	23	DRS
6	IRS	15	IRS	24	IRS
7	CRS	16	DRS	25	DRS
8	IRS	17	DRS	26	DRS
9	DRS	18	DRS	27	CRS

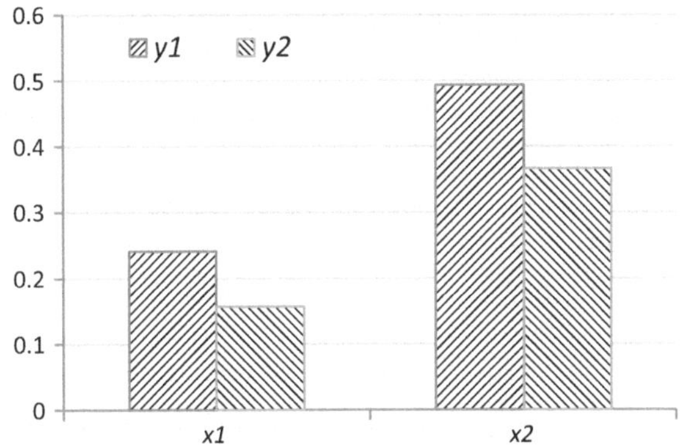

Fig. 2 Correlation between input and output parameters

5 Conclusions and Future Trends

This paper exhibits the different classic models of DEA. The aim here to review different aspects of traditional DEA models, viz. CRS, VRS, and their limitations, etc. However, the applications of DEA are immense and it is hard to cover all in a paper. In recent, multiplier restriction models, multilevel modeling, fuzzy logic, and stochastic models, etc., are in trends among researchers and economists. A lot of research still indispensable in the future for betterment.

References

1. Charnes A, Cooper WW, Rhodes E (1978) Measuring the efficiency of decision making units. Eur J Oper Res 2:429–444
2. Kumar V, Kumar N, Ghosh S, Singh K (2014) Indian thermal power plant challenges and remedies via application of modified data envelopment analysis. Int Trans Oper Res 00:1–23

3. Ghosh S, Yadav VK, Mehta G, Birajdar R (2017) Evaluation of Indian power sector reform strategies and improvement direction though DEA. In: 2017 IEEE power & energy society general meeting, pp 1–5
4. Laplante AE, Paradi JC (2014) Evaluation of Bank Branch Growth Potential Using Data Envelopment Analysis. Omega. https://doi.org/10.1016/j.omega.2014.10.009
5. Nazarko J, Šaparauskas J (2014) Application of DEA method in efficiency evaluation of public higher education institutions. Technol Econ Dev Econ 20(1):25–44
6. Kohl S, Schoenfelder J, Fügener A, Brunner JO (2018) The use of data envelopment analysis (DEA) in healthcare with a focus on hospitals. Health Care Manag Sci
7. Djordjević DP, Vujošević M, Martić M (2015) Measuring Efficiency Of Football Teams By Multi-Stage DEA Model. Tech Gaz 22(3):763–770
8. Soheilirad S, Govindan K, Mardani A, Zavadskas EK, Nilashi M, Zakuan N. (2017) Application of data envelopment analysis models in supply chain management : a systematic review and meta-analysis. Ann Oper Res
9. Toma E, Dobre C, Dona I, Cofas E (2015) DEA applicability in assessment of agriculture efficiency on areas with similar geographically patterns. Agric Agric Sci Proc 6:704–711
10. Ghosh S, Yadav VK, Mukherjee V (2017) Evaluation of cumulative impact of partial shading and aerosols on different PV array topologies through combined Shannon's entropy and DEA. Energy, pp 765–775
11. Ghosh S, Yadav VK, Mukherjee V, Yadav P (2017) Evaluation of relative impact of aerosols on photovoltaic cells through combined Shannon's entropy and data envelopment analysis (DEA). Renew Energy 105:344–353
12. Podinovski VV, Thanassoulis E (2007) Improving discrimination in data envelopment analysis: some practical suggestions. J Product Anal 28(1/2):117–126
13. Charnes A, Cooper WW, Golany B, Seiford L, Stutz J (1985) Foundations of data envelopment analysis for Pareto-Koopmans efficient empirical production functions. J Econom 30(1–2):91–107
14. Sexton TR, Silkman RH, Hogan AJ (1986) Data envelopment analysis : Critique and extensions. In: Silkman RH (ed), Measuring efficiency: an assesrment of data envelopment analysis. New Directions for Program Evaluation, San Francisco, no. 32, pp 73–105
15. Li XB, Reeves GR (1999) Multiple criteria approach to data envelopment analysis. Eur J Oper Res 115(3):507–517
16. Tone K (2001) A slacks-based measure of efficiency in data envelopment analysis. Eur J Oper Res 130:498–509
17. Tulkens H, Deprins D (1984) Measuring labor-efficiency in post offices. In: Marchand M, Pestieau P, Tulkens H (eds), The performance of public enterprises: concepts and measurement chap. 10. Amsterdam, North-Holland, pp 243–268
18. Andersen P, Christian N (1993) A procedure for ranking efficient units in data envelopment analysis. Manage Sci 39(10):1261–1264
19. Cooper WW, Seiford LM, Zhu J (2004) Handbook on data envelopment analysis
20. Ray SC (2004) Data envelopment analysis: theory and techniques for economics and operations research
21. Kuah CT, Wong KY, Behrouzi F (2010) A review on data envelopment analysis (DEA). In: 2010 Fourth Asia Int. Conf. Math. Model. Comput. Simul., pp 168–173
22. Banker RD, Charnes A, Cooper WW (1984) Some models for estimating technical and scale inefficiencies in data envelopment analysis. Manage Sci 30(9):1078–1092
23. Wu J, Sun J, Liang L (2012) DEA cross-efficiency aggregation method based upon Shannon entropy. Int J Prod Res 50(23):6726–6736
24. Golany B, Roll Y (1989) An application procedure for DEA. Omega Int J Manage Sci 17(3):237–250

Designing of Adaptive Depth Control for Autonomous Underwater Vehicle Using Type-2 Fuzzy Logic Controller

Devjani Bhattacharya and C. Puttamadappa

Abstract Adaptive depth control for an autonomous underwater vehicle is presented in this research. The response of an underwater system is moderate compared to air monitoring systems. The slow response is due to the speed constraints and high density of water. Establishing an accurate control technique for AUV is a difficult task due to the nonlinearity of hydrodynamic elements. The depth control analysis was done using PID control and type-2 fuzzy logic controller (Type-2 FLC). Type-2 FLC is used to tune the PID controller. The dynamic functionalities of an AUV are determined by using six degrees of freedom differential equations of motion keeping fixed earth as reference. Modeling and analysis of AUV are presented in the research and simulation of the model is done using MATLAB. Precise results for depth analysis are obtained using type-2 FLC.

Keywords Autonomous underwater vehicle · PID controller · Fuzzy logic controller

1 Introduction

Underwater vehicles are gaining popularity due to increasing research on underwater operations. Many researchers have concentrated their inclinations on the development of underwater vehicles which allows them to investigate the sea operations from the surface. These vehicles are employed to carry out various exercises such as surveillance under the sea, ocean data sampling and observation. Irrespective of their operating characteristics, i.e., either controlled by cable or autonomous vehicles, it is essential to come up with different control strategies to obtain aspired vehicle movements [14]. Underwater surroundings instigate various complexities in

D. Bhattacharya · C. Puttamadappa (✉)
Dayananda Sagar University, 560068 Bengaluru, Karnataka, India
e-mail: puttamadappa@gmail.com

D. Bhattacharya
e-mail: devjani.banerjee@gmail.com

M. N. Favorskaya et al. (eds.), *Innovations in Electrical and Electronic Engineering*, Lecture Notes in Electrical Engineering 661, https://doi.org/10.1007/978-981-15-4692-1_35

controlling and navigation strategies and make communication of underwater vehicles a problematic task. The important distinction between an autonomous underwater vehicle (AUV) and remotely operated vehicle (ROV) is that ROV is associated with a command station using a tethered cable. The tethered cable secures the power supply data communication signals allowing the operators to track, control and access the vehicle information. AUV is provided with a battery as power supply and SONAR to accomplish the given assignment eliminating the need of operator, which continually requires complex control strategies. It must be sufficiently straightforward to make sure there is efficient online execution of various control strategies. Still, simultaneously, the control strategies need to cope up with abrupt vehicle or territorial association [2]. In recent times, AUV is turning out to be increasingly more well known since it can work and traverse through utmost depths, which is helpful for offshore applications. While simultaneously, it is a cost-effective fit for carrying out surveillance and disposal operations in the NAVY. The concept of AUV consists of two vehicle groups such as flight vehicles and hovering vehicles. Flight vehicles are widely applied in related search operations, location tracking and object monitoring, whereas hovering vehicles find its usage in the investigation and manual work-related projects [2]. Different control strategies were developed depending upon the mathematical model of an underwater vehicle. Besides, due to the nonlinear behavior of the marine system due to fluid–vehicle interaction, it is not desirable to develop an accurate and predictive model for the underwater system. In an underwater vehicle, the operation of navigation, guidance, and control (NGC) possess multiple states of sophistication based on various autonomy levels [3]. The navigation system is one of the important elements in the control structure. The navigation framework gives an estimation of the position, velocity vehicle behavior corresponding to system inertia which is placed at the control station. Widely used algorithm to accomplish the navigation system is the Kalman filter (KF) algorithm [5]. KF is an estimating algorithm which is factually ideal corresponding to a quadratic error function which enables the operator to approximate the vehicle state. One more important component of the control structure shown in Fig. 1 is the control system. It consists of a cluster of algorithms required for balancing the vehicle state to provide instructions to the vehicle to act according to the commands or instructions initiated in the path planning system. Controlling an underwater vehicle is not an easy task as they are extremely nonlinear, and the nonlinearity affects the communication of a marine system with the physical surroundings [6].

Underwater systems are used substantially in marine engineering, wherein underwater vehicles are widely used in marine research and expedition of oil and natural gas

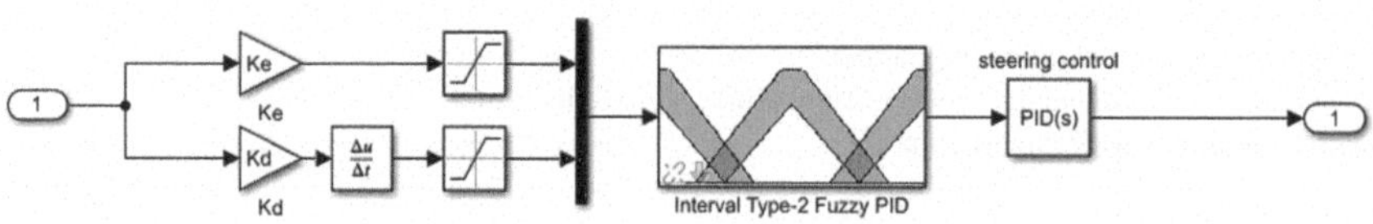

Fig. 1 Integration of FLC with PID controller

[1]. There are basically two types of underwater systems which are remotely operated vehicles (ROV) and autonomous underwater vehicles (AUV). The important distinction between an AUV and ROV is that ROV is associated with a command station using a tethered cable. The tethered cable secures the power supply data communication signals allowing the operators to track, control and access the vehicle information. AUV is provided with a battery as power supply and SONAR to accomplish the given assignment eliminating the need of operator, which continually requires complex control strategies [3]. Empirical modeling, along with a superior controller design, was developed to secure the dynamic functionality of a newly developed ROV [6]. The model was designed for adaptive depth control of ROV systems and is based on single input FLC. The designed framework created for SIFLC was tuned using a PSO algorithm.

There are various techniques proposed to develop suitable control architecture to improve the performance of the underwater system [7]. Conventional controllers like PID controllers are used extensively in developing a robust control methodology to accelerate static and dynamic functionality for the underwater system. A control strategy for underwater systems using a fuzzy logic controller was proposed by [12]. FLC uses simple rhetorical variables instead of numerical factors, which does not need a well-characterized mathematical model. An adaptive design employing FLC for depth controlling in AUV is presented by [4]. The dynamic functionalities of an AUV are determined to employ six degrees of differential equations keeping body and earth fixed as reference. A prototype designed using reduced order technique was developed for a diving approach using depth plane dynamics and a proper dual loop controlling technique. The proposed model is executed by integrating FLC with a phase lead dynamic compensator to obtain required tracking point and favorable performance under various disruptions which occur in marine territories [11]. A survey of the latest technological advancements in underwater navigation systems is presented by [10]. The study also focuses on future scope and challenges associated with it. Enhancements in sensor technologies for marine navigation have enabled vast research on AUV. This review focuses on technological improvements in stochastic algorithms developed especially for underwater navigation systems. Future difficulties in submerged vehicle navigation paths are reviewed in this paper, including an ideal survey, terrestrial estimation and mid-water navigation. Application of FLC in controlling pitch and depth of an underwater vehicle is discussed by [13]. The FLC discussed in this paper has six degrees of control system to measure speed and depth of an underwater vehicle.

In this work, a design for autonomous underwater vehicles is presented using a fuzzy logic controller (FLC). The response of an underwater system is moderate compared to air monitoring systems. The slow response is due to the speed constraints and high density of water. Establishing an accurate control technique for AUV is a difficult task due to the nonlinearity of hydrodynamic elements [8]. To solve the problem of nonlinearity, it is essential to design a stable, secure and performance-oriented control strategy to enhance static and dynamic functioning of the system. There are various techniques available to develop suitable control architecture to improve the performance of the underwater system [10]. Conventional controllers like PID controllers are used extensively in developing a robust control methodology to accelerate static and dynamic functionality of the system. But, in basic PID controller, essential parameters are tuned manually by a proficient controller. The techniques used in PID controllers are effective only when the mathematical model of the system is known [11]. A new technology known as fuzzy logic controller (FLC) is designed for AUV, which has high-quality precision. FLC uses simple rhetorical variables instead of numerical factors, which does not need a well-characterized mathematical model. FLC is an essential technique employed to solve the problems of nonlinearity and to reduce the complexity of the system [9].

2 Design, Analysis and Modeling

Controlling an underwater vehicle is not an easy task as they are extremely nonlinear, and the nonlinearity affects the communication of a marine system with the physical surroundings. In this work, a design of depth and speed control for autonomous underwater vehicles is presented using a type-2 fuzzy logic controller (FLC). Type-2 FLC is used to tune the PID controller. The dynamic functionalities of an AUV are determined by using six degrees of freedom differential equations of motion keeping fixed earth as reference.

2.1 Depth Subsystem Modeling

The analysis and modeling of the system for depth analysis are described below. The modeling of the system is done by keeping (X-Y plane) as reference.

The motion equations for heave and pitch are:

$$m(w - u_o q) = Z \tag{1}$$

$$I_y q = M \tag{2}$$

where I_y is the moment of inertia of the body along y axis, '*m*' is the mass of the body, '*q*' is the component of angular velocity vector for along axis, '*w*' is the weight of the body, 'u_o' is the state of motion of body at time *t*.

The heave external force *Z* and pitch moment *M* are made up of mass containing hydrodynamics, deflection of elevator plane, linear damping and the cross flow drag. Along with these properties, there is a righting moment due to distance between center of mass and buoyancy.

$BG_z = z_G - z_B$. *There is positive buoyancy of the vehicle* $\Delta B = B - mg$ *that acts in* Z *– axis.*

$$Z = Z_{\dot{w}}\dot{w} + Z_{\dot{q}}\dot{q} + Z_w w + Z_q q + Z_\delta \delta_s + \Delta B \tag{3}$$

$$\begin{aligned} M &= M_{\dot{w}} w + M_{\dot{q}}\dot{q} + M_w w + M_q q - \dot{m}g(z_G - z_B)\sin(\theta) + M_\delta \delta_s \\ M &= M_{\dot{w}}\dot{w} + M_{\dot{q}}\dot{q} + M_w w + M_q q - mgBG_Z\theta + M_\delta \delta_s \end{aligned} \tag{4}$$

where $B =$ Buoyancy, $g =$ acceleration due to gravity, $\delta_s =$ angular displacement of the control surface, $\theta =$ pitch angle, $m =$ mass of the body, $Z =$ external force.

From the analysis of kinematics in x-z plane, assuming a small pitch angle,

$$\theta = q \tag{5}$$

$$\dot{z} = -\theta u_o + w \tag{6}$$

By substituting Eqs. 3 and 4 into Eqs. 1 and 2, similarly, combining 5 and 6, we obtain a state space representation with state variables $w(t)$, $q(t)$, $\Theta(t)$ and $z(t)$;

$$\begin{bmatrix} \dot{w} \\ \dot{q} \\ \dot{\theta} \\ \dot{z} \end{bmatrix} = \begin{bmatrix} C_{11} & C_{12} & C_{13} & 0 \\ C_{21} & C_{22} & C_{23} & 0 \\ 0 & 1 & 0 & 0 \\ 1 & 0 & -u_0 & 0 \end{bmatrix} \begin{bmatrix} w \\ q \\ \theta \\ z \end{bmatrix} + \begin{bmatrix} d_1 \\ d_2 \\ 0 \\ 0 \end{bmatrix} \delta_s + \begin{bmatrix} e_1 \\ e_2 \\ 0 \\ 0 \end{bmatrix} \Delta B \tag{7}$$

Considering value of C_{21}w as constant and heave velocity is not fluctuating, the linear model shown in Eq. 7 is reduced to:

$$\begin{bmatrix} \dot{q} \\ \dot{\theta} \\ \dot{z} \end{bmatrix} = \begin{bmatrix} C_{22} & C_{23} & 0 \\ 1 & 0 & 0 \\ 0 & -u_0 & 0 \end{bmatrix} \begin{bmatrix} q \\ \theta \\ z \end{bmatrix} + \begin{bmatrix} d_2 \\ 0 \\ 0 \end{bmatrix} \delta_s + \begin{bmatrix} C_{21}w + e_2\Delta B \\ 0 \\ 0 \end{bmatrix} \tag{8}$$

The pitch dynamics is defined by following equations

$$\theta = q \tag{9}$$

$$q = C_{22}q + C_{23}\theta + d_2\delta_s + C_b \tag{10}$$

$$C_b = C_{21}w + e_2\Delta B \tag{11}$$

The modeling of the subsystem for depth analysis is determined in Eq. 11, assuming that AUV is positively buoyant.

2.2 Modeling of AUV

The analysis and modeling of an AUV is based on Newton Euler equations. Two coordinate points are taken into account as reference. The body of the vehicle fixed frame is considered to be placed in the body of the vehicle and the coordinate points are measured with a fixed frame as reference known as inertial frame. This consists of six coordinates for determining velocity (three translational velocity and three rotational velocity) along X, Y and Z, respectively.

The fixed frame is entitled as

Vector $V = [uvwpqr]^T$, $where\, v_1 = [uvw]^T$ are $the\, translational\, velocities\, along\, X, Y, Z\, direction\, respectively$

Where, u = surge, v = sway and w = heave velocity.

$$V_2 = [pqr]^T\; are\, the\, rotational\, velocities$$

Where, p = roll, q = pitch and r = yaw motions.

Similarly, the fixed frame with earth as reference contains six coordinates. These coordinates represent the orientation and position of the AUV. The earth fixed frame is given by:

$$Vector\, \eta = [xyz\phi\theta\psi]^T\; where\, \eta_1 = [xyz]^T\; (position\, coordinates)\, and$$
$$\eta_2 = [\phi\theta\psi]^T\; are\, the\, rotational\, coordinates$$

The initialization point of the body fixed coordinate is assumed to be the center of mass of an AUV and the translational motion is determined by Newton's second law $F = ma$

The rotational movement of the AUV is monitored by using Euler's mathematical relation:

$$M_c = I_c\dot{w} + w \times I_c, \quad where\, F\, and\, M_c\, are\, external\, forces\, and\, moments$$

According to Newton's and Euler's mathematical equalization, the six degrees of motion equation for AUV are reframed using coordinates defined for body fixed frame.

$$m\left[\dot{u} - vr + wq - x_g\left(q^2 + r^2\right) + y_g(pq - \dot{r}) + z_g(pr + \dot{q})\right] = \sum x_{ext}$$
$$m\left[\dot{v} - wp + ur - y_g\left(r^2 + p^2\right) + z_g(qr - \dot{p}) + x_g(qp + \dot{r})\right] = \sum Y_{ext}$$
$$m\left[\dot{w} - uq + vp - z_g\left(p^2 + q^2\right) + x_g(rp - \dot{q}) + y_g(rp + \dot{p})\right] = \sum z_{ext}$$

2.3 Design of PID Controller

In this methodology, PID controller is used to measure speed and response of the AUV. PIC controller consists of two loops: an inner loop (to measure the pitch of the vehicle) and an outer loop (to measure the depth of the vehicle). The transfer function of an inner loop is of second-order, and the response is fast, whereas the reaction for the outer loop is slow. To obtain an adequate output, corresponding to response time and control, PID controller techniques are implemented for the inner loop as well as for outer loop.

2.4 Fuzzy Logic Controller

FLC is used to synchronize the voltage generated across the controller, for excess energy dissipation and for charging and discharging of energy-storing batteries. The FLC consists of three parts which are fuzzification, interference engine and defuzzification. One of the advantages of using a FLC instead of conventional controller is that it calculative less acute and it enhances imprecise mathematical modeling. FLC consists of semantic variables as inputs and produces a specific precise output after defuzzification. A technique of center of gravity is employed for the process of defuzzification of the generated output. In this work, a fuzzy logic controller is used to tune PID controller. A control variable is defined for error analysis: error (e) which is the difference between estimated depth and achieved depth) and this is estimated after applying input and error change (Δe). In this study, type-2 fuzzy logic controller is implemented. Type-2 FLC consists of a rule base, fuzzifier, interference engine and defuzzifier. It also consists of interval kind of fuzzy sets to notify input and output of an AUV. It is used to simplify the computation of the controller. Initially, the input values are subjected for fuzzification and singleton fuzzification is advantageous due to its reliability and its proneness toward embedded processors (Fig. 2).

In this research, FLC is integrated with PID controller to track speed of AUV and to measure depth and pitch of the vehicle. The block diagram for integration is shown in Fig. 3. The systematic modeling of type-2 FLC is a difficult task as the output cannot be determined in a closed form because of KM-type reduction in FLC. The proposed approach will reduce the parameter dependency which needs to be tuned to achieve desired results.

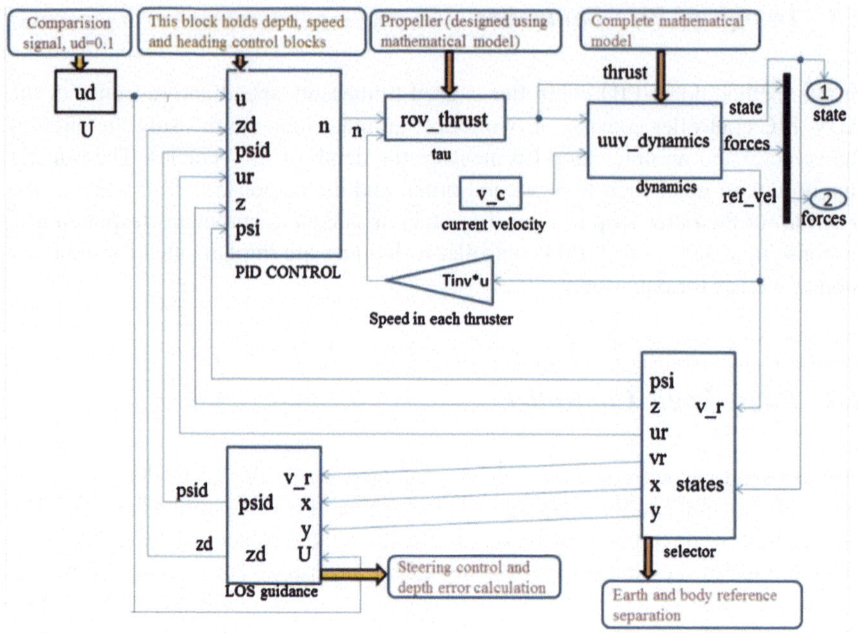

Fig. 2 Simulink model for AUV

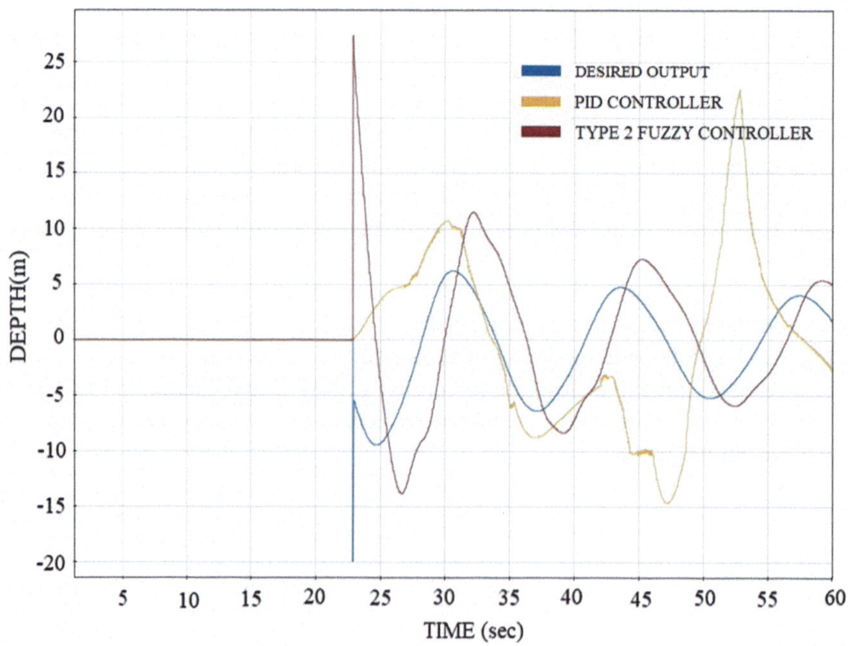

Fig. 3 Depth analysis using PID controller and Type-2 FLC

3 Simulation Results

3.1 Simulink Model for AUV

See Fig. 2.

3.2 Analysis of Depth Control for AUV for Positive and Negative

See Fig. 3.

3.3 Output of Vehicle Path Determined by Fuzzy-PID Controller

See Fig. 4.

The three- and four-way points were used to analyze the stability of the vehicle. To check the vehicle stability, we add one more way point (four-way points) to check its stability.

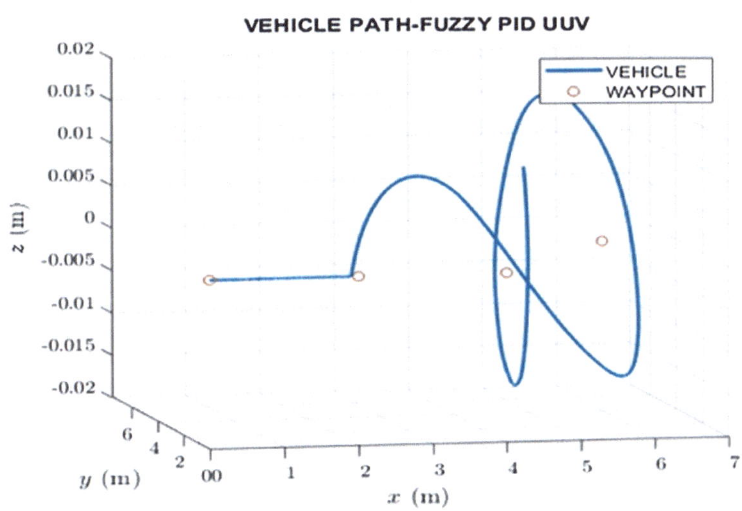

Fig. 4 Vehicle path determined by Fuzzy-PID controller for four-way points

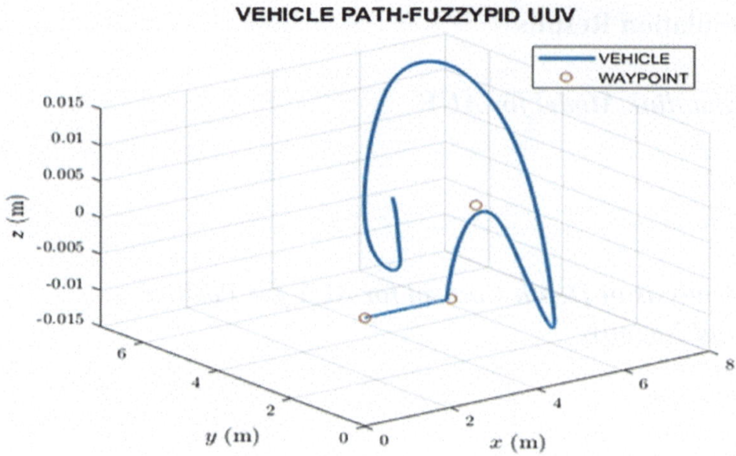

Fig. 5 Vehicle path determined by Fuzzy-PID controller for three-way points

4 Conclusions

The main aim of this study was to design an effective depth control technique for autonomous underwater vehicles. In the research, analysis and modeling of AUV was done for depth analysis using PID controller. Type-2 fuzzy logic controller was integrated with PID controller for tuning the controller. FLC was implemented to track the speed of AUV and to measure depth and pitch of the vehicle. From the simulated results as shown in Fig. 5, it is shown that the precise depth control was obtained using type-2 FLC compared to output using PID controller. The vehicle path is also analyzed by using three-way points and four-way points. For future scope, pitch control analysis can be done using the proposed research.

References

1. Roberts G (2008) N Trends in marine control systems. Annual rev control 32(2):263–269
2. Vervoort JHAM (2009) Modeling and control of an unmanned underwater vehicle. Master Traineesh. Rep, 5–15
3. Abdullah AM, Zakaria NI, Jalil KAA, Othman N, Dahalan WM, Hamid H, Kamro MIM (2018) Review of the control system for an unmanned underwater remotely operated vehicle. In: Engineering applications for new materials and technologies, Springer, Cham, pp 609–631
4. Nag A, Patel SS, Akbar SA (2013) Fuzzy logic based depth control of an autonomous underwater vehicle. In: International mutli-conference on automation, computing, communication, control and compressed sensing (iMac4s), IEEE, pp 117–123
5. Aras MSM, Abdullah SS, Othman SYB, Sulaiman M, Basar MF, Zambri MKM, Kamarudin MN (2016) Fuzzy logic controller for depth control of underwater remotely operated vehicle. J Theoretical Appl Inf Technol 91(2)

6. Khodayari MH, Balochian S (2015) Modeling and control of autonomous underwater vehicle (AUV) in heading and depth attitude via self-adaptive fuzzy PID controller. J Mar Sci Technol 20(3):559–578
7. Smith SM, Rae GJS, Anderson DT (1993) Applications of fuzzy logic to the control of an autonomous underwater vehicle. In: [Proceedings 1993] Second IEEE international conference on fuzzy systems, pp 1099–1106
8. Ishaque K, Abdullah SS, Ayob SM, Salam Z (2011) A simplified approach to design fuzzy logic controller for an underwater vehicle. Ocean Eng 38(1):271–284
9. Aruna D, Sri G, RajaRao PVV, Rama Mohana Rao K, Raghuram (2017) Int J Innovative Res Electr, Electron, Instrum Control Eng. ISSN (Online) 2321–2004, ISSN (Print) 2321–5526
10. Blain M, Lemieux S, Houde R (2003) Implementation of a ROV navigation system using acoustic/Doppler sensors and Kalman filtering. In: Oceans celebrating the past. teaming toward the future (IEEE Cat. No. 03CH37492) vol 3, pp 1255–1260
11. Kinsey JC, Eustice RM, Whitcomb LL (2006) A survey of underwater vehicle navigation: Recent advances and new challenges. IFAC Conference of Manoeuvring and Control of Marine Craft 88:1–12
12. Geng Y, Sousa J (2010) Hybrid derivative-free EKF for USBL/INS tightly-coupled integration in AUV. In: IEEE International conference on autonomous and intelligent systems, AIS, pp 1–6
13. Sname T (1950) Nomenclature for treating the motion of a submerged body through a fluid. The Society of Naval Architects and Marine Engineers, Technical and Research Bulletin No 1–5
14. Rúa S, Vásquez RE (2016) Development of a low-level control system for the ROV Visor3. International Journal of Navigation and Observation

Improvement in Maintenance Practices for Enhancing the Reliability of SMS Mould Oscillators

Rajul Misra, Kshitij Shinghal, and Amit Saxena

Abstract In the continuous casting of steel, the continuously cast strand withdrawn from the mould consists of a thin outer shell surrounding a molten core (because only a thin layer of steel adjacent to the walls of the mould undergoes a solidification process). There is friction between the shell and the mould where the shell has a tendency to stick to the mould. The shell is relatively weak because it is very thin and at very high temperature. Utmost care is required to limit the stress which is applied to the shell in order to withdraw the strand from the mould. Excessive stress will cause the shell to rupture. Rupture of the shell inside the mould is bad for the surface quality of the strand while rupture outside of the mould will result in a breakout, i.e. an escape of the molten core. Therefore, mould oscillators are provided to prevent the sticking and reduce the friction-resistant force between the solidified shell and the mould castings. These oscillators are very critical for continuous casting of steel. Generally, an oscillator comprises of a motor drive which is coupled to a reducing gearbox where the gear output shaft is coupled to an eccentric rotating shaft for oscillation of the mould. This eccentric rotating shaft is simply supported on two bearings. Problem in oscillator directly affects the production and quality of the steel. The SMS maintenance team of JSPL was facing chronic problems in all of its six mould oscillators. The common problem was the abnormal sound from the gearbox. More often, the coupling bolts of gearbox and mould oscillators were getting sheared off. Sometimes, the oscillator bearings were getting damaged. In addition to it, the vibration level on DE side of motor was high. Due to poor reliability of mould oscillators, the steel's productivity and quality were getting affected. But a small change in the maintenance strategy of mould oscillators minimized their vibration

R. Misra
Department of Electrical Engineering, Moradabad Institute of Technology, Moradabad, U.P, India
e-mail: rajulmisra71@gmail.com

K. Shinghal · A. Saxena (✉)
Department of Electronics & Communication Engineering, Moradabad Institute of Technology, Moradabad, U.P, India
e-mail: amitssaksena@gmail.com

K. Shinghal
e-mail: kshinghal@gmail.com

M. N. Favorskaya et al. (eds.), *Innovations in Electrical and Electronic Engineering*, Lecture Notes in Electrical Engineering 661, https://doi.org/10.1007/978-981-15-4692-1_36

levels and reduced the noise intensity from their gearboxes. Since last eight months, the reliability of mould oscillators has improved a lot and all are running without any breakdowns.

Keywords Mould oscillator · Vibration test · SMS mould · Fault detection

1 Introduction

Continuous casting, also called strand casting, is the process whereby molten metal is solidified into a "semi-finished" billet, bloom or slab for subsequent rolling in the finishing mills.

Molten metal is tapped into the ladle from furnaces. From the ladle, the hot metal is transferred to tundish [1, 3]. Metal is drained from the tundish through shroud into the top of an open-base copper mould. Refer to Fig. 1 which shows the schematic diagram of strand casting.

The mould is water-cooled to solidify the hot metal directly in contact with it.

In the mould, a thin shell of metal next to the mould walls solidifies while the bulk of metal within the walls of the strand is still molten. There is friction between the shell and the mould where the shell has a tendency to stick to the mould. The shell is relatively weak because it is very thin. Rupture of the shell inside the mould is bad for the surface quality of the strand while rupture outside of the mould will result in a breakout, i.e. an escape of the molten core. Therefore, to withdraw the strand from

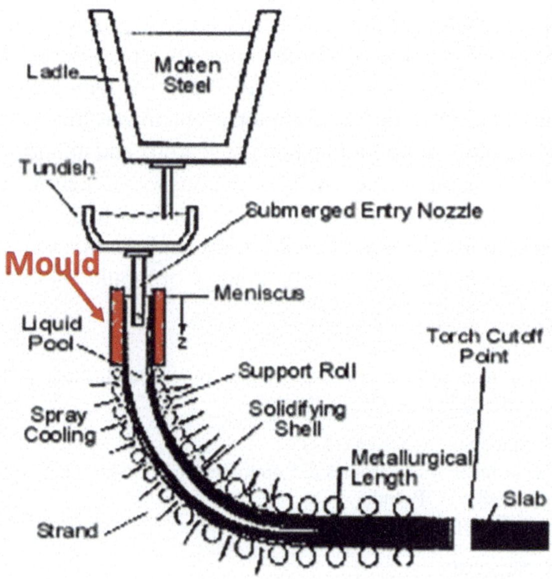

Fig. 1 Schematic diagram of strand casting

Motor
Rating-11KW
RPM-992
VVF Drive

Gearbox
Single stage

1 = Motor NDE Bearing 2 = Motor DE Bearing 3 = Gearbox input bearing
4 = Gearbox output bearing 5 = Oscillator Bearing Block - 1 6 = Oscillator Bearing Block - 2

Fig. 2 Mould oscillator

the mould, the stress (on the strand) is applied in a very controlled way otherwise the excessive stress will cause the shell to rupture [2, 5].

Therefore, the mould is oscillated in a controlled manner with the help of a mould oscillator. This prevents the sticking of the shell to the mould. Besides, it also reduces the friction between the shell and the mould for smooth strand casting. Generally, an oscillator comprises of a motor drive which is coupled to a reducing gearbox where the gear output shaft is coupled to an eccentric rotating shaft for oscillation of the mould [4, 6, 7]. This eccentric rotating shaft is simply supported on two bearings. Figure 2 illustrates the mould oscillator assembly.

Therefore, the mould oscillator is very critical for strand casting of steel [8]. Problem in an oscillator directly affects the production and surface quality of the steel [9–11].

2 Problem Description

Six mould oscillators were newly installed. After running for six months, problems started to surface. Bolts of coupling sheared off in three oscillators. In some cases, the motor and gearbox bearings got damaged. In one oscillator, the gearbox base bolts were broken. When checked, it was found that five out of the six oscillators were having high vibration levels. There was high-intensity abnormal sound from almost all the gearboxes. In three oscillators, the temperature of gearbox output bearing was relatively much higher than the other similar bearings. Besides, there was abnormal sound from bearing blocks of three oscillators. Due to frequent problems in the oscillators, their reliability had decreased. Production and surface quality of strands were getting affected.

3 Impact of the Problem

- Frequent problems mean poor reliability of the oscillators which will ultimately lead to production loss.
- Increased vibration level in oscillator will create more stress on the shell of strand casting. More stress will rupture the shell which will affect the surface quality of the strand casting. More importantly, rupture of shell outside the mould may result in a breakout, i.e. an escape of the molten core. Thus, the dimensional quality of the strand casting will be poor resulting in huge production loss on account of rejections.
- Maintenance cost will increase on account of frequent repairs and replacements.
- Breakdowns result in asset loss to the company on account of damage to bearings, couplings and other vital components.
- Frequent and unplanned maintenance activities on urgent basis demoralize the maintenance team.

4 Problem Analysis

Since all the six mould oscillators were having problems, it was certain that some common factors were responsible. The case had to be studied thoroughly so that meaningful solution to the problem could be provided.

To start with, the existing health status of all the mould oscillators were evaluated through a series of activities as mentioned in Table 1.

From the preliminary study, it was found that the possible reasons responsible for abnormality in the mould oscillators were:

(a) Misalignment
(b) Weak structural base
(c) Lubrication issue

Table 1 Activities to assess the existing health status of mould oscillators

S. No.	Activity	Observations
1	Vibration measurement	In five oscillators, vibration level was high on motor bearings. In four oscillators, vibration was relatively high at gearbox input bearing
2	Temperature measurement	In three oscillators, temperature at motor DE bearing and at gearbox input bearing was high
3	Visual inspection	In four oscillators, jerks observed at gearbox output bearing and at oscillator bearing blocks
4	Sound monitoring	In almost all the oscillators, there was abnormal sound from gearbox bearings. In three oscillators, abnormal sound observed at motor DE bearings. Besides, there was abnormal sound from the bearing blocks of three mould oscillators

(d) Eccentric force
(e) Bearing deterioration
(f) Bearing loose in housing.

Vibration signature analysis of motor and gearbox bearings indicated that misalignment and weak base were the prime reasons for the abnormality. Refer to the Figs. 3, 4 and 5 where bearing vibration signature of three oscillators has been

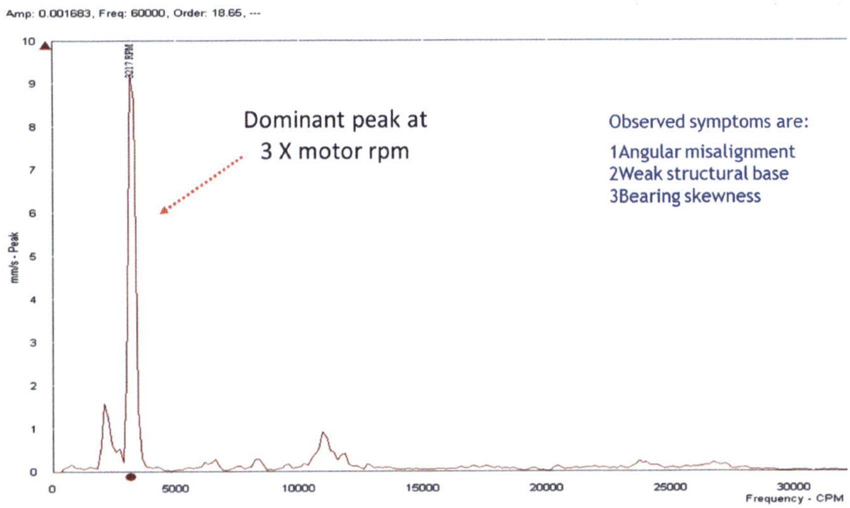

Fig. 3 Vibration signature of motor DE bearing of mould oscillator # 3

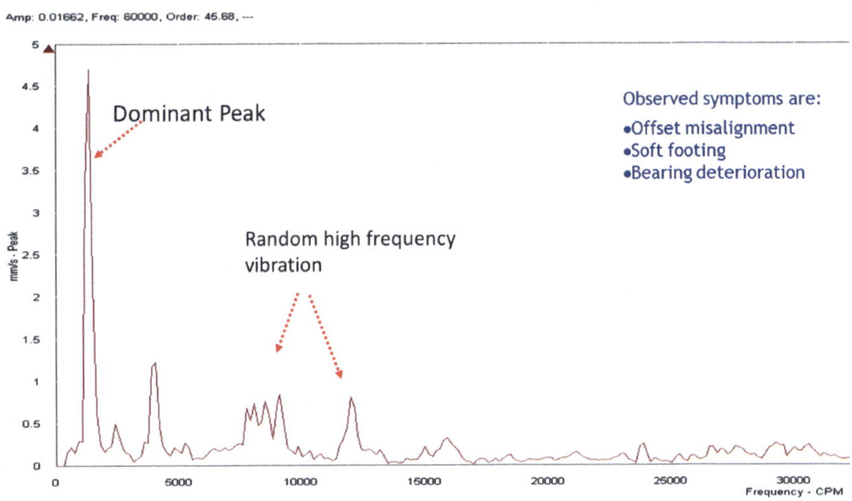

Fig. 4 Vibration signature of motor DE bearing of mould oscillator # 6

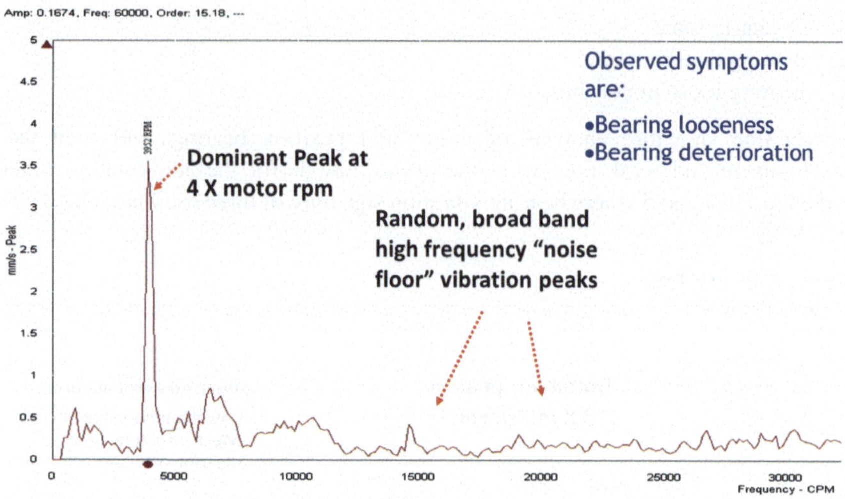

Fig. 5 Vibration signature of gearbox input bearing of mould oscillator # 5

shown. Observed symptoms have been given in each graph.

It is important to note that the reasons like bearing looseness in housing, increased bearing temperature and bearing deterioration were the cascading effects of the prime reasons.

In almost all the oscillators, there was high-intensity abnormal sound from the gearboxes. Reasons could be coupling misalignment, poor gear teeth meshing conditions, bearing loose in housing and bearing deterioration. But there were no peaks in the vibration signature of gearbox bearings which could be related to gear meshing problems. So, gear teeth meshing was proper.

Physical jerks could be observed at the gearbox output bearing. There was an abnormal sound from the oscillator bearing blocks which could be correlated to eccentric force owing to cam arrangement for oscillatory motion. However, the sound was prominent in three oscillators only. The possible reasons could be coupling misalignment, bearing looseness in housing, bearing deterioration and poor lubrication.

During physical inspection, it was observed that all the oscillators were mounted on a common channel base with vertical supports missing at three points. The structural base appeared weak.

5 Root Cause of the Problem

There were many possible reasons for the chronic problems in oscillators. But misalignment and poor base strength were the prime reasons. However, the other reasons were also duly considered with the same seriousness.

Following steps were taken to determine the root cause(s) of the oscillator problems:

Step # 1 Vibration characteristics of structural base were studied to assess the adequateness of base rigidity. All the oscillators were mounted on a common channel base. But at some points, it appeared that the vertical supports were missing. Vibration measurement of the base was done.

It was found that the base vibration readings of mould oscillator # 6 were very high compared to the base vibration readings of mould oscillator # 5. Vibration reading at point # E was four times greater than the vibration reading at point # A. Refer to Figs. 6 and 7 which show the comparative vibration readings on the base of two oscillators.

Fig. 6 Base vibration (oscillator # 5)

Fig. 7 Base vibration (oscillator # 6)

Table 2 Vertical alignment readings (Initial)

	Angularity (mm/100 mm)	Offset (mm)
Permissible limit (1000 RPM)	0.09	0.12
Alignment readings found	**0.69**	**0.77**
Alignment status	**NOT OK**	**NOT OK**
Correction	**0.7 mm** shim to be added to motor front feet	**2.5 mm** shim to be added to motor rear feet

Table 3 Horizontal alignment readings (initial)

	Angularity (mm/100 mm)	Offset (mm)
Permissible limit (1000 RPM)	0.09	0.12
Alignment readings found	**0.54**	**0.14**
Alignment status	**NOT OK**	**NOT OK**
Correction	**1.5 mm** shifting of motor at its front feet towards north	**3.0 mm** shifting of motor at its rear feet towards north

On physical inspection of base, it was found that wherever the vertical supports were missing, there the vibration readings were high. Here, the base of oscillator # 6 was not supported from beneath.

Step # 2 Laser technology was employed to check the coupling alignment tolerances of all the oscillators. It was found that all the oscillators were misaligned. Tables 2 and 3 show the alignment readings between motor and gearbox for oscillator # 3. It could be seen that oscillator # 3 was grossly misaligned. Similarly, the alignment readings between gearbox output bearing and oscillator bearing block were found high.

It was observed that most of the oscillators were offset misaligned. This meant that the level of motor base and gearbox base was not in proper level with respect to the base level of oscillator bearing blocks. Therefore, vertical misalignment was to be corrected in common for all the oscillators.

Step # 3 It was found that the bearings which showed high temperature were the bearings of oscillators which were highly misaligned. So, misalignment could be the reason for temperature build-up at the bearings and alignment correction could resolve the problem of heating.

Step # 4 On gearbox output bearing, physical jerks could be observed. Since cyclic eccentric force was inherent in the system due to oscillatory motion, the gearbox output shaft bearing and the oscillator bearing blocks were the most affected. Therefore, jerks on these bearings could be felt. Intensity of jerks was high in the oscillators where coupling was grossly misaligned. Therefore, it could be the repeated jerks which were responsible for creeping misalignment in the system.

6 Corrective Actions

6.1 Base Strengthening

Based on the base vibration readings, relatively weak points were identified on the structural base. Base strengthening was done by providing vertical supports wherever required. The structural base was integrated by abridging the gaps in the base structure by providing ribs and by welding.

6.2 Inspection of Bearings and Suitable Corrective Actions

The bearings which showed the symptoms of looseness and deterioration were inspected. One motor bearing was replaced. In two bearings, re-lubrication was done.

6.3 Alignment Correction

All the oscillators were aligned one-by-one with laser technology. First, the alignment correction was done for the coupling between the gearbox output shaft and the oscillator bearing block bearing. Then, the alignment correction was done for coupling between motor and gearbox input bearing. At two instances, base grinding was done on gearbox side. In one case, the base hole was enlarged to accommodate the lateral shifting of motor.

7 Results

Vibration level on motor bearings is reduced from 9.8 to 2.5 mm/s. Similarly, the readings on gearbox input bearings decreased from 5.9 to 1.8 mm/s.

Overall vibration level structural base was found in the range of 25–50 microns (μ).

For last one year, all the six oscillators are running without any breakdowns. Their reliability and availability have increased.

8 Preventive Measures

- All the six oscillators have been included in the predictive schedule of vibration, temperature and sound monitoring. The vibration data collected are trended and analysed through advanced vibration software to detect the budding problem in the system so that timely corrective action could be taken.
- Thermal imaging of motor, gearbox and oscillator bearing blocks being done on monthly basis to detect the temperature build-up.
- In case of increase in abnormal sound from gearbox or increase in vibration level, the alignment of coupling between the gearbox output shaft and oscillator bearing block is checked through laser technology. Since the gearbox output bearing experiences the maximum jerk due to eccentric force of oscillation, there is always a chance of lateral shift of gearbox and alignment getting disturbed.
- In case of increase in vibration level at motor bearings only, thermal profile study and bearing signature analysis are done. Suitable corrective actions are taken based on the thermal status, vibration level and bearing sound.
- Lubrication of bearings and oscillator unit are being done through preventive maintenance schedule.

9 Conclusions

When there is a recurrent problem in a system, then its maintenance strategy needs to be reviewed. The possible reasons for the problem need to be discussed, and root cause analysis of the problem must be done. Once the root cause is known, solution to the problem becomes easy. Emphasis must be given to formulate and implement a preventive strategy to avoid the recurrence of similar or potential problems in the system. With systematic maintenance approach and effective condition monitoring techniques, the recurrent problems will be permanently eliminated.

References

1. Haoyang T, Wei P, Hua H, H. Min (2017) Experimental study on sound level calculation of dry transformer via vibration velocity method. In: 2017 1st International conference on electrical materials and power equipment (ICEMPE), Xi'an, pp 410–413. https://doi.org/10.1109/icempe.2017.7982115
2. Islam M, Islam R, Sebastian T (2011) Noise and vibration characteristics of permanent magnet synchronous motors using electromagnetic and structural analyses. In: 2011 IEEE Energy conversion congress and exposition, Phoenix, AZ, pp. 3399–3405. https://doi.org/10.1109/ecce.2011.6064228
3. Islam MS, Islam R, Sebastian T (2014) Noise and vibration characteristics of permanent-magnet synchronous motors using electromagnetic and structural analyses. In IEEE transactions on industry applications, vol 50, no 5, pp 3214–3222. https://doi.org/10.1109/tia.2014.2305767

4. Taylor JI (1900) The vibration analysis handbook: A practical guide for solving rotating machinery problems. Ipp books
5. Fahi F, Gardonio P (2006) Sound & structural vibration. Academic Press
6. Ginsberg JH Mechanical & structural vibrations, published by Wiley
7. Brandt A (2010) Noise and vibration analysis, published by Wiley
8. Ray SK (2003) Surface quality of steel products. Allied Publishers
9. Cartmell M (2003) Modern practice in stress and vibration analysis. Trans Tech Publications
10. Brumbach ME, Clade JA (2012) Industrial maintenance, Cengage learning
11. Bloch HP (1998) Improving machinery reliability, Gulf Professional Publishing

Test Bench of Automotive Component of an Electric Vehicle for Electrical Parameter Measurement

Sonu Raut, Prema Daigavane, and M. B. Shaikh

Abstract In this paper, a novel and economical test bench of automotive component of an electric vehicle (EV) has been developed to measure the electrical parameters. This test bench provides the flexible testing and monitoring system to ensure the quality and traceability on the production line. This paper briefly describes the hardware and software implementation of the test bench. The automotive component is subjected to different electrical parameters tests such as low voltage short circuit test, high voltage insulation test and low voltage connection test. The tests parameters are displayed on supervisory control and data acquisition (SCADA) and human–machine interface (HMI) screen.

Keywords EV · SCADA · HMI · PLC · Servo motor · ETB component

1 Introduction

The electric vehicles are predicted to be the next disruptive market force for transportation and technology. The primary reasons for the introduction of EV's into the market is the concern over greenhouse gas emissions and their contribution to global warming. The EV has significantly proved to be cleaner and safer for the environment. Also, EV's are 75% efficient at turning input energy into the moving energy. Due to the increased demand for EV's in the market, it has encouraged the automakers to get better at making them by using newer ideas and innovations. This can be only achieved by improving the technology and materials used to make EV.

S. Raut (✉)
G H Raisoni College of Engineering, Nagpur, India
e-mail: raut_sonu.ghrcemtechips@raisoni.net

P. Daigavane
Department of Electrical Engineering, G H Raisoni College of Engineering, Nagpur, India
e-mail: prema.daigavane@raisoni.net

M. B. Shaikh
Industrial Project Head, SAS Automation, Pune, India
e-mail: marketing@sasengineering.in

M. N. Favorskaya et al. (eds.), *Innovations in Electrical and Electronic Engineering*, Lecture Notes in Electrical Engineering 661, https://doi.org/10.1007/978-981-15-4692-1_37

Hence, to provide the same quality testing and monitoring capabilities for the EV, there is a need for the platform where the components are to be tested before its serial production.

Therefore, a novel and economical test bench of automotive component of an EV has been developed to measure the electrical parameters. The test bench ensures the quality and traceability on the production line by providing the flexible testing and monitoring system. The test bench is equipped with hardware assembly to hold the automotive component in stable condition at the time of test conduction. The automotive component which is to be tested is the electrical test bench (ETB) cover. The different tests are carried out on the ETB cover. The tests can be performed in two modes either autotest or manual test. The autotest is performed in SCADA, whereas the manual test is performed in HMI. The SCADA and HMI interfaces with programmable logic controller (PLC) through communication protocols. The PLC modules are fed by the real-time information from sensors through analog and digital cards.

2 Hardware Design and Development

2.1 *Fixture Assembly*

The fixture assembly is bolted to the test bed. It is provided with an assembly to hold the component in place at the time of test conduction (Fig. 1).

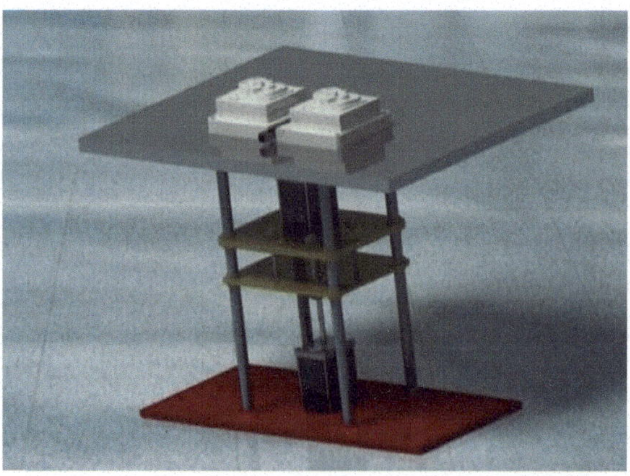

Fig. 1 Isometric view of fixture assembly

Fig. 2 ETB component holding arrangement

2.2 ETB Component Holding Arrangement

This arrangement includes the motorized mechanism, which consist of two servo motors, one rotating shaft, four guiding pipes for lifting movement and two sheet metal plates. A servo motor is a rotary machine. It is used for controlling of the angular position, velocity and acceleration (Fig. 2).

2.3 Fixture

The fixture is capable of holding the two automotive components at a time. The fixture is provided with the slots for connections at respective position to the component (Fig. 3).

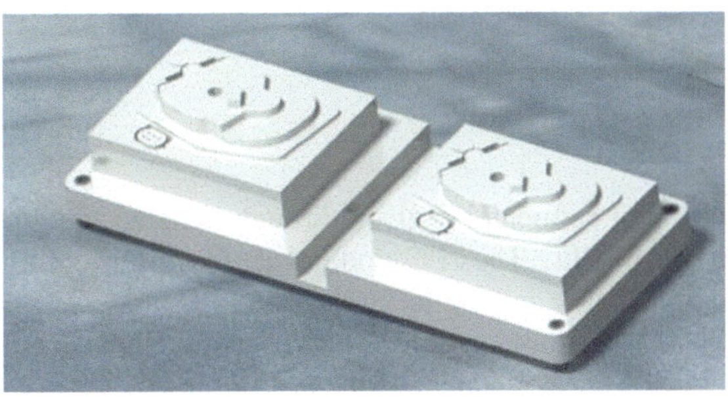

Fig. 3 Isometric view of the fixture

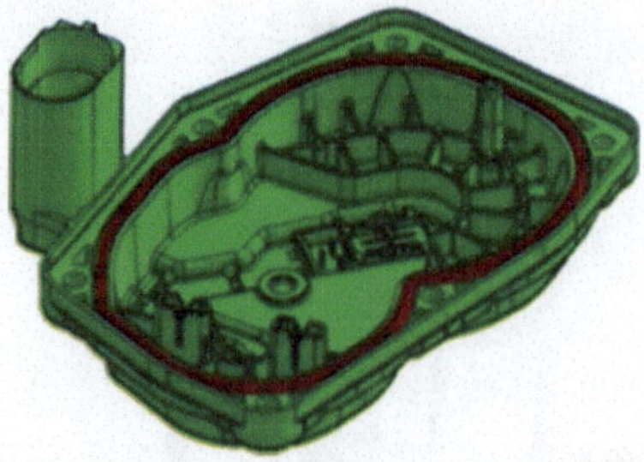

Fig. 4 ETB cover

2.4 ETB Cover

ETB cover is the automotive component which is to be tested against different electrical parameters. The connector area of the component consists of the ten terminals on which the electrical tests are to be done. The tests are done on different sets of combination of the terminals (Fig. 4).

3 Software Development

When the automotive component is placed on the fixture, the operation of the fixture assembly of test bench gets started. The start command is given through SCADA. After receiving the start command to servo drive, it sets the operation to the servo motor. The servo motor lifts the rotating shaft and clamper upwards, then stops at its desired position and rotates the shaft to 90 °. As soon as the clamper clamps the automotive component, the electrical parameter testing gets started. The tests can be performed in two modes either autotest or manual test. The autotest is performed in SCADA, whereas the manual test is performed in HMI (Fig. 5).

3.1 Operation in SCADA

First, the autostart input command is given through SCADA, and then the fetched input data interfaces with PLC. The PLC modules are fed by the real-time information from sensors through analog and digital cards. The PLC holds the command which is fetched from SCADA and uses it according to the operation cycle. Then, the data is switched to insulation tester meter, here, the making and braking of the contractors occurs. Also, different electrical parameters are tested. Then, the occurred result and output data is transferred to PLC. The PLC then displays the output data on the SCADA screen (Figs. 6 and 7).

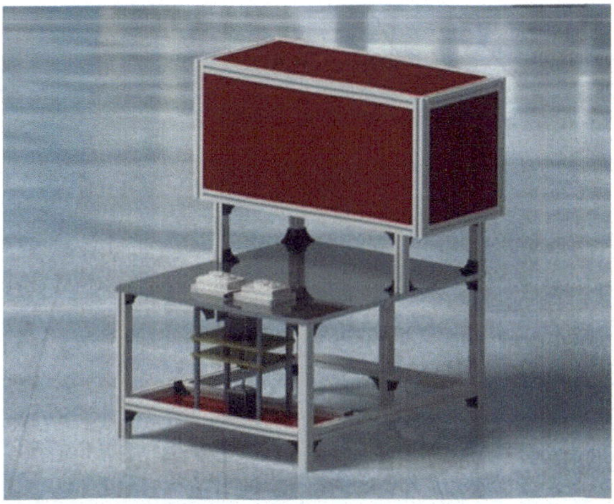

Fig. 5 Test bench for electrical parameter measurement

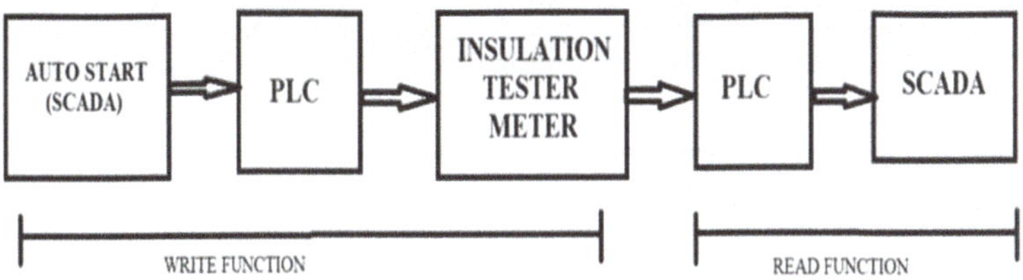

Fig. 6 Flowchart of autotest

3.2 Operation in HMI

The manual selection of the required test is done in HMI. The operation of the test to be carried out is similar to that of SCADA. The final results are displayed on the HMI screen (Figs. 8 and 9).

4 Measured and Calculated Parameters

The following are the electrical tests to be carried out on the ETB component:

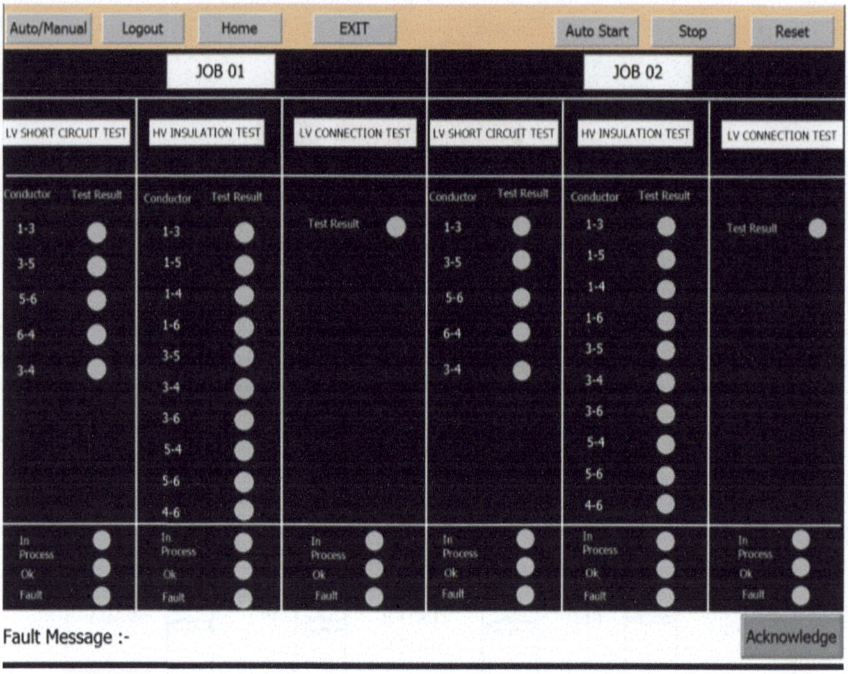

Fig. 7 SCADA screen

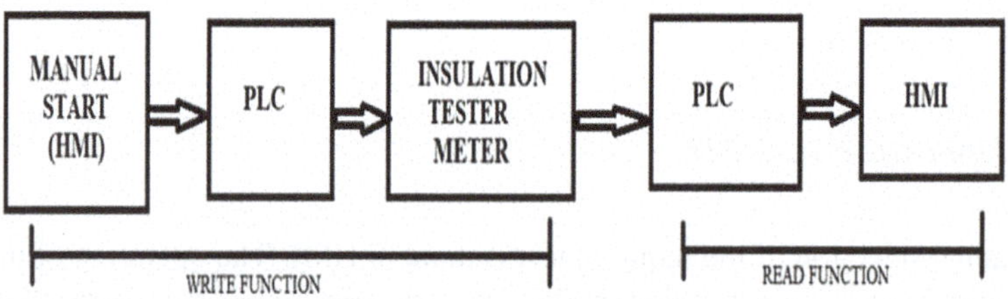

Fig. 8 Flowchart of manual test

4.1 Low Voltage Short Circuit Test

- The acceptance criteria: when the current flow for the given time within the specified limits then the resistance offered should be more than 5 MOhm ($>5M\Omega =$ O.K). If the condition is satisfied, then the automotive component is correct (O.K), otherwise, the automotive component is faulted (not O.K) (Fig. 10) (Table 1).
- Implementation notes
- The terminals should be electrically isolated.
- No repeating of tests after the test is not O.K.
- The test starts when the test voltage is reached.

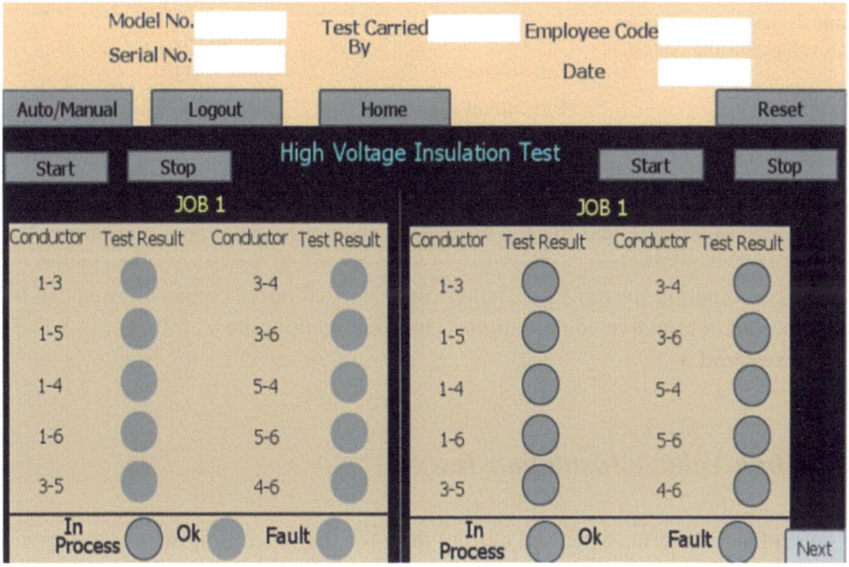

Fig. 9 HMI screen

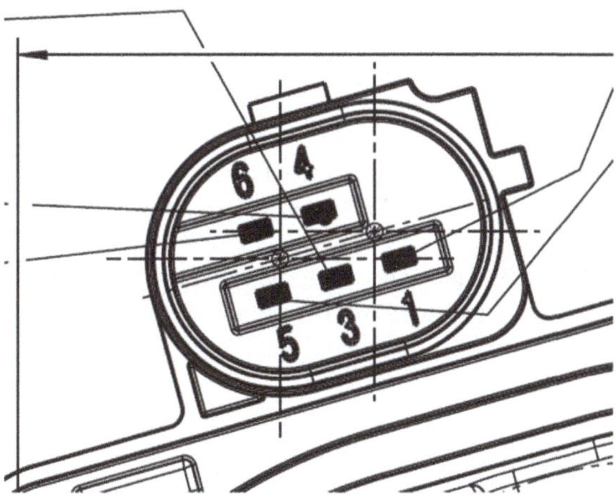

Fig. 10 Terminals of ETB cover

Table 1 Test parameters for low voltage short circuit test

Test parameter	Rating
Test voltage	5 V
Test current	Max. 1 μA
Release time	<100 ms
Test time	>500 ms

Table 2 Test parameters for high voltage insulation test

Test parameter	Rating
Test voltage	1000 V
Test current	Max. 100μA
Test time	1 s

- Conductors located next to each other must have a different potential during the test.
- Test combination (terminals): The test to be carried out in connector area can be connected in total five combinations. The combinations are as follows: 1–3, 3–5, 5–6, 6–4 and 3–4.

4.2 *High Voltage Insulation Test*

- Acceptance criteria: when the current flow for the given time within the specified limits then the resistance offered should be equal to or more than 10MOhm ($\geq$10MOhm = O.K). If the condition is satisfied, then the automotive component is correct (O.K), otherwise, the automotive component is faulted (not O.K) (Table 2).
- Implementation notes:
- All conductors are connected against all others.
- No repeating of tests after the test is not O.K.
- The test starts when the test voltage is reached.
- Conductors located next to each other must have a different potential during the test.
- Test combination (terminals): The test to be carried out in connector area can be connected in total ten combinations. The combinations are as follows: 1–3, 1–5, 1–4, 1–6, 3–5, 3–4, 3–6, 5–4, 5–6 and 4–6.

4.3 *Low Voltage Connection Test*

- Acceptance criteria: When the current flow for the given time within the specified limits then the voltage offered should be less than 1 V (<1 V = O.K). This voltage is possibly depending on equipment and material. If the condition is satisfied, then the automotive component is correct (O.K), otherwise, the automotive component is faulted (not O.K) (Table 3) (Fig. 11).
- Implementation notes
- All conductors are series connected.
- No repeating of tests after the test is not O.K.
- The test starts when the test voltage is reached

Table 3 Test parameters for low voltage connection test

Test parameter	Rating
Test current	5A
Test time	2 s

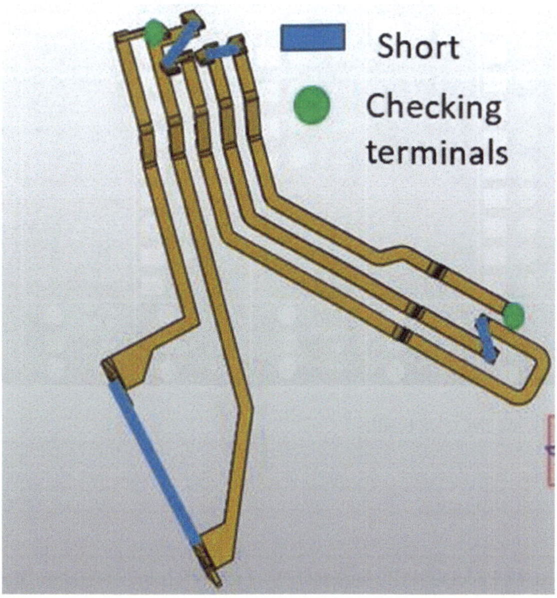

Fig. 11 Testing concept

5 Results and Conclusions

In this paper, the hardware and software development of test bench for electrical parameter measurement of the automotive component is introduced in detailed. The tests are carried either automatically or manually through SCADA and HMI. Also, the different types of tests which are performed on the automotive component are studied briefly. These tests results helped in the identification of the correct automotive component. Further, this test bench can be used at industrial level for the accurate examination of the different components or parts of the electric vehicles (Fig. 12).

Fig. 12 Screen showing the tests results

References

1. Marra F, Sacchetti D, Pedersen AB, Andersen PB, Larsen E Implementation of an electric vehicle test bed controlled by a virtual power plant for contributing to regulating power reserves. In: 2013 World electric vehicle symposium and exhibition
2. Izhar T, Ali M, Sohaib M, Nazir A Development of a motor test bench to measure electrical/mechanical parameters. In: 2017 International conference on energy conservation and efficiency
3. Gaushell DJ, Darlington HT (1987) Supervisory control and data acquisition. IEEE, 75
4. Shimada A, Kishiwada Y Arimura M (2006) AC Servo motor position sensorless control using mechanical springs. In: 9th IEEE International workshop on advanced motion control
5. Saini HS, Daruwala RD Human machine interface in internet of things system. In: 2016 International conference on computing communication control and automation
6. Isik MF, Haboglu MR, Yanmaz H (2016) Monitoring and control of PLC Based motion control systems via device-net. In: 16th International power electronics and motion control conference and exposition 2014. Vol 2(5), pp 99–110

Measurement of Torque Using Leakage Flux for Induction Motors in Electric Vehicles by Non-invasive Method

Shweta S Hooli, Anusha Vadde, Krishnan Manickavasagam, and Govind R. Kadambi

Abstract This paper proposes a non-invasive method of predicting health monitoring of a three-phase Squirrel Cage Induction Motor (SCIM). A 15 kW SCIM is modelled using Finite Element Analysis (FEA) for performing magnetic analysis to design a restraining coil. Since, torque developed depends on magnitude of flux, an electronic system is designed to display the torque. Health condition of SCIM is predicted based on torque magnitude. A typical implementation of the proposed scheme on 15 kW SCIM revealed a satisfactory correlation between the experimental and simulation results. The proposed approach aids to monitor the SCIM in Electric Vehicles (EVs) without physical intervention.

Keywords Finite element analysis (FEA) · Squirrel Cage Induction Motor (SCIM) · Finite element method (FEM) · Totally enclosed fan cool (TEFC)

1 Introduction

Induction motors find their widespread utility across various system applications in the industries. The robustness, simple construction, efficiency and reliability have made the Squirrel Cage Induction Motor (SCIM) commonly used drives. Many studies are being carried out over decades to improvise its efficiency and reduce losses.

S. S Hooli (✉) · A. Vadde · K. Manickavasagam · G. R. Kadambi
Department of Electrical Engineering, Ramaiah University of Applied Sciences, Bangalore, Karnataka, India
e-mail: shwetahooli99@gmail.com

A. Vadde
e-mail: anushav.ee.et@msruas.ac.in

K. Manickavasagam
e-mail: manickavasagam.ee.et@msruas.ac.in

M. N. Favorskaya et al. (eds.), *Innovations in Electrical and Electronic Engineering*, Lecture Notes in Electrical Engineering 661, https://doi.org/10.1007/978-981-15-4692-1_38

The magnetic field analysis with Finite Element Method (FEM) is used in the practical design of electrical machines [1]. To detect stator winding fault, rotor faults, rotor slot harmonics, and speed leakage, flux sensor coils are used in [2]. An external coil is placed on the frame of an induction motor to identify the air gap harmonics. Prediction of rotor speed of an induction motor from the e.m.f of an external coil using spectral analysis techniques is presented in [3]. A frequency-domain detection of stator winding faults is performed by comparing the stator current and the stray flux using the spectrum analysis for a normal induction machine in static as well as in dynamic conditions of with and without torque [4].

A novel harmonic current injection method to optimize air gap and yoke flux density distribution for multiphase motor are discussed in [5]. The stray flux in the vicinity of an electrical machine is attributed to the magnetic state of the machine and can be affected by the presence of a fault in the machine. Analysis of magnetic flux distribution of SCIM is complicated and hence simple models are required to analyze the effect of an internal fault on the stray flux [6].

Field analysis is used in [7] to provide an accurate evaluation of magnetic field distribution and mechanical performance of the induction motor. A new air-gap flux-based technique is proposed in [8] for Turn-to-Turn Fault (TTF) detection and identification of defective region in inverter-fed induction motors. Stray load losses in the end-region, leakage fluxes and losses, are investigated for a high-power cage induction motor. The study is performed at locked rotor, no-load and rated load conditions using a 3-D FEM approach [9]. A model is presented in [10] to capture magnetic stray fields outside the machine to reduce the electromagnetic signature.

Ship propulsion, aircraft, health care centers and other specific applications insist on the suppression of the effect of electromagnetic signatures outside the electrical machines. The International Agency for Research on Cancer (IARC) reviewed the health effects of static, electric fields, magnetic fields and concluded that Power Frequency Magnetic Field (PFMF) is possibly carcinogenic to humans. Exposure of PFMF) above 0.4 μT (4mG) increases the chance of childhood cancer [6]. Calculation of stator end-winding leakage inductance and influence of the excitation frequency of the stator current on the leakage inductance is explained in [11]. A survey of existing broken rotor bar fault detection techniques, classification based on fault signature, various monitoring conditions and signal processing techniques for the detection process is discussed in [12]. The field quantities in a longitudinal plane along the axis of an Axial Flux Machine (AFM) are calculated using both the FEM and the Fourier approach in [13]. A novel sensorless method for the induction motor control is proposed in [14]. Electric vehicles are having the potential to reduce the threat of global warming and expected to become a viable alternative to traditional vehicles [15]. At the same time, exposure of PFMF on human beings [6] is also a major concern and leads to a necessity to carry out to predict the leakage flux inside and outside of the SCIM. A method for estimating the torque from the measurement of the stator current without mechanical sensors is presented in [16]. From the spectrogram of the stator current, slip and rotational speed are estimated in stationary or operating conditions. In the proposed method, the only restrained coil is inserted in the fins of motor to capture leakage flux from the casing of SCIM. As a novel approach an electronics circuit is designed with restrained coil used to estimate the torque without current measurement. This arrangement can be utilized with any electric motor with slight modification in programming of microcontroller.

From the literature review presented above, it is inferred that most of the authors have concentrated only on the estimation of the stray magnetic flux and its interference with the nearby particles. In this paper, a method is proposed for reduction of leakage flux at casing of SCIM by placing a restraining coil on its fins. Impact on electric field and magnetic field distribution inside and outside of the SCIM by placing a restraining coil in the fins has been analyzed.

In this paper, leakage flux inside and outside of 15 kW SCIM is predicted using FEA. A restraining coil is designed to generate induced e.m.f from the flux external casing of SCIM.

The main objectives of the proposed method are as follows:

- To analyze the performance of a SCIM using FEM
- To analyze both electric and magnetic flux distribution inside and outside the SCIM with restraining coil
- To compute the torque from flux by providing restraining coil on the surface of fins
- To predict the health condition of SCIM without physical intervention.

2 Block Diagram

Induction motor is an important component integrated with commercially available equipment and industrial processes. Many research studies are available in the literature on monitoring of induction motors to ensure a high degree of reliability. Induction motor applications are found in many industries and include a wide range of rated power. The leakage flux outside SCIM can be utilized to generate e.m.f with the help of restraining coil. Figure 1 indicates the proposed approach to arrive at the specifications of restraining coil.

The designed restraining coil is analyzed by placing it around the fins of SCIM using FEA simulation.

3 Determination of Leakage Flux Using FEA

The main focus of this paper is to monitor the motor by the computation of torque. To compute the torque, prediction of magnetic flux at varying distances external to the casing of SCIM is required. Design specifications of a 15 kW SCIM are given in Table 1. Frequency, current and number of turns are the variable inputs for FEA analysis. The number of turns and current input for different load conditions of SCIM is calculated using dimensions and specifications of SCIM. The material assigned for different parts of SCIM for the analysis is given in Table 2.

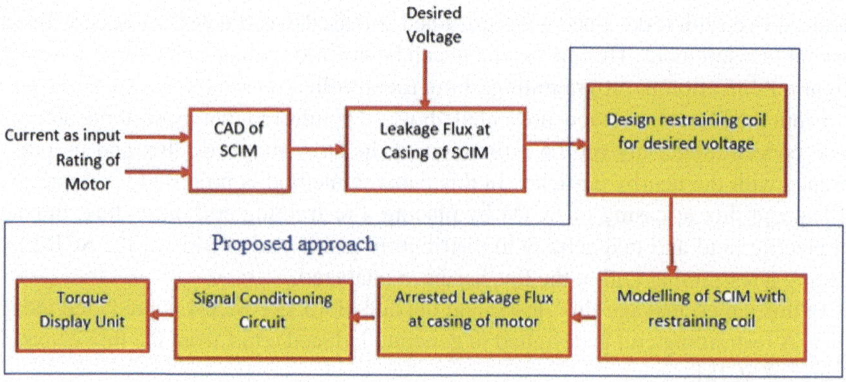

Fig. 1 Block diagram representation of the proposed approach

Table 1 Specifications of the 15 kW Induction Motor

Type	M2BAX160MLB4 (IE1)
Voltage	415 V
Rated current	28.18 A
No-load current	11.72 A
Power	15 kW
Pole	4 no.
Rated speed	1461.8 rpm
No-load speed	1498 rpm
Efficiency	90%
Power factor	0.822
Torque	97.98 Nm
Ambient temperature	25 °C
Stator temperature rise	70 °C
Temperature tolerance class	B
Insulation class	F

In this paper, closed type rotor bar is considered for the modelling and analysis of SCIM. The flux distribution in open type and closed type of rotor is almost the same [8]. Leakage flux distribution inside and outside of SCIM using FEA can be obtained for any given specifications of SCIM.

4 Design of Restraining Coil

The stator coil, restraining coil on the fin and outer frame of the SCIM is imitating the transformer action. SCIM stator induced emf E_1 and induced emf in restraining

Table 2 Material assigned for the Motor Parts

Motor parts	Materials	Electrical conductivity (S/m)	Relative permeability
Rotor bars and stator slots	Copper	5.998 × 107	1
Rotor	Aluminum 6061	3.72 × 107	gh1
Stator	Steel AISI 4340	1.6 × 106	30
Casing	Low carbon steel	5.51 × 106	1797.9241
Outer environment	Air	0	1

coil around the fin is E_2 is written as

$$E_1 = 4.44\, f\, \text{Ø}_1\, N_1 \tag{1}$$

$$E_2 = 4.44\, f\, \text{Ø}_2\, N_2 \tag{2}$$

Using the turns ratio:

$$\frac{\text{Ø}_1}{\text{Ø}_2} = \frac{N_2}{N_1} \tag{3}$$

Equation (3), Ø_1—average stator flux inside the SCIM is 0.3975 wb. Ø_2—average stator leakage flux around the fin of SCIM is 0.3975 wb. N_1—number of turns per phase in stator of SCIM is 108. This is obtained from datasheet of 15 kW SCIM. N_2—total number of turns around the 32 fins of SCIM. Using (3), N_2 is calculated as 68.54 turns for 32 fins.

SCIM is used with distributed winding in the stator. Location of the restating coil around all the fins is not possible due to space-constrained between fins. Even though space is available, if the restraining coil is placed around all the fins, the motor will exhibit transformer property. The main aim of the work is to extract leakage flux for displaying torque. From the simulation, it is assumed that six restraining coils are placed around the fins at every 60° along the circumference of SCIM. From the known values, the number of turns required for one fin (68.54 turns/32 fins) is 2.14 turns per fin.

5 Analysis of Magnetic Field Distribution in SCIM

Modelling of SCIM involves designing stator slots, rotor bars and shaft assembly. SCIM is modelled based on the design specifications mentioned in Table 1. A two-dimensional view of 15 kW SCIM with stator slots and closed type rotor bar is presented in Fig. 2. Simulation model of SCIM is developed with and without restraining coil placed on fins to measure leakage flux inside and outside of the casing of motor.

The magnetic leakage flux distribution inside and outside of SCIM at a section of 100 mm without and with restraining coil at one of the phase of '*A*' is represented in Figs. 3 and 4 respectively.

Simulation was also carried out with the placement of a restraining coil on all the 32 fins of the chosen SCIM with a rating of 15 kW. The magnetic leakage flux distribution with and without the restraining coil on all the fins at a section corresponding to an incremental axial length of '180 mm' is shown in Fig. 5.

Simulation of leakage flux distribution was carried out at points, which are exterior to the rim of the casing, but still coinciding with the plane of the rim of SCIM. The points of observation for simulation were on circles, which are external to the casing. The radial distances of separation between the circles of observation measured with respect to the rim of the casing of SCIM are at 6, 12, 18, 24, 50 and 100 mm. The leakage flux was averaged over 100 points measured on the circles of observation. The results of the leakage flux distribution on exterior circles with respect to rim of the casing were shown in Fig. 6. It can be observed that the leakage flux was reduced by placing a restraining coil placed over all fins of SCIM.

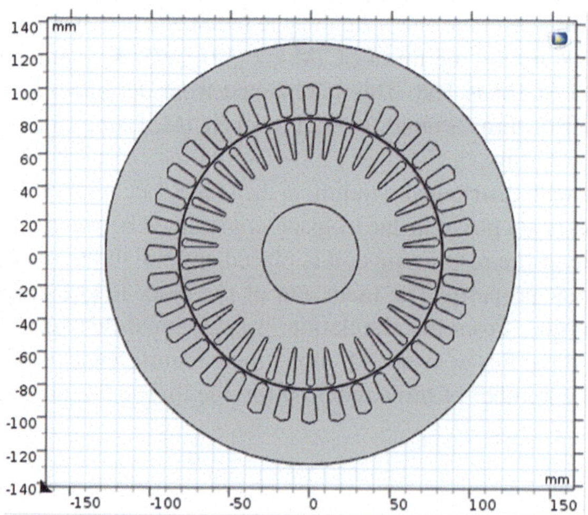

Fig. 2 2-Dimensional model of 15 kW SCIM

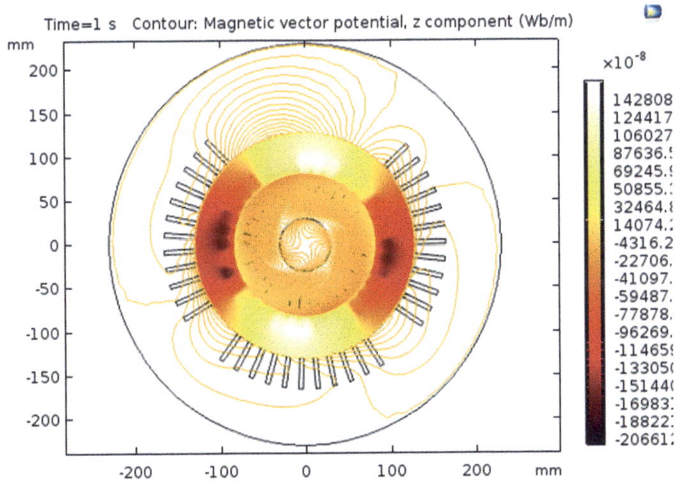

Fig. 3 Leakage flux distribution inside and outside of SCIM without restraining coil at a section of 100 mm

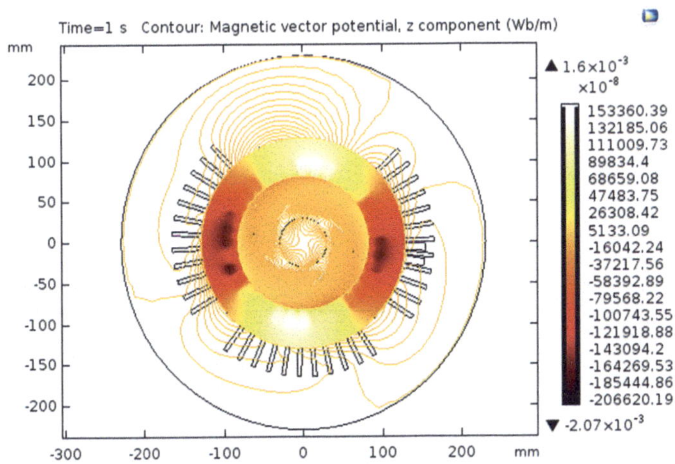

Fig. 4 Leakage flux distribution outside the SCIM with single restraining coil at a section of 100 mm

The leakage magnetic flux distribution at any distance external to casing is obtained by analyzing average flux at a hundred points on different distances away from SCIM. From this observation, fractional reduction in leakage flux in percentage is

$$\text{Avg}\,\varnothing_{\text{reduction}} = \frac{\text{Avg}\,\varnothing_{\text{WRC}} - \text{Avg}\varnothing_{\text{WORC}}}{\text{Avg}\,\varnothing_{\text{WORC}}} * 100 \tag{4}$$

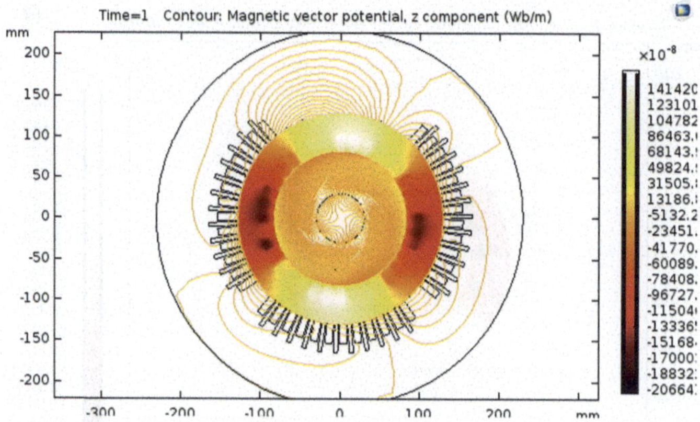

Fig. 5 Leakage flux distribution outside the SCIM with restraining coil placed over all fins at a section of 100 mm

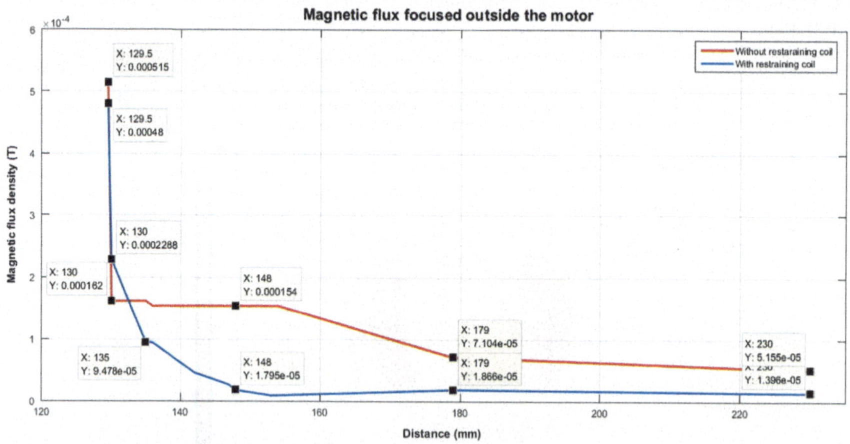

Fig. 6 Leakage flux distribution outside the SCIM at considered distances with and without restraining coil on all fins

where

Avg $\text{Ø}_{\text{reduction}}$	Average leakage flux reduction
Avg Ø_{WRC}	Average leakage flux with restraining coil
Avg Ø_{WORC}	Average Leakage flux without restraining coil

From (4), leakage flux is reduced by a factor of 73% by placing a restraining coil on all the 32 fins of the motor.

6 Analysis of Electric Field Distribution in SCIM

In the previous section, the focus was on the distribution of flux both inside and outside the casing of SCIM. In this section, the emphasis is on the impact of placement of restraining coil on electric field distribution inside the motor. The induced voltage on the casing of the motor has phase displacement, which is similar to supply voltage. The induced voltage outside the SCIM is shown in Fig. 7a. The desired or specified voltage distribution is illustrated in Fig. 7b. In this research study, induced voltage from *R*, *Y* and *B* phase at each fin is obtained as 0.4 V as shown in Fig. 8b and each phase have six fins carrying six restraining coils. Therefore, induced e.m.f obtained in each phase is computed as 2.4 V (6×0.4). The restraining coil is designed to generate 3 V. But through simulation, the voltage at the terminals of restrained coil is 2.4 V. The difference in the design and simulated value voltage at the restrained coil can possibly be attributed to the leakage flux present after inserting restraining coil also.

The amount of induced voltage at the terminals of restraining coil depends upon magnitude of leakage flux. In this study, restraining coil is designed for an induced voltage of 3 V. The corresponding electric field is shown in Fig. 8a. The same design procedure is followed for an induced voltage of 6 V to analyze the variation of the electric field inside the motor. When the induced voltage at the terminals of restraining coil is increased to 6 V, the electric field inside the motor is also increased as shown in Fig. 8b. The lowest value of electric field norm is considered to calculate the impact of electric field on SCIM from Fig. 8a, b.

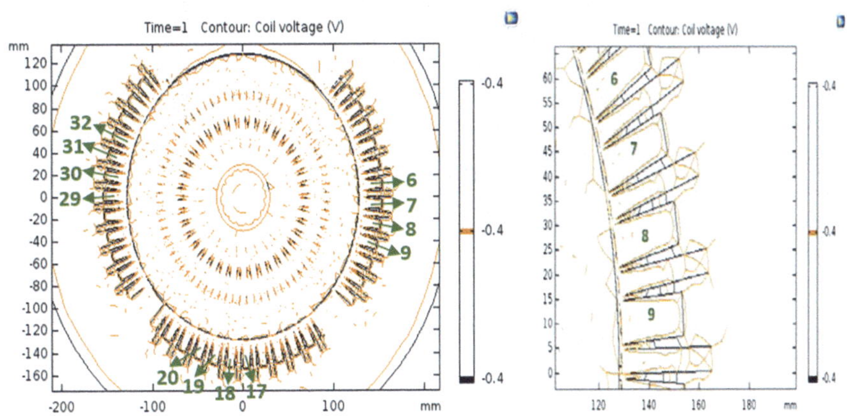

(a) (b)

Fig. 7 **a** Stator and rotor induced voltage inside and outside the SCIM **b** Induced voltage at casing of the SCIM

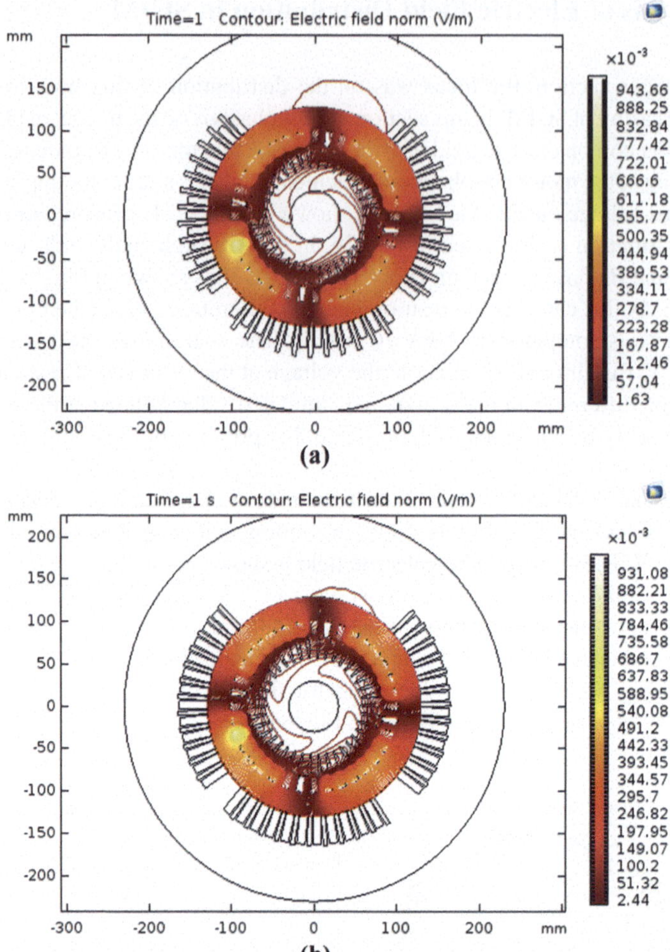

Fig. 8 **a** Electric field norm for designed model of 3 V **b** Electric field norm for designed model of 6 V

Increment in electric field norm (V/m):

$$E = \left(\frac{2.44 - 1.63}{2.44}\right) * 100 = 33.196\,\%$$

When the induced voltage is 6 V the electric field inside the motor is increased to 2.44 V/m from 1.63 V compared to the induced voltage of 3 V. When the induced voltage is 6 V, there is a possibility of restraining coil extracting power from the supply due to transformer action. Hence, the voltage extracted by the restraining coil is restricted up to 3 V only.

7 Electronic System for Torque Measurement

In real-time, measurement of torque is obtained from the speed is measured by tachometer using mathematical equations. According to Faraday's Laws of induction, the amount of leakage flux is proportional to load current and load current is function of torque. Hence, by capturing leakage flux, speed and torque can be measured and displayed directly using an electronic system designed as given in Fig. 9.

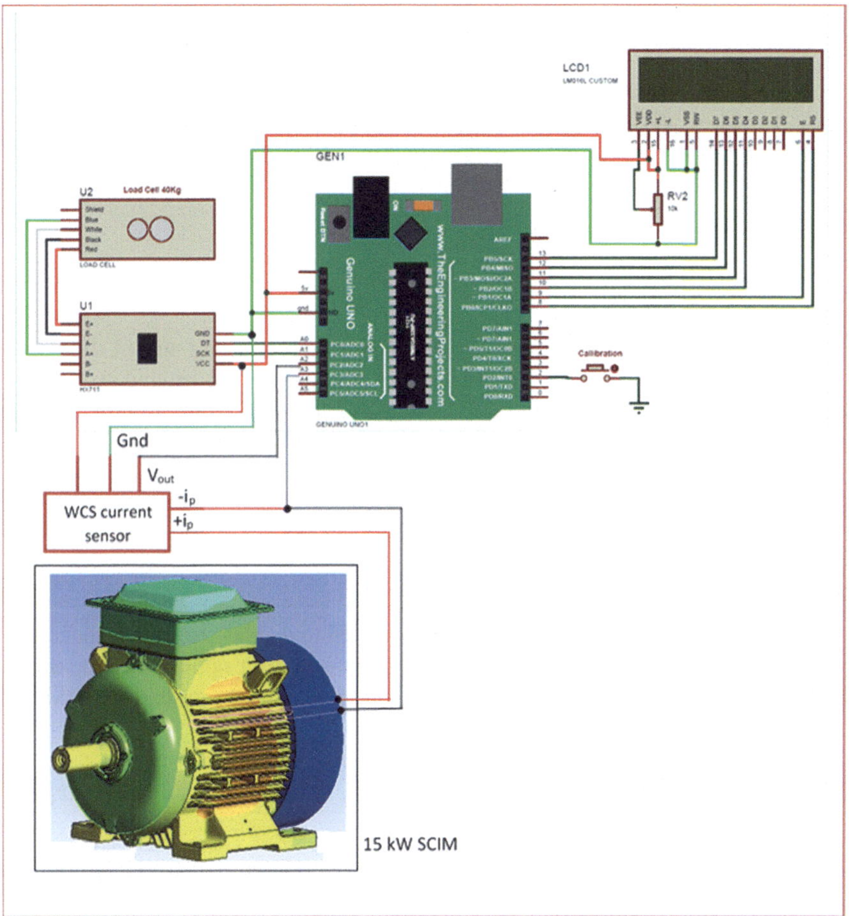

Fig. 9 Electronic system for torque display

7.1 WCS Current Sensor

The current sensor is connected in series to the restraint coil, to measure the current and generate a corresponding voltage. The generated voltage signal is given to the controller for processing. The sensor must relate to power supply and ground. Output terminal is configured as shown in Fig. 9.

7.2 Load Cell

The load cell (U2) is used to capture the amount of load connected to the motor. The physical quantity is converted to voltage by load cell. Output of load cell is connected to amplifier HX711 (U1) to amplify the signal.

7.3 Geniuno UNO

The Geniuno Uno is Ardinuo ATmega328P board with 6 analog and 14 digital i/o pins operating with 5 V. It has onboard power-up facility with ground connection and reset option. In this circuitry three analog and six digital pins are utilized. The program is written to process the inputs for obtaining speed and torque as output.

7.4 LCD for Torque

LM016L is a 16 × 2 liquid crystal display mounted to display the output speed and torque value. It has eight digital input, power up and reset option. The measured torque and speed are in Table 3 using the leakage flux from the casing of SCIM.

Table 3 Measured Torque and Speed Values

Sl.no	Induced EMF (V)	Torque (Nm)	Speed (RPM)
1	0.38	97.27	1461
2	0.285	54.24	1471
3	0.19	24.34	1460
4	0.095	5.92	1490
5	0.076	3.99	1492

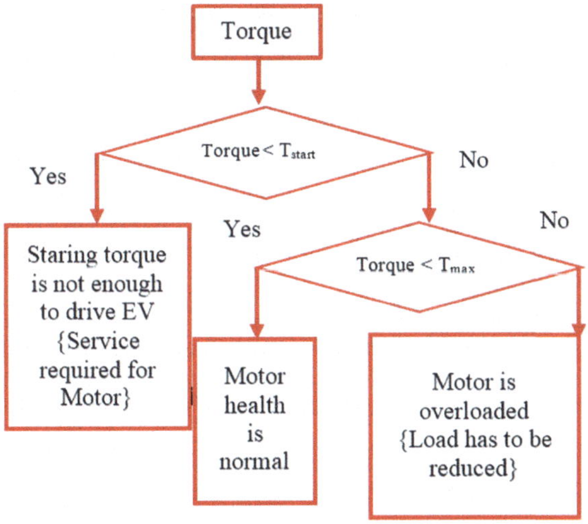

Fig. 10 Decision making on health of SCIM based on torque

8 Health Monitoring of Induction Motor in EV

The knowledge of this torque reveals failures or malfunctions and thus can be exploited in monitoring systems for preventing unscheduled maintenance shutdowns or to determine the dynamic performance and the efficiency of the energy conversion process [16]. Direct measurement of speed and torque can be displayed while the motor is running. Addition of restraining coil is used to predict the speed and torque without mathematical calculations. The speed and torque can be used to find health of the SCIM by knowing the standard value of starting and maximum torque. The health condition is predicted as if the starting torque is less than the specified value then SCIM required service. If the torque exceeds the maximum torque value than the specified value, then SCIM load is maximum. The specified values of starting and maximum torque of SCIM in EV must be provided by the manufacturer to the consumer. Health monitoring of SCIM using torque is given by Fig. 10 and the decision can be displayed if the programming is added. This is the future scope of the work.

9 Conclusions

Magnetic analysis of a 15 kW, SCIM is performed with and without the restraining coil using COMSOL software. The analysis of leakage flux inside and around the SCIM using is presented in this paper. This research leads to predict the harmful effect of leakage flux on human beings when SCIM is used as drives in EVs. The simulation results prove that a 73% reduction in leakage flux outside the SCIM when restraining coils is placed on all the fins. Restraining coil is designed for displaying the torque using electronic system, when the SCIM is in operation. Torque is measured from leakage flux of SCIM using an electronic system. Health condition of motor is predicted by comparing the magnitude of starting torque and maximum torque of the SCIM with torque displayed. The health condition can also be displayed using intelligence techniques in the future. This will reduce the burden on the physical human intervention of monitoring of SCIM in EV.

Acknowledgements The authors would like to express their profound gratitude to Dr. M. Arulantham, Dean, Faculty of Engineering and Technology, Dr. K.M. Sharath Kumar, Deputy Director, Directorate of sponsor research and the management of M.S. Ramaiah University of Applied Sciences for their continuous encouragement and kind support for providing the facility to carry out the research.

References

1. Goktas T et al (2016) Comprehensive analysis of magnet defect fault monitoring through leakage flux. IEEE Trans Magn 53(4):1–10
2. Goktas T, Zafarani M, Lee KW, Akin B, Sculley T (2017) Comprehensive analysis of magnet defect fault monitoring through leakage flux. IEEE Trans Magn 53(4):1–10
3. Ertan HB Keysan O (2009) External search coil as a means of measuring rotor speed of an induction motor. In: International symposium on advanced electromechanical motion systems & electric drives joint symposium, pp. 1–6
4. Henao H, Capolino GA, Martis G (2003) On the stray flux analysis for the detection of the three-phase induction machine faults. In: IAS annual meeting on conference record of the industry applications conference, pp. 1368–1373
5. Kong W, Qu R, Kang M, Huang J, Jing L (2017) Air-gap and yoke flux density optimization for multiphase induction motor based on novel harmonic current injection method. IEEE Trans Ind Appl 53(3):2140–2148
6. Koroglu S, Adam AA, Umurkan N, Gulez K (2009) Leakage magnetic flux density in the vicinity of induction motor during operation. Electr Eng (Archiv fur Elektrotechnik) 91(1):15–21
7. Elkasabgy NM, Eastham AR, Dawson GE (1992) Detection of broken bars in the cage rotor on an induction machine. IEEE Trans Ind Appl 28(1):165–171
8. Dehghan H, Haghjoo F, Cruz SM (2018) A flux-based differential technique for turn-to-turn fault detection and defective region identification in line-connected and inverter-fed induction motors. IEEE Trans Energy Convers 33(4):1876–1885
9. Cheaytani J et al (2015) End-region leakage fluxes and losses analysis of cage induction motors using 3-d finite-element method. IEEE Trans Magn 51(3):1–4
10. Le Coat G et al (1999) Electromagnetic signature of induction machines. IEEE Trans Energy Conver 14(3):628–632
11. Lin R, Arkkio A (2009) Calculation and analysis of stator end-winding leakage inductance of an induction machine. IEEE Trans Magn 45(4)
12. Eastham JF, Balchin MF, Rodger D (1981) Measurement and calculation of rotor-core flux densities in axial flux linear induction motors. IEE Proc 128, Pt. B, No. 6
13. Sangsefidi Y, Ziaeinejad S, Nabi HP, Shoulaie A (2014) Induction motor control based on approximate stator flux. IET Power Electron 7(11):2765–2777

14. Farasat M, Trzynadlowski AM, Fadali MS (2014) Efficiency improved sensorless control scheme for electric vehicle induction motors. IET Electr Syst Transp 4(4):122–131
15. Silva WL, Lima AMN, Oliveira A (2014) A method for measuring torque of squirrel-cage induction motors without any mechanical sensor. IEEE Trans Instrum Measur 64(5):1223–1231
16. Chen J, Yang S (2011) Application of Proteus software in MCU teaching. In: 2011 second international conference on mechanic automation and control engineering, Hohhot, pp 6359–6362

Fuzzy Logic-Based Intelligent PID Controller for Speed Control of Linear Internal Combustion Engine

K. Sandeep Rao, V. N. Siva Praneeth, and Y. V. Pavan Kumar

Abstract Internal combustion (IC) engines are widely used in many applications such as automobiles, light aircraft systems, ships/boats, and diesel generators. In order to effectively cater the needs of all these traditional as well as modern applications such as hybrid electric vehicles, the engine speed should be controlled smoothly and rapidly. Usually, proportional-cum integral-cum derivative (PID)-based controller is used for controlling the speed according to the requirement. However, there are some limitations exist with the conventional offline tuning methods available for PID gain parameter design such as restricted to nonlinear changes and poor disturbance rejection due to having constant gain values irrespective of system condition. Besides, the identification of suitable tuning method for a given system is another tedious job. By keeping all these points in view, this paper proposes an intelligent and online tuning method for PID gain parameters using fuzzy logic theory. Instead of tuning the gain parameter values offline, the proposed fuzzy-PID (FPID) tunes the PID gain values online, i.e., the PID gain values are continuously updated according to the system condition. So, this improves the controller ability to reject the disturbances. The system is modeled and simulated using MATLAB/Simulink software. The results proved that the proposed FPID controller provides better transient and steady-state speed response of the IC engine under all conditions.

Keywords Fuzzy-PID · IC engine control · Parameter tuning · PID controller

K. Sandeep Rao · V. N. Siva Praneeth · Y. V. Pavan Kumar (✉)
School of Electronics Engineering, Vellore Institute of Technology—Andhra Pradesh (VIT-AP) University, Amaravati 522237, India
e-mail: pavankumar.yv@vitap.ac.in

K. Sandeep Rao
e-mail: sandeep.18bec7094@vitap.ac.in

V. N. Siva Praneeth
e-mail: praneeth.18bev7022@vitap.ac.in

M. N. Favorskaya et al. (eds.), *Innovations in Electrical and Electronic Engineering*, Lecture Notes in Electrical Engineering 661, https://doi.org/10.1007/978-981-15-4692-1_39

1 Introduction

Linearized internal combustion engine (LICE) generates power by burning the fossil fuels such as petrol, gas, and oil . LICE should generate the required power irrespective of the disturbances and provide the required speed. However, the control over the speed is very crucial in order to manage the speed according the required output. In this process of control, there are different combinations of conventional controllers available such as PI, P, PD, and PID. [1]. A suitable controller is selected depending upon the system behavior and necessity. Among all the controllers, PID controller is the most commonly used controller [2]. The efficacy of the controller depends upon the suitable selection of its gain parameters. There are many traditional procedures which are existing to design the PID controller gain parameters, but these traditional methods are "offline" design methods. The controller gain parameters are initially tuned, and then, the controller is put into the system operation. So, these offline methods are limited with respect to the following points [3, 4].

- As the controller gain parameters are designed prior to its operation in real-time, the controller becomes unaware of practical disturbances. Hence, these methods possess very poor disturbance rejection capacity.
- The controller cannot suffice for variable inputs (change of user inputs) due to the retention of constant controller gain parameters, which were designed for one input. Further, these methods are applicable to only linear variations of the error.
- Most of the traditional methods are applicable for either open-loop or closed-loop systems. The methods applicable for both types of systems are very limited.
- Another key limitation is, even though there are many methods, the method identified as best to tune the gain parameters of one given plant, cannot be generalized, i.e., it may not work same for another plant. So, the best suited method has to be identified for each and every practical system. Hence, the searching of suitable tuning method for a given plant is a key job in overall system control operation.

The considered application in this paper, i.e., LICE, is inherently nonlinear in nature [4, 5]. So, the traditional PID controller design methods cannot give better results for LICE control, which needs some intelligent and "online" tuning approaches. To use the traditional tuning methods given in [6–9], it approximates the system behavior to linear. On the other hand, artificial intelligent-based online controllers such as fuzzy logic controller that uses fuzzy theory for process control were discussed in [10–13]. However, these methods have to be improved to obtain superior time-domain transient response parameters to operate the considered systems more robustly with respect to the disturbance conditions. Besides, these methods were not validated under real-time disturbances.

By keeping all these points in view, this paper proposes an intelligent and online tuning method for PID gain parameters using fuzzy logic theory. Instead of tuning the gain parameters values offline, the proposed fuzzy-PID (FPID) tunes the PID gain values online, i.e., the PID gain values are continuously updated according to the system condition. Hence, this proposed FPID possesses the advantages of both fuzzy controller and the traditional PID controller, and thereby, improves the controller ability to reject the disturbances.

2 LICE System Description and Controller Modeling

2.1 System Model Description

LICE is the one of the heat engines, which is used in many automobile applications, power generation systems, etc. Its plant model can be derived as a linear model by using the four-cylinder or six-cylinder configuration [14]. This paper uses six-cylinder configuration for developing the system model, whose schematic is shown in Fig. 1.

For the system model development, the nonlinear IC engine parts shown in Fig. 1 are approximated to a linear model by assuming the following conditions. The total block diagram is shown in Fig. 2.

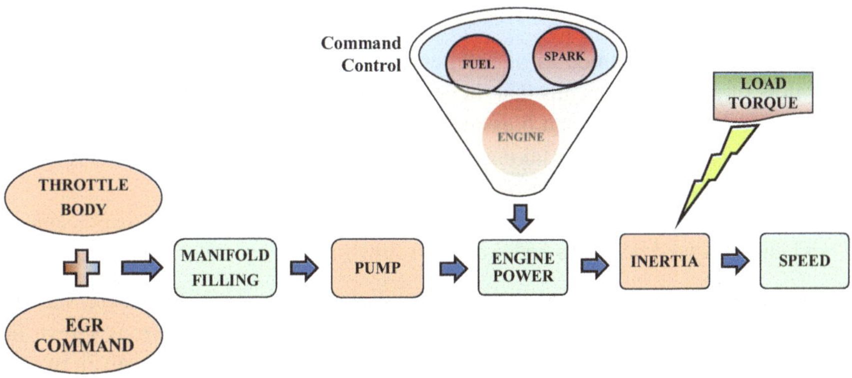

Fig. 1 Schematic model of nonlinear combustion engine system

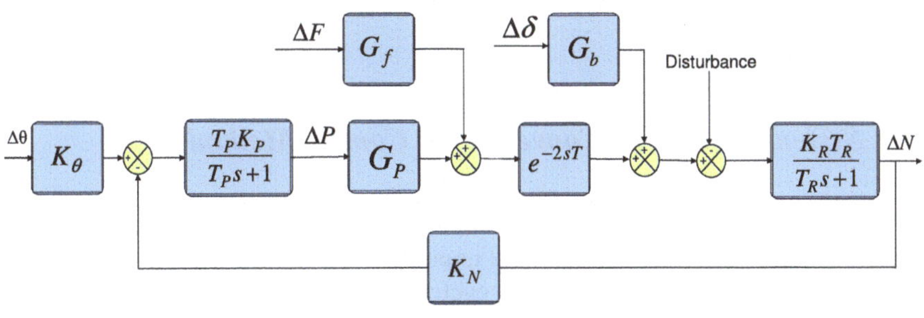

Fig. 2 Block diagram of LICE system

Table 1 Model parameters of LICE system

S. no	Parameter	Variable	Values	Units
1	Gain of spark advance on torque	G_i	13.372	ft-lbf/psi
2	Gain of delayed fuel on torque	G_s	36.61	ft-lbf/lbm
3	Manifold time constant	T_P	0.215	s
4	Rotating inertia time constant	T_R	3.988	s
5	Gain to adjust polar term of inertia	K_R	67.21	rpm/ft-lbf-s
6	Function of gas constant	K_P	0.7768	lbf-h/lbm-in^2-s
7	Pumping feedback gain	K_N	0.98	lbm/rpm-min
8	Exponential delay time	T	0.0333	s

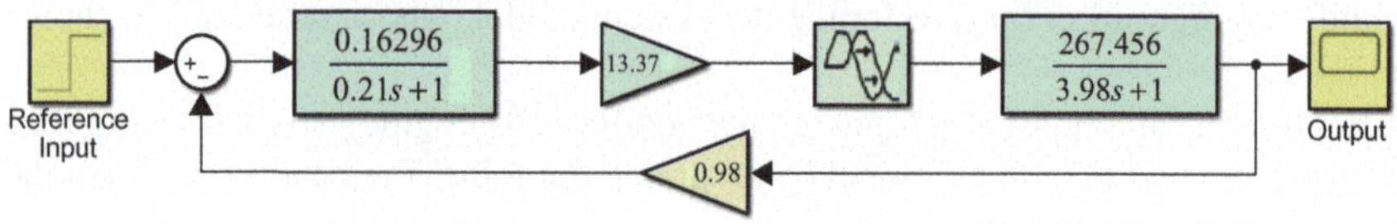

Fig. 3 LICE system simulation model for the controller development

- Throttle body is assumed as one-dimensional, adiabatic, and reversible process of flow of gas, which means that there is no loss or transfer of heat.
- In manifold filling, uniform pressure between throttle and all valves at constant temperature is assumed. By taking these considerations, a linear first order relation is obtained between the manifold pressure to flow rates.
- The engine is treated as pump. By doing this, the product of engine displacement and engine speed becomes equal to mass flow rate.

The modeling specifications are given in Table 1. By substituting these values, block diagram shown in Fig. 2 is converted to the modeling diagram as shown in Fig. 3, from which, the total closed-loop transfer function is derived as given in (1).

$$\frac{759.997s + 3578.052}{0.83s^2 + 766.077s + 3579.052} \tag{1}$$

2.2 System Control by Conventional Control Methods

There are many conventional tuning methods available for tuning the PID controller gain parameters [15]. Among these, the important methods are open-loop transient response (OLTR) methods and ultimate cycles (UC) methods that are further classified as shown in Fig. 4. These methods are sufficient for the effective design of most of the linear systems. However, there are certain limitations as explained follows.

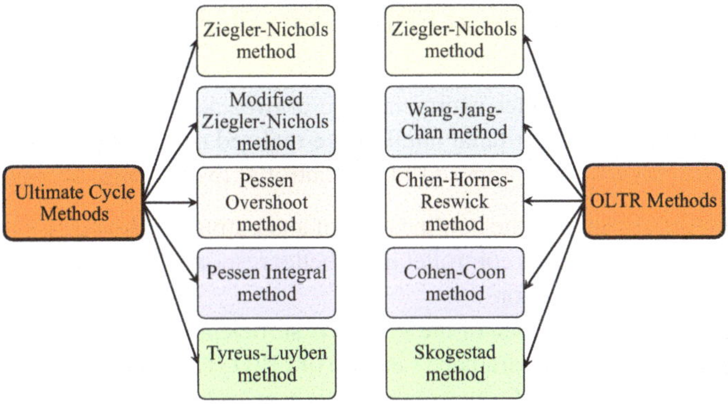

Fig. 4 Different OLTR and UC methods to tune controller parameters

OLTR methods are the fundamental methods developed for PID controller gain parameters' tuning. These are simple and base to the next-generation methods, which draws an open-loop curve of the system to calculate the transient response index parameters such as stationary gain, dead time, time constant, and slope of the curve. Using these parameters, the controller gains will be calculated as per the relations given in [15]. These methods are simple to implement but only useful to estimate proportional gain and integral time, but cannot be used for the calculation of derivative time. Another major limitation is the procedure of estimation, where the controller parameters are estimated using open-loop response rather than closed loop. Hence, the OLTR methods cannot be used where there is a strict necessity of PID controller [16].

Another group of methods for PID tuning is UC methods, where closed-loop response is used for the controller gain parameter tuning instead of open-loop response used in OLTR methods. So, UC methods may work better than the OLTR methods. In the UC methods, ultimate cycles or sustained oscillations have to be obtained by adjusting the closed-loop gain value of the system [17]. This gain value at which the ultimate cycles exist is called "critical gain." The time period of the oscillation needs to be calculated, and by using these critical gain and critical time period, PID controller gain parameters are calculated as per [15]. With respect to the literature mentioned in [15–18], among various UC group methods, Ziegler–Nichols (UZPID) closed-loop method gave better output for different systems. So, this is considered to validate the proposed FLPID in this paper. The major limitation of UC methods is that all types of systems may not produce sustained oscillations, for which UC methods cannot be used.

3 Description of Proposed FPID Controller

The proposed FPID controller overcomes the drawbacks of the conventional PID controller with the help of artificial intelligence-realized using fuzzy logic concepts. The main advantages of the proposed FPID controller over UZPID are [19, 20],

- FPID controller can map the nonlinear input and outputs in the range of membership functions. Hence, this controller gives better response for LICE as it possesses inherent nonlinear behavior.
- The behavioral performance of the system is sufficient in generation of membership functions and other parameters in case of FPID. There is no specific need for mathematical modeling of a system which is required in case of PID controller.

Various parts of the controller realization are shown in Fig. 5. In the development of proposed FPID, there are three main steps involved. The first step is fuzzification process, where the domain of the input/output data is transformed, i.e., the crisp data is converted into fuzzy set data. In this transformation, each input/output will be converted into its own membership function, which graphically represents the given data. Different types of membership functions such as triangular, trapezoidal, Gaussian are used. Triangular membership function is used for the implementation of the proposed FPID as it is the most commonly used one for periodic data sets. The second step is developing the fuzzy interface structure (FIS). In this FIS, the input and outputs are converted into membership functions. The inputs and outputs are related using set of mapping rules, which replicate how the decisions are taken by the humans, i.e., for a certain action what type of response need to be given. Fuzzy logic controller's efficiency depends on establishment of these rules with the correct relation of inputs and outputs. Table 2 shows the rules implemented for the proposed FPID controller, and Table 3 shows their input and output ranges identified.

The third step is defuzzification, which is the inverse process of the fuzzification. In this process, the fuzzy set values are converted into the output values. There are various methods for converting the rage of output values into a single output value, such as centroid method, bisector method, and middle of the maximum technique. As shown in Fig. 6, the proposed FPID controller is implemented using one input variable

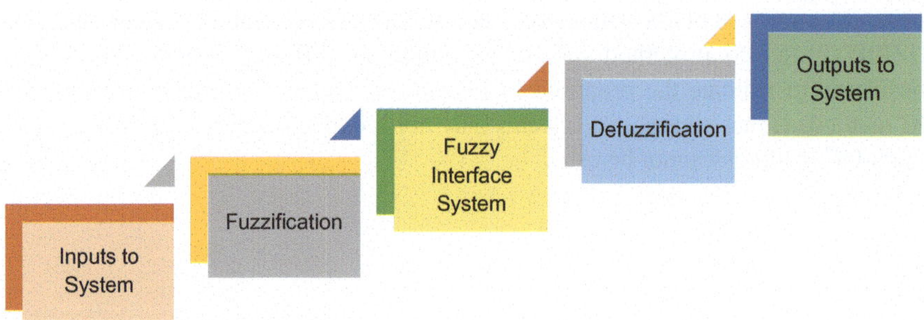

Fig. 5 Steps involved in implementing fuzzy logic controller

Table 2 Rules for relating the inputs and outputs variables

Input	Output		
	K_P	K_I	K_D
MPV	K_{P1}	K_{I1}	K_{D1}
NPV	K_{P2}	K_{I2}	K_{D2}
ZEV	K_{P3}	K_{I3}	K_{D2}
NNV	K_{P4}	K_{I3}	K_{D3}
MNV	K_{P5}	K_{I1}	K_{D3}

Table 3 Input and output ranges considered for generating FIS rules

Membership functions		Data ranges	
Input variable	Deviation	MPV	[0.1 0.2 0.3]
		NPV	[0 0.1 0.2]
		ZEV	[−0.1 0 0.1]
		NNV	[−0.2 −0.1 0]
		MNV	[−0.3 −0.2 −0.1]
Output variables	K_P	K_{P1}	[0.0025 0.003 0.0035]
		K_{P2}	[0.008 0.009 0.01]
		K_{P3}	[0.0038 0.0048 0.0058]
		K_{P4}	[0.0012 0.00220.0032]
		K_{P5}	[0.008 0.009 0.01]
	K_I	K_{I1}	[0.04 0.045 0.05]
		K_{I2}	[0.05 0.06 0.07]
		K_{I3}	[0.02 0.03 0.04]
	K_D	K_{D1}	[0.02 0.03 0.04]
		K_{D2}	[0.03 0.04 0.05]
		K_{D3}	[0.045 0.05 0.055]

(i.e., error variable) and three output variables (i.e., PID gain values—K_P, K_I, K_D) with the use of triangular membership function. Whenever a disturbance is injected into the system, the response of the system undergoes underdamped in nature. The deviation from desired response is divided into segments, namely maximum positive value (MPV), normal positive value (NPV), zero value (ZEV), i.e., normal value, normal negative value (NNV), and maximum negative value (MNV) as shown in Fig. 7. The corresponding controller gain values are also represented in ranges. All

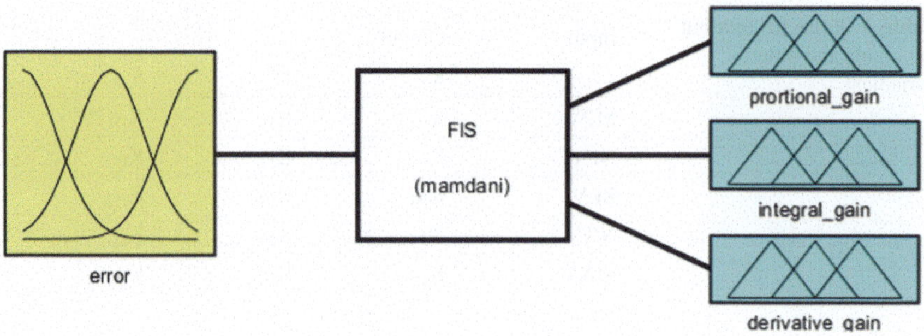

Fig. 6 Inputs and outputs for generating FIS of proposed FPID controller

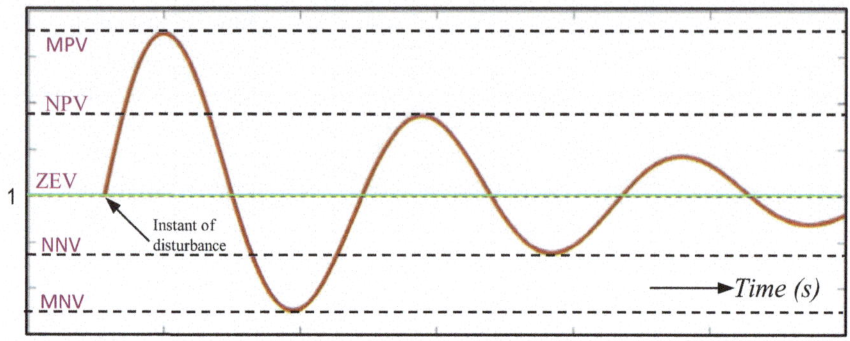

Fig. 7 Response of the system when subjected to a disturbance

the ranges of input/output with respect to the proposed segments are given in Table 3. The overall simulation model including conventional and proposed models is shown in Fig. 8.

4 Simulation Results

As the main objective of the proposed FPID controller is to achieve superior performance compared to the conventional PID, especially during disturbances, various test cases referred in Table 4 are implemented to validate the FPID. As mentioned in Sect. 2.2, conventional UZPID, which is providing better system characteristics among the conventional methods, is implemented to validate the proposed FPID method. The PID gain values obtained by UZPID method are $K_P = 0.0699$, $K_I = 0.1856$, and $K_D = 0.006$. The efficacy of conventional UZPID controller and the proposed FPID controller for the speed control of LICE system are compared with respect to the system behavior obtained during transient-state and steady-state with

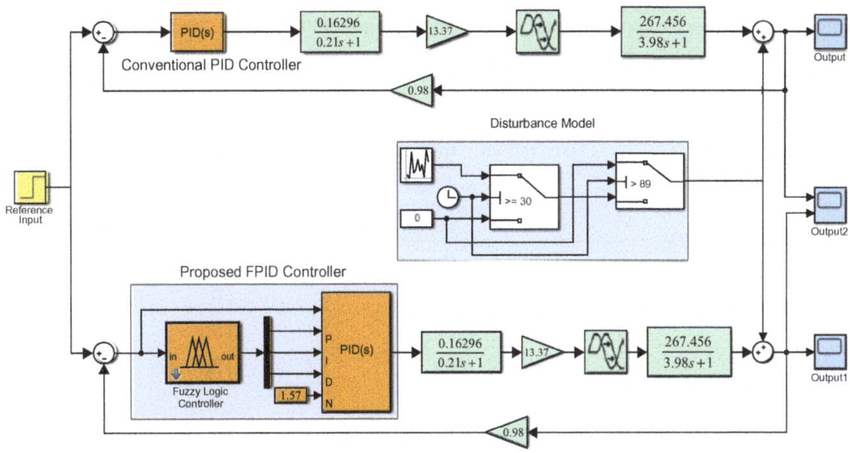

Fig. 8 Simulink model of LICE system with conventional and proposed controller

Table 4 Test cases implemented for LICE system analysis

Test case no.	Disturbance (or variation) condition
1	Ideal condition (zero variation)
2	Positive step variation with 20% of output
3	Negative step variation with 20% of output
4	Repeating sequence interpolated variation
5	Uniform random number variation
6	Sinusoidal variation
7	Band-limited white noise variation
8	Repeating stair sequence variation

disturbances. The cumulative quantitative results with respect to various time-domain performance indices are given in Tables 5 and 6.

The comparative simulation results are shown from Figs. 9, 10, 11, 12, 13, 14, 15, 16, 17, and 18. These results include both time-domain transient analysis as well

Table 5 Performance of conventional UZPID and proposed FPID during transient state

Method	Raise time (in s)	Peak time (in s)	Delay time (in s)	Settling time (in s)	Peak overshoot (in %)
Conventional UZPID	0.321	1.45	1.627	3.203	66.34 (oscillatory)
Proposed FPID	1.18	5.13	1.663	8.082	10.48 (smooth)

Table 6 Performance of conventional UZPID and proposed FPID during steady state

S. no	Type of disturbance	Time (sec)	Conventional UZPID method		Proposed FPID method		Superior method
			% Peak overshoot	% Peak undershoot	% Peak overshoot	% Peak undershoot	
1	Step disturbance of positive 20%	30	20	17.4	20	1.1	FPID
		45	0	0	0	0	
		60	0	0	0	0	
		75	0	0	0	0	
		89	17.4	20	1.1	20	
2	Step disturbance of negative 20%	30	17.4	20	1.1	20	FPID
		45	0	0	0	0	
		60	0	0	0	0	
		75	0	0	0	0	
		89	20	17.4	20	1.1	
3	Repeating sequence interpolated disturbance	30	10	6.7	10	1.1	FPID
		45	0	0	0	0.5	
		60	0	0	0	0	
		75	0	0	0	0	
		89	6.7	10	1.1	10	
4	Uniform random number disturbance	30	8.2	5.8	8.2	0.7	FPID
		45	0	0	0.5	0	
		60	10	6.7	10	1.1	
		75	17.4	20	1.1	20	
		89	5	2.3	5	0.3	
5	Sinusoidal disturbance	30	2.5	0.12	2.5	0.12	FPID
		45	0	0	0	0.7	
		60	0	0	0.7	0	
		75	0	0	0	0.7	
		89	6.7	10	1.1	10	
6	Band-limited white noise disturbance	30	2.3	5	0.3	5	FPID
		45	18	15.6	18	0.9	
		60	6.2	9	0.8	9	
		75	2.3	5	0.3	5	
		89	3	2.2	3	0	
7	Repeating sequence stair disturbance	30	20	17.4	20	1.1	FPID
		45	5.8	8.6	0.8	9	
		60	3	2.2	3	0	
		75	2	1.8	2	0	

(continued)

Table 6 (continued)

S. no	Type of disturbance	Time (sec)	Conventional UZPID method		Proposed FPID method		Superior method
			% Peak overshoot	% Peak undershoot	% Peak overshoot	% Peak undershoot	
		89	15.6	18	0.9	18	

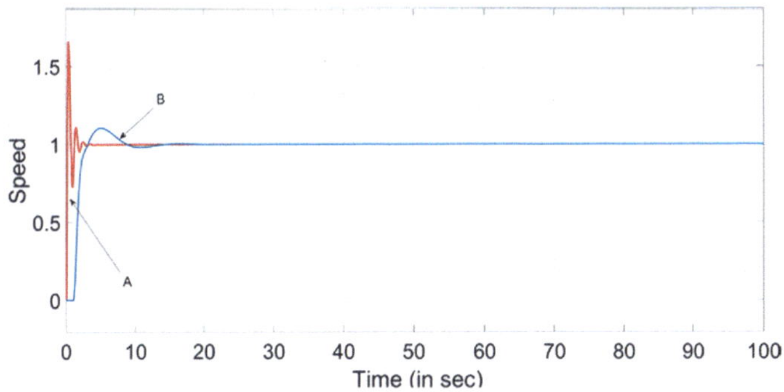

Fig. 9 LICE response with conventional UZPID and proposed FPID for test case-1

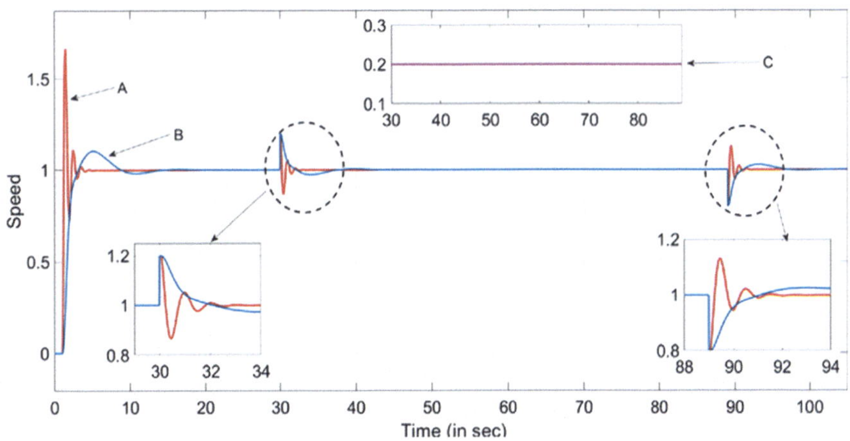

Fig. 10 LICE response with conventional UZPID and proposed FPID for test case-2

as frequency-domain stability analysis. From Figs. 9, 10, 11, 12, 13, 14, 15, and 16, various plots of the response are represented as given follows.

- Plot 'A' represents the response of the system with conventional UZPID controller.
- Plot 'B' represents the response of the system with proposed FPID controller.
- Plot 'C' represents the characteristics of disturbance injected into the system.

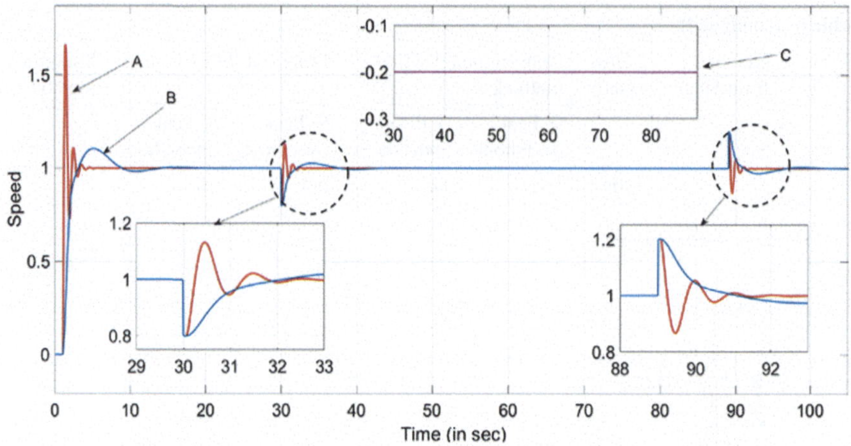

Fig. 11 LICE response with conventional UZPID and proposed FPID for test case-3

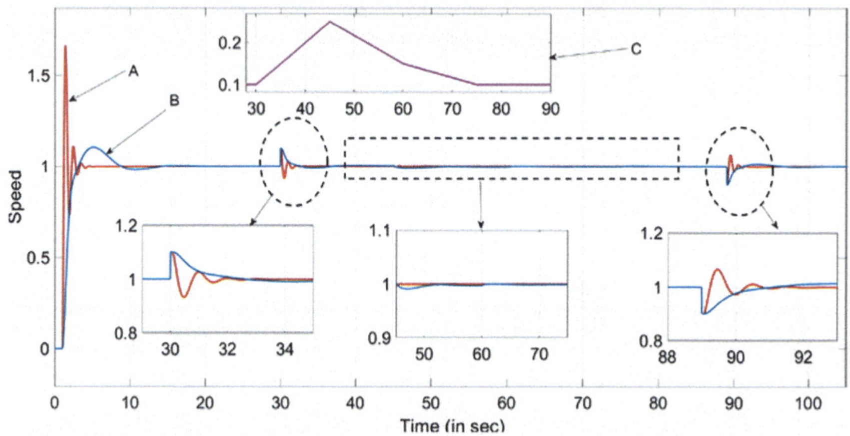

Fig. 12 LICE response with conventional UZPID and proposed FPID for test case-4

Figure 9 shows the response when there is no disturbance. Different types of disturbances with sample time 15 s are injected at 30 s and removed at 89 s into the system in its steady state. Figures 10 and 11 show the simulation results obtained for test case-2 and test case-3, respectively. At the time when the disturbance is injected, both the conventional UZPID controller and proposed FPID controller are rectifying the disturbance, but UZPID controller leads to oscillatory transient behavior high peak-over and undershoot when compared to proposed FPID controller. Similarly, repeating sequence, uniform loads, and sinusoidal current drawn loads are applied in test cases 4–6. The corresponding results are shown in Figs. 12, 13, and 14. In test cases 7 and 8, band-limited white noise and repeating sequence stair type of disturbances are injected into the system. The corresponding results are shown in

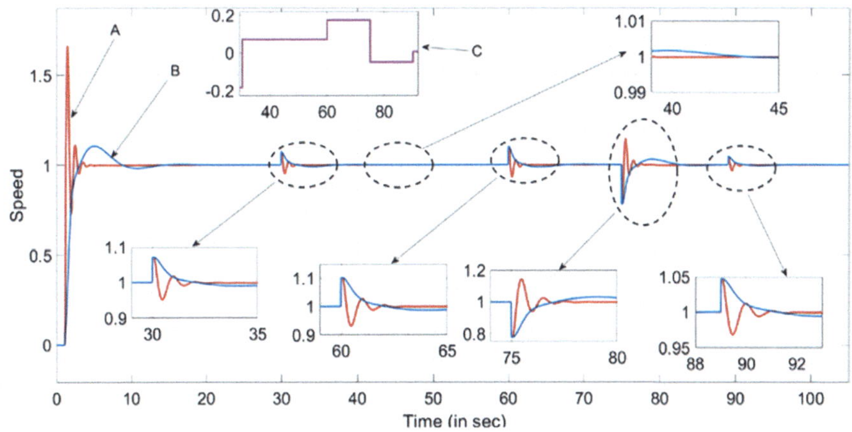

Fig. 13 LICE response with conventional UZPID and proposed FPID for test case-5

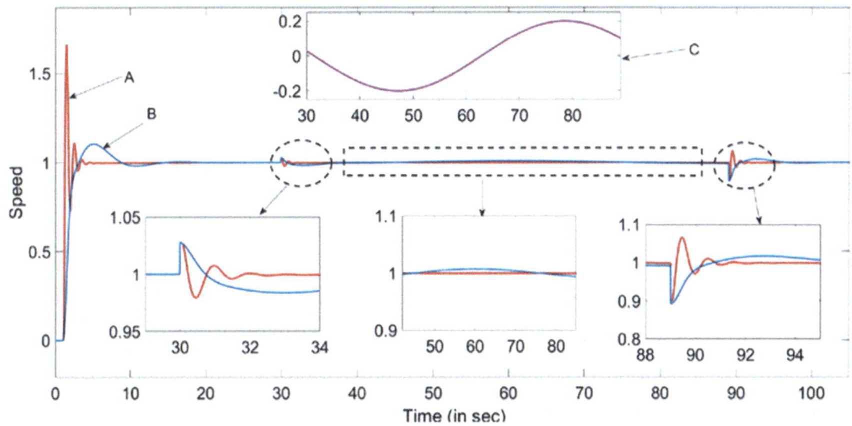

Fig. 14 LICE response with conventional UZPID and proposed FPID for test case-6

Figs. 15 and 16. The disturbances are introduced at 30 s and removed at 89 s. In all these cases, the proposed FPID controller is providing smooth response with less transient time in addressing the disturbance when compared to conventional UZPID controller.

From the stability plots of UZPID and FPID as in Figs. 17 and 18, the phase margin for UZPID is calculated as 99.9°, and phase margin for FPID is 138°. As the phase margin for FPID is more than that of UZPID, the proposed FPID controller gives more stability to the system than the conventional UZPID controller. Further, from Table 5, it can be observed that the peak overshoot is more in case of UZPID (66.34%) when compared to FPID (10.48%). Similarly, from Table 6, it is clearly

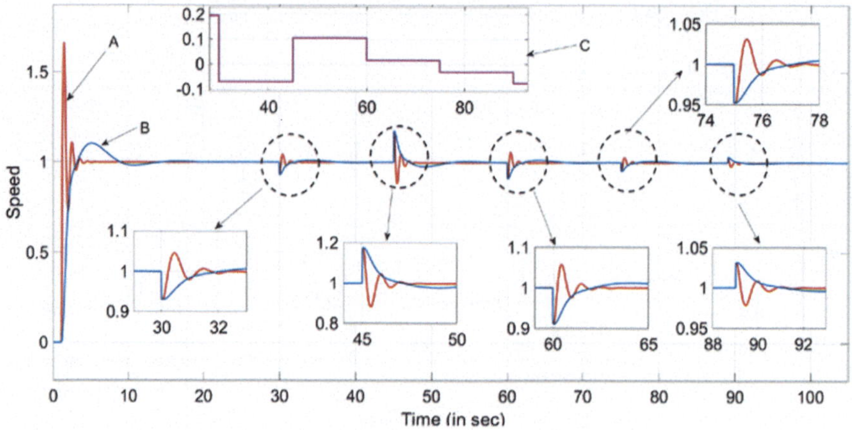

Fig. 15 LICE response with conventional UZPID and proposed FPID for test case-7

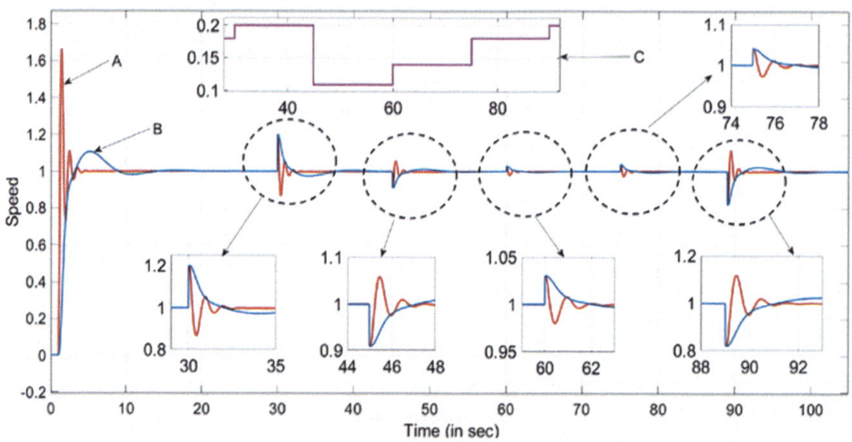

Fig. 16 LICE response with conventional UZPID and proposed FPID for test case-8

seen that the proposed FPID controller provides better response to the system in all disturbances when compared to the conventional UZPID controller.

5 Conclusions

The constraints with the conventional PID controller for controlling the speed of LICE such as poor disturbance rejection, poor transient response, and poor stability margins are addressed and satisfactorily improved in this paper with the proposed fuzzy-PID

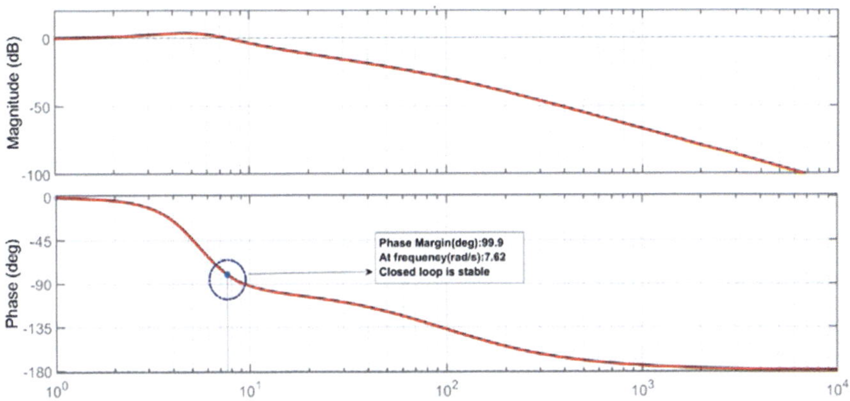

Fig. 17 Frequency response with conventional UZPID controller for test case-1

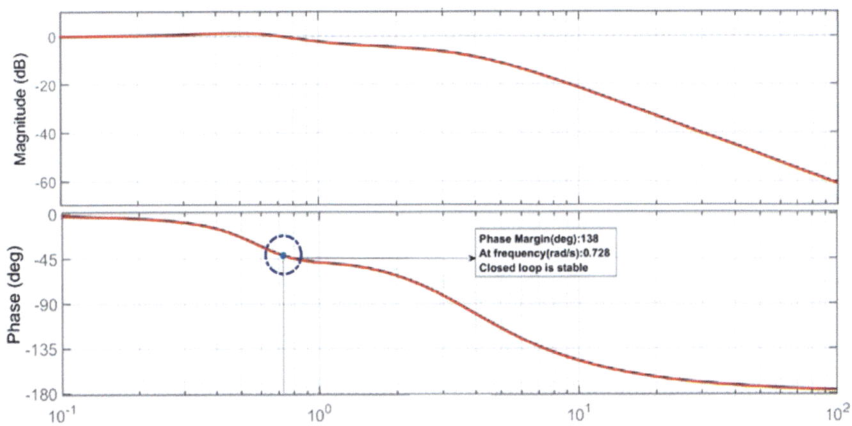

Fig. 18 Frequency response with proposed FPID controller for test case-1

(FPID) controller. The cumulative achievements of the proposed controller are given as follows.

- The proposed FPID provides better closed-loop performance and thereby reduces the oscillatory behavior in transient state.
- The proposed FPID improves key time-domain performance indices, i.e., peak overshoot and undershoot. The reduction of these values protects the loads connected to the LICE.
- The proposed FPID controller has commendably reduced the effect of various disturbances on the system behavior when compared to the conventional controller.

- Besides, the proposed FPID controller produces more stability margins when compared to the conventional controller, and thereby, enhances the system stability.

Hence, it could be concluded that the proposed FPID controller is a better controller to control the speed of a LICE smoothly according to the user requirements.

References

1. Boiko IM (2014) Application of second-order sliding-mode control algorithms in continuous cycling tests for PID tuning. European control conference. Strasbourg, France
2. Srinivas P, Lakshmi KV, Kumar VN (2014) A comparison of PID controller tuning methods for three tank level process. Int J Adv Res Electr Electron Instrum Eng 3(1):6810–6820
3. Sambariya DK, Prasad R, Birla D (2015) Design and performance analysis of PID based controller for SMIB power system using firefly algorithm. In: 2nd international conference on recent advances in engineering & computational sciences. Chandigarh, India
4. Arun VKS, Subramaniam U, Padmanaban S, Bhaskar MS, Almakhles D (2019) Investigation for performances comparison PI, adaptive PI, fuzzy speed control induction motor for centrifugal pumping application. In: IEEE 13th international conference on compatibility, power electronics and power engineering. Sonderborg, Denmark
5. Wang X, Wang P, Fei W, Chen H (2019) Stability and trajectory control for post-impact vehicle based on fuzzy PID algorithm. In: Chinese control conference. Guangzhou, China
6. Long G, Yang Y, Xiaolin R, Jianbo C, Mei T, Guihong Z, Ziyun J, Han Q (2019) Design of air source heat pump intelligent temperature control system based on adaptive fuzzy PID control. In: Chinese control conference. Guangzhou, China
7. Live HJ, Wang HB, Zhu XM, Shen ZH, Chen JY (2017) Simulation research of fuzzy auto-tuning PID controller based on matlab. In: International conference on computer technology, electronics and communication (ICCTEC). Dalian, China
8. Manikandan R, Arulprakash A, Arulmozhiyal R (2015) Design of equivalent fuzzy PID controller from the conventional PID controller. In: International conference on control, instrumentation, communication and computational technologies. Kumaracoil, India, (2015)
9. Ataei M, Ehsan SS (2008) Sliding mode PID-controller design for robot manipulators by using fuzzy tuning approach. 27th Chinese Control Conference. Kunming, China
10. Li S, Ma C, Fu W, Li Z, Tao S (2019) Design of temperature control system for high-precision position sensor calibration thermostat based on Fuzzy-PID compensation. IEEE Innovative Smart Grid Technologies—Asia. ISGT Asia, Chengdu, China
11. Khan AA, Nishkam R (2006) Fuzzy PID controller: design, tuning and comparison with conventional PID controller. In: IEEE international conference on engineering of intelligent systems. Islamabad, Pakistan
12. Nguyen BK, Kyoung KA (2006) Position control of shape memory alloy actuators by using self tuning fuzzy PID controller. In: 1st IEEE conference on industrial electronics and applications. Singapore
13. Murthy BV, Kumar YVP, Kumari UVR (2012) Fuzzy logic intelligent controlling concepts in industrial furnace temperature process control. In: IEEE international conference on advanced communication control and computing technologies (ICACCCT). Ramanathapuram, India
14. Cook JA, Barry KP (1998) Modeling of an internal combustion engine for control analysis. IEEE Control Syst Mag 8(4):20–26
15. Kumar YVP, Kiran KMNS, Sadhu Y, Raju K.P (2013) Online attitude controlling of longitudinal autopilot for general aviation aircraft using artificial neural networks. Nirma University International Conference on Engineering (NUiCONE), Ahmedabad, India

16. Rao PVGK, Subramanyam MV, Satyaprasad K (2014) Study on PID controller design and performance based on tuning techniques. In: International conference on control, instrumentation, communication and computational technologies, Kanyakumari, India
17. Kumar YVP, Ravikumar B (2014) Real time and high fidelity controller design for hardware in the loop (HIL) testing of flight attitude control. In: International conference on control, instrumentation, communication and computational technologies. Kanyakumari, India
18. Xian X, Wang Q (2017) Speed control of hydraulic elevator by using PID controller and self-tuning fuzzy PID controller. 32nd Youth Academic Annual Conference of Chinese Association of Automation. YAC, Hefei, China
19. Lebukan DEP, Wardana ANI, Effendy N (2019) Implementation of plant-wide PI-fuzzy controller in tennessee eastman process. International Seminar on Application for Technology of Information and Communication, Semarang, Indonesia
20. Hongxia Y, Yang C, Zhao Q (2019) Division of working variable domain fuzzy PID control for valley electric phase change heat storage. IEEE Innovative Smart Grid Technologies—Asia. Chengdu, China

Security Implementations in IoT Using Digital Signature

Purnima Gupta, Amit Sinha, Prabhat Kr. Srivastava, Ashwin Perti, and Aswani Kumar Singh

Abstract In recent advancements, different types of embedded IoT devices are connected with the wired or wireless network and continuously access the internet for communication. Cybercriminals are finding vulnerabilities on IoT devices and compromise them to launch massive attacks (e.g. DDoS, Spamming, MITM, RFID Skimming) to destroy the network. IoT devices having default authentication credentials are an easy target. To avoid cybercriminals, we need a more sophisticated authentication mechanism embedded with existing security measures. Digital signature has become an integral part of IoT to restrict illegal users. The digital signature verification process is a time-consuming operation and not advisable in IoT however we can minimize the time of verification through some optimization schemes. This paper summarizes all the existing digital signature implementation aspects in IoT and states their technological properties as well as features and loopholes.

Keywords Digital signature · IoT · DSA · ECDSA · EdDSA · BLS · RSA

P. Gupta
School of IT, IMS, Noida, India
e-mail: purnimaa018@gmail.com

A. Sinha · A. Perti (✉)
Department of IT, ABES Engineering College, Ghaziabad, India
e-mail: ashwinperti@abes.ac.in

A. Sinha
e-mail: amit.sinha@abes.ac.in

P. Kr. Srivastava
Department of Computer Science & Engineering, Dr. KNMIET, Modinagar, Ghaziabad, India
e-mail: sri_prabhat@rediffmail.com

A. K. Singh
Soft-Tech Development Solution, DDU, Bharwalia Buzurg, UP, India

M. N. Favorskaya et al. (eds.), *Innovations in Electrical and Electronic Engineering*, Lecture Notes in Electrical Engineering 661, https://doi.org/10.1007/978-981-15-4692-1_40

1 Introduction

Kevin Ashton introduced the term Internet of things (IoT) in 1982 to establish an interface between human beings and virtually interconnected devices [1]. IoT is a network of interconnected sensors and smart network devices having different communication protocols and communication medium. IoT may include various types of networks topologies having different types of communication technologies like Worldwide Interoperability for Microwave Access (WiMax), Low Power Wireless Personal Area Network (LoWPAN), Near Field Communication (NFC), Radio Frequency Identification (RFID), Wireless Sensor Network (WSN), Wireless Fidelity (Wi-Fi). It is assumed that IoT will grow up to 50 billion devices by 2020 that will further grow up to 125 billion devices by 2030 [2]. Various applications like Smart Health monitoring, Smart Home appliances, Smart City and Smart Power Grid have millions of devices and generate a massive volume of data. Sensors are used to monitor various physical conditions like temperature, sound, pressure, light, speed, etc. Fig. 2 shows the basic structure of IoT.

IoT devices use the internet for communication hence vulnerable to external threats. Cybercriminals can attack the IoT device and use them as a bot. Further, these devices can be used as botnets to destroy communication of a victim network. Botnets are major threats for IoT. Botnets are nothing but compromised internet-connected IoT devices. Attackers can use these compromised devices as a botnet to launch DDoS attacks. Aidra, Bashlite, Linux/IRCTelnet, Hajime, Linux, Wifatch, Brickerbot and Mirai are some popular IoT botnets. Hajime is one of the IoT botnets who target IoT devices via Telnet and gains access by brute-forcing default credentials. Although defaults credentials of devices like routers must be changed before use, however, this could be ignored. These devices are an easy target for cybercriminals. TLS, SSL and Digital Signature Certificate standards can be used for security from such threats [3–6].

Digital signatures are gaining legal acceptance over the traditional hand-written signatures. it works on public-key cryptography which is designed to protect the genuineness of a digital document. Digital signature schemes have been proposed to get overcome the security problems in authentication, confidentiality, and Integrity related problems in IoT. Many digital signature schemes have been proposed earlier like RSA, DSS, and Al-Gamal digital signature schemes. Using a digital signature for a secure authentication process would certainly prevent the IoT devices to be compromised easily. Figure 1 represents the base model of the digital signature algorithm. The digital signature is proof that the coming message or command is from an authentic source [7–11] (Fig. 2).

This paper has been divided into eight sections. Section 1 represents the basic structure of the IoT and the requirement for the digital signature has been found for security-related issues. Section 2 explains the related work done previously by the researchers to implement digital signature algorithms like RSA, DSA, EdDSA, Short Signatures (BLS) including their algorithms. Section 3 represents the latest digital signature implementation with its technical introductions. Section 4 demonstrates the

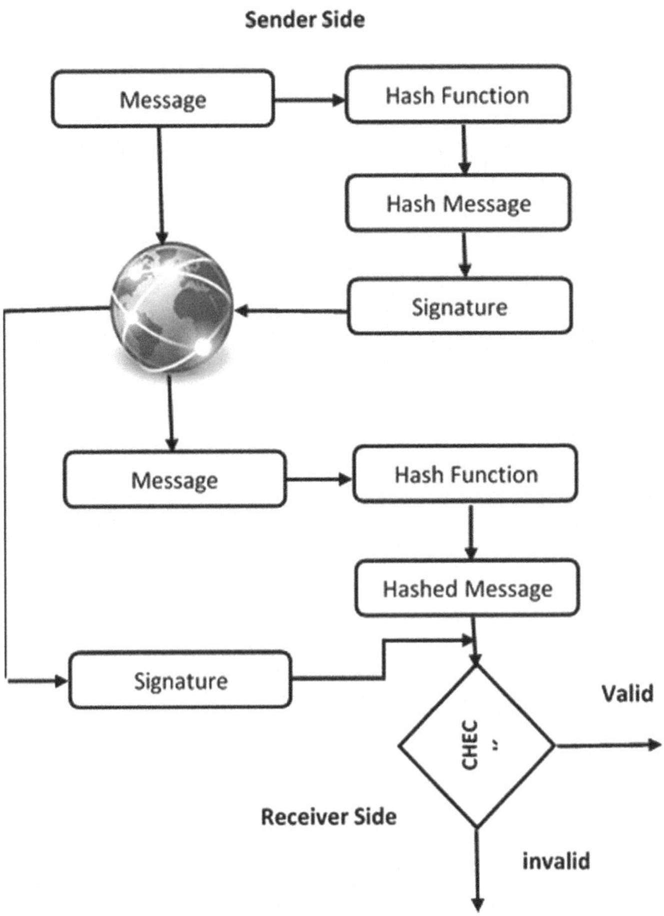

Fig. 1 Digital signature model

problem statement. Sections 5 and 6 represent the proposed solution and performance analysis details respectively. Section 7 summarizes the findings of the paper, as well as Sect. 8, which holds the conclusion about this survey paper.

2 Related Work

2.1 RSA Based Signature Scheme

The method which uses public-key cryptography, developed by Rivest, Shamir and Adelman in 1977 is known as RSA. A digital signature scheme with public-key

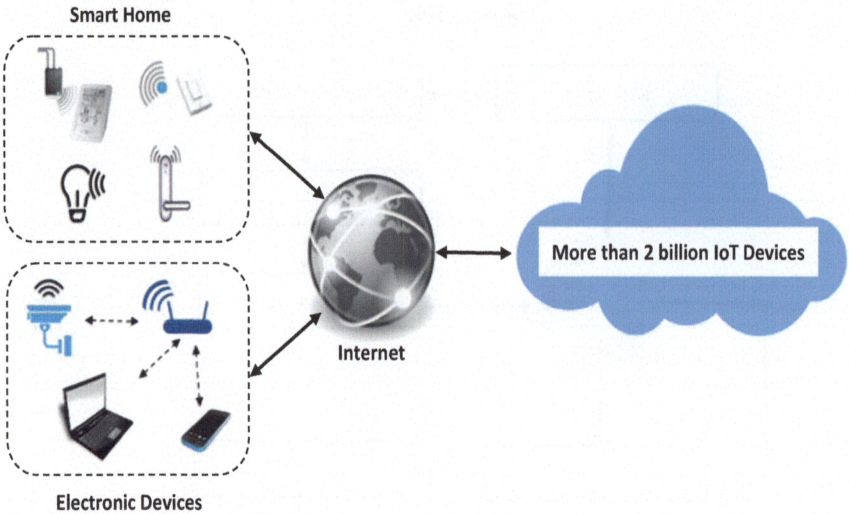

Fig. 2 Internet of things (IoT) basic structure

cryptography has been introduced in 1976 by Whitfield Diffie and Martin Hellman. Sender's private key is used to create an RSA digital signature. RSA based digital signature scheme has two basic steps [12].

- Signature generation through the private key of the signer.
- Message verification through the corresponding public key.

A signature generated by the first step will always verify the second step. Old RSA implementations are vulnerable to some attacks. Dan Boneh and others identified an attack on RSA. This attack was based on doing faulty calculations. The private key can be obtained from the faulty signatures by injecting random faults in RSA calculation. For protection from the Fault Based Attacks, Mihir Bellare and Phillip Rogaway suggested two provably secure scheme for RSA in 1996.

- The Optimal Asymmetric Encryption Padding (RSA-OAEP)
- The Probabilistic Signature Scheme (RSA-PSS)

Encryption and decryption functions are inverse of each other for RSA public/private key triple (e, d, n).

$$\text{Public key } f(m) = m^e \bmod n$$
$$\text{Private key } g(m) = m^d \bmod n$$

Then,

$$f(g(m)) = m \text{ and } g(f(m)) = m$$

RSA-PSS is now able to handle Fault Based Attacks by encoding random values in the signature. Those random values could not be extracted from the signatures. Coron and Mandal have also proved that it is impossible to get the private key through fault-based attacks [13].

2.2 *Digital Signature Algorithm (DSA)*

A DSA digital signature is generated using some set of parameters like data that requires to be signed, private key 'x', secret key 'k' per message, and hash function. The same digital signature can be verified using the same domain parameters and public key 'y', data to be verified and the same hash [14].

Components of Global Public-Key

- 'P'- A prime number of n bits where n is a multiple of 64 and $512 \leq n \leq 1024$
- 'q'- A 160-bit prime factor of p–1
- 'g'- $= h(p-1)/q \bmod p$, where h is any integer with $1 < h < p-1$, such that $(h(p-1)/q \bmod p) > 1$

Private Key
'x' is a random or pseudorandom integer with $0 < x < q$.

Public Key
'y' $= g^x \bmod p$

Secret Key
A random or pseudorandom integer with $0 < k < q$.

Sign

$$\begin{aligned} r &= \left(g^k \bmod p\right) \bmod q \\ s &= \left[k^{-1}(H(M) = xr)\right] \bmod q \\ \text{Signature} &= (r, s) \end{aligned}$$

Verify

$$\begin{aligned} w &= \left(s'\right)^{-1} \bmod q \\ u1 &= \left[H\left(M'\right)w\right] \bmod q \\ u2 &= \left(r'\right) w \bmod q \\ v &= \left[\left(g^{u1} y^{u2}\right) \bmod p\right] \bmod q \\ \text{Test}: v &= r' \end{aligned}$$

2.3 Elliptic Curve El-Gamal Digital Signature Scheme

Taher El-Gamal was the first mathematician who proposed a public-key cryptosystem based on a discrete logarithm problem (DLP). ECDSA is a variant of RSA and DSA. It offers the small key size as well as significant improvements in the generation and verification timings with the small requirement of bandwidth, processing capability, and storage space. This digital signature scheme has the same form as the public key and private key.

Key Generation

- Entity T selects a random integer k_a from the interval [1, n–1] as a private key.
- the public key is computed as $T = k_a G$.
- The public key is stored on the public key server.

Signing Scheme

- Random integer k is selected from the interval [1, n–1].
- Computes $R = kG = (x_R, y_R)$ where, $r = x_R$ MOD n, if $r = 0$ then go to step 1.
- Compute $e = h(M)$, where h is a hash function $\{0, 1\}^* \rightarrow F_n$.
- Computes $s = k^{-1}(e + rk_T)$ mod n; if then go to step 1 (R, s) is the signature message.

Verifying Scheme

- Verify that s is an integer in [1, n–1] and $R = (x_R, y_R) \in E(F_q)$.
- Compute $V_1 =$ sR.
- Compute $V_2 = h(M)\, G + rT$, where $r = x_R$
- Accept only if $V_1 = V_2$.

Elliptic curve ciphers require less computation power, communication bandwidth memory and require a high level of mechanical abstraction for abstraction. Cryptographic implementation in smaller chip sizes is highly required the ECDSA to reduces the heating problem in chips and power consumption [15].

2.4 Edwards-Curve Digital Signature Algorithm (EdDSA)

This signature scheme is a variant of Schnorr's signature system with (possibly twisted) Edwards curves which provides high performance on a variety of platforms. EdDSA does not require the use of a unique random number for each signature. It is more effective for side-channel attacks and hash function collisions do not break this system hence it is collision resilience [16].

Key Setup

- Compute $H(k) = (h_0, h_1, \ldots h_{2b\text{-}1})$.
- $a = (h_0, \ldots, h_{b\text{-}1})$.
- $b = (h_b, \ldots, h_{2b\text{-}1})$.
- Compute public key $A = aB$.

Signature Generation

- Compute ephemeral key $r = H\ (b, M)$.
- Compute ephemeral public key $R = rB$.
- Compute $h = H\ (R, A, M)$ and convert to integer.
- Compute $S = (r + ha) \bmod 1$.
- Signature pair is (R, S).

Verification Stage

- Compute $h = \text{hash}(R + A + M) \bmod q$
- Compute $P1 = s{*}G$
- Compute $P2 = R + h{*}A$
- Check if $P1 == P2$

The signing and verification process of EdDSA is simpler, faster and more secure than ECDSA. Unlike ECDSA the EdDSA signatures do not provide a way to recover the signer's public key from the signature and the message [17].

2.5 *Short Signature Scheme*

A short signature is about half the size of the DSA signature with the same level of security and promisingly used in the low-bandwidth communication environment. RSA, DSA, and ECDSA provide relatively long signature which is undesirable [19, 20]. The short signature scheme provides 160-bit signature length which is the smallest among signature sizes produced by RSA, DSA, and ECDSA. BLS based short signature scheme has been proposed by Boneh et al. in 2001 [18]. Bilinear pairing '*e*' and elliptic curve group elements as signatures are the components of BLS signature [21].

Key Generation
Select a number x as your private key in the range [0, r-1]. The public key shared is g^x.

Sign
Given the private key x and some message m, the signature is given by

$(H(m))x$.

Table 1 Advantages and disadvantages of some widely used digital signatures [22]

Signature scheme	Advantage	Disadvantage	Signing speed
RSA	• Fearless key distribution • Large networks have a smaller number of keys	• Slow operation • High computation cost • Vulnerable to multiplicative attacks	Slow
DSA	• Short signature length • Lower signature computation time • Less storage requirement	• Signature verification must have complicated remainder operators	Moderate
ECDSA	• No application performance issues • Fast signing and verifying process • Support for national information protection standards	• A chance of error that makes it possible to select a private key value and identical signatures for different documents can be obtained • Requires a random number per signature	Moderate
EdDSA	• High speed • High performance • Independence of the random number generator	• Can be hacked by large quantum computers	Fast
BLS [23]	• Short signatures • Better performance • Simplified computing • No need for random number • Many signature blocks can be combined into a single signature	• Pairing is hard and not efficient • Security proof is hard	Fast

Verify
To verify that the signature is valid, we check that

$$e((H(m))x,\ g) = e(H(m),\ gx),\ \text{given}\ gx,\ g.$$

The major drawbacks of the short signature scheme (BLS) are pairing is hard as well as inefficient and security proof generation is also harder than other algorithms (Table 1).

3 Digital Signature Schemes in IoT

Digital signatures are getting used in e-commerce, banking, software systems, and an effective technique to verify authenticity and non-repudiation of the message. All these sectors have versatile computing devices connected through a network, need

Table 2 Different digital signature schemes implemented in IoT for security

Digital signature scheme	Year	Proposed by	Technique used
Shortened Complex Digital Signature Scheme (SCDSA) [24]	2018	Mughal, M. A., Luo, X., Ullah, A., Ullah, S., & Mahmood, Z.	A lightweight Shortened Complex Digital Signature Algorithm (SCDSA) for providing secure communication between smart devices in human-centered IoT
Lamport Signature Scheme [25]	2018	Abdullah, G. M., Mehmood, Q., & Khan, C. B. A.	Lamport signature scheme, which is quantum-resistant, for authentication of data transmission and its feasibility in IoT devices.
Proxy Blind ECDS Algorithm [26]	2018	Harini N, Kamakshi Devisetti R N, Aruna D	Elliptic Curve Cryptography (EEC) based on proxy blind signing procedure
Cloud-Based Digital Signature Application [27]	2018	Sahar A. El-Rahman, DaniyahAldawsari, OmaimahAlrashed, Ghadeer Alsubaie	a digital signature mobile applicationwhere it provides a cloud-based digital signature with high security to sustain with the growth of IoT.
Elliptic Curve Digital Signature Algorithm [28]	2016	B. Sindhu, R. M. Noorullah	Used the standards of the Elliptic Curve digital signature scheme.

a very strong authentication scheme to verify the identity proof and genuineness of transmitted data. Many digital signature schemes have been proposed by researchers to mitigate the security-related vulnerabilities in IoT.

In Table 2, we have illustrated some newest digital signatures with their technical summary. These digital signature techniques certainly deliver us new improved algorithms that provide better security than earlier algorithms. Still, these digital signature algorithms need to be verified for their hidden drawbacks to present an improved and completely secure Internet of Things network. This improvement is also necessary to make its users fearless about security-related risks in IoT.

4 Problem Statement

Secure communication is the key requirement between two communicating IoT devices. The owner of the IoT network may need to upgrade the firmware of IoT devices to deploy some security patches. It is to be ensured that data transmission is to be done within a secure and authenticated environment.

In many cases, cybercriminals may compromise the communicating device and use them as a lot to launch future attacks to harm a specific target system. Patching or executing a malicious program can lead to severe damages or losses if the device is deployed in a critical environment such as smart power grids, nuclear facilities, traffic management systems or flight management systems. So, this is highly required to verify the authenticity of the receiver before transferring the confidential data or installing the important patches.

There are several digital signature schemes such as DSA, ECDSA, EdDSA but the problem with these cryptographic signatures they are very much insecure and vulnerable to be broken by a quantum computer [10]. There should be a stronger, complex and unbreakable algorithm for the digital signature scheme so that the security of IoT devices can be ensured.

5 Proposed Solution

An existing digital signature system must not vulnerable to attackers. An attacker can launch Fault Based Attacks and Bleichenbacher Attack in old RSA based digital signature schemes to obtain the private key. RSA-PSS may be a solution to this attack. It has been proven that the signature generated by RSA-PSS could not let the attacker extract the private key. Performance-related issues can be minimized using batch processing and multithreading approaches.

ECDSA has less impact over EdDSA and is a newer technology than EdDSA. EdDSA provides high performance on platforms and does not need a random number for each signature. It is enough strong against side-channel attacks. EdDSA provides collision resilience, meaning that hash-function collisions do not break this system. Table 1 presents that EdDSA is the best select digital signature scheme on various factors like performance, complexity, and speed. EdDSA has minimum disadvantages among DSA, RSA, ECDSA, BLS based short signature schemes. This digital signature scheme is desirable in the IoT network for the protection of the authenticity of messages. The only drawback of the EdDSA scheme found that it can be hacked through large quantum computers.

6 Performance Analysis

DSS is considered to be stronger than El-Gamal since in this scheme the secret number k is harder to obtain from r because of the reduction mod q. The verification step in DSS is also faster than the corresponding step in El-Gamal, since there are fewer modular exponentiations to perform, and this is an important practical consideration. Performance analysis shows the RSA has the worst performance regarding signature generation time (7.8 ms) and BLS has the worst performance in the signature verification phase (0.8.6 ms). The outcome of this paper can affect the selection

Table 3 Digital Signature Schemes characteristics

Signature scheme	Key size (bits)	Signature size (bytes)	Signature generation time (ms)	Signature verification time (ms)
RSA	1024	128	7.8	0.4
DSA	1024	384	3.5	4.5
ECDSA	160	64	3.1	8.2
EdDSA	256	512	0.08	0.16
BLS	100	20	2.2	8.6

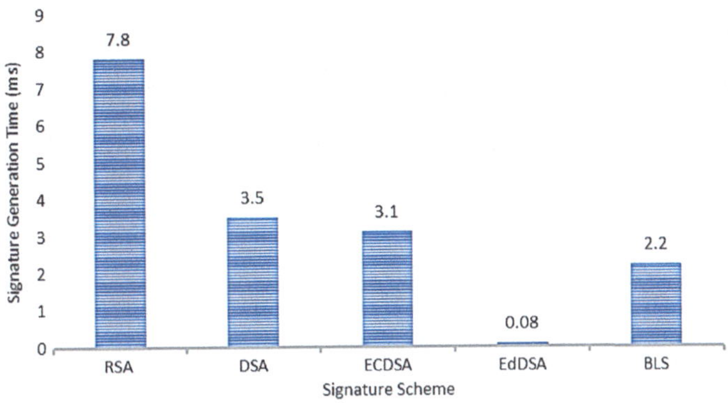

Fig. 3 Digital signature generation time comparison

of the digital signature scheme of the sophisticated software network system seeking security-related issues (Table 3).

Performance analysis has been done on various factors like key size, signature size, signature generation time, and signature verification time. EdDSA has the minimum signature generation time (0.08 ms) and signature verification time (0.16 ms) having 256 bits of key size and 512 bytes of signature size. Figure 3 demonstrates the chart of signature generation time comparison and Fig. 4 displays the signature verification time comparison for all five selected algorithms.

7 Conclusions

The meticulous survey of the different digital signature schemes reflects their advantages in terms of performance, correctness, complexity, suitability in IoT platform and stability against security threats found as a major challenge for IoT network. Performance improvement is the key challenge in implementing digital signature in IoT. This survey opens up the need for a detailed study of practical threats and related

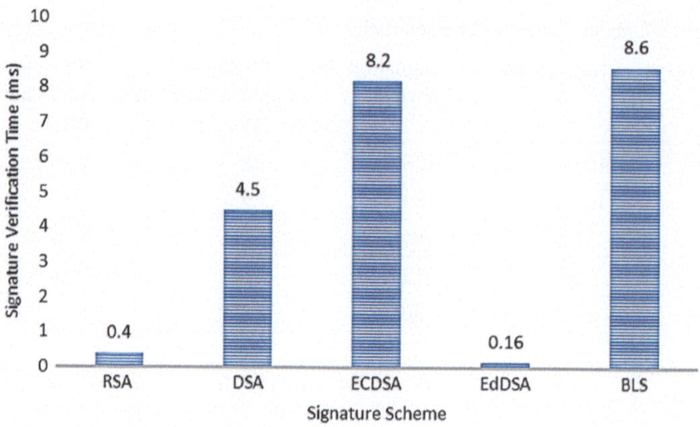

Fig. 4 Digital signature verification time comparison

analysis. This can help create a generic, concrete and usable authentication scheme based on the digital signature.

8 Summary

Authors have analyzed different types of digital signature schemes like EdDSA, RSA based digital signature, Lamport Signature Scheme, and Secure Proxy Blind ECDS. The working technique, advantages, security strength with limitation and drawbacks have been discussed briefly. A few problems of existing digital signature schemes for securing the IoT network have been identified like performance, complexity, and storage related. Authors find EdDSA has the top performance regarding the signature generation and verification processes. EdDSA can be adopted for IoT as the optimized digital signature scheme. RSA signature can be selected if the signature generation is getting performed on the client-side. This will certainly affect the server performance as the signature verification process takes minimal time among all five selected and benchmarked digital signatures.

References

1. Ashton K (2009) That 'internet of things' thing. RFID J 22(7):97–114
2. Hajkowicz SA et al (2016) Tomorrow's digitally enabled workforce: megatrends and scenarios for jobs and employment in Australia over the coming twenty years. Australian Policy Online
3. Zhang W (2010) Integrated security framework for secure web services. In: 2010 third international symposium on intelligent information technology and security informatics. IEEE
4. Geer D (2003) Taking steps to secure web services. Computer 36(10):14–16
5. Stallings W et al (2012) Computer security: principles and practice.: Pearson Education, Upper Saddle River, NJ, USA

6. Hiltgen A, Kramp T, Weigold T (2006) Secure internet banking authentication. IEEE Secur Priv 4(2):21–29
7. Kardi A, Zagrouba R, Alqahtani M (2018) Performance evaluation of RSA and elliptic curve cryptography in wireless sensor networks. In: 2018 21st Saudi computer society national computer conference (NCC). IEEE
8. Hassan, R, Toheed Q (2010) Asymmetric-key cryptography for Contiki. MS thesis
9. Wong, CK, Lam, SS (1998) Digital signatures for flows and multicasts. In: Proceedings sixth international conference on network protocols (Cat. No. 98TB100256). IEEE
10. Rabah K (2005) Elliptic curve ElGamal encryption and signature schemes. Inf Technol J 4(3):299–306
11. Stallings W (2013) Digital signature algorithms. Cryptologia 37(4):311–327
12. Kenneth L (2013) Digital signatures using RSA. mathematical sciences UMass Lowell [Online]. Available: http://faculty.uml.edu/klevasseur/math/RSA_Signatures/RSA_Signatures.pdf
13. Coron J-S, Mandal A (2009) PSS is secure against random fault attacks. In: International conference on the theory and application of cryptology and information security. Springer, Berlin, Heidelberg
14. Barker EB (2009) Digital signature standard (DSS). No. Federal Inf. Process. Stds. (NIST FIPS)-186-3
15. Imem AA (2015) Comparison and evaluation of digital signature schemes employed in NDN network. arXiv preprint arXiv:1508.00184
16. Nakov S (2018) EdDSA and Ed25519 [Online]. Available: https://cryptobook.nakov.com/digital-signatures/eddsa-and-ed25519
17. Samwel N et al (2018) Breaking ed25519 in wolfssl. In: Cryptographers' track at the RSA conference. Springer, Cham
18. Boneh D, Lynn B, Shacham H (2004) Short signatures from the Weil pairing. J Cryptol 17(4):297–319
19. Sethi A, Vijay S, Saini JP (2018) Analytical Model for Secure Pairing in ad hoc Network. In: 2018 IEEE 3rd international conference on computing, communication and security (ICCCS). IEEE
20. Tso R, Okamoto T, Okamoto E (2009) Efficient short signatures from pairing. In: 2009 Sixth international conference on information technology: new generations. IEEE
21. Srikar V (2018). The BLS signature scheme-a short intro [Online]. Available: https://medium.com/chain-intelligence/the-bls-signature-scheme-a-short-intro-801c723afffa
22. Zhanna L (2018). A guide to digital signature algorithms [Online]. Available: https://dzone.com/articles/digital-signature-1
23. Stepan (2018) BLS signatures: better than Schnorr. Crypto Advance [Online}. Available: https://medium.com/cryptoadvance/bls-signatures-better-than-schnorr-5a7fe30ea716
24. Mughal, MA et al (2018) A lightweight digital signature based security scheme for human-centered Internet of Things. IEEE Access 6:31630–31643
25. Abdullah, GM, Mehmood, Q, Khan, CBA (2018) Adoption of Lamport signature scheme to implement digital signatures in IoT. In: 2018 International conference on computing, mathematics and engineering technologies (iCoMET). IEEE
26. Harini N, Kamakshi D, Aruna D (2018) Secure proxy blind ECDS algorithm for IoT. Int J Pure Appl Mathe 118(7):437–445
27. El-Rahman, SA, et al (2018) A secure cloud based digital signature application for IoT. Int J E-Ser Mobile Appl (IJESMA) 10(3):42–60
28. Sindhu B, Noorullah RM (2016). Secure elliptic curve digital signature algorithm for internet of things. Global J Comput Sci Technol

Low Power and Highly Reliable 8-Bit Carry Select Adder

Pritty and Mansi Jhamb

Abstract In today's enhanced technological advancements, a carry select adder performs an indispensable integral part of the complex processing of data. It exhibits promising results in minimizing cost and power consumption. The logic operation of it has simulated over 90 nm CMOS technology of spice tool at 0.8 V. These works present a Carry select adder using multiplier and adder circuits. In proposed designs, extra circuitry is eliminated and leads to a reduction in 43.995% of power consumption, 93.912% of time delay, and 97.515% of energy. The transistor count of the proposed carry select adder is 61.788% less than that of the conventional carry select adder. The power, delay, energy of it has been compared with state-of-art carry select adder. This advanced circuit of adder shows highly efficient and reliable performance.

Keywords Adder · Carry select adder · Multiplexer · Ripple carry adder · XOR gate

1 Introduction

The development of CMOS technology continues to drastically minimize power consumption, time delay, and transistor count of integrated circuits (ICs) [1]. The enhanced bit counts have increased the transistor counts, power, and decreased speed. The power dissipation has further three types as under Static power (P_{stat}), Short circuit power (P_{short}), Dynamic power (P_{dyn}).

$$P_{\text{total}} = P_{\text{stat}} + P_{\text{short}} + P_{\text{dyn}} \tag{1}$$

Pritty (✉) · M. Jhamb
USIC&T, GGSIPU, Dwarka, New Delhi, India
e-mail: pritty.14516490019@ipu.ac.in

M. Jhamb
e-mail: mansi.jhamb@ipu.ac.in

M. N. Favorskaya et al. (eds.), *Innovations in Electrical and Electronic Engineering*, Lecture Notes in Electrical Engineering 661, https://doi.org/10.1007/978-981-15-4692-1_41

where P_{total} is the total power consumed in a particular design.

The static power is dissipated because of unnecessary leakage current when the transistor is switched off. However, dynamic power has dissipated when parasitic components are switched from on to off or off to on. During the current flowing from VDD to VSS, short circuit power takes place [2].

The propagation delay is calculated using the following formula:

$$T_{prop} = (N - 1) * (T_{carry} + T_{sum}) \quad (2)$$

T_{prop} is the total propagation delay
N is number of bits
T_{carry} is a time delay for output carry of adder
T_{sum} is a time delay for output sum of adder [3]

The calculation of energy (E) is done using the product of power and time delay. The equation is as follows:

$$E = P * T_{prop} \quad (3)$$

Therefore, the reduction of power consumption, energy, and delay is obligatory for a circuit operation [4]. Addition, accumulation, and multiplication are the necessary basic functioning of various digital signal processing (DSP) systems. The addition is the essential logic operation for DSP, control system, and digital system [5]. Full adder applications such as Voting systems, ALU, multiplication, and graphics are related to complex arithmetic computation. The wide range of adder circuits has used for addition, such as carry look-ahead adder, ripple carry adder (RCA), (carry skip) CS adder, carry select adder (CSLA). The simplest, fastest design of adders is RCA and CSLA that have significantly decreased the time delay [6]. For sum operation 2 XOR gates and for carry operation 3 AND, 1 OR GATES are used to operate for the conventional adder. In contrast, in RCA, the sum operation has performed by full adder blocks [7]. The carry propagates from one full adder block to another to generate the final carry output. The n bit full adder blocks are used to perform the N bit RCA operations [8]. CSLA can be designed using various methods. First method, It was designed using RCA, 5 bit BEC, and Multiplexer (MUX) circuits [9]. It can also be designed using RCA, modified BEC (4 bit and 3 bit), and 8:4 MUX or 6:3 MUX [10]. Area-efficient CSLA, high speed, and low-power CSLA are few examples of CSLA that has designed as per need of application [11]. The second method, it can be architecture using RCA, AND gate, XOR gate and MUX [12]. The third method, it can also be designed utilizing RCA, D-latch and MUX blocks [13]. Other methods for CSLA are using parallel prefix adder [14], asynchronous CSLA, and synchronous CSLA [15]. The proposed CSLA is designed using three RCA of 4 bit each and five 2:1 MUX circuits. Three transistor XOR gates and multiplexers have applied for the schematic design of the full adder. 2:1 MUX block has been drawn using pass transistor logic/transmission gates [16]. The area, transistor count, and time delay have further reduced by eliminating extra circuitry. It can be designed for different requirements, i.e., High-speed, performance, highly efficient, low power, etc. for different applications [17–19].

The organization of the remaining paper is as follows. Section 2 covers information related to state-of-art methods of CSLA design. Section 3 includes the implementation of the proposed CSLA. It also contains schematics and layout of adder, CSLA, and MUX. Section 4 of the paper includes tables, results, and comparison graphs of it. Section 5 covers the conclusion of the article.

2 State-of-Art Methods

2.1 CSLA Using BEC

The design of CSLA generates multiple carries at different stages, and final sum output is generated by carry using as select line. CSLA uses the pair of RCA. Binary to excess-1 converter (BEC) and one RCA with multiplexer is utilized for design Fig. 1. Many CSLA design using this technique is proposed such as design of the area and power-efficient modified carry select adder [8], low power area-efficient carry select adder [11], area-efficient carry select adder [10]. In all these papers, BEC or XOR gate has changed. In [8], Each BEC used 4 bit and NOT, AND, XOR gates used for it. The logical expressions of binary to excess convertor (4-bit) are:

$$X_0 = (B_0)' \text{ and } X_1 = B_0 \oplus B_1 \tag{4}$$

$$X_2 = B_2 \oplus (B_0 + B_1) \text{ and } X_3 = B_3 \oplus (B_0 + B_1 + B_2) \tag{5}$$

This paper gives the area reduction, power reduction; delay overhead (each value in percentage). Its drawback that it does not gives an idea related to transistor count, power delay product (PDP/energy). In [10], BEC has modified by using 3-bit BEC,

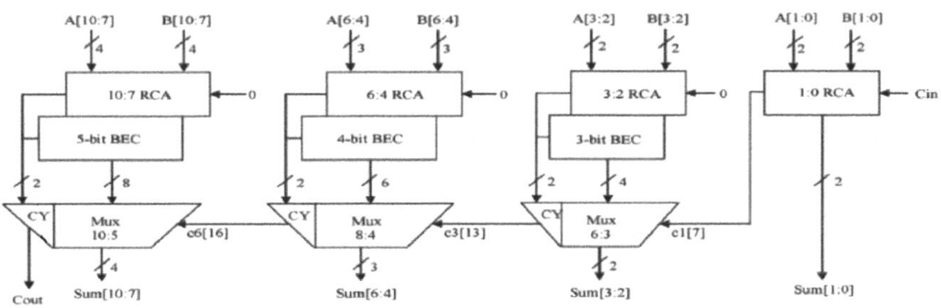

Fig. 1 Block diagram for CSLA using BEC [8]

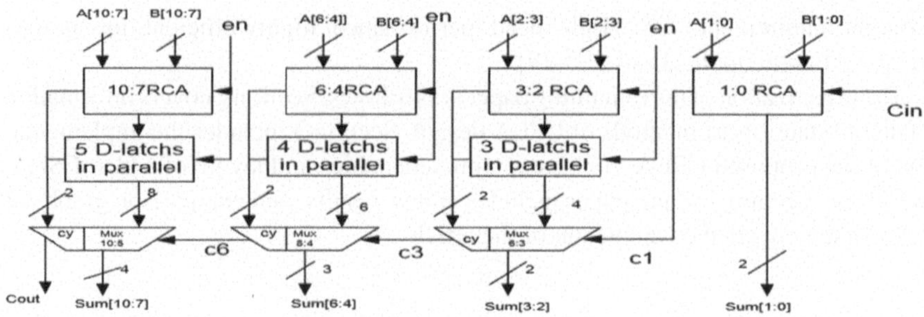

Fig. 2 Block diagram for CSLA using D-latch [13]

4-bit BEC. Transistor count in [10], it is above 250 transistors for 16-bit CSLA and 196 for 8-bit CSLA. Similarly, the CSLA design of [9] does not give the exact value of transistor count and leads to a reduction in power, delay, PDP/energy.

In [11], CSLA is designed using full adder, BEC (XOR, AND, NOT gates) and multiplexer. This design used approx. 295 transistor counts for 16-bit CSLA and 188 transistor counts for 8 bit CSLA. Overall, this technique leads to reduce transistor count, power, delay but up to an extent only.

2.2 *CSLA Design Using D-Latch*

In this CSLA circuitry, BEC has replaced by D-Latch. *N* bit D-latch has required for N bit CSLA design. An enable (en) input is used in D-latch and RCA for the calculation of outputs.

When en = 0, then RCA calculates outputs for carry-input zero. When en = 1, then RCA calculates outputs for carry-input one. D-latch has used for storing first operation output Fig. 2. Multiplexer has used for comparison of RCA output for carry-input = 1 and D-latch store data. It leads to a reduction in delay, power consumption, PDP, and area.

3 Implementation of CSLA

The full adder circuit can explain as: the adding of two inputs with Carry-in gives the Sum and Carry, where

$$S = (A \oplus B) \oplus \text{Cin} \tag{6}$$

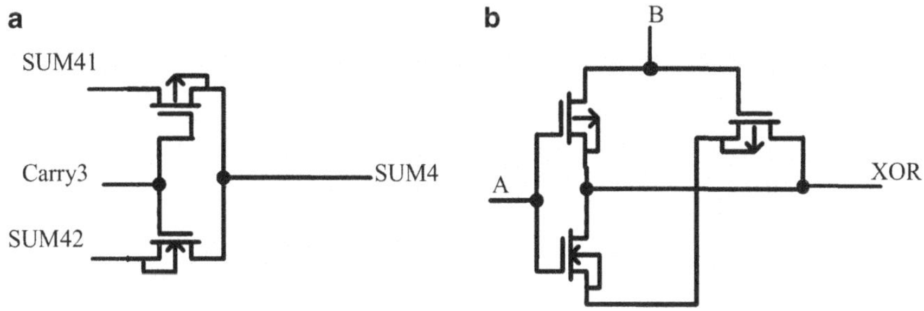

Fig. 3 **a** XOR gate **b** multiplexer

$$\text{Cout} = A.B + \text{Cin}\,(A \oplus B) \quad (7)$$

The 1-bit adder has designed by XOR and multiplexer circuits. Five multiplexer circuits are designed using two transistors only. The primary way to design the 1-bit adder by XOR gate and multiplexer has displayed in Fig. 4. The gate and multiplexer XOR circuits are building blocks in full adder design.

The arithmetic addition operation of the adder leans on the performance of the XOR, multiplexer circuits. The design must have minimal transistors used to implement the XOR block for low power consumption Fig. 3. The MUX circuit of the multiplexer has adopted in the proposed design to generate optimized and carried output sum Fig. 3. The transmission gates have been used as multiplexers 2: 1. The transmission gate has benefits for the CSLA: it accelerates the carry/sum as a buffer by the side of the sum/carry series/blocks. Also, it provides a better oscillation of the output voltage.

The adder design, which utilized 2 XOR gates and a multiplexer circuit, requires eight transistors. Choose a wide-width ratio of transistors. When A and B both inputs are low or zero, the circuit produces low logic or zero as output. When the inputs A is zero and B is one, then the transistor MS2, P is on, MS1, N and MS1, P is off. They provide high logic or one in XOR output. When the input A equals to high, and B low, MS2, P is disabled, MS1, N is activated, and MS1, P leads to provide the output as one or high. When A is one and B is one, then the circuit provides zero as an exit. Then, this circuit works like an XOR circuit. The complete summing circuit of 8 transistors has been designed using the 2 XOR circuits and a 2: 1 multiplexer Fig. 4. In this circuit, the summation output has generated by two XOR gates and the output has produced by a 2:1 multiplexer block or a MUX circuit of transmission gates. Extra circuitry has eliminated from RCA3, which leads to a reduction in the reduction in 12 transistors.

It provides better performance than the circuits of the previous full adder and gets less use of the area due to the decrease in the number of transistor counts. The addition of 2 or more binary (1 or 0) numbers can be done using RCA. RCA is designed by connecting 1-bit cascaded adders with the C1 of the 1-bit adder linked to the C0 carry-out of the previous adder design and so on. Figure 5 depicts the combination

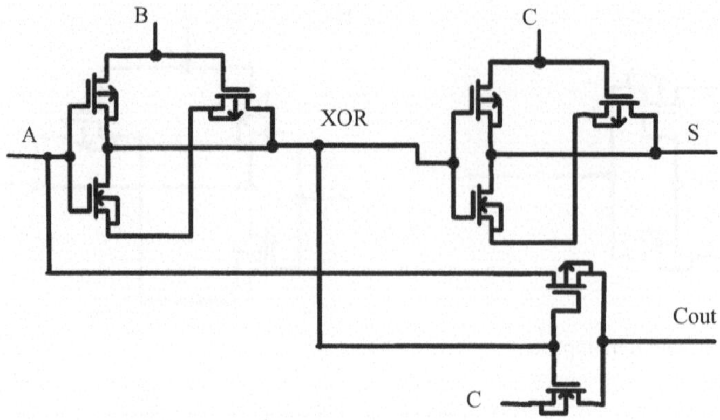

Fig. 4 Design of 1-bit adder

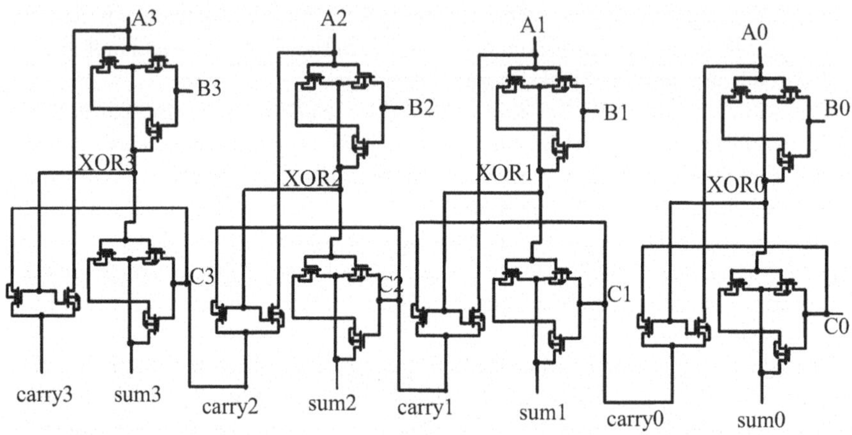

Fig. 5 Design for 4-bit RCA

of 4 1-bit adder circuits (FA) which result in sum and carry output of it. The RCA has been implemented using a whole summing cell with eight transistors based on the transistor logic. An XOR gate and 2:1 multiplexer are used to form 1-bit adder as presented in Fig. 8. The sum (A ⊕ B ⊕ Cin) is produced by two XOR circuits. Examining the truth table of the adder using Boolean algebra expression summarized that if XOR of *A* and *B* are correct, then Carry is complement of Sum. When XOR of *A* and *B* are low, Carry is equal to *A* input (or *B* input) and sum is equal to low. The carry select adder formed of one MUX circuit and two RCA in Fig. 6. The addition of *A* and *B* of n bits in the carry selection adder has done with two RCA, to obtain the outputs twice, once with the consideration that the carry-in is equal to zero and the other considering that will be equal to one. The two results have obtained, final

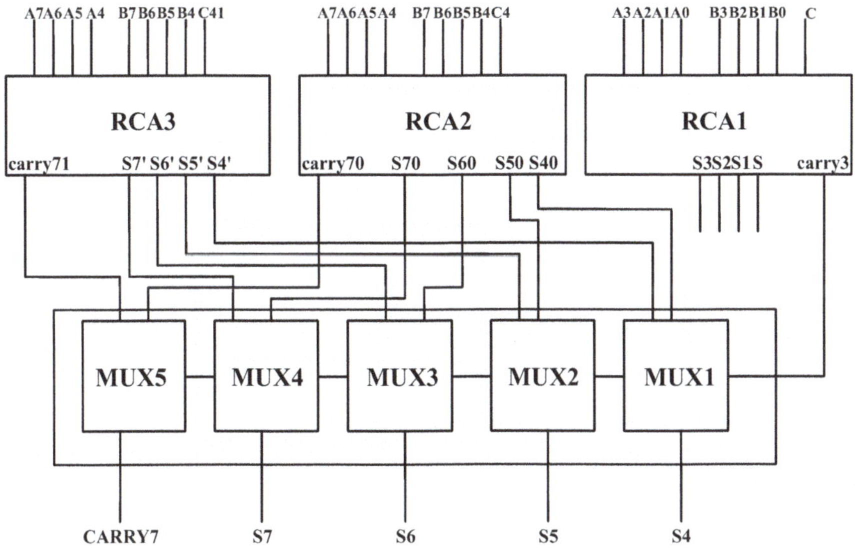

Fig. 6 Block diagram for 8-bit CSLA

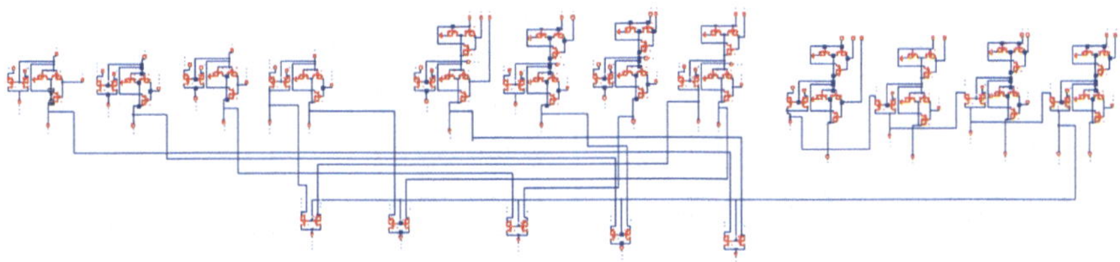

Fig. 7 Schematic of 8-bit modified CSLA

carry has selected, as well as the final sum, with the multiplexer once the final carry has calculated Fig. 7.

4 Result and Discussion

The schematic diagrams and layouts of each CSA block have been designed in the spice schematic editor and L-edit. Subsequently, schematic diagram and layout of CSA can be formed by integrating the individual blocks as presented in Figs. 7, 8, 9, 10, and 11. The optimized scheme has simulated using the 90 nm CMOS technology in the spice tool. The scheme has been designed and simulated. The output waveform of CSLA has been displayed in Figs. 12 and 13.

It shows the final result of C7 S7 S6 S5 S4 S3 S2 S1 S0. Table 1 shows the

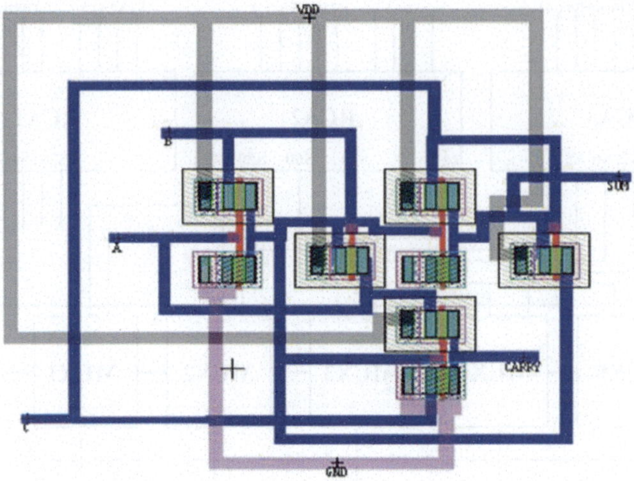

Fig. 8 Full adder layout

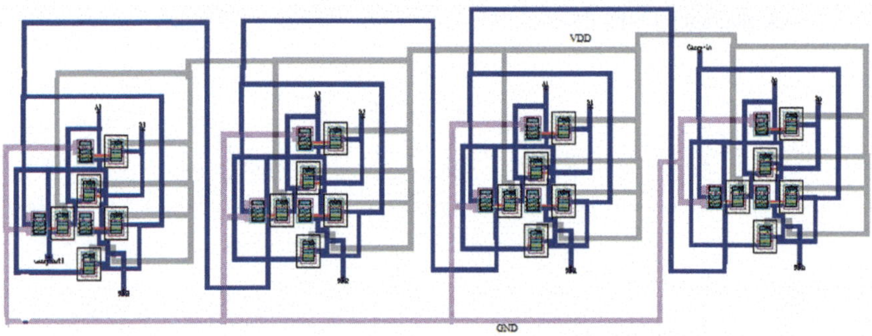

Fig. 9 Layout of RCA

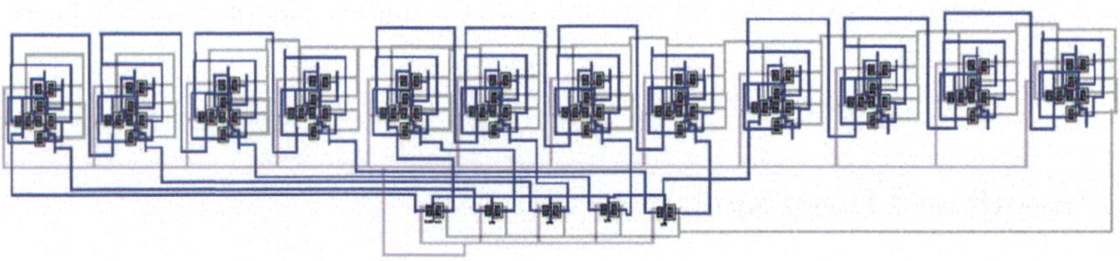

Fig. 10 Layout of 8-bit CSLA

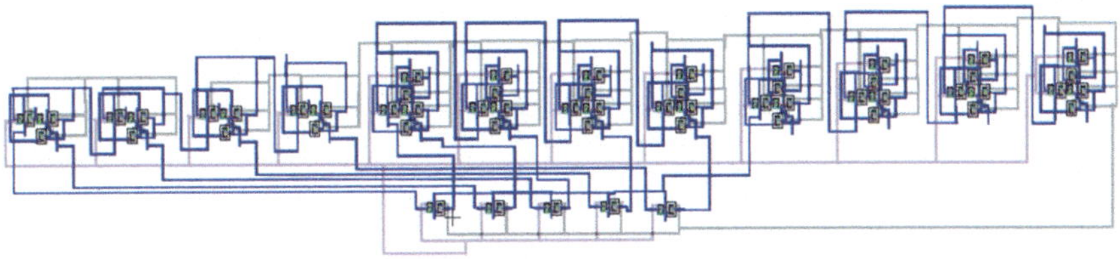

Fig. 11 Layout of modified CSLA

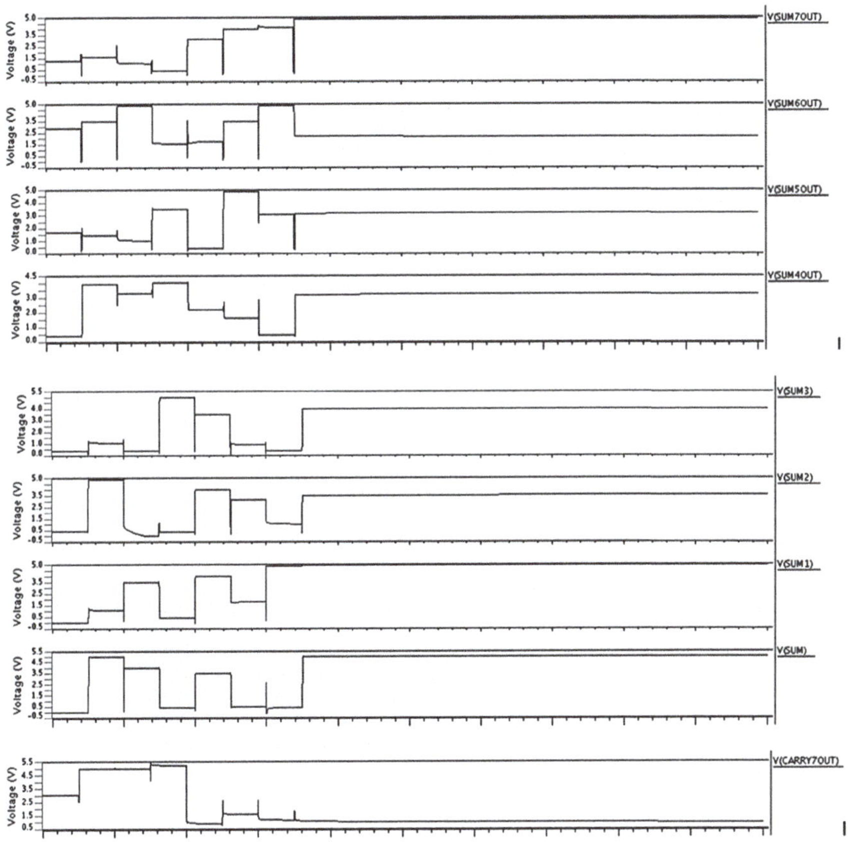

Fig. 12 Output waveform for 8-bit CSLA

mathematical and Figs. 14, 15, 16, and 17 shows graphical presentation of power, delay, PDP, and transistor count of multiplexer, adder, RCA, CSLA, modified CSLA. Modified CSLA is compared with previous CSLA designs using power consumption, propagation, energy parameters. These comparisons have shown in Table 2 mathematically, and Figs. 15, 16, and 17 graphically.

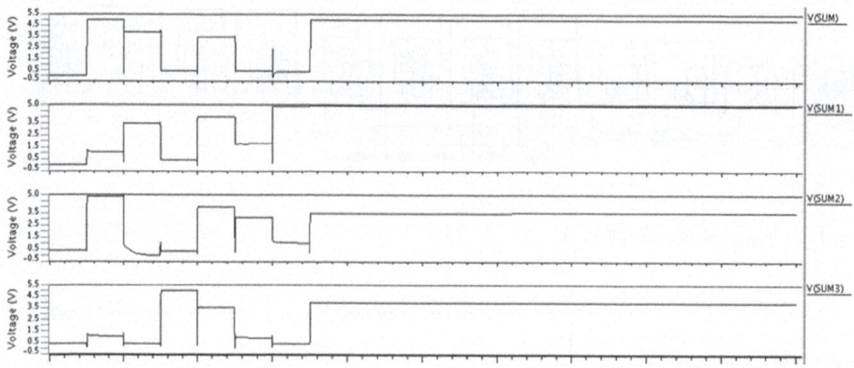

Fig. 13 Output waveform for 4-bit RCA1

Table 1 Measured parameters of implemented designs

Circuits	Number of transistors	Power (in μW)	Time delay (in ps)	PDP (energy) (in fJ)
MUX	2	0.000191	30.262	0.005
Adder (1 bit)	8	0.0801	0.050	0.004
RCA (4 bit)	32	15.977	244.38	3.904
CSLA (8 bit)	106	84.7518	94.197	7.983
Modified CSLA (8 bit)	94	72.5271	103.158	7.481

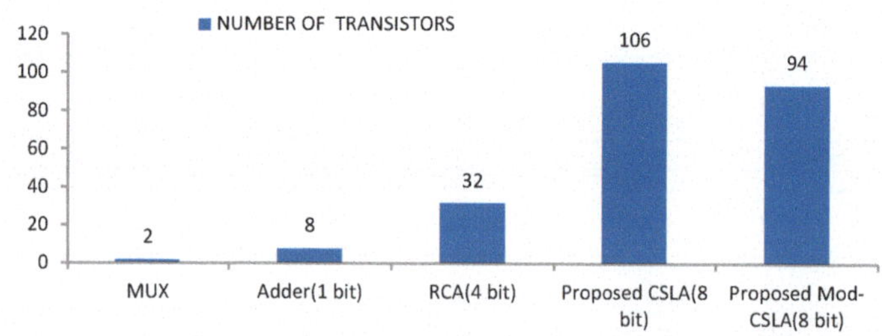

Fig. 14 Transistor count of implemented designs

5 Conclusions

The state-of-art CSLA had utilized either two blocks of RCA using 1-bit adder (8 transistors) or one block of RCA with the help of 1-bit adder (8 transistors) and binary to excess one convertor or D-latch. This paper had simulated the carry select adder using a 1-bit adder (8 transistors) and carry select adder further modified by removing XOR gates used in RCA3 block. The results of carry evaluation block

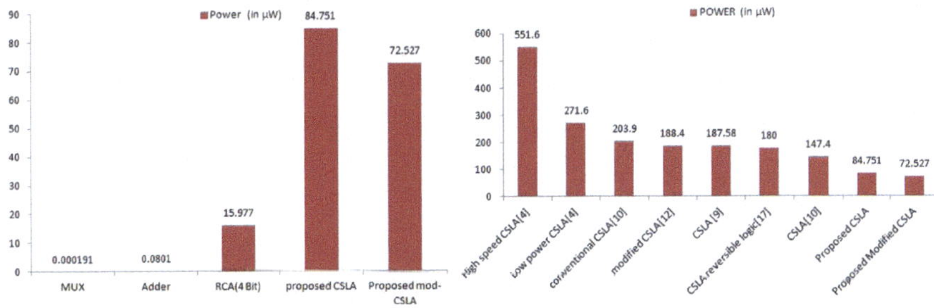

Fig. 15 Power comparison of implemented designs and different CSLAs

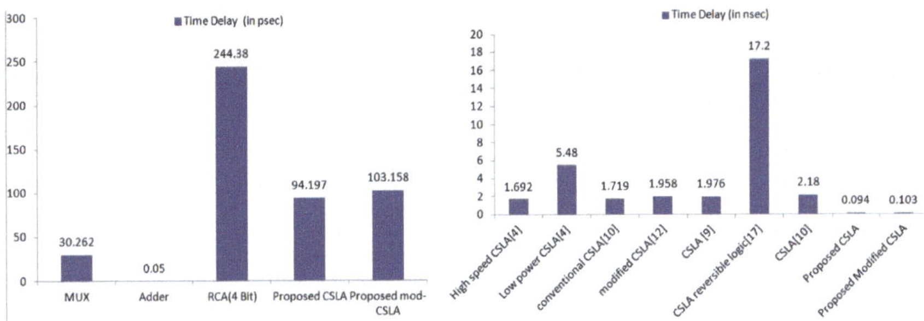

Fig. 16 Time delay comparison of implemented designs and different CSLAs

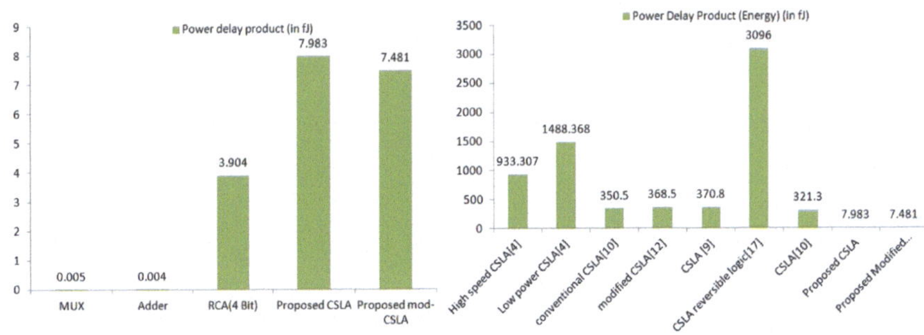

Fig. 17 Power delay product comparison of implemented designs and different CSLAs

with carry-in one have derived from carry-in 0 with an effectively designed multiplexer circuit. The proposed CSLA has better performance than state-of-art CSLA in case of delay, transistor count, and energy. The output results depicted 72.5271 μW average power and 103.158 ps average delays for proposed CSLA. The CSLA design works with improved speed when XOR circuit from RCA3 block is eliminated. The proposed circuit presented minimal Transistor count with reduce in PDP and power. At VDD = 1 V, CSLA depicted value of PDP 7.481 fJ and power 72.5271 μW.

Table 2 Comparison table for different CSLA designs

Circuits	Number of bits in each input	Power (in μW)	Time delay (in ns)	PDP (Energy) (in fJ)
High-speed CSLA [3]	8	551.6	1.692	933.307
Low-power CSLA [3]	8	271.6	5.48	1488.368
conventional CSLA [9]	8	203.9	1.719	350.5
modified CSLA [11]	8	188.4	1.958	368.5
CSLA [8]	8	187.58	1.976	370.8
CSLA using reversible logic [17]	8	180	17.2	3096
CSLA [9]	8	147.4	2.18	321.3
Proposed CSLA	8	84.751	0.094	7.983
Proposed Modified CSLA	8	72.527	0.103	7.481

The proposed work also compared with state-of-art CSLA designs. It has observed 43.995% reduction in power dissipation, 93.912% reduction in propagation delay, and 97.515% reduction in PDP/energy. The no of transistor or transistor count has also minimized as compared to previous designs of CSLA. This design of CSLA leads to highly reliable and efficient performance.

References

1. Rajendra Prasad KV et al (2013) Layout design of 4-bit ripple carry adder using NOR and NAND logic. IJVES 4(09157)
2. Kumar KR et al (2014) A novel 3 transistor XOR gate based full adder design for VLSI application. ICRACVES-2014, 2349-0020
3. Hiremath Y (2014) A novel 8-bit carry select adder using 180 nm CMOS process technology. IJEERT, 2(6):187–194
4. Rawat K et al (2002) A low power and reduced area carry select adder. In: The 2002 45th midwest symposium on circuits and systems. MWSCAS-2002. ISSN 0-7803-7523-8/02
5. Uma R et al (2001) Area, delay and power comparison of adder topologies. VLSICS 3(1)
6. Sathyabhama B et al (2015) Area and power-efficient carry select adder using 8T full adder. IEEE ICCSP. 978-1-4799-8081-9/15
7. Maurya KAK et al (2017) Design and implementation of 32-bit adders using various full adders. i-PACT 987-1-5090-5682-8/17
8. Prakash M et al (2015) Design of low power 8-bit carry select adder. IJSEAS 1(6)
9. Singh S, Kumar D (2011) Design of area and power-efficient modified carry select adder. Int J Comput App 33(3)

10. Singh G, Goel C (2014) Design of low power and efficient carry select adder using 3-T XOR gate. Hindawi Publishing Corporation, 2014 (564613)
11. Singh G et al (2014) Area efficient carry select adder using cadence tools. IJETT 10(10)
12. Saranya K (2013) Low power and area-efficient carry select adder. IJSCE 2(6)
13. Shanigarapu L et al (2013) Low power and high speed carry select adder. In J Sci Res Pub 3(8)
14. Hebbar AR et al (2018) Design of high speed carry select adder using modified parallel prefix adder, ICACC 1877–0509
15. Balasubramanian P (2017) Asynchronous carry select adder. Eng Sci Tech Internet J 20:1066–1074
16. Pant D et al (2012) CMOS design of 2:1 multiplexer using complementary pass transistor logic. Nat Conf, ADVICE
17. Maity S et al (2012) Design and implementation of low power high performance carry skip adder. IJEAT 1(4)
18. Valinataj M, Mohammadnezhad A, Nurmi J (2018) A low-cost high-speed self-checking carry select adder with multiple-fault detection. Microelectron J 81:16–27
19. Nam M, Choi Y, Cho K (2018) High-speed and energy efficient carry select adder (CSLA) dominated by carry generation logic. Microelectron J 79:70–78

SAGNAC-Error Calculation in Two-Way Satellite Time and Frequency Transfer

Vimla Sharma, Charu Prabha, and Mirza Mohammad Zaheer

Abstract TWSTFT techniques are employed for time interval measurements of two different atomic clocks or timescales located far apart. There are various methods of time and frequency transfer between timing labs. Among all-time transfer methods, TWSTFT is very precise time transfer method for geographically apart but visible from common GEO satellite stations. In this paper, we have discussed the errors included in the process of TWSTFT and also derived and calculated the mathematical value of the SAGNAC effect delay and the ionospheric delay. Among all the delays, SAGNAC effect delay and the ionospheric delay are considered to be the most complicated in terms of calculation and are responsible for the increased inaccuracy in TWSTFT.

Keywords TWSTFT · SAGNAC · Synchronization

1 Introduction

India has build-up its own regional satellite navigation system "Navigation with Indian Constellation (NavIC)". The heart of any navigation system is a precise time measurement system. More precise are measurements of time, more accuracy will be provided to the user in terms of navigation. To cater to the time measurement requirement, Indian Space Research Organization (ISRO) has established two IRNSS Network Timing facilities IRNWT-I and IRNWT-II with the associated TWSTFT facility at geographically apart locations i.e. Bangalore and Lucknow, respectively. IRNWT facility maintains the navigation time and metrological time. A similar

V. Sharma (✉) · C. Prabha (✉) · M. M. Zaheer
Department of Space, ISTRAC/ISRO, Lucknow 226021, India
e-mail: vimla_sharma@istrac.gov.in

C. Prabha
e-mail: charuprabha@istrac.gov.in

M. M. Zaheer
e-mail: mirza@istrac.gov.in

M. N. Favorskaya et al. (eds.), *Innovations in Electrical and Electronic Engineering*, Lecture Notes in Electrical Engineering 661, https://doi.org/10.1007/978-981-15-4692-1_42

type of TWSTFT facility is also established at NPL Delhi, India. IRNWT facilities maintain time interval measurements between both IRNWT timescales, i.e. IRNWT-I, IRNWT-II and NPL Delhi timescales using these TWSTFT facilities [1].

Time transfer techniques are required to maintain the traceability between the timing labs geographically separated apart. Time transfer is a method used to compare time scales or frequency standards which are separated geographically. Time transfer is needed for many reasons. Instantaneous measurements of widely separated clocks can be made to monitor the performance of time scales through the intermediary of the clocks being compared. One can also be interested in making comparisons of advanced frequency standards, such as caesium fountains, and/or checking the long-term stability of such standards. Time transfer is done using the following commonly known methods namely:

1. Common view time transfer
2. All in view time transfer
3. TWSTFT

2 TWSTFT

Among the above three methods, TWSTFT has been accomplished to be the most precise method for time and frequency transmission as the accuracy in this prescript is better than 1 ns.

TWSTFT utilize geostationary spacecraft as a relay station and delivers stable and precise time transfer. It is the most accurate method for time transfer as almost all the delays cancel out due to analogy. The only disadvantage of this method is the increase in cost as both the ground stations require transmission and receiving equipment. BIPM (International Bureau of Weights and Measures) near Paris, France apply these techniques in coordination with other national laboratories to collate atomic timescale standards for the calculation of the International Atomic Time Scales (TAI) and Universal Coordinated Time (UTC) [2].

The working principle of TWSTFT is very simple. Both the TWSTFT terminals simultaneously transmit signals at the same moment to a geostationary satellite. These signals are transmitted back by the geostationary satellite and received at the TWSTFT facility on the ground. Then the time elapsed between the time of signal transmission and the time of signal reception is calculated at both the TWSTFT stations separately. Both the stations require spread spectrum modems to exchange information through communication satellites. At the transmitting station, 1 PPS (one pulse/second) signal is superimposed into an intermediate frequency (IF), up converted into radio frequency (RF), amplified and then transmitted to the spacecraft. Over receiving station, radio signal (RF) is therewith amplified, down-converted to an intermediate frequency (IF) and demodulated by the modem in order to produce a received 1 PPS. Thus, the time interval between the local 1 PPS and received 1 PPS is measured at each station [2, 7] through Time Interval Counter (Fig. 1).

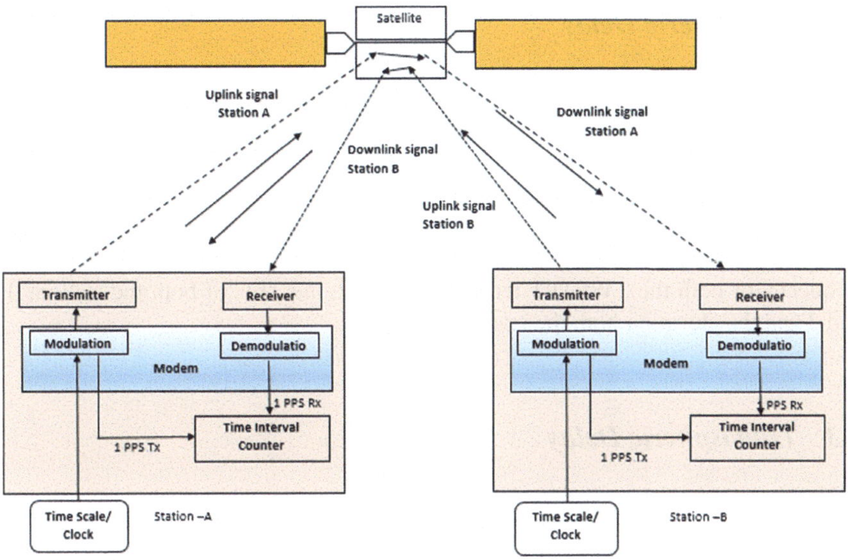

Fig. 1 Block diagram of TWSTFT [2]

The uncertainty in TWSTFT measurement is because of some atmospheric effects and delays which are [2, 3, 7]:

2.1 *SAGNAC Effect*

As explained by the experiment conducted by French scientist Georges SAGNAC in 1913, due to the diurnal motion of the satellite, the uplink geometric path becomes unequal to the downlink geometric path. This is known as the 'SAGNAC effect'. The TWSTFT terminal on the ground and the spacecraft both orbit around the equatorial rotation axis of the planet earth. This movement of both spacecraft and planet earth takes place when the signals are transmitted from TWSTFT terminal to the spacecraft and back over TWSTFT. It is proportional to the equatorial projection of the area of the four-sided plane figure (a quadrangle), whose vertices are centre of the earth and the positions of the stations on the surface of the earth also the position of the satellite with respect to the earth's surface. Thus, it is a function of the positions of stations 1 and 2 and the satellite.

2.2 *Ionospheric Delay*

Ionospheric delay (which occurs when a signal enters the ionospheric layer) is dependent upon both the uplink and downlink frequencies, the TEC (total electron content) and the elevation. The total electron content is defined as the total number of electrons integrated between two points, along a tube of one meter squared cross-section, i.e. the electron columnar number density and is expressed in electrons/m^2. While calculating the ionospheric delay, one needs to know the uplink frequency and downlink frequency of both the TWSTFT terminals 1 and 2, the TEC of both the stations 1 and 2 and the elevation both the stations 1 and 2.

2.3 *Tropospheric Delay*

Other than ionospheric layer, tropospheric layer also affects the propagation of the signal to and from the satellite. But the tropospheric delay is mostly less than 10 ps and therefore the delay caused due to the tropospheric layer is negligible. It is frequency independent (negligible) at the frequencies used for satellite communication.

2.4 *Satellite Error*

The error in the transponder of the satellite or in any internal subsystem of the satellite can also cause uncertainty. Therefore, this error should be known before the satellite is launched.

2.5 *Earth Station Delay*

The errors in the ground equipment have to be determined in each station. For example, the difference between the transmitting and receiving section inclusive the D/C and U/C converters, Modem, Power amplifiers, Antenna , cables connected, etc. The equipment delay closely depends upon temperature.

3 Calculation of Errors

Out of all the causes of inaccuracy in time synchronization through TWSTFT, the SAGNAC effect delay is the most complicated one in terms of calculation. This is because of its dependence on unusual factors. The SAGNAC effect delay calculation requires a geometric analysis of positions of the stations, earth's geometric centre position and the geometric satellite positions. The other factors affecting the accuracy of TWSTFT such as the earth station delay and the satellite error could be easily measured. Also, the tropospheric delay doesn't have much effect on TWSTFT.

3.1 SAGNAC Effect Delay Calculation

If the clock pulse is transmitted from station *A* to station *B*, then the SAGNAC delay in the time is given as [3, 6, 7]

$$= (2\omega \, \mathrm{Ar}/c2) \quad (1)$$

where

ω: Earth angular velocity in radian/second
c: Specified light's speed in meter/second

'Ar' is the area prescribed by the projections onto the equatorial plane by the lines connecting the earth's centre and the satellite to the two earth stations.

3.2 Calculation of 'Ar'

In a Geocentric coordinate system, the earth's centre is considered as the origin. Let the following variables denote (Fig. 2):-

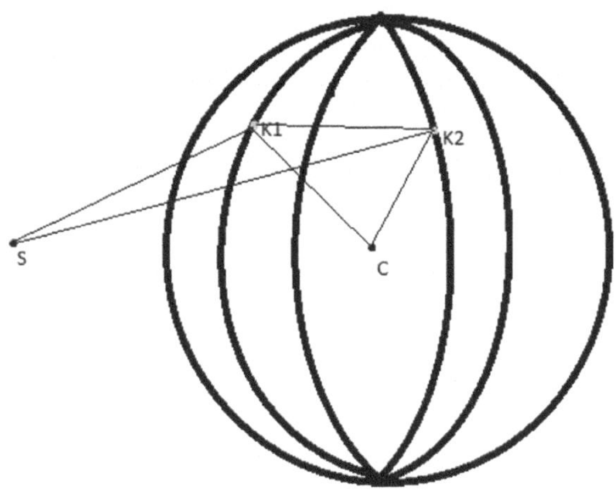

Fig. 2 Location of TWSTFT terminals K1 and K2 on the earth's surface

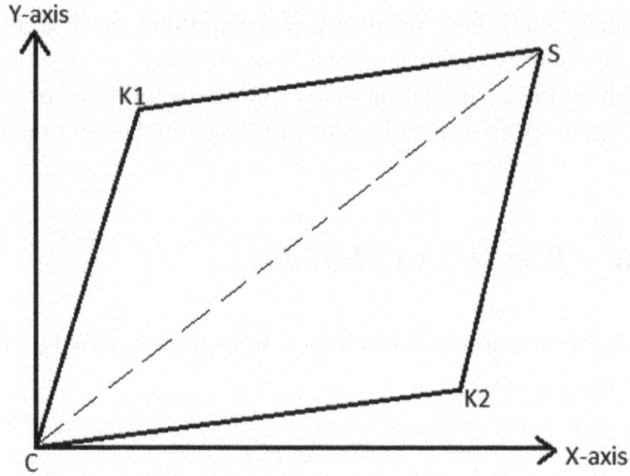

Fig. 3 Projection of area of triangles SK1K2 and CK1K2 on the equatorial plane

XA: Geocentric *x*-coordinate of station *A*
YA: Geocentric *y*-coordinate of station *A*
XB: Geocentric *x*-coordinate of station *B*
YB: Geocentric *y*-coordinate of station *B*
Xs: Geocentric *x*-coordinate of satellite *S*
Ys: Geocentric *y*-coordinate of satellite *S*

Now, as the satellite used in the method of TWSTFT is geostationary, both points *C* and *S* would lie on the equatorial plane [6].

Assuming the Equatorial plane to be the *XY* plane and the earth's centre at the origin, the projection of the area covered by the triangles SK1K2 and CK1K2 on the equatorial plane is shown in the following figure. The area of the quadrilateral CK1SK2 is the required value of 'Ar' (Fig. 3).

The area of Quadrilateral CK1SK2 is the sum of the area of the triangles CK1S and CK2S. Now the area of the triangles could be calculated by the method of cross product of two vectors.

Position vector of K_1: $X_1a_x + Y_1a_y$
Position vector of K_2: $X_2a_x + Y_2a_y$
Position vector of S: $X_sa_x + Y_say$

(As it is projection on equatorial, therefore, the *Z* value from equatorial plane will be considered as zero)

Area of quadrilateral CK_ASK_B

$$= \tfrac{1}{2}|CK_A \times CS| + \tfrac{1}{2}|CS \times CK_B|$$
$$= \tfrac{1}{2}\left|\left(X_Aa_x + Y_Aa_y\right)x\left(X_sa_x + Y_sa_y\right)\right| + \tfrac{1}{2}\left|\left(X_sa_x + Y_sa_y\right)x\left(X_2a_x + Y_2a_y\right)\right|$$
$$= \tfrac{1}{2}[X_s(Y_B - Y_A) + Y_s(X_A - X_B)]$$

Hence, the value of 'Ar' is given as

$$\text{Ar} = \tfrac{1}{2}[X_s(Y_B - Y_A) + Y_s(X_A - X_B)] \tag{2}$$

Therefore, the SAGNAC error between station A and station B can be formulated from Eqs. (1) and (2) as

$$SC_{AB} = \omega[X_s(Y_B - Y_A) + Y_s(X_A - X_B)]/c^2$$

Converting the Geocentric coordinates to Geodetic Coordinates [4, 5],

$$X_A = \left\{\begin{array}{l} a\,\text{Cos}\left[\tan^{-1}\left\{(1-f)\tan\left[\text{LAT geodetic}(K_A)\right]\right\}\right] \\ +\ \text{HT}(K_1)\text{Cos}\left[\text{LAT geodetic}(K_A)\right] \end{array}\right\}\text{Cos}\,[\text{LOT}\,(K_A)]$$

$$Y_A = \left\{\begin{array}{l} a\text{Cos}\left[\tan^{-1}\left\{(1-f)\tan\left[LAT\ \ \text{geodet}(K_A)\right]\right\}\right] \\ +\ \text{HT}(K_1)\text{Cos}\left[\text{LAT geodet}(K_A)\right] \end{array}\right\}\text{Sin}\,[\text{LOT}(K_A)]$$

$$X_B = \left\{\begin{array}{l} a\,\text{Cos}\left[\tan^{-1}\left\{(1-f)\tan\left[\text{LAT geodetic}(K_B)\right]\right\}\right] \\ +\ HT\,(K_2)\,\text{Cos}\left[\text{LAT geodetic}(K_B)\right] \end{array}\right\}\text{Cos}[\text{LOT}(K_B)]$$

$$Y_B = \left\{\begin{array}{l} a\text{Cos}[\tan^{-1}\{1-f)\tan[LAT\ geodetic(K_B)]\}] \\ +\ HT(K_2)\text{Cos}[LAT\ geodetic(K_B)] \end{array}\right\}\text{Sin}\,[LOT(K_B)]$$

$$X_s = R\,\text{Cos}\,[\text{LAT}(S)]\text{Cos}\,[\text{LOT}(S)]$$

$$Y_s = R\,\text{Cos}\,[\text{LAT}(S)]\text{Sin}\,[\text{LOT}(S)]$$

where

c: Specified light's speed = 299792458 m/s
ω: Earth angular velocity = 7.2921 * 10-5 rad/s
f: Flattening of the planet Earth ellipsoid = 1/298.257222
a: Planet earth semi-major axis or equatorial radius = 61378137 m
R: Radius of Satellite's orbit = 42164000 m
LAT (K): Latitude of the TWSTFT terminal (radian)
LOT (K): Longitude of the TWSTFT terminal (radian)
HT (K): Altitude of TWSTFT terminal (meter).
For geostationary satellites LAT(s) = 0° N

4 Results

Calculation results of SAGNAC error for the TWSTFT stations placed anywhere, i.e. station *A* and station *B*.

Location of station *A*:

LA geodetic (1) = 26.91° E
HT (1) = 6378165.34 m
LAT (1) = 80.956° N
LOT (s) = 55° E
B. Location of station *B*
LA geodetic (2) = 12.89° N
HT (2) = 6378836 m
LAT (2) = 77.37° E
LOT (s) = 55° E

Substituting the values, we get-
SC_{AB} = 89.44 ns

It may appear that SAGNAC error will be cancelled out in TWSTFT but SAGNAC error depends upon the path travelled by the signal. Error due SAGNAC effect will differ in polarity for signal travelling from TWSTFT terminal *A* to TWSTFT terminal *B* as compared to signal travelling from TWSTFT terminal *B* to TWSTFT terminal *A*.

5 Conclusions

TWSTFT is a two-way time transfer method that makes a geostationary spacecraft work as a relay station. This method provides stable and accurate time transfer. Periodic time transfers are required to maintain traceability between IRNWT time scales. All errors contributing in the process are to be precisely calculated and accounted for. It is seen that among all contributing errors SAGNAC effect plays a significant role. In a system with an accuracy of the order of 1 ns, error of values around 89.44 ns becomes crucial. The precise coordinates of the TWSTFT terminal enable the fine computation of the corrections due to the SAGNAC effect at each site.

References

1. An in-house developed Timescale for NavIC PTF
2. Hanson DW (1989) The operational use of two-way satellite time and frequency transfer employing pseudorandom noise code. In: 43rd Annual symposium on frequency control, pp 174–178
3. Sun H, Li Y, Yang H, Ding X (2013) Effect of satellite motion on TWSTFT. ICCNCE-13

4. Hedgley Jr, DR, Edwards C (1976) An exact transformation from geocentric to geodetic coordinates for non-zero altitude. NASA Technical Report, R-458
5. . Conversion of geocentric to geodetic latitude—mathworks.com https://www.mathworks.com/help/aeroblks/geocentrictogeodeticlatitude.html
6. Soler T, Eisemann DW (1994) Determination of look angles to geostationary communication satellite. JSE 120(3):115–116
7. (08/2015) The operation use of two-way satellite time and frequency transfer employing pseudorandom noise codes. Recommendation ITU-R TF. 1153–4, 2015

Evaluation of Sector-Based Routing Approach in Mobile Ad Hoc Networks

Palanisamy Vigneshwaran and Satkunarajah Suthaharan

Abstract Infrastructure and ad hoc network architectures are the two major architectures that can use wireless networks. Unlike infrastructure architecture using the centralized base station, mobile ad hoc network (MANET) is an infrastructure-less architecture which can be connected without any centralized control. In MANET, a mobile device can be connected with another device within the appropriate radio range that is typically 30 MHz to 5 GHz. Therefore, MANET is considered as a self-organizing network where each node acts as a participant and a router. Sector-based routing is a method in MANET which can be used to separate the positions of every node in a network topology. In addition, a specific node called server node can be assigned among the nodes in each sector. The server node is responsible to reserve the path resources and to maintain the information for its associated nodes and links. The performance of sector-based routing depends on the number of nodes available in the sector. Further, the routing approach in sector-based MANET can be considered in various scenarios using server nodes. Our research in this paper is to evaluate the efficiency of the sector-based routing in MANET considering several scenarios. Further, we address the variation of the performance while the number of nodes changes with several routing methods in sector-based MANET. In this paper, we investigate the detailed simulation experiments on a different number of nodes and various scenarios of sector-based routing in MANET. Our findings are as follows. (1) The average collision rate is significantly lower in sector-based routing when compared to MANET. (2) The resource efficiency in sector-based routing is considerably higher. (3) Unlike in a small number of nodes, the higher number of connections can be allowed in a high number of nodes distributed in a sector.

Keywords MANET · Sector-based routing · Server node · Resource efficiency

---

P. Vigneshwaran (✉) · S. Suthaharan
Department of Physical Science, Faculty of Applied Science, University of Jaffna, Vavuniya Campus, Jaffna, Sri Lanka
e-mail: p.vickey22@gmail.com

S. Suthaharan
e-mail: suthaharan@vau.jfn.ac.lk

M. N. Favorskaya et al. (eds.), *Innovations in Electrical and Electronic Engineering*, Lecture Notes in Electrical Engineering 661, https://doi.org/10.1007/978-981-15-4692-1_43

1 Introduction

The wireless networks of the mobile nodes can be classified into two types such as infrastructure architecture and ad hoc network architecture [1, 2]. In infrastructure architecture, mobile devices can be connected with a wireless network where a mobile node communicates with a geographically fixed base station. In this network, the wireless link is dedicatedly connected between the mobile node and the centralized base station [3, 4]. On the other hand, mobile ad hoc network (MANET) is an infrastructure-less multi-hop mobile network in which each node communicates with each other either directly or indirectly as existing in the typical communication networks [5]. Particularly, the device can directly communicate with each other devices within the radio frequency range of 30 MHz to 5 GHz; in contrast, intermediate nodes can be deployed for the communication where the nodes are located out of the radio range. In addition, MANET is the wireless ad hoc network in which each node has an independence to move freely in any direction; hence, a new node can join and an existing node can leave the network at any time without any limitations [6, 7]. Therefore, the MANET is the infrastructure-less and self-organizing wireless networks mainly used for mobile nodes, and each node in a network acts as a participant as well as router. Furthermore, people and vehicle with hand-held wireless devices can be interconnected in areas where proper communication infrastructures are unavailable [8]. Due to the mobility of the nodes, there are some essential characteristics incorporated with the MANET. Firstly, mobility of the nodes causes a change of the topology constantly, and thus, it follows the dynamic and autonomous network topology. Further, this topology exists only for a small amount of time until the next movement of the devices. The main drawback of this characteristic is unlike protocols used in wired networks, different and compatible protocols must be used for the MANET. Bandwidth usage is another significant constraint of MANET in which nodes can be connected using wireless links, and thus, the lower capacity of bandwidth can be achieved when compared with the wired transmission. Similarly, storage capacity is also limited in MANET.

Figures 1 and 2 show the MANET with four mobile devices and several dynamic links. In Fig. 1, node A and node B are considered as source and destination, respectively. At time $t = t1$, node A and node B are connected through an intermediate node C. At time t=t2, similarly, source node A and destination node B can be connected through two intermediate nodes which are node D and node C as shown in Fig. 2. Therefore, MANET always follows the dynamic topology in which all nodes and links can be changed the location frequently. Particularly, routing and path discovery are the essential parts of the MANET as the routing can be performed similarly to the network topology dynamically [9].

All the routing solutions have been proposed in order to make sure the efficient data traffic between source and destination. Position-based routing is a novel approach which has been proposed for quality of service and received significant attention in the distributed environment used with the server architecture [10]. This approach is based on the geographical locations of the mobile nodes [11]. In this approach, the

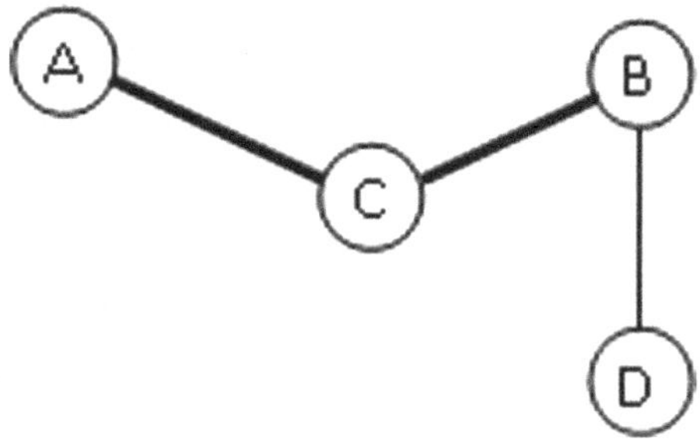

Fig. 1 Dynamic network topology at time $t = t1$ in MANET

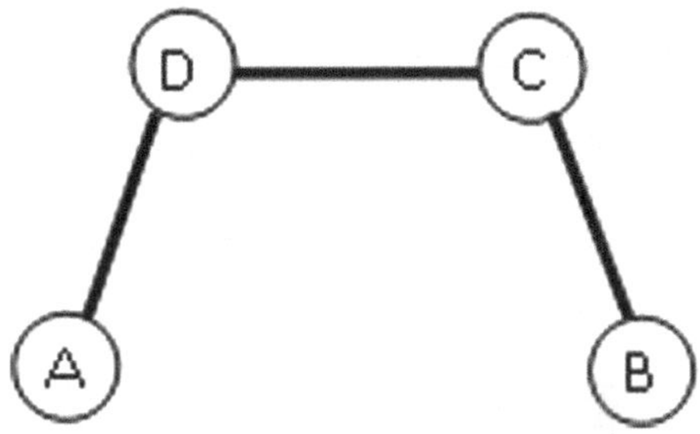

Fig. 2 Dynamic network topology at time $t = t2$ in MANET

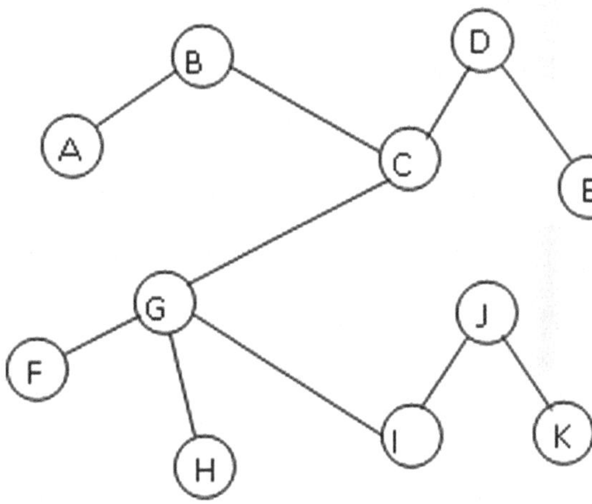

Fig. 3 Dynamic network topology in traditional MANET

ad hoc network is divided into sectors. Therefore, this approach can also be called as sector-based routing. Each sector contains various numbers of nodes as the network topology follows the dynamic network environment. Each divided sector is monitored and controlled by a node called server node. The server node is responsible to maintain the information regards to the associated nodes and links. Further, reservation of resources and multiple path selection are the main responsibilities of a server node in a sector.

Eleven mobile nodes and ten wireless links are shown in Figs. 3 and 4. In Fig. 3, a dynamic network topology just before the sectorization is shown. To implement

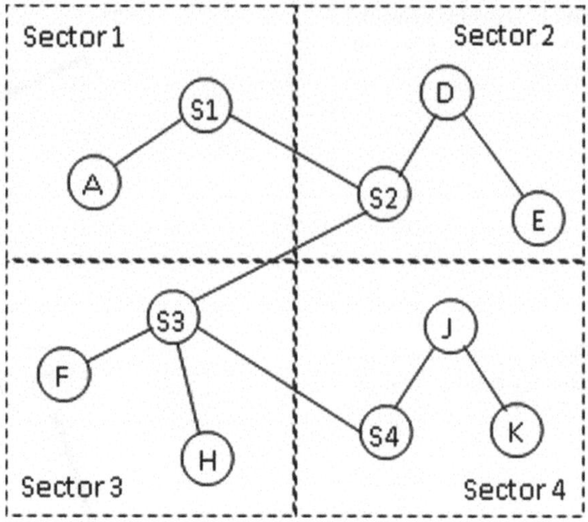

Fig. 4 Dynamic network topology in sector-based MANET

the sector-based MANET, ten nodes shown in Fig. 3 are sectorized in four regions shown in Fig. 4. All the server nodes S1, S2, S3, and S4 shown in Fig. 4 contain the information of the corresponding nodes and the links. The performance of sector-based routing in MANET depends on the number of nodes it contains. To ensure the method, a certain number of nodes can be handled by the server node of a particular sector. In addition, routing approach in sector-based MANET can be realized in various scenarios using server nodes of each sector. To the best of our knowledge, this study has not been addressed in an appropriate way. In this paper, we investigate the efficiency of the sector-based routing in MANET considering several scenarios. Further, we address the variation of the performance while the number of nodes changes with several routing methods in sector-based MANET.

This paper is organized as Sect. 2 presents of related work on MANET. In Sect. 3, the methodology of sector-based MANET is explained. Section 4 evaluates the performance study. Finally, the conclusion is discussed in Sect. 5.

2 Related Work

In this section, we discuss the relevant study carried out by other researchers. In [12], author describes the wireless routing protocol (WRP) as one of the path-finding algorithms which have the shortest path algorithm. There is some information used to estimate the shortest path such as length and hops from the source node to the destination node. Distance table, routing table, link cost table, and message retransmission list are the four aspects where each node can consist of routing. Dynamic source routing (DSR) is explained in [13]. It is particularly proposed to use multi-hop wireless ad hoc networks of mobile nodes. Two stages in DSR are considered, and those

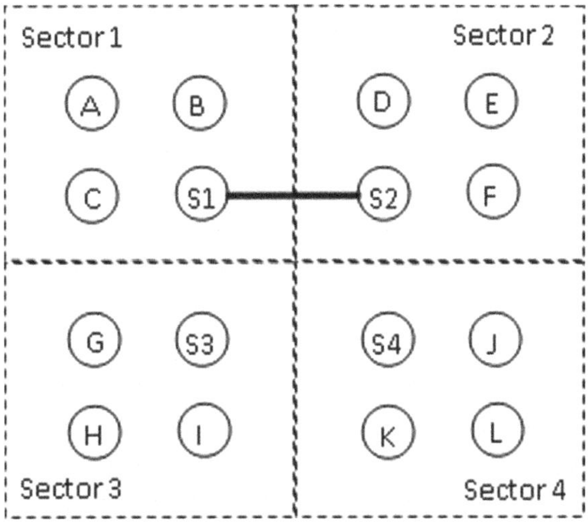

Fig. 5 Sector-based mobile ad hoc network topology in which source and destination are the Server nodes (1-hop)

are route discovery and route maintenance. In the route discovery process, a control packet travels from source to destination to discover the paths and to arrange the path for sending original data, whereas, in the route maintenance stage, the entire paths are maintained to ensure the successful transmission of the data. In [14], the author explains the features of zone routing protocol (ZRP) for the reconfigurable wireless networks. In this approach, zones are predefined based on the availability of network nodes; thus, the number of hops is calculated. Each node contains routing information about its associated nodes which are located in a particular zone. This is because of reducing the network overhead caused by proactive routing protocols. Authors in [15] used the arbitrary waypoint mobility model to calculate the length of the path in MANET. In this study, they found path duration and MANET design parameters as specific features. Protocol for routing in interest-defined mesh enclaves (PRIME) for combined routing in MANETs has been addressed and tested in [16]. In this literature, meshes are initiated and disabled by the occurrence of interest in individual destination nodes. In [17], the authors describe two aspects such as query delay and data availability in order to balance the incompatible features using data replication techniques.

3 Methodology

In this section, several scenarios in sector-based mobile ad hoc network are shown in Figs. 5, 6, 7, and 8 with a dynamic network topology of four sectors, sixteen nodes, and some dynamic links (solid lines show the particular dynamic links for the illustration). These dynamic links can be changed frequently, because of the real-time mobility of the nodes. For illustration, it is assumed that sixteen nodes are equally

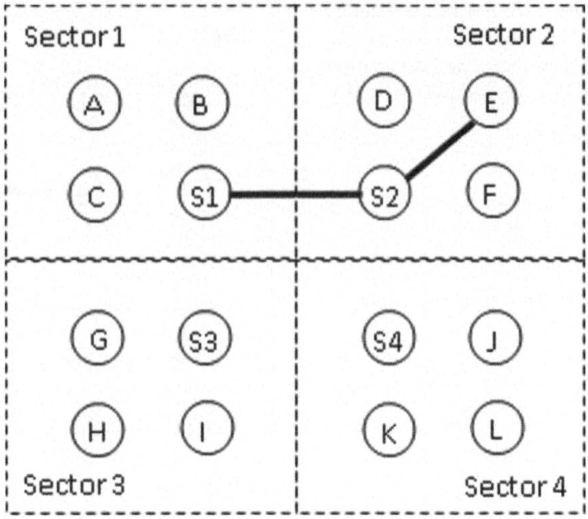

Fig. 6 Sector-based mobile ad hoc network topology in which source is a Server node (2-hop)

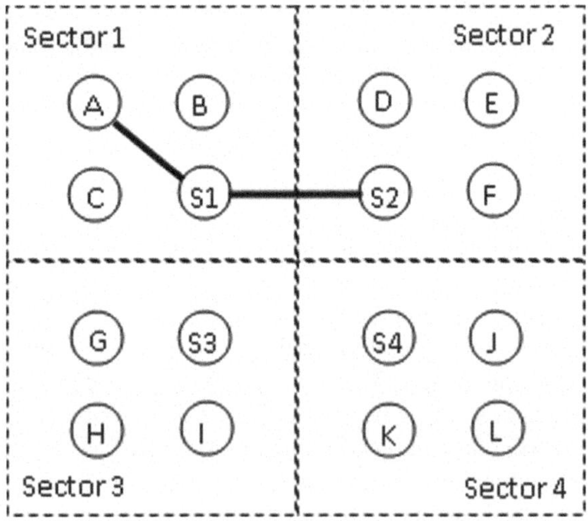

Fig. 7 Sector-based mobile ad hoc network topology in which destination is a Server node (2-hop)

distributed in four sectors (the number of nodes in a sector may be varied based on the mobility of the nodes). Since each node has the capability to move anywhere at any time, a node's sector to sector mobility also considered. Specifically, one node in each sector will be selected as a server node which has the correspondence of all nodes in the particular sector. Thus, the server node will have the details (say, configuration details) of all nodes in a sector. Since the configuration details are saved in the server nodes, the signal can be propagated to the destination in an efficient manner. It is important to note that, before the mobility of the entire nodes, server node will be reassigned as a normal mobile node. Similarly, after the mobility, a new node will be assigned as a server for each sector. In Figs. 5, 6, 7, and 8, server node S1 is

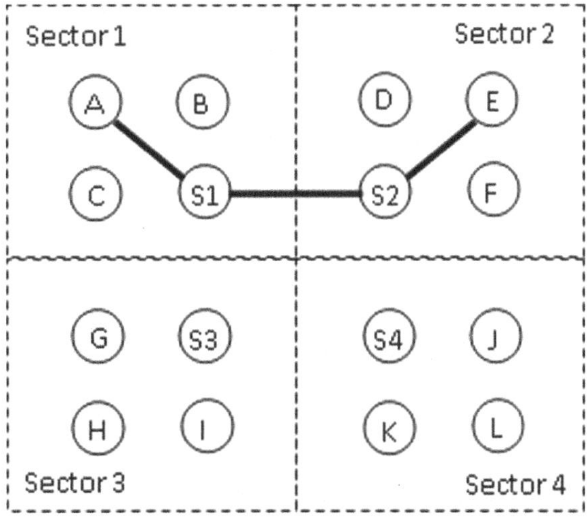

Fig. 8 Sector-based mobile ad hoc network topology in which two Server nodes are the intermediate nodes and those are not a source node and a destination node (3-hop)

corresponding to node A, node B, and node C in Sector 1. During a small amount of time of mobility, Server 1 can be connected to the other three corresponding nodes. Similarly, all other server nodes S2, S3, and S4 are connected and corresponding to their nodes in Sector 2, Sector 3, and Sector 4, respectively. In addition to that S1, S2, S3, and S4 are interconnected to represent their sectors (not all the connected links are shown in the figures). This is because all the nodes in all sectors should go through their corresponding server nodes to ensure the sector-based routing.

In this paper, we address four methods to support sector-based MANET as shown in Figs. 5, 6, 7, and 8. We consider the same topology to illustrate the methods. In Fig. 5, server node S1 from sector 1 and server node S2 from sector 2 are considered as a source node and destination node. There are no intermediate nodes used to transmit the data as S1 and S2 are the server nodes and able to connect directly from sector 1 to sector 2, and hence, the number of hop from source to destination is 1. In the second method (shown in Fig. 6), we consider S1 in sector 1 and E in sector 2 as a source and destination nodes, respectively. In order to implement the sector-based MANET, destination node E must have the communication to its associated server node, which is S2. Therefore, source node S1 can transmit the data to node E through server node S2 in sector 2. Particularly, the intermediate node of this method will be a server node and the number of hops is 2. Similarly, two hops and an intermediate server node S1 in sector 1 are considered in the third method as shown in Fig. 7. In this method, node A and node S2 are the source and destination nodes, respectively. To establish a connection from node A to server node S2, node A should communicate with the server node S1 which is in sector 1. Finally, node A and node E are considered as source and destination nodes shown in 8. In this method, both the nodes are non-server nodes. Both nodes should communicate with their corresponding server nodes S1 and S2. Therefore, the transmission is enabled

```
Input: G = (V, E), V :mobile nodes, E : wireless links, Sr : Server nodes, Request i = (S, D)
Output: Finding wireless path between two nodes.
Step-1
Assign all mobile nodes in to n sectors.
Step-2
Assign a server node for each sector.
Step-3
Find wireless path between source S and destination D.
if bandwidth for a wireless path is available then
  if S = Server (SrS) and D = Server (SrD) then
    Allow the transmission.
  else
    if S ! = Server (SrS) and D = Server (SrD) then
      Identify the corresponding server (SrS) for the source node.
      Assign SrS as an intermediate node between S and D.
      Allow the transmission.
    else
      if S = Server (SrS) and D ! = Server (SrD) then
        Identify the corresponding server (SrD) for the destination node
        Assign SrD as an intermediate node between S and D
        Allow the transmission
      else
        if S! = Server (SrS) and D ! = Server (SrD) then
          Identify the corresponding servers (SrS) and (SrD) for the source and destination nodes respectively
          Assign SrS and SrD as intermediate nodes between S and D
          Allow the transmission
        end if
      end if
    end if
  end if
else
  Block the transmission
end if
```

Algorithm 1 Assign wireless link between two nodes in MANET using sector-based routing.

A–S1, S1–S2, and S2–E. Thus, two intermediate nodes S1 and S2 are used with 3 hops.

To implement the methods illustrated above, we propose an algorithm shown in Algorithm 1. The input of this algorithm is graph, nodes, links, source, destination, and server node. Aim of this algorithm is to find the wireless path between two nodes. Four methods illustrated above are incorporated into the algorithm.

Table 1 Resource efficiency achieved in traditional MANET

No. of nodes	Resource efficiency (b/s/Hz)
5	0.000390042
10	0.000126662
15	0.000206905
20	0.000101785
25	0.000200225
30	0.000132417

4 Performance Study

Simulation of the traditional routing technique of MANET and the sector-based routing in MANET is considered in this performance study. The evaluation of the sector-based routing in MANET in several criteria such as (a) average collision rate, (b) resource efficiency, (c) transmission speed, and (d) number of connections admitted is measured in this study. Source node and destination node of each connection follow a uniform distribution. Dynamic network topology with various numbers of mobile nodes such as 5, 10, 15, 20, 25, and 30 is taken for this study. Totally ten bandwidth resources using wavelength division multiplexing method are considered. Transmission between two mobile nodes follows Poisson distribution and holding time of requests follows an exponential distribution with unit mean. Connection establishment is performed in a dynamic network environment. Each experiment is evaluated with a large number of connection request arrivals.

In this study, several mobile nodes are taken into consideration. This is because to properly observe the fine behaviour of traditional and sector-based MANETs, while increasing the number of nodes. Hence, more connectivity of mobile nodes in MANET can be provided. In Fig. 9, average collision rate obtained in sector-based MANET is compared with the average collision rate obtained in traditional MANET. Similarly, with the same number of nodes resource efficiency in sector-based MANET and resource efficiency in traditional MANET are compared as shown in Fig. 10. Due to the values obtained in resource efficiency in traditional MANET are smaller, a table containing such values is shown in Table 1. In Fig. 11, the transmission speed of sector-based MANET is compared against the transmission speed of the traditional MANET. Further, the comparison of the number of connections admitted in the sector-based MANET and the number of connections admitted in traditional MANET are shown in Fig. 12. This comparison is particularly performed only for 3-hops from the source node to the destination node, because of including typically two server nodes in each sector. In addition, we compare the number of connections admitted in sector-based MANET individually for 1-hop (server-server), 2-hop (source-server-destination), and 3-hop (source-server-server-destination). This is because to observe the patterns of admitted connections in various hops, while the number of nodes is varied.

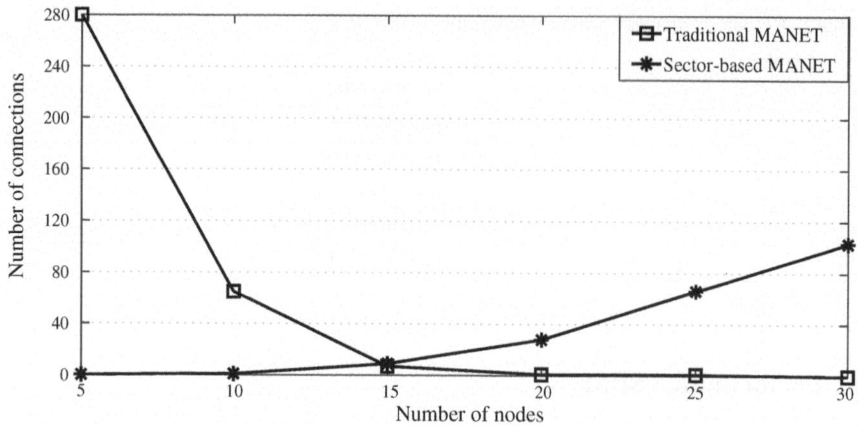

Fig. 9 Performance of sector-based MANET and traditional MANET in terms of average collision rate

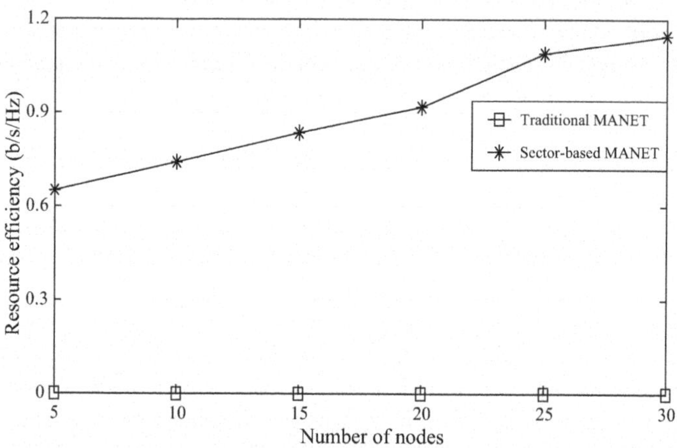

Fig. 10 Performance of sector-based MANET and traditional MANET in terms of resource efficiency

This performance study is considered in twofold. Firstly, the comparisons of sector-based MANET against traditional MANET and secondly, the admission of connections in a different number of hops in sector-based MANET are considered. The reduction of the average collision rate in sector-based MANET is shown in Fig. 9. In this comparison, it can be observed that the reductions are 0.1107, 0.1892, 0.1848, 0.201, 0.0025, and 0.0007 in 5, 10, 15, 20, 25, and 30 nodes, respectively. It is noted that the significant reductions have been achieved in less number of nodes. In Fig. 10, resource efficiency in sector-based MANET outperforms the notable increase compared with traditional MANET. Particularly, this difference can be observed in 25 and 30 nodes. The resource efficiency of traditional MANET is very low

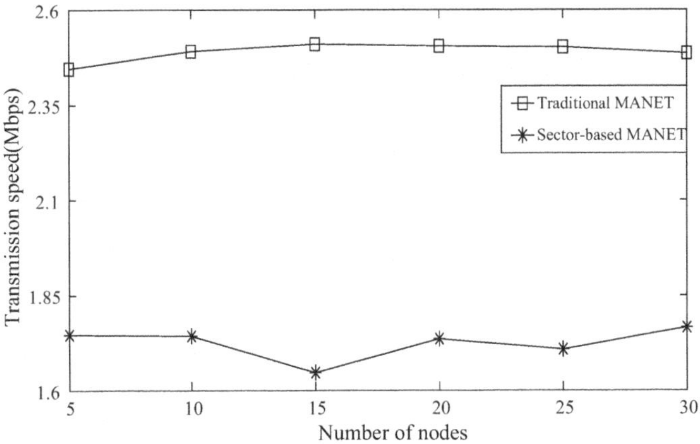

Fig. 11 Performance of sector-based MANET and traditional MANET in terms of transmission speed

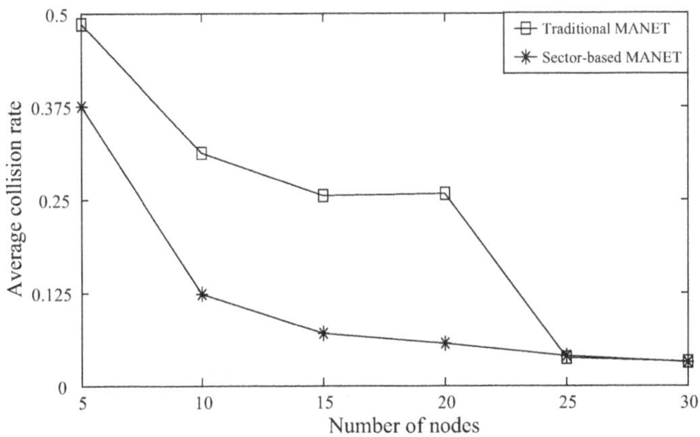

Fig. 12 Performance of sector-based MANET and traditional MANET in terms of number of connections allowed

and near to zero as shown in Table 1. Further, the performance of the transmission speed of traditional MANET is realized in all number of nodes when compared to transmission speed achieved in sector-based MANET shown in Fig. 11. However, the maximum transmission speed achieved by the traditional MANET and the sector-based MANET are 2.5092 Mbps and 1.6467 Mbps, respectively, at the number of nodes 15; hence, these transmission speeds are comparable. In Fig. 12, unlike in the low number of nodes like 5 and 10, more number of connections can be admitted in the high number of nodes (20, 25, and 30) using sector-based MANET. In contrast, a small number of nodes can be deployed to admit more connections in traditional MANET. It can be observed that for the number of nodes 15, both the

Table 2 Number of connections admitted in various hops in sector-based MANET

No. of nodes	1-hop	2-hop	3-hop
5	104	4	0
10	528	41	1
15	987	108	9
20	1430	236	28
25	1796	425	66
30	2162	550	103

traditional MANET and the sector-based MANET are performing approximately the same value. Secondly, various numbers of hops are compared in sector-based MANET shown in Table 2. In this comparison, a high amount of connections can be admitted in 1-hop when compared to 2-hop and 3-hop, using sector-based MANET. This amount of connection increases along with the number of nodes in all 1-hop, 2-hop, and 3-hop. Specifically, 1-hop (server node to server node) connection has the highest amount of connection at the number of nodes 30.

5 Conclusions

In this paper, we addressed various scenarios on sector-based routing in mobile ad hoc networks (MANET) and the behaviour of sector-based routing when the number of nodes in a sector varied. We evaluated the performance in various aspects such as average collision rate, resource efficiency, transmission speed, and the number of connections. Firstly, we compared sector-based MANET against the traditional MANET in our study. In this comparison, significant changes have been observed in the performance of sector-based MANET. Specifically, the average collision rate, resource efficiency, and the number of connections admitted in sector-based MANET outperform when compared it against traditional MANET. Further, the admission of connections in the different number of hops in sector-based MANET is considered. In this comparison, it is clearly observed that more number of connections can be admitted when a sector contains more nodes. Even though a high number of connections can be admitted, when the hop number is equivalent to one, that is server-to-server connection between two sectors.

References

1. Schiller JH (2003) Mobile communications. Pearson Education, New York
2. Marti S, Giuli T, Lai K, Baker M (2000) Mitigating routing misbehavior in mobile ad hoc networks. In: ACM MobiCom
3. Sushma Ghode D, Shilpa Ghode D (2013) A novel approach for energy efficient routing algorithm For ad hoc network. Int J Pure Appl Res Eng Technol
4. Singh G, Singh J (2012) MANET: a study of challenges and routing principles. Int J Adv Innov Res 1(1)
5. Saeed NH, Abbod MF, AlRaweshid H (2012) Manet routing protocols taxonomy. In: Proceedings of international conference on future communication networks (ICFCN). IEEE Computer Society, pp 123–128
6. Royer EM (1999) A review of current routing protocols for adhoc mobile wireless networks. Int J Personal
7. Royer EM, Toh CK (1999) A review of current routing protocols for ad hoc mobile wireless networks. IEEE Personal Conununications
8. IEEE Std. 802.11 (1997) Wireless LAN medium access control (MAC) and physical layer (PHY) specifications
9. Vas SR, Castaneda R (2000) Simulation-based perfonnance evaluation of routing protocols for mobile ad-hoc networks. Mob Networks Appl 3:179–189
10. Erbas F, Garcia JF, Jobmann K (2004) Position-based QoS routing in mobile ad hoc networks: problem statement and a novel approach. In: IEEE performance, computing and communications conference, pp 619–623
11. Mauve M, Widmer L, Hartenstein H (2001) A survey on position-based routing in mobile ad-hoc networks. IEEE Network
12. Murthy S, Garcia-Luna-Aceves JJ (1996) An efficient routing protocol for wireless networks. Mob Networks Appl 1(2):183–197
13. Royer EM (1999) A review of current routing protocols for adhoc mobile wireless networks. Int J Pers Commun 6(2)
14. Haas ZJ, Pearlman MR, Samar P (2002) The zone routing protocol (ZRP) for ad hoc networks. draft-ietfmanet-zone-zrp-04. txt
15. Namuduri K, Pendse R (2012) Analytical estimation of path duration in mobile ad hoc networks. IEEE Sen J 12(6)
16. Garcia Luna Aceves JJ, Menchaca Mendez R (2011) PRIME: an interest-driven approach to integrated unicast and multicast routing in MANETs. IEEE/ACM Trans Network 19(6)
17. Zhang Y, Yin L, Zhao J, Cao G (2012) Balancing the trade-offs between query delay and data availability in MANETs. IEEE Trans Parallel Distrib Syst 23(4):643–650

Printed Slot Wideband Rectangular Dielectric Resonator Antenna

Sovan Mohanty and Baibaswata Mohapatra

Abstract In this paper, a high performance wideband dielectric resonator antenna is investigated. It is observed that wide bandwidth and multi-band operations can be realized with a hybrid structure by amalgamating wide slot patch antenna and dielectric resonator antenna which are resonating at nearby frequencies. Further wide bandwidth can be realized by reducing the dielectric constant of the substrate to 3.38 and by using thick and relatively bigger dielectric material of dimensions $23.5 \times 36.25 \times 9.8\ \text{mm}^3$, $\varepsilon_r = 10.2$ as the radiating structure. Here, the width of the slot is being enhanced to $19.5 \times 32\ \text{mm}^2$ and an open-ended micro-strip feed line with fork-shaped opening is placed within the slot to provide impedance matching.

Keywords Dielectric resonator antenna (DRA) · Dielectric waveguide model (DWM) · Finite element method (FEM)

1 Introduction

Modern technology demands high speed communication, computation and control with high sensitivity, selectivity, and fidelity. High speed complex circuitry demands wide bandwidth, minimum internal noise, and highly precise measurement techniques. It is quite challenging to understand the impact of dynamic engineering environment and design necessary circuits to cope up with challenges like nonlinearity, inhomogeneity and distributed parameters analysis, etc. Apart from this the designed circuit has to survive in a real-world environment like external radio interference, static electricity, lightning, power line interference, etc. Cellular system

S. Mohanty (✉)
Department of Electronics and Communication Engineering, SRMS College of Engineering and Technology, Bareilly, U.P, India
e-mail: mohanty.sovan@gmail.com

B. Mohapatra
School of Electrical, Electronics and Communication Engineering, Galgotias University, Greater Noida, U.P, India
e-mail: bmohapatra9@gmail.com

M. N. Favorskaya et al. (eds.), *Innovations in Electrical and Electronic Engineering*, Lecture Notes in Electrical Engineering 661, https://doi.org/10.1007/978-981-15-4692-1_44

including upcoming 5G technology, high speed microprocessor, and medical electronics employ microwave and millimeter wave devices at their core. At this range of operation dielectric becomes an efficient candidate for energy transmission and radiation rather than normal conductor like Ag, Au, Cu, etc. [1]. DRA has high radiation efficiency, high gain, and high directivity, strong control over bandwidth and physical structure, minimum tolerance error, high noise handling capability and can be easily integrated with the latest technology [2]. Bandwidth enhancement technique mainly concern with impedance bandwidth of the DRA which can be defined as the range of frequency over which antenna is perfectly matched with the feeding element such that VSWR lies less than 2.0 or 1.5 [3]. The factors which are responsible to improve impedance bandwidth are the impedance matching networks. Matching network should be placed closed to the radiating structure [4]. Here, higher impedance bandwidth can be realized by (i) increasing the thickness of the substrate [5], (ii) using dielectric material with low permittivity in the substrate [6], (iii) increasing the physical size of the DRA, (iv) reducing the Q-factor of the radiator [7], (v) using hybrid structure in which two resonators are coupled in such way that their frequency response lies close to each other [8], (vi) using larger air gap between the feed and the DRA [9], (vii) right positioning of the open-ended feed line within the slots [10], (viii) using stacked configuration [11], (ix) exciting multimode rather than using lower-order mode only [12], (x) Controlling the size of the slot [13] and (xi) using antenna array structure [14].

Printed slot wideband dielectric resonator antenna is designed and discussed in this paper. This structure is a hybrid design based on the effective resonance of two resonators formed by patch geometry and RDRA. It provides multi-band and wide-band response from 2.75 to 3.26 GHz resonating at 3 and 3.92–7.68 GHz resonating at 4.4 GHz. This antenna is a suitable candidate for commercial applications like imaging systems, spectrum sensing, and indoor and outdoor wideband-based communication systems.

2 Method of Analysis

A high degree of control over the bandwidth is possible by controlling the aspect ratio of the DRA. Aspect ratio can be effectively controlled in rectangular dielectric resonator antenna rather than in any other structure. Dimension of the RDRA can be determined by using dielectric waveguide model (DWM) which is based on both Marcatili's and effective dielectric constant (EDC) approximation [15]. It is used to calculate the resonant frequency and quality factor of the RDRA. Following equations describe the sinusoidal variations of the field within the dielectric and exponentially decaying far field [16].

$$k_x^2 + k_y^2 + k_z^2 = \varepsilon_r k_0^2 \tag{1}$$

$$k_z \tan\left(\frac{k_z d}{2}\right) = \sqrt{k_0^2(\varepsilon_r - 1) - k_z^2} \tag{2}$$

$$k_x = \frac{m\pi}{a} \tag{3}$$

$$k_y = \frac{n\pi}{b} \tag{4}$$

$$F = \frac{2\pi a f_0}{c}\sqrt{\varepsilon_r} \tag{5}$$

$$f_{\text{GHz}} = \frac{15F}{a_{\text{cm}}\pi\sqrt{\varepsilon_r}} \tag{6}$$

where ε_r relative dielectric permittivity, a, d, b = length, width and height of rectangular resonator, k_0 = free space wave number, k_{x0}, k_{y0} = decay constant of field along x and y direction, m, n = number of extrema along x and y direction of wave number inside the waveguide.

3 Antenna Design

Figures 1, 2a, b and 3 show printed slot wideband RDRA. The DRA is placed over the etched portion of the ground plane to take the advantage of finite ground plane. Generally, narrow slot results in the formation of narrowband device. Here, bandwidth can be improved by proper interaction of fork-shaped micro-strip transmission line, comparatively wide slot and DRA with high dielectric constant. The proposed

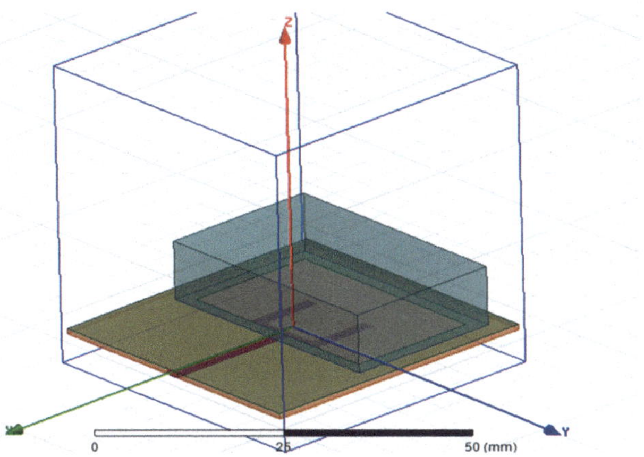

Fig. 1 Structure of printed slot RDRA

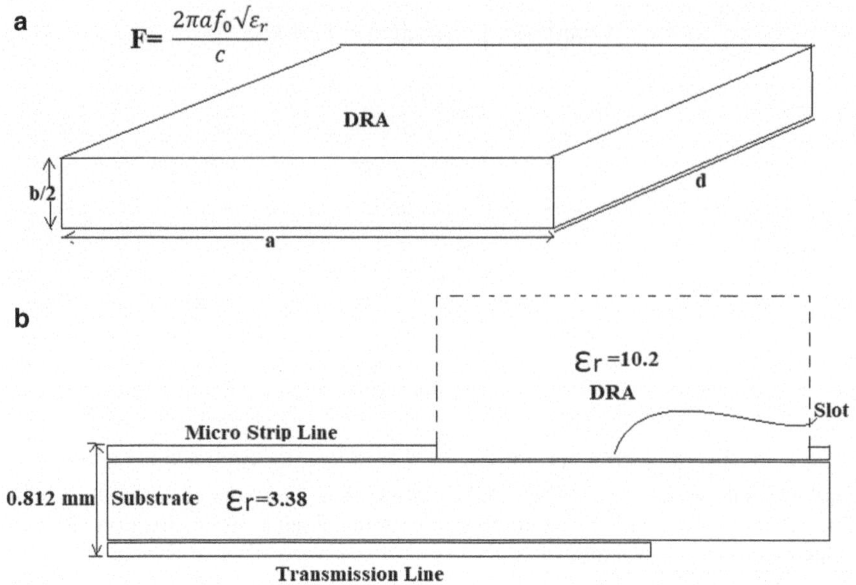

Fig. 2 **a** Geometry of the RDRA, **b** side view geometry of the printed RDRA over the substrate

DRAs patch and feeding geometry is shown in Fig. 3. The radiating structure is having a dimension of 0.47λ × 0.72λ × 0.19λ at 6 GHz. It uses Arlon AD 1000 TM as dielectric with permittivity 10.2 and loss tangent $\tan\delta = 0.003$. Antenna is placed over a substrate made up of Arlon 25 N^{TM} with $\varepsilon_r = 3.38$ and $\tan\delta = 0.002$. It has dimension of 0.87λ × 0.84λ × 0.016λ at 6 GHz. Again thicker dielectric size is preferred to reduce the inductive effect and to improve the impedance bandwidth. Top of the substrate is coated with perfectly electrical conductor to reduce the physical height of the antenna by half based on image theory [1]. A slot of dimension 0.39λ × 0.64λ is etched and DRA is placed exactly over the slotted portion as shown in Fig. 3. The presence of slot over half wavelength increases numbers of modes which lead to certain spurious distortions which is observed at the input characteristics of the antenna.

4 Results and Discussion

The simulation tool used in this work is Ansys HFSS 14.0 which is based on finite element method (FEM). Figure 4 shows the plot of return loss w.r.t frequency from 2 to 8 GHz. The 10 dB return loss is found to be over the frequency range of 2.75 to 3.26 GHz with bandwidth 18.36%. Another band is found to be from 3.92 to 7.58 GHz with bandwidth 63.65%. Return loss response shows multiple nulls which

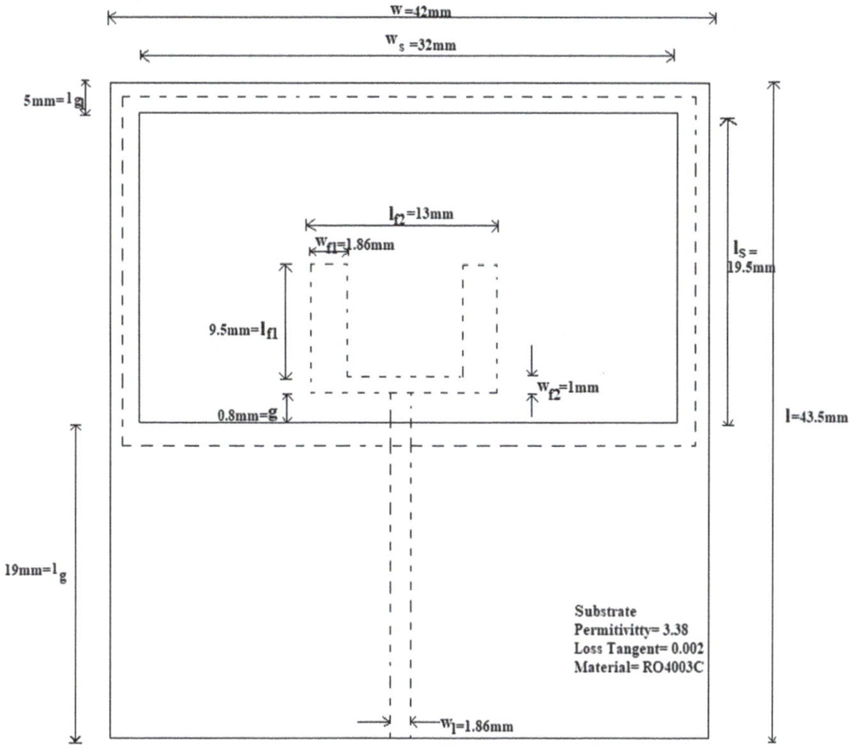

Fig. 3 Top view geometry of the printed RDRA over the substrate

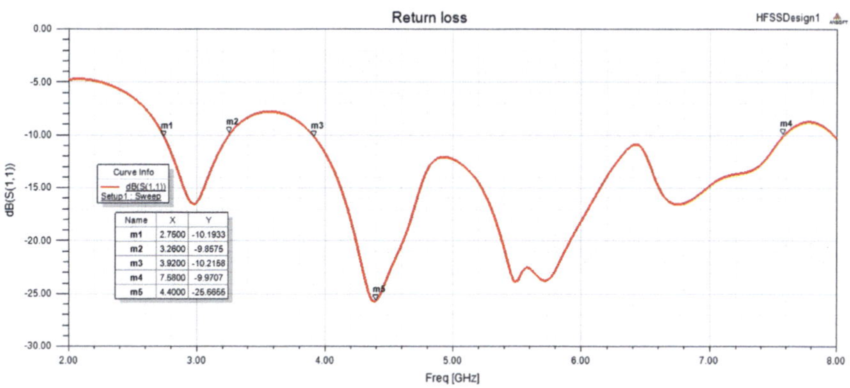

Fig. 4 Reflection coefficient plot of rectangular DRA

implies this antenna operates at several resonant mode due to interaction with fork-shaped micro-strip transmission line, etched slot RDRA and coupling between them. Geometrical parameters of the fork shape and types of the feeding structure basically determine the impedance bandwidth.

Figure 5 shows radiation characteristics of the RDRA w.r.t the impedance bandwidth. The electric and magnetic field along *yz* and *xy* plane when $\phi = 90°$ & $\phi = 0°$, respectively, with $\theta = 90°$ at the 6 GHz frequency (Fig. 6).

Figures 7 and 8 show the radiation pattern of the antenna in E and H plane at the operating frequency of 6 GHz. Higher level of cross-polarization occurs due to the

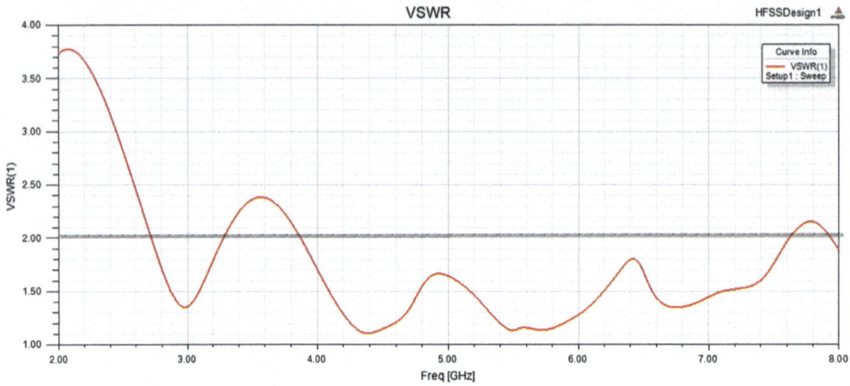

Fig. 5 VSWR versus frequency

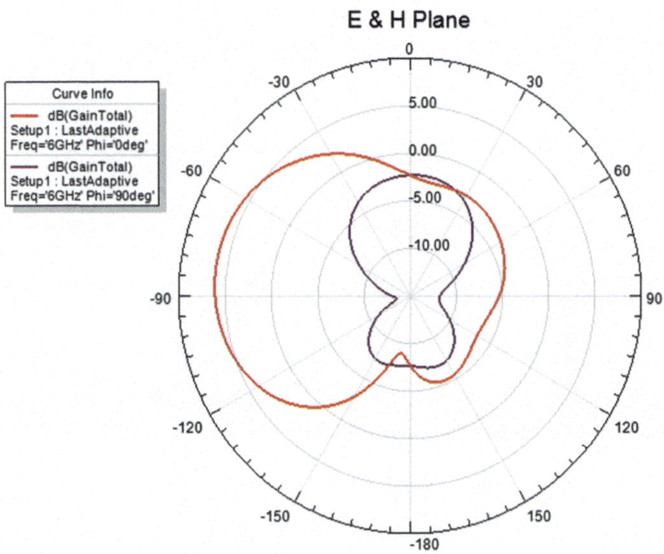

Fig. 6 E and H field along *yz* and *xy* plane

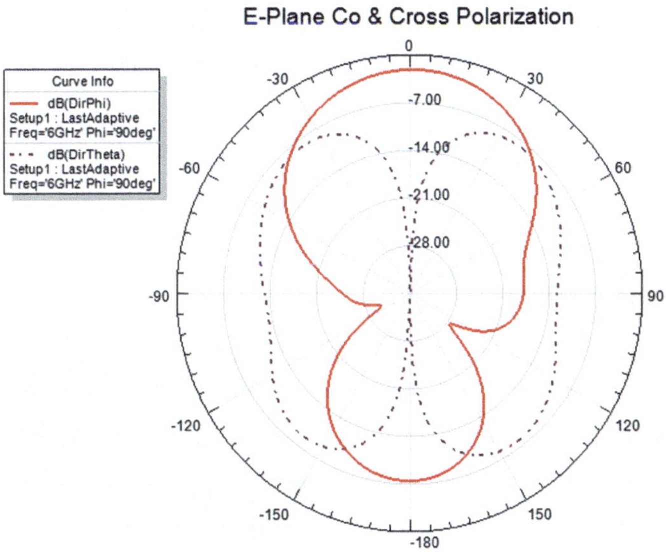

Fig. 7 E-plane co- and cross-polarization

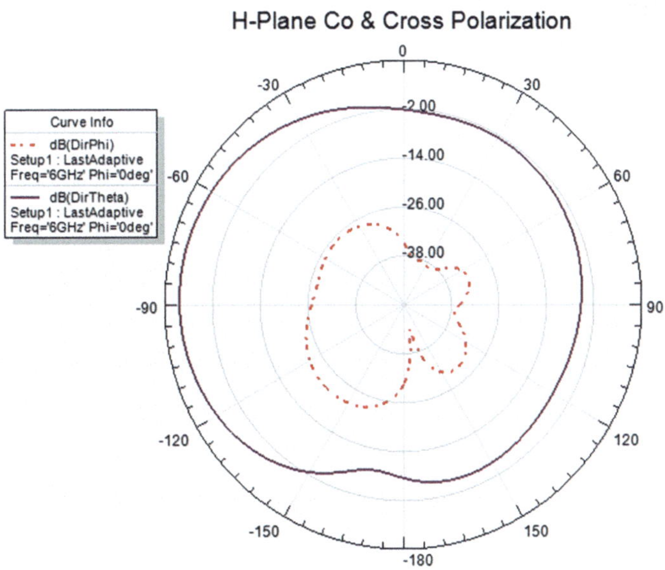

Fig. 8 H-plane co- and cross-polarization

presence of wider slot and amalgamation of RDRA, wide slotted patch, finite ground plane and fork-shaped transmission line which inherently depends upon operating frequency.

Figure 9 shows 3D plot of gain in dB along the spherical co-ordinate system. It is observed that maximum gain occurs in the direction of $\phi = 180°$ (Fig. 10).

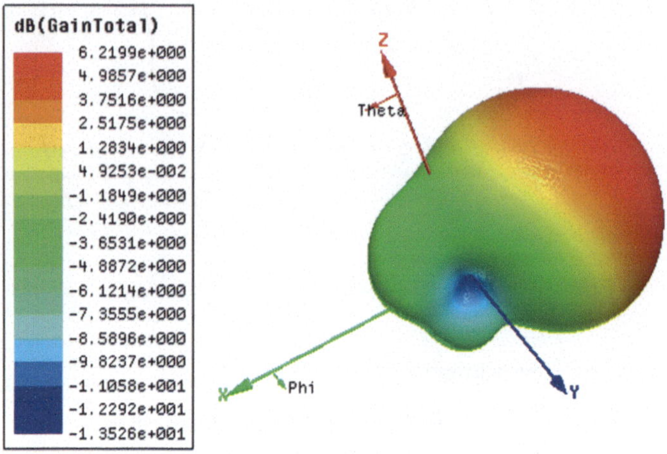

Fig. 9 3D gain plot of RDRA

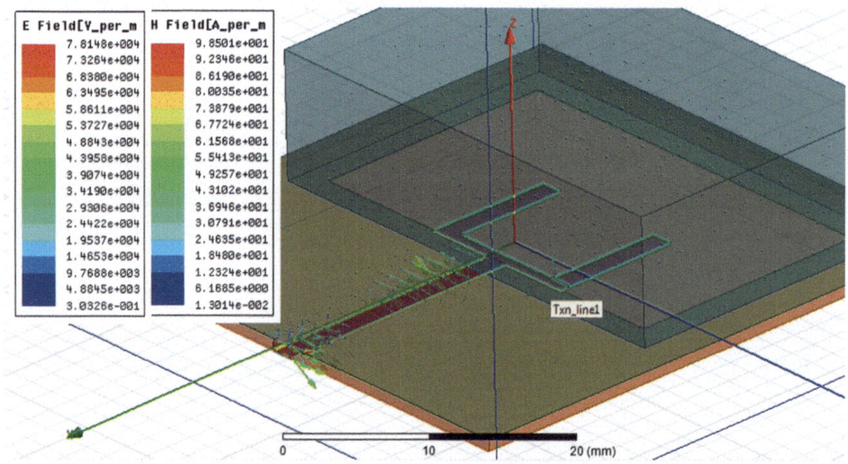

Fig. 10 E and H field in the proposed RDRA

5 Conclusions

In this paper, a novel printed slot wideband rectangular dielectric antenna has been proposed. In this design, wide bandwidth can be realized by employing a hybrid structure which is resonating to nearby frequencies. The proposed antenna provides stabilized impedance bandwidth of 18.36% from 2.75 to 3.26 GHz and 63.65% from 3.92 to 7.58 GHz with stabilized gain of 5 dB. This wideband antenna is suitable for imaging systems, spectrum sensing, indoor and outdoor wideband-based communication systems.

References

1. Harrington RF (2001) Time harmonic electromagnetic fields. Wiley-IEEE Press
2. Petosa A, Ittipiboon A, Antar YMM, Roscoe D, Cuhahi M (1998) Recent advances in dielectric resonator antenna technology. IEEE Antenna Propag Mag 40(3)
3. Pozar DM (1986) Microstrip antennas. IEEE Proc 80:79–91
4. Luk KM, Lee MT, Leung KW, Yung EKN (1999) Technique for improving coupling between microstripline and dielectric resonator antenna. IEEE Electron Lett 35(5):357–358
5. Petosa A, Ittipiboon A, Cuhaci M, Larose R (1999) Micro-strip fed stacked dielectric resonator antenna. In: Antenna technology and applied electromagnetics symposium ANTEM'96', Ottawa, Canada, pp 705–708
6. Garg R, Bhartia P, Bahl I, Ittipiboon A (2001) Microstrip antenna design handbook. Artech House, Canton, MA
7. Yaduvanshi RS, Parthasarathy H (2016) Rectangular: dielectric resonator antennas theory and design. Springer, New Delhi. ISBN 978-81-322-2499-0, ISBN 978-81-322-2500-3 (eBook). https://doi.org/10.1007/978-81-322-2500-3. Heidelberg, New York Dordrecht, London © Springer, India (2016)
8. Balanis CA (2005) Antenna theory analysis and design, 3rd edn
9. Shum SM, Luk KM (1995) Stacked annular ring dielectric resonator antenna excited by axi-symmetric co-axial probe. IEEE Trans Antennas Propag 43(8):889–892
10. Fan Z, Antar YMM (1997) Slot-coupled DR antenna for dual frequency operation. IEEE Trans Antennas Propag 45(2):306–308
11. Wang F, Zhang C, Sun H, Xiao Y (1999) Ultra-wideband dielectric resonator antenna design based on multilayer form. Int J Antenna Propag. Article ID 4391474
12. Li X, Yang Y, Gao F, Ma H, Shi X (2019) A compact dielectric resonator antenna excited by a planar monopole patch for wideband applications. Int J Antenna Propag. Article ID 9734781
13. Ittipiboon A, Mongia R, Antar YMM, Bhartia P, Cuhaci M (1993) Aperture fed rectangular and triangular dielectric resonators for use as magnetic dipole antennas. Electron Lett 29(23):2001–2002
14. Petosa A, Mongia RK, Ittipiboon A, Wright JS (1995) Investigation on a micro strip-fed series array of dielectric resonator antennas. IEEE Electron Lett 31(16):1306–1307
15. Long SA, McAllister MW, Shen LC (1983) The resonant cylindrical dielectric cavity antenna. IEEE Trans Antennas Propag AP-31:406–412
16. Luk KM, Leung KW (2003) Dielectric resonator antennas. Electron Electr Eng Res Stud

Design and Analysis of Different Full Adder Cells Using New Technologies

Nandhaiahgari Dinesh Kumar, Somineni Rajendra Prasad, Chilukuri Raja Kumari, and Challa Dhanunjaya Naidu

Abstract CMOS transistors are most widely used for the design of computerized circuits, when scaling down the nanometer technology these devices faces the short channel effects and causes I-V characteristics to depart from the traditional MOSFETs, so the researchers have developed the other transistors technologies like CNTFET and GNRFET. Carbon nanotube field-effect transistor is one of the optimistic technologies and it is a three-terminal transistor similar to MOSFET. The semiconducting channel between the two terminals called source and drain comprises of the nanotube which is made of carbon. Graphene nano-ribbon field-effect transistor is the most optimistic technology here the semiconducting channel is made of graphene. When contrasted with barrel-shaped CNTFETs, GNRFETs can be prepared in situ process, transfer-free, and silicon compatible, thus have no passage-related and alignment problems as faced in CNTFET devices. This paper presents different 1-bit Full Adder Cells(FACs) like TG MUX-based FAC (TGM), MN MUX-based FAC (MNM), proposed TG Modified MUX-based FAC (TGMM) and another proposed MN Modified MUX-based FAC (MNMM) are designed using different technologies like CNTFET and GNRFET at 16 nm technology with a supply voltage of 0.85 V and simulation is done by using Synopsys HSPICE tool and the proposed designs are best when compared to the TGM and MNM FACs in terms of Static and Dynamic powers Dissipations and Delay.

Keywords Full adder cell(FAC) · TG modified mux · MN modified mux · CNTFET · GNRFET · Power dissipation · Delay and HSPICE tool

---

N. Dinesh Kumar (✉) · S. Rajendra Prasad · C. Raja Kumari · C. Dhanunjaya Naidu
Department of Electronics and Communication Engineering, Vallurupalli Nageswara Rao Vignana Jyothi Institute of Engineering and Technology (VNR VJIET), Hyderabad, Telangana, India
e-mail: dineshkumarnandhaiahgari@gmail.com

M. N. Favorskaya et al. (eds.), *Innovations in Electrical and Electronic Engineering*, Lecture Notes in Electrical Engineering 661, https://doi.org/10.1007/978-981-15-4692-1_45

1 Introduction

The booming demand for low power in very large-scale integration (VLSI) can be seen at distinctive block levels, such as process technology, circuit, structure, and the layout level. The FACs are the fundamental building blocks of the ALU circuits. The different FACs are designed by using the multiple threshold voltages for different transistors in the design [1]. Majority function or threshold function gives the output as logic '1' when more number of inputs are '1' based on the logic level the multiple threshold voltages are obtained. The diameter of the CNT is controlled by varying the chirality vector based on the diameter the threshold voltage could be varied [2]. GNRs can be prepared in situ process, transfer-free and silicon compatible, thus have no passage-related and alignment problems as faced by CNTFET devices [3].

The TGM FAC [4] is designed by using the 25 transistors, MNM FAC [5] is designed by using the 26 transistors, proposed TGMM FAC is designed only by using the 23 transistors and another proposed MNMM FAC is designed only by using the 18 transistors. This paper is structured as follows Sect. 2 presents the Review of CMOS, CNTFET and GNRFET. Section 3 tells about the design of different FACs.

Section 4 shows the comparative analysis of different FACs. Section 5 is done with the conclusion of the paper.

2 Review of CMOS, CNTFET and GNRFET

The CMOS transistor is the basic and most widely used for the design of computerized circuits, As nanometer of the transistor is decreasing, chip thickness and working frequency are enlarged, making power utilization in battery-worked handy devices a significant concern. CNTFET shows a considerable advantage over the CMOS. Both P-type and N-type CNTFET transistors have the analogous mobility. Sizing of transistors in complex advanced devices based on CNTFETs would be simple compared to CMOS devices. Considering the properties like electrical characteristics, for example, ballistic transport and low Off-Current of CNTFETs empowers rapid and low power circuit structures [2].

In CNTFETs, the edge potential of the transistor is relying upon the width of the CNTs by varying the chirality vector the threshold voltage can be altered. CNTs play a vital role in the design of the integrated circuits in the future due to their uncommon electrical and mechanical properties as compared with CMOS devices. CNT transistors are divided into two types of single wall (SWCNT) and multi-wall (MWCNT) [6]. Single wall CNTFET consists of one barrel whereas the multi-wall has more than one barrel. The structure of the CNTFET is shown in Fig. 1. The carbon nanotubes materials are quiescent devices to substitute si in the low power.

Channel length of a CNTFETs could be decreased to five nanometers before tunneling temperature brings an undesirable drop in currents. Diameter of a CNTFET is similar to the arrangement of the amount of tubes [7].Graphene is a two-dimensional layer of carbon atoms having 2D honeycomb lattice. The most attractive

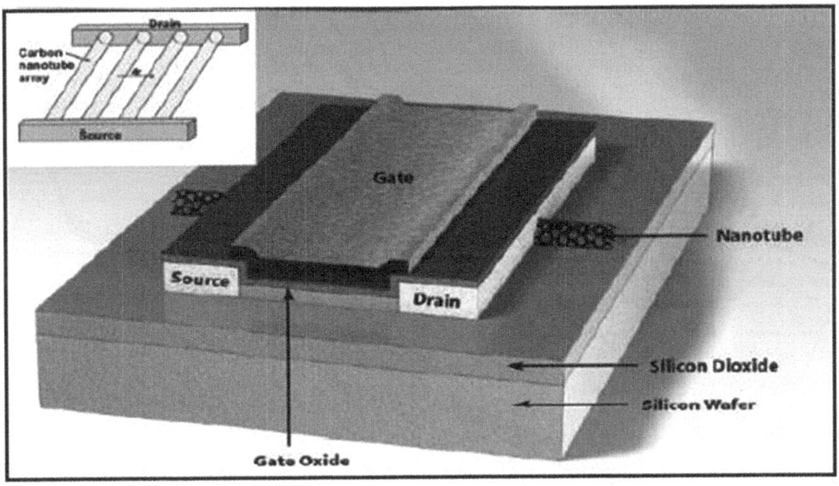

Fig. 1 CNTFET transistor structure

feature of graphene is high carrier mobility which renders faster switching times and GNRs are ultra-thin sheets of graphene support one-dimensional ballistic transport of charge carriers. Graphene is categorized into two types monolayer, bilayer or multilayer graphene. In the bilayer graphene band-gap can be induced to make it semiconducting.

GNRFET has emerged to replace the traditional silicon transistors in scaled technologies [8]. Since these devices are very small and the effect of course variation on the circuit performance is very large. The most researched are carbon nanotube FETs (CNTFETs) and Graphene nano-ribbon FETs (GNRFETs), When contrasted to barrel-shaped CNTs, GNRs can be prepared in situ process, transfer-free and silicon compatible, thus have no passage-related and alignment problems as faced by CNTFET devices. The GNRFET transistor structure is shown in Fig. 2.

In GNRFET number of dimmer lines determine the width. There are two sorts of GNRFETs Schottky barrier (SB) GNRFET and MOS like GNRFET [9]. SB type uses metal to shape contacts along these lines keeping away from fundamental extra doping. In any case, the SB type GNRFETs offer particularly less Ion to Ioff proportion when contrasted and the MOS-type GNRFET.

3 Design of Different FAC's Using New Technologies

This section discusses the design of different 1-bit FACs like TG MUX-based FAC (TGM), MN MUX-based FAC (MNM), proposed TG Modified MUX-based FAC (TGMM) and another proposed MN Modified MUX-based FAC (MNMM) are designed using different technologies like CNTFET and GNRFET.

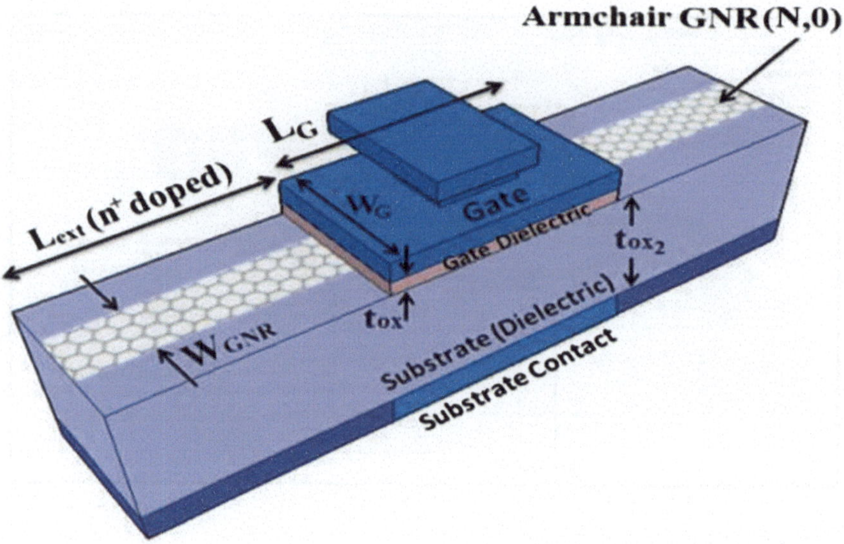

Fig. 2 GNRFET transistor structure

3.1 TG MUX-Based FAC (TGM)

The TGM 1-bit FAC is designed using the CNTFET and GNRFET technologies. Here the FAC consists of transmission gates. The circuitry consists of different levels, in the first level NOT logic operation, the second level AND and OR logic operation to them A, B inputs are given, the third level will be having XOR/XNOR logic operation and finally fourth level MUX's logic operation, to the MUX's logic C is given as input and output from the MUX's are SUM and CARRY. The CNTFET design parameters are pitch: 6 nm, tubes: 3, Kox: 16, Hox: 4 nm, Lg: 16 nm, Cb: 100 nm and Efo: 0.6.

The TGM 1-bit FAC is designed by using the GNRFET technology with the GNRFET transistors. The different design parameters for GNRFET technology are nRib: 6, N: 6, Tox: 0.6 nm, dop: 0.001, L: 16 nm and Tox2:20 nm. The total numbers of transistors used for the design are 25. The TGM 1-bit FAC in GNRFET technology is shown in Fig. 3.

3.2 Proposed TG Modified MUX-Based FAC (TGMM)

The block diagram describes the implementation of the FAC, the block diagram is shown in Fig. 4. The FAC is implemented in different levels, in the first level NOT logic operation, the second level AND and OR logic operation, the third level will be

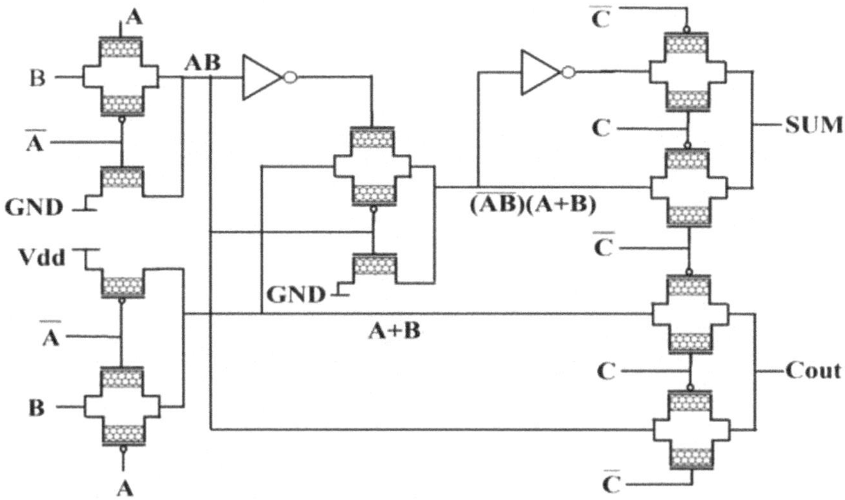

Fig. 3 TGM 1-bit FAC in GNRFET technology

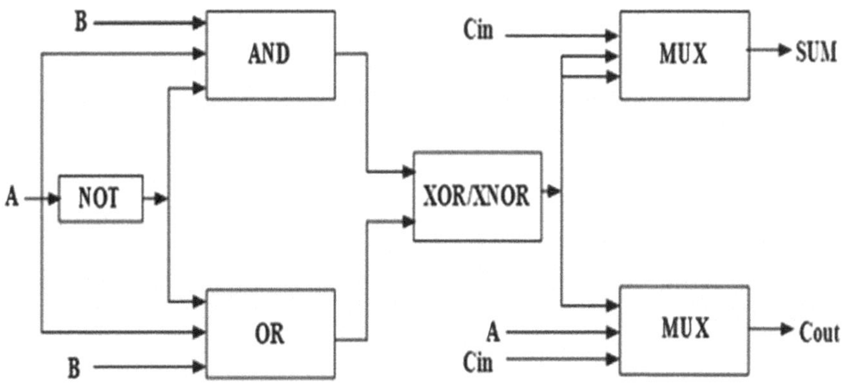

Fig. 4 Block diagram of proposed TGMM 1-bit FAC

having XOR/XNOR logic operation and finally fourth level MUX's logic operation. In this proposed design the modification is made in the MUX design.

The proposed TGMM 1-bit FAC is designed using the CNTFET and GNRFET technologies. Here the FAC is implemented by transmission gates. The circuitry consists of different levels, in the first level NOT logic operation, the second level AND and OR logic operation to them A, B inputs are given, the third level will be having XOR/XNOR logic operation and finally fourth level MUX's logic operation, to the MUX's logic C is given as input and output from the MUX's are SUM and CARRY. Design parameters for the Proposed TGMM CNTFET 1-bit FAC are pitch:6 nm, tubes:3, Kox:16, Hox:4 nm, Lg:16 nm, Cb:100 nm and Efo:0.6.

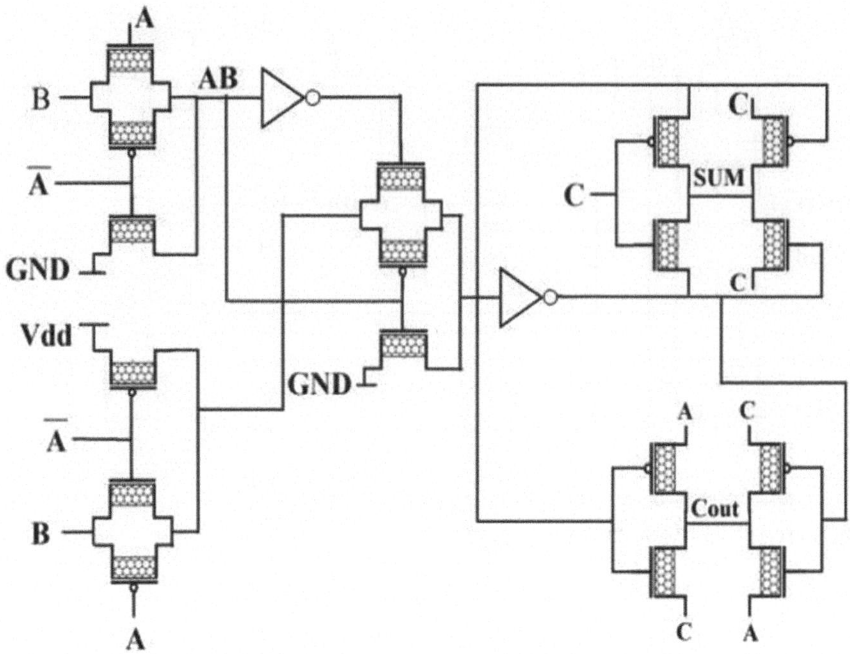

Fig. 5 Proposed TGMM 1-bit FAC in GNRFET technology

The proposed TGMM 1-bit FAC is designed by GNRFET technology with the GNRFET transistors and the design parameters are nRib: 6, N: 6, Tox:0.6 nm, dop:0.001, L:16 nm and Tox2:20 nm. The total numbers of transistors used for the design are 23. The proposed TGMM 1-bit FAC in GNRFET technology is shown in Fig. 5.

3.3 MN MUX-Based FAC (MNM)

The MNM 1-bit FAC is designed using the CNTFET and GNRFET technologies, The MNM 1-bit FAC is shown in the below Fig. 6. The operation of the MNM 1-bit FAC has majority function or threshold function it gives output as logic '1' when more number of inputs are '1' based on the logic level the multiple threshold voltages are obtained. The circuitry consists of different levels, in the first level majority logic operation to them A, B inputs are given, the second level NAND logic

operation, the third level will be having XOR/XNOR logic operation and finally fourth level MUX's logic operation, to the MUX's logic, A and C is given as input and output from the MUX's are SUM and CARRY. Here the total numbers of transistors used for the circuit design are 26. The CNTFET design parameters are pitch:6 nm, tubes:3, Kox:16, Hox:4 nm, Lg:16 nm, Cb:100 nm and Efo:0.6.

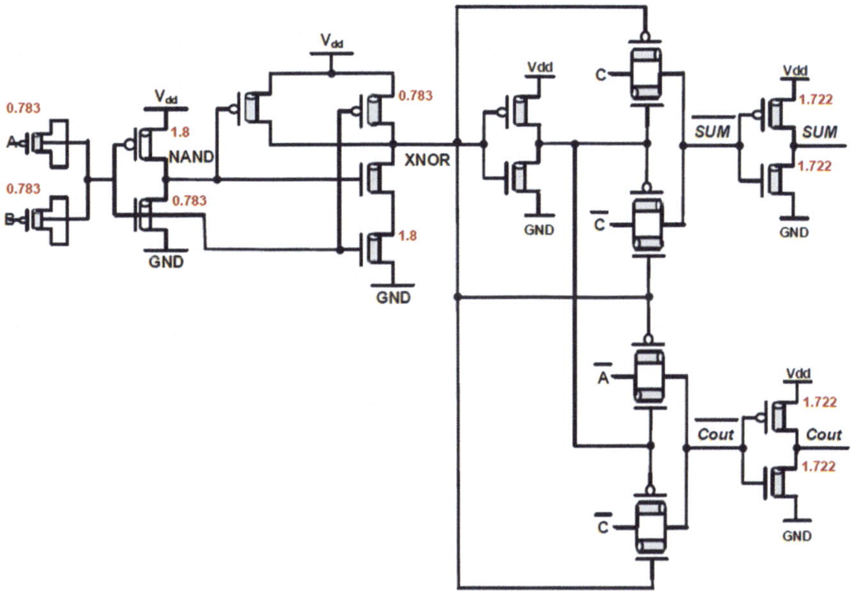

Fig. 6 MNM 1-bit FAC in CNTFET technology

The MNM 1-bit FAC is designed by GNRFET technology with the GNRFET transistor. The different design parameters for GNRFET technology are nRib:6, N:6, Tox:0.6 nm, dop:0.001, L:16 nm and Tox2:20 nm.

3.4 *Proposed MN Modified MUX-Based FAC (MNMM)*

The block diagram describes the implementation of the Proposed MNMM 1-bit FAC, the block diagram is shown in Fig. 7. The operation of the proposed FAC has the majority function or threshold function it gives output as logic '1' when more number

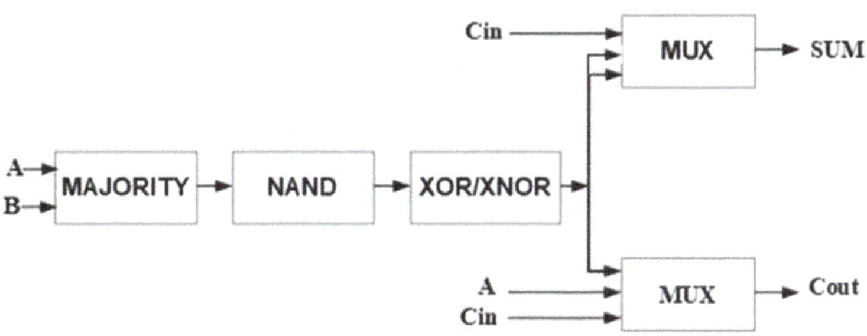

Fig. 7 Block diagram of proposed MNM 1-bit FAC

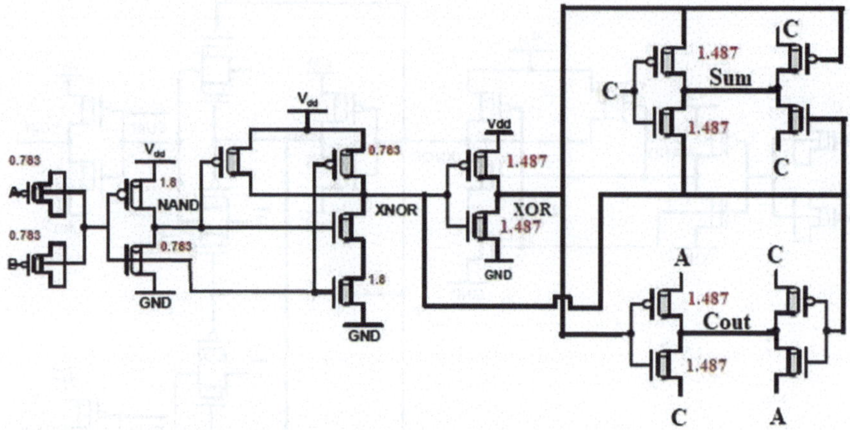

Fig. 8 Proposed MNM 1-bit FAC in CNTFET technology

of inputs are '1' based on the logic level the multiple threshold voltages are obtained. The first level is majority logic operation, the second level NAND logic operation; the third level will be having XOR/XNOR logic operation and finally fourth level MUX's logic operation. In this proposed design the modification is made in the MUX design.

The MNM 1-bit FAC is designed using the CNTFET and GNRFET technologies. The Proposed MNM CNTFET 1-bit FAC is shown in the Fig. 8. The operation of the FAC have the first level is majority logic operation to which A, B inputs are given, the second level NAND logic operation, the third level will be having XOR/XNOR logic operation and finally fourth level MUX's logic operation, A and C is given as input and output from the MUXs are SUM and CARRY. Here the total numbers of transistors used for the circuit design are 18. The CNTFET design parameters are pitch:6 nm, tubes:3, Kox:16, Hox:4 nm, Lg:16 nm, Cb:100 nm and Efo:0.6.

The proposed MNM 1-bit FAC is designed by GNRFET technology with the GNRFET transistor and the design parameters for GNRFET technology are nRib:6, N:6, Tox:0.6 nm, dop:0.001, L:16 nm and Tox2:20 nm.

4 Comparative Analysis of Different 1-Bit FAC's

This section discusses the comparative analysis of different 1-bit FACs like TG MUX-based FAC (TGM), MN MUX-based FAC (MNM), proposed TG Modified MUX-based FAC (TGMM) and another proposed MN Modified MUX-based FAC (MNMM) are designed using different technologies like CNTFET and GNRFET. These designs are simulated using Synopsys HSPICE tool at 16 nm technologies with

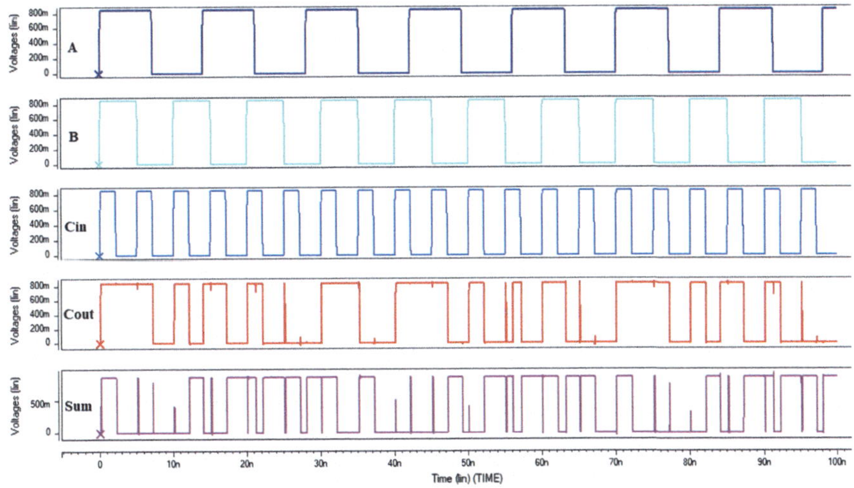

Fig. 9 Simulation waveform for TGM GNRFET 1-bit FAC

0.85 V supply and the performance parameter values of Propagation Delay, dynamic and static powers are tabulated in Table 1.

4.1 TG MUX-Based FAC (TGM)

The analysis of TGM 1-bit FAC with CNTFET and GNRFET technologies which are discussed in Sect. 3 are simulated and analyzed. Simulation results of TGM GNRFET 1-bit FAC are shown in Fig. 9. The operation of the circuit is obtained as per the functionality of the FAC truth table.

4.2 Proposed TG Modified MUX-Based FAC (TGMM)

The analysis of proposed TGMM 1-bit FAC with CNTFET and GNRFET technologies which are discussed in Sect. 3 are simulated and analyzed. Simulation results of the proposed TGMM GNRFET 1-bit FAC are shown in Fig. 10. The operation of the circuit is obtained as per the functionality of the FAC truth table.

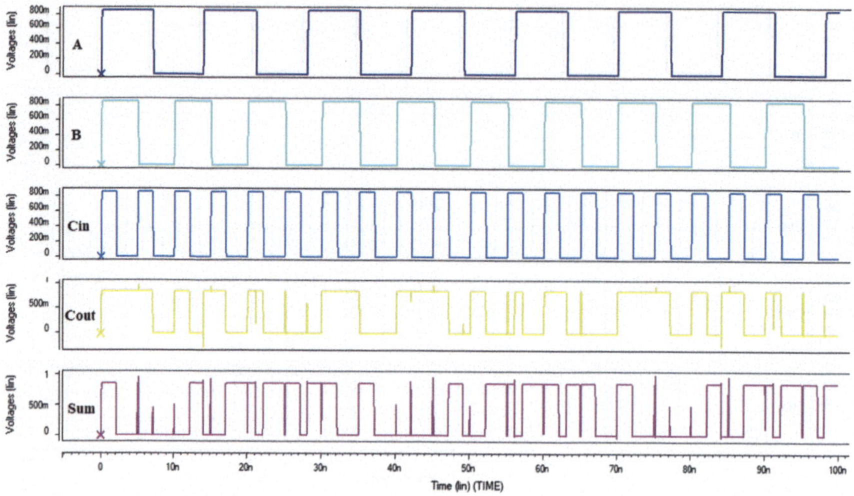

Fig. 10 Simulation waveform for proposed TGMM GNRFET 1-bit FAC

4.3 MN MUX-Based FAC (MNM)

The analysis of MNM 1-bit FAC with CNTFET and GNRFET technologies which are discussed in Sect. 3 are simulated and analyzed. Simulation results of MNM CNTFET 1-bit FAC are shown in Fig. 11. The operation of the circuit is obtained as per the functionality of the FAC truth table.

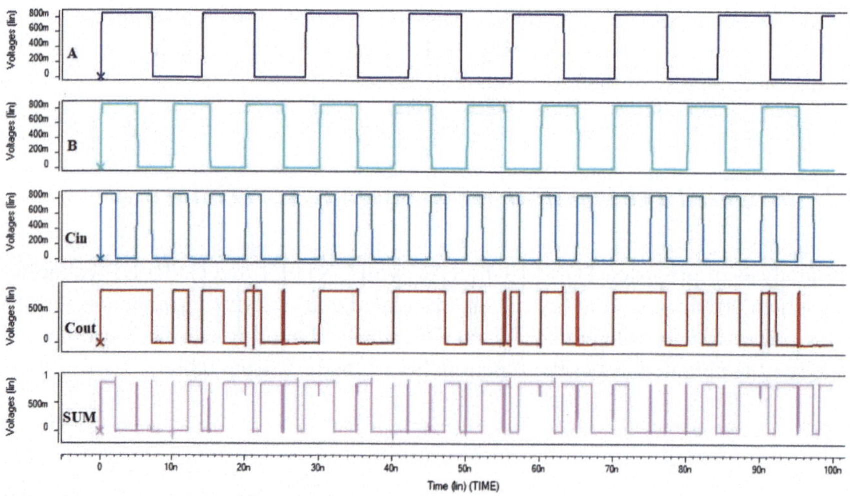

Fig. 11 Simulation waveform for MNM CNTFET 1-bit FAC

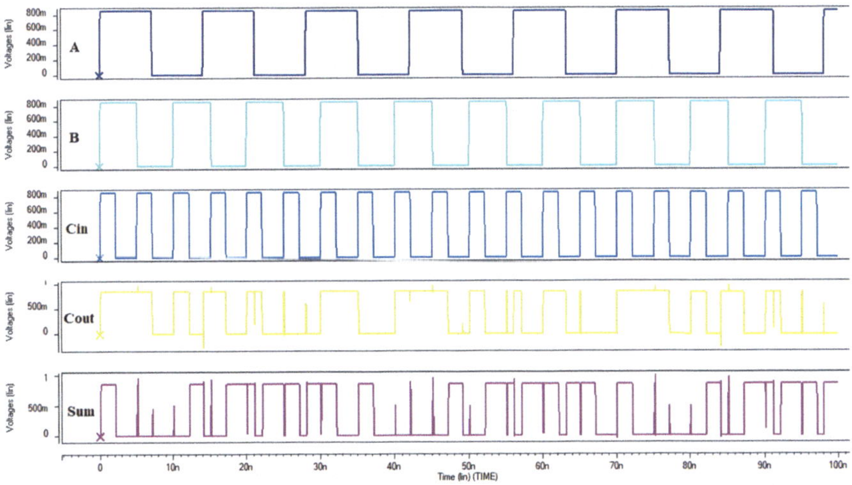

Fig. 12 Simulation waveform for proposed MNMM GNRFET 1-bit FAC

4.4 Proposed MN Modified MUX-Based FAC (MNMM)

The analysis of MNMM 1-bit FAC with CNTFET and GNRFET technologies which are discussed in Sect. 3 are simulated and analyzed. Simulation results of MNM CNTFET 1-bit FAC are shown in Fig. 12. The operation of the circuit is obtained as per the functionality of the FAC truth table.

From Table 1, TGM GNRFET FAC has 5% better speed performance than TGM CNTFET FAC. Dynamic Power Dissipation of TGM GNRFET FAC is 28% less than Dynamic Power Dissipation of TGM CNTFET FAC. Static Power Dissipation of TGM GNRFET FAC is 25% less than TGM CNTFET FAC, whereas TGM CNTFET FAC having more Power Dissipation than TGM GNRFET FAC.

TGMM GNRFET FAC has a 27% better speed performance than TGMM CNTFET FAC. Dynamic Power Dissipation of TGMM GNRFET FAC is 17% less than Dynamic Power Dissipation of TGMM CNTFET FAC. Static Power Dissipation of TGMM GNRFET FAC is 19% less than TGMM CNTFET FAC Static Power Dissipation.

MNM GNRFET FAC has a 99% better speed performance than the MNM CNTFET FAC. Dynamic Power Dissipation of MNM GNRFET FAC is 11% less than the MNM CNTFET FAC. Static power Dissipation of MNM GNRFET FAC is 5% less than that of the MNM CNTFET FAC, From the overall comparison the MNM GNRFET based FAC is a better one.

MNMM GNRFET FAC has a 99% better speed performance than the MNMM CNTFET FAC. Dynamic Power Dissipation of MNMM GNRFET FAC is 7% less than the MNMM CNTFET FAC. Static power Dissipation of MNMM GNRFET FAC

Table 1 Comparative analysis of different FACs in new technologies

Parameters	TGM		TGMM		MNM		MNMM	
	CNTFET	GNRFET	CNTFET	GNRFET	CNTFET	GNRFET	CNTFET	GNRFET
Propagat ion delay (ns)	0.0224	0.0212	0.0210	0.0153	2.1069	0.0179	1.3286	0.01182
Dynamic power (nw)	32.933	23.430	19.963	16.435	66.928	59.235	27.097	25.021
Static power (nw)	33.657	25.061	20.967	16.909	61.381	58.138	26.558	23.432

is 11% less than that of the MNMM CNTFET FAC, From the overall comparison, the MNMM GNRFET based FAC is a better one.

Finally from the table it is clear that TGMM GNRFET FAC is better than TGM GNRFET FAC and MNMM GNRFET FAC is better than MNM GNRFET FAC. GNRFET technology is better when compared to CNTFET technology.

5 Conclusions

This paper presents the Design and Comparative Analysis of different FACs using CNTFET and GNRFET Technologies and the design is simulated using the Synopsys HSPICE tool at 16 nm technology with a supply voltage of 0.85 V. The performances of all these designs are compared in terms of Propagation Delay, Static Power Dissipation and Dynamic Power Dissipation. The TGM 1-bit FAC is analyzed in CNTFET and GNRFET technologies. The TGM 1-bit FAC is designed by using 25 transistors whereas in the TGMM FAC which is newly designed by using only 23 number of transistors. The TGMM FAC is designed by modifying the MUX logic circuitry. The MNM 1-bit FAC is analyzed in the CNTFET and GNRFET technology and it is designed by using 26 transistors. The MNMM 1-bit FAC is analyzed in the CNTFET and GNRFET technologies and it is designed by using only 18 number of transistors. Thus, proposed designs TGMM and MNMM FACs are best in terms of power dissipations, delay, and the transistor count.

References

1. Maleknejad M, Mohammadi S, Mirhosseini SM, Navi K, Naji HR, Hosseinzadeh M (2018) A low-power high-speed hybrid multi-threshold full adder design in CNFET technology. Springer Science+Business Media, LLC, part of Springer Nature
2. Ghanatghestani MM, Ghavami B, Salehpour H (2017) A CNFET full adder cell design for high-speed arithmetic units. Turkish J Electr Eng Comput Sci 25:2399–2509
3. Deng J, Wong HSP (2006) A circuit-compatible SPICE model for enhancement mode carbon nanotube field effect transistors. Int Conf Simul Semicond Process Devices, 166–169
4. Somineni RP, Sai YP, Leela SN (2015) Low leakage CNTFET full adders. In: IEEE 2015 global conference on communication technologies (GCCT), pp 174–179, 23–24 Apr 2015
5. Somineni RP, Jaweed SM (2017) Design of low power multiplier using CNTFET. In: 2017 IEEE 7th international advance computing conference (IACC-2017), Hyderabad, pp 556–559, 05–07 Jan 2017
6. Amini-Valashani M, Ayat M, Mirzakuchaki S (2018) Design and analysis of a novel low-power and energy-efficient 18T hybrid full adder. Microelectron J 74:49–59
7. Stanford University CNFET Model. http://nano.stanford.edu/model.php?id=26
8. Chen YY, Sangai A, Gholipour M, Chen D Graphene nano-ribbon field-effect transistors as future low-power devices. In: Symposium on low power electronics and design
9. Mishra M, Singh RS, Imran A (2017) Performance optimization of GNRFET Inverter at 32 nm technology node. ScienceDirect Mater Today Proc 4:10607–10611

An Efficient Hardware Architecture for Deblocking Filter in HEVC

P. Kopperundevi and M. Surya Prakash

Abstract This paper proposes a new hardware architecture for deblocking filter in a high efficiency video coding (HEVC) system. The proposed hardware is designed by using mixed pipelined and parallel processing architectures. The pixels are processed in the stream of two blocks of 4×32 samples in which edge filters are applied vertically in a parallel fashion for the processing of luma and chroma samples. These pixels are transposed and reprocessed through the vertical filter for horizontal filtering in a pipelined fashion. Finally, the filtered block will be transposed back to the original direction. The proposed filter is implemented using Verilog HDL, and the design is synthesized using the GPDK 90 nm technology library. Experimental results show that the proposed deblocking filter architecture achieves similar or up to two times higher throughput compared with the existing architectures while occupying a moderate chip area and consuming relatively low logic power. The proposed architecture supports the real-time deblocking filter operation of $4k \times 2k$ @60 fps under the clock frequency of 125 MHz with a gate count of 110K.

Keywords Deblocking filter · Application-specific integrated circuit · Hardware architecture

1 Introduction

HEVC is an international standard for video compression, developed by a working group of ISO/IEC Moving Picture Experts Group (MPEG) and ITU-T Video Coding Experts Group (VCEG), jointly published as ISO/IEC 23008-2 and ITU-T Recommendation H.265 in April 2013 [1]. It is stated that the performance of HEVC is identical as that of former H.264/AVC with 50% reduced bit rate [2]. This reduction in bit rate is mainly due to the introduction of more flexible and adaptive basic cod-

P. Kopperundevi (✉) · M. Surya Prakash
Department of Electronics and Communication Engineering, National Institute of Technology Calicut, Calicut, Kerala, India
e-mail: suryaprakash@nitc.ac.in

M. N. Favorskaya et al. (eds.), *Innovations in Electrical and Electronic Engineering*,
Lecture Notes in Electrical Engineering 661, https://doi.org/10.1007/978-981-15-4692-1_46

ing units (CUs). In H.264/AVC, a picture is divided into 16 x 16 sample fixed size macroblocks, but in HEVC, a picture is divided into 16 × 16, 32 × 32, or 64 × 64 samples coding units. Subsequently, these CUs can be divided into predictive units (PUs) and act as a root of the quadtree transform. Each transform quadtree child node defines a transform unit (TU). The video compression efficiency achieved in HEVC standard is not a result of any single feature but rather a combination of several encoding tools. One of those tools is the deblocking filter (DF). The DF is intended to reduce the blocking artifacts due to block-based coding. The computational complexity of HEVC is less when compared to the H.264/AVC. It consumes one-fifth of the overall computational complexity of the HEVC video codec. Some consumer electronics for video applications require high-definition quality videos to be processed with low power dissipation and high-speed specifications. To meet these needs, it is essential that the algorithms related to HEVC standard are implemented on hardware by developing computationally efficient architectures. Therefore, the development of an efficient architecture on application-specific integrated circuit (ASIC) and field-programmable gate array (FPGA) is of great importance before making commercial prototypes. The hardware implementations of DF by making use of parallel processing, pipelining, and memory reuse techniques in real-time applications have a higher demand for reducing power dissipation and execution time. In this work, an efficient DF architecture is designed to filter video frames by partitioning each frame into the 32 × 32 block, and then, filtering operation is performed on its multiple edges without considering the previous blocks. The deblocking architecture is inspired by an unified cross unit mentioned in [3]. In this, each block is filtered in two steps: (i) Parallel edge filtering is performed on eight edges of 4 × 8 adjacent multiple sample grid blocks in one sub block of 8 × 32. (ii) Multiple 8 × 32 sub block filtering operation in a 32 × 32 block. The largest coding unit (LCU) 64 × 64 block is processed based on quarter-LCU. The proposed design is synthesized using Xilinx 12.5 at 125 MHz and implemented in ASIC using the 90 nm library, and the RTL simulations of deblocking architecture are verified using ModelSim SE 6.5b. The proposed hardware ensures that it is capable of decoding 60 full UHD video frames per second. The proposed work results in a moderate area as well as power due to the hybrid parallel and pipeline processing architecture and also architecture-independent optimizations of equations.

The rest of this paper is organized as follows: Overview of DF algorithm is presented in Sect. 2. Section 3 provides information related to conventional architectures of the DF in HEVC. Proposed DF hardware architecture is presented in Sect. 4. Implementation results and comparisons with the existing techniques are presented in Sect. 5.

2 Deblocking Filter Algorithm

The deblocking filter is applied to all samples adjacent to a PU or TU boundaries which are aligned on an 8 × 8 sample grid except at picture boundary. Unlike in H. 264 /AVC where deblocking filter is applied on 4 x 4 sample grid, in HEVC deblock-

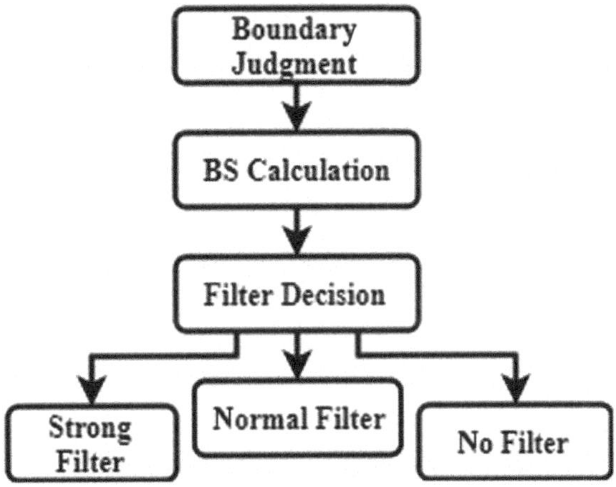

Fig. 1 Processing stages of a deblocking filter

ing filter is applied on vertical boundaries first followed by horizontal boundaries on 8 x 8 sample grid. Figure 1 illustrates the overall deblocking filter processing steps. Thus, it processes through the stages of boundary judgement, boundary strength (BS) calculation, and finally filtering decision and operation.

2.1 *Boundary Judgement*

The initial step in deblocking filter is to proceed with whether the current boundary belongs to a boundary of PU, TU, CU. The filtering should not be applied to the current boundary if either of them exits. Figure 3(b) shows the horizontal boundary to be filtered, as the vertical boundary is analogous.

2.2 *Boundary Strength and Filtering Parameters*

In HEVC, calculation of boundary strength reflects how strong the filtering is needed for the boundary. The value of BS is an integer ranging from 0 to 2, and it depends on transform parameter, motion vector (MV), reference frame, and coding information. The filtering operation is performed when BS > 0 and BS > 1 for luma and chroma samples, respectively [4]. Table 1 shows the different criteria for evaluating the boundary strength values of the edges required for filtering. Thresholds values of β and tc are decided from the predefined table, according to the BS value and the quantization parameter (QP) of P and Q blocks.

Table 1 Boundary strength calculation

Condition	BS value
Any one of the P and Q blocks is an intra block	2
At least one of the P and Q blocks has transform coefficients	1
Absolute difference between the motion vectors(MVs) that belongs to the P and Q blocks is more are equal to one integer luma sample	1
Motion prediction in the P and Q blocks indicates different reference picture or total number of reference motion vectors is different	1
Others	0

2.3 Filtering Decisions and Operation

Filtering of boundary is not required when estimation of BS value is zero. Deblocking filtering is performed for block boundaries with an associated boundary strength value greater than zero and for which Eq. (1) holds.

$$\begin{aligned} &|p20 - 2p10 + p00| + |p23 - 2p13 + p03| + \\ &|q20 - 2q10 + q00| + |q23 - 2q13 + q03| < \beta \end{aligned} \tag{1}$$

The filtering is processed further based on the value of tc and β, those are threshold values obtained based on quantization parameter (QP). HEVC has two modes of deblocking filter, i.e., strong filtering mode and normal filtering mode. The deblocking filter switches between the strong and the normal filtering mode on the basis of the signal characteristics for each block boundary of 4×8 samples.

3 Related Works

Few works have been reported on developing the hardware architecture for deblocking filter. For HEVC, as used in H.264/AVC, the deblocking filter (DF) is applied to 8×8 blocks instead of 4×4 blocks. Sequential dependency occurs with H.264/AVC block edge filtering as the block size is 4×4. This is the major difference between H.264/AVC and HEVC deblocking filters. It is therefore possible to achieve more parallelism with HEVC deblocking filters [3]. As a consequence, several efforts have recently centered on developing effective parallel architecture for HEVC filter deblocking [4–11]. In Ozcan et al. [6], the first HEVC DF hardware uses two data paths in parallel. In which, it performs edge filtering decisions and operations simultaneously in order to increase its performance.

It was found that the architecture achieves the throughput of 1920 × 1080 @30 fps at 86 Mhz when implemented by ASIC 90nm CMOS technology with a gate count of 16.4K (without including on-chip memory). In [7], the hardware architecture for deblocking filter based on interleaved processing is presented. The design is the outcome of the reduction in intervening data storage of pixel blocks which can process 4k @30 fps. In [8], a new parallel zigzag order processing scheme is used to enhance the parallelism by dissolving the data dependency between adjacent filtering operations. This architecture supports 1080p HD processing at 60fps. Further, Shen et al. [9] presented an efficient architecture that processes a block of 32 × 32. In order to possess efficient handling of data access for both vertical and horizontal filtering, it uses memory interlaced technique, and it is a four-stage pipelined architecture. The hardware was synthesized in 0.13 μm technology, with a gate count of 75K. It achieves a throughput of 4k × 2k @30 fps at 200 MHz clock frequency. A four-stage pipeline architecture was introduced by Shen et al. [12] to filter a 64 × 64 block. It has been synthesized in 65 nm technology with a gate count of 103.3 K operating at 200 MHz. Finally, Hsu et al. [10] proposed a six-stage pipelined two-line deblocking filter with high processing throughput and low latency, in which it is capable of decoding 4k × 2k resolution at 60 fps under the clock frequency of 100 MHz with a gate count of 466.5K. It is evident from the literature that the design of high processing technology is of great significance. Hence, an efficient architecture is proposed in this work in terms of throughput.

4 Proposed Hardware Architecture

The proposed hardware architecture for the deblocking filter is shown in Fig. 2. It consists of an input-output buffer for storing boundary parameters, filtering parameters, and input-output samples. Block memory is used for storing the transposed

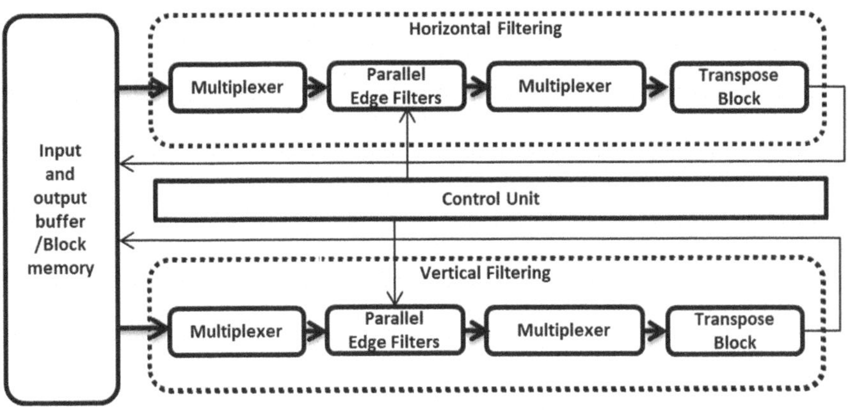

Fig. 2 Proposed hardware architecture of deblocking filter

block pixels, eight edge filters in which it filters the samples parallely, multiplexer to select the edges at particular timing instant.

The proposed architecture is designed based on the fact that a video frame, i.e, an image is partitioned into 32 × 32 blocks. Each 32 × 32 block is first divided into four rows of 8 × 32 blocks. Each 8 × 32 block contains two rows of 4 x 32 blocks. Two adjacent 4 × 8 samples are selected as an input to each of the edge filters from the block of 8 × 32 samples via multiplexer. Then, these samples are filtered horizontally by taking eight parallel edge filters. Since one adjacent 4 × 8 block filtering is independent of other edge filtering, filtering operations are performed using parallel edge filters. Eight edges present in 8 × 32 block are filtered using edge filters and then transposed using transpose block, and output is stored in block memory. The same procedure is followed for filtering the four rows of 8 × 32 block for horizontal filtering in a pipelined manner. Unified cross unit [3] approach is used in order to avoid an extra edge filter and also to reduce the delay. For horizontal (vertical edge) filtering and vertical (horizontal edge) filtering, separate units of edge filters are used. The proposed architecture processes a complete LCU in the form of quarter-LCU. The subsequent sections explain individual models of deblocking filter.

4.1 *Input and Output Buffer*

Horizontal and vertical filters are provided with the information about filtering parameters, boundary parameters as well as filtering samples from the input-output buffer. These parameters are sent synchronously with respect to the clock for each 4 × 8 sample to the edge filters.

4.2 *Edge Filter*

The adjacent samples of size 4 × 8 from two rows of 8 × 32 sub block in 32 × 32 block are the input to the edge filters in which it performs parallel filtering operation on the entire 32 × 32 block. The filtering parameters and boundary strength values are sent to edge filters synchronously with reference to clock from the buffer. The edge filters are designed by using the HEVC standard equations that are subjected to architecture-independent optimizations, such as equations that facilitate the reuse of the same operations multiple times. Further, multiplication operations are replaced with shift-add operations, and since parallel edge filters are employed, these architecture-independent optimizations along with these shift-add operations will lead to low chip area.

The flowchart explains the operation of edge filter that is applicable to filter the vertical edge is shown in Fig. 3a. The filtering operation for the horizontal edge is analogous. Edge filter operation will take place only if BS > 0 is satisfied. Edge

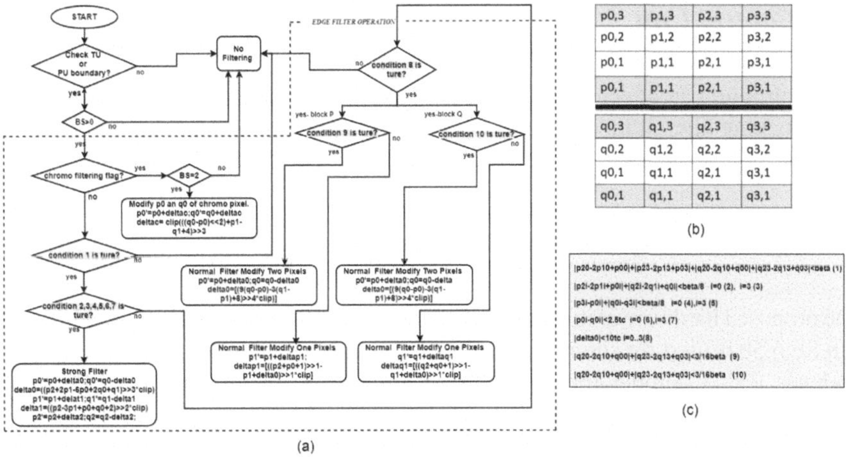

Fig. 3 Edge filter operation (**a**). Example for two 4 × 4 with block with horizontal edge (**b**). Conditions 1–7 for edge filter (**c**) operation

filter performs the chroma filtering block when the following two conditions are met. (i) The particular boundary should belong to chroma block. (ii) The value of BS should be equal to two. Otherwise, further conditions have to be evaluated as shown in Fig. 3c. Specifically, condition (1) examines the pixels at first and fourth rows/columns of an edge to decide whether the filtering operations need to continue. If (1) is satisfied, as well as (2)–(7), strong filtering will be applied to that particular edge. Furthermore, if (1) is satisfied, but (2)–(7) are not, (8) is checked, if it is satisfied, it implies to apply normal filtering. Finally, (9) and (10) are examined to perform operations in the second level of pixels, i.e., to decide whether to modify two pixels or one pixel at each P and Q block, respectively. If any of those conditions fails, it leads to no filtering operation.

4.3 Transpose Block

Once the edge filter operation is completed (vertical egde), the transpose operation of these samples is performed by considering 4 × 8 block as input via multiplexer and obtain 8 × 4 as output. The output of transpose block is stored in block memory of size 32 × 32. The same transpose block operation is carried out at the output of vertical filter (horizontal edge) block to get back to the original orientation. Two such transpose blocks are used to transpose 32 × 32 blocks in terms of eight 4 × 8 samples as input at regular instances of time.

4.4 Control Unit

The control unit architecture is relatively simply due to the distinct data flow nature of deblocking filter. It monitors the edge filter operations of deblocking filter.

5 Implementation Result and Comparison

The proposed hardware architecture is implemented using Verilog HDL, and Modelsim SE 6.5b is used for simulation purpose. Gate-level synthesis is performed using Cadence RTL Compiler for which 90 nm technology libraries are used. Physical design is performed using Cadence SOC Encounter. The proposed design consists of a mechanism to decode 64×64 LCU in the form of quarter-LCU. The proposed DF hardware architecture achieved low power and moderate area due to architecture-independent optimizations of standard HEVC equations, such as equations reformulations to facilitate the reuse of many operations using the same operators, replacing multiplications with add/shift operations, and also hardware-specific optimizations, such as determining the sizing of registers and operators, the number of datapaths in parallel, balancing of operating stages, and scheduling. Table 2 shows the performance comparison of our hardware architecture with existing architectures, and the postlayout result for the proposed deblocking filter is shown in Fig. 4.

When compared to the architectures proposed by Ozcan et al. [6] and Le et al. [8], the throughput of our architecture is high due to the reduction in the number of cycles/LCU of our architecture. However, the architecture of Oczan et al. [6] is designed in such a way that it can process only luma samples with a resolution of 1080p @30 fps. Furthermore, the proposed design supports identical resolution

Table 2 Comparison with previous designs

	Ozcan et al. [13]	Fang et al. [4]	Le et al. [5]	Shen et al. [3]	Hsu et al. [14]	Proposed work
Process (nm)	90	90	65	130	90	90
Technology	ASIC	ASIC	ASIC	ASIC	ASIC	ASIC
Gate count (NAND2 x1)	16.4K	30K	41.77K	75K	30K	110.3K
Frequency (MHz)	108/86	141.5	226	28	100	125
Resolution	1080p	4k × 2k	1080p	4k × 2k	4k × 2k	4k × 2k
LCU size	64 × 64	16 × 16	32 × 32	32 × 32	64 × 64	32 × 32
Throughput (fps)	30	30	60	30	60	60

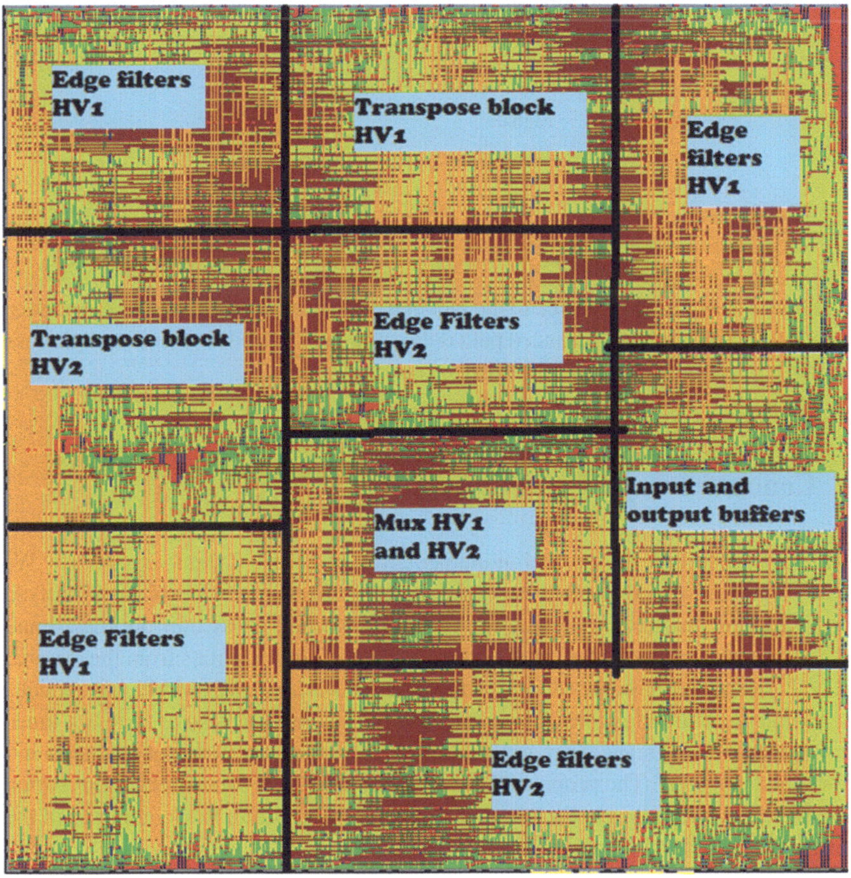

Fig. 4 Postlayout result of deblocking filter

as well as throughput as that of architecture given in [10] with moderate increase in hardware resource utilization. This is due to the fact that our design consists of parallel as well as pipelined architecture. The design in [7, 9] occupies lesser amount of area but much reduced throughput compared with our proposed DF.

Table 3 shows the results of power estimation. Power analysis for the proposed architecture is estimated using Xilinx Xpower analyzer tool where the design is mapped to virtex 6 FPGA board with a speed grade of −3. A DF hardware power consumption can be split into three main categories: logic power, clock power, and signal power. Logic power is the amount of power dissipated in the parts where computations take place. Clock power is due to clock tree used in the FPGA. Signal power is the power that is dissipated between logic blocks in routing tracks. Our architecture consumes up to three times low logic power when compared with architectures pre-

Table 3 Power consumption results of the proposed architecture

Category	Power (mW)
Clock	46
Logic	3
Signals	5
IOs	2
Total	56

sented in [6, 14, 15]. This is due to the fact that our hardware architecture is designed in such a way that it has a balanced path delay, and thus, it leads to minimum glitches in circuit, ultimately leads to low power consumption.

6 Conclusions and Future Works

This paper describes design and implementation of a HEVC deblocking filter. We have employed a methodology to design an optimized hardware by using mixed pipelined and parallel processing architectures, unified cross unit approach, and also architecture-independent and hardware-specific optimizations from the original HEVC standard equations. Simulation results are verified with MATLAB models. Experimental result illustrates that our architecture achieves similar or up to two times throughput than the state or art architecture, with moderate amount of chip area and power consumption. The proposed DF can be used for real-time applications which demands decoding of 4k × 2k resolution at 60fps. The future work concerns in the design of area-efficient architecture.

References

1. ITU-T and ISO/IEC (2013) High efficiency video coding, document H.265 and ISO/IEC 23 008-2:2013, ITU-T Recommendation, Geneva, Switzerland
2. Sullivan GJ, Ohm JR, Han WJ, Wiegand T (2012) Overview of the high efficiency video coding (HEVC) standard. IEEE Trans Circ System Video Technol 22(12):1649–1668
3. Shen W, Shang Q, Shen S, Fan Y, Zeng X (2013) A high-throughput VLSI architecture for deblocking filter in HEVC. In: 2013 IEEE international symposium on circuits and systems (ISCAS2013). IEEE, pp 673-676
4. Fang CC, Chen IW, Chang TS (2015) A hardware-efficient deblocking filter design for HEVC. In: 2015 IEEE international symposium on circuits and systems (ISCAS), pp 1786–1789
5. Le HHN, Bae J (2014) High-throughput parallel architecture for H. 265/HEVC deblocking filter. J Inf Sci Eng 30(2):281–294
6. Cheng W, Fan Y, Lu Y, Jin Y, Zeng X (2015) A high-throughput HEVC deblocking filter VLSI architecture for 8k × 4k application. In: 2015 IEEE international symposium on circuits and systems (ISCAS). IEEE, pp 605–608

7. Norkin A, Bjontegaard G, Fuldseth A, Narroschke M, Ikeda M, Andersson K, Zhou M, Van der Auwera G (2012) HEVC deblocking filter. IEEE Trans Circ Syst Video Technol 22(12):1746–1754
8. Ostermann J et al (2004) Video coding with H.264/AVC: tools performance and complexity. IEEE Circ Syst Mag 4(1):7–28
9. Srinivasarao BKN, Chakrabarti I, Ahmad MN (2015) High-speed low-power very-large-scale integration architecture for dual-standard deblocking filter. IET Circ Dev Syst 9(5):377–383
10. Peesapati R, Das S, Baldev S, Ahamed SR (2017) Design of streaming deblocking filter for HEVC decoder. IEEE Trans Consum Electron 63(3):1–9
11. Li M, Zhou J, Zhou D, Peng X, Goto S (2012) De-blocking filter design for HEVC and H.264/AVC. In: Proceedings of Advanced Multimedia Information Processing, pp 273–284
12. Shen W, Fan Y, Bai Y, Huang L, Shang Q, Liu C, Zeng X (2016) A combined deblocking filter and SAO hardware architecture for HEVC. IEEE Trans Multimedia 18(6):1022–1033
13. Ozcan E, Adibelli Y, Hamzaoglu I (2013) A high performance deblocking filter hardware for high efficiency video coding. IEEE Trans Consum Electron 59(3):714–720
14. Hsu PK, Shen CA (2016) The VLSI architecture of a highly efficient deblocking filter for HEVC systems. IEEE Trans Circ Syst Video Technol 27(5):1091–1103
15. Parlak M, Hamzaoglu I (2008) Low power H. 264 deblocking filter hardware implementations. IEEE Trans Consum Electron 54(2):808–816

Differentiator-Based Universal TAM Filter Topology Using Operational Floating Current Conveyors

Ashmin Gangal, Varun Bhanoo, and Neeta Pandey

Abstract This work puts forward a universal transadmittance mode (TAM) filter topology based on a differentiator approach using an analog building block known as operational floating current conveyor (OFCC). The proposed topology uses three OFCC blocks, two grounded capacitances, and three grounded resistances. This circuit has a number of advantages which make it highly suitable for a number of applications. All the passive elements of the circuit are grounded which make it highly favorable from fabrication point of view. The circuit has high input as well as high output impedance. The proper impedances of the circuit make it suitable for interfacing between voltage mode and current mode circuits. The pole frequency of the circuit is electronically tunable using MOS transistor-based implementation of grounded resistances. SPICE simulations using 0.5 μm technology parameters from MOSIS (AGILENT) are provided.

Keywords Electronically tunable · OFCC · Transadmittance filter · Universal filter

1 Introduction

In continuous time signal processing, analog filters play an important part of partial or total suppression of any undesirable portion from a signal, and therefore, it is a field of continuous research. There are different modes of analog filters, namely voltage mode (VM), current mode (CM), transadmittance mode (TAM), and transimpedance mode (TIM). Among all these filters, TAM filters have significant importance. TAM filters accept voltage input and deliver the filtered current output. This

A. Gangal (✉) · V. Bhanoo · N. Pandey
Department of Electronics and Communication, Delhi Technological University, New Delhi, India
e-mail: ashmingangal@gmail.com

V. Bhanoo
e-mail: varunbhanoo@gmail.com

N. Pandey
e-mail: n66pandey@rediffmail.com

M. N. Favorskaya et al. (eds.), *Innovations in Electrical and Electronic Engineering*, Lecture Notes in Electrical Engineering 661, https://doi.org/10.1007/978-981-15-4692-1_47

property of TAM filters is used in interfacing circuits between voltage mode and current mode. This conversion from voltage to current can also be combined with signal processing feature in this stage itself for efficient usage in later stages. One of the other applications of TAM filters is the receiver baseband block used in radio systems [1]. There are a number of TAM filters [1–22] reported previously in the literature. All of the reported filters are realized using the following active blocks: second-generation current conveyor (CCII) [1], differential difference current conveyor (DDCC) [2], operational floating current conveyor (OFCC) [3], voltage differencing transconductance amplifier (VDTA) [4, 12, 13, 15, 17, 21], current controlled current conveyor transconductance amplifier (CCCCTA) [5, 14], multiple output second-generation current controlled conveyor (MOCCCII) [6, 9], differential voltage current conveyor transconductance amplifier (DVCCTA) [7], current differencing transconductance amplifier (CDTA) [8], current conveyor transconductance amplifier (CCTA) [10], digitally current controlled differential voltage current conveyor (DCCDVCC) [11], current feedback operational amplifier (CFOA) [16], dual output-operational transconductance amplifier (DO-OTA) [18], operational transconductance amplifier (OTA) [19], third-generation current conveyor (CCIII) [20], and operational amplifier (OP-AMP) [22].

Following are the features of previously reported TAM filters:

- The filters reported in [1, 4, 11–14, 16, 18, 20–22] provide less than or equal to three filtered responses.
- More than three ABB are used in [11, 16, 18].
- Floating components are used in [1, 2, 8, 10, 11, 13–16, 20–22] making the circuit unfavorable for fabrication purposes.
- Proper input and output impedance are not achieved in [1–3, 8, 10, 14–16, 19, 20, 22].

TAM filters reported [1–22] in the recent past use either a differentiator-based approach or an integrator-based approach. Differentiator-based filters are useful in applications such as biomedical instrumentation [23] and linear control systems [24] and thus, a few differentiators-based TAM filters are reported previously [1, 14, 22] but have limitations such as improper impedances, limited availability of filtered outputs, and existence of floating elements in the circuit. The proposed work focuses on a differentiator-based approach having advantages such as proper impedances employing only grounded elements providing all the five responses along with the feature of electronic tunability of the pole frequencies.

This paper has six sections including the introduction. Section 2 describes the analog building block (ABB) used, i.e., OFCC. Section 3 presents an OFCC-based differentiator used later to obtain the proposed universal filter. Section 4 describes the proposed OFCC-based universal TAM Filter. Section 5 describes the simulation results. Conclusions drawn are discussed in Sect. 6.

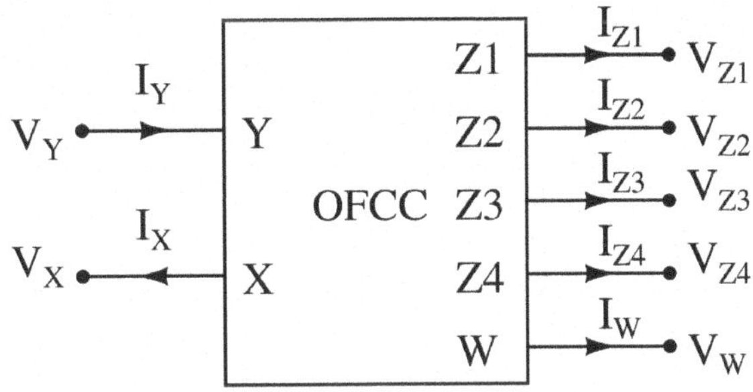

Fig. 1 OFCC symbolic notation

2 The OFCC

OFCC [25–27] is an ABB having a total of seven terminals out of which, two are input terminals and rest five are output terminals. Its symbolic notation is shown in Fig. 1.

The two input terminals of the OFCC block, *X* and *Y*, have low and high input impedance, respectively, making it appropriate for current and voltage inputs, respectively. Output port *W* is a low output impedance port having voltage equal to the multiplication of open-loop transconductance gain Z_t and the current at port *X*. All other output ports have high output impedance where $Z1$ and $Z2$ copy the current of port *W* with no phase shift, while $Z3$ and $Z4$ copy the current of port *W* with a 180° phase shift. The port relationships of the block denoted by terminal equations are given in (1).

$$\begin{bmatrix} I_Y \\ V_X \\ V_W \\ I_{Z1} \\ I_{Z2} \\ I_{Z3} \\ I_{Z4} \end{bmatrix} = \begin{bmatrix} 0 & 0 & 0 & 0 & 0 & 0 & 0 \\ 1 & 0 & 0 & 0 & 0 & 0 & 0 \\ 0 & Z_t & 0 & 0 & 0 & 0 & 0 \\ 0 & 0 & 1 & 0 & 0 & 0 & 0 \\ 0 & 0 & 1 & 0 & 0 & 0 & 0 \\ 0 & 0 & -1 & 0 & 0 & 0 & 0 \\ 0 & 0 & -1 & 0 & 0 & 0 & 0 \end{bmatrix} \begin{bmatrix} V_Y \\ I_X \\ I_W \\ V_{Z1} \\ V_{Z2} \\ V_{Z3} \\ V_{Z4} \end{bmatrix} \quad (1)$$

3 Differentiator Using OFCC

Differentiator is a vital building block which is useful in signal conditioning applications and to design higher-order filters depending upon the application. OFCC-based differentiator is shown in Fig. 2. It consists of an OFCC block, a resistance, and a capacitance. The functionality of the circuit is described in Eq. (2).

Fig. 2 Differentiator using OFCC

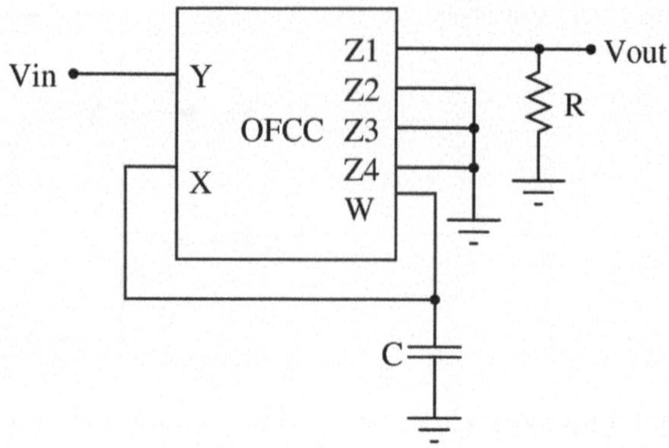

$$\frac{V_{\text{out}}}{V_{\text{in}}} = sRC \tag{2}$$

4 Proposed Filter

The proposed filter is designed using three OFCC blocks, out of which the first OFCC block acts as the gain stage cascaded with two OFCC-based differentiators which are described in Sect. 3. The filter is depicted in Fig. 3 and is capable of producing all the five responses—low pass (LP), high pass (HP), band pass (BP), band reject (BR), and all pass (AP) response whose transfer functions are given in (3)–(7), respectively, where the denominator $D(s)$ is described in (8).

$$\frac{I_{\text{LP}}}{V_{\text{in}}} = -\frac{\frac{1}{R_1 R_2 R_3 C_1 C_2}}{D(s)} \tag{3}$$

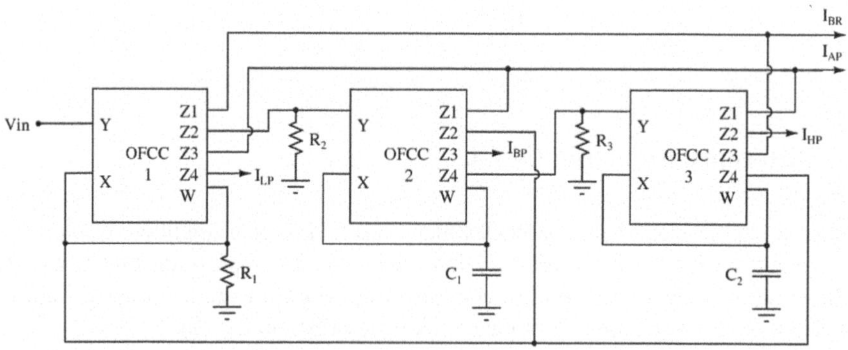

Fig. 3 Proposed universal TAM filter

$$\frac{I_{\text{HP}}}{V_{\text{in}}} = -\frac{s^2/R_1}{D(s)} \tag{4}$$

$$\frac{I_{\text{BP}}}{V_{\text{in}}} = -\frac{\frac{s}{R_1 R_3 C_2}}{D(s)} \tag{5}$$

$$\frac{I_{\text{BR}}}{V_{\text{in}}} = \frac{\frac{s^2}{R_1} + \frac{1}{R_1 R_2 R_3 C_1 C_2}}{D(s)} \tag{6}$$

$$\frac{I_{\text{AP}}}{V_{\text{in}}} = -\frac{\frac{s^2}{R_1} - \frac{s}{R_1 R_3 C_2} + \frac{1}{R_1 R_2 R_3 C_1 C_2}}{D(s)} \tag{7}$$

$$D(s) = s^2 + \frac{s}{R_3 C_2} + \frac{1}{R_2 R_3 C_1 C_2} \tag{8}$$

The filter is characterized by the pole frequency (ω_o), bandwidth (BW), and the quality factor (Q_o) as described in (9)–(11), respectively.

$$\omega_o = \frac{1}{\sqrt{R_2 R_3 C_1 C_2}} \tag{9}$$

$$\text{BW} = \frac{1}{R_3 C_2} \tag{10}$$

$$Q_o = \sqrt{\frac{R_3 C_2}{R_2 C_1}} \tag{11}$$

The grounded resistances in the circuit can be realized using MOS-based resistance [28]. Electronic tunability in the proposed circuit is performed by changing the gate voltage of the MOS-based resistances, which results in the change of the cut-off frequency of the filter responses.

5 Simulation Results

The proposed universal filter is designed using CMOS implementation of OFCC [29] from MOSIS (AGILENT) 0.5 μm technology parameters. The schematic is depicted in Fig. 4, and the aspect ratio of the transistors is taken from [29]. V_{DD} and V_{SS} in the CMOS implementation are considered as ±1.5 V and the bias voltages (V_{b1} and V_{b2}) are considered as ±0.8 V. Simulations are performed using SPICE.

The AC analysis is done using the resistance values as $R_1 = R_2 = R_3 = 1$ kΩ and the capacitance values as $C_1 = C_2 = 10$ nF. The magnitude responses of the LP, HP,

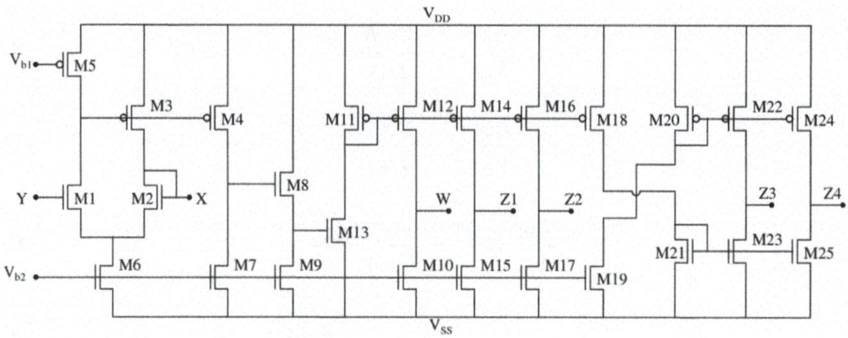

Fig. 4 CMOS implementation of OFCC

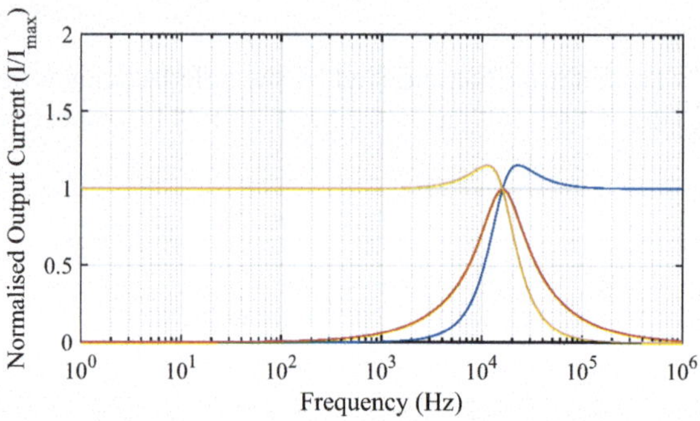

Fig. 5 LP, HP, and BP response

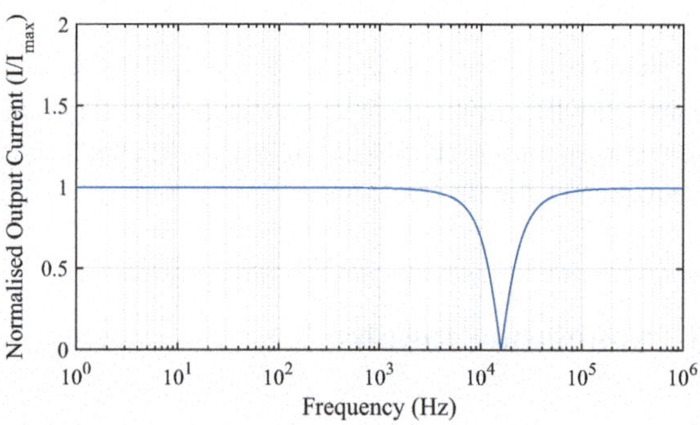

Fig. 6 BR response

and BP are shown in Fig. 5. The magnitude response of BR filter is shown in Fig. 6. The AP response is shown in Fig. 7.

To illustrate the electronic tunability pole frequencies of the filter, resistances (R_2 and R_3) are varied by changing the voltages in the MOS transistor-based resistance in a low pass response whose results have been shown graphically in Fig. 8. The simulation results of the electronic tunability experiment are given in Fig. 9.

Frequency spectrum analysis is done to check the efficiency of the proposed filter. It is carried out using a mixed frequency voltage input of amplitude 50 mV and

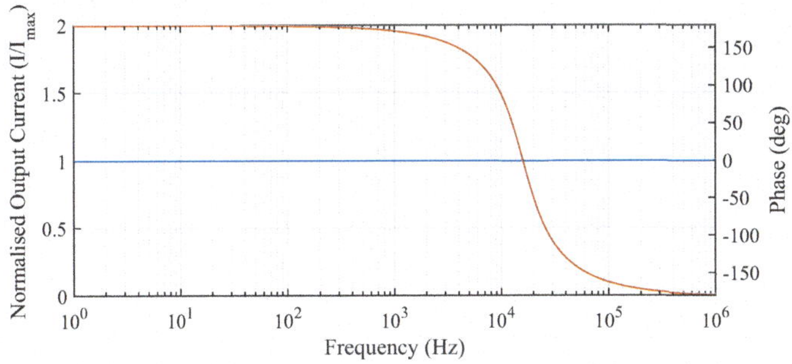

Fig. 7 AP response

Fig. 8 Electronic tunability of frequency of the filter

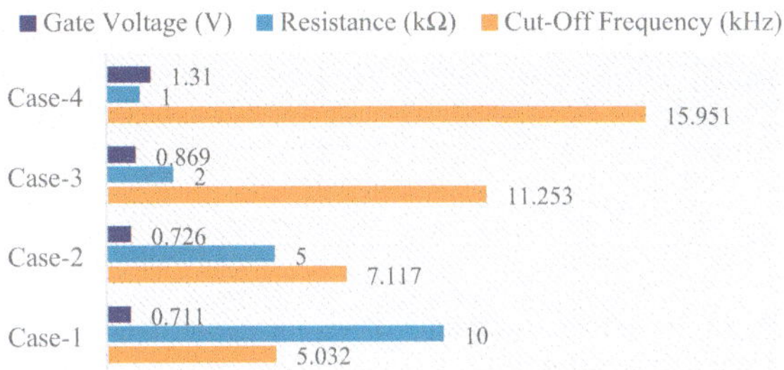

Fig. 9 Variation in cut-off frequency of a low pass response with resistance values

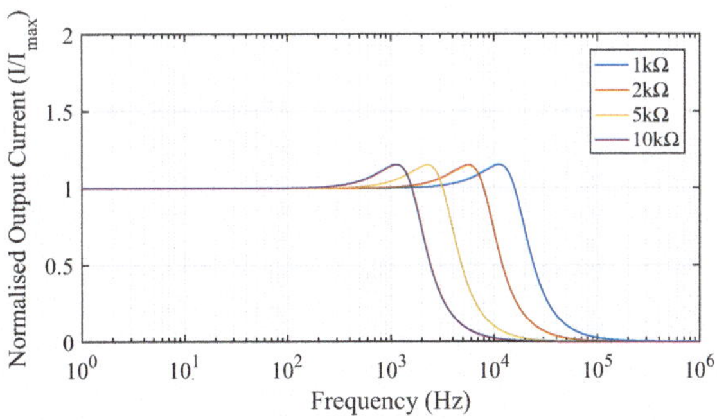

having frequencies 2, 20, and 100 kHz. The frequency spectrum of all the output currents is depicted in Fig. 10. As observed in Fig. 10, all the responses adhere to the expected frequency spectrums.

To check the effect of mismatch in passive components, statistical analysis is performed using Monte Carlo statistical analysis using the resistance nominal values as $R_1 = R_2 = R_3 = 1\ \text{k}\Omega$ and the capacitance nominal values as $C_1 = 20$ nF, $C_2 = 10$ nF. To study this effect, HP response has been simulated with 5% Gaussian deviation in all the resistances, in all the capacitances and in both the resistances and capacitances present in the circuit whose frequency response after 40 runs are depicted in Figs. 11, 12, and 13, respectively.

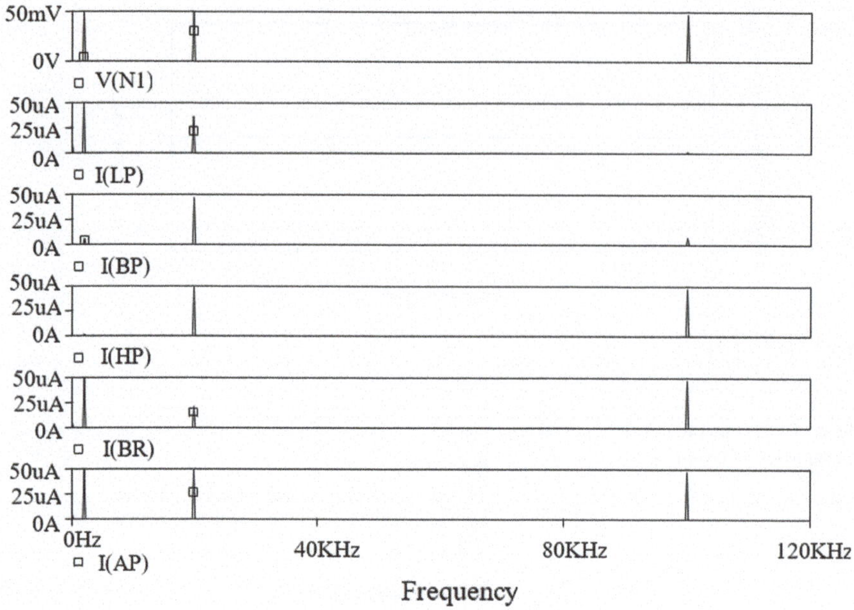

Fig. 10 Frequency spectrum analysis for mixed frequency voltage input along with all output filtered response currents

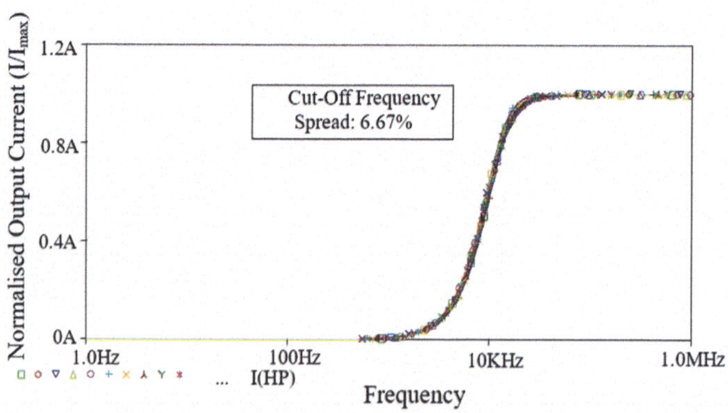

Fig. 11 Simulated HP response with 5% variation in all resistances

6 Conclusions

This paper proposes a universal TAM filter topology based on a differentiator approach using OFCC as the ABB. It provides all the five filtered responses as current outputs. All the passive elements of the circuit are grounded which makes it highly favorable from fabrication point of view by reducing the parasitic capacitances

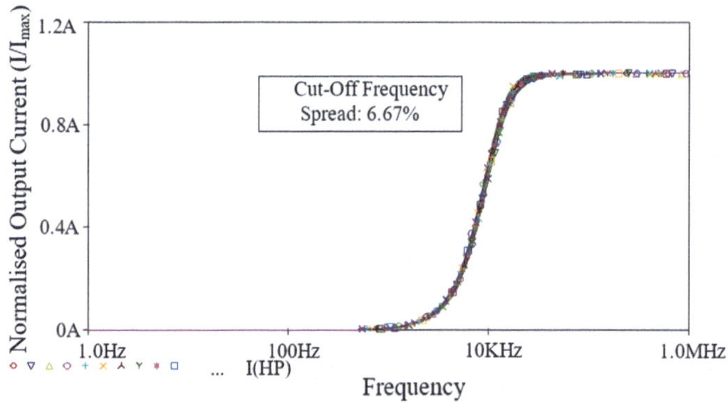

Fig. 12 Simulated HP response with 5% variation in all capacitances

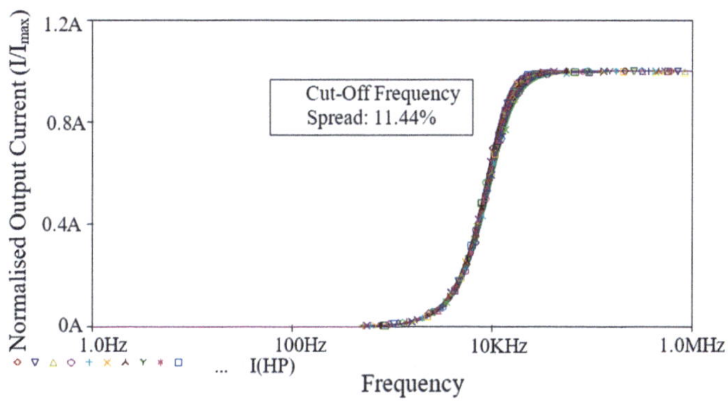

Fig. 13 Simulated HP response with 5% variation in all resistances and capacitances

that make the circuit more favorable for integrated circuit (IC) implementations. The pole frequency of the circuit is electronically tunable using MOS transistor-based implementation of grounded resistances. The circuit has high input and high output impedance. The proper impedances of the circuit make it suitable for interfacing between VM and CM circuits. The functionality of the proposed filter has been verified by SPICE simulations.

References

1. Toker A, ÇiÇekoglu O, Özcan S, Kuntman H (2001) High-output-impedance transadmittance type continuous-time multifunction filter with minimum active elements. Int J Electron 88(10):1085–1091
2. Liao WB, Gu JC (2011) SIMO type universal mixed-mode biquadratic filter

3. Pandey N, Nand D, Khan Z (2014) Operational floating current conveyor-based single-input multiple-output transadmittance mode filter. Ara J Sci Eng 39(11):7991–8000
4. Yeşil A, Kaçar F (2013) Electronically tunable resistorless mixed mode biquad filters. Radioengineering 22(4):1016–1025
5. Chen HP, Yang WS (2017) Electronically tunable current controlled current conveyor transconductance amplifier-based mixed-mode biquadratic filter with resistorless and grounded capacitors. Appl Sci 7(3):244
6. Paul SK, Pandey N, Bhattacharyya A (2009) Current controlled conveyor based transadmittance mode universal filter. In: 2009 IEEE symposium on industrial electronics & applications, vol. 2. IEEE, pp 764–767
7. Singh SV (2019) High input impedance trans-admittance mode biquad universal filter employing DVCCTAs and grounded passive elements. Indian J Pure Appl Phys (IJPAP) 57(1):52–62
8. Shah NA, Quadri M, Iqbal SZ (2007) CDTA based universal transadmittance filter. Analog Integr Circ Sig Process 52(1–2):65–69
9. Pandey N, Paul SK (2010) SIMO transadmittance mode Active-C universal filter. Circ Syst 1(02):54
10. Tomar R (2015) A new trans-admittance-mode biquad filter suitable for low voltage operation. Int J Eng 28(12):1738–1745
11. Beg P, Maheshwari S, Siddiqi MA (2013) Digitally controlled fully differential voltage-and transadmittance-mode biquadratic filter. IET Circ Dev Syst 7(4):193–203
12. Prasad D, Srivastava M, Bhaskar DR (2014) Transadmittance type universal current-mode biquad filter using VDTAs. Int Sch Res No
13. Shankar C, Singh SV (2016) Single VDTA based multifunction transadmittance mode biquad filter. Int J Eng Technol (IJET) 7(6):2180–2188
14. Singh SV, Maheshwari S, Mohan J, Chauhan DS (2009) An electronically tunable SIMO biquad filter using CCCCTA. In: International conference on contemporary computing. Springer, Berlin, pp 544–555
15. Shankar CHANDRA, Singh SV (2015) A new trans-admittance mode biquad filter using MO-VDTA. WSEAS Trans Circ Syst 14:8–18
16. Shah NA, Iqbal SZ, Parveen B (2004) SITO high output impedance transadmittance filter using FTFNs. Analog Integr Circ Sig Process 40(1):87–89
17. Gupta G, Singh SV, Bhooshan SV (2015) VDTA based electronically tunable voltage-mode and trans-admittance biquad filter. Circ Syst 6(03):93
18. Ibrahim MA, Kuntman H (2003) A novel transadmittance-type KHN-biquad employing DO-OTA with only two grounded capacitors. WSEAS Trans Circ Syst 2:400–403
19. Senani R (2005) New OTA-C universal current-mode/trans-admittance biquads. IEICE Electron Express 2(1):8–13
20. Çam U (2005) A new transadmittance type first-order all pass filter employing single third generation current conveyor. Analog Integr Circ Sig Process 43(1):97–99
21. Shankar C, Singh SV (2015) A low voltage operable VDTA based biquad filter realizing band pass and high pass filtering functions in trans-admittance-mode. In: International conference on computing, communication & automation. IEEE, pp 1288–1293
22. Shah NA, Iqbal SZ, Parveen B (2005) Lowpass and bandpass transadmittance filter using operational amplifier pole. AEU-Int J Electron Commun 59(7):410–412
23. Northrop RB (2003) Analysis and application of analog electronic circuits to biomedical instrumentation. CRC press
24. d'Azzo JJ, Houpis CD (1995) Linear control system analysis and design: conventional and modern. McGraw-Hill Higher Education
25. Toumazou C, Payne A (1991) Operational floating conveyor. Electron Lett 27(8):651–652
26. Khan AA, Al-Turaigi MA, El-Ela MA (1994) Operational floating current conveyor: characteristics, modelling and applications. In: Conference proceedings. 10th anniversary. IMTC/94. Advanced technologies in I & M. 1994 IEEE instrumentation and measurement technology conference (Cat. No. 94CH3424-9). IEEE, pp 788–791

27. Ghallab YH, El-Ela MA, Elsaid MH (1999) Operational floating current conveyor: characteristics, modelling and experimental results. In: ICM'99. Proceedings. Eleventh international conference on microelectronics (IEEE Cat. No. 99EX388). IEEE, pp 59–62
28. Wang Z (1990) 2-MOSFET transresistor with extremely low distortion for output reaching supply voltages. Electron Lett 26(13):951–952
29. Hassan HM, Soliman AM (2005) Novel CMOS realizations of the operational floating conveyor and applications. J Circ Syst Comput 14(06):1113–1143

Information Hiding Using Private Content-Based Random Encryption and Data Reduction for RGB Images

Amit Kumar Shakya, Ayushman Ramola, and Anurag Vidyarthi

Abstract In this research work, we have successfully reduced the content present in red-green-blue (RGB) image for secure transmission, for this purpose, we have used a private content-based image encryption (PCBIE), model. The PCBIE model works on three basic steps image encryption followed by first-order statistical parameter calculation for the encrypted image and original image and finally a comparison of the obtained parameters with the original image parameters. Through this methodology, one can easily prevent the original identity of the image. Here, we have taken four different categories of images and performed encryption on them, in the result, we have obtained that identity of the image is totally hidden and one can identify the original image only on the basis of the parameter value. The first-order statistical parameter includes the mean, variance, and standard deviation.

Keywords Private content · Encryption · Mean · Variance · Standard deviation

1 Introduction

Data robbery is one of the potential threats in the current World, as trillions of gigabits of data are generated every day [1]. Today, with the advancement of technology, everything has become online and this has activated a new group of cyber robbers that are engaged in the activity of cyber mugging and unethical hacking [2]. Several incidents are reported daily where someone confidential data is robbed for the purpose of ransom. Nowadays, several organizations are engaged in creating advanced software's so that problems related to data theft can be stopped or prevented. In this research work, we have been motivated by such advanced software creators who are creating software to prevent data from digital robbery we have developed an algorithm

A. K. Shakya (✉) · A. Ramola
Department of Electronics and Communication Engineering, Sant Longowal Institute of Engineering and Technology, Sangrur, Punjab, India
e-mail: xlamitshakya.gate2014@ieee.org

A. Vidyarthi
Graphic Era (Deemed to Be University), Dehradun, Uttarakhand, India

M. N. Favorskaya et al. (eds.), *Innovations in Electrical and Electronic Engineering*, Lecture Notes in Electrical Engineering 661, https://doi.org/10.1007/978-981-15-4692-1_48

so that our data (images) can be easily hidden and does not reveal their original identity. They can only be identified on the basis of their first-order statistical parameters, and finally, a comparison between the original and encrypted images is performed only on the basis of these parameters. Several scientists have been working on the scheme of data hiding through private content-based image query [3] and advanced private content-based image query [4]. Besides this data hiding techniques include spatial domain data hiding [5], prediction error expansions [6], transfer domain data hiding [7, 8], steganalysis [9], DWT-based data hiding [10], LSB coding [11], digital watermarking [12], etc., in literature, some image compression based on region of interest is also considered as an data hiding scheme [13, 14]. In this research work, we have first all created encrypted images from the original images, followed by plotting of histograms for original and encrypted image, and finally, a comparison is made in between statistical parameters for original and encrypted images.

2 Mathematical Relation of First-Order Statistical Parameters

First-order statistical parameters include mean, variance, and standard deviation the mathematical representations of the formulas are listed in Eq. (1–3), where N is the total number of pixels within an image, μ represents the mean value, x is the statistical parameter, e.g., for an 8-bit image it will have 2^8 values.

2.1 *Mean (μ)*

It is the measure of the central tendency of a probability distribution of the image pixels widely distributed within an image.

$$\mu = \frac{1}{N} \sum_{\text{first}=0}^{\text{last}=N-1} x_{[\text{first}]-[\text{last}]} \tag{1}$$

2.2 *Variance $(\sigma)^2$*

It is the mean difference between the average values of the image pixel with individual pixel value.

$$\sigma^2 = \frac{1}{N-1}\sum_{\text{first}=0}^{\text{last}=N-1}\left(x_{[\text{first}]-[\text{last}]} - \mu\right)^2 \tag{2}$$

2.3 *Standard Deviation (σ)*

It is the representation of how pixel values are spread within an image.

$$\sigma = \sqrt[2]{\frac{1}{N-1}\sum_{\text{first}=0}^{\text{last}=N-1}\left(x_{[\text{first}]-[\text{last}]} - \mu\right)^2} \tag{3}$$

These parameters are calculated for the original image and the encrypted image and later compared with each other and similarity in terms of numerical values is obtained from the image pixels.

Flowchart of the proposed methodology

See Fig. 1.

3 Experimental Results

In this experiment, we have taken six different RGB images which are shown in Fig. 2. The images include Lena and Mandril which are taken from Digital Image Processing, Third Edition, Gonzalez and Woods [15]. The Red Planet, Jupiter, Asteroid and Moon are the courtesy of European Space Agency ESA [16].

Now, the images are converted into their encrypted version by arranging pixels in the original image in a random manner so that every pixel gets at a new coordinate location. The obtained image represents a noisy picture which does not reveal the original identity of the image. Thus, for the original image, their encrypted versions can be used and can be transmitted to a new place without revealing the original identity of the image. Thus, for secure transmission purposes, these images can be used. The encrypted version of the images is shown in Fig. 3.

Now, the histogram plot for both original and encrypted images is plotted together where the original image contains a total number of pixels and the encrypted image contains only "Ten" percent pixels. The histogram for the original and encrypted image both are shown in Fig. 4. Later, the first-order statistical parameters are computed for both original and encrypted images. The parameters may attain the same value for both original and encrypted image, respectively, the change in the statistical parameter values can be obtained from Tables 1 to 2.

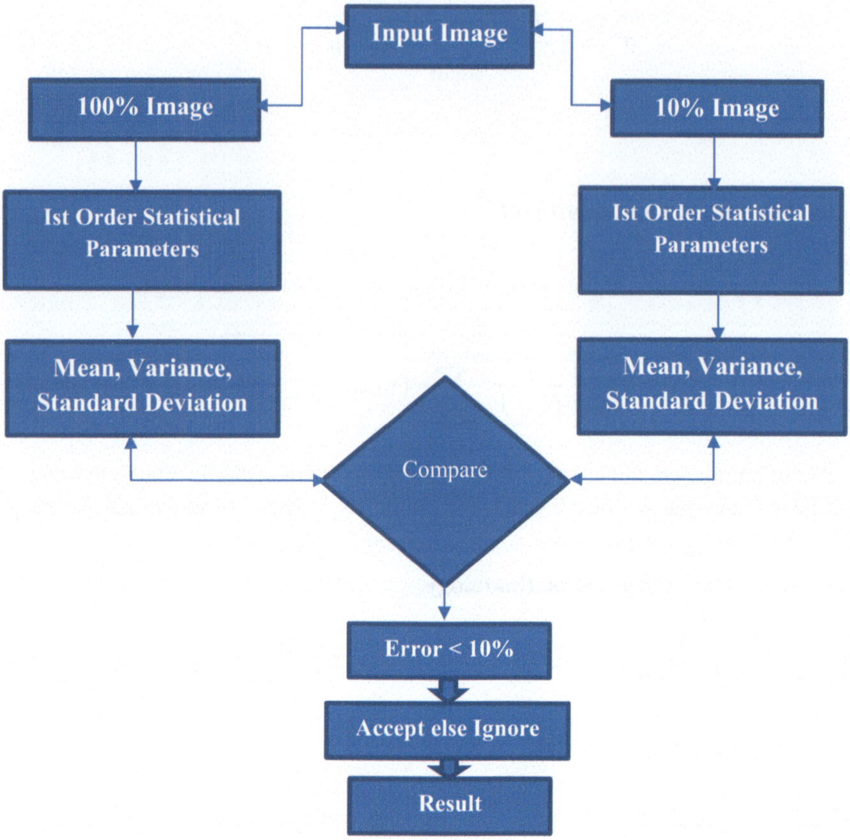

Fig. 1 Flowchart of the proposed methodology

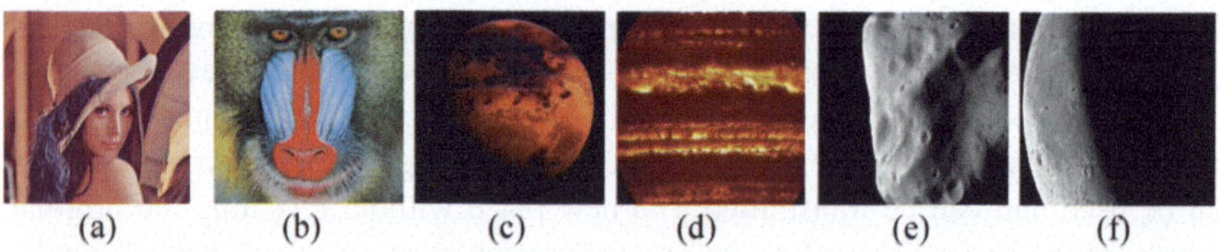

Fig. 2 **a** Lena (Color 512), **b** Mandril (Color 512), **c** Red Planet, **d** Jupiter, **e** Asteroid, **f** Moon

Now, the comparison of the first-order statistical parameters is performed to obtain the numerical similarity in the first-order statistical parameters.

Tables 1 and 2 represent the statistical parameter values for both original and encrypted images which show similarity in the feature value.

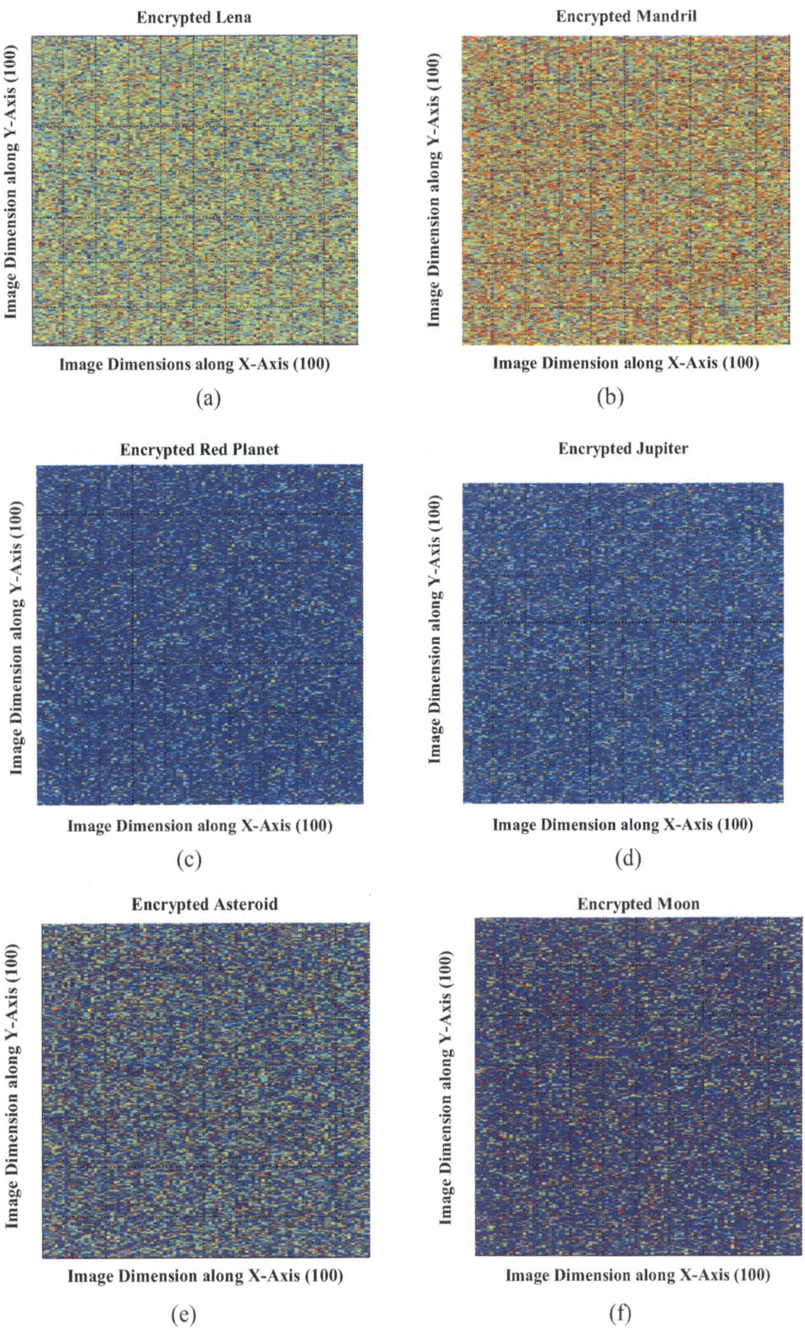

Fig. 3 **a** Encrypted Lena, **b** encrypted Mandril, **c** encrypted Red Planet, **d** encrypted Jupiter, **e** encrypted Asteroid, **f** encrypted Moon

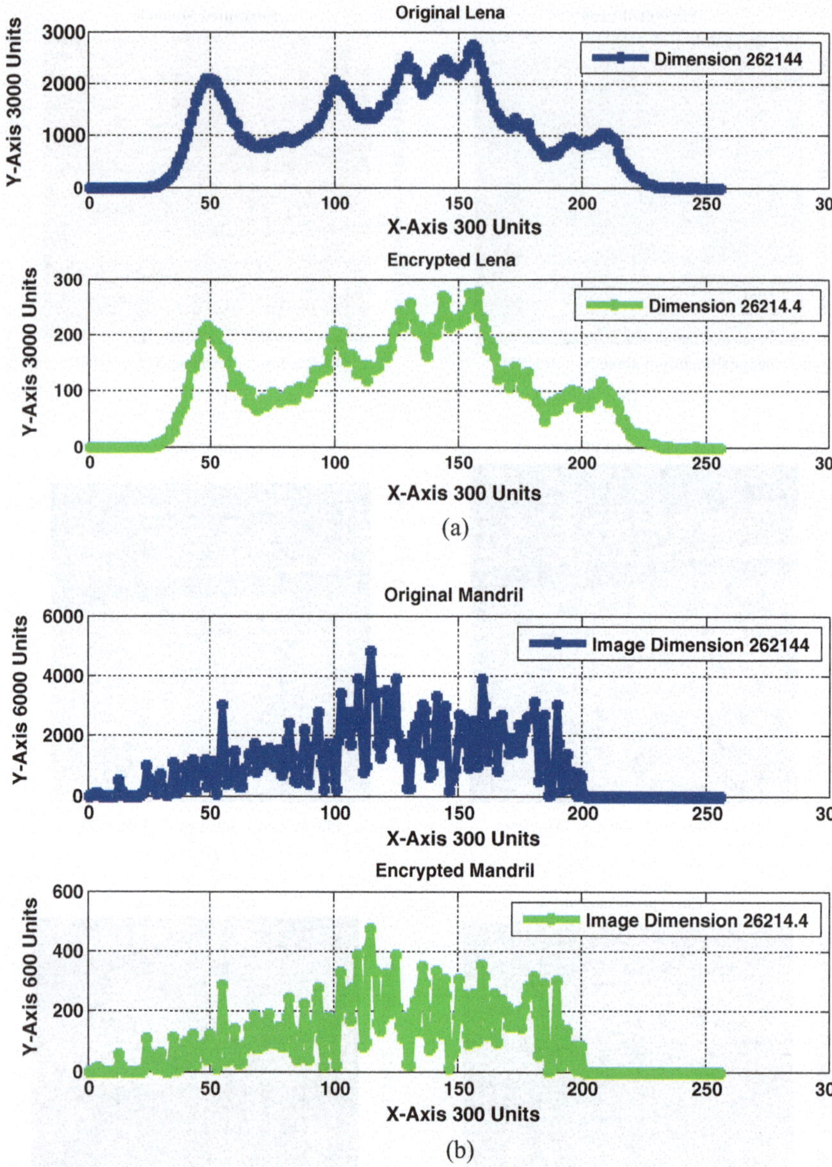

Fig. 4 Histogram signature plot for the original and encrypted version of **a** Lena (Color 512), **b** Mandril (Color 512), **c** Red Planet, **d** Jupiter, **e** Asteroid, **f** Moon

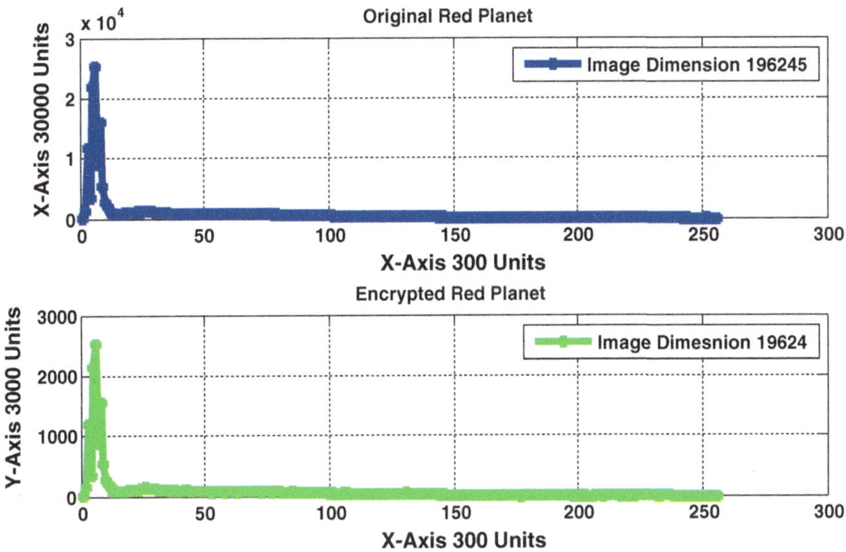

(c)

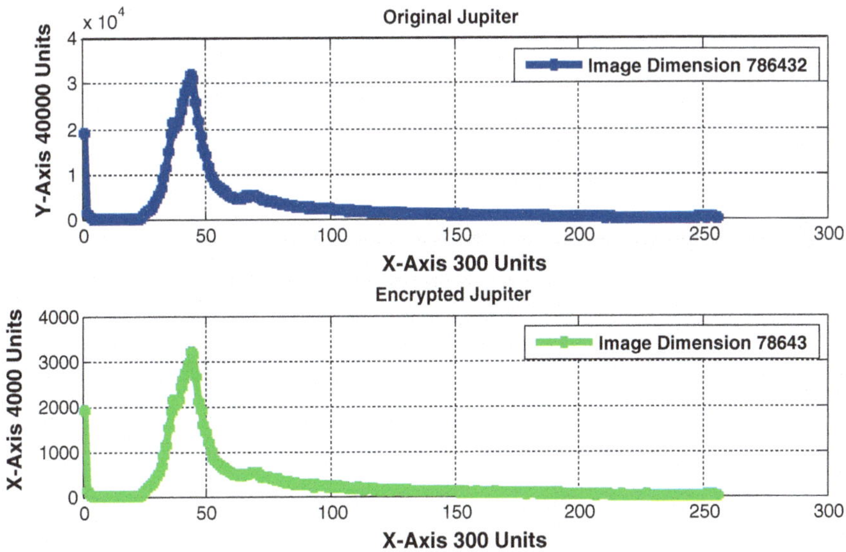

(d)

Fig. 4 (continued)

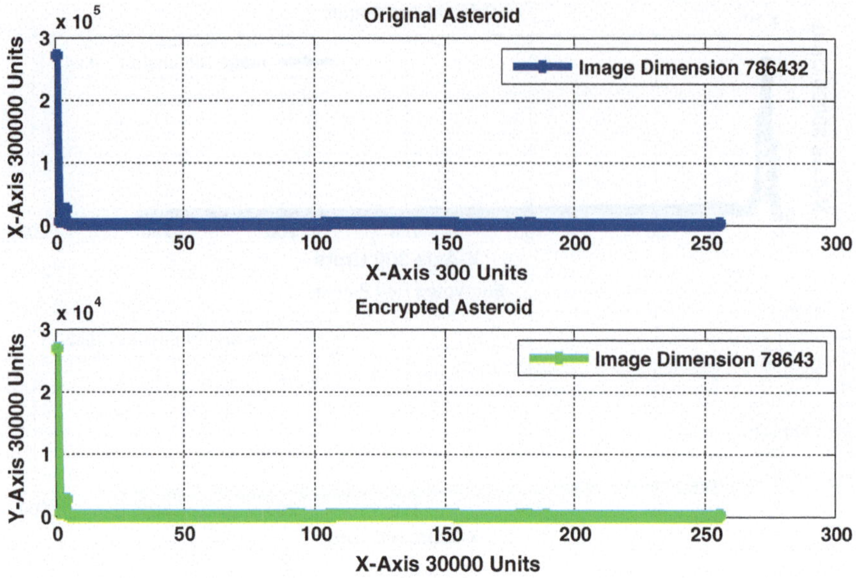

(e)

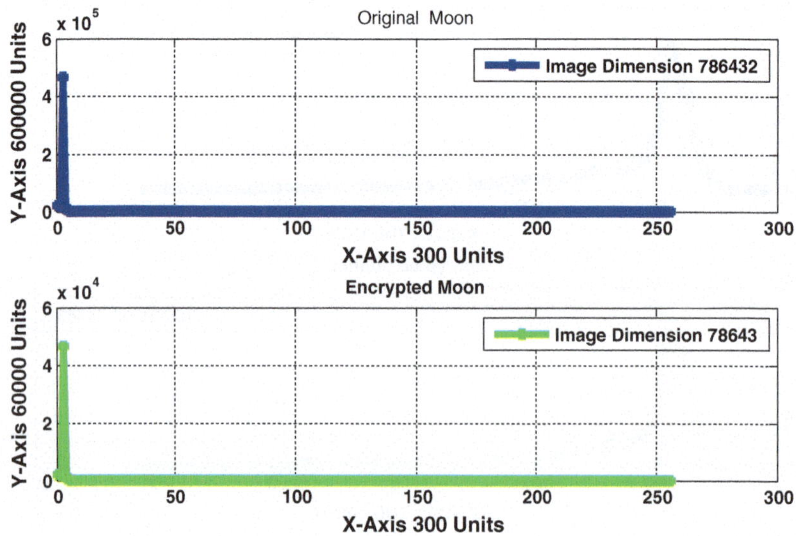

(f)

Fig. 4 (continued)

Table 1 First-order statistical parameters for original image

S.No	Image	Total pixels	Mean	Variance	Standard deviation
1	Lena	262,144	124.043	10.7258	3.27502
2	Mandril	262,144	122.091	10.4112	3.22663
3	Red Planet	196,245	35.8831	8.78709	2.96430
4	Jupiter	786,432	64.4904	8.04031	2.83554
5	Asteroid	786,432	73.4906	10.4829	3.23773
6	Moon	786,432	48.3714	8.71656	2.95238

Table 2 First-order statistical parameters for the encrypted image

S. No	Image	Total pixels	Mean	Variance	Standard deviation
1	Lena	26,214	123.996	10.7193	3.27403
2	Mandril	26,214	122.156	10.4164	3.22744
3	Red Planet	19,624	36.002	8.80579	2.97622
4	Jupiter	78,643	64.4339	8.0198	2.83192
5	Asteroid	78,643	73.6827	10.4732	3.23623
6	Moon	78,643	48.4295	8.72467	2.95375

4 Conclusions

The outcome of this research work is the development of a new methodology for data hiding. The prevention of data robbery is also prevented through this approach. The methodology is easy and the encryption is easy. The limitation of this approach is if two or more images develop the same type of statistical features, then it may get difficult to identify the original image.

Acknowledgements The authors are thankful to the European Space Agency (ESA) for the satellite images used in this research work.

References

1. Kettimuthu R, Liu Z, Wheeler D, Foster I, Heitmann K, Cappello F (2018) Transferring a petabyte in a day. Elsevier Future Gener Comput Syst 88(1):191–198
2. Ibarra J, Jahankhani H, Kendzierskyj S (2019) Cyber-physical attacks and the value of healthcare data: facing an era of cyber extortion and organised crime, pp 115–137
3. Shakya A, Ramola A, Kandwal A, Chamoli V (2018) Privacy-preserving random permutation of image pixels enciphered model from cyber attacks for covert operations. In: Communications in computer and information science I end, vol. 905. Springer, Singapore, Dehradun
4. Shakya A, Ramola A, Pokhariya H, Kandwal A (2019) Fusion of IoT and machine learning approach to prevent confidential data from digital crimes and cyber mugging for covert

transmission. In: Lecture notes in electrical engineering. Springer, Singapore, Sikkim, pp 563–579
5. Fangjun H, Xiaochao Q, Joong K, Jiwu H (2016) Reversible data hiding in JPEG images. IEEE Trans Circ Syst Video Technol 26(9):1610–1621
6. Hu X, Zhang W, Li X, Yu N (2015) Minimum rate prediction and optimized histograms modification for reversible data hiding. IEEE Trans Inf Forensics Secur 10(3):653–664
7. Ramola A, Shakya A (2017) Private content-based image query system using statistical properties. In: IEEE international conference on innovations in control, communication and information systems (ICICCI), vol. 1. Noida, pp 1–6
8. Wang L, Pan Z, Zhu R (2017) A novel reversible data hiding scheme using SMVQ prediction index and multi-layer embedding. Springer Multimed Tools Appl 76(I):26225–26248
9. Xia Z, Wang X, Sun X, Liu Q, Xiong N (2016) Steganalysis of LSB matching using differences between nonadjacent pixels. Springer Multimed Tools Appl 75(4):1947–1962
10. Weng C-Y (2019) DWT-based reversible information hiding scheme using prediction-error-expansion in multimedia images. Springer Peer-to-Peer Network Appl 1–10
11. Papadopoulos N, Psannis K (2018) Sequential multiple LSB methods and real-time data hiding: variations for visual cryptography ciphers. Springer J Real-Time Image Process 14(1):75–86
12. Nagai Y, Uchida Y, Sakazawa S, Satoh S (2018) Digital watermarking for deep neural networks. Springer Int J Multimed Inf Retrieval 7(1):3–16
13. Tang Z, Xu S, Yao H, Qin C, Zhang X (2019) Reversible data hiding with differential compression in encrypted image. Springer Multimed Tools Appl 78(8):9691–9715
14. Shakya A, Ramola A, Pandey D (2017) Polygonal region of interest-based compression of DICOM images. In: IEEE international conference on computing, communication, and automation (ICCCA), vol. 1. Noida, pp 1035–1040 (2017)
15. Gonzalez R, Woods R (2008) Digital image processing, 3rd edn. Pearson
16. Observatory E Breakthrough watch and the European Southern Observatory achieve “first light” on the upgraded planet-finding instrument to search for Earth-like planets in the nearest star system. In: ESA. Available at: https://www.eso.org/public/news/eso1911/

An Algorithm Design for Anomaly Detection in Thermal Images

Chhavi Mishra, T. Bagyammal, and Latha Parameswaran

Abstract Anomaly detection is useful in diverse domains including fault detection system, health monitoring, intrusion detection, fraud detection, emotion recognition, cancer detection, animal rescue, detecting ecosystem disturbances, and event detection in sensor networks. Thermal image is a widely used night vision technology. Anomaly detection using thermal image features has been proposed in this work. Three major classes of features, namely textural features, color features, and shape features, have been extracted. Correlation model has been used for detecting anomalies. Thermal image of perishable objects has been analyzed, and the evaluation result confirms the hypothesis. It is found that using a set of features while using correlation as similarity measure the achieved average recall is 76.06%.

Keywords Thermal images · Anomaly detection · Correlation

1 Introduction

Thermal imaging method is useful to capture radiation of the objects and create the image out of it. Thermal camera is basically a heat sensor that captures and record small differences in temperature [1]. Thermal imaging is a means of convalescing visibility of objects in a dark environment by finding out the objects infrared energy and creating an image based on that information. In those environments (background) where objects cannot be easily differentiated, thermal camera easily detects them and they appear as distinctive objects in a thermal image; because objects are

C. Mishra (✉) · T. Bagyammal · L. Parameswaran
Department of Computer Science and Engineering, Amrita School of Engineering, Amrita Vishwa Vidyapeetham, Coimbatore, India
e-mail: cb.en.p2cse18003@cb.students.amrita.edu

T. Bagyammal
e-mail: t_bagyammal@cb.amrita.edu

L. Parameswaran
e-mail: p_latha@cb.amrita.edu

M. N. Favorskaya et al. (eds.), *Innovations in Electrical and Electronic Engineering*, Lecture Notes in Electrical Engineering 661, https://doi.org/10.1007/978-981-15-4692-1_49

seldom at the alike temperature as other objects around them. One of the biggest advantages is that it can see through smoke, fog, haze, and also, it can penetrate solid obstructer to some extent to capture images [2]. Thermal cameras have been in notice mainly for wildlife and armed forces applications. Over the time, resolution and image quality of thermal camera have increased which opens wide use in different fields such as manufacturing industry and medical industry. [2]. Out of many applications of thermal imaging, our focus is anomaly detection which is useful in different domains, such as fault detection, gas leakage detection, system health monitoring, cancer detection, emotion recognition, intrusion detection, fraud detection, animal rescue, event detection in sensor networks, and detecting ecosystem disturbances [9].

2 Related Works

Many researchers have worked on thermal images for a variety of applications. A few are listed in this section. In [1], the authors have given description of thermal camera and its ability to capture images. Thermal images are usually grayscale in nature where objects in white are hot and cold are black objects. Depth of gray indicates the variations between the hot and cold. In [2], the authors have described how thermal images can be used for identifying the objects in a complex background. When observed through a thermal imaging camera, warm objects stand out fine in opposition to cool backgrounds; human and other warm-blooded animals become effortlessly noticeable in the environment, be it day or night. When an object is hidden behind some obstacle or at the back of another object, change detection techniques based on visual imagery fails. The authors have developed an object detection method using the record of heat signature. Average filter and median filter were used to remove noise from thermal images. Otsu's threshold algorithm was used to extract the required segment. Outer boundary of the thermograph was detected using Canny edge detector. Viola–Jones object detection framework for matching the sample input with the features stored in the database was done [2]. In [3], the authors have described a model for anomaly detection in medical images using correlation. Anomalies from electrocardiograph (ECG) signal had been detected using threshold.

They have proposed a similarity-based function between signal attributes for the correlation, to consolidate the previously detected anomalies. Correlation algorithm that detects the different relation between the medical image anomalies has been proposed. Unnoticed hidden defects may cause the failure of the component which may lead to colossal damage to life and property [4]. In [4], algorithm to detect detects in metals using image processing-based concepts using IR images have been proposed. Images containing two types of defects, namely holes in pulsed thermography images and cracks in lock-in thermography, are considered. The proposed method consists of (1) Canny edge detection which includes the following steps: (i) video-to-frame conversion, (ii) preprocessing to reduce noise, iii) principal component thermography, (iv) Segmentation of defects, and (2) thresholding-based defect detection method uses the following steps: (i) histogram equalization, (ii) morphological opening and background subtraction, (iii) thresholding. Metals with flat bottom holes defects are detected by Canny edge detector-based methodology. Defects such as cracks that occur during welding are detected using thresholding-based methodology. In [5], algorithm to detect breast cancer in thermal images at an early stage has been proposed. The image is classified based on the features extracted from the region of interest to normal or abnormal using a deep neural network (DNN) classifier. Uncertainty in the output of DNN is further analyzed using support vector machine (SVM) classifier.

In [6], authors have proposed a method to distinguish orange fruit in sweet citrus tree using thermal images. Steps performed are (i) temperature to grayscale image conversion, (ii) contrast enhancement with a fuzzy intensification operator, (iii) image segmentation using minimum fuzzy divergence value and fruit detection using circle Hough transform. The proposed method gives better result for recognition of isolated fruit when compared to grouped fruits. In [7], authors proposed an algorithm for pedestrian detection in thermal infrared images with limited annotations. The main idea is to adapt the plenty of color images associated with bounding box annotations to the thermal domain for training the pedestrian detector. Algorithms used were VGG16 and RNN. In [8], authors have proposed an algorithm for segmentation of foot from infrared image using Otsu thresholding and morphological operations under three different image acquisition setups. Proposed algorithm was analyzed using the three measures such as Jaccard index, false positive rate, and false negative rate in comparison with the result of manual segmentation. From the segmentation done under each imaging conditions, it was found that, morphological operations combined with Otsu thresholding performed well at par with the ground truth or the manually segmented image. In [9], thermal image processing algorithms for damage detection of dual-phase high-performance steel material and damage analysis have been discussed. First-order and second-order derive features of the image were studied. Second-order derivative gave successful result, and the correlation coefficient trend was used to identify the amount of damage. The higher is the coefficient, the lowest is the damage. In [10], the authors presented how to detect human using infrared image processing technique. Combination of the pixel-gradient and body parts processing and three-phase classification process (head modeling, human modeling, and classifier) has been proposed to lessen the false detection. The presented algorithm has been tested on thermal images taken in a real environment. Some limitations of the model have been reported, as problem with detection groups of the overlapped people, and identification of some poses are not included in the dataset. In [11], the authors have proposed an algorithm for scene change detection by analyzing the linear relationship that exists between the corresponding regions saliency and wavelet features. The quality of thermal infrared radiation and imagery creates certain challenges to image analysis algorithms. Thus, it can be inferred that study and analysis of thermal imagery are a potential area of research.

Given a set of thermal images of the same scene object taken at varying time interval, objective of the proposed work is to detect changes or anomalies in objects. In order to detect changes, feature extraction methods have been applied. Three major classes of features, namely textural features, color features, and shape features, have been extracted. Correlation model has been used for detecting anomalies.

3 Dataset Description

3.1 Image Acquisition

For preparing a dataset, thermal images are captured using fluke thermal camera. The camera version is Tis40 which is used to create a dataset.

3.2 Dataset Generation

The dataset is collected in a controlled environment; the dataset collected consists of thermal images of different objects which has an element of heat and has characteristics of temperature variation. For experimentation, thermal images of perishable items (objects) have been collected as shown in Fig. 1. Thermal images were obtained at varying time stamps. Thus, each image has a time stamp of different time sequence such as apple has a time range starting from time 1 to time 2 on various dates. A dataset consisting of 3000 + thermal images has been constructed for this study of anomaly detection.

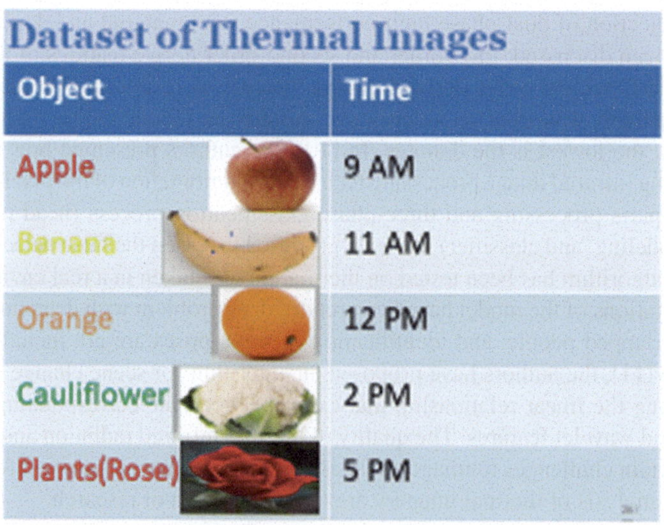

Fig. 1 Sample dataset at different time stamps

In Fig. 2, first five images of apple are taken at 5 pm and the apple is 5 days older is shown, and in next five images are of first day images that are taken at 10 am. Similarly, images are taken on various date and time. Thermal image of banana and flower is shown in Figs. 3 and 4, respectively. Figure 5 shows the grayscale thermal image of few objects from the dataset. Table 1 shows grayscale thermal image and normal image of objects at varying date/time intervals.

4 Proposed Anomaly Detection Algorithm

4.1 *Architecture*

Architecture of the proposed work on anomaly detection with an objective to detect if the image has anomalies is shown in Fig. 6. The proposed algorithm has steps as given below:

Algorithm:

(1) Capture the heat emission of various objects as thermal images at various date/time intervals.
(2) Preprocess the thermal images and extract texture, color, and shape signature features.
(3) Detect anomalies in thermal images using correlation model.
(4) Validate results with ground truth values.

4.2 *Feature Extraction*

Feature extraction entails reducing the amount of resources required to describe a large set of data. Features extracted in this proposed algorithm include textural features, color features, and shape features.

4.2.1 Textural Features

i. Graylevel difference statistics (GLDS)

GLDS [7] is based on the estimation of the probability density $P(i)$ of image pixel pairs at a given distance $\mathrm{d} = (\mathrm{d}x, \mathrm{d}y)$ having a certain absolute graylevel difference value i. Statistical measures computed on the basis of GLDS are given in Eqs. 1, 2, 3, and 4.

Mean: Mean indicates the average value in a GLDS matrix.

$$\mathrm{Mean}(\mu) = \sum_{\mathrm{i}=1}^{\mathrm{m}} i P(i) \tag{1}$$

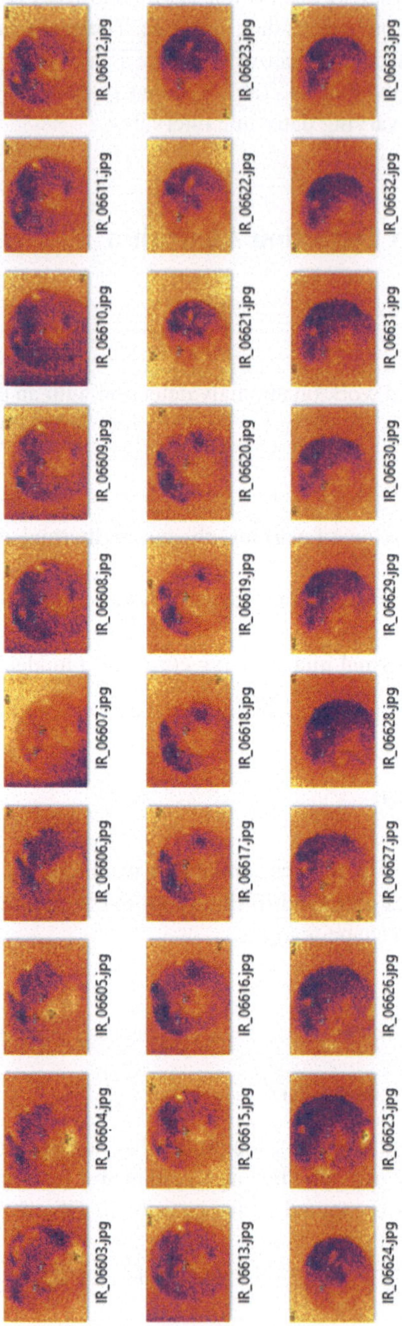

Fig. 2 Thermal images of apple

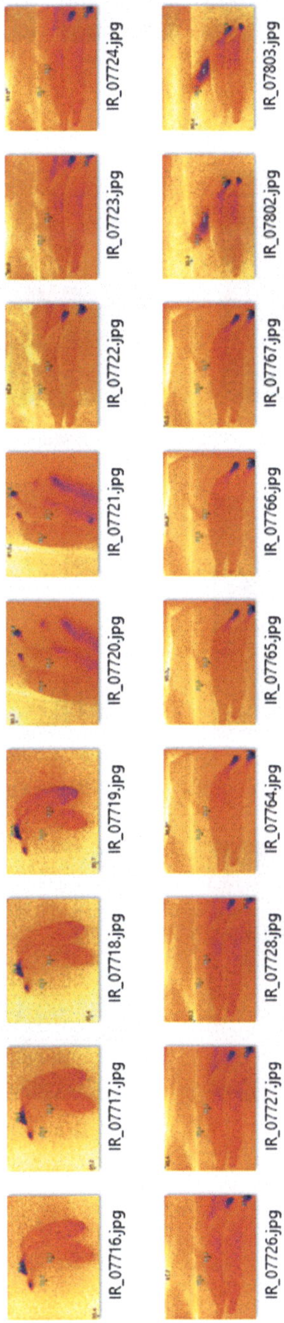

Fig. 3 Thermal images of banana

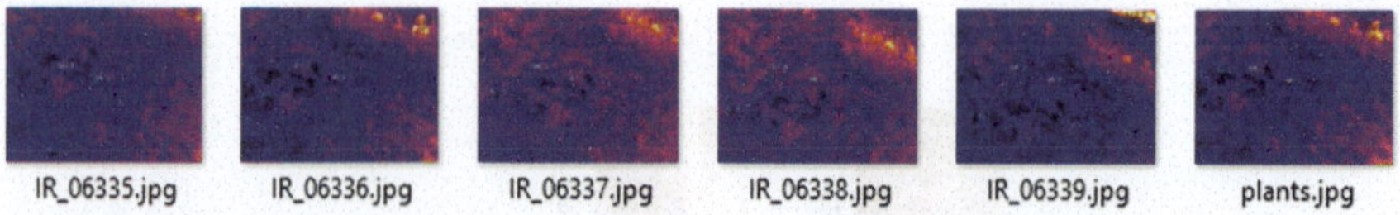

Fig. 4 Thermal images of flowers

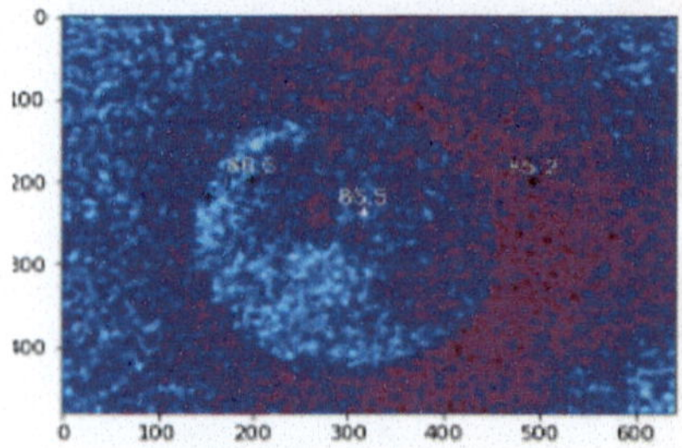

Apple gray scale 1st image

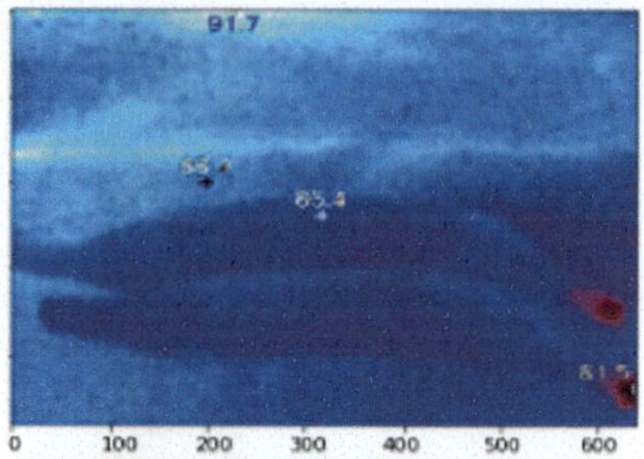

Banana gray scale 1st image

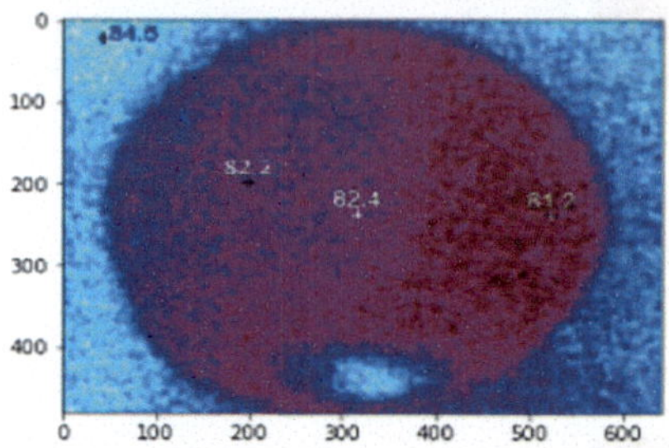

Apple gray scale 2nd Image

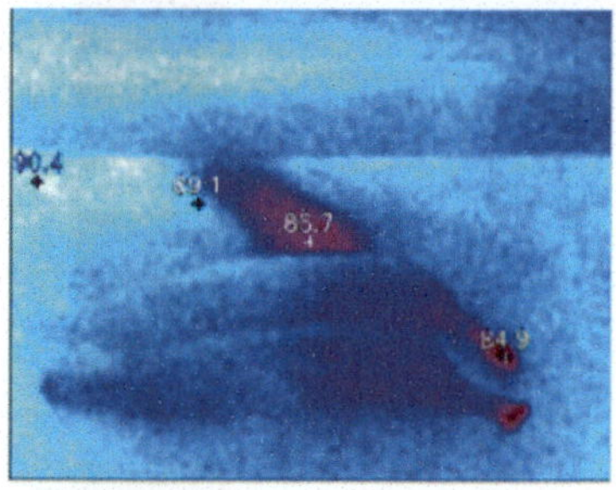

Banana gray scale 2nd image

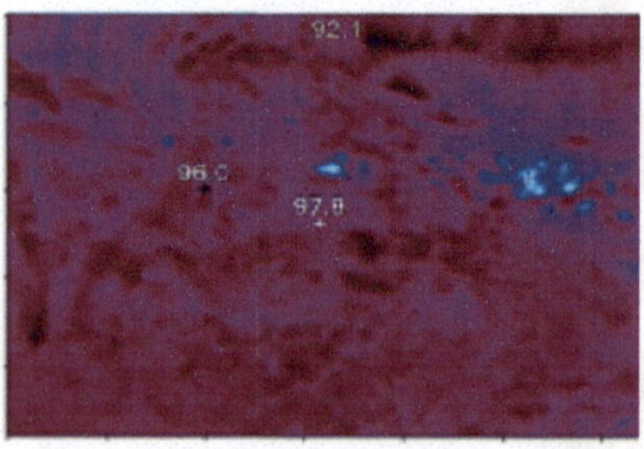

Rose gray scale image

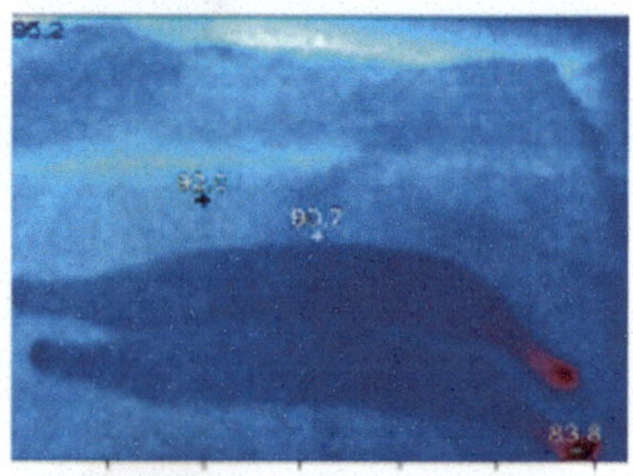

Banana gray scale3rd image

Fig. 5 Grayscale thermal images of objects

Table 1 Table of different objects with time

29th July 4:39 PM	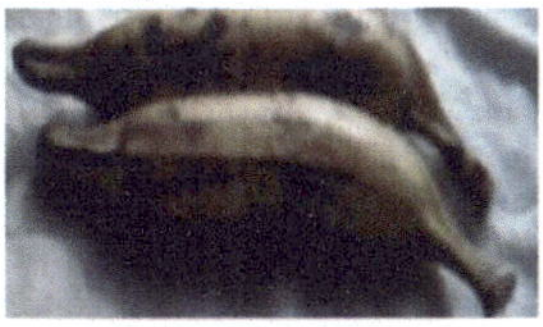	
30th July 1:44 PM	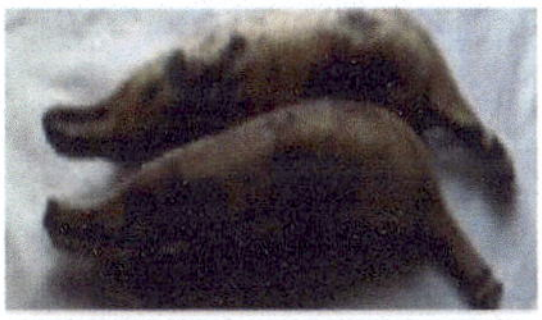	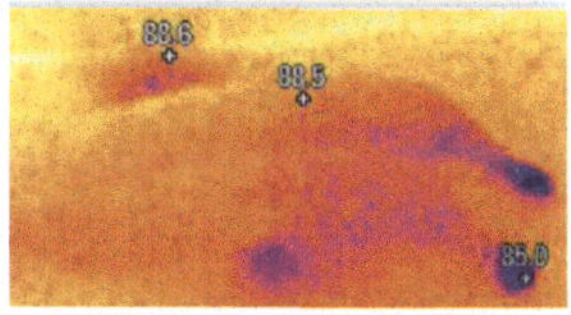
31st July 11:50 AM	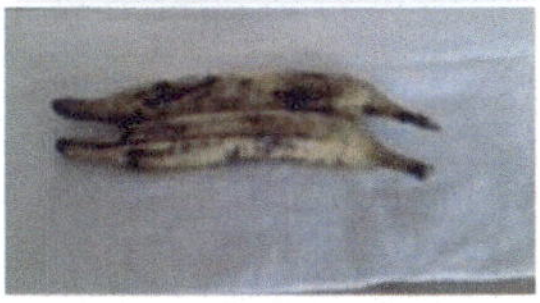	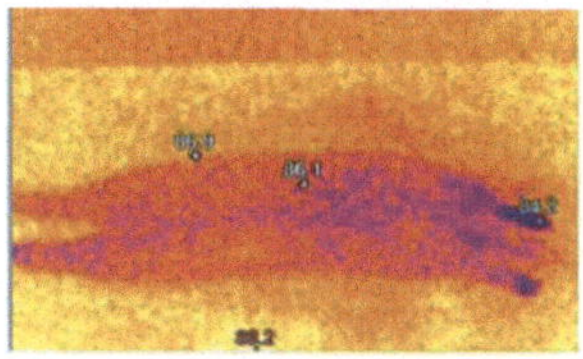
2nd August 4:13 PM		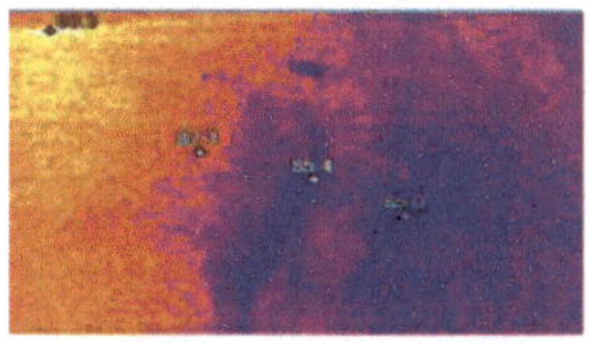
29th July 3:47		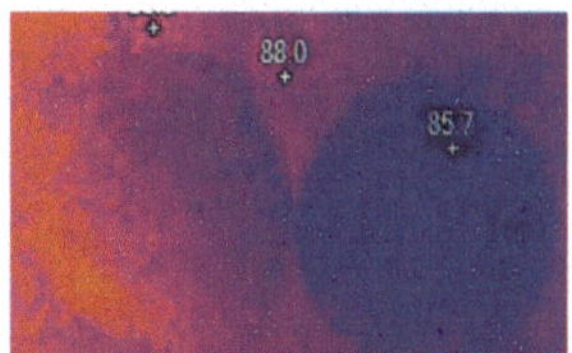
29th July 4:38PM		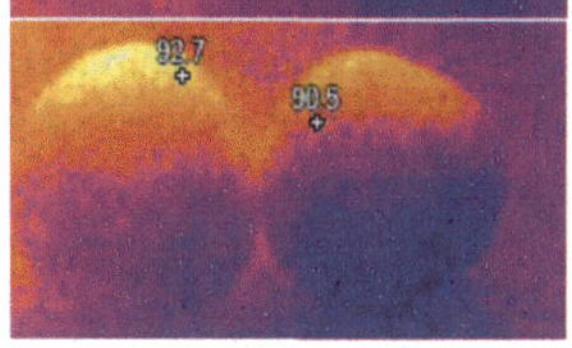

Contrast: Contrast is the difference between high illumination and low illumination value.

$$\text{Contrast} = \sum_{i=1}^{m} i\,P(i) \tag{2}$$

Variance: Variance measures how far a set of data is spread out. Variance is the average of the squared distances from each point to the mean. Zero variance indicates

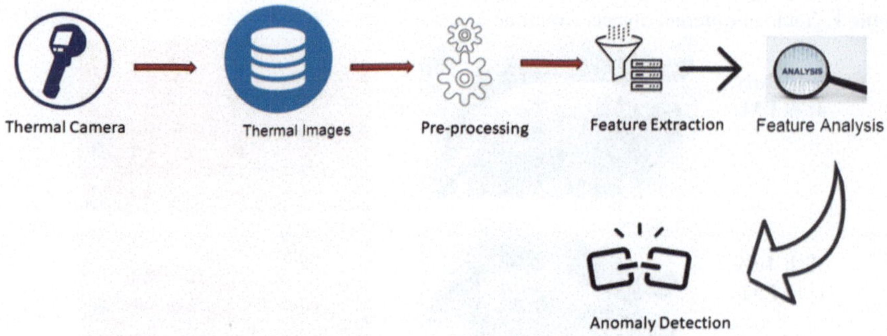

Fig. 6 Architecture of the proposed algorithm

that all the data values are identical.

$$\text{Variance} = \sum_{i=1}^{m} (P(i) - \mu)^2 P(i) \tag{3}$$

where μ denotes the average value.

Entropy: Entropy measures information content. It shows the randomness of intensity distribution. The higher the entropy, the less energy is available in the image.

$$\text{Entropy} = \sum_{\mathrm{i}=1}^{\mathrm{m}} \mathrm{P(i)log_2P(i)} \tag{4}$$

ii. Graylevel co-occurrence matrix (GLCM)

Texture can also be analyzed using GLCM [8]. It is a statistical method to examine texture that considers the spatial relationship of pixels. The GLCM dimension is n × n matrix where n is number of gray levels in the image. GLCM features used are contrast, variance, entropy, and energy which is given in Eqs. 5, 6, 7, and 8.

Contrast: It shows the amount of local variations present in the image. For uniform images, this value is zero; as variation increases, this value also increases.

$$\text{Contrast} = \sum_{i,j=0}^{N-1} P_{ij}(i - j)^2 \tag{5}$$

Variance: Variance is defined as the sum of the difference between intensity of the center pixel and its neighborhood. It is also known as second central moment.

$$\text{Variance} = \sum_{i,j=0}^{N-1} P_{ij}(i - \mu)^2 \tag{6}$$

Entropy: Entropy is a measure of information content. It measures the randomness of intensity distribution.

$$\text{Entropy} = \sum_{i,j=0}^{N-1} -ln\left(P_{ij}\right)P_{ij} \tag{7}$$

Energy: The texture energy measure (TEM) is computed by summing the absolute values in a local neighborhood.

$$\text{Energy} = \sum_{i,j=0}^{N-1}\left(P_{ij}^{2}\right) \tag{8}$$

iii. Wavelet transform

In addition to extracting features in textural domain, frequency domain features have been extracted. Wavelet features are useful as they have inbuilt capability to de-noise image. Wavelet transform decomposes the low-frequency content of the image into different levels which helps in feature extraction of noisy data [9]. The output of the wavelet transform is wavelet coefficients, and from the obtained coefficients mean, variance, standard deviation, skewness, kurtosis are computed as features of the images.

iv. principal component analysis (PCA)

PCA features help to analyze the image geometry and extract its geometrical features. First “r” eigenvalues of the image are extracted to form the feature vector. A matrix is computed that summarizes how the variables in the given image are related to one another. The main idea of (PCA) is to reduce the dimensionality of a dataset consisting of many variables correlated to each other. The principal components are the eigenvectors of a covariance matrix, and hence, they are orthogonal [10]. In this proposed work, the first few eigenvalues are extracted as features for analysis.

4.2.2 Color Features

Color moments are measures that characterize color distribution in an image. The first-order, second-order, and the third-order color moments have been proved to be capable and successful in representing color distributions of images [11]. Also, color moments compactly represent features compared to other color features [11]. If the value of the ith color channel at the jth image pixel is p_{ij}, then the color moments used are calculated based on Eqs. 9, 10, 11, and 12.

Mean: Gives the average color in the image.

$$E_i = \frac{1}{N}\sum_{J=1}^{N} P_{ij} \tag{9}$$

Standard Deviation: This is the second color moment, obtained by taking the square root of the variance of the color distribution.

$$\delta_i = \sqrt{\frac{1}{N}\left(\sum_{j=1}^{N}(P_{ij} - E_i)^2\right)} \tag{10}$$

Skewness: It is used to measure the amount of asymmetric content in the color distribution.

$$s_i = \sqrt{\frac{1}{N}\left(\sum_{j=1}^{N}(P_{ij} - E_i)^3\right)} \tag{11}$$

Kurtosis: It is a used to measure how extreme the tails are in comparison with the normal distribution

$$\text{Kurt}_i = E\left[\left(\frac{P_{ij} - E_i}{\delta_i}\right)^4\right] \tag{12}$$

4.2.3 Shape Features

Fourier descriptors are a technique used in image processing to characterize the boundary shape of a image segment. The first few terms in a Fourier series provide the foundation of this descriptor. The global and local features of the shape can be obtained using a few coefficients of the Fourier descriptors. Centroid distance gives both global and local features of the shape signatures. Fourier descriptor extracts contours by applying fast Fourier transform (FFT) [9]. Fourier descriptors are used for contour representation of object in object detection module [13].

Texture, color, and shape features extracted from a sample thermal image are shown in Table 2.

4.3 Algorithm to Find Anomaly

In this proposed work, mathematical model, correlation model, has been used to detect changes between the given images from the features extracted. Correlation

Table 2 Texture, color, and shape feature of a sample thermal image

Mean	Entropy	Variance	Contrast	AM	Max_prob	Homogeneity
114.1	-1.9×10^8	$6. \times 10^{11}$	33,897,566	844,383	201	114.11
112.7	-1.9×10^8	$6. \times 10^{11}$	33,701,178	745,755	250	37,664
97.4	-1.6×10^8	$7. \times 10^{11}$	29,750,506	326,123	255	29,946
105.8	-1.8×10^8	$7. \times 10^{11}$	3,213,371	511,201	255	36,472
97.4	-1.6×10^8	$7. \times 10^{11}$	29,750,506	326,123	255	29,946
113.37	-1.9×10^8	6.66×10^{11}	33,824,542	644,877	253	32,075
107.07	-1.8×10^8	6.95×10^{11}	32,390,750	608,565	255	37,636
113.73	-1.9×10^8	6.64×10^{11}	33,906,060	769,691	215	35,410
113.37	-1.9×10^8	6.66×10^{11}	33,824,542	644,877	253	32,075
112.44	-1.9×10^8	6.71×10^{11}	33,663,828	645,929	255	36,019
114.6	-1.9×10^8	6.58×10^{11}	33,950,380	813,825	201	32,887
113.3	-1.9×10^8	6.66×10^{11}	33,764,128	718,499	196	35,479
107.53	-1.8×10^8	6.94×10^{11}	32,536,342	700,991	246	45,914
111.7	-1.9×10^8	6.75×10^{11}	33,452,104	820,449	247	38,474

GLCM mean	GLCM variance	GLCM skew	GLCM Kusrtosis
403	89,023,077	2.8×10^8	1.54×10^{12}
13068	76,684,784	3.34×10^8	1.45×10^{12}
−6079	52,768,247	90,998,165	1.49×10^{12}
−4312	60,904,912	1.49×10^8	1.50×10^{12}
7025	78,917,735	2.94×10^8	1.56×10^{12}
403	89,023,077	2.8×10^8	1.54×10^{12}
−6079	52,768,247	90,998,165	1.49×10^{12}
8252	79,494,284	4.21×10^8	1.48×10^{12}
−45076	68,597,856	1.13×10^8	1.68×10^{12}
−2303	80,195,895	1.27×10^8	1.55×10^{12}
−7647	75,359,077	1.36×10^8	1.54×10^{12}
−979	87,215,733	3.85×10^8	1.61×10^{12}
−4252	79,972,768	3.41×10^8	1.68×10^{12}
−5817	65,569,357	2.36×10^8	1.51×10^{12}

shows how strongly the two variables x and y are related to each other, and it is represented by r; it ranges from –1 to +1 [12]. Correlation is computed as given in Eq. 13.

$$r = \frac{N\sum XY - (\sum X)(\sum Y)}{\sqrt{\left[N\sum X^2 - (\sum X)^2\right]\left[N\sum Y^2 - (\sum Y)^2\right]}} \tag{13}$$

where

N = Number of pairs of values
ΣXY = Sum of products of X and Y value
ΣX = Sum of X values
ΣY = Sum of Y values
ΣX^2 = Sum of squared X values
ΣY^2 = Sum of squared Y values

If the correlation is high it can be inferred that there is no changes in the given pair of images, if correlation is low it may be inferred that there may be changes in the given image pairs.

5 Experimental Results

The features extracted include color features, shape features, and texture features for all the images. Correlation has been computed between the ground truth image and the received test image(s). Based on the correlation, a threshold has been determined experimentally and all images which have a correlation less than the threshold have been identified as images which may contain anomalies or changes compared to the original, ideal ground truth image. Correlation of a few sample images in the dataset with ground truth image is shown in Table 3. Table 4 shows correlation of a few banana images in the less anomalous dataset with ground truth image values of highly anomalous data.

Table 3 Correlation of a few sample images in the dataset with ground truth image

ACA	ACB	ACC	ACD	ACE	ACF	ACG
0.999	0.999	0.999	0.743	0.966	0.986	0.947
0.999	0.999	0.999	0.736	0.962	0.984	0.942
0.999	0.999	0.999	0.738	0.963	0.985	0.944
0.999	0.999	0.999	0.739	0.963	0.985	0.944
0.999	0.999	0.999	0.738	0.963	0.985	0.944
0.998	0.998	0.998	0.712	0.952	0.977	0.931
0.999	0.999	0.999	0.737	0.969	0.984	0.944
0.999	0.999	0.999	0.737	0.963	0.984	0.943
0.999	0.999	0.999	0.746	0.967	0.987	0.949
0.999	0.999	0.999	0.742	0.965	0.986	0.946
0.999	0.999	0.999	0.772	0.966	0.946	0.948
0.999	0.999	0.999	0.756	0.971	0.989	0.953

Table 4 Correlation of a few banana images in the less anomalous dataset with ground truth image values of highly anomalous data

0.991	0.985	0.534	0.983	0.544	0.991	0.544
0.459	0.429	0.993	0.421	0.999	0.573	0.999
0.982	0.990	0.295	0.991	0.301	0.953	0.300
0.458	0.429	0.989	0.422	0.999	0.574	0.999
0.996	0.997	0.479	0.997	0.486	0.994	0.486
0.455	0.425	0.985	0.417	0.991	0.568	0.991
0.973	0.981	0.247	0.383	0.252	0.936	0.252
0.458	0.430	0.990	0.422	0.999	0.574	0.999
0.975	0.984	0.261	0.985	0.266	0.941	0.266
0.455	0.426	0.984	0.418	0.993	0.570	0.992

5.1 *Anomaly Detection Using Threshold*

If the computed correlation is greater than the defined threshold (T), then the data is normal otherwise the data is said to have anomaly. A correlation model has been built with the ground truth image and is shown in Fig. 7, where x-axis and y-axis correspond to the number of images and the correlation coefficient value, respectively.

Figure 8 presents the correlation values for images of each of the class. Empirically, the value of threshold has been fixed at $T = 0.8$. Based on the observation made with the correlation model and the ground truth images, the value of threshold has been set to 0.8. Figure 9 shows the correlation graph for banana object with Fourier descriptor feature. Result for the anomaly detection is given in Table 6. Based on the observation and experimentation, an average recall of 76.06% has been obtained based on the features extracted and recall is defined in Eq. 14 [6].

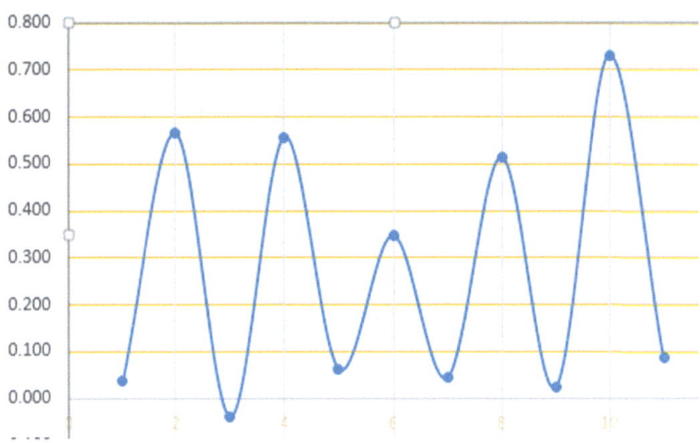

Fig. 7 Correlation graph

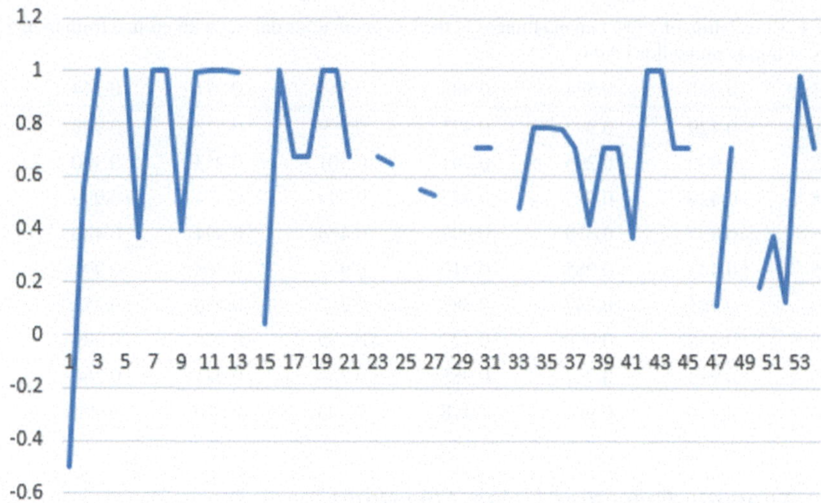

Fig. 8 Correlation graph for banana object

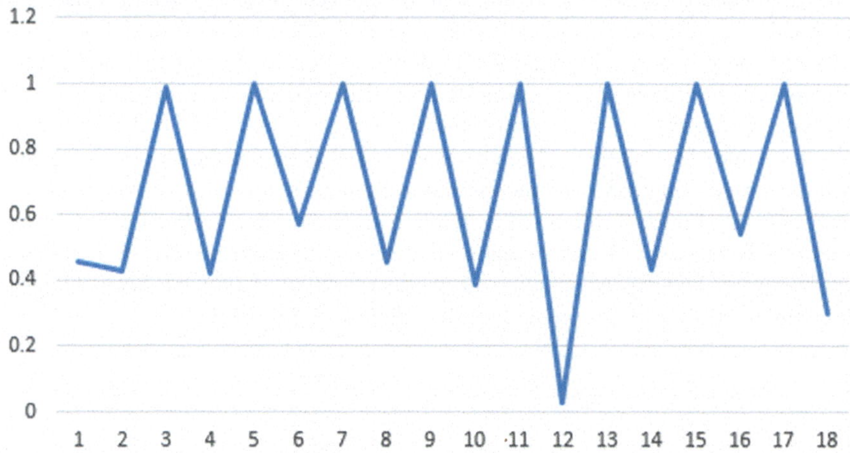

Fig. 9 Correlation graph for banana with Fourier descriptor feature

Table 6 Summary of the anomaly detection result

Image type	Number of images	Ground truth	Number of images identified as anomaly
Apple	660	350	240
Banana	550	240	310
Flowers	2000	120	60

$$Recall = \frac{|\{anamoly\ objects\}^\wedge\{identifiedanamoly\ objects\}|}{|\{anamoly\ objects\}|} \tag{14}$$

6 Conclusions

This thesis has aimed to detect anomalies in objects using its thermal images. Changes of the object at varying time/date interval have been studied with the objectives to detect if the objects (in images) have anomaly. An average recall of 76.06% has been achieved. Deep learning architecture can also be studied to detect changes in the given image dataset, in order to improve computation time and find exact location of anomaly.

References

1. Ansari S, Salankar S (2017) An overview on thermal image processing. In: Proc. 2nd International conference research in intelligent and computing in engineering, 117–120
2. Bhartee AK, Srivastava KM, Sharma T, Scholar UG, GCET GN (2017) Object identification using thermal image processing. Int J Eng Sci 11400
3. Mohamed MB, Makhlouf AM, Fakhfakh A (2018) Correlation for efficient anomaly detection in medical environment. In: 2018 14th International wireless communications & mobile computing conference (IWCMC), IEEE, 548–553
4. Selvathi D, Hannah Nithilla I, Akshaya N (2019) Image processing techniques for defect detection in metals using thermal images. In: 2019 3rd International conference on trends in electronics and informatics (ICOEI), IEEE, 939–944
5. Soliman OO, Sweilam NH, Shawky DM (2018) Automatic breast cancer detection using digital thermal images. In: 2018 9th Cairo International biomedical engineering conference (CIBEC), IEEE, 110–113
6. Pedraza ILA, Diaz JFA, Pinto RM, Becker M, Tronco ML (2019) Sweet citrus fruit detection in thermal images using fuzzy image processing. In: IEEE Colombian conference on applications in computational intelligence, Springer, Cham, 182–193
7. Guo T, Huynh CP, Solh M (2019) Domain-adaptive pedestrian detection in thermal images. In: 2019 IEEE International conference on image processing (ICIP), IEEE, 1660–1664
8. Sudha BG, Umadevi V, Shivaram JM (2017) Thermal image acquisition and segmentation of human foot. In: 2017 4th International conference on signal processing and integrated networks (SPIN), IEEE, 80–85
9. Vandone A (2011) Algorithms for infrared image processing. 9–29
10. Budzan S (2015) Human detection in thermal images using low-level features measurement automation monitoring. 61
11. Thirumurthy B, Parameswaran L, Vaiapury K (2018) Visual-based change detection in scene regions using statistical-based approaches. J Electron Imaging 27(5):051217
12. Huang Y, Han X, Tian X, Zhao Z, Zhao J, Hao D (2010) Texture analysis of ultrasonic liver images based on spatial domain methods. In: 2010 3rd International congress on image and signal processing, vol 2. IEEE, pp 562–565
13. Çevik T, Alshaykha AMA, Çevik N (2016) Performance analysis of GLCM-based classification on Wavelet Transform-compressed fingerprint images. In: 2016 Sixth international conference on digital information and communication technology and its applications (DICTAP), 131–135

14. Gonzales RC, Richard E (2002) Woods. Digital image process
15. Mudrova M, Procházka A (2005) Principal component analysis in image processing. In: Proceedings of the MATLAB technical computing conference, Prague
16. Afifi AJ, Ashour WM (2012) Image retrieval based on content using color feature. Int Scholarly Res Not
17. YenYen K, Yen EK, Johnston RG (1996) The ineffectiveness of the correlation coefficient for image comparisons
18. Parameswaran L (2013) A hybrid method for object identification and event detection in video. In: 2013 Fourth national conference on computer vision, pattern recognition, Image Processing and Graphics (NCVPRIPG), IEEE, 1–4

AIDS Detection Using Genomics Signal Processing Techniques on DNA

Darshankumar C. Dalwadi, Vatsal Shah, Harsh Navadiya, and Yagnik Mehta

Abstract AIDS is a most dangerous disease that is developed in the person having HIV virus. HIV stands for human immunodeficiency virus. Lakhs of people suffer from these diseases, and ultimately, they died due to AIDS. So, it is necessary to detect these diseases in an early stage. Genomics signal processing deals with advance research in genetics. So, by applying various GSP techniques, it becomes easier to predict one of the most hazardous diseases AIDS. In this paper, we have represented the binary mapping of the raw genomic data to convert into digital data and on applying the fast Fourier algorithm as well as the discrete wavelet transform in our algorithm to predict abnormalities present in the coding region of DNA of the gene of HIV infected cell. The purpose of this research is to provide an accurate prediction of HIV infected cell to the researcher's so that the life efficiency of any patient suffering from these deadly disease increases. We have implemented the algorithm on MATLAB 2015a which consists of signal processing toolbox. The proposed algorithm is tested for several DNA sequences present in normal genes as well as HIV infected genes of Homo sapiens chromosomes available in the National Center of Biotechnology Information (NCBI) database.

Keywords Fast Fourier transform · Genomic signal processing · DNA · AIDS and its causes · Discrete wavelet transforms · Binary mapping · NCBI database

1 Introduction

Fast Fourier transform is an algorithm that reduces the complexity of computing N-point discrete Fourier transform of the discrete sequences or binary sequences. The discrete Fourier transform converts the discrete sequences or binary sequences into frequency domain. This method is useful in many fields, but the computation of this method by using the equation directly becomes slow. So, these transforms compute the DFT very faster. FFT computes DFT matrix as a product of zero factors. The

D. C. Dalwadi (✉) · V. Shah · H. Navadiya · Y. Mehta
Birla Vishvakarma Mahavidyalaya Engineering College, V. V. Nagar, Gujarat 388120, India
e-mail: darshan.dalwadi@bvmengineering.ac.in

M. N. Favorskaya et al. (eds.), *Innovations in Electrical and Electronic Engineering*, Lecture Notes in Electrical Engineering 661, https://doi.org/10.1007/978-981-15-4692-1_50

fast Fourier transform is much more accurate then discrete Fourier transform. The equation of the fast Fourier transform is

$$\sum_{n=0}^{N/2-1} a_n^{\text{even}}\, \mathrm{e}^{-2\pi jnk/(N/2)} + \mathrm{e}^{-j2\pi k/N} \sum_{n=0}^{N/2-1} a_n^{\text{odd}}\, \mathrm{e}^{-2\pi jnk/(N/2)} \tag{1}$$

where N stands the number of points in FFT and is known as twiddle factor. Fast Fourier transform is used in various applications like recording, sampling of the data, fast algorithm for discrete sine and cosine transform, fast Chebyshev approximation, solving difference equations and computation of the isotopic distribution.

Genomic signal processing can be defined as the analysis, processing, and use of genomic signals to gain biological knowledge, and the translation of that knowledge into the system-based application that can be used to diagnose and treat genetic disease [1]. The purpose of the genomic signal processing is to combine the principles and techniques of signal processing with the understanding of the genomics to detect, classify, control, statistical and dynamical modeling of the gene networks. It is a fundamental discipline that brings the model-based analysis of the gene. Application of genomic signal processing is tissue classification and discovery of the signaling pathway both based on the expressed macromolecule phenotype of the cell. The accomplishment of these applications is done by signal processing techniques such as fast Fourier transform, discrete wavelet transform, and discrete Fourier transform.

DNA was first discovered by Francis Crick and James Watson at the Cavendish Laboratory at Cambridge University. It is a double-helical structure comprises four nucleic acids, Adenine (A), Cytosine (C), Guanine (G), Thymine (T) a deoxyribose and phosphate group. The Adenine is connected with Thymine by hydrogen bond, and the Guanine is connected with Cytosine by a hydrogen bond. Both polynucleotides of DNA contain some biological information in it. This information is replicated when the polynucleotides get separated. RNA is produced by DNA through the transcription process. DNA usually occurs as linear chromosomes in Homo sapiens. The set of chromosomes makes up the human genome. The genome has 3 billion base pairs of DNA. The biological information that is carried by DNA occurs in a piece of sequences of DNA called genes. Transmission of the genetic information occurs via complementary base pairing. The helical structure of the DNA is shown in Fig. 1. DNA molecules store the digital information that shapes the genetic scheme of living organisms [2]. By understanding the structure and properties of DNA, one can predict the genetic diseases.

HIV is a virus that targets our immunity system cells which protects our body from diseases. As soon as it targets to the immunity system, our immunity of the body to resist the diseases decreases. If the HIV is not detected for the longer, it causes the disease known as AIDS. AIDS stands for acquired immune deficiency syndrome. Symptoms of these hazardous diseases are fever, weight losses rapidly, profuse night sweats, extreme and unexplained tiredness, prolonged swelling of the lymph glands in the armpits, groin, or neck, diarrhea that lasts for more than a week, sores of

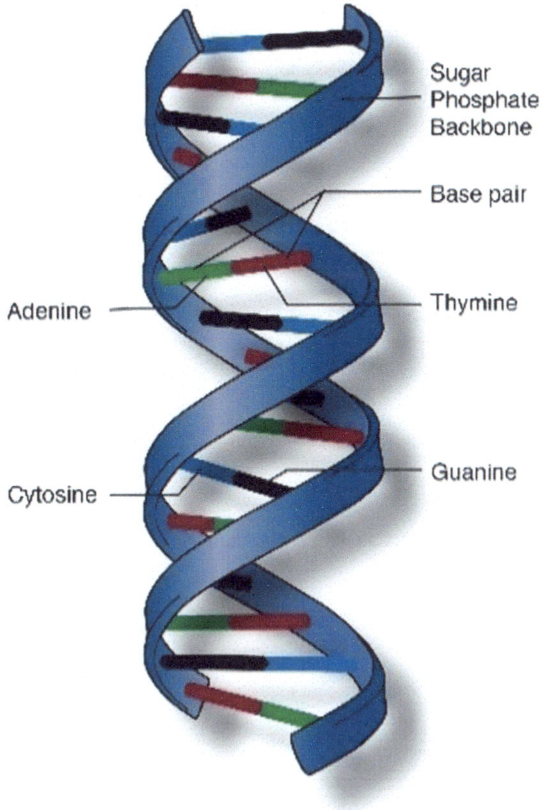

Fig. 1 Helical structure of DNA [3]

the mouth, anus, or genitals, pneumonia, red, brown, pink, or purplish blotches on or under the skin or inside the mouth, nose, or eyelids, memory loss, depression, and other neurologic disorders. HIV virus is a major component of the AIDS. So, it is necessary to understand how these HIV virus spread the virus spreads through various sources like blood, breast milk, semen, rectal fluids, and vaginal fluids. So, one should take care of these following things. HIV is not transmitted through water, air, shaking hands, mosquito bite, etc., when HIV infected cells get interact with the body, then it damages some cells of brain, kidney, GI tract, kidney, as well as oral keratinocytes and epithelial cells. The infected cells grow rapidly, and as a result, it completely destroys our immunity system, and as a result, the person suffers from the various diseases mentioned above. Figure 2 shows the interaction of these HIV infected cells with the CD4 cell which is responsible for the protection of the body from the various diseases (Figure 3).

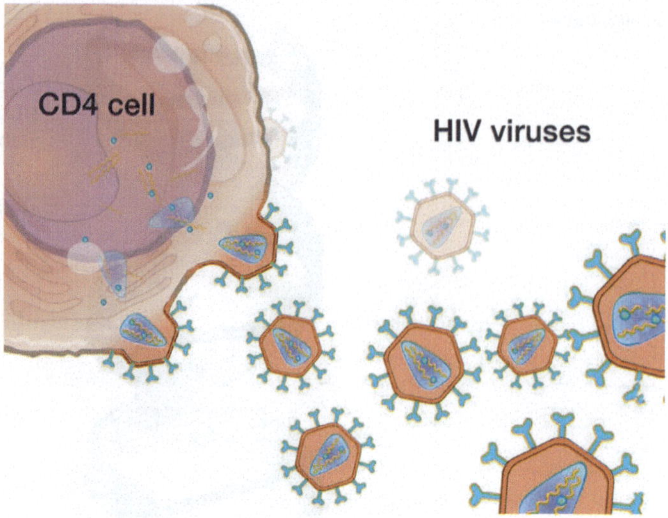

Fig. 2 HIV interaction with normal cell [4]

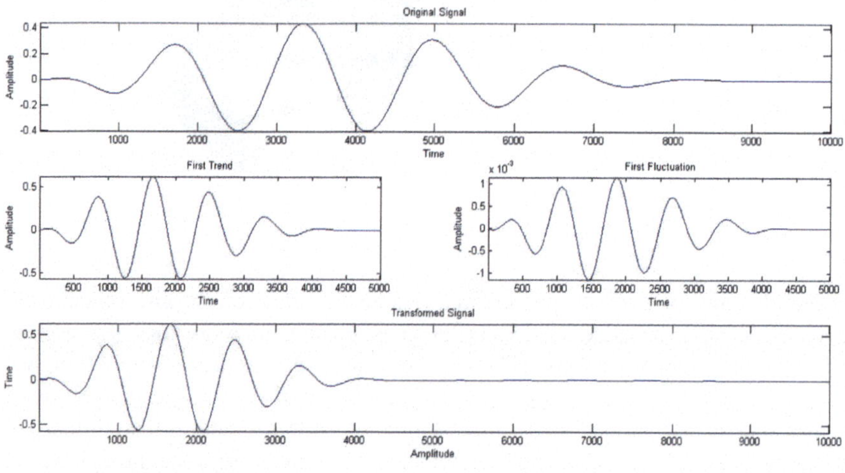

Fig. 3 Haar wavelet transform [5]

A wavelet is a mathematical technique that is used to abstract particular data from many types of data. Discrete wavelet transform means the transform which divides the signals into two orthogonal sets of the wavelets. The first discrete wavelet transform was invented by Alfred Haar. If an input has 2^n numbers, the Haar wavelet transforms may be considered to pair up the input values, storing their differences and passing the sum. This process is repeated recursively to prove the next scale which leads to 2^n-1 differences and a final sum. Now, another wavelet transform is most commonly used, and it was discovered by mathematician Ingrid Daubechies in 1988. This wavelet transform is known as Daubechies wavelet transform. This formulation is based on the use of recurrence relation to generating progressively finer discrete wavelet samplings of an implicit mother wavelet function, where each resolution is twice that of the previous scale. We have used here Haar wavelet transform, and the formula of the Haar wavelet transform is

$$y_n = H_n x_n \tag{2}$$

where: H_n is a Haar Matrix
y_n is an output
$x_{n\text{ is}}$ an input signal.

The NCBI stands for National Center for Biotechnology Information which consists of information regarding genes genomic data and biomedical information. It consists of genes not limited up to humans but all organisms. It also consists of information regarding protein chains as well as gene expression, taxonomy, and sequence analysis. One can easily access the genomic data in order to predict the disease. The front end of the NCBI Web site is shown in Fig. 4.

2 Literature Review

Prediction of cancer cell using DSP technique was proposed by Satapathi et al. [6]. They have described the DFT algorithm on the coding region of the DNA in order to predict the cancerous cell. DFT means discrete Fourier transform, it means to apply the discrete time Fourier transform at particular sample of the signal, so here, the author has first of all mapped the bases of the DNA (A, C, G, T) to the binary, it means the A and T are converted to binary value 0, and C and G are converted to binary sequence 1. After that Discrete Fourier Transform is applied to the binary sequence. Then, by plotting the power spectrum density (PSD) on DFT of cancer cell and normal cell, they have predicted cancer cell. They have applied the IIR filter to denoise the PSD signal. If the ratio of mean amplitude and mean frequency is greater than 1, then the cell is normal, otherwise it is cancer cell.

DWT-based cancer identification using EIIP was proposed by Chakraborty and Gupta [7]. They have mapped the DNA bases (A, C, G, T) to the 0.1260, 0.1335, 0.0806, 0.1340. This method is known as the EIIP mapping method. Then, they have applied the discrete wavelet transform to the signal. Discrete wavelet transform is unique algorithm in order to do feature extraction of the signal and to find out the statistic of the signal. If the ratio of mean amplitude of the signal and standard deviation of the signal is greater than 1, then normal cell, otherwise it is cancer cell. Figure 3 shows the discrete wavelet analysis of signal. And different level of decomposition of the signal is shown in Fig. 3. In our research, we have done binary

Fig. 4 NCBI Web site [5]

mapping of the DNA sequence which converts into digital signal. On applying fast Fourier transform and discrete wavelet transform on binary coded DNA sequence, we have predicted the HIV infected gene. We have use the FFT because it computes faster than DFT ultimately which improves our research more accurate.

Table 1 Binary mapping of DNA

DNA components	Binary equivalent	Decimal value
Adenine	00	0
Cytosine	01	1
Guanine	10	2
Thymine	11	3

3 Binary Mapping of DNA Sequence

As we know that DNA consists of four chemical elements, they are A (Adenine), C (Cytosine), G (Guanine), and T (thymine). So, binary mapping means mapping of these elements into the binary one. This is one of the most important elements in our research. Table 1 shows the binary mapping of DNA. By using these binary mapping concepts, we have predicted the cancer cell. We have mapped the DNA sequence to its binary equivalent using MATLAB 2015a.

4 Our Methodology of Research

- The method that we have applied in the prediction of the HIV infected cell is as follows:
- First, take a DNA sequence of HIV infected gene as well as the normal gene from the database available on the NCBI Web site.
- Then, apply the binary mapping into the DNA sequence.
- Thereafter, find the fast Fourier transform of the converted binary sequence for analysis.
- Find out the discrete wavelet transform of the resultant binary coded DNA sequence for statistical analysis.
- Determine the ratio of the median of amplitude () by standard deviation (S) of the binary DNA sequence
- If is greater than 3.5, then the predicted cells are HIV infected cell.
- If is lesser than 3.5, then the predicted cells are normal cell.

5 Flowchart of Research Process

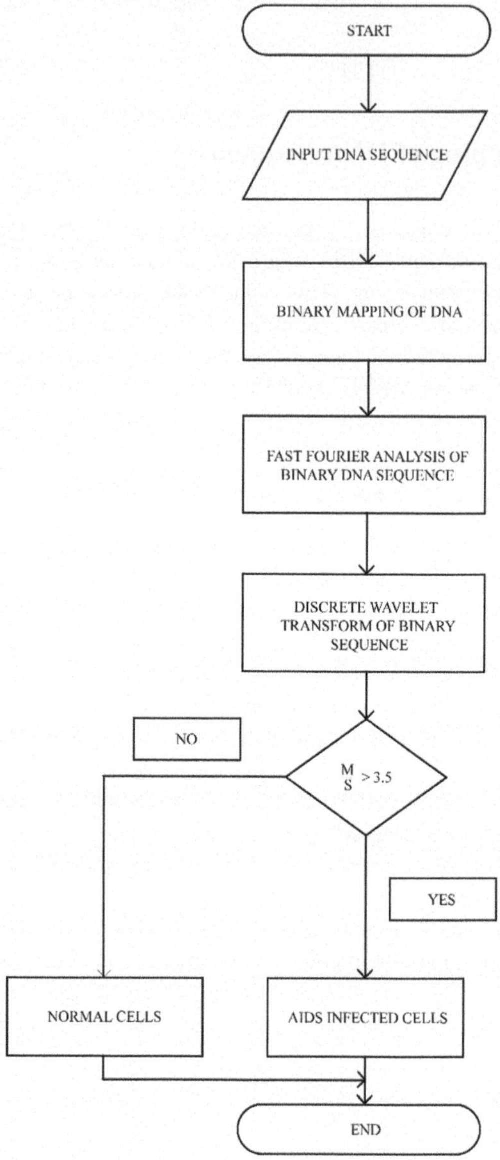

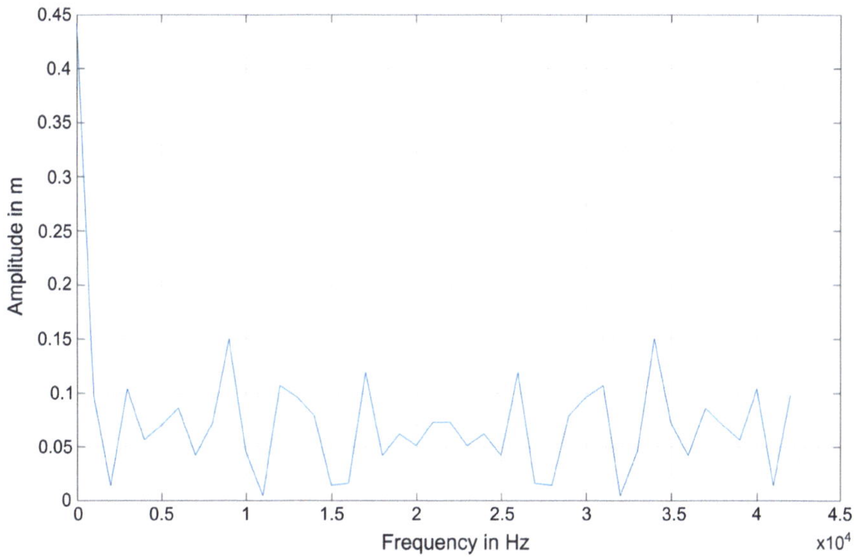

Fig. 5 CNNTB HIV infected gene

6 Fast Fourier Analysis of the Normal Cell and HIV Infected Cell

From these results, we have concluded that there are many peaks in the waveform of the HIV infected gene, while there are fewer peaks in the normal gene, this means that the HIV infected gene contains more noise compared to the normal gene (Figs. 5, 6, 7, 8, 9 and 10).

7 Statistical Analysis of the Normal Gene and HIV Infected Gene

From the statistical analysis, we can say that if the ratio median of signal by standard deviation is greater than 3.5, then the predicted cell is HIV infected gene. Otherwise, the predicted cell is normal gene. Tables 2 and 3 justify the same.

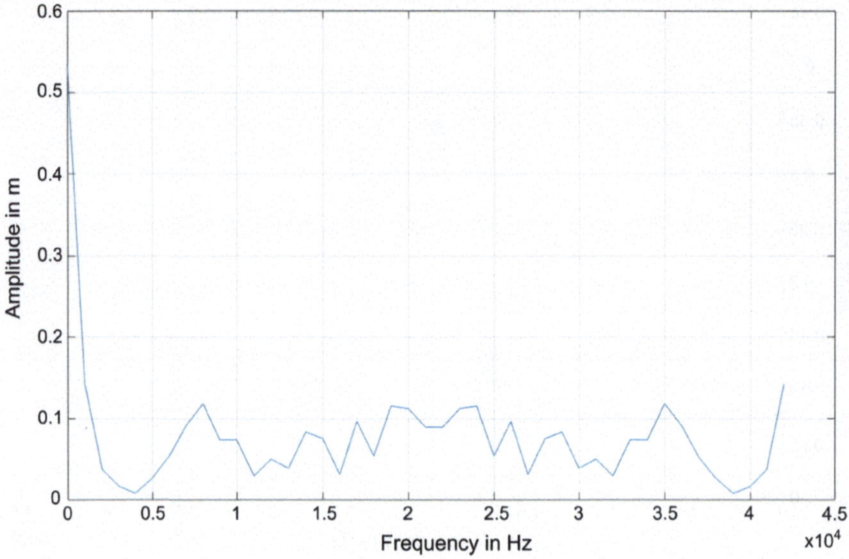

Fig. 6 HBA1 normal gene

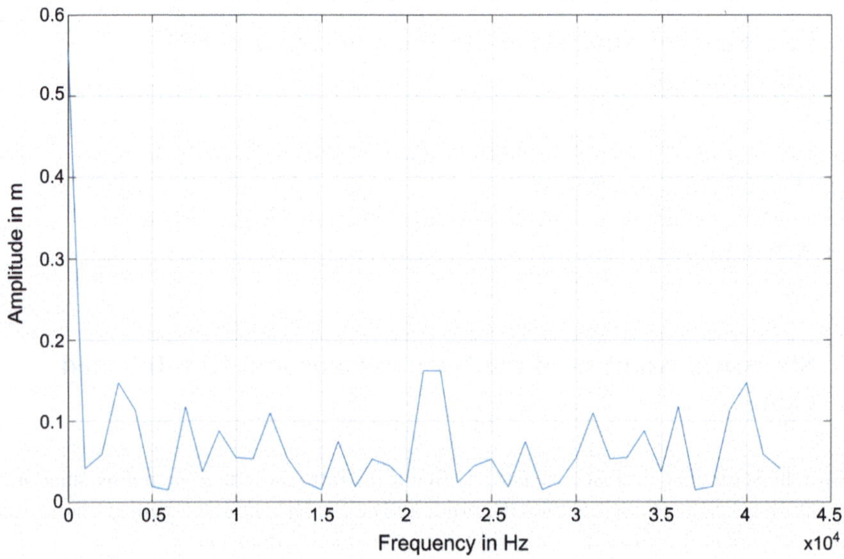

Fig. 7 HLA-DRB HIV infected gene

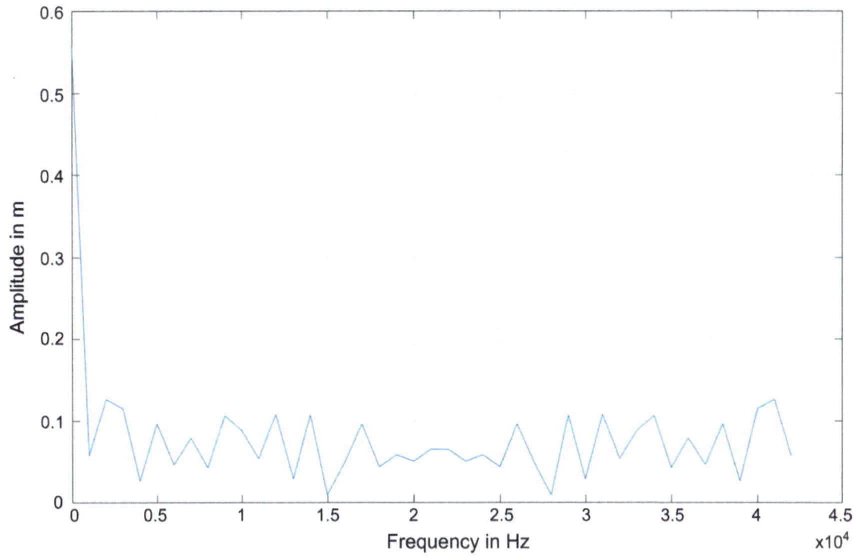

Fig. 8 HBG1 normal gene

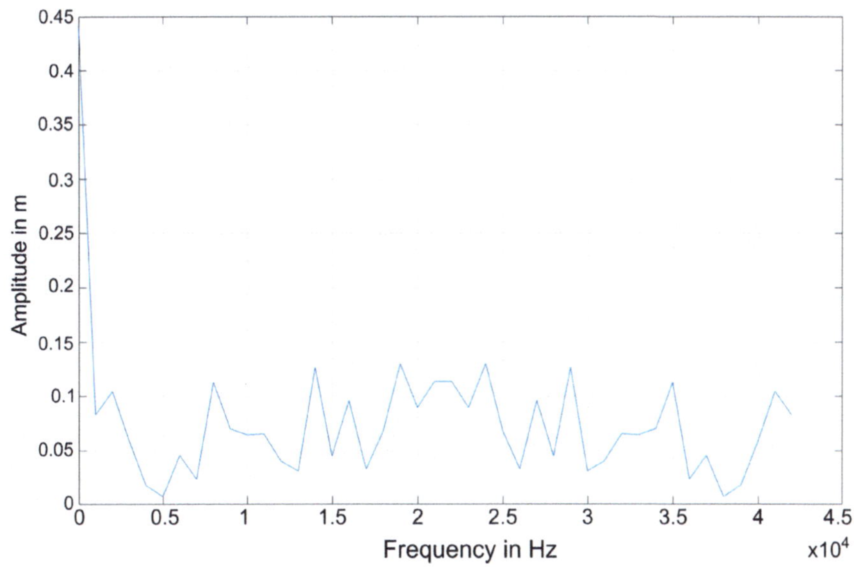

Fig. 9 HLA-C HIV infected gene

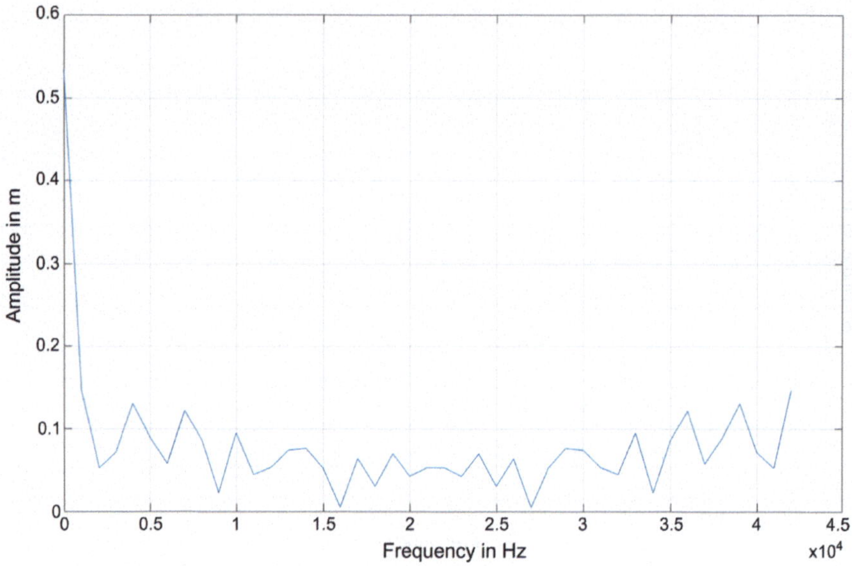

Fig. 10 Globin normal gene

Table 2 HIV infected gene

Sr. No	Name of the gene	Median (M)	Standard deviation (S)	$\frac{M}{S}$
1	CTNNB1	2.5	0.6435	3.885
2	HLA B	2.5	0.5672	4.407
3	HLA-C	2.5	0.5478	4.5637
4	HLA-DRB1	3	0.8258	3.6328
5	MTOR	2.25	0.6173	3.6449

Table 3 Normal gene

Sr. No:	Name of gene	Median (M)	Standard deviation (S)	$\frac{M}{S}$
1	HBA1	2.5	0.7973	3.135
2	HBA2	2.5	0.7811	3.200
3	HBG1	2	0.922	2.169
4	HBG2	2.25	0.7912	2.843
5	GATA	2	0.7033	2.8437
6	GLOBIN	2	0.911	2.195

8 Conclusions

In the present work, an effective algorithm has been developed in MATLAB which consists of signal processing toolbox. And the most important is the mapping of the DNA sequences, and the mapping technique that is used here is binary mapping. A combination of the fast Fourier-based spectral analysis and wavelet-based techniques was found to be more accurate due to their attractive properties, such as local feature extraction time-frequency domain representation, multi-resolution, scalability, denoising, and also compressing of big sample data or sequences of data. Fast Fourier analysis technique has been applied to raw genomic data for detection of the location of exon region of DNA. So in the future, it has a great scope in early diagnosis and prognosis of AIDS. We experimentally found that if we find the ratio of change in mean by standard deviation of the HIV infected cell, then the value of that ratio is equal to either normal cell's ratio or cancer cell's ratio. This was the reason that the HIV infected cells are not detected, so in our algorithm, we have determined the ratio of median of the signal by the standard deviation. So, in future, there will be great scope of developing effective drugs for curing these diseases.

References

1. Wang H-Q, Wong H-S, Huang D-S, Shu J (2007) Extracting gene regulation information for cancer classification. Pattern Recognit
2. Naeem SM, Eldosoky MA, Mabrouk MS (2017) Detecting genetic variants breast cancer using different power spectrum methods, pp 147–153
3. https://www.genome.gov/genetics-glossary/Double-Helix
4. https://www.aboutkidshealth.ca/Article?contentid=910&language=English
5. https://ch.mathworks.com/matlabcentral/fileexchange/38061-haar-wavelet-transformhttps://ncbi.nlm.nih.gov
6. Satapathi GN, Srihari P, Jyothi A, Lavanya S (2013) Prediction of cancer cell using digital signal processing. In: International conference on communication and signal processing, vol. 23, pp. 149–153
7. Chakraborty S, Gupta V (2016) DWT based cancer identification using EIIP. In: 2016 second international conference on computational intelligence and communication technology, pp 718–723

8 Conclusions

In the present work, an effective algorithm has been developed in MATLAB which [illegible]

Semantic Segmentation-Based Image Inpainting Detection

Nitish Kumar and Toshanlal Meenpal

Abstract Image manipulation detection has evolved as a very challenging problem in the field of multimedia forensics. Object removal from the image is one of the most used manipulation operations in the image. Object removal accomplished by exemplar-based image inpainting makes image visually pleasing and physically plausible without any noticeable trace. So, detection of inpainted region has always been a challenge for researchers. In this paper, we have proposed a semantic segmentation-based approach to detect the inpainted region in an image. This is a deep learning-based approach which uses fully convolutional network. We have achieved a remarkable result in the detection of inpainted region. For training and validation of our results, we have created inpainted image database. We have evaluated our results on different performance metrics.

Keywords Inpainting detection · Forgery detection · Image inpainting · Image forensics

1 Introduction

Nowadays, with the increase in usage of mobile phones, computer and Internet, capturing and sharing images of each moment have become one of the important parts of everyone's life. People are using a lot of image editing tools for beautification of photos like cropping, brightness changing, colour enhancement, background replacement, adding or removing any object from the image. So, we can define image editing or image manipulation as any change done on digital images with the help of any software [15]. These image manipulation techniques can be used unethically

N. Kumar (✉) · T. Meenpal
Electronics and Communication Engineering Department, National Institute of Technology Raipur, Raipur, Chhatisgarh, India
e-mail: nkumar.phd2018.etc@nitrr.ac.in

T. Meenpal
e-mail: tmeenpal.etc@nitrr.ac.in

M. N. Favorskaya et al. (eds.), *Innovations in Electrical and Electronic Engineering*, Lecture Notes in Electrical Engineering 661, https://doi.org/10.1007/978-981-15-4692-1_51

by forger to deliver fake information. Image forgery is a part of image manipulation which focuses on changing the graphic content of image like morphing, compositing, enhancing, etc. Compositing is a sub-category of image forgery which is rephrased as image tampering [13]. Image tampering is used either for adding a new object in the image or for removing an object or some part of an image. Always there is an original image behind a tampered image. Tampered part of image can be easily identified by visually comparing the tampered image to the original, whereas it is very difficult to identify the tampered region without using the original image. Blind forensic has always been challenging when the tampering clues are visually imperceptible.

Tampering can be done on any kind of images whether it is real-world image or computer-generated. Adding, removing or modifying some objects from images are the main target of image tampering. Image tampering can be categorized based on the source and composition of the new content in tampered region. They are classified as copy-move, cut-paste and erase-fill. Copy-move [3] can be done by copying the region from the image and pasting at the region to be tampered in the same image. It is one of the most common method of object removal from an image. Cut-paste [9] can be achieved if the new content at the tampered region is entirely from the different source of image. Image splicing is most commonly used word in case of cut-paste [7]. Erase-fill [11] can be achieved by removing the object and filling the region by elementary patches from the untampered part of image. Image inpainting is the most commonly used terminology for erase-fill.

Image inpainting [10] is one of the effective image editing technique used for restoring damaged or missing parts of image using known information. Image inpainting makes the target region visually pleasing. Some of the common application of image inpainting are scratch removal, crack removal, missing blocks recovery, object removal, etc. But when inpainting can be exploited for unethical motives, it is difficult to analyse whether the image is original or with modified image content. Figure 1 shows the example of modified image with the help of image inpainting without any visible clues of tampering. To deal with such kind of problem inpainting forensics is required for image investigation. Two important forensic tasks need to

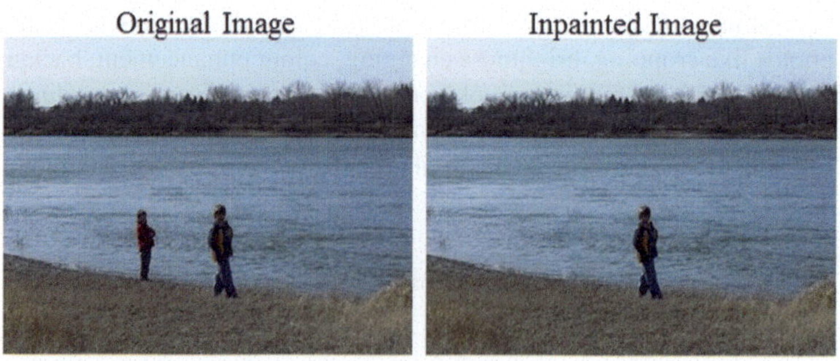

Fig. 1 Image Inpainting example in case of object removal

be examined: the given image has been inpainted or not; if yes, we need to find the region of inpainting in the image.

In this paper, we have presented a deep learning model for image inpainting detection. In our proposed model, we have used a fully convolutional network which detects the inpainted and uninpainted mask of given image. Other sections of this paper are arranged as follows. Section 2 explains the previous work done on image inpainting and its detection. Section 3 explains the inpainting database and proposed methodology for inpainting detection. We have explained our results in Sect. 4 and in Sect. 5 concluded our work.

2 Related Work

Image inpainting has been proposed for the content correction or restoring damaged area in an image [5]. Image inpainting can be used for removing unwanted object in case of image tampering. Neighbouring texture is being used for filling the holes. Based on region filling approach, image inpainting has been categorized into two groups. First group is diffusion-based inpainting in which local structure is being smoothly propagated from exterior of the holes to interior [1]. This approach is well suited for filling small holes and keeps the shape of structure present in neighbourhood. However, this method does not perform well in case of recovering larger areas. Second group is exemplar-based inpainting in which patches have been copied from the known part of the image and stitched together to fill the holes [4]. This approach is called as block duplication approach which is similar to copy-move where we perform region duplication. Block duplication is visually more imperceptible as compared to region duplication.

Since the improved inpainting technique leaves behind no perceptual artefacts, inpainting forensic has become challenging field for researcher. Earlier blind inpainting detection was first proposed by Wu et al. [14] which was based on zero-connectivity labelling (ZCL) and fuzzy membership. ZCL was used in suspicious region to provide matching degree of the blocks and fuzzy membership had been computed to identify tampered region. The main drawback of this method was manual selection of suspicious region. Chang et al. [2] proposed a two-stage searching algorithm for accelerating search of suspicious patches to overcome the above drawback. False-alarm in this case had been filtered by multi-region relations. Because of similar patches detection accuracy was still limited. This work was further improved [8] using central pixel mapping for patch search and false-alarm patches had been removed by fragment splicing detection.

Since convolutional neural network (CNN) has been used widely in the field of image processing and computer vision and achieved a great success, researchers have started adopting CNN in the field of forensics for better performance. Zhu et al. [17] had proposed CNN-based architecture for inpainting forensics for the first time. Encoder-decoder-based CNN network had been used, and for better training of CNN, a label matrix had been used. For calculating the effective loss, weighted cross entropy had been used. Experimental results had achieved better performance. Motivated with these works we have aimed to design an effective detection algorithm for exemplar-based inpainting.

3 Methodology

For better understanding of our work on inpainting forensic, first we have briefly explained the exemplar-based image inpainting. Using this technique, we have created inpainted image database. Then we have explained the proposed architecture of deep learning-based inpainting detection.

3.1 Exemplar-Based Image Inpainting

Image inpainting is a method used for effectively repairing damaged areas, scratches or missing region in an image. Goal of this method is to make inpainted image visually pleasing and physically plausible as much as possible. Different approaches have been used to achieve better inpainting result like diffusion-based approach, exemplar-based approach, etc. Since exemplar-based approach typically gives better inpainting results, here we are going to use this approach. This approach inpaint the region by inwardly propagating the patches from the known region into unknown region, so we also use the name patch-based approach. Hence, our work is focused on patch-based inpainting forensics.

Here we have explained in Fig. 2 stepwise patch-based image inpainting which has been proposed by Criminisi et al. [4]. Figure 2a represents the original image with Ω as the unknown region, $\partial\Omega$ as its contour and Φ as the whole known region. In Fig. 2b, suppose p is having the highest priority in the interior of $\partial\Omega$ and image patch Ψ_p has been selected as the patch to be inpainted which is centred at p. Now in the known region Φ similar patch has to be searched by using any recognized patch matching rule. Image patch Ψ_q has been found out as the most matching patch of Ψ_p as in Fig. 2c. At last we need to replace the patch region Ψ_p with the region of Ψ_q and the unknown region has been updated as shown in Fig. 2d. Hence, the unknown region has been filled by iteratively performing the above steps.

3.2 Inpainted Image Database

Inpainted image database is required for performing inpainting forensics task. Since there is no specific database available for inpainting detection, we have used the above methodology to create database for inpainting. Here we have used MIT Places database [16] for randomly selecting 2000 images of different category having sizes

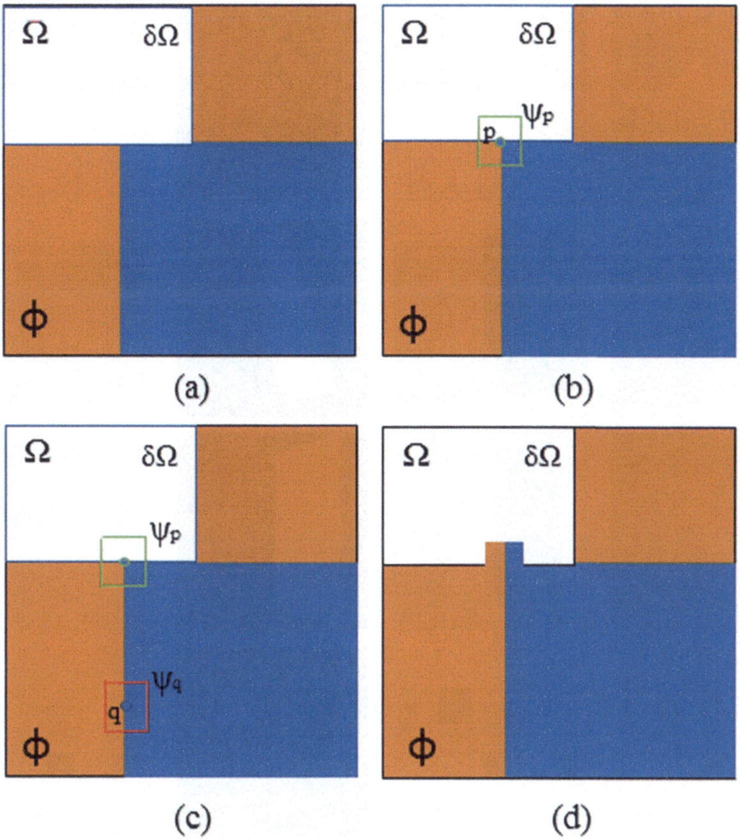

Fig. 2 Steps of exemplar-based image inpainting: original image with unknown region Ω (**a**), selection of target patch Ψ_p centred at p in the interior of $\partial\Omega$ (**b**), Searching of most likely patch Ψ_q for Ψ_p (**c**), Region filling of Ψ_p and updating the contour $\partial\Omega$ (**d**)

256×256. We have removed the objects of irregular shapes and different sizes from the image by manually selecting the boundary to create the target region for inpainting. Sizes of the removed objects are varying from 2 to 40% of image size.

We have shown some of the images of database created for image inpainting in Fig. 3. Figure 3a is the original image, Fig. 3b is the image after removing object from it. Black region in the image shows the target region to be filled by inpainting. Figure 3c is the mask of removed object and Fig. 3d is the final inpainted image. During inpainting detection this mask image will act as ground truth for validation of detected inpainted region.

Fig. 3 Image inpainting for object removal: original image (**a**), cropped image after object removal (**b**), mask of removed object (**c**), inpainted image (**d**)

3.3 Inpainting Detection

To deal with inpainting detection task, we have to come across blind inpainting detection. In blind detection task, we do not have any information regarding size, shape or location of inpainted region neither we are having the original image for comparing. Inpainting decision on either image is inpainted or not can be taken completely based on image itself. Suppose we have been given an image M of width and height ($W \times H$) for inpainting forensics. Our aim is to divide pixels m_{ij} of input image M into two classes inpainted and uninpainted pixels. To check whether the pixels of image M are inpainted or not, we create another matrix N of same size as M which represent the probability that each pixels of image M is inpainted. Values of each element n_{ij} of matrix N vary from [0,1]. We can say matrix N as inpainting probability matrix (IPM). We can define IPM as:

$$N = F(M) \tag{1}$$

where F is the unknown function to be learned from training.

We can form $1 - N$ as another matrix of probability having pixels of M, those are uninpainted. So, IPM can be used to estimate two outputs from the given image, and hence, we can predict the inpainted region in the image.

3.4 CNN-based Inpainting Detection

We have used convolutional neural network (CNN)-based architecture for the inpainting forensics. CNN architecture is used for learning features of image content instead of manipulation information. Since image inpainting mainly depends on image content and it is a global operation. A label matrix X is formed by assigning a class label for every pixel of input image M which has been used for training the database. Label matrix X can be defined as

$$X_{i,j} = \begin{cases} 1 & (i, j) \in \Omega \\ 0 & \text{elsewhere} \end{cases} \tag{2}$$

where Ω represents the inpainted region of input image. We will use this label matrix along with input image samples (M, X) of database for training of CNN.

We have designed the fully convolutional network (FCN)-based architecture which is based on convolution and deconvolution. In this model, some of the representation used are: convolutional layers as Conv2D, max pooling layers as MaxPooling2D and deconvolution layers as Conv2DTranspose. This architecture is made up of 20 layers as shown in Fig. 4. Here, the first 5 blocks consists of first 13 layers and these are part of VGG model [12]. We have the input image of size $224 \times 224 \times 3$ to the model and binary image of same size is output of this model.

In the first layer of the model, we have applied convolution with kernel size $3 \times 3 \times 3$. 64 kernels have been used in this layer. We get output of first layer as 64 feature maps having size 224×224. We have used rectified linear units (Relus) as the activation function in all the layers except in the last layer where we have used softmax function. In second layer, we are having input as $224 \times 224 \times 64$ and we have applied the convolution with same number of kernel and we get the output of same size as layer 1. After layer 2 we apply the max-pooling with 2×2 window size and step size of 2. After max-pooling we get output as $112 \times 112 \times 64$. Two Conv2D layers along with 1 MaxPooling2D have been marked as block 1 of VGG model. Similarly, we have marked the block 2 to block 5 of VGG model. After that we have applied 2 Conv2D layers and we have 4096 feature maps of size 7×7. Since

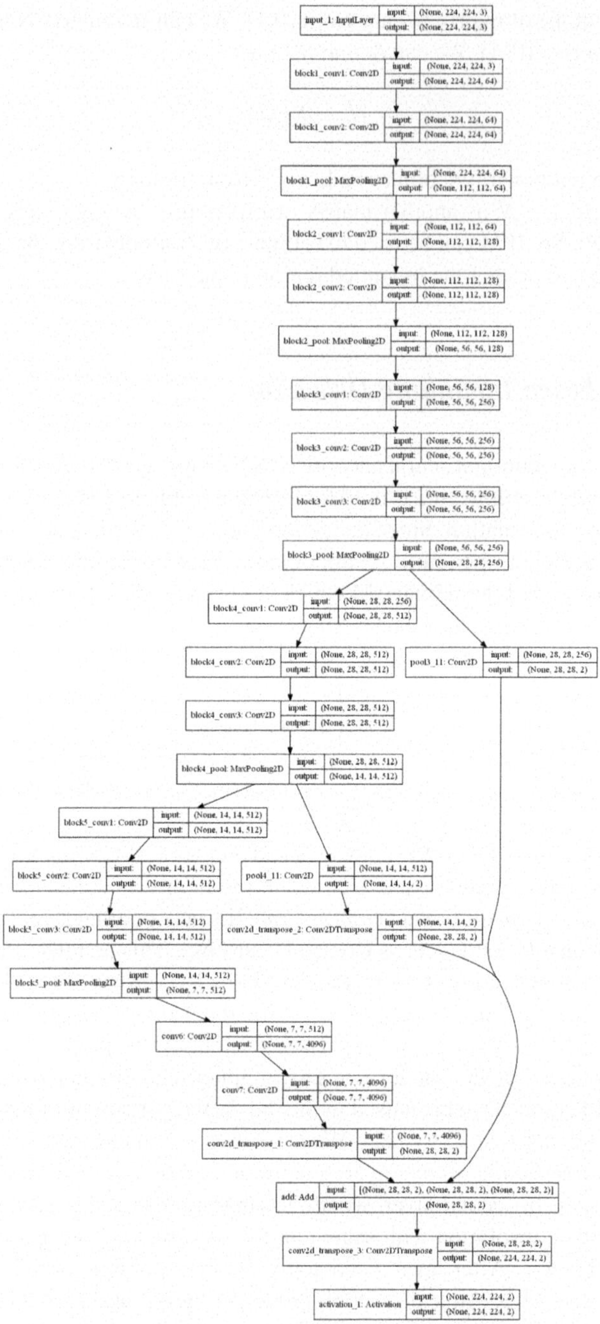

Fig. 4 Proposed inpainting detection architecture based on fully convolutional network

our expected output is 2 binary image of same size as input, we have performed the upsampling operation. We have used the deconvolution operation which helps in upsampling. Deconvolution operation has been represented as Conv2dTranspose in our proposed model. After deconvolution we have 2 feature maps of size 28 × 28.

In this model, we have also used 2 skip connection which is similar to residual neural network [6]. There are some information which is acquired from the initial layers and is required for reconstruction during upsampling. With the help of skipping layers, information of primary layers is fed to the later layers. In skip connection, same input can be fed to 2 different layers. Pool 3 and pool 4 layers have been taken from block 3 and block 4, respectively, as skipping layers and made up of same size as the output of deconvolution layer stated above. All these feature maps of same size obtained from three different layers have been added. And at the last layer we used the deconvolution to make our output of 2 layers with size 224 × 224. In the two-output layer, one will represent the binary inpainted mask and another will be uninpainted binary mask.

4 Experiments and Results

4.1 Training Databse

We have created inpainted image database of 2000 colour images of size 256 × 256. Inpainted region in the database is of irregular shape and size varying from 2 to 40% of image size. Out of 2000 images, we have randomly selected 1600 images for training and 400 images for testing. Since we have taken the input of proposed model is 224 × 224 so first we resized our database images to this size and then applied for training. We have formed the ground truth label matrix of utilized tampered region for each inpainted image. Weighted cross entropy has been used as a loss function as the inpainted region is comparatively small with respect to uninpainted region in the training samples. Stochastic gradient decent has been used for iteratively updating the CNN parameters. We have a set of training parameters of stochastic gradient decent. The learning rate has been taken as e^{-2}. The momentum value is 0.9. The weight decay has been fixed to 5^{-4}. The batch size is taken as 8.

4.2 Testing and Results

We have randomly selected 400 colour images from the database for the validation of our proposed architecture. These images have not been considered for training. We have achieved better detection results on irregular shapes as compared to existing detection approach [2]. Performance of detection by our proposed model has been shown in Fig. 5, where first column represents the original images, second column

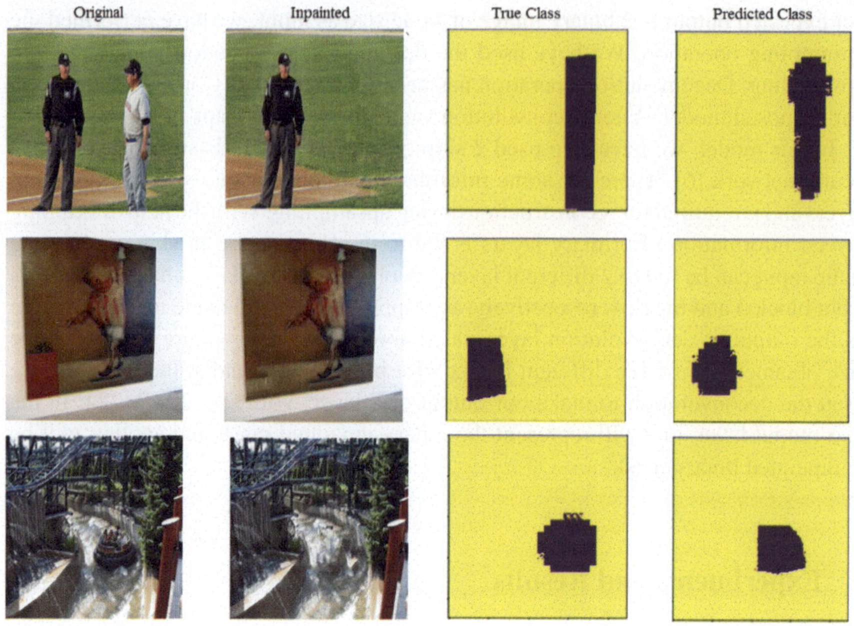

Fig. 5 Inpainting detection result: original images (1st column), inpainted images (2nd col.), prediction results (3rd col.), reference masks (4th col.)

represents inpainted images, third column represents detected inpainted region and fourth column represents the reference mask. As we can see, the inpainted region is large, and still we have achieved better detection results.

Our proposed inpainting detection results have been evaluated based on some of the standard performance metrics such as precision, recall, F_1 score and intersection over union (IoU). These parameters have been calculated based on true positive (TP), false positive (FP) and false negative (FN). These parameters have been defined as follows:

1. **Precision (P)**: It is defined as the ratio of number of TP to the sum of number of TP and FP as explained in Eq. 3.

$$P = \left(\frac{\text{TP}}{\text{TP} + \text{FP}}\right) \tag{3}$$

2. **Recall (R)**: It is defined as the ratio of number of TP to the sum of number of TP and FN as explained in Eq. 4.

$$R = \left(\frac{\text{TP}}{\text{TP} + \text{FN}}\right) \tag{4}$$

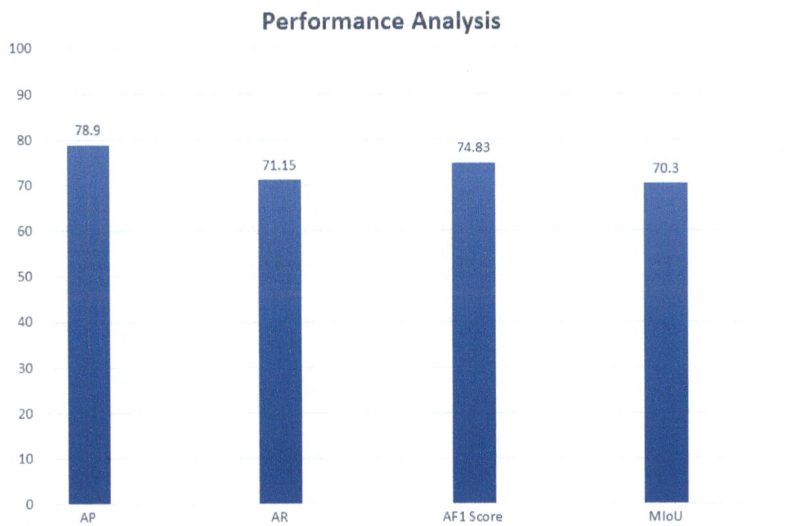

Fig. 6 Performance analysis based on average precision (AP), average recall (AR), average F1 score (AF1) and mean intersection over union (MIoU)

3. ***F1* Score** (F_1): It is defined as the harmonic mean of P and R taking both into account. It can be calculated as product of 2 times of P and R divided by sum of P and R as explained in Eq. 5.

$$F_1 = 2.(\frac{P \cdot R}{P + R}) \tag{5}$$

4. **Intersection over union (IoU)**: It is a metric defined to calculate the percentage overlap between the predicted output and target mask. It is calculated as the intersection of number of pixels of predicted mask and target mask to the union of number of pixels of both the masks.

We have two labels as the output of proposed model; first label represents the inpainted region and the second label represents the uninpainted region. Since we have two different output labels, we have calculated the average precision (AP), average recall (AR), average F1 score (AF1) and mean IoU on the tested result. We have achieved a remarkable result as 78.9% of AP, 71.15% of AR, 74.83% of AF1 and 70.3% of mean IoU as shown in Fig. 6. In this analysis, IoU is very important metric which shows the similarity between the detected mask and target mask and we have achieved the target result quite satisfactorily.

5 Conclusions

Image inpainting is one of the image manipulation technique. This paper has presented the new approach towards the detection of image inpainting. We have proposed the semantic segmentation-based CNN model for the detection of target inpainted region. For performing training and validation of proposed model, we have created inpainted image database and tested our model having irregular shaped inpainted region. We have achieved a remarkable accuracy in detection measured by some standard metrics like 78.9% of average precision, 74.83% of average $F1$ score and 70.3% of mean average intersection over union. As a future work, this model can be trained to make it robust for other manipulation operation as well like copy-move and splicing.

References

1. Bertalmio M, Sapiro G, Caselles V, Ballester C (2000) Image inpainting. In: Proceedings of the 27th annual conference on computer graphics and interactive techniques. ACM Press/Addison-Wesley Publishing Co., New York, pp 417–424
2. Chang IC, Yu JC, Chang CC (2013) A forgery detection algorithm for exemplar-based inpainting images using multi-region relation. Image Vis Comput 31(1):57–71
3. Christlein V, Riess C, Jordan J, Riess C, Angelopoulou E (2012) An evaluation of popular copy-move forgery detection approaches. IEEE Trans Inform Forens Secur 7(6):1841–1854
4. Criminisi A, Pérez P, Toyama K (2004) Region filling and object removal by exemplar-based image inpainting. IEEE Trans Image Process 13(9):1200–1212
5. Guillemot C, Le Meur O (2013) Image inpainting: overview and recent advances. IEEE Sig Process Mag 31(1):127–144
6. He K, Zhang X, Ren S, Sun J (2016) Deep residual learning for image recognition. In: Proceedings of the IEEE conference on computer vision and pattern recognition, pp 770–778
7. He Z, Lu W, Sun W, Huang J (2012) Digital image splicing detection based on markov features in dct and dwt domain. Pattern Recogn 45(12):4292–4299
8. Liang Z, Yang G, Ding X, Li L (2015) An efficient forgery detection algorithm for object removal by exemplar-based image inpainting. J Vis Commun Image Represent 30:75–85
9. Ng TT, Chang SF, Sun Q (2004) Blind detection of photomontage using higher order statistics. In: 2004 IEEE international symposium on circuits and systems (IEEE Cat. No. 04CH37512). vol 5. IEEE, pp V–V
10. Ružić T, Pižurica A (2014) Context-aware patch-based image inpainting using markov random field modeling. IEEE Trans Image Process 24(1):444–456
11. Schetinger V, Oliveira MM, da Silva R, Carvalho TJ (2017) Humans are easily fooled by digital images. Comput Graph 68:142–151
12. Simonyan K, Zisserman A (2014) Very deep convolutional networks for large-scale image recognition. arXiv preprint arXiv:1409.1556
13. Wang W, Dong J, Tan T (2009) A survey of passive image tampering detection. In: International workshop on digital watermarking. Springer, Berlin, pp 308–322
14. Wu Q, Sun SJ, Zhu W, Li GH, Tu D (2008) Detection of digital doctoring in exemplar-based inpainted images. In: 2008 international conference on machine learning and cybernetics, vol 3. IEEE, pp 1222–1226
15. Zheng L, Zhang Y, Thing VL (2019) A survey on image tampering and its detection in real-world photos. J Vis Commun Image Represent 58:380–399

16. Zhou B, Lapedriza A, Khosla A, Oliva A, Torralba A (2017) Places: a 10 million image database for scene recognition. IEEE Trans Pattern Anal Mach Intell
17. Zhu X, Qian Y, Zhao X, Sun B, Sun Y (2018) A deep learning approach to patch-based image inpainting forensics. Signal Process Image Commun 67:90–99

An Enhanced Depth Approximation Model for Haze Removal Using Single Image

Suresh Raikwar and Shashikala Tapaswi

Abstract Contrast of images which are captured in poor weather (i.e., hazy, foggy, rainy and cloudy) degrades due to optical and physical properties of light and atmospheric particles. This degradation lessens the performance of computer vision assisted systems. The weight of degradation at a pixel depends upon depth of the pixel from camera. Therefore, accurate depth estimation of a pixel is essential to improve quality of image. This paper presents color attenuation prior-based depth approximation model to approximate depth of a pixel from the camera using a single degraded image. The proposed method observed that the depth of a pixel from the camera is directly proportional to the difference of the saturation from the sum of brightness and hue. Visual quality and quantitative metrics are used to compare results of the proposed method with prominent existing methods in literature.

Keywords Fog · Haze · Atmospheric light · Atmospheric scattering · Transmission

1 Introduction

Performance of surveillance systems, object recognition in outdoor systems, intelligent transportation systems, traffic monitoring using outdoor vision systems is highly interrupted by poor weather. Computer vision assisted systems work effectively only if input is noiseless. Image captured in mediocre weather degrades due to absorption and scattering of light by the particles present in atmosphere (such as fog, haze, and smoke), which fade color and contrast of the captured image. Camera receives sum of reflected light and scattered light as radiance of a pixel [2]. It is observed that

S. Raikwar (✉)
GLA University, Mathura, Mathura, Uttar Pradesh, India
e-mail: suresh.raikwar@gla.ac.in

S. Tapaswi
ABV-Indian Institute of Information Technology and Management, Gwalior, Madhya Pradesh, India

M. N. Favorskaya et al. (eds.), *Innovations in Electrical and Electronic Engineering*, Lecture Notes in Electrical Engineering 661, https://doi.org/10.1007/978-981-15-4692-1_52

amount of scattered light is more than absorbed light [28]. Thus, removal of fog or haze (dehazing) is highly essential. Main task of dehazing is to improve quality of degraded image using atmospheric scattering model [2, 21]. The quantity of degradation at a scene point depends upon depth of the scene point from camera. [2].

Histogram equalization and contrast stretching are generally used to enhance degraded images earlier. However, these methods focus on improvement of brightness and contrast, which make these methods unable to produce ideal dehazed images [4, 5, 19].

Methods in [2, 22, 23, 29] are based on atmospheric scattering model and generate quality results. However, the performance of these methods depends upon filtering approach. Due to inappropriate filtering or inexact depth estimation, these methods do not restore degraded edges.

It is observed that the depth of a pixel from the camera is directly proportional to the difference of the saturation from the sum of brightness and hue. Therefore, an enhanced model to approximate depth of each pixel is proposed. Original edges are preserved, and degraded edges are restored by the proposed method. Visual quality of results which are obtained by the proposed method is validated qualitatively and quantitatively.

Work in literature is presented in Sect. 2. Introduction to atmospheric scattering model and problem formulation is presented in Sect. 3. Mathematical foundation of the proposed work is discussed in Sect. 4. In Sect. 5, the proposed work is modeled mathematically. The process to restore original clear day image from hazy image is given in Sect. 6. Result analysis based on visual quality and quantitative metrics is discussed in Sect. 7. In Sect. 8, conclusion is presented.

2 Related Work

Concentration of haze changes with unknown depth of each scene point, which makes dehazing a tiring task. Objective of dehazing is to recover unknown scene depth using degraded image. Dehazing is mainly classified into three categories based on type of input; (1) extra information based [14, 20], (2) multiple images based [11–13, 17, 18], and (3) single image based [1, 2, 21, 22, 29].

Extra information-based methods need some additional input like cues of depth, which are obtained through different camera positioning [14, 20]. Due to extra need, these methods are not good for real-time applications. Multiple image-based methods [17, 18] require more than one image of a scene, which are taken at varying degree of polarization. However, these methods need extra hardware, which increases hardware cost and other expenses.

Thus, single image-based methods [1, 2, 21, 22, 29] have been proposed, which solves problem of dehazing by putting many constraints. Performance of these methods depends upon strong assumptions and priors. In [21], a method to maximize local contrast is proposed, which is based on an observation that haze free images have more contrast than hazy images. However, this method produces blocky artifacts due to window-based filtering. In [1], it is assumed that transmission and local surface shading are not correlated. However, method in [1] performs poorly in dense haze.

A most prominent work is proposed in [2]. In [2], it has been observed that one of the color channel of outdoor haze free images has very low intensity in non-sky region, which is used to estimate transmission. However, this method performs incorrectly in the presence of sky region, or an object brighter than atmospheric light.

More fast method is presented in [29], which estimates scene depth to recover scene transmission. This method is fast and handles sky regions to certain extent. However, it loses a few edges and does not recover degraded edges in certain hazy conditions due to inaccurate estimated depth.

3 Problem Formulation

Color and contrast of images change due to scattering of light reflected from scene point by atmospheric particles. Type, size, orientation, and distribution of particles decide severity of scattering [10]. Figure 1 shows the process of image formation in outdoor environment. A light beam reflected from surface of an object is attenuated due to atmospheric scattering. Camera receives fraction of non-attenuated light (direct attenuation) and attenuated light (airlight), which is described by atmospheric scattering model [10, 13]. Mathematical expression of atmospheric scattering model is given in Eq. 1.

$$I_2(y) = \underbrace{I_1(y) * \mathrm{Tr}(y)}_{\text{Direct Attenuation}} + \underbrace{\mathrm{Ar} * (1 - \mathrm{Tr}(y))}_{\text{Airlight}} \tag{1}$$

where y is position (usually coordinate) of a scene point, intensity of degraded image at position y is $I_2(y)$, intensity of original clear day image at position y is $I_1(y)$, Ar is atmospheric light, transmission at position y is represented by $\mathrm{Tr}(y)$, and given by Eq. 2.

$$\mathrm{Tr}(y) = e^{-\gamma \mathrm{Dep}(y)} \tag{2}$$

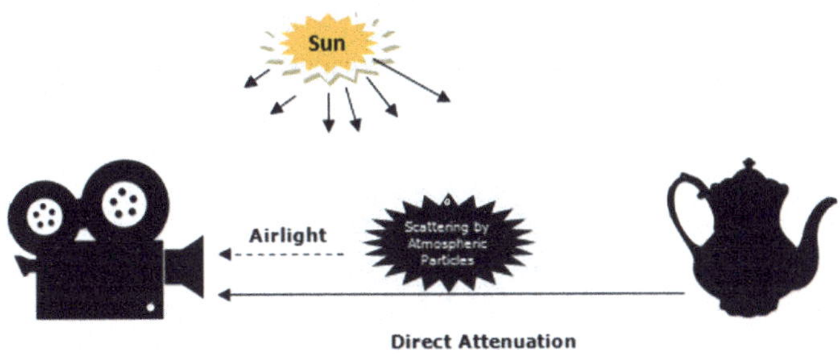

Fig. 1 Image formation based on optical model

where scattering coefficient is γ, Dep(y) is depth of a pixel at position y from camera. If atmosphere contains homogeneous and little particles in size, then γ and Ar will be constants. Approximation of $I_1(y)$, Tr(y), and A from single degraded image $I_2(y)$ is the main goal of dehazing. If Dep(y) is known, then Tr(y) can be obtained using Eq. 2. Obtained Tr(y) can recover image $I_1(y)$ using Eq. 1.

4 Mathematical Foundation of the Proposed Model

Information about depth of a scene point is not contained in single image. Thus, dehazing is challenging due to unknown depth. To restore clear day image, depth recovery is essential. The proposed work is influenced by color attenuation prior [29] to approximate depth. According to [29], the depth of a pixel increases with increased difference of saturation and brightness at same pixel. The results produced by [29] are good; however, this method loses existing edges due to lack of accuracy in depth estimation. This inspired the proposed work to introduce an enhanced depth approximation model. Method in [29] works in HSV color space which can be best described by following transformation equations [7].

$$H(y) = \begin{cases} \frac{1}{6} * \left(\frac{g'(y)-b'(y)}{\delta(y)}\right), & \text{if } r'(y) = \mathrm{MAX}(y) \\ \frac{1}{6} * \left(2 + \frac{b'(y)-r'(y)}{\delta(y)}\right), & \text{if } g'(y) = \mathrm{MAX}(y) \\ \frac{1}{6} * \left(4 + \frac{r'(y)-g'(y)}{\delta(y)}\right), & \text{if } b'(y) = \mathrm{MAX}(y) \end{cases} \tag{3}$$

$$S(y) = \frac{\delta(y)}{\mathrm{MAX}(y)}, \qquad V(y) = \mathrm{MAX}(y) \tag{4}$$

$$\mathrm{MAX}(y) = \max(r'(y), g'(y), b'(y)), \quad MIN(y) = \min(r'(y), g'(y), b'(y))$$

where $[r'(y), g'(y), b'(y)]$ are normalize triplets representing r, g, b color intensities of input image at location y, and $\delta(y) = \mathrm{MAX}(y) - \mathrm{MIN}(y)$. Saturation, brightness, and hue of input image at location y are represented by $S(y)$, $V(y)$, and $H(y)$, respectively.

Equation 1 represents that original clear day image $I_1(y)$ is degraded due to, (1) multiplicative reduced transmission Tr(y) and (2) additive airlight $Ar(1 - \mathrm{Tr}(y))$. It is proved that the transmission Tr(y) depends upon depth $Dep(y)$ of each scene point. Thus, the degradation increases with depth. Therefore, additive airlight increases with depth due to which $MAX(y)$ and MIN(y) start approaching to airlight. At long distance, transmission will be zero and airlight will be maximum. Therefore, Tr(y) and MIN(y) will be almost same. Thus, $\delta(y)$ will decrease with depth and can be expressed as a function of depth Dep(y) as shown in Fig. 2

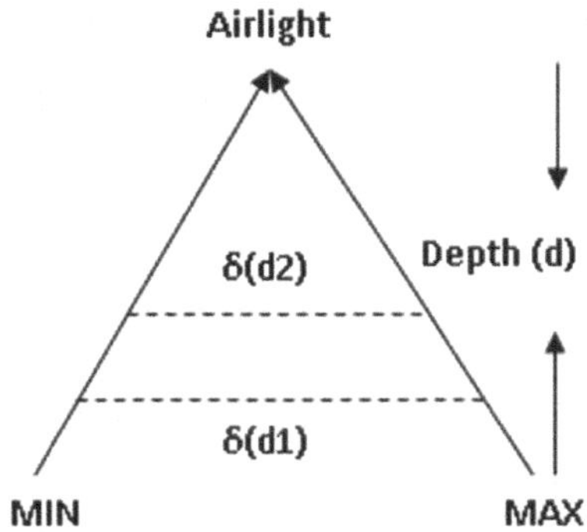

Fig. 2 $\delta(y)$ as function of depth Dep(y)

Figure 2 shows $\delta(y)$ as a function of depth Dep(y). Two solid lines represent values of MIN(y) and MAX(y) at varying level of depth Dep(y). Difference of MAX(y) and MIN(y) is the value of $\delta(y)$ and represented by dashed lines. It can be observed that increased airlight causes increase in MAX(y) and MIN(y) intensity. However, at very long distance MAX(y) and MIN(y) will be almost equal due to reduced transmission and increased airlight. Thus, $\delta(y)$ decreases with depth which causes saturation $S(y)$ to decrease and brightness $V(y)$ to increase. Therefore, [29] considered difference of brightness and saturation as a function of depth.

However, it can be observed from Eq. 3 that hue $H(y)$ increases with depth due to reduction in $\delta(y)$. This implies that hue $H(y)$ and brightness $V(y)$ are positively correlated with depth while saturation $S(y)$ is negatively correlated.

Consider objects which are not too close to camera and not too far from camera. For these objects, maximum difference of $V(y) - S(y)$ will be one, which is not true estimation of depth by [29]. Thus, difference of brightness and saturation is unable to estimate true depth. However, the value of $H(y) + V(y) - S(y)$ will be more for same objects. Therefore, the model in [29] and the proposed model are combined to estimate the scene depth more accurately. This combination can be best represented by a combination of hue, brightness, and saturation as:

$$\text{depth}(y)\alpha(\text{hue}(y) + \text{brightness}(y) - \text{saturation}(y)) \tag{5}$$

5 Mathematical Modeling

Mathematical model of the proposed enhanced depth approximation model is described as:

$$D(y) = (c_1 + c_2 B(y) + c_3 H(y) - c_4 S(y))/\alpha + \epsilon(y) \tag{6}$$

where $D(y)$ is depth, $B(y)$ is brightness, $H(y)$ is hue, $S(y)$ is saturation, c_1, c_2, c_3 and c_4 are linear coefficients, α is used to normalization constant and $\epsilon(y)$ is random image to represent random error of the model.

5.1 Computation of Linear Coefficients

Linear coefficients are computed with an objective to improve structural similarity index ($ssim$), which measures variance of mixed effect of contrast, structure and luminance of two images [16, 26].

$$ssim(j,k) = l(j,k) * c(j,k) * s(j,k) \tag{7}$$

where $l(j,k)$ is luminance, $c(j,k)$ is contrast, $s(j,k)$ defines structure of two images j and k. Formulas to compute the value of $l(j,k)$, $c(j,k)$, and $s(j,k)$ are defined as:

$$l(j,k) = \frac{2\mu_j\mu_k + C_1}{\mu_j{}^2 + \mu_k{}^2 + C_1}, \quad c(j,k) = \frac{2\sigma_j\sigma_k + C_2}{\sigma_j{}^2 + \sigma_k{}^2 + C_2} \quad s(j,k) = \frac{\sigma_{jk} + C_3}{\sigma_j\sigma_k + C_3} \tag{8}$$

where μ_j, μ_k are local mean of images j and k, respectively, σ_j, σ_k are standard deviations of images j and k, respectively, σ_{jk} is cross covariance. The value of $ssim(j,k) = 1$ is possible, if and only if $l(j,k) = 1$, $c(j,k) = 1$, and $s(j,k) = 1$. Thus equating $l(j,k) = 1$, $c(j,k) = 1$, and $s(j,k) = 1$.

$$(\mu_j - \mu_k)^2 = 0, \quad (\sigma_j - \sigma_k)^2 = 0, \quad \sigma_j\sigma_k = \sigma_{jk} \tag{9}$$

It can be inferred from Eq. 9 that $ssim$ between two images will be high if squared difference of those images is low. Thus, linear coefficients are computed such that squared difference(error) is minimized. Therefore, ordinary least square estimation (OLS) is used to compute linear coefficients. Generalized regression for Eq. 6 can be expressed as:

$$D_i(y) = \left(\frac{c_1}{\alpha} + \frac{c_2}{\alpha}B_i(y) + \frac{c_3}{\alpha}H_i(y) - \frac{c_4}{\alpha}S_i(y)\right) + \epsilon_i(y) \tag{10}$$

where $D_i(y)$ dependent variable of regression and represents random depth of ith sample. Brightness, hue, and saturation of ith sample are represented by $B_i(y)$, $H_i(y)$, and $S_i(y)$, respectively. Random error of ith sample is $\epsilon_i(y)$.

It is assumed that the random error $\epsilon_i(y)$ is based on normal distribution with σ^2 variance. If $\frac{c_1}{\alpha}, \frac{c_2}{\alpha}, \frac{c_3}{\alpha}$ and $\frac{c_4}{\alpha}$ are replaced by $\beta_0, \beta_1, \beta_2$ and β_3, respectively in Eq. 10, then

$$D_i(y) = (\beta_0 + \beta_1 B_i(y) + \beta_2 H_i(y) - \beta_3 S_i(y)) + \epsilon_i(y) \tag{11}$$

where $\beta_i \leq 1$ for $(i = 0, 1, 2, 3)$. Using Eq. 11, the sum of square of errors s is given by following equation.

$$s = \sum_{k=1}^{\text{num}} (D_i(y) - (\beta_0 + \beta_1 B_i(y) + \beta_2 H_i(y) - \beta_3 S_i(y)))^2 \tag{12}$$

where num are number of samples used in OLS. The following equations are derived by equating partial differentiation of Eq. 12 to zero (i.e., $\frac{\partial s}{\partial \beta_0} = 0$, $\frac{\partial s}{\partial \beta_1} = 0$, $\frac{\partial s}{\partial \beta_2} = 0$ and $\frac{\partial s}{\partial \beta_3} = 0$).

$$\sum_{k=1}^{\text{num}} D_i(y) = \text{num}\beta_0 + \beta_1 \sum_{k=1}^{\text{num}} B_i(y) + \beta_2 \sum_{k=1}^{\text{num}} H_i(y) - \beta_3 \sum_{k=1}^{\text{num}} S_i(y) \tag{13}$$

$$\begin{aligned} \sum_{k=1}^{\text{num}} D_i(y) * B_i(y) = \beta_0 \sum_{k=1}^{\text{num}} B_i(y) + \beta_1 \sum_{k=1}^{\text{num}} B_i(y)^2 \\ + \beta_2 \sum_{k=1}^{\text{num}} H_i(y) * B(y) - \beta_3 \sum_{k=1}^{\text{num}} S_i(y) * B_i(y) \end{aligned} \tag{14}$$

$$\begin{aligned} \sum_{k=1}^{\text{num}} D_i(y) * H_i(y) = \beta_0 \sum_{k=1}^{\text{num}} H_i(y) + \beta_1 \sum_{k=1}^{\text{num}} B_i(y) * H_i(y) \\ + \beta_2 \sum_{k=1}^{\text{num}} H_i(y)^2 - \beta_3 \sum_{k=1}^{\text{num}} S_i(y) * H_i(y) \end{aligned} \tag{15}$$

$$\begin{aligned} \sum_{k=1}^{\text{num}} D_i(y) * S_i(y) = \beta_0 \sum_{k=1}^{\text{num}} S_i(y) + \beta_1 \sum_{k=1}^{\text{num}} B_i(y) * S_i(y) \\ + \beta_2 \sum_{k=1}^{\text{num}} H_i(y) * S_i(y) - \beta_3 \sum_{k=1}^{\text{num}} S_i(y)^2 \end{aligned} \tag{16}$$

Solution of Eqs. 13, 14, 15, and 16 gives values of β_0, β_1, β_2 and β_3. Linear coefficients c_1, c_2, c_3 and c_4 can be obtained if α is known. As discussed that α is used to normalize depth. Thus, it can be obtained as:

$$\alpha = \begin{cases} 1 & if \quad d(y) \leq 1 \\ d_{\max} & \text{otherwise} \end{cases} \tag{17}$$

where $d_{\max}$ is maximum of scene depth $D(y)$.

5.2 *Data Preparation for Regression Analysis Using Ordinary Least Square Estimation Method*

Ground truth of the depth is unavailable due to constraint of nature (Depth of an object in a scene may change over time). Thus, 200 images captured in fine weather of outdoor scenes (mountains, animals, trees, etc.) are used to prepare sample space. For each sample image, depth $D_i(y)$ is obtained randomly using Gaussian distribution with parameters $\mu = 0, \sigma^2 = 0.5$. Experiments have been conducted to select value of σ^2 to bring proper diversity in the depth map. It is found that $\sigma^2 = 0.5$ is sufficient to bring proper diversity. Atmospheric light Ar is randomly obtained using uniform standard distribution. Hazy images with respect to each clear day image are prepared using Eq. 2. Curve of Eq. 11 is fitted on this sample space using Eqs. 13, 14, 15, and 16. This curve fitting gives values of linear coefficients ($c1 = 0.0122$, $c2 = 0.9592$, $c3 = 0.9839$, and $c4 = 0.7743$).

6 Properties of the Enhanced Depth Approximation Model and Scene Restoration

6.1 *Edge Preserving Property*

Existing edges are preserved and degraded edges are recovered by enhanced depth approximation model. Gradient of Eq. 6 will be as in Eq. 18 [9].

$$\Delta D(y) = c_2 \Delta B(y) + c_3 \Delta H(y) - c_4 \Delta S(y) + \Delta \epsilon \tag{18}$$

where $\Delta\epsilon = 0$ according to principle of OLS Eq. 18 proves that gradients of $D(y)$ depend upon gradient of $B(y)$, $H(y)$, and $S(y)$, which indicates presence of edge in $D(y)$ if and only if there is an edge in $B(y)$, or $H(y)$, or $S(y)$. Figure 3 shows effect of hue on color attenuation prior. It can be noticed from Fig. 3f, g that edges preserved by the proposed model are more accurate. Peak signal-to-noise ratio($psnr$) obtained by the proposed model proves its accuracy.

6.2 *White Regions Handling*

Due to additive airlight, amount of whiteness increases in degraded image. Thus, differentiation of real white objects from atmospheric light becomes difficult. The proposed method may approximate wrong transmission in the presence of white objects in the scene. Presence of white objects results in increased brightness, low saturation, and moderate hue.

Fig. 3 Hazy image (**a**), Ground truth of clear day image (**b**), Edges in Fig. 3a (**c**), Edges in Fig. 3b (**d**), Approximated depth using method in [29] (**e**), Edges in Fig. 3e ($psnr = 62.85$) (**f**), Approximated depth obtained by the proposed method (**g**), Edges in Fig. 3f ($psnr = 63.71$) (**h**)

In [2, 29], the problem of white region is solved by assuming that pixels are locally at same depth. Minimum filter is used by these methods to refine depth map locally. However, minimum filters results in loss of existing edges as shown in Fig. 3f. Therefore, median filter is used by the proposed method, which solves problem of white object upto a level and preserve existing edges. Thus, the refined transmission is expressed as:

$$T_r(y) = \text{med}_{x \epsilon \omega_r(y)} \text{Tr}(x) \tag{19}$$

where $T_r(y)$ is refined transmission, $\omega_{r(y)}$ is window of size $y \times y$, and $\text{Tr}(x)$ is approximated transmission. To reduce blocking artifacts which are introduced due to window-based operation, refined depth map is further smoothed using guided filter [3].

6.3 *Restoration of $I_1(y)$*

Atmospheric light A^c in each c color channel is estimated using method of [2], where $c\epsilon(R, G, B)$. The proposed method further takes minimum of atmospheric light of each color channel as global atmospheric light, which is given by Eq. 20.

$$Ar = \min_{c\epsilon(R,G,B)} (A^c) \tag{20}$$

where min is a function to compute minimum of given values, Ar is global atmospheric light. Equation 6 is used to approximate scene depth $D(y)$, transmission is obtained using Eq. 2 and refined using Eq. 19. Once Ar and $Tr(y)$ are obtained then

Eq. 1 can be used to restore image $I_1(y)$. The value of γ is critical in restoration of $I_1(y)$. Low value of γ generates residual haze and its high value increases dehazing level, which darkens the dehazed image. Thus, the proper value of γ is vital for restoration. The proposed work considered $\gamma = 1$.

7 Experimental Analysis

The MATLAB version R2014a is used to implement the proposed method. Waterloo IVC dehazed image data set(WID) [6] is used to verify effectiveness of the proposed method. This data set consists of 25 hazy images of outdoor scenes. Results obtained by the proposed method are compared with method in [2, 8, 22, 23, 25, 29].

7.1 Qualitative Evaluation

Figure 4 shows comparison of the results based on visual quality. Figure 4b shows that method of [2] generates artifacts near depth discontinuities and color distortion in sky region. Figure 4c shows the results obtained in [23]. It can be observed from Fig. 4c that results are dark due to overestimation of transmission. Results obtained in [8] are promising as shown in Fig. 4d. However, method of [8] is time consuming due to regularization. Method of [22] obtains better results, which are shown in Fig. 4e. However, this method produces wrong results in dense haze. Method of [29] is fast. However, this method wrongly estimates depth in the presence of white object as shown in Fig. 4f. Method in [25] is fast in comparison with other methods. However, results obtained by method in [25] are not visually pleasant as shown in Fig. 4g. The

(a) (b) (c) (d) (e) (f) (g) (h)

Fig. 4 Comparison of visual quality Hazy image (**a**), [2] (**b**), [23] (**c**), [8] (**d**), [22] (**e**), [29] (**f**), [25] (**g**), and proposed method (**h**)

(a) Hazy Image (b) $\gamma = 1$ (c) $\gamma = 1.2$ (d) $\gamma = 1.5$

Fig. 5 Effect of varying γ on restored images. Hazy image (**a**), are restored images (**b–d**)

proposed method obtains better visual results as shown in Fig. 4h. It restores natural colors in sky region as well as in non-sky region.

Furthermore, Fig. 5 shows the effect of varying γ on restored images. Image shown in Fig. 5a is restored with varying values of $\gamma = [1, 1.2, 1.5]$ using the proposed method. Restored images are shown in Fig. 5b–d. It can be observed that darkness of restored images increases with increased value of γ. Thus, adaptive γ is essential for accurate dehazing.

7.2 *Quantitative Evaluation*

The proposed method is validated using quantitative metrics to measure strength on the basis of restored edges, structure, and texture. Metrics e and $\bar{r}$ are computed, which quantifies the strength of the proposed method to restore and preserve edges [24, 27] using WID data set. Increasing values of e and $\bar{r}$ indicate improved quality of results.

Obtained values of metrics e and $\bar{r}$ for images shown in Fig. 6 are given in Tables 1 and 2. The proposed method performs well in comparison with method in [2, 8, 22, 23, 29] on the basis of obtained values of e and $\bar{r}$.

(a) (b) (c) (d) (e) (f)

Fig. 6 Images used to compute parameters e and $\bar{r}$ as given in Tables 1 and 2, respectively

Table 1 Comparison on the basis of e of images shown in Fig. 6 [15]

e	[23]	[2]	[22]	[29]	Proposed method
(a)	1.11	0.47	0.51	0.62	1.52
(b)	0.30	0.10	0.80	0.13	0.19
(c)	4.47	3.15	2.88	2.74	3.12
(d)	0.99	2.12	2.73	1.58	1.87
(e)	1.55	2.08	2.33	1.01	2.28
(f)	0.79	0.72	0.75	0.95	1.74
Avg.	1.53	1.44	1.55	1.18	1.79

Table 2 Comparison on the basis of $\overline{r}$ of images shown in Fig. 6 [15]

r	[23]	[2]	[22]	[29]	Proposed method
(a)	1.79	1.08	1.15	1.47	1.48
(b)	1.41	0.95	0.96	1.14	1.15
(c)	2.46	1.84	1.50	1.82	3.98
(d)	1.58	3.11	3.76	2.53	4.44
(e)	2.27	2.83	3.01	2.23	3.34
(f)	1.93	1.50	1.50	2.50	2.10
Avg.	1.91	1.89	1.98	1.95	2.75

8 Conclusions

An enhanced depth approximation model has been proposed. The proposed depth approximation model is based on an observation that depth of a pixel from the camera is directly proportional to the difference of the saturation from sum of brightness and hue. Transmission obtained by the proposed method is further refined using local median filtering, which helps in preserving existing edges. Accuracy of the proposed model is proved on the basis of visual quantity and qualitative metrics. However, the proposed method is based on homogeneous scattering of light. Thus, this issues will be part of future work.

References

1. Fattal R (2008) Single image dehazing. In: Proceedings of ACM SIGGRAPH, pp 72:1–72:9
2. He K, Sun J, Tang X (2011) Single image haze removal using dark channel prior. IEEE Trans Pattern Anal Mach Intell 33(12):2341–2353
3. He K, Sun J, Tang X (2012) Guided image filtering. IEEE Transaction on Pattern Analysis and Machine Intelligence 35(6):1397–1409

4. Kim JY, Kim LS, Hwang SH (2001) An advanced contrast enhancement using partially overlapped sub-block histogram equalization. IEEE Trans Circ Syst Video Technol 11(4):475–484
5. Kim TK, Paik JK, Kang BS (1998) Contrast enhancement system using spatially adaptive histogram equalization with temporal fltering. IEEE Trans Consum Electron 44(1):82–87
6. Ma K, Liu W, Wang Z (2015) Perceptual evaluation of single image dehazing algorithms. In: IEEE International Conference on Image Processing, Quebec City, QC, Canada, pp 3600–3604, Sept 2015. https://doi.org/10.1109/ICIP.2015.7351475
7. MathWorks T, Color space conversion (2017), https://in.mathworks.com/help/vision/ref/colorspaceconversion.html
8. Meng G, Wang Y, Duan J, Xiang S, Pan C (2013) Efficient image dehazing with boundary constraint and contextual regularization. In: IEEE international conference on computer vision, pp 617–624
9. Mi Z, Zhou H, Zheng Y, Wang M (2016) Single image dehazing via multi-scale gradient domain contrast enhancement. IET Image Process 10(3):206–214. https://doi.org/10.1049/iet-ipr.2015.0112
10. Narasimhan SG (2004) Models and algorithms for vision through the atmosphere. Ph.D. thesis, New York, NY, USA
11. Narasimhan SG, Nayar SK (2000) Chromatic framework for vision in bad weather. In: IEEE conference on computer vision and pattern recognition, vol 1, pp 598–605
12. Narasimhan SG, Nayar SK (2003) Contrast restoration of weather degraded images. IEEE Trans Pattern Anal Mach Intell 25(6):713–724
13. Nayar SK, Narasimhan SG (1999) Vision in bad weather. IEEE Conf Compu Vis 2:820–827
14. Nayar SK, Narasimhan SG (2003) Interactive deweathering of an image using physical models. In: IEEE workshop on color and photometric methods in computer vision in conjunction with IEEE conference on computer vision, Oct 2003
15. Raikwar SC, Tapaswi S (2017) An improved linear depth model for single image fog removal. Multimedia Tools Appl 77(15):19719–19744
16. Raikwar SC, Tapaswi S (2018) Tight lower bound on transmission for single image dehazing. The Visual Computer. https://doi.org/10.1007/s00371-018-1596-5
17. Schechner YY, Narasimhan SG, Nayar SK (2001) Instant dehazing of images using polarization. IEEE Conf Comput Vis Pattern Recogn 1:325–332
18. Shwartz S, Namer E, Schechner YY (2006) Blind haze separation. IEEE Conf Comput Vis Pattern Recogn 2:1984–1991
19. Stark JA (2000) Adaptive image contrast enhancement using generalizations of histogram equalization. IEEE Trans Image Process 9(5):889–896
20. Tan K, Oakley JP (2000) Enhancement of color images in poor visibility conditions. IEEE Conf Image Process 2:788–791
21. Tan R (2008) Visibility in bad weather from a single image. In: IEEE conference on computer vision and pattern recognition, pp 24–26
22. Tang K, Yang J, Wang J (2014) Investigating haze-relevant features in a learning framework for image dehazing. In: IEEE international conference on computer vision and pattern recognition, pp. 2995–3002
23. Tarel JP, Hautière N (2009) Fast visibility restoration from a single color or gray level image. In: Proceedings of IEEE international conference on computer vision. pp 2201–2208, Sept 2009
24. Wang R, Li R, Sun H (2016) Haze removal based on multiple scattering model with superpixel algorithm. J Signal Process 127(C):24–36
25. Wang W, Yuan X, Wu X, Liu Y (2017) Fast image dehazing method based on linear transformation. IEEE Trans Multimedia 19(6):1142–1155. https://doi.org/10.1109/TMM.2017.2652069
26. Wang Z, Bovik AC, Sheikh HR, Simoncelli EP (2004) Image qualifty assessment: from error visibility to structural similarity. IEEE Trans Image Proces 13(4):600–612
27. Xu Y, Wen J, Fei L, Zhang Z (2015) Review of video and image defogging algorithms and related studies on image restoration and enhancement. IEEE Access 4:165–188

28. Zhang YQ, Ding Y, Xiao JS, Liu J, Guo Z (2012) Visibility enhancement using an image filtering approach. EURASIP J Adv Signal Process 2012(1):220–225
29. Zhu Q, Mai J, Shao L (2015) A fast single image haze removal algorithm using color attenuation prior. IEEE Trans Image Process 24(11):3522–3533

Energy-Efficient and Secured Mobile Agent Itinerary Approach in Wireless Sensor Network

Shivani Chaudhary, Umesh Kumar, and Sapna Gambhir

Abstract In wireless sensor network, mobile agent (MA) paradigm is the substitute which has many advantages over client–server paradigm. As energy consumption is the major issue for research in WSNs, MA-based paradigm has been a boon in resolving this. In spite of sending data from all nodes to sink, mobile agent itself migrates to all the sensor nodes and collects data which results in minimized energy and bandwidth consumption. But for better efficiency, it is required to plan the itinerary in a best way so that MA can collect data from source nodes efficiently. Many algorithms have been given by researchers which consider all sensor nodes to be visited by MA which is incompetent, and none of the algorithm considers the detection of malicious node and preventing mobile agent to migrate to malicious nodes. Proposed algorithm, first stores the device signature of all intended participating sensor nodes, cluster formation, and eventually by authenticating the nodes based on fingerprints an itinerary for MA will be designed. Cluster head nodes are traversed excluding MNs in CHs itinerary, and data will be aggregated from benign nodes only.

Keywords Mobile agent · Wireless sensor network · Clustering · Device fingerprinting · Itinerary planning · Authentication

S. Chaudhary · U. Kumar (✉) · S. Gambhir
J.C. Bose University of Science and Technology, YMCA, Faridabad, Haryana, India
e-mail: umesh554@gmail.com

S. Chaudhary
e-mail: shivanichaudhary522@gmail.com

S. Gambhir
e-mail: sapnagambhir@gmail.com

M. N. Favorskaya et al. (eds.), *Innovations in Electrical and Electronic Engineering*, Lecture Notes in Electrical Engineering 661, https://doi.org/10.1007/978-981-15-4692-1_53

1 Introduction

Wireless sensor network (WSN) is one of the emerging technologies for collecting data in different real-time sensitive applications at high level. A device which is tiny in size, low powered, and at low cost is used in WSNs for sensing the physical or environmental conditions which is known as sensor nodes (SNs). These nodes then communicate the sensed data to the sink and sink to the intended user or client at remote site according to the application's requirement [1].

After sensing, SNs need to send data to sink, for that purpose a traditional client–server based model is earlier used which is implemented as: each sensor node in the network via multi-hop routing sends the complete sensed data to the sink. But it has many drawbacks, and hence, it is considered as inefficient in many ways. Sensor nodes are resource constrained device as limited battery, and due to human unreachable deployed networks, it is difficult to replace the batteries. This is the reason that managing power somehow in SNs is the important issue.

Another major drawback is in large-scale network, when large amount of data is processed to sink over a wireless link. In this, due to shared bandwidth, there is possibility that low bandwidth links may not handle the enormous data transmission [2].

To counter these problems another technology, mobile agent-based paradigm has come to light and accepted by many researchers for limiting the data transmission by each sensor node and preventing the critical resources like battery and bandwidth. In MA-based paradigm instead of each sensor nodes sending data to sink, a processing element migrates to each node for the phenomena of data gathering. This processing element is mobile agent, and it is a software entity or a program encoded in a data packet to perform a specific function and roams for gathering data after processing at nodes [3].

Energy-efficient agent migration is of major concern but besides that another serious problem is the security of MAs against compromised or malicious nodes in the dynamic network. In mobile agent-based model, each mobile agent has been providing an execution environment by the sensor nodes in the particular itinerary. Unknowingly, when sensor nodes are getting compromised by any internal or external malicious node, that malicious adversary node can severely spoil the execution of an agent by accessing the security keys and actual information can be affected by modifying or corrupting mobile agent's state information or code, by preventing the agent's processing code execution, or denying agent service requests [4].

In order to prevent the intrusion of such malicious node, integrity protection and symmetric key authentication only are not enough, reliability of nodes should be determined and avoid the compromised nodes during MA itinerary planning or the agent's visit to source nodes. For this, device fingerprinting technique [5] is used for the authentication of nodes and only allowing those authenticated nodes to be visited by the agent for collecting the data [6].

Main technical contributions of this paper are as:

1. Fingerprint-based framework for evaluating the authentication of nodes.
2. Cluster head selection on the basis of density impact factor and clusters formation.
3. An energy-efficient and security-aware itinerary planning approach based on fingerprints and cluster heads.

The result of proposed approach with the existing local closest first approach using performance parameters is compared.

2 Related Work

In wireless sensor network, for MA-based paradigm discovering an efficient itinerary or route for MA migration is the major research area for the efficient data collection from large number of source nodes or sensor nodes. Extensive research has been done in recent years for the same. The selection of set of SNs and sequence to be visited in an itinerary plays very crucial part and affects noticeably the accuracy and quality of data aggregation. Various proposed algorithms related to mobile agent itinerary planning [4, 7] are studied.

Chen et al. [8] have given routing protocols in wireless sensor networks and conclude that in wireless sensor networks, client–server model is the most commonly used for data aggregation from sensor nodes, in which every node transmits the sensed data itself to sink node via multi-hop. Due to the reason that data transmission takes place in huge amount to the sink, this model becomes inefficient with the growing network size. Then mobile agent-based model has been proposed by many authors as the energy-efficient solution. But it has been noticed that in this model to plan the route or itinerary for agent is the challenging issue.

First solution for this issue was given by Akkaya et al. [9] and Qi et al. [10]. Two heuristic algorithms have been proposed; in LCF, an agent has been dispatched by the sink and then agent looks for the next node which can be selected on the criteria that distance between the current node and next node to be selected must be the shortest and so on for remaining nodes. Whereas GCF works as initially mobile agent dispatched from sink and checks for the next sensor node closest to the sink every time. GCF chooses the source nodes in its itinerary based on global network distance matrix not based on local location of nodes.

An algorithm almost similar to LCF has been designed by Kwon et al. [11] called (MADD) mobile agent-based directed diffusion algorithm which is almost similar to LCF but differs only when selecting the first node after sink, it is done in such a way that chose the farthest node as the initial one after the sink instead of considering the nearest one.

In [12], Wu et al. have given an approach for determining the itinerary for a mobile agent based on genetic algorithm. To design a static itinerary using GA, global knowledge of network topology must be known prior. In terms of energy consumption, LCF and GCF can be defeated by this genetic algorithm in performance. Approaches in

[10] and [12] perform adequately for normal size networks, but for the maintenance of global topology information, each node has to report its state to sink which leads GA to suffer loads of communication overhead.

Konstantopoulos et al. [13] and Venetis et al. [14] proposed two efficient approaches: first as (IEMF) itinerary energy minimum for first-source selection and second itinerary energy minimum algorithm (IEMA). IEMF employs as after sink it selects the first-source node in itinerary as the one with the lowest estimated energy cost of its subsequent route among other routes and then uses the LCF approach for further remaining route planning. When IEMF gets iterated k times for energy efficiency enhancement, it is then considered as IEMA approach. In general, IMEF signifies the prominence of selecting the first node to be visited. Energy costs of different alternatives are estimated for the selection of first node, and decision is made on the basis of minimum energy cost. IEMA iterations are performed on the basis of choosing the next node with optimal energy from remaining nodes.

Mobile agent paradigm is major research area but matter in this is planning the migration protocol, for mobile agents. Initially, two algorithms for itinerary planning problem are proposed in [15] local closest first (LCF) and global closest first (GCF) algorithms. In LCF, the MA begins from the sink node and moves to the closest or nearest SN based on the current location of the MA. In GCF, mobile agent selects the next node which is most nearer to the sink node. LCF gives poor performance as it considers the far nodes to be visited in last. On the other hand, GCF incurs very long routes.

A number of researchers have worked and proposed many protocols to solve this problem, but all protocols proposed that MAs visit all SNs, and this leads to consumption of more energy. But in MAPA protocol [16], routes are planned for MA which includes only CHs in the itinerary. By planning itinerary for MAs among CHs based on minimum spanning tree, the CHs located near to each other are added to the same itinerary. MAPA protocol gives better performance as compared to the protocols proposed previously as it consumes less energy and takes less time for MAs data collecting tasks. In this protocol, consumption of energy is minimized and it also consumes lesser time as data is collected from CH's only.

Authors in [17] have presented the problem of static itinerary-based agent migration protocol in WSN. ETMAM has presented an integrated solution for reliable agent migration within network. In ETMAM, next node is selected based on both energy and trust. In addition, ETMAM uses agent to reduce itinerary length and trusted node on basis of trust value. It finally comes out with the trust and energy-aware routing protocol for mobile agents.

Authors in [18] dealt with the issue of MA itinerary planning in WSN and briefly reviewed the various itinerary planning approaches, i.e., both for the single agent and for the multi-agent networks. In this approach, previous node asks for information packet from all neighbors, calculates information gain, and then selects the subsequent node of MA migration. The proposed approach dealt with two important aspects of MA-based paradigm: accuracy and information gain and is therefore advantageous over the existing approaches in terms of task duration and energy usage.

3 Proposed Work

In this section, a mobile agent itinerary planning algorithm for data aggregation in wireless sensor network which is secured and energy aware is proposed. Proposed idea aims to avoid the internal attacks such as denial-of-service (DOS) attack, selective forwarding attack, flooding, and blackhole attack by planning an itinerary. This, for mobile agent will also preserve the resources of sensor nodes, energy is one of them which is a major constraint for wireless sensor network.

In order to reduce the energy consumption of sensor nodes as when mobile agent migrates to all the nodes in network, then each and every node has to be active and itinerary is formed using each node; therefore, there is wastage of energy which can be rectified using the MA migration only to the cluster heads so that lifetime of network can be prolonged. For that purpose firstly, we divide our network into clusters and later on with the help of routing mobile agent only to the cluster heads network life may be extended up to an extent. As wireless sensor networks are prone to attacks [19], mostly to internal attacks. Internal attacks are those when any legitimate node acts illicit and make use of compromised nodes for forwarding the data to wrong destination or can drop the packets which lead to the disruption of network. To avoid the participation of compromised nodes in itinerary and data aggregation, we use the authentication of sensor nodes based on fingerprinting of devices which is extracted using the physical parameters of sensor nodes, that is, MAC addresses, received signal strength, clock skew which are known as unique signature of the device or node.

The procedure for designing an MA itinerary which is based on cluster heads and fingerprints-based authentication is as at the starting of nodes deployment in network we will first extract the node signature which are defined as the unique features of nodes. When signatures of all the nodes are stored in a set, it is known as fingerprint set, which is basically can be understood as registrations of nodes are done and fingerprints are forwarded to sink node. After this, on the basis of impact factor density and remaining battery of nodes cluster heads are elected from the network. Then each node by calculating and comparing transmission range of nodes with cluster head's transmission range. Now when we are done with the clustering, further itinerary is planned for MA's migration to aggregate the data in an efficient way. In this part, when MA is dispatched to the first cluster head using LCF approach, it first verifies for its signature if matching with the registered fingerprints. And if it is, then MA is migrated through this authenticated node, and CH notifies its cluster members to send the sensed data and their signature to identify the illicit node or authenticated node. While in the other case, if signatures of individual nodes does not matched, then node is considered as compromised or malicious node, and MA excludes that malicious CH and does not follow that node in itinerary. Eventually when MA has reached all the CHs in network, it begins migrating back to the sink and collecting data from CHS that are involved in itinerary at the time of discovering the route.

In this way, MA itinerary which is formed through authenticated cluster heads only and bypassing the compromised nodes is used. Also data is being collected through registered nodes only. Following flowchart is demonstrating the flow of proposed work step by step in the next section.

3.1 Phases of Proposed Work

In the following scenarios, different phases of the proposed algorithms have been presented with the help of diagrams. First we would have network topology of sensor nodes deployed in the network monitoring area, and sink is located at center of this monitoring region. Our proposed approach for MA itinerary planning consists of four phases as Phase 1: Fingerprint extraction and storing at sink node, Phase 2: Cluster head selection and cluster formation, Phase 3: Fingerprint and clustering-based MA itinerary planning, and Phase 4: Data aggregation by MA.

A network model assumed for this work, a wireless sensor network having '*i*' number of sensor nodes distributed at random in a square area and sink at the center of monitoring network area. Sink and SNs both have specific transmission range but sink is more powerful and intelligent than SNs. Also, sink is equipped with more computational capabilities and energy. Network model is shown in Fig. 1. In this model, sink is supposed to be the more powerful and has the needed information regarding SNs, that is, their position with the coordinates with the help of GPS or localization phenomenon [20].

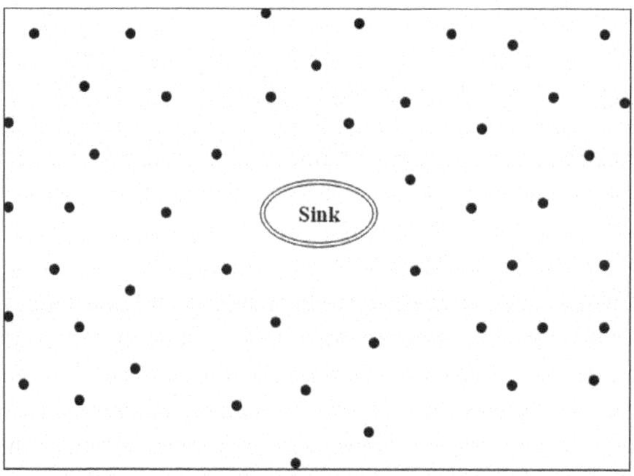

Fig. 1 Network topology of i SNs and sink node

3.2 Fingerprint Generation of SNs and Stored at Sink

After setting up network topology, next task for sink to have all unique physical characteristics of sensor nodes. And with the help of all node's physical properties which are known as sensor node signature sink node will generate a set of all node's signature which is called as device fingerprints and noted as FSN. This fingerprint will be used by the mobile agent in further process for comparing the fingerprints so that only benign or trusted node can participate in the itinerary planning which is our sole purpose.

Input: S_n - Set of Sensor nodes
GN - Gateway or Sink node, (base station)
MA - Mobile Agent
Output: F_{SN} – set of Fingerprints stored at GN
Begin:
1. Repeat for all in S_n
2. Extract
   2.1 Measured Clock skew
   2.2 RSSD
   2.3 MAC address
3. End loop
4. Forward extracted Device fingerprint F_{SN} to GN
5. GN stores F_{SN} in its cache memory & forward to the application server

Algorithm 1: Physical features extraction from sensor nodes

In Algorithm 1, different physical characteristics like SNs Mac address, received signal strength distribution (RSSD), clock skew (difference between clock running speeds of SNs) can be used as unique features of an individual node.

Using these device signatures, we can create a unique id or fingerprint for all the SNs so that these nodes can be identified as the one which are intentionally deployed by user for an application. The output of Algorithm 1 is as FSN will be extracted and stored at the sink node cache and forwarded to the user application server also for later use (Fig. 2).

3.3 Cluster Head Selection and Cluster Formation

In this part, main idea is to calculate the remaining energy of all the nodes and distribute density impact factor of all SNs to other nodes. When there is i number of nodes, each SN will receive $(i - 1)$ impact factor from others excluding itself. After comparing the battery of each node with threshold say 15% and computing the density of each node i with the help of Eq. 1, the node with the maximum density will be elected as the cluster head.

Density impact factor can be calculated as [21]

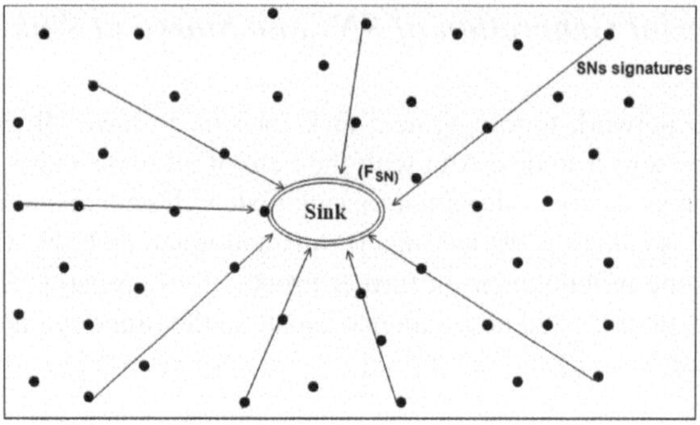

Fig. 2 Fingerprints extraction of each SN and stored at sink

$$Ijk = e^{\frac{-\left(H_k^j - 2\right)^2}{2\sigma^2}} \tag{1}$$

where H_k^j is the estimation of hop count between nodes j and k and parameter σ determines how strongly nodes impact each other. Formula for hop count estimation is given in [22] as

$$H_k^j = \frac{d(k-1,k)}{T_{\max}} \tag{2}$$

In Eq. (2), d is the distance between k and $(k - 1)$th nodes and $T_{\max}$ is the transmission range of each sensor node. In Algorithm 2, first we will mention how many cluster heads we are going to select in the network area that is N_c. We are first calculating the impact factor density of each node with the help of Eq. (1). After that we have to check the remaining energy of each SN by comparing it with the threshold, if node's energy is less than $T_{h,}$ then this node is not the part of further work and will be eliminated and added to the set faulty nodes and nodes with the energy greater than T_h will be added to the set $SN_{left.}$ After that we select node i from group SN_{left} which will be having the maximum density impact factor value and elected as cluster head and will be added to the set $S_{ch;}$ after selection of cluster head nodes, we include all the nodes whose distance from cluster head is less than its transmission range $T_{max,}$ that is, $(l_{ch}, i) < T_{max}$ and take out these nodes from remaining SN_{left} group. In this way as shown in Fig. 3, cluster heads selected and clusters have been formed.

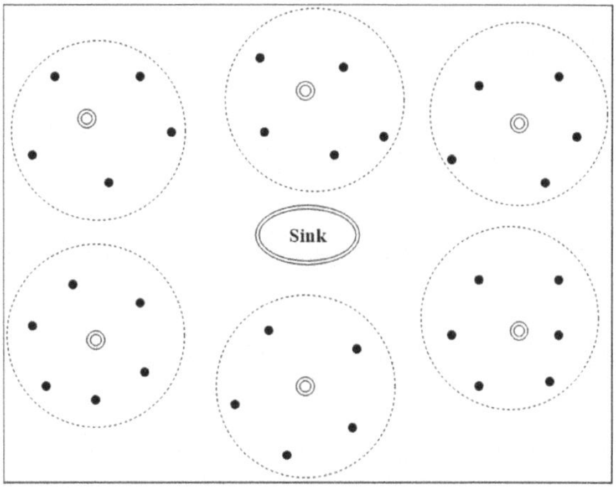

Fig. 3 Cluster head selection and cluster formation

Input: S_n - set of Sensor nodes
GN - Gateway or Sink node, **MA** - Mobile Agent
Output: S_{ch}, S_{Group}
Parameters: d_i : density factor, l_{ij} : distance between i and j nodes

1. Repeat for each node i in S_n do
   $d_i \leftarrow 0$
2. end loop
3. For all $i \in S_n$ do
   3.1. For all $j \in S_n$ do
      3.1.1. Calculate d_{ij} using(1)
      3.1.2. $d_i = d_i + d_{ij}$
   3.2. end for
4. end for
5. for all i in S_n, compare remaining energy (E_i) of nodes
   5.1. If (E_i < Threshold(T_h))
      CH_{faulty} <- CH_{faulty} + i
      $S_{n(left)}$ <- S_n – i
6. For all i in $S_{n(left)}$ do
   6.1. If (max d_i)
      6.1.1. Select i as CH
         $S_{n(left)}$ <- $S_{n(left)}$ – CH
         S_{CH} <- S_{CH} + CH
      6.1.2. For i in Sn(left)
         Calculate (L_{CH} , i)
      6.1.3. if (L_{CH} , i) < T_{max} , then
         $S_{n(left)}$ <- $S_{n(left)}$ – i
         S_{group} <- S_{group} + i
   6.2. End if
7. End for

Algorithm 2: Optimized clustering and cluster head selection

This process will be continued to select the cluster heads and clustering again and again for all SNs until any of the node left from being the part of a cluster.

From Algorithm 2, we will get output as two groups S_{CH} and S_{group}, that is, cluster head nodes and cluster members associated. These will be used in Algorithm 3 for planning the cluster head itinerary planning. Clustering reduces the number of nodes taking part in communication, ensures scalability for large number of nodes, and reduces the communication overhead for both single hop and multi-hop.

3.4 Fingerprinting and Cluster Head-Based Secured MA Itinerary Planning

In this section, after selecting cluster heads and forming clusters based on the remaining battery greater than threshold and density of nodes as shown in Fig. 4. In the process of itinerary planning just like LCF, we first chose the cluster head node in the range of sink to begin the itinerary with min(u, v) where v is used for all cluster head nodes and in beginning sink is the u node. Algorithm 3 is used to plan the itinerary among CHs with the help of weight of cluster heads. Sink calculates weight of each CH as it has the location coordinates of each node in the network. Formula from [23] can be used for computing weights between CHs.

$$W = \alpha * H_C Hi^{(CHj)} + (1 - \alpha) * (H_s^C Hi + H_s^{(CHj)}) \quad (3)$$

where $0 < \alpha <= 1$

To generate an itinerary H_{CHi}^{CHj} is the hop count estimation between jth and ith cluster head nodes. Hop count estimation between sink is represented by H_{CHi}^{CHj}.

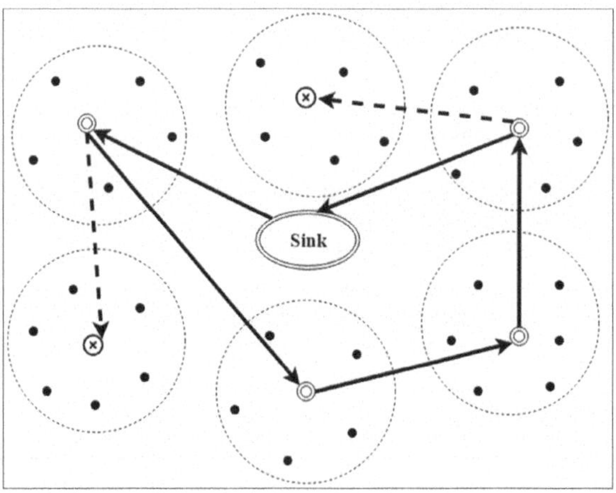

Fig. 4 Fingerprint and cluster head-based MA itinerary

Input: S_{CH} - set of Cluster head nodes
S_{group} - set of associated cluster head members
GN - Gateway or Sink node
MA - Mobile Agent
F_{SN} - Set of extracted device fingerprints (from algorithm 1)
Output: F_{SN} - set of Fingerprints stored at GN
Parameters: P_{CH} - set of CH noes participating in path or itinerary to be followed
S_{MN} - set of malicious nodes
1.Do
  1.1 $P_{CH} \leftarrow GN$
  1.2 $S_{MN} \leftarrow 0$
2.While ($u \in PCH$, $v \in SCH$) do
  Calculate min W (u, v)
3.If (signature (v)) $\in F_{SN}$
  3.1. P_{CH} <- P_{CH} + v
  3.2. Else
    S_{MN} <- S_{MN} + v
4.Do S_{CH} <- S_{CH} – v
5.End while

Algorithm 3: Fingerprinting and cluster head-based secured MA itinerary planning

In Algorithm 3, we begin from the sink node, then with the help of weight function in Eq. 3 find the first cluster head to be visited by MA which has minimum cost. Thereafter, the CH node with the minimum cost is matched with the fingerprints stored in FSN at sink. If node has same physical features or node signature as in FSN then it is considered as the next CH node to be in participating in itinerary; otherwise, it is known to be as malicious node and excluded from the itinerary. And further this process gets executed until all the CH nodes will be traversed by MA in itinerary. All the nodes which are included in itinerary are added in PCH.

In Fig. 4, it is shown first CH is selected on the basis of minimum cost and fingerprint matching and then MA is dispatched to this path and further path is constructed in the same way until all the CHs covered.

3.5 Optimized Data Aggregation and MA Migration

After forming clusters and itineraries planned among CHs based on fingerprints, sink thereafter dispatches MA to CHs in the planned itinerary. Firstly, when MA reached the CH, CH sends a packet to notify its cluster members for sending the sensed data and their device signatures to the cluster heads in their range and check the signatures sent by SNs are matching to FSN or not. If node seems like fingerprints not matching, data is not accepted by the CH and dropped else data is collected at

CH from all its ranging nodes. Subsequently, when MA reached to last CH it starts collecting the data from all the CHs in its itinerary back to the sink using MST [23]. In this way, data is aggregated in secured manner from legitimate nodes only, less energy of SNs consumed by MA. Eventually data collected by MA from CH, and that collected sensed data of all sensor nodes is forwarded to the sink and further to the user. Optimization of itinerary in proposed itinerary planning is such, this approach ensures bypassing malicious or compromised CHs in itinerary, resource preservation as only few CHs are participating in itinerary which saves the energy consumption of all nodes and leads to longer life of network. Data is being collected by authenticated sensor nodes only so there is no chance of flooding attacks by compromised nodes.

Input: P_{CH} - set of CH noes participating in itinerary
S_{CH} - set of Cluster head nodes
S_{group} - set of associated cluster head members
GN - Gateway or Sink node
MA - Mobile Agent
F_{SN} – Set of extracted device fingerprints(from algorithm 1)
Output: trusted data aggregated at sink node
Parameters - P_{CH} : set of CH noes participating in path or itinerary to be followed
S_{MN} : set of malicious nodes
MA <- 0

1. For each CH node i in P_{CH}
   1.1. Send hello packet to associated CH group members
2. For each node j in S_{group}
   2.1. Forward (sensed data$_j$ & signature$_j$) to CH$_i$
   2.2. If (signature$_j \in F_{SN}$)
      2.2.1. MA = MA+ data$_j$
   2.3. End if
3. End for
4. Forward MA to next CH in P_{CH}
5. End For

Algorithm 4: Optimized and secured data aggregation

In Algorithm 4, data is aggregated from benign or registered nodes only and itinerary is also formed among authenticated cluster head nodes. Initially, when MA migrates to the cluster head node in the planned itinerary, it notifies all its cluster members to send the sensed data and their signatures (Fig. 5).

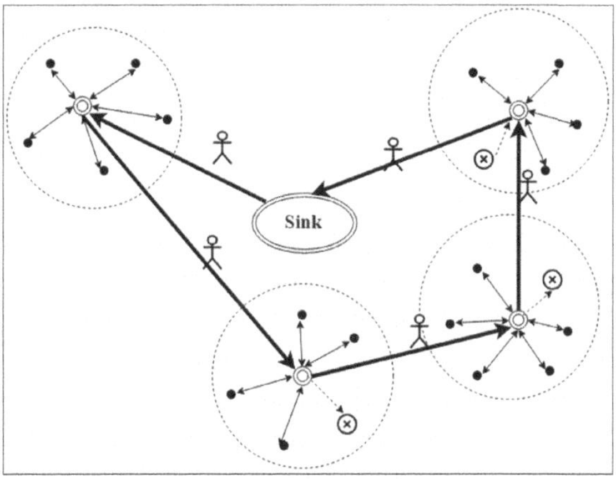

Fig. 5 Data aggregation from authenticated nodes

And after getting signatures, it matches with the pre-extracted fingerprints if those are being matched only, then it will accept the data from that sensor node and append the collected data to MA. This process is continued until MA reached to the sink node following the planned itinerary. While in other case, it rejects the incoming data from that malicious node and considers that node as malicious node. In this way, data is aggregated from secured or benign nodes only and mobile agent migrates only through authenticated cluster head nodes which ensure the secured and energy-efficient planning of MA migration route or itinerary.

4 Simulation and Results

In the proposed algorithm, we have formulated the solution for mobile agent itinerary which consider the security and energy for planning the itinerary for mobile agent when it is going to collect the data from sensor nodes and delivering back to the sink node. In our proposed work, we use ns2 (network simulator 2) for simulating the proposed algorithm with the help of network animator. Finally, following the itinerary which is secured and energy-efficient MA finally delivers the aggregated data from all cluster heads to the sink node. MA is dispatched by the sink and returns back to the sink after collecting the data in shown in Fig. 6.

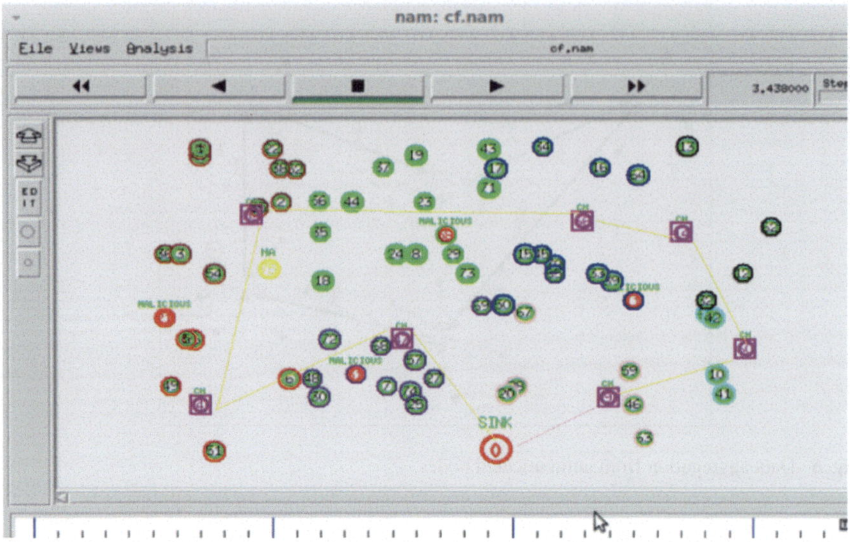

Fig. 6 Simulation of secured MA itinerary

4.1 Result Analysis

In Fig. 7, the comparison of existing algorithm and proposed algorithm is shown by using energy consumption parameter. It is clear that in proposed algorithm energy consumption of all the nodes is lesser than the existing algorithm LCF.

In Fig. 8, the comparison of existing algorithm and proposed algorithm is shown by using network lifetime parameter. It is clear that in proposed algorithm as energy consumption of nodes is lesser so lifetime of network is increased as compared to existing algorithm.

As we are using mobile agent here for gathering data network traffic is reduced and less overhead in the network. In existing algorithm, MA migrates to each SN therefore comparatively more overhead. In the proposed algorithm, there is less overhead as mobile agent visiting only cluster head nodes shown in Fig. 7.

Figure 9 shows that malicious nodes have been detected and avoided to form itinerary and collect data from CH nodes in the proposed algorithm while in existing algorithm even not a single MN is detected.

Table 1 shows the comparison of existing algorithm and proposed algorithm by using three important parameters like energy consumption, network lifetime, overhead, itinerary length, execution time, security.

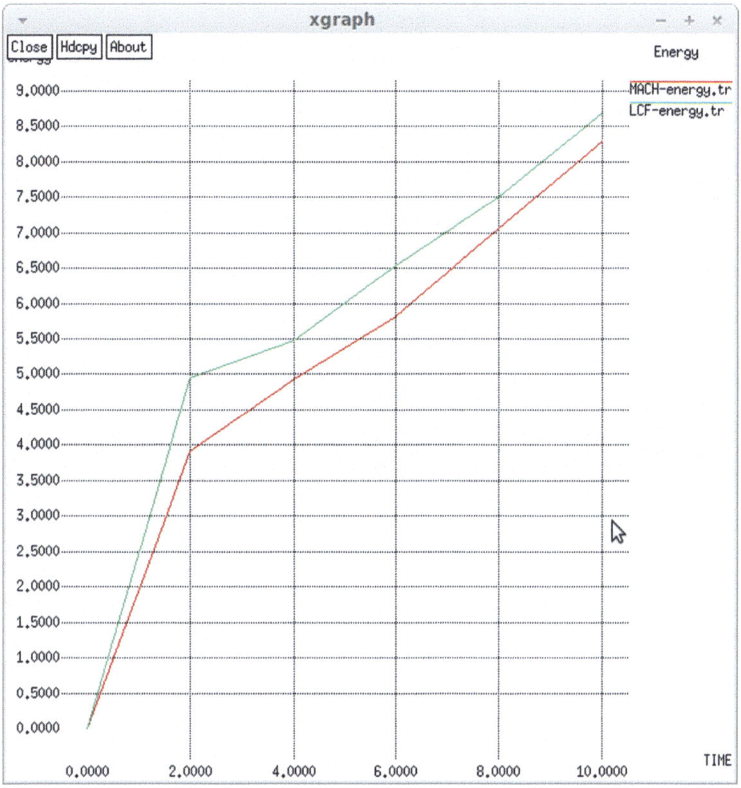

Fig. 7 Comparison of energy consumption

5 Conclusions

Our experiments based on NS2 for ordinary existing algorithm and proposed algorithm classifying network into clusters and by fingerprints matching can detect the malicious nodes which results in energy-efficient and reliable mobile agent migration in the network. In the proposed work energy and reliability of nodes, both have been considered as the criteria for selecting the next node for the migration of agent.

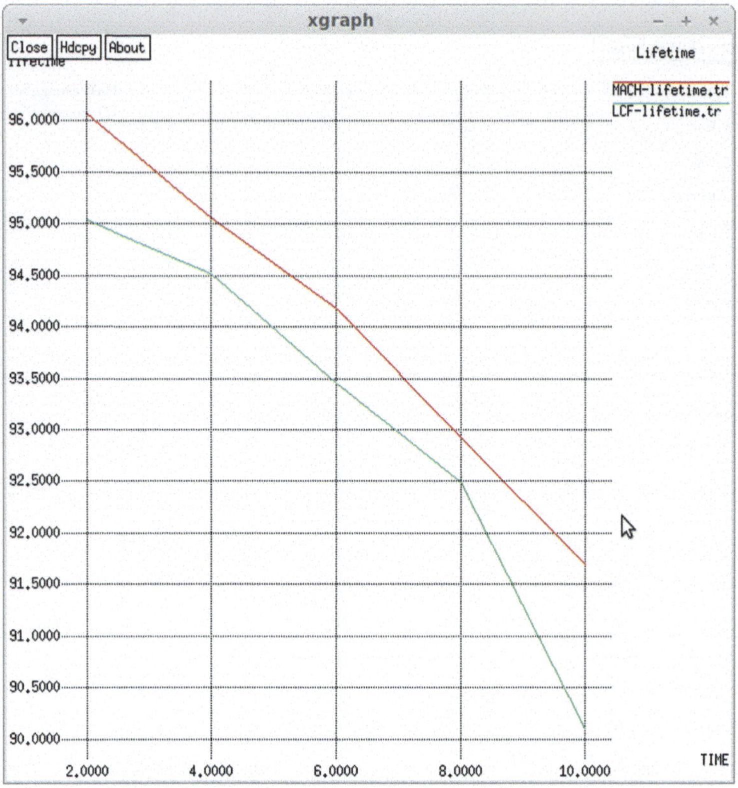

Fig. 8 Comparison of lifetime of network

It is also notable that with the energy, various attacks can be prevented to be held by using the fingerprint authentication of nodes. In the presence of malicious nodes in the network, our proposed algorithm performs better than the existing algorithms.

Henceforth, we conclude that the proposed work in this thesis ensures the energy efficient and reliable migration of agent through authenticated nodes. Simulation and implementation results found are acceptable and adequate to consider our proposed solution being capable of increasing the network lifetime and reliable transfer of mobile agent.

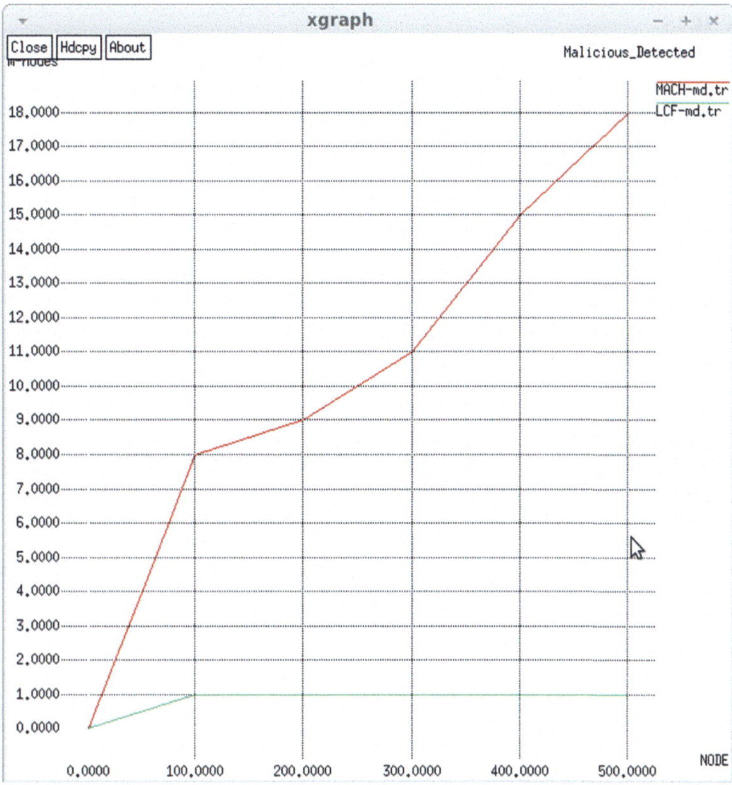

Fig. 9 Comparison using MN node detected and avoided

Table 1 Comparison of existing algorithm and proposed algorithm

Parameter	Existing algorithm (LCF)	Proposed approach
Energy consumption	High	Low
Network lifetime	Moderate	Moderate high
Overhead	Moderate	Very low
Itinerary length	Large	Very short
Execution time	High	Moderate low
Security	No	Security aware

References

1. Jayram BG, Ashoka DV (2013) Merits and demerits of existing energy efficient data gathering techniques for wireless sensor networks. Int J Comput Appl 66(9)
2. Cheng C-T, Tse Chi K, Francis CM, Lau (2010) A delay-aware data collection network structure for wireless sensor networks. IEEE Sens J 11(3):699–710
3. Vukasinovic I, Babovic Z, Rakocevic G (2012) A survey on the use of mobile agents in wireless sensor networks. In: 2012 IEEE international conference on industrial technology 2012. IEEE, Greece, pp 271–277
4. Chen X, Makki K, Yen K, Pissinou N (2009) Sensor network security: a survey. IEEE Commun Surv Tutor 11(2):52–73
5. Kumar U, Gambhir S (2018) Device fingerprint and Mobile agent based authentication technique in wireless networks. Int J Fut Gen Comm Netw 11(3): 33–48
6. Mei X, Liu D, Sun K, Xu D (2013) On feasibility of fingerprinting wireless sensor nodes using physical properties. In: 2013 IEEE 27th international symposium on parallel and distributed processing. IEEE, pp 1112–1121
7. Akyildiz IF, Melodia T, Chowdury KR (2007) Wireless multimedia sensor networks: a survey. IEEE Wirel Commun 14(6):32–39
8. Chen M, Kwon T, Yuan Y, Leung VC (2006) Mobile agent based wireless sensor networks. J Comput 1(1):4–21
9. Akkaya K, Younis M (2005) A survey on routing protocols for wireless sensor networks. Ad hoc Netw 3(3):325–349
10. Qi H, Wang F (2001) Optimal itinerary analysis for mobile agents in ad hoc wireless sensor networks. Proc IEEE 18(5):147–153
11. Chen M, Kwon T, Yuan Y, Choi Y, Leung VC (2006) Mobile agent-based directed diffusion in wireless sensor networks. EURASIP J Adv Signal Process 2007, 1–13
12. Wu Q., Rao NS, Barhen J, Iyenger SS, Vaishnavi VK, Qi H, Chakrabarty K (2004) On computing mobile agent routes for data fusion in distributed sensor networks. IEEE Trans Knowl Data Eng 16(6):740–753
13. Konstantopoulos C, Mpitziopoulos A, Gavalas D, Pantziou G (2009) Effective determination of mobile agent itineraries for data aggregation on sensor networks. IEEE Trans Knowl Data Eng 22(12):1679–1693
14. Venetis IE, Gavalas D, Pantziou GE, Konstantopoulos C (2018) Mobile agents-based data aggregation in WSNs: benchmarking itinerary planning approaches. Wirel Netw 24(6):2111–2132
15. Qadori HQ, Zulkarnain ZA, Hanapi ZM, Subramaniam S (2017) Multi-mobile agent itinerary planning algorithms for data gathering in wireless sensor networks: a review paper. Int J Distrib Sens Netw 13(1):1550147716684841
16. El Fissaoui M, Beni-Hssane A, Saadi M (2018) Multi-mobile agent itinerary planning-based energy and fault aware data aggregation in wireless sensor networks. EURASIP J Wirel Commun Netw 92(1)
17. Gupta GP, Misra M, Garg K (2014) Energy and trust aware mobile agent migration protocol for data aggregation in wireless sensor networks. J Netw Comput Appl 41:300–311
18. Lohani D, Varma S (2015) Dynamic mobile agent itinerary planning for collaborative processing in wireless sensor networks. In: 2015 annual IEEE India conference (INDICON). IEEE, pp 1–5
19. Brar RS, Arora H (2013) Mobile agent security issue in wireless sensor networks. Int J 3(1)
20. Alrajeh N, Bashir A, Shams MB (2013) Localization techniques in wireless sensor networks. Int J Distrib Sens Netw 9(6):304628
21. Chen M, Gonzalez S, Zhang Y, Leung VC (2009) Multi-agent itinerary planning for wireless sensor networks. In: International conference on heterogeneous networking for quality, reliability, security and robustness. Springer, Berlin, Heidelberg, pp 584–597

22. Chen M, Cai W, Gonzalez S, Leung VC (2010) Balanced itinerary planning for multiple mobile agents in wireless sensor networks. In: International conference on ad hoc networks. Springer, Berlin, Heidelberg, pp 416–428
23. Aşçı M, İleri CU, Dağdeviren O (2017) An energy-efficient capacitated minimum spanning tree algorithm for topology control in wireless sensor networks. In: 2017 25th signal processing and communications applications conference (SIU). IEEE, pp. 1–4

A New One Time Password Generation Method for Financial Transactions with Randomness Analysis

Varun Shukla, Arpit Mishra, and Shubhi Agarwal

Abstract All of us know that one time passwords (OTPs) are a part of life in today's scenario. We have to use them for authentication in many applications such as financial transactions, e-commerce websites, and in various government and nongovernment websites, etc. The strength of any OTP is uniqueness and it must not be predictable in any case. The method of OTP generation should be simple but safe enough to avoid any vulnerability. So a simple and unique OTP generation mechanism using RSA (Rivest, Shamir, Adleman) based string is presented in this paper with complete randomness analysis which also solves the associated problem of Syndicate bank's token based OTP procedure.

Keywords Authentication · OTP (One time Password) · Prime factorization · RSA (Rivest, Shamir, Adleman) · TFA (Two-factor Authentication)

1 Introduction

Authentication is a very important goal of cryptography [1]. In a peer to peer application, users (or entities) are supposed to get verified by different methods. Login-password mechanisms, fingerprint and retina detection, all are used for authentication but OTPs have dominated all the other authentication mechanisms because of many advantages [2]. One needs not to remember OTP as they are dynamic in nature and we don't need to store OTPs as well. Syndicate bank is a well known public sector bank in India and it is one of the major commercial banks in India. For internet banking for its users, Syndicate bank has introduced TFA (Two Factor Authentication) mechanism. In TFA, users are supposed to download and install RSA-Secure ID application software on their computers provided by the bank [3]. The step by step procedure is shown in Fig. 1.

V. Shukla (✉) · A. Mishra · S. Agarwal
Department of Electronics & Communication, Pranveer Singh Institute of Technology, Kanpur, Uttar Pradesh 209305, India
e-mail: varun.shuklaa@gmail.com

M. N. Favorskaya et al. (eds.), *Innovations in Electrical and Electronic Engineering*, Lecture Notes in Electrical Engineering 661, https://doi.org/10.1007/978-981-15-4692-1_54

Fig. 1 Showing the steps involved in TFA mechanism

Step 1:Select Two Factor Authentication
Step 2:Download RSA Secure ID MSI File
Step 3:Install RSA Secure ID Application
Step 4:Import RSA Secure ID Token
Step 5:Generate PIN/Usage of RSA Secure ID Token

The system requirements are Windows operating system for computers and for smart phones, it is Android 2.1 and above, Apple IOS 4.3 and above, windows phone 7.5 and above. Figure 2 shows the token [3].

First-time users need to generate a 4 digit numerical PIN (Personal Identification Number) which will be used along with RSA-token for authentication. A message appears on the screen and users are asked to do so. The message is shown in Fig. 3 [3].

PIN with token will be the password for users for any financial transactions. For token generation, one needs to run the software. The generated token is timer-based but there are many problems in this which makes it so complicated. First problem is that apart from RSA-token, users have to generate a 4 digit PIN which increases complications. Second, it is a difficult process. A person who is not so tech-savvy will feel difficult to use it. Third problem which makes it very difficult to use is that if RSA-Secure ID application software token is corrupted (due to any reason) or lost or stolen then you have to contact your branch and put a request to re-allow you to use it. It takes some days to respond and during this period users can't use internet banking. It makes it very troublesome specifically for common users who are not so

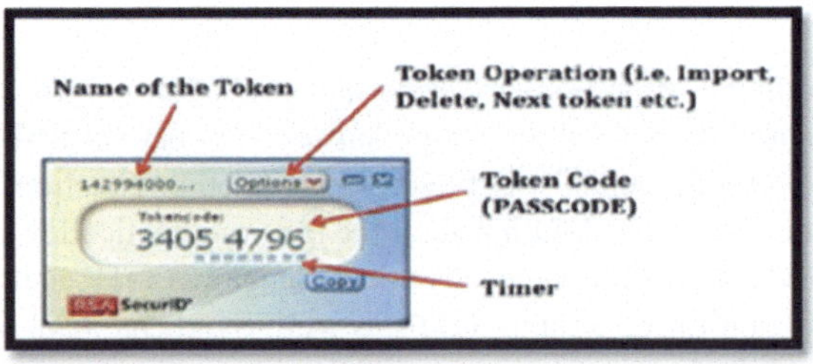

Fig. 2 Showing the generated token [3]

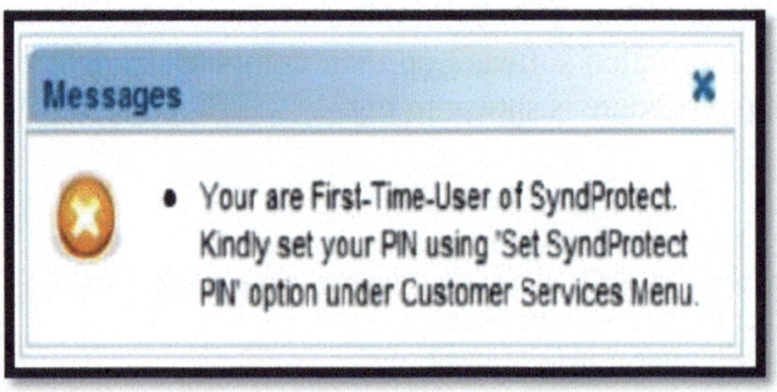

Fig. 3 Showing the necessity of PIN for first-time users [3]

tech-savvy. The rest of this paper is organized as follows: In Sect. 2, we discuss the proposed method. The tabular and graphical analysis is provided in Sect. 3. Security analysis and advantages are given in Sect. 4. Conclusion and future scope is given in Sect. 5.

2 Proposed Method

Our method utilizes RSA for cipher string generation which is based on prime factorization problem [4]. RSA is the most reliable encryption algorithm known so far. One needs to select two prime numbers p and q. These prime numbers must be large enough for security [5]. We have a composite number $N = pq$ and it is a public parameter. $\Phi(N)$ is the Euler totient function and it is given as $\Phi(N) = (p-1)(q-1)$. The public key is denoted by e and it can be freely selected but must be coprime with $\Phi(N)$. The private key d is selected in such a way so that $ed \mod \Phi(N) = 1$. The extended Euclidean algorithm can be used to find d. If m is our message or plain text and c represent ciphertext then $c = m^e \mod n$ and for decryption we use $m = c^d \mod n$. Our plain text is "session OTP". We have taken small plain text for illustration point of view but one can take any plain text. We have converted the plain text into its corresponding ASCII equivalent. We have added space (ASCII equivalent 032) in ASCII equivalent of m in order to make a segment of size 16. The implementation results are as shown in below Table 1.

Now we have to generate the HMAC string and we utilize the ASCII value of m as a key [6]. HMAC (Hash-based Message Authentication Code) generation requires an initiator text or input string and it is "Hello, let us initiate a session" in our case. This initiator text is a public parameter. We use SHA (Secure Hash Algorithm) variants such as SHA1, SHA224, SHA256, SHA384 and SHA512. The string length of SHA1, SHA224, SHA256, SHA384 and SHA512 is 40, 56, 64, 96 and 128 hex digits respectively. The first eight odd-numbered numeric values are selected for the

Table 1 Showing the values of all parameters along with the cipher text

Parameters	Values
m	Session OTP
ASCII of m	115101115115105111110032079084080032032032032032
p	232104590928947159522 2517
q	188077839016534762671 4757
N	43653729887733178634103465010581209766811025833 69
$\Phi(N)$	43653729887733178634103422992338215218618806460 96
e	$2^{16}+1$
d	64451148266430602020779822218573412035240485728 1
c	02829432693507137649490074515336350306203560098 70

formation of corresponding OTP. The HMAC strings generated from the ingredients along with the OTPs are shown in Table 2.

3 Tabular and Graphical Analysis

Now we provide tabular and graphical analysis of all the above generated OTP as shown in Table 2. The reliability and strength of any OTP is the associated randomness. Our OTP contains numerals (or symbols) from 0 to 9. We represent the occurrence percentage of symbols for all SHA variants used (Table 3a–e). We also represent the occurrence percentage in graphical format for a better understanding of the readers of this paper (Fig. 4a–e).

From Table 3a–e and Fig. 4a–e, it is clear that the occurrence of all the symbols is extremely random which makes it very difficult for intruders to predict. For example, the occurrence of 0 is 15.3846% in SHA1, 20% in SHA256 but 0% in SHA224. Similarly the occurrence of 7 is 11.5384 in SHA1, 8.8888 in SHA256 but 20.5128 in SHA224 which is comparatively quite high.

4 Security Analysis and Advantages

4.1 Secure Key Due to Prime Factorization

A key is required for HMAC string and OTP. This key is our ASCII equivalent of m secured by RSA encryption. RSA is the most reliable method so far as it is based on the prime factorization problem. To find prime factors of a composite number (subject to the condition that p and q. are large enough) is not possible in computationally feasible time.

4.2 Prime Range Can Be Increased

We have selected the prime number p. and q. in the range 2^{80}. to 2^{81} for illustration purpose only. One can select bigger prime numbers by increasing the range. It will increase the security level.

Table 2 Showing the HMAC string and the OTPs for various SHA versions

Initiator text	Key	Algo.	HMAC string	OTP
Hello, let us initiate a session	115101115 1151051111 1003207908 4080032032 032032032	SHA1	f75229e0bf06a35d9a1c8a81c2ce9a901824707c	72903981
		SHA224	7c73b75f52b09ffd3058a77a297946 73848f19dba9ee0e27c8066044	73529087
		SHA256	0532070940e757ad08586003f4f034 819a51ada4e8dddfc86b865250df676a63	03004778
		SHA384	7acd25f57fd9db3a460bc11813e90f 6e7c330190af54e70bf8dc3c2e9b9e 69aa3235043d7a9c55af74a74b4ca16135fb	75736183
		SHA512	0c30aa9e080586ede8b5a3e453fc524fcc0ea9c8 12226c1e9692b3dedd95dfdd0944cc087de52a26 8af631a15aa9fb2fae2cd78ead65a742b3e5e351165bbe6c	00008835

Table 3 Representing occurrence percentage of symbols in different SHA versions

S. No.	Algo.	Sym.	Freq.	Occurrence percentage
a				
1	SHA1	0	4	15.3846
2		1	3	11.5384
3		2	4	15.3846
4		3	1	3.8461
5		4	1	3.8461
6		5	2	7.6923
7		6	1	3.8461
8		7	3	11.5384
9		8	3	11.5384
10		9	4	15.3846
b				
1	SHA224	0	5	0.0000
2		1	1	2.5641
3		2	3	7.6923
4		3	3	7.6923
5		4	4	10.2564
6		5	3	7.6923
7		6	3	7.6923
8		7	8	20.5128
9		8	4	10.2564
10		9	5	12.8205
c				
1	SHA256	0	9	20.0000
2		1	2	4.4444
3		2	2	4.4444
4		3	4	8.8888
5		4	4	8.8888
6		5	6	13.3333
7		6	6	13.3333
8		7	4	8.8888
9		8	6	13.3333
10		9	2	4.4444

(continued)

Table 3 (continued)

S. No.	Algo.	Sym.	Freq.	Occurrence percentage
d				
1	SHA384	0	6	10.5263
2		1	6	10.5263
3		2	3	5.2631
4		3	9	15.7894
5		4	6	10.5263
6		5	7	12.2807
7		6	4	7.0175
8		7	7	12.2807
9		8	2	3.5087
10		9	7	12.2807
e				
1	SHA512	0	7	9.8591
2		1	6	8.4507
3		2	10	14.0845
4		3	7	9.8591
5		4	5	7.0422
6		5	11	15.4929
7		6	8	11.2676
8		7	3	4.2253
9		8	7	9.8591
10		9	7	9.8591

4.3 Primality Test Can Be Changed

We have selected the Miller Rabin primality test but one can select other tests like Solovay-Strassen or Fermat etc. The reason is that all the above-mentioned tests provide probable primes and there may be a case of pseudo pme. So the user can select the algorithm of his own choice.

4.4 Hash-Based Security

Hash-based string generations enjoy the advantages of collision resistance, one-way trapdoor, etc [7]. We can't have two different inputs having the same hash. We use HMAC string generation so it is not vulnerable to length based attacks also [8].

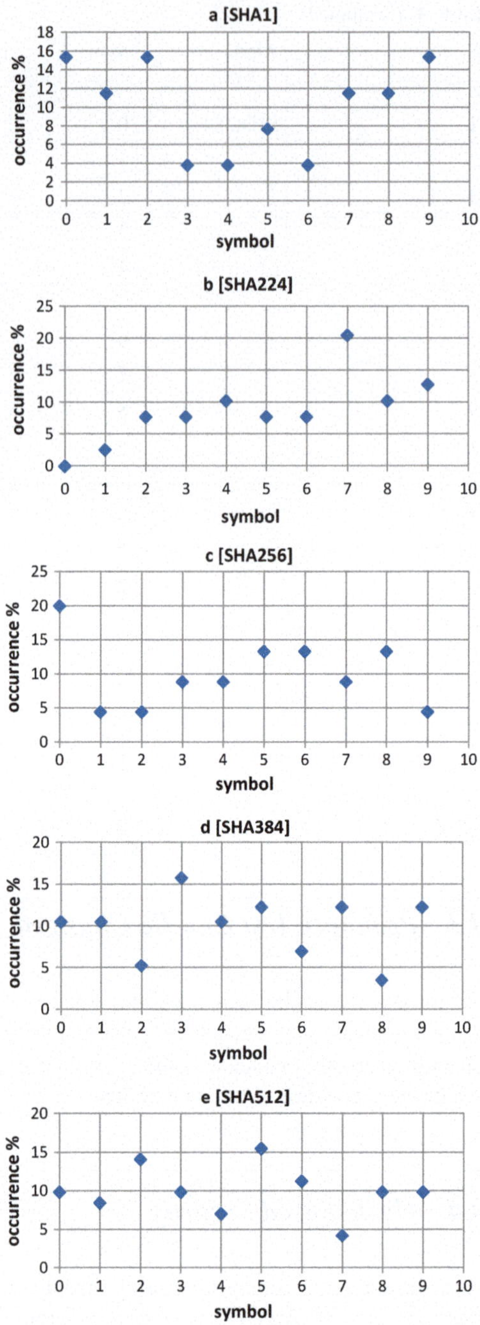

Fig. 4 **a–e** Representing occurrence of symbols graphically in various SHA versions

4.5 Algorithm (SHA) Can Be Changed Any Time

We have used SHA variants for OTP generation but other algorithms like RIPEMD128 or MD5 (Message Digest 5) can also be used. More options will always enhance the difficulty level for intruders.

4.6 Avoidance of Brute Force

An intruder who wishes to apply brute force has two options. First one is that he tries to get ciphertext but it is not possible because it is encrypted and an intruder has to break hard problems which are not possible in computationally feasible time. Second option is that intruder predicts the same HMAC string which is not feasible because to guess 56 hex digits (taking SHA224 as an example) where any digit can have any value from 0 to f in hexadecimal notation is a computationally very expensive because intruder has to search millions of millions combinations [9].

4.7 Variations in OTP Can Be Possible

We have selected the first eight odd number numeric values for OTP generation but one can select even number values also. The generation of alphanumeric OTPs is also possible by the same method. It makes it impossible for intruders to make any prediction and enhances the security of data communication.

4.8 Kerckhoffs's Principle Is Satisfied

Dutch cryptographer Auguste Kerckhoff had given a very important cryptographic principle [10]. According to it, any cryptographic mechanism should be secure if the key is secure and everything else is public knowledge. It holds true in our case because if an intruder wants to develop the HMAC string, he needs the key but the key has been securely transmitted to the receiver using the hardness of prime factorization.

4.9 SMS Sending of OTP Is not Required

In financial transactions, we need to send the OTP to the client's mobile and usually the medium is SMS (Short Message Service). Researchers have shown that SMS sending of OTP is vulnerable to various security attacks [11, 12]. In the proposed method, user himself generates the required OTP for the session and it is verified by the bank. So our proposed method resists the associated vulnerability of SMS sending of OTP.

4.10 Acts as a Dynamic or Session-Based OTP

A major problem associated with the traditional password is that they need to be stored in a system. It poses vulnerability and occupies space too. Since the password generated from the proposed method is session-based, there is no need to store them. So if the security of a particular system is in danger, the security of the password is not compromised [13].

5 Conclusions and Future Scopes

We have proposed a unique method for OTP generation using the strength of prime factorization problem. Users can generate their own parameters and increase the level of security whenever they want. The proposed method can be used as an improvement over the Syndicate bank's TFA mechanism. Brute force analysis is computationally infeasible for the proposed method and Kerckhoff's principle is also satisfied. Since users generate the OTP, the proposed method solves the problem of OTP transportation or the problem of sending OTP via SMS. The future scope of the paper is very good because we have implemented SHA variants only but one can implement other algorithms like MD5, RIPEMD 160, etc.

References

1. Shukla V, Chaturvedi A, Srivastava N (2019) A secure stop and wait communication protocol for disturbed networks. Wirel Pers Commun 110(2):861–872. https://doi.org/10.1007/s11277-019-06760-w
2. Shukla V, Chaturvedi A, Srivastava N (2019) Nanotechnology and cryptographic protocols: issues and possible solutions. Nanomater Energy 8(1):78–83. https://doi.org/10.1680/jnaen.18.00006
3. Syndicate Bank RSA SecurID Software Token. Available at https://syndicate-bank-rsa-securid-software-toke.software.informer.com/. Accessed 28 Jan 2020
4. Guan J, Li W, Ouyang B (2018) On rank factorizations and factor prime factorizations for multivariate polynomial matrices. J Syst Sci Complexity 31(6):1647–1658. https://doi.org/10.1007/s11424-018-7446-8
5. Zhou X, Tang X (2011) Research and implementation of RSA algorithm for encryption and decryption. In Proceedings of 2011 6th international forum on strategic technology, pp 1118–1121. https://doi.org/10.1109/ifost.2011.6021216
6. Gilbert H, Handschuh H (2003) Security analysis of SHA-256 and sisters. In International workshop on selected areas in cryptography, selected areas in cryptography, (part of the lecture

notes in computer science book series (LNCS, vol 3006)), pp 175–193. https://doi.org/10.1007/978-3-540-24654-1_13

7. Andreeva E, Preneel B (2008) A three-property-secure hash function. In International workshop on selected areas in cryptography (Part of the lecture notes in computer science book series (LNCS, vol. 5381)), pp 228–244. https://doi.org/10.1007/978-3-642-04159-4_15
8. Tiwari H, Asawa K (2012) A secure and efficient cryptographic hash function based on NewFORK-256. Egypt Inf J 13(3):199–208. https://doi.org/10.1016/j.eij.2012.08.003
9. Shukla V, Chaturvedi A, Srivastava N (2019) Authentication aspects of dynamic routing protocols: associated problem & proposed solution. Int J Recent Technol Eng 8(2):412–419. https://doi.org/10.35940/ijrte.B1503.078219
10. Petitcolas FAP (2011) Kerckhoffs' principle. Encyclopedia of cryptography and security, pp 663–706. https://doi.org/10.1007/978-1-4419-5906-5_487
11. Mulliner C, Borgaonkar R, Stewin P, Seifert J-P (2013) SMS-based one-time passwords: attacks and defense. International conference on detection of intrusions and malware, and vulnerability assessment, DIMVA 2013 (Part of the lecture notes in computer science book series (LNCS, vol 7967)), pp 150–159. https://doi.org/10.1007/978-3-642-39235-1_9
12. Hamdare S, Nagpurkar V, Mittal J (2014) Securing SMS based one time password technique from man in the middle attack. Int J Eng Trends Technol (IJETT) 11(3):154–158. https://doi.org/10.14445/22315381/IJETT-V11P230
13. Chang B, Li Y, Wang Q, Zhu WT, Deng RH (2018) Making a good thing better: enhancing password/PIN-based user authentication with smartwatch. Cybersecurity 1(7):1–13. https://doi.org/10.1186/s42400-018-0009-4

Performance Analysis of DCW-Based Improved TCP Variant for Safety Message Broadcast in VANETs

Poonam Verma, Somendra Prakash Singh, and Neeta Singh

Abstract Considering highly dynamic nature of a vehicular network, timely dissemination of safety messages is required by vehicles to form and maintain an ad hoc network. As CAMs are periodically broadcasted, an ample amount of CAMs on the channel can lead to collision, thus leaving no scope for EDM dissemination. To cope up with this issue, optimal safety message dissemination is essential for a vehicular network to perform efficiently. The paper proposes a DCW-based improved TCP variant for safety message broadcast over the network. The number of backlogged messages at the nearest RSU is considered as the governing factor for dynamically adapting the CW size at the vehicle's end. CW size plays an important role in managing broadcast of safety messages in VANETs. Round trip time is considered as the key parameter for estimating the number of safety messages at the nearest RSU. Simulation is performed considering three different vehicle's transmission range (200–500 m). It is worth noting that, the proposed I-TCP variant outperforms the existing fixed CW schemes in terms of congestion probability for all the values of transmission range. Moreover, it can be evident that, for fixed CW size, lower transmission range should be preferred over the larger transmission range.

Keywords Safety messages · I-TCP variant · RTT · Backlogged messages · Collision probability · CW size

---

P. Verma (✉)
Thapar Institute of Engineering & Technology, Patiala, India
e-mail: poonam.verma@thapar.edu

S. P. Singh · N. Singh
Gautam Buddha University, Greater Noida, India

M. N. Favorskaya et al. (eds.), *Innovations in Electrical and Electronic Engineering*, Lecture Notes in Electrical Engineering 661, https://doi.org/10.1007/978-981-15-4692-1_55

1 Introduction

Traditionally, VANET can be portrayed into two applications, user-related application and safety-related application [1]. Safety-related application incorporates two sorts of messages: periodic broadcast messages (Beacon message or CAMs: cooperative awareness messages) and accident messages (EDMs: event-driven messages). Out of seven dedicated short-range communication (DSRC) channels, only one control channel (CCH) of 10 MHz is dedicated for the use of safety applications [2]. Moreover, safety-related applications are sensitive to message loss and delay. As CAM messages are periodically broadcast at a rate of 10 Hz, CCH is often being occupied by these messages only, leaving no scope for EDM messages. Nevertheless, CAMs carry the core status information of vehicles contributing toward the network and managing their broadcast is also of great importance. Many of VANET's systems use a common channel assessment procedure for both the safety-related messages (CAMs and EDMs), without considering the fact that both the messages have different characteristics and need to be handled likewise [3, 4].

Taking into account this issue, the paper presents an improved-transmission control protocol variant (I-TCP) model. Conventional TCP was meant for wired networks. On wired networks, there are fewer chances of high delay and congestion because of outer parameters. On the contrary, wireless connections have a serious issue of variable and high delay with significant bit error rate (BER). Due to these characteristics of wireless networks, traditional TCP begins to perform inadequately, and hence, to manage the issues of wireless connections, research was made in the field of TCP, and changes were done by the prerequisites to enhance the TCP performance. Variants named Tahoe, Reno, New Reno, Vegas, and SACK and numerous more came into existence [5–7].

Considering this issue, a modified TCP variant is required to cope up with the highly dynamic nature of VANET. This paper proposes a dynamic contention window (DCW)-based I-TCP variant for safety message broadcast. The proposed I-TCP variant is based on estimating the number of backlogged messages at the nearest RSU for dynamically adapting CW size at the vehicle's end. As the backlogged messages at the RSU are the result of safety message dissemination by vehicle and their propagation throughout the channel, backlogged safety message can be considered as the key parameter for evaluating the load over the channel. This is because more the number of backlogged packets, more are the chances of the collision on the channel, and therefore choosing the CW size at the vehicle's end according to the backlogged safety message at the nearest RSU can control collision on the channel along with providing optimal management of safety message broadcast.

Rest of the paper is structured as follows: Sect. 2 presents some of the related research work. Section 3 presents some of the key features of our proposed I-TCP variant for the safety message of VANETs. In Sect. 4, the proposed I-TCP is validated by extensive simulations. Section 5 concludes and summarizes the paper.

2 Related Work

A good amount of research has been done in the area of VANET for optimizing the performance of safety application (e.g., CAMs and EDMs dissemination). Using queueing theory, Yousefi et al. [8] studied connectivity issues within vehicles. In 2013, Mughal et al. [9] concluded that for practical purpose, using power control strategies and transmission control procedure for determining the prerequisites of imagined reference point subordinate safety applications, which is not feasible. In 2013, Nagy et al. [10] presented a single server model for IP multimedia subsystem network. However, the M/M/1 queueing model being used in [10] is not practically suitable for real-time applications. In 2013 [11], Aly et al. proposed a gateway integrating VANETs, WSNs, and IMSs, presuming all the substances in the system are known to each other. In 2014 [12], Abd-Elrahaman et al. presented an analytical model using genetic routing algorithm, but the durability and reliability of the V2V connections were neglected. Without considering IMS processing time, Marot et al. [13] calculated the number of hops and time needed to diffuse the information in 2015. In 2015 [14], Qiong et al. proposed an analytical model based on dynamic CW for safety messages, but the principle of back-off time freezing was not considered in the model. In 2016 [15], Marot et al. computed RTT without considering message loss within the network. In 2017 [16], Khalifa et al. analyzed the performance of TCP variants for data collection in a smart power grid. In 2017 [17], Verma et al. analyzed network performance by introducing the concept of time-out for CAM messages wherein based on time-ot, recent CAMs were prioritized over expired CAM. In 2018 [18], Verma et al. presented a one-dimensional Markovian model for IEEE 802.11p MAC for safety messages. All these models considered either CAM or EDM.

It can be noted from the previous work that vehicular networks are likely to get congested when the CW size is fixed, due to ample number of CAMs on the CCH. This perhaps necessitates the need for adapting the CW size in accordance with the network requirements. Considering the above points, this paper considers backlogged packet as the governing parameter for improving the network performance in terms of collision probability, by applying DCW-based Improved-TCP Variant. Detailed elaboration of the proposed model is presented in the next section.

3 Proposed I-TCP Variant

The provision of acknowledgment is not feasible for safety messages in VANET (due to their broadcast nature), and hence, the paper assumes the nearest RSU as the destination for the broadcasted messages. The paper also assumes that the vehicles arrive according to the Poisson process with a rate λ_a. The initial CW size for performing the simulation is considered to be 128. The min and max speeds for the vehicle are V_{min} and V_{max} with mean μ and standard deviation σ. The inter-distance of the vehicles at any given time is exponentially distributed which is given by λ [15]. A detailed network architecture for VANET applications using the IP multimedia subsystem (IMS) as a service controller in the USN environment is considered [15]. Based on the above assumptions, this section describes an approximate analytic model for VANET. The notations used in the paper are listed below.

Notations	Description
N_b	Backlogged messages at the nearest RSU
B	Vehicle density
T	Stationary probability that a node broadcast safety message in a chosen time slot
Th	Message transmission delay at vehicle's end
Tc	Message generation time at RSU
Tt	Message transmission time
RTT	Round trip time; delay experienced by a safety message from being transmitted from the source node to receiving its response from the core network
TP/RTTmin	Minimum RTT from source node to the nearest RSU
Tq	Core network's application execution time
Tcscf	CSCF server's processing time
T ehr	EHR server's processing time
Tasca	AS\CA server's processing time
Tps	PS server' processing time
Ttdb	TDB server' processing time
Contacts	Contacts server' processing time

For reducing the number of backlogged messages at any RSU, merely decreasing the transmission rate of safety messages (CAMs, in particular) is not an appropriate solution, as decreasing the CAM frequency can lead to an inadequately updated network (as high CAM rates are required to keep the network like VANET up-to-date). Apparently, direct increase or decrease in CAM broadcast rate can affect the network performance to a great extent. This is because the message that encounters collision due to direct adaptation of CAM is lost and cannot be retransmitted, due to lack of acknowledgment for CAMs in VANET (as mentioned in the above section). Keeping in mind this issue, CW size required at the channel assessment process is varied without altering the collision probability of safety messages. CW size affects the waiting time of safety messages in the queue of a vehicle but eventually is secured from getting lost due to a collision on the channel, followed by their broadcast.

RTT is the key feature of the proposed scheme as it plays an important role in evaluating the performance of any application within a communication network. For VANET, RTT indicates the time taken by an application starting from the source vehicle, until receiving a response from the core network. This eventually helps to evaluate the number of backlogged packets (N_b) at the RSU, as here in our case the nearest RSU acts as the core network. RTT is calculated in two parts: firstly, the paper calculates the time taken by VANET application in the vehicle-to-vehicle communication and then the average processing time taken by IP multimedia subsystem is calculated. For estimating the number of backlogged safety messages (N_b), the following mathematical expressions are evaluated using [16],

$$\mathrm{N_b} = \frac{\mathrm{CW}}{\mathrm{RTT}}(\mathrm{RTT} - \mathrm{RTT_{min}}) \tag{1}$$

where

$$\mathrm{RTT} = \mathrm{Tp} + \mathrm{Tq} \tag{2}$$

$$\mathrm{Tp} = E(\mathrm{no.of\ hops}) * T_h \tag{3}$$

$$\mathrm{Tq} = \mathrm{Tc} + \mathrm{Tcscf} + \mathrm{Tasca} + \mathrm{Tehr} + \mathrm{Ttdb} + \mathrm{Tps} + \mathrm{Tcontacts} + \mathrm{Tt} \tag{4}$$

$$|CW| = \begin{cases} CW - (CW/4), & N_b < A1 \\ CW - (CW/8), & A1 < N_b < A2 \\ CW - (CW/16), & A2 < N_b < A3 \\ CW, & A3 < N_b < A4 \\ CW + (CW/16), & A4 < N_b < A5 \\ CW + (CW/8), & A5 < N_b < A6 \\ CW + (CW/4) & A6 > N_b \end{cases} \tag{5}$$

Depending upon the N_b estimation, the proposed I-TCP variant dynamically chooses the CW size in accordance with the threshold limits derived in Eq. (5).

For safety message broadcast, parameters like throughput, packet, delivery ratio cannot be calculated in VANETs. Therefore, the collision probability is important for performance evaluation of safety message broadcast in VANETs. Let us denote the probability of collision Pc experienced by a safety packet being transmitted on the channel given by [18].

$$Pc = 1 - (1 + 2\beta R\tau)e^{-2\beta R\tau} \tag{6}$$

Pseudocode developed for the proposed-based improved-TCP variant is stated below. The algorithm is validated by MATLAB simulation.

Input: N_b $Threshold$ $(A1 - A6)$, $CW_{Pre-Defined}$, β
Output: CW
Parameter Limitation: N_b $Threshold$ $(A1 - A6) \in [20, 120\ messages]$
$CW \in [15, 1023]$
$CW_{Pre-Defined} = 128$
$\beta \in [0.01, 0.1]$
Initially: $CW = CW_{Pre-Defined}$
if Nb $< A1$ **then**
$CW = CW - (CW/4)$
else if $A1 < Nb < A2$ **then**
$CW = CW - (CW/8)$
else if $A2 < Nb < A3$ **then**
$CW = CW - (CW/16)$
else if $A3 < Nb < A4$ **then**
$CW = CW$
else if $A4 < Nb < A5$ **then**
$CW = CW + (CW/16)$
else if $A5 < Nb < A6$ **then**
$CW = CW + (CW/8)$
else if $A6 > Nb$ **then**
$CW + (CW/4)$
end if

Algorithm 1. Improved-TCP variant algorithm for safety messages

4 Model Validation and Performance Evaluation

The author presents an I-TCP variant for safety message broadcast over the vehicular network. For a vehicular network, contention window (CW) size is an important parameter affecting the performance of the network. Based on this, to manage the backlogged messages at the nearest RSU, CW is dynamically varied without degrading the network performance in terms of collision probability. This eventually helps in securing safety messages at vehicle's end without allowing their loss over the channel due to a collision. It is also proven from the simulation results that the overall collision probability of the network is improved, even when the CW size is decreased.

Due to dynamic nature of the vehicular network, timely dissemination of safety messages is highly required, to maintain the essence of the network and keep it up-to-date. At the same time, merely disseminating safety messages without assuring minimum collision at the channel is not advisable. Therefore, this paper uses the concept of backlogged messages at the nearest RSU (i.e., receiver end) to adapt CW size in order to minimize collision across the channel. Keeping this in mind, a lesser number of backlogged safety messages are required to assure less collision over the channel. Therefore, the proposed model adapts the CW size in accordance with N_b, assuring lower message loss over the channel due to a high transmission rate and/or fixed CW size.

For the proposed I-TCP variant, initially, CW is kept at a nominal pre-defined value, so as to minimize the number of backlogged messages (obtained by iterative simulation of the proposed model). Depending upon this value, N_b thresholds are defined for further adaptation of CW, varying from 20 (lower limit) to 120 (upper limit). Every time, N_b is estimated based on RTT value expressed in Eq. (1), the system checks for the pre-defined N_b threshold limits. This helps in choosing appropriate CW depending upon the range within which the calculated current N_b lies. Hence, based on these threshold limits, defined initially, I-TCP variant chooses the CW size dynamically over the range of 15–1023, as expressed in Eq. (5). This helps in reducing the number of backlogged safety messages, thus improving network performance. Also, when the N_b size is smaller than the affordable limits, the model decreases the CW size, allowing more safety messages over the channel. This perhaps improves the overall performance of the network without accelerating the probability of collision due to an overloaded channel.

For designing the simulation framework of the presented model, parameter consideration is listed in Table 1. The simulation scenario is created taking a road of length 3 km. Three different transmission ranges of 200, 350 and 500 m are considered for vehicles participating in the network. As per IEEE 802.11p standards, safety message are broadcast over CCH having a bandwidth of 10 MHz. Vehicles are assumed to be moving at a speed of 100 km/hr (considering highway scenario). The size of the contention window lies between 15 and 1023 for safety messages, as standardized by IEEE 802.11p. The paper has designed an additional algorithm to limit the CW size of safety messages to 1023. The initial CW size for performing the simulation is kept at 128. The paper has considered six different threshold levels for choosing appropriate CW size, in order to improve the performance of the network. N_b threshold values

Table 1 Simulation parameters

Parameters	Value
Road length (L)	3 km
Transmission range (R)	200, 350, 500 m
Number of channel	1
Queue capacity	∞
Bandwidth	10 MHz
Vehicle density ($\beta = \lambda \backslash L$)	0.01–0.1
CW size	[15–1023]
CW pre-defined	128
Vehicle speed	100 km/h
N_b threshold (A1–A6)	20–120 safety messages

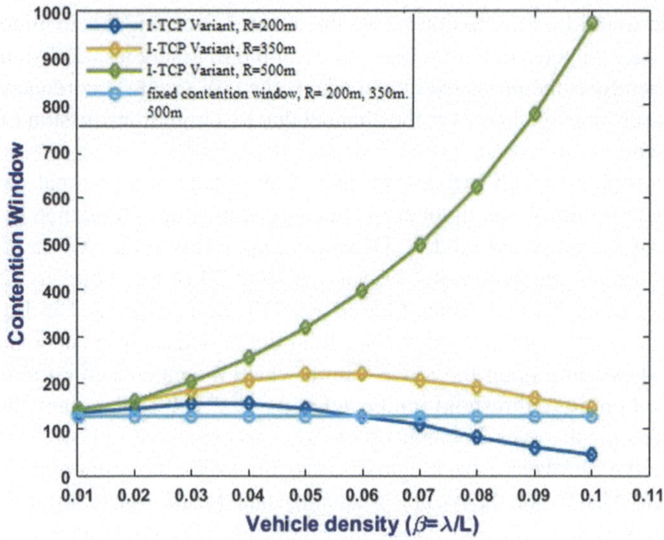

Fig. 1 Comparative plot of CW versus vehicle density

lie between 20 and 120. This value is determined by performing iterative simulation of our proposed work.

Figure 1 depicts the dynamic values of CW selected by the proposed model over three different transmission ranges varying from 200 to 500 m. Different R values are considered to elaborate the impact of R over the RTT, i.e., larger the range less will be the RTT and vice versa. The CW size for our proposed model dynamically changes with the vehicle density. Keeping the CW size fixed over the range of vehicle density increases the overall delay of the network. This is because, with an increase in vehicle density, the number of vehicle contending for channel assessment will gradually increase. For a transmission range of 200 and 300 m, the CW size decreases with increasing vehicular density. This is because, for lower values of *R*, the RTT time increases, as the number of hops required by a safety message to reach the nearest RSU increases. This eventually reduces the number of backlogged safety message s at the nearest RSU. Therefore, the proposed model gets the opportunity to allow more safety messages over the channel by reducing the CW size, in order to improve the network performance.

On the contrary, it can be clearly seen that, for our proposed work at $R = 500$ m, the CW size increases for larger vehicle density. This is because, for a wider range of transmission, a lesser number of hops are required by a safety message to reach the nearest RSU, thus reducing the RTT time. This leads to more number of safety messages at the nearest RSU, thus increasing the chances of the collision over the channel. Hence, the system requires to reduce the number of safety messages over the channel, in order to reduce N_b at the nearest RSU. To deal with the situation, the model allows to adapt the CW size in accordance with the N_b threshold defined. This apparently leads to increasing the CW size, in order to lower the N_b at the nearest RSU, thus reducing any further chances of the collision over the CCH. Moreover, larger CW size facilitates each vehicle to select CW over a wider range. This also lowers the collision at the channel, which is the result due to the same random back-off number selection.

To evaluate the performance of our proposed work, a comparative view between fixed and dynamic CW in terms of congestion probability is elaborated in Fig. 2. The simulation is performed over three different transmission ranges varying from 200 to 500 m so as to clearly depict the influence of transmission range over RTT. Figure 2 shows that the proposed DCW selection scheme outperforms the existing fixed CW in terms of congestion probability. Considering the dynamic CW shown in Fig. 1, it can be clearly seen that the CW size has a great impact on the collision probability. And therefore Figs. 1 and 2 are mutually dependent on each other, as increase in CW size lowers the collision probability and vice versa.

It is worth noting that for $R = 200$ m, the collision probability increases marginally at higher vehicular density as CW chosen for at vehicular density is smaller (due to less number of N_b at the RSU). It can be proven from the simulation graph that at $R = 500$, the network performance is best, as negligible collision probability is experienced even at the highest vehicular density. This is because a higher value of CW is over for this transmission range, due to more number of safety messages at the nearest RSU.

Moreover, Fig. 2 depicts that for fixed CW size, lower transmission range should be preferred over larger transmission range, as the collision probability increases with

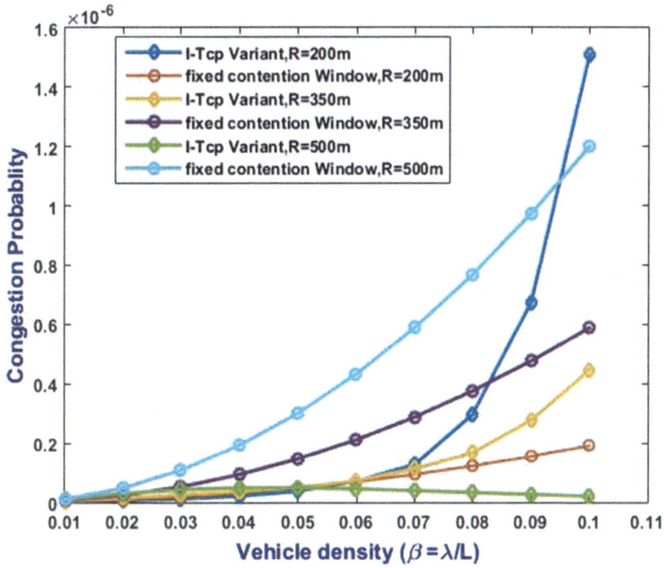

Fig. 2 Comparative plot of congestion probability versus vehicle density

Table 2 Simulation comparison: fixed CW versus I-TCP at $\beta = 0.07$

S. No.	Range (m)	P_C (10^{-6})		Contention window	
		Fixed CW	I-TCP	Fixed CW	I-TCP
1.	200	0.1	0.12	128	115
2.	350	0.3	0.1	128	200
3.	500	0.6	0.07	128	490

an increase in the transmission range. This is because more is the transmission range, lesser would be the number of intermediate hops, thus reducing the RTT time at the nearest RSU. This eventually leads to more numbers of safety messages over CCH. At this situation, if the system does not have the provision of CW adaptation, the collision probability will gradually increase with an increase in vehicular density, thus degrading the network performance to a great extent. Thus, to maintain the network performance, dynamic CW is required, in order to adapt to the network condition. A tabular comparison between the simulation results obtained from fixed CW and I-TCP is provided in Table 2.

5 Conclusions

The paper proposes a DCW scheme-based I-TCP Variant for safety message broadcast in VANETs. The paper considers the number of backlogged safety messages at the road-side unit (RSU) as the governing factor for dynamically adapting the CW size at the vehicle's end (required at the time of broadcast of safety messages). For estimating N_b, the paper considers round trip time (RTT) as the key parameter at the nearest RSU. Simulation is performed considering three different vehicle's transmission ranges, varying from 200 to 500 m. It is worth noting that, the proposed I-TCP variant outperforms the existing fixed CW schemes in terms of congestion probability for all the values of transmission range. Also, it can be evident from the simulation results that, for fixed CW size, lower transmission range should be preferred over the larger transmission range.

References

1. IEEE 1609.3/D5.0 (2010) Trial-use standard for wireless access in vehicular environments (WAVE)—networking services. In: IEEE vehicular technology society
2. IEEE Std. 802.11-2007 (2007) IEEE standard for wireless LAN medium access control (MAC) and physical layer (PHY) specifications
3. Casetti C, Gerla M, Mascolo S, Sanadidi MY, Wang R (2002) TCP Westwood: end-to-end congestion control for wired/wireless networks. Wirel Netw 8(5):467–479

4. Ha S, Rhee I, Xu L (2008) CUBIC: a new TCP-friendly high-speed TCP variant. ACM SIGOPS Oper Syst Rev 42(5):64–74
5. Ding L, Wang X, Xu Y, Zhang W (2008) Improve throughput of TCP-Vegas in multihop ad hoc networks. Comput Commun 31(10):2581–2588
6. De Vendictis A, Baiocchi A, Bonacci M (2003) Analysis and enhancement of TCP Vegas congestion control in a mixed TCP Vegas and TCP Reno network scenario. Perf Eval 53(3–4):225–253
7. Brakmo LS, Peterson LL (1995) TCP Vegas: end to end congestion avoidance on a global internet. IEEE J Sel Areas Commun 13(8):1465–1480
8. Yousefi S, Altman E, El-Azouzi R, Fathy M (2008) Analytical model for connectivity in vehicular ad hoc networks. IEEE Trans Veh Technol 57(6):3341–3356
9. Mughal BM, Wagan AA, Hasbullah H (2010) Efficient congestion control in VANET for safety messaging. Int Symp Inf Technol (ITSim) 2:654–659
10. Nagy L, Tombal J, Novotny V (2013) Proposal of a queueing model for simulation of advanced telecommunication services over IMS architecture. In: 36th international conference on telecommunications and signal processing (TSP), pp 326–330
11. Aly M, Quintero A (2013) A presence-based architecture for a gateway to integrate vehicular ad-hoc networks (VANETs), IP multimedia subsystems (IMS) and wireless sensor networks (WSNs). In: 9th international wireless communications and mobile computing conference (IWCMC), pp 1648–1653
12. Abd-Elrahman, E, Said AM, Toukabri T, Afifi H, Marot M (2014) Assisting V2V failure recovery using device-to-device communications. Wirel Days (WD), 1–3
13. Marot M, Said AM, Afifi H (2015) On the maximal shortest path in a connected component in V2V. Perf Eval 94(1):25–42
14. Yang Q, Xing S, Xia I, Shen L (2015) Modelling and performance analysis of dynamic CW scheme for periodic broadcast in vehicular ad hoc networks. IET Commun 9(11):1347–1354
15. Said AM, Marot M, Ibrahim AW, Afifi H (2016) Modeling interactive real-time applications in VANETs with performance evaluation. Comput Netw 104(1):66–78
16. Khalifa, Shaban KB, Abdrabou A, Bilh A, Naik K (2017) Modelling and performance analysis of TCP variants for data collection in smart power grids. Comput Commun 103(1):39–48
17. Verma P, Singh N (2017) Performance evaluation and validation of session-expiration scheduling technique for queued CAMs in vehicular ad-hoc networks. Int J Veh Inf Commun Syst 3(2):98–114
18. Verma P, Singh N, Sharma M (2018) Modelling a vehicle-ID-based IEEE 802.11 OCB MAC scheme for periodic broadcast in vehicular networks. IET Commun 12(19):2401–2407

A Secure Steganographic Technique for Medical Images

Phanindra Dheeraj Varma, Shikhar Mishra, and Ankita Meenpal

Abstract Communication and storage of Private/Secret information have become important parts of our employment or personal life, after the inception of the Information age. The development in digital technologies and the Internet caused many breakthroughs in the realm of data communication. Unfortunately, this advancement is also offering endless opportunities for attacking the transmission media and steal confidential data. Therefore, secure transmission of data being robust to attacks is required and this objective has gained the attention of many researchers. The technique of hiding information in another information is called steganography. When digital images are used for concealing the secret information, it is called Image Steganography. An important challenge in designing a steganographic scheme is to keep up a good trade-off between features such as robustness, security and imperceptibility. Most of the methods proposed in the literature lack either a feature or the trade-off between them. The proposed method combines image encryption, Lifting Scheme based Wavelet Transform (LS-WT), Discrete Cosine Transform (DCT), and Singular Value Decomposition (SVD) for providing the trade-off between features. Peak Signal–Noise Ratio (PSNR), Normalized Cross-Correlation (NCC), and Structural Similarity Index (SSIM) are used to evaluate the performance of the scheme.

Keywords Steganography · Imperceptibility · Steganalysis · Logistic chaotic map · Wavelet transform · DCT · SVD

P. D. Varma (✉) · S. Mishra · A. Meenpal
National Institute of Technology, Raipur 492010, India
e-mail: dheeraj.vinukonda@gmail.com

S. Mishra
e-mail: shikharmishra27@gmail.com

A. Meenpal
e-mail: ameenpal.phd2017.etc@nitrr.ac.in

M. N. Favorskaya et al. (eds.), *Innovations in Electrical and Electronic Engineering*, Lecture Notes in Electrical Engineering 661, https://doi.org/10.1007/978-981-15-4692-1_56

1 Introduction

Today, advancements in digital communication and innovations in internet technologies have a great impact on everyday life. The transformation of information into digital form and its transfer has been boosted on a very large scale. The breakthroughs occurring are readily accessible to adapt for a specified application. However, the downside is that this easy access provides limitless opportunities for the attacks which aim at stealing private information, particularly in medicine. Hence, the need for the development of techniques to secure confidential data is of higher importance.

The data security systems are classified into two types, Cryptography and Data hiding [1]. Cryptographic schemes transform meaningful private data into a meaningless form called ciphertext. They use an algorithm with a secret key for this purpose, which is called encryption. The secret key is needed at the receiver to extract original private data from the transmitted ciphertext. This is called decryption and it is impossible without the knowledge of key. As cryptography provides security by converting the data into a meaningless form, there is always a suspicion about the presence of data. This suspicion is avoided by data hiding systems. They embed the private data in a carrier, which can be any digital media file.

Information hiding systems are two types: Steganography and Watermarking. Steganography aims at hiding the private information in the carrier, known as the cover file. It is based on the assumption that the visibility of feature confirms the presence of information, so the goal is to completely hide it [2]. The attacks that occur on a steganographic system to retrieve the secret data are called steganalysis. Cryptography and steganography have the same ultimate goal, but their methodologies differ. Steganography doesn't change the format of data as cryptography does. It just hides the confidential data in a cover file, with an aim to make no visible changes in it. More security is provided when cryptography collaborates with the steganographic system [3]. Watermarking is similar to steganography, but it embeds the secret data to provide authentication and restricts unauthorized use.

The basic process in image steganography happens in two stages, embedding and extraction. Embedding occurs in the transmitter side, in which the secret image is added into a cover image such that no visible changes occur in it. The resulting image is called stego image, which is transmitted through the channel of communication. During embedding, there are some properties to be considered to make the scheme better. The most important property is imperceptibility, which determines the strength of the system in concealing the secret data in cover file in a way to be not recognized by the naked eye, or statistical analysis. Another important feature is the robustness, which is the capability of security system to be valid even if a data retrieval attack hits the stego image. Also, the system should aim at transmitting the maximum amount of secret information using minimum cover data. Extraction occurs at the receiver side, in which the embedded secret image is pulled out from the stego image.

The rest of this paper is organized as follows: Sect. 2 briefly discusses the existing work in the literature, Sect. 3 briefly describes the methods used in the proposed algorithm, Sect. 4 provides complete details about the implementation steps of algorithm, Sect. 5 presents the results of experiments and comparisons, and Sect. 6 gives the conclusion.

2 Related Work

The image steganographic methods are classified in many ways based on references like the domain of embedding, nature of retrieval and the dimension of the cover image. With respect to the embedding domain, there are two types of steganographic techniques: Spatial domain and Transform domain.

2.1 Spatial Domain Steganography

When the secret data are embedded directly into the pixel values of the cover image, it comes under the spatial domain steganography. Least Significant Bit (LSB) embedding [4, 5] is the most common technique among spatial domain methods. It just embeds the secret data bits in the LSBs of the cover image pixels. This provides strong imperceptibility, but steganalysis is easier in this case, thus, less robustness. Sutaone et al. [6], proposed the LSB embedding at random locations according to a secret key. This key also determines the order in which secret data are embedded. The dimension of the cover image is selected according to that of secret image. Rajendran et al. [7] generated a chaotic sequence using logistic map function and the embedding is carried out in LSBs according to that sequence. Methods based on Pixel Value Differencing (PVD) [8, 9] and Pixel Intensity Modulation (PIM) [10] are proposed in literature. Hybrid techniques are also proposed combining LSB and PVD [11, 12].

2.2 Transform Domain Steganography

When the secret data are embedded into the transformed frequency domain coefficients of the pixel values of cover image, the method comes under transform domain steganography. Saidi et al., proposed a method [13] that uses Discrete Cosine Transform (DCT) subbands for embedding and a chaotic map to decide the embedding positions. Discrete Wavelet Transform (DWT) based methods [14] are more famous in recent time due to the multi-resolution properties of wavelets. Kumar and Kumar employed a DWT based method [15] to embed secret into the detail coefficients. They employed a secret key computation and a block matching concept between cover and secret, to keep the imperceptibility. Some embedding schemes employed two or more transformations to the images, which appeared to be more robust than single transform techniques. Experiments prove the predominance of these hybrid techniques [16]. Factorizing methods like Singular Value Decomposition (SVD) are also employed in combination with wavelet transform [17] in these hybrid approaches.

This paper proposes a secure steganographic method in the transform domain. The embedding is done in the WT coefficients, by further decomposing them using DCT and SVD [18]. The hiding of data is carried out in the singular values of the DCT decomposition. This offers better imperceptibility. The method adds more security to the data by enciphering the secret image using a key generated from Logistic Chaotic Map function and Linear Feedback Shift Register (LFSR) [19], before embedding.

3 Preliminaries

In this section, the methods used in the proposed scheme are briefly reviewed.

3.1 Logistic Chaotic Map

Logistic chaotic map generates pseudo-random sequence with numbers in an uncertain manner. It is described mathematically with the following function.

$$a_{n+1} = t \times a_n \times (1 - a_n) \tag{1}$$

In the above equation, a_n denotes the population at a given generation n, t denotes the growth rate or fertility rate with the value in between 0 and 4. The chaotic map produced with initial value a_0 in range (0, 1) and t in the range [3.56995, 4] is neither periodic nor convergent, but largely chaotic.

3.2 Linear Feedback Shift Register

The Linear Feedback Shift Register (LFSR) is another way of creating a random sequence. An 8-bit LFSR generates the random sequence by performing XOR operation between some of the bits of an initial byte called Seed. The result is appended at the LSB end and one left shift operation is performed, discarding the MSB. This process is carried out iteratively to generate the sequence, as shown in Fig. 1.

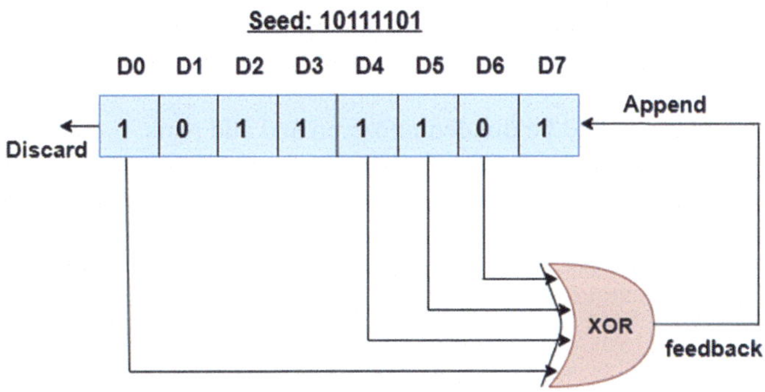

Fig. 1 Linear feedback shift register

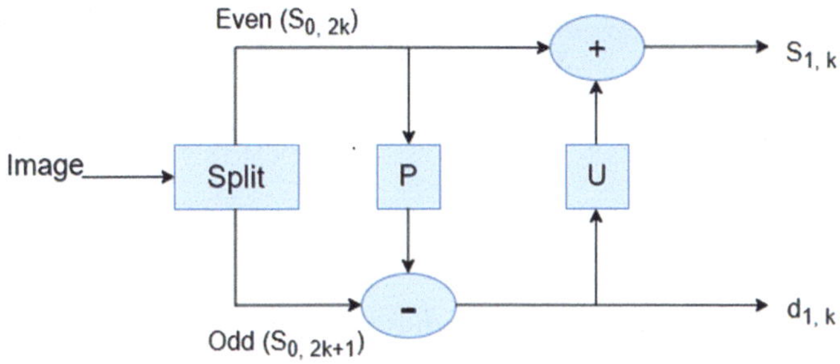

Fig. 2 Lifting scheme of Haar wavelet

3.3 *Lifting Scheme Based Wavelet Transform*

Lifting scheme helps to construct wavelets easily, with calculations in the spatial domain. It is also a faster way of implementing wavelets. We consider the Haar wavelet, which disintegrates the image into four frequency sub-bands: LL, HL, LH, and HH. Among these, LL band consists of low-frequency or approximate coefficients and all other bands comprise of high-frequency or detail coefficients. Embedding in the approximate band offers robustness, but image quality degrades, whereas in the detail band provides good image quality and less robustness. This wavelet is implemented in images by averages and differences between pairs of pixels.

$$S_{1,k} = (S_{0,2k+1} + S_{0,2k})/2 \tag{2}$$

$$d_{1,k} = S_{0,2k+1} - S_{0,2k} \tag{3}$$

This wavelet is implemented in three steps using lifting scheme, as shown in Fig. 2. The three steps are as follows:

1. Split: The image should be divided into even and odd parts.
2. Predict: The odd components are predicted from linear combination of even components. The odd samples are replaced by difference between them and the predictions.

$$d_{1,k} = S_{0,2k+1} - S_{0,2k} \tag{4}$$

3. Update: The even samples are updated by adding them with linear combination of predicted samples from the predict step.

$$S_{1,k} = d_{1,k}/2 + S_{0,2k} \tag{5}$$

The inverse transformation is implemented by simply reversing Eqs. (4) and (5).

$$S_{0,2k} = S_{1,k} - d_{1,k}/2 \tag{6}$$

$$S_{0,2k+1} = d_{1,k} + S_{0,2k} \tag{7}$$

3.4 Discrete Cosine Transform

DCT decomposes an image into low, medium and high-frequency regions. Low-frequency represents coarse information and high-frequency represents edge information. Embedding in low-frequency band degrades the visual quality of image and in high-frequency band makes the stego image less robust. The forward 2D DCT is defined by the following equation.

$$J(a,b) = s(a)s(b)\sum_{l=0}^{N-1}\sum_{m=0}^{N-1} f(l,m)\cos\left[\frac{(2l+1)\pi a}{2N}\right]\cos\left[\frac{(2m+1)\pi b}{2N}\right]$$

$$s(a) = s(b) = \begin{cases} \frac{1}{\sqrt{N}},\ ifa = 0, b = 0 \\ \frac{2}{\sqrt{N}},\ ifa \neq 0, b \neq 0 \end{cases} \tag{8}$$

3.5 *Singular Value Decomposition*

SVD is a factorization method for matrices, based on the concept of eigen decomposition. It factorizes a matrix M, with dimension $x \times y$, into three matrices as in the following equation.

$$SVD(M) = USV^* \tag{9}$$

Here, U denotes a left unitary matrix of dimension $x \times x$, V denotes a right unitary matrix of size $y \times y$ and $V*$ is the conjugate transpose of V. The columns of U and V are called left and right singular vectors of M, respectively.

$$UU^* = I_x, VV^* = I_y \tag{10}$$

S denotes a diagonal matrix of size $x \times y$, with only non-negative numbers. The diagonal elements are called singular values of M and are in descending order. Slight changes in the singular values doesn't disrupt the visual quality of image, hence keeping the imperceptibility good. Due to this reason, the embedding can be done into every singular value, thus providing good capacity.

4 Methodology

In this section, the proposed methodology is described step-wise, in terms of two phases: Embedding and Extraction. The encryption of the secret image is carried out before the embedding process. We shall mention some important details which will help for a better understanding of these sections.

- Dimension of secret medical image: 256 × 256.
- Dimension of cover image used: 512 × 512.
- Length of the keys generated using Logistic Map and LFSR = 256 * 256 = 65,536 samples.
- The birth rate (t) and initial value (a_0) in Logistic Map function and Seed byte in LFSR are treated as secret keys.

4.1 *Embedding*

In this stage, the secret image is encrypted and embedded into the cover image after successive decompositions using lifting Haar WT, DCT, and SVD, thus resulting in stego image. The total process of embedding is discussed in the following steps.

1. $K1$ and $K2$ are generated using logistic chaotic map and LFSR. Secret key is generated by xor operation between them.

$$\text{Key} = K1 \oplus K2 \tag{11}$$

2. Secret image is enciphered with the key generated.

$$\text{Encrypted Image}(I) = \text{Secret Image} \oplus \text{Key} \tag{12}$$

3. Cover image is divided into non-overlapping blocks, each (Cb_i) of dimension 8 × 8. So, in total, $\frac{512\times512}{8\times8} = 4096$ blocks are obtained.
4. Cover image blocks (Cb_i) are subjected to 1-level lifting Haar WT. This results in four sub-bands: LL, LH, HL and HH for each block. The size of each sub-band is 4×4.
5. 2D-DCT of encrypted image (I) is calculated. This gives a coefficient matrix (M) of same dimension, i.e. 256×256.

$$M = DCT(I) \tag{13}$$

6. DCT matrix (M) is divided into 4×4 sized non-overlapping blocks. This division gives $\frac{256\times256}{4\times4} = 4096$ blocks (MB_i).
7. Embedding is chosen to be done in LL sub-band of cover image. So, 2D-DCT of LL bands of cover image blocks (LL_i of Cb_i) are computed. This gives a coefficient matrix (C_i) for each LL_i. The dimension of each D_i is 4×4.

$$C_i = DCT(LL_i) \tag{14}$$

8. SVD is performed on each MB_i and C_i. It goes as follows.

$$SVD(MB_i) = US_i \times SS_i \times VS_i \tag{15}$$

$$SVD(C_i) = UC_i \times SC_i \times VC_i \tag{16}$$

9. This is the embedding step. Here, a scaling factor α is chosen. Embedding is done as follows.

$$S_i = SC_i + (\alpha \times SS_i) \quad (17)$$

10. Inverse SVD is performed using S_i, UC_i and VC_i to compute the modified coefficient matrix (C_i').

$$SVD^{-1}(UC_i, S_i, VC_i) = UC_i \times S_i \times VC_i = C_i^{'} \quad (18)$$

11. Inverse 2D-DCT is performed on C_i' to get the modified sub-band LL'.

$$LL^{'} = DCT^{-1}\left(C_i^{'}\right) \quad (19)$$

12. Inverse lifting Haar WT is performed on sub-bands LL', LH, HL and HH, to output the blocks of stego image (Sb_i).
13. The stego image blocks (Sb_i) are chained together to form the stego image (S).

The complete embedding stage as a flowchart is represented in Fig. 3.

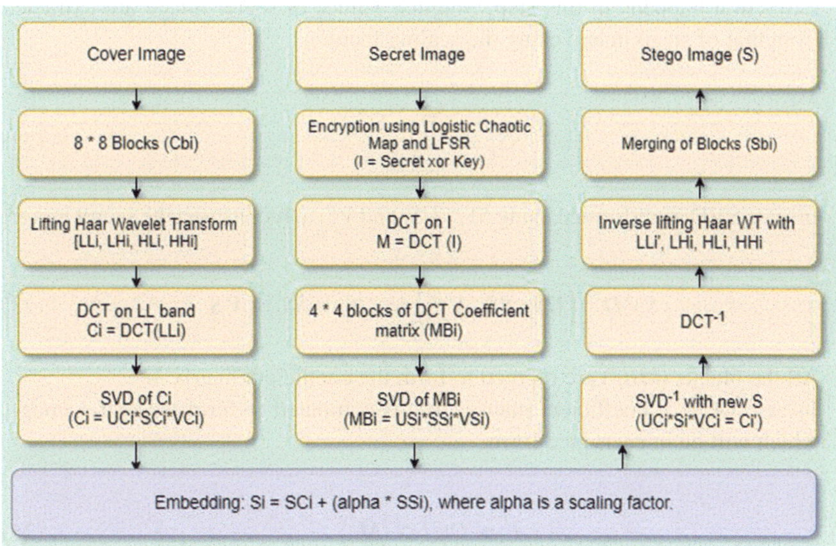

Fig. 3 Embedding procedure

4.2 Extraction

After stego image is formed in the embedding stage, it is transferred through a network to a receiver. At the destination, the secret image is to be extracted out from the received stego image. This process of extraction is discussed in detail in the following series of steps.

1. Received stego image is disintegrated into nonoverlapping blocks of dimension 8×8 (Rb_i).
2. These blocks (Rb_i) are subjected to 1-level lifting Haar WT, resulting in 4 sub-bands: LL_r, LH_r, HL_r and HH_r.
3. 2D-DCT of LL_r is computed to give the coefficient matrix Di_r.

$$Di_r = DCT(LL_r) \tag{20}$$

4. Singular value matrix (S_i') is computed using Di_r, UC_i and VC_i.

$$S_i' = UC_i^{-1} \times Di_r \times UC_i^{-1} \tag{21}$$

5. Extraction happens in this step. Singular values of secret image are extracted from that of stego image using the scaling factor α.

$$SS_i' = \frac{(S_i' - SC_i)}{\alpha} \tag{22}$$

6. Inverse SVD is performed using SS_i', US_i and VS_i to reconstruct the secret image.

$$SVD^{-1}\left(US_i, SS_i', VS_i\right) = US_i \times SS_i' \times VS_i \tag{23}$$

7. All the blocks (MB_i') are merged to form the coefficient matrix M'.
8. Inverse DCT of coefficient matrix (M') is computed to form the secret image, which will be in encrypted form.

$$I' = DCT^{-1}\left(M'\right) \tag{24}$$

9. The encrypted image (I') is decrypted using the secret key. This will result the required secret image.

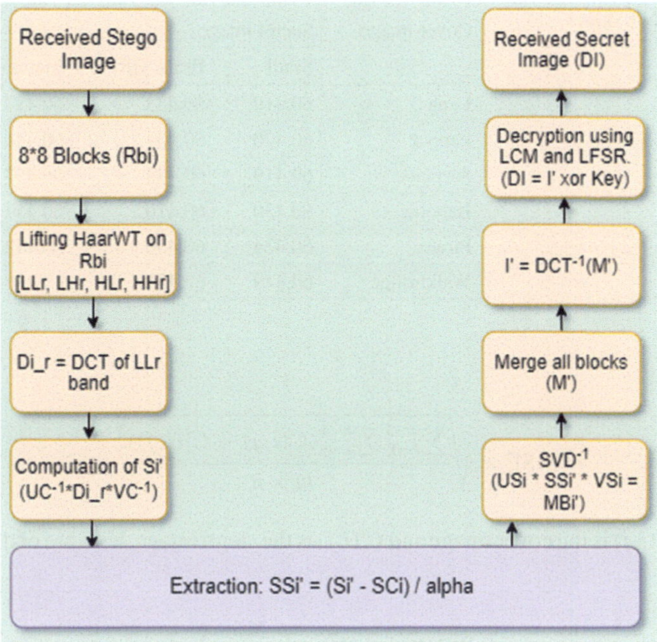

Fig. 4 Extraction procedure

$$\text{Decrypted Image } (DI) = I' \oplus \text{Key} \tag{25}$$

All the process of extraction phase is represented in a flowchart in Fig. 4.

5 Results and Discussion

The performance of the method proposed is evaluated using the following metrics.

5.1 *Peak Signal to Noise Ratio*

PSNR indicates the quality of stego image, by measuring the Mean Square Error (MSE) between cover and stego images.

Table 1 PSNR values

Cover image	Secret image		
	Skull	Brain MRI	Mammogram
Lena	60.419	60.433	60.437
Pepper	60.420	60.434	60.441
Airplane	60.419	60.434	60.437
Baboon	60.430	60.410	60.434
Pirate	60.424	60.450	60.442
Walkbridge	60.419	60.433	60.438

$$\text{MSE} = \sqrt{\frac{\sum_{i=0}^{m-1}\sum_{j=0}^{n-1}(C(i,j) - C'(i,j))^2}{m \times n}} \tag{26}$$

Here, $C(i,j)$ is the cover image and $C'(i,j)$ is the stego image, both are of dimension $m \times n$.

$$\mathbf{PSNR} = 10\mathbf{log}_{10}\left(\frac{255^2}{\mathbf{MSE}}\right) \tag{27}$$

Higher the PSNR, better the imperceptibility and hence, better stego image quality. Table 1 presents the PSNR values obtained between specified cover images and their respective stego images formed by embedding the mentioned secret image.

5.2 Normalized Cross-Correlation

NCC gives the similarity between secret image and its extracted version. If NCC = 1, then they are exactly similar, if 0, no similarity, and if 1, completely opposite images. It is computed by the following formula.

$$\text{NCC} = \frac{\sum_{i=1}^{M}\sum_{j=1}^{N}\left[s_{i,j} - \mu_s\right]\left[s'_{i,j} - \mu'_s\right]}{\sqrt{\sum_{i=1}^{M}\sum_{j=1}^{N}\left[s_{i,j} - \mu_s\right]^2}\sqrt{\sum_{i=1}^{M}\sum_{j=1}^{N}\left[s'_{i,j} - \mu'_s\right]^2}} \tag{28}$$

In the above equation, $s(i, j)$ and $s'(i, j)$ are the pixel values of secret and its extracted version at locations i, j, respectively. μs and $\mu' s$ are the mean values of pixels of the secret and extracted images.

Table 2 gives the NCC coefficient values evaluated between the original secret images and extracted versions from the specified cover images.

Table 2 NCC values

Cover image	Secret image		
	Skull	Brain MRI	Mammogram
Lena	0.981	0.991	0.958
Pepper	0.982	0.991	0.954
Airplane	0.981	0.990	0.957
Baboon	0.979	0.991	0.958
Pirate	0.980	0.992	0.958
Walkbridge	0.979	0.990	0.955

5.3 *Structural Similarity Index*

SSI indicates the similarity between the structural content of cover and stego images. It is evaluated as follows.

$$\mathrm{SSIM}(x, y) = \frac{(2m_x m_y + c_1)(2s_{xy} + c_2)}{\left(m_x^2 + m_y^2 + c_1\right)\left(s_x^2 + s_y^2 + c_2\right)} \tag{29}$$

Here, m_x, m_y are the means, s_x, s_y are the variances of x and y, s_{xy} is the covariance of x and y, and c_1, c_2 are constants depending on the dynamic range of pixels. The Structural content similarity index (SSIM) for all the secret images experimented using the mentioned cover images in the above tables, is reported to be approximately 0.99.

Figures 5 and 6 show results obtained during the simulation of the proposed steganographic algorithm. The dimensions of cover images and secret images used for the experimentation are all 512 × 512 and 256 × 256, respectively. So, the Embedding Capacity (EC) is fixed. The value of the embedding capacity is evaluated as follows.

$$\begin{aligned}\mathrm{EC(bits/pixel)} &= (256 \times 256)/(512 \times 512) \\ &= 1/4 = 0.25 \text{ bits/pixel or bpp.}\end{aligned} \tag{30}$$

5.4 *Resistance Against Steganalysis*

Steganalysis can detect the presence of secret data inside the stego image and hence, it can nullify the ultimate aim of the steganographic system. So, the method should be strong enough to avoid any suspicion about the presence of hidden data. Steganalysis can be either a visual attack or a statistical one. From the high PSNR values obtained,

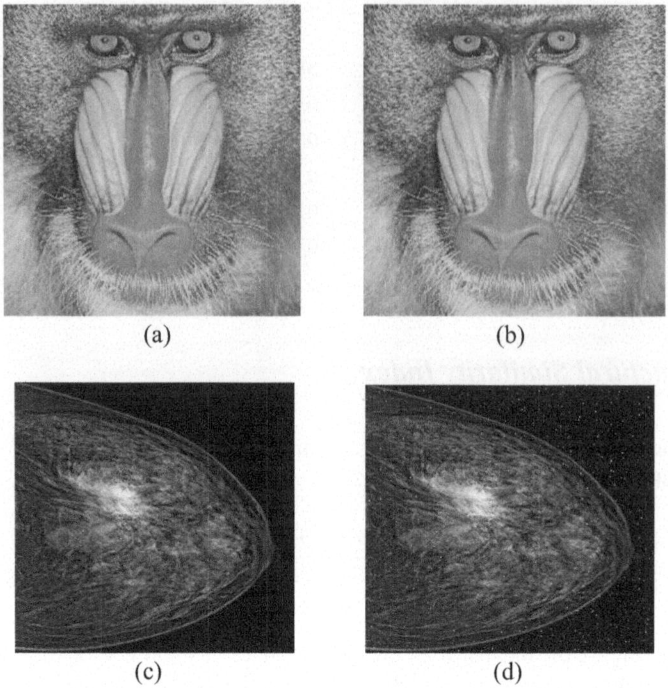

Fig. 5 Mammogram embedded into baboon image **a** cover image, **b** stego image, **c** secret image, **d** extracted secret image

it can be inferred that the algorithm is robust to visual attacks. Still, the statistical steganalysis can detect the presence of confidential data.

So, to evaluate the security of the algorithm against statistical analysis, machine learning models are implemented on the features obtained from the wavelet analysis. As the method implements lifting Haar WT and DCT, and embeds the secret data in low-frequency sub-bands, the same coefficients are analyzed for the purpose of steganalysis.

Using lifting Haar WT on a dataset comprising 500 cover images and 500 stego images, the approximation coefficients are generated up to level-3, and their statistics like mean, variance, skewness, kurtosis, mean absolute deviation, and geometric mean are collected. The same features are also generated for DCT coefficients evaluated from wavelet coefficients of level-1. This data is used for training machine learning models such as Support Vector Machine, Random Forest classifier, and AdaBoost classifier, to differentiate between cover and stego images.

Table 3 represents the metrics obtained by different machine learning models on the above-generated dataset.

The precision scores obtained represent the inability of machine learning algorithms inaccurate classification between cover and stego images. So, the steganographic method proves to be secure.

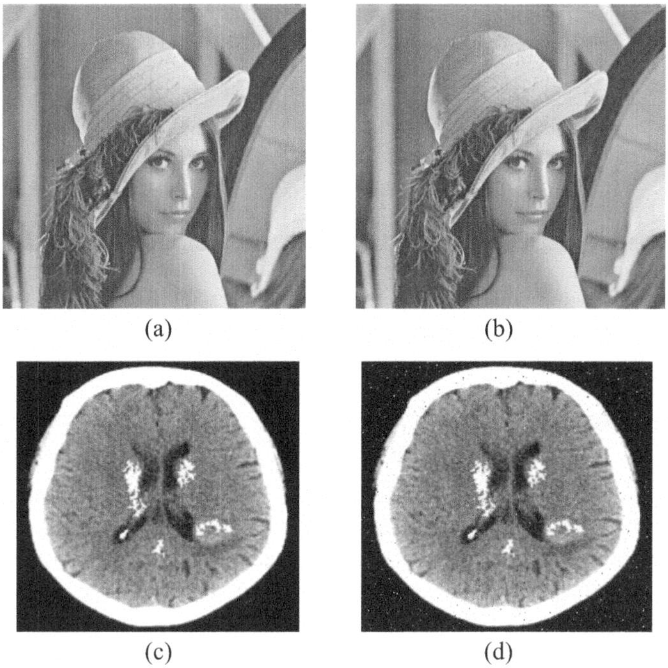

Fig. 6 Brain MRI embedded into lena image **a** cover image, **b** stego image, **c** secret image, **d** extracted secret image

Table 3 Evaluation of ML models for steganalysis

ML algorithm	Accuracy	Precision
Linear SVM	45.5	44.5
SVM (Gaussian kernel)	40.0	36.4
Random forest	13.5	13.9
AdaBoost	29.5	26.9
Decision tree	13.5	16.5

5.5 Histogram Analysis

Steganalysis can detect the presence of hidden data using histograms. If there is a noticeable difference between the histograms of cover and stego images, then there is a high probability of suspicion. Figure 7 represents the histograms plotted for the distribution of the Pepper cover image and its stego image. The two of them appear to be mostly similar visually.

These histograms are compared using correlation analysis, in which the correlation coefficient obtained for the two histograms is approximately 0.99.

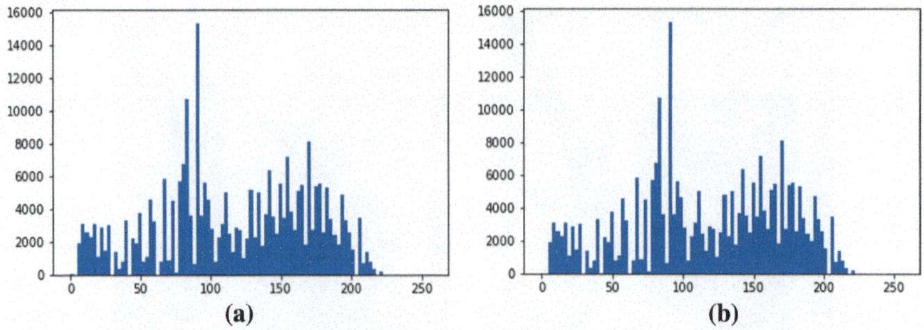

Fig. 7 Histogram of **a** cover image, **b** stego image

5.6 Computational Complexity

The time required for the total process to complete usually denotes the computational complexity. The total process in the algorithm comprises of two phases: Embedding and Extraction. The proposed method is implemented on a computer with Core i3 Gen-6 processor of 2 GHz base speed and 8 GB physical memory. The simulation is totally carried out in Python version 3.6, using Jupyter notebook. The embedding phase took 3.45 s to hide a brain MRI image in Lena cover image, and the extraction phase took 0.63 s.

Figure 8 shows a pie chart with different processes in the embedding phase and their constituent percentage in the time taken.

Fig. 8 Computational time (in percentage) in the Embedding phase

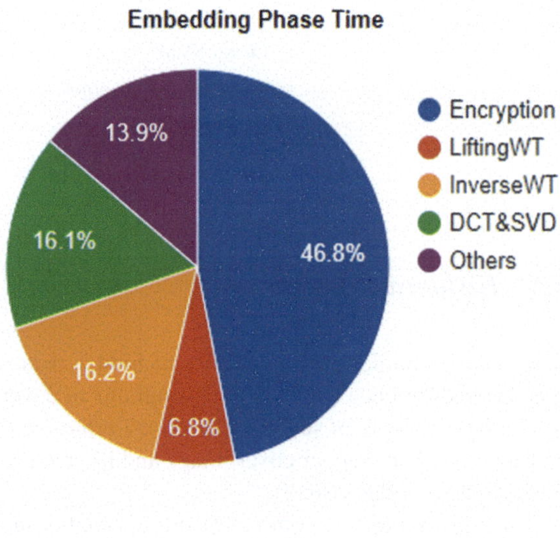

Table 4 Comparison of the method with other schemes (PSNR (dB))

Method	Cover image				
	Pepper	Baboon	Airplane	Lena	Average
Wu et al. [20]	43.56	43.54	43.53	43.54	43.54
Chang et al. [21]	40.96	40.93	40.73	40.97	40.89
Kanan et al. [22]	45.13	45.12	45.18	45.12	45.14
Arun Kumar et al.	49.67	49.26	50.02	50.12	49.77
Ours	60.42	60.43	60.41	60.41	60.42

5.7 *Comparison with Other Methods*

The comparison of the presented algorithm is carried out with some other techniques existing in the literature, in terms of PSNR. The following table represents the comparison.

The PSNR values in Table 4 shows that the proposed method is performing better than others and it can produce a stego image with better visual quality.

6 Conclusions

A high-secure steganographic method is proposed in this paper, which combines Lifting Haar WT, DCT and SVD. As, the embedding is in the singular values, the quality of the image is not disturbed and hence, stego image will be of great quality. This prevents the visual attack steganalysis. The embedding is done after the decomposition of a frequency sub-band. This makes statistical steganalysis difficult too. The encryption of secret image using Logistic chaotic map and LFSR, adds more security to the private data. As discussed in results, it also takes less amount of time to perform the embedding and extraction process. The high PSNR values suggest that the probability of suspicion about hidden data is minimum.

References

1. Kadhim IJ, Premaratne P, Vial PJ, Halloran B (2019) Comprehensive survey of image steganography: techniques, evaluations, and trends in future research. Neurocomputing 335:299–326
2. Cheddad A, Condell J, Curran K, Kevitt PM (2010) Digital image steganography: survey and analysis of current methods. Signal Process 90(3):727–752
3. Anees A, Siddiqui AM, Ahmed J, Hussain I (2014) A technique for digital steganography using chaotic maps. Nonlinear Dyn 75(4):807–816
4. Chandramouli R, Memon N (2001) Analysis of LSB based image steganography techniques. In: IEEE international conference on image processing, pp 1019–1022

5. Jung KH, Yoo KY (2015) Steganographic method based on interpolation and LSB substitution of digital images. Multimedia Tools Appl 74(6):2143–2155
6. Sutaone MS, Khandare MV (2008) Image based steganography using LSB insertion. In: IET international conference on wireless, mobile and multimedia networks, pp 146–151. https://doi.org/10.1049/cp:20080166
7. Rajendran S, Doraipandian M (2017) Chaotic map based random image steganography using LSB technique. Int J Netw Sec 19(4):93–598. https://doi.org/10.6633/ijns.201707.19(4).12
8. Pan F, Li J, Yang X (2011) Image steganography method based on PVD and modulus function. In: International conference on electronics, communications and control, pp 282–284. https://doi.org/10.1109/icecc.2011.6067590
9. Wu T (2003) A steganographic method for images by pixel value differencing, 24(9–10):1613–1626
10. Das S, Sharma S, Bakshi S, Mukherjee I (2018) A framework for pixel intensity modulation-based image steganography. Adv Intell Syst Comput 563:3–14. https://doi.org/10.1007/978-981-10-6872-0_1
11. Swain G (2016) A steganographic method combining LSB substitution and PVD in a block. Proc Comput Sci 39–44. https://doi.org/10.1016/j.procs.2016.05.174
12. Kalita M, Tuithung T (2016)A novel steganographic method using 8-neighboring PVD (8nPVD) and LSB substitution. In: International conference on systems, signals and image processing. https://doi.org/10.1109/IWSSIP.2016.7502756
13. Saidi M, Hermassi H, Rhouma R, Belghith S A new adaptive image steganography scheme based on DCT and chaotic map. Multimedia Tools Appl 76:13493–13510, 2017. https://doi.org/10.1007/s11042-016-3722-6
14. Divya V, Sasirekha N (2016) High capacity steganography technique based on wavelet transform. In: Online international conference on green engineering technologies.. https://doi.org/10.1109/get.2016.7916612
15. Kumar V, Kumar D (2017) A modified DWT-based image steganography technique. Multimedia Tools Appl 1–30. https://doi.org/10.1007/s11042-017-4947-8
16. 1Rohit T, Surekha B (2018) A color image steganography in hybrid FRT–DWT domain. J Inf Sec Appl 40:92–102
17. Singh S, Singh R, Siddiqui TJ Singular value decomposition based image steganography using Integer wavelet transform. Adv. Signal Process Intell Recogn Syst 201:593–601. https://doi.org/10.1007/978-3-319-28658-7_50
18. Arunkumar S, Subramaniyaswamy V, Vijayakumar V, Chilamkurti N, Logesh R (2019) SVD based robust image steganographic scheme using RIWT and DCT for secure transmission of medical images. Measurement 139:426–437. doi:https://doi.org/10.1016/j.measurement.2019.02.069
19. Rohith S, Hari Bhat KN, Sharma AN (2014) Image encryption and decryption using chaotic key sequence generated by sequence of logistic map and sequence of states of linear feedback shift register. In: International conference on advances in electronics computers and communications. https://doi.org/10.1109/icaecc.2014.7002404
20. Wu CC, Kao SJ, Hwang MS (2011) A high quality image sharing with steganography and adaptive authentication scheme. J Syst Softw 84(12):2196–2207
21. Chang CC, Hsieh YP, Lin CH (2008) Sharing secrets in stego images with authentication. Pattern Recogn 41(10):3130–3137
22. Kanan HR, Nazeri B (2014) A novel image steganography scheme with high embedding capacity and tunable visual image quality based on a genetic algorithm. Expert Syst Appl 41(14):6123–30

Analysis of Energy Consumption of Energy Models in Wireless Sensor Networks

Indrani Das, Rabindra Nath Shaw, and Sanjoy Das

Abstract In the wireless sensor networks, energy consumption is a very crucial issue because they work through the limited battery power of individual nodes. Here, we have analyzed various energy models on the performance. Different routing protocols are used on various node densities. In WSN, if nodes battery limit once exhausted the ongoing message exchanges for any kind of work get completely discontinued. A sensor node in WSN uses battery power when they are sending, receiving, and listening ongoing communication in transmitting, receiving, idle, and sleep mode. The total energy is consumed for different activities, like routing protocols message exchanges, MAC layer operations as well as other network layers. Any kind of message exchanges required a certain amount of power. Therefore, battery life is a major concern in the network. The maximum power consumption is observed when nodes receive and transmit packets from one node to another. So, proper utilization of battery of individual nodes is always desirable in this network. The individual node battery level determines the overall network performance and lifetime. In this paper, we have used various energy models. The AODV and DSR routing protocols are used for evaluating their performance in terms of energy consumption. Our simulation results show that AODV routing protocol used less battery power as compared to DSR in all energy models.

Keywords Nodes · Transmit mode · Receive mode · Energy consumption · Network lifetime

---

I. Das
Department of Computer Science, Assam University, Silchar, Assam, India
e-mail: indranidas2000@gmail.com

R. N. Shaw
School of Electrical Electronics & Communication Engineering, Galgotias University, Greater Noida, India
e-mail: r.n.s@ieee.org

S. Das (✉)
Department of Computer Science, Indira Gandhi National Tribal University, Regional Campus Manipur, Imphal, India
e-mail: sdas.jnu@gmail.com

M. N. Favorskaya et al. (eds.), *Innovations in Electrical and Electronic Engineering*, Lecture Notes in Electrical Engineering 661, https://doi.org/10.1007/978-981-15-4692-1_57

1 Introduction

In the wireless sensor networks, with the help of energy models energy of individual node is well conserved. The sensor nodes are deployed randomly or certain fashion for information acquisition [7]. These nodes collect data and transmit them to a data concentrator which analyzes and processes data.

A sensor node can work in transmit, receive, idle, and sleep mode. The WSN's various network layers perform various activities and exchanges various control packets to fulfill the network requirements. But the sensor network does not have external source of power supply, and nodes are operated with limited battery power. So, identifying suitable energy models is very essential to increase lifetime of the network. This depends on the battery life of individual node. So, all node participation should be avoided in the sensor network [8].

In this paper, AODV and DSR routing protocols are used to understand the energy consumption by individual nodes in different energy models. This study helps in understanding the overall network life span with various energy models. Through simulation as well as other research works show that sensor node consumed major portion of battery power during message transmission and receive operations.

The paper is organized as follows: Sect. 1 included the introduction of WSNs, and literature review is discussed in Sect. 2. The methodology and experiment analysis are discussed, and observation of the performance of AODV and DSR is thoroughly studied as shown in Sect. 3. The paper is concluded in Sect. 4.

2 Literature Review

A lot of research works has addressed the energy consumption issues of sensor nodes. Also, the analysis of various routing protocols is studied by different research communities. In this section, we have included literature concerning to our problem specifically work focused on energy models.

In [6], authors have discussed the energy consumption for MANET. They have considered various activities of network layers, application layer, and MAC layer. The DSDV, DSR, and AODV routing protocols are used for analyze. The result shows the AODV routing protocol consumed lesser energy than DSR and DSDV protocols.

In [4], authors have proposed a linear discriminant packet flow analysis system (LDPA) algorithm to analyze energy consumption and detect the attack, and the results analysis show the effectiveness of the LDPA algorithm.

The authors[11] focused on service quality as well as survival time of WSNs. A novel shuffled frog leaping algorithm is proposed. The energy consumption in the physical layer is mathematically modeled and applied optimization function of energy consumption. The total energy consumption in sending message and transmission power of WSN are considered. In terms of survival of individual sensor nodes, error frame, and network lifetime, the proposed algorithm is performing better as compared to others.

In [1], authors develop energy consumption models, which considered energy consumption due to data packets, control packets, and re-transmission occur in the network. The simulation results show that their model is better than the existing one.

In WSN, network lifetime is determined by battery life of individual node. So, all node participation should be avoided in the sensor network. The opportunistic routing and geographic random forwarding technique are used for reducing energy consumption [8]. With the help of sleep mode operation, network lifetime is improved as well as reduced the energy consumption.

In [9], authors proposed a mathematical framework for energy model to evaluate the energy consumption. The IEEE 802.15.4, DMAC with radio frequency identification (RFID) impulse is used for performance analysis. The mathematical results obtained show RFID impulse performs better than others in terms of energy consumption.

In this paper, authors [10] proposed a temperature-based analytical battery model. This is a software-based approach, which calculates the state of charge, and the voltage level of batteries of nodes in WSNs. The results show that without disturbing the tasks performed by nodes this model can be embedded to compute the lifetime of batteries.

In [5], a series of experiments conducted for energy consumption of IEEE 802.11 based wireless network is studied. The energy consumption is modeled as linear equation with send, receive, and discard broadcast and point-to-point data packets of various sizes.

In [2], authors have investigated three scenarios based on existing energy models. Authors used Mica, MicaZ, and generic energy models along with AODV and DYMO routing protocols. The experiment results show that both the routing protocols consumed less energy in MicaMote energy model.

3 Methodology and Experimental Analysis

During various operations performed by different layers of a network consumed certain amount of battery power. Sensor node may operate in transmit, receive, idle, and sleep modes. If a node is always active in the network consumed lots of energy, so sometimes based on the need to save the battery power, they are switching in different mode. Doing this, they prolong the battery power and increase the lifetime of the network. It is observed from studies and experiments that during transmit and receive mode operations, nodes are consuming maximum power as compared to other network activities. Our analysis considered transmits and receives mode operations with various energy models and AODV and DSR routing protocols. The number of node and simulation time is fixed throughout the simulation. The AODV and DSR routing protocols are used for performance analysis done in Qualnet simulator. The simulation parameters and their values used in the experiments are given in Table 1.

Table 1 Simulation parameters and their values [3]

Parameter	Values
Simulation time	30 min
Number of node	20
Data rate	2 Mbps
Packet reception model	PHY802.11b
Network area	1500 × 1500
Number of channel	1
Path loss model	Two ray
Battery model	Residual life estimator
Antenna model	Omnidirectional
Radio type	802.11b
Battery type	Duracell AA(MX-1500)
Routing protocol	AODV and DSR
Energy model	Mica-Motes, Generic, MicaZ
Mobility model	Random waypoint
Pause time	1 s
Minimum speed of nodes	0 mps
Maximum speed of nodes	10 mps

A. **Energy Consumption in Transmit Mode**

The total energy (power) consumed (in mWh) by radio interface in the transmission mode.

In Fig. 1, the results of energy consumption are computed based on various energy models, namely generic, MicaMote, and MicaZ for AODV routing protocol.

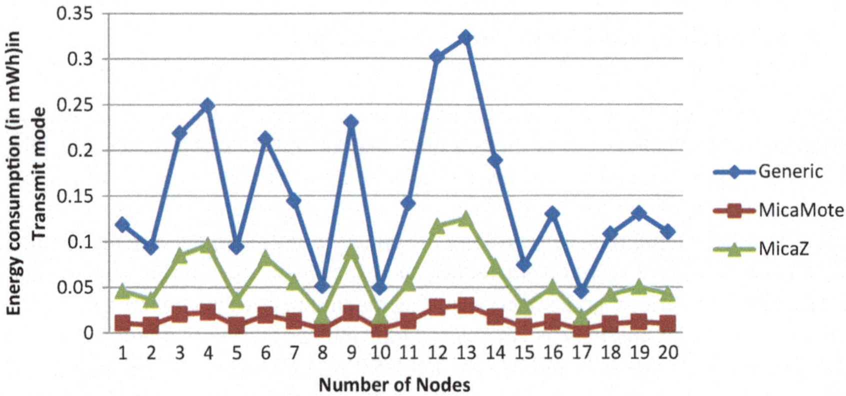

Fig. 1 Energy consumed for AODV routing protocol

The results show that energy consumption is lowest in MicaMote energy model irrespective of the number of nodes.

In Fig. 2, the results of energy consumption are computed based on various energy models, namely generic, MicaMote, and MicaZ for DSR routing protocol. The results show that energy consumption is lowest in MicaMote energy model irrespective of number of nodes. Further, for some of the nodes, MicaZ also gives the same results but energy consumption is too high for generic model.

B. Energy Consumption in Receive Mode

Total energy (power) consumed (in mWh) by radio interface in reception mode.

In Fig. 3, the results of energy consumption during receive mode operation are computed based on various energy models, namely generic, MicaMote, and MicaZ

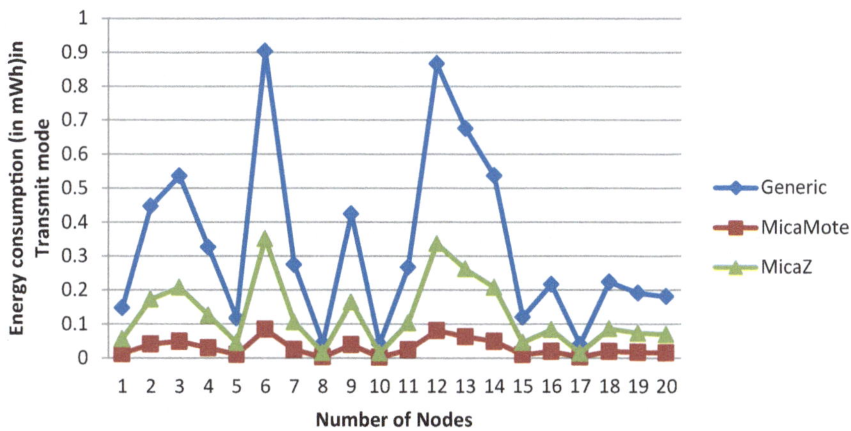

Fig. 2 Energy consumed for DSR routing protocol

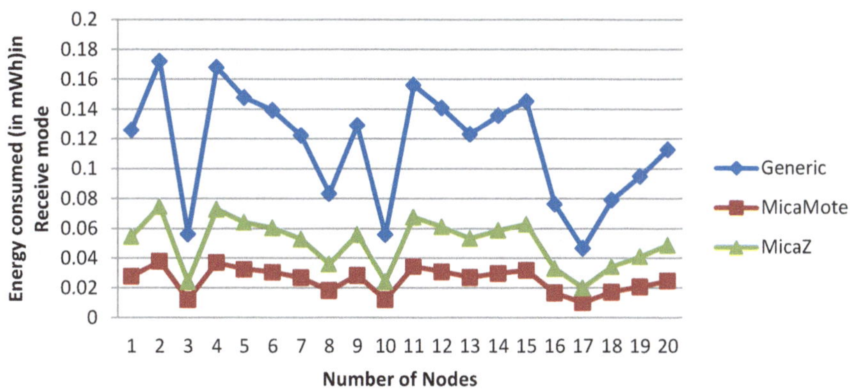

Fig. 3 Energy consumed for AODV routing protocol

for AODV routing protocol. The results show that energy consumption is lowest in MicaMote energy model irrespective of the number of nodes.

In Fig. 4, the results of energy consumption during receive mode operation are computed based on various energy models for DSR routing protocol. The results show that energy consumption is lowest in MicaMote energy model irrespective of the number of nodes. Also, this is observed that MicaZ is also giving better results.

C. Energy Consumption comparison of routing protocols in Transmit Mode

In Fig. 5, the results of energy consumption during transmit mode are compared for the AODV and DSR routing protocols for generic energy model. The results show that energy consumption is lower in AODV routing protocol.

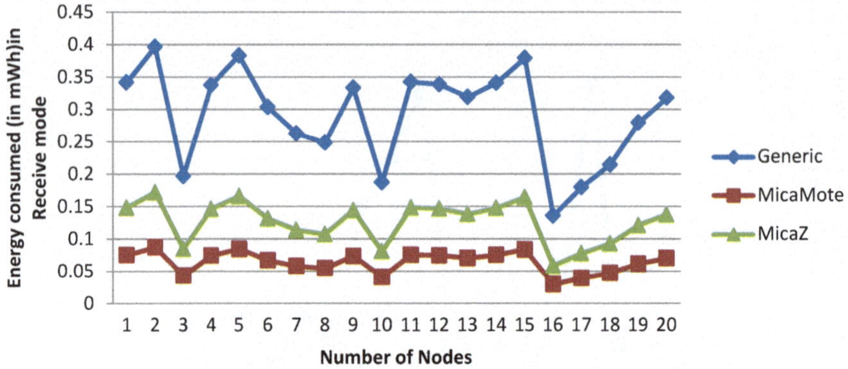

Fig. 4 Energy consumed for DSR routing protocol

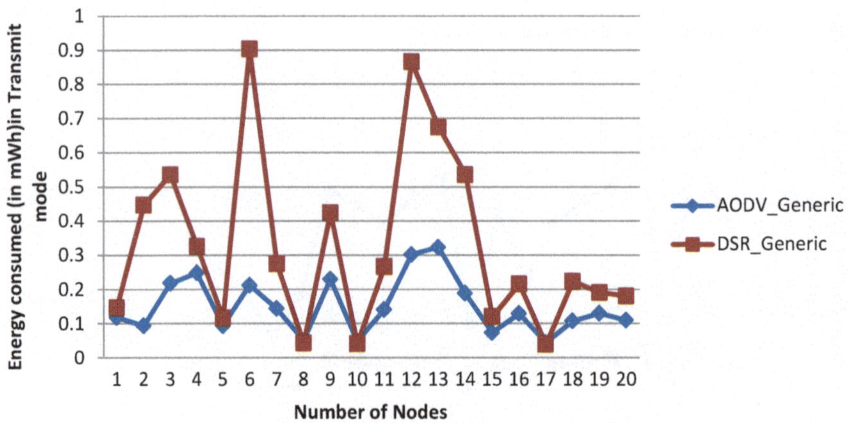

Fig. 5 Energy consumed during transmit mode for AODV versus DSR routing protocol with generic energy model

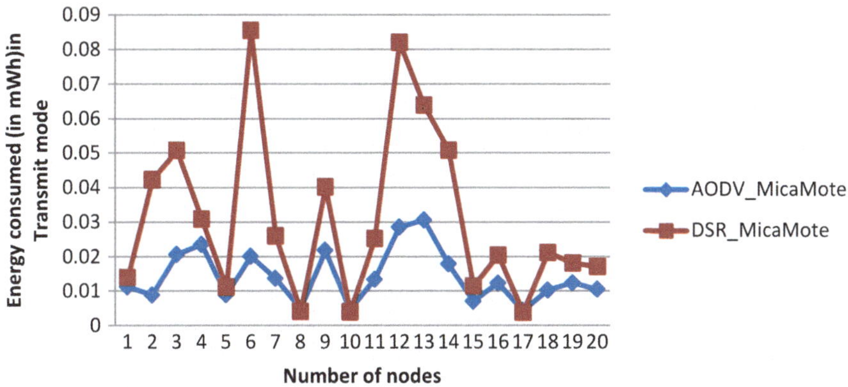

Fig. 6 Energy consumed during the transmit mode for AODV versus DSR routing protocols with MicaMote energy model

In Fig. 6, the results of energy consumption computed for transmit mode operation are compared for MicaMote energy model. The results show that energy consumption is lower for the AODV routing protocol during transmit mode operation.

In Fig. 7, the results of energy consumption during transmit mode operation are compared for the AODV and DSR routing protocols for MicaZ energy model. The results show that energy consumption is lower in AODV routing protocol but for some nodes DSR routing protocol gives the similar results.

D. **Energy Consumption comparison of routing protocols in Receive Mode**

In Fig. 8, the results of energy consumption during operation during receive mode are compared for the AODV and DSR routing protocols with the generic energy model.

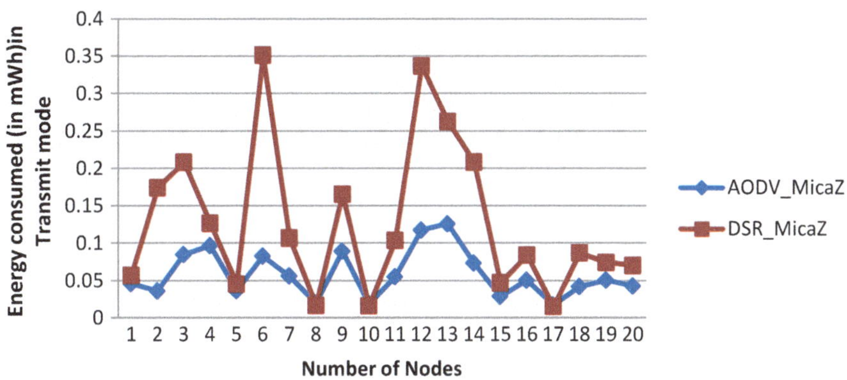

Fig. 7 Energy consumed during transmit mode for AODV versus DSR routing protocol with MicaZ energy model

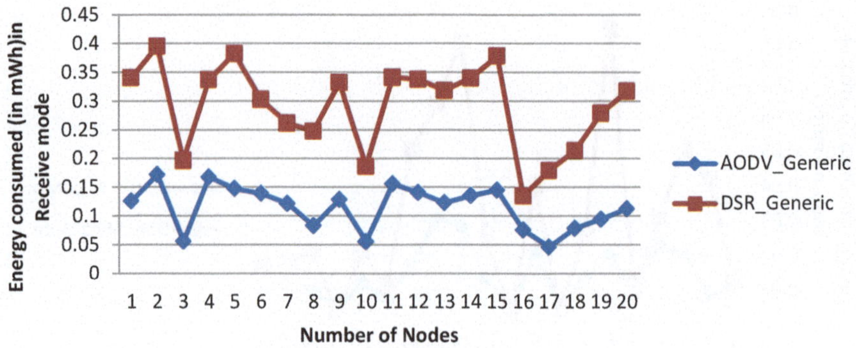

Fig. 8 Energy consumption during receive mode operation for AODV versus DSR routing protocol with generic energy model

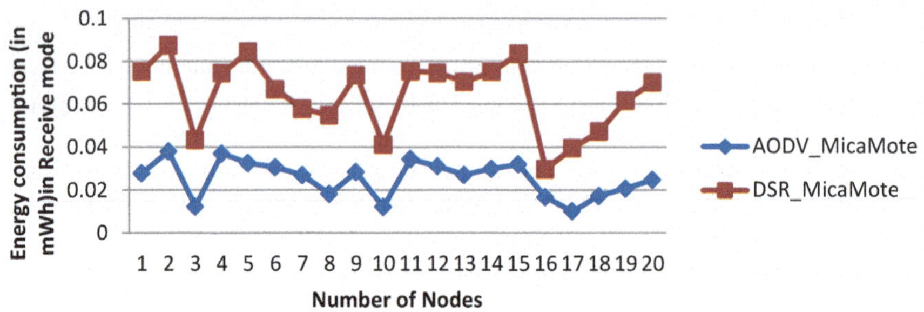

Fig. 9 Consumption during receive mode operation for AODV versus DSR routing protocol with MicaMote energy model

The results show that energy consumption is lower in AODV routing protocol but for some nodes DSR routing protocol gives the similar results.

In Fig. 9, the results of energy consumed in transmit mode are compared for the AODV and DSR routing protocols for MicaMote energy model. The results show that energy consumption is lower in AODV routing protocol.

In Fig. 10, the results of the energy consumption during the transmit mode are compared for the AODV and DSR routing protocols for MicaZ energy model. The results show that energy consumption is lower in AODV routing protocol.

4 Conclusions

In this paper, the energy consumption is analyzed for AODV and DSR routing protocols with various energy models analyzed. The thorough simulation results show that almost in all the cases, AODV routing protocol outperforms than DSR in terms

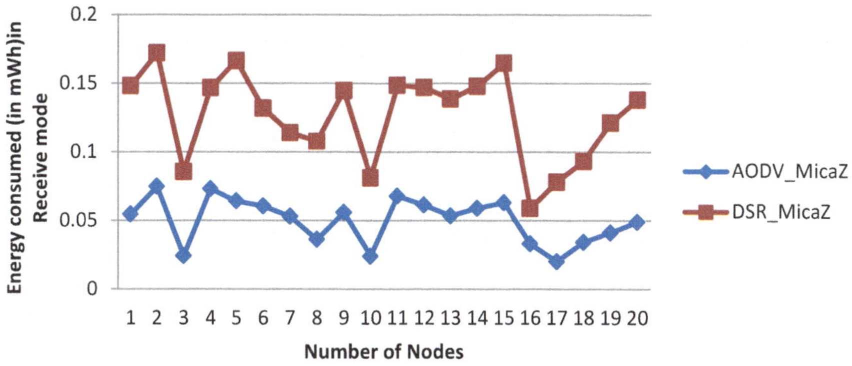

Fig. 10 Consumption during receive mode operation for AODV versus DSR routing protocol with MicaZ energy model

of energy consumption. We have shown energy consumption only in transmit and receive mode operation of sensor nodes. In transmit and receive mode operation of nodes, AODV required lesser energy as compared to DSR routing protocol. So, the network longevity will be higher when the networks operate with AODV routing protocol. In future, these energy models can be analyzed along with other battery models, MAC protocol operations, and higher node density.

References

1. Anjum Farooq, Mouchtaris Petros (2006) Security for wireless ad hoc networks. Secur Wireless Ad Hoc Netw 5(11):1–247. https://doi.org/10.1002/9780470118474
2. Asheer S, Kumar S (2015) Performance of wireless sensor network under various energy models, routing protocols and node mobility conditions. In: Proceedings of 2014 IEEE international conference on advanced communication, control and computing technologies, ICACCCT 2014, vol 978, pp 813–817. https://doi.org/10.1109/ICACCCT.2014.7019205
3. Das Indrani, Shaw Rabindra Nath, Das Sanjoy (2020) Analysis of energy consumption in dynamic mobile ad hoc networks. In: Jain L, Tsihrintzis G, Balas V, Sharma D (eds) Data communication and networks, advances in intelligent systems and computing. Springer, Singapore, pp 235–243
4. Divya M, Kavitha KR, Jaya Krishna AP (2019) Energy efficient data transmission in wireless sensor network using cross site leaping algorithm. Int J Recent Technol Eng 7(53):515–521
5. Feeney LM, Nilsson M (2001) Investigating the energy consumption of a wireless network interface in an ad hoc networking environment. Proc - IEEE INFOCOM 3:1548–1557. https://doi.org/10.1109/infcom.2001.916651
6. Heni M (2012) Energy consumption model in ad hoc mobile network. Int J Comput Netw Commun 4(3):207–217. https://doi.org/10.5121/ijcnc.2012.4314
7. Lozneanu D, Gheorghe P, Miron EL (n.d.) Energy model of sensor nodes in WSN. Tech Sci Appl Math, 33–38
8. Pradeebaa G, Lavanis N (2016) Network lifetime improvement using routing algorithm with sleep mode in wireless sensor network. In: Proceedings of the 2016 IEEE International conference on wireless communications, signal processing and networking, WiSPNET 2016, vol 1, 1572–1575. https://doi.org/10.1109/WiSPNET.2016.7566401

9. Ram M, Kumar S (2014) Analytical energy consumption model for MAC protocols in wireless sensor networks. In: 2014 International conference on signal processing and integrated networks, SPIN 2014, pp 444–447. https://doi.org/10.1109/spin.2014.6776994
10. Rodrigues LM, Carlos M, Gerson B, Francisco V, Paulo P (2017) Estimating the lifetime of wireless sensor network nodes through the use of embedded analytical battery models. J Sens Actuator Netw 6(2). https://doi.org/10.3390/jsan6020008
11. Zhou C, Ming W, Weiqing Q, Zhengqiu L (2018) A wireless sensor network model considering energy consumption balance. Math Probl Eng. https://doi.org/10.1155/2018/8592821

Predictive Data Analysis for Energy Management of a Smart Factory Leading to Sustainability

Milan Kumar, V. M. Shenbagaraman, Rabindra Nath Shaw, and Ankush Ghosh

Abstract Smart manufacturing and data analytics can help in manufacturing sector, where the current scenarios transmission and data analysis from across the plant create manufacturing intelligence; it can be used to have a good impact across all side of operations. Predictive analysis helps that in a lot of ways in smart manufacturing. In this work, authors have tried to take values of a smart manufacturing company and have done predictive analysis to take out the energy consumption based on historical data.

Keywords Smart manufacturing · Deep learning · Data augmentation · Statistical analysis · Energy consumption

1 Introduction

The pace of transformation in very fast nowadays. In the nineteenth century, the whole manufacturing was dominated by Stream, while twentieth century was all about electricity and now what we are talking about in manufacturing is all about data and only data [1–3]. Internet of things abbreviated as IoT is such a revolution that makes us collect the data from all kinds of machines. Now we have even machine-to-machine communications (M2M) which are helping manufacturing engineers, production engineers, and plant managers to feed with lots and lots of data. The information collected is not only helping us to improve the productivity but also helping to improve downtime which is bringing better disciplines in a factory.

M. Kumar · V. M. Shenbagaraman
SRM Institute of Science and Technology, Chennai, India

R. N. Shaw
Galgotias University, Noida, India

A. Ghosh (✉)
The Neotia University, Bishnupur, West Bengal, India
e-mail: ankushghosh@gmail.com

M. N. Favorskaya et al. (eds.), *Innovations in Electrical and Electronic Engineering*, Lecture Notes in Electrical Engineering 661, https://doi.org/10.1007/978-981-15-4692-1_58

Advancement of big data analytics [4, 5] now the speed to market is significantly improved as in the background it is eliminating error, reducing lead times, boosting manufacturing. Data can be collected from structured sources as well as unstructured sources, and such evolution of high data is a big challenge for any industry to manage which is often un-noticed. With the further evolution of the cloud technology, you can store the data infinitely and that is how the new term is "Big Data," But, in this case, the biggest challenge is not only to collect data but to process it and that is how data analytics comes into picture. This gives us great insight, benefitting manufacturing companies across the world.

Large auto makers like Volkswagen, Tesla have started using big data to spot patters which are helping to reduce the quality issues moving into zero-defect scenarios. The companies can easily predict the manufacturing downtimes and even predict it in advance. Companies like Microsoft, Amazon, and Google are getting bigger and bigger as companies are using their cloud storage to enhance data capabilities. The conventional energy resources are getting exhausted day by day. Energy saving is therefore has become a mandatory criteria for all factories [6–9].

Today, machine learning and artificial intelligence enables manufacturers to process petabytes of data in actual scenario that can be useful to engineers.

2 Data Analytics

It is the way to make and understand the pattern of data leading us to make the insights. With the right set of technologies like R, Python, you are able to make the data more insightful for you enabling us to take decision [10, 11]. It allows for the recognition of imperfections and more precisely than a human subject matter specialist. In the past, it was used to understand the historic set of data and try to learn it from it and try to put some manual intervention how to overcome that business challenge. But, now the historic data is used to derive a machine learning pattern based on the statistical models which help us to predict. Traditionally, a enormous amount of data is already present in the manufacturing sector. Predictive maintenance and asset performance optimizer management are acquiring of data management and analytic strategies focused on operational excellence. Previously, data analytics was used to understand the past and current status of a target, like predicting the chances of machine defects/the probable frequency of machine stoppages.

3 Data's Role in Smart Factories

Nowadays, data is an extremely important asset for manufacturers; as per Economist magazine, "Data is a New Oil." It plays a vital role in smart factory to see into the data and make meaning out of it. In fact, we can easily say it is the foundation of all Industry 4.0 and digital manufacturing lead. The key role of data analytics in smart factories is to built connection between data being generated and to focus on creating new measures which may not be immediately obvious in the past. If the manufacturer can minimize production shutdowns or downtimes through forecast failure itself is the huge benefit of data analytics. It can enhance production competitive advantage via efficiencies, cost savings, quality leading to top-line growth.

In current scenarios, where item and utility are commoditized day by day, cost and quality are become important major advantage that data collection can be easily enabled and triggered via big data analytics. A smart factory is a exact environment where management and staff can optimize their work, machinery and equipment helps to improve the processes in a company. In assembly, there are multiple sources of information which can be connected to the Internet of things (IoT) via sensors and other SCADA devices [11, 12]. The industrial Internet of things (IIoT) [12, 13] are made up of sensor nodes that capture manufacturing data on processes and inventory and transmitted through a network that connects to the cloud data storages.

It uses this information and helps identify based on data patterns shows the anomalies. In the supply chain process, it extends the quality, logistics, assembly, machining and product lifecycle management. Smart factories enable high optimized performance that uses advanced IT; data analytics plays a key role in how those plant operates.

Manufacturers are facing less development cycles, product cycles, stringent budgets, and high supply chain diversity. But, it needed strong data search platforms which can monitor and manage production lines. Data can provide critical functions like security, capability to analyze it within no time. Big data analytics, ML techniques can be applied on old data to find the meaningful information in data.

4 Basic Elements of Data Analytics

The new phrase like cloud, cloud computing, edge computing, data lake are often heard new terms which are main elements to enable any data analytics. Data ingestion occurs through plant sensors embedded, or even now, we can retrofit in production equipment such as SCADA machines, assembly machines, conveyor belts, co-bots, robots. In addition to data acquisition hardware, other key enablers are data acquisition, data visualization, analytics platform, and monitoring software.

Three steps are involved in data analytics process: applying analytics, data acquisition, and data visualization. Data acquisition is connecting with production machines, often through edge devices or other types of h/w, and if the machine is not IoT enabled, then you have to retrofit the sensors then capturing and importing data which is called data ingestion. At Step 2, it applies analytics using advanced statics modeling or ML s/w like R, Python. The last step is result of an application or visualization layer, like dashboards which display data about uptime or throughput, quality. The main h/w components include IoT sensors and actual located on sensor nodes, fix equipment, and combined data to send to host that is require to process and storage capacity in the cloud or on area in the hosts. Programmable logic controllers (PLCs) are the main components in data collection. Programmable logic controllers (PLCs) enable machinery and bots to be monitored.

The data can be seen in programmable logic controllers (PLCs) doesn't stay long as it is immediate. Therefore, the data can be extracted and saved in the corner server or a cloud. The edge server remains close to the production line to temporarily save data and do some data preprocessing.

5 Emerging Application Areas of Data Analytics

Data analytics is helping manufacturers improve efficiency, productivity, streamlining assembly operations. In the past, a machine would simply grind to a halt, hindering production for days or even weeks before big data analytics, factories had few signs of equipment malfunction. Today, using IoT-enabled sensors, IIoT and data analytics, production engineers can measure equipment temperature, heat, sound, vibrations to understand what "normal" operations look and feel like. Big Data analytics also enables manufacturers and companies to track the amount of time it takes products to move through different areas of the assembly line, value stream points out.

Today's industry and manufacturing landscape demands we run lean manufacturing, because consumers desire their products faster and much cheaper as it used to be. On the other hand, our industry faces a skilled manpower and talent shortage, so expanding our team is not a sustainable solution and most wise solution.

Technologies like artificial intelligence, machine learning and deep learning, can not only help engineers more quickly and accurately identify right statistical patterns and trends from historical data, but they can learn and adapt as new data comes making it more predictive. Advance analytics can merge previously unavailable data from various business systems so that assembly lines can get smarter enabling us to take the decisions well in advance.

IoT-enabled products delivered to individual customers and business customers can send data back to the companies, helping improve product or services, and open up new revenue streams and models never before thought off or explored, all enabled by data. Main benefits of smart factories are cost reduction, quality improvement, improving efficiencies; this will enhance in the future how the data can generate and grow new business opportunities and new revenue models.

6 Data and Study Area

Data analytics with predictive analysis is shown here. It shows how to predict energy consumption for a certain period of time. To do this, data analytics and machine learning are used here. It will predict how much energy (KWh) will be used when

Date	runTime	OAT	P1VFDOutput	P2VFDOutput	month	Day_of_w	vacation	kWh
2/18/2018	0	33.81698113	79.70628931	0	2	6	0	593.2075
2/19/2018	0	36.45763889	74.99340278	0	2	0	0	1611
2/20/2018	0	35.61666667	77.59340278	0	2	1	0	1628
2/21/2018	0	23.44305556	51.61944444	41.90833333	2	2	0	1729.5
2/22/2018	0	28.175	0	90.39201389	2	3	0	1641
2/23/2018	0	36.24583333	0	80.94652778	2	4	0	1207.5
2/24/2018	0	32.06597222	0	83.13854167	2	5	0	598.5
2/25/2018	0	31.77708333	0	84.48472222	2	6	0	638.5
2/26/2018	0	36.74097222	0	68.23263889	2	0	0	1624.5
2/27/2018	0	46.46875	0	57.34097222	2	1	0	1546
2/28/2018	0	38.15208333	27.93715278	23.46145833	2	2	0	1537.5
3/1/2018	0	35.15902778	79.39513889	0	3	3	0	1592.5
3/2/2018	0	36.29930556	67.87638889	0	3	4	0	1597.5
3/3/2018	0	36.25	53.68402778	0	3	5	0	446
3/4/2018	0	35.77665505	54.70034843	0	3	6	0	491.7073
3/5/2018	0	32.44652778	83.91701389	0	3	0	0	1665
3/6/2018	0	31.41944444	85.61840278	0	3	1	0	1659
3/7/2018	0	24.64652778	51.24618056	41.88784722	3	2	0	1692.5
3/8/2018	0	25.03402778	0	93.73784722	3	3	0	1668
3/9/2018	0	26.43402778	0	89.0875	3	4	0	1556.5
3/10/2018	0	28.78055556	0	71.91666667	3	5	0	556
3/11/2018	0	32.76606498	0	44.1566787	3	6	0	560.9242

Fig. 1 Energy consumption dataset

machines run inside a smart manufacturing industry [14–16]. This forecast will help to prepare the proposed budget (Fig. 1).

This is a historical data of a smart manufacturing industry. Here, two machines runs and their power consumption unit in KWh is given. To calculate this, Keras [15] is used to predict the KWh of energy which will be spent when specific inputs are given.

7 Framework and Prediction Model

A deep learning regression model is created with the Keras/TensorFlow library and compiled for a target variable KWh per day. The final model is saved with a root mean squared error (RSME) value of approximately 40 KWh after training 1500 epochs (cycles) on the data set. The last part of this report is an example (and possible strategy) of how an energy manager may be able to utilize a model in a simple python script in predicting electrical energy consumption.

The process started with extracting one year's worth of time series data recorded on 5 min intervals from the machine automation system. Preprocessing the data was done in Excel, and then Python was used to resample the data to compute daily mean values in the Pandas computing library (Fig. 2).

Then, it is divided into two data, i.e., training and testing, for further calculation (Fig. 3).

```
: X = dfTrain.drop(['kWh'],1)
  Y = dfTrain['kWh']

: offset = int(X.shape[0] * 0.7)
  X_train, Y_train = X[:offset], Y[:offset]
  X_test, Y_test = X[offset:], Y[offset:]
```

Fig. 2 Data resample program

```
: model = Sequential()
  model.add(Dense(60, input_dim=7, kernel_initializer='normal', activation='relu'))
  model.add(Dense(55, kernel_initializer='normal', activation='relu'))
  model.add(Dense(50, kernel_initializer='normal', activation='relu'))
  model.add(Dense(45, kernel_initializer='normal', activation='relu'))
  model.add(Dense(30, kernel_initializer='normal', activation='relu'))
  model.add(Dense(20, kernel_initializer='normal', activation='relu'))
  model.add(Dense(1, kernel_initializer='normal'))
  model.summary()
  model.compile(loss='mse', optimizer='adam', metrics=[rmse])

  Model: "sequential_1"
  _________________________________________________________________
  Layer (type)                 Output Shape              Param #
  =================================================================
  dense_1 (Dense)              (None, 60)                480
  _________________________________________________________________
  dense_2 (Dense)              (None, 55)                3355
  _________________________________________________________________
  dense_3 (Dense)              (None, 50)                2800
  _________________________________________________________________
  dense_4 (Dense)              (None, 45)                2295
  _________________________________________________________________
  dense_5 (Dense)              (None, 30)                1380
  _________________________________________________________________
  dense_6 (Dense)              (None, 20)                620
  _________________________________________________________________
  dense_7 (Dense)              (None, 1)                 21
  =================================================================
```

Fig. 3 Data processing program

Then, a sequential model is created with the help of Keras/TensorFlow to perform the predictive analysis. Subsequently, the program is run for 1500 epochs to train the model and which will improve the accuracy (Fig. 4).

The epoch is run to train the dataset and increase the accuracy.

```
In [7]: history = model.fit(X_train, Y_train, epochs=1500, batch_size=1, verbose=2)

        Epoch 1/1500
         - 2s - loss: 560347.0809 - rmse: 570.2821
        Epoch 2/1500
         - 0s - loss: 134732.7105 - rmse: 312.4636
        Epoch 3/1500
         - 1s - loss: 116291.1931 - rmse: 281.4558
        Epoch 4/1500
         - 1s - loss: 110181.1986 - rmse: 261.0042
        Epoch 5/1500
         - 0s - loss: 103795.6129 - rmse: 260.4303
        Epoch 6/1500
         - 1s - loss: 102719.3804 - rmse: 253.0388
        Epoch 7/1500
         - 0s - loss: 99835.5238 - rmse: 247.1861
        Epoch 8/1500
         - 0s - loss: 87888.4128 - rmse: 233.9145
        Epoch 9/1500
         - 0s - loss: 94877.5023 - rmse: 246.1768
        Epoch 10/1500
```

Fig. 4 Data training program

8 Results and Discussions

The result can be depicted from the graph plotted between KWh versus epoch. The graph clearly shows that how the loss is decreasing, and the accuracy is increasing (Fig. 5).

```
plt.plot(history.history['rmse'])
plt.title("kWh RSME Vs Epoch")
plt.show()
```

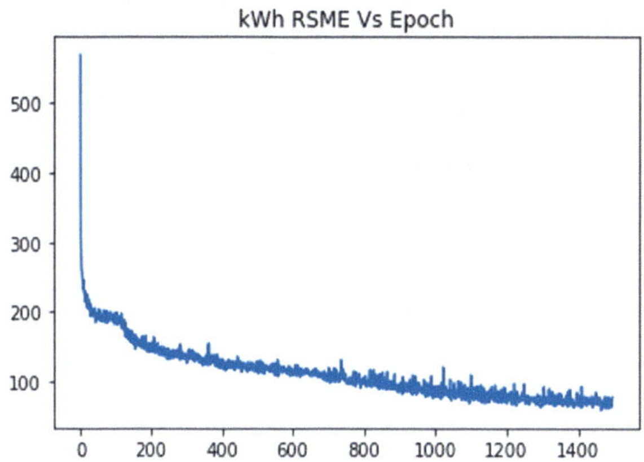

Fig. 5 Graph between KWh versus epoch

```
In [9]: model.predict(X_test)

Out[9]: array([[1054.3888 ],
               [1192.5333 ],
               [ 482.58896],
               [ 407.71115],
               [1015.0923 ],
               [ 972.8841 ],
               [ 985.09064],
               [1137.5066 ],
               [1170.0199 ],
               [1169.9825 ],
               [ 698.3013 ],
               [1043.8987 ],
               [ 965.6087 ],
               [ 803.42786],
               [ 892.33075],
               [1069.0815 ],
               [ 441.18106],
               [ 460.13577],
               [1064.4578 ],

In [10]: Y_test

Out[10]: Date
         2018-11-01    1092.499998
         2018-11-02    1068.000000
         2018-11-03     413.500000
```

Fig. 6 Predictive analysis result

The result can be predicted from the test dataset to make predictive analysis of the sample data. The predictive analysis result is shown in the table (Fig 6).

Here, we have predicted over the test dataset and compared with the original dataset, which passed the accuracy test. From the predictive result, energy saving can be calculated. Assuming the machine averages 40% load during unoccupied hours (108 h/week), then having the chiller shutoff during unoccupied hours can save:

$$\text{KWh Savings Potential} = (4\ \text{weeks/month})^{*}\ (108\ \text{h/week})^{*}$$
$$(140\text{ns}^{*}\ 0.7\ \text{kW/Ton}\ ^{*}\ 40\%) = 16,900\ \text{kWh potential monthly savings.}$$

9 Conclusions

Smart manufacturing is coming in a big way in manufacturing sector to process the real-time transmission and analysis of data from across the factory creates manufacturing intelligence, and to make prediction. Here, in this work, a predictive analysis is done on energy consumption in a smart factory. The prediction have been done over the test dataset and compared with the original data to substantiate our result. Thus, our predictive analysis has proven the accuracy test. The predictive analysis result helps that in a lot of ways in smart manufacturing as well as to prepare advanced budget of the company.

References

1. Kusiak A (2018) Smart manufacturing. Int J Prod Res 56(1–2):508–517
2. Lade P, Ghosh R, Srinivasan S (2017) Manufacturing analytics and industrial internet of things. IEEE Intell Syst 32(3):74–79
3. Tao, F Qi Q, Liu A, Kusiak, A (2018) Data-driven smart manufacturing. J Manuf Syst
4. Kusiak A (2017) Smart manufacturing must embrace big data. Nature 544(7648):23
5. Hu H, Wen Y, Chua TS, Li X (2014) Toward scalable systems for big data analytics: a technology tutorial. IEEE Access 2:652–687
6. Shaw RN, Walde P, Ghosh A (2020) Review and analysis of photovoltaic arrays with different configuration system in partial shadowing condition. Int J Adv Sci Technol 29(9):2945–2956
7. Shaw RN, Walde P, Ghosh A (2020) Enhancement of power and performance of 9x4 PV arrays by a novel arrangement with shade dispersion. Test Eng Manage 13136–13146, ISSN: 0193–4120
8. Shaw RN, Walde P, Ghosh A (2019) Effects of solar irradiance on load sharing of integrated photovoltaic system with IEEE standard bus network. Int J Eng Adv Technol 9(1). October 2019.
9. Shaw RN, Walde P, Ghosh A, A new model to enhance the power and performances of 4x4 PV arrays with puzzle shade dispersion. Int J Innovative Technol Exploring Eng 8(12)
10. Casado R, Younas M (2015) Emerging trends and technologies in big data processing. Concurr Comput: Pract Exper 27(8):2078–2091
11. Hopkins S, Kalaimannan E (2019) Towards establishing a security engineered SCADA framework. J Cyber Secur Technol 1–13. https://doi.org/10.1080/23742917.2019.1590920
12. Perwej Yusuf, Aboughaly M, Bedine K, Harb H (2019) An extended review on internet of things (IoT) and its promising applications. 7:8–22
13. Yaqoob I, Hashem I, Ahmed A, Kazmi S (2019) Internet of things forensics: Recent advances, taxonomy, requirements, and open challenges. Future Gener Comput Syst 92:265–275
14. Andreas Sepp (2016) Predictive analytics. https://doi.org/10.13140/rg.2.2.28803.96802
15. Wang J, Ma Y, Zhang L, Gao RX, Wu D (2018) Deep learning for smart manufacturing: methods and applications. J Manuf Syst
16. Zhao R, Yan R, Chen Z, Mao K, Wang P, Gao RX (2015) Deep learning and its applications to machine health monitoring: a survey. J Latex Class Files 14(8):1
17. Das I, Shaw RN, Das S (2020) Analysis of energy consumption in dynamic mobile ad hoc networks. 1049

Location-Based and Multipath Routing Performance Analysis for Energy Consumption in Wireless Sensor Networks

Indrani Das, Rabindra Nath Shaw, and Sanjoy Das

Abstract The wireless sensor networks sensor nodes are operated through battery power; for every particular activity, amount of energy is used. The node lifetime also depends on battery life. The longer the battery life leads to a longer life of the network. So, energy utilization is crucial. In this paper, we have used location-based routing protocol and multipath routing protocol for analyzing the consumption of energy. The principal objective of this work is to analyze the performance of different categories of routing protocols in terms of energy utilization. So that appropriate routing protocol may be adapted for future communication. The total energy is consumed for various activities, like routing protocols message exchanges, MAC layer operations as well as other network layers. Any kind of message exchanges required a certain amount of power. We have analyzed power consumption for individual nodes. The AOMDV and LAR routing protocols are used to evaluate their performance in terms of energy consumption. This is evident from the result that the AOMDV routing protocol utilizes lesser battery power as compared to protocol.

Keywords Sensor node · AOMDV · LAR · Transmit mode · Receive mode · Energy consumption · Network lifetime

---

I. Das
Department of Computer Science, Assam University, Silchar, Assam, India
e-mail: indranidas2000@gmail.com

R. N. Shaw
School of Electrical Electronics & Communication Engineering, Galgotias University, Greater Noida, India
e-mail: r.n.s@ieee.org

S. Das (✉)
Department of Computer Science, Indira Gandhi National Tribal University, Regional Campus Manipur, Imphal, India
e-mail: sdas.jnu@gmail.com

M. N. Favorskaya et al. (eds.), *Innovations in Electrical and Electronic Engineering*, Lecture Notes in Electrical Engineering 661, https://doi.org/10.1007/978-981-15-4692-1_59

1 Introduction

Wireless sensor networks (WSNs), sensor nodes are operated through limited battery power. The conservation of energy in WSNs is a major issue nowadays. Day by day, applications of sensor networks increase by leap and bound. The sensor nodes are deployed on particular applications. Nowadays, for monitoring a smart home, agriculture fields, animal monitoring, health care monitoring, industrial monitoring, environmental monitoring, etc., different kinds of sensors are used. All these applications need the continuous transmission of data from the source to sink for control. Due to the limited battery power and computational power, optimum uses of sensor nodes are required. Overutilization of sensor nodes may discharge higher power, and the battery power level gets exhausted quickly. This creates a fragmentation in the network, and ongoing transmission gets hampered. The objective is to how to save the battery power and increase the life of sensor nodes is a major concern.

A sensor node uses battery power in a different mode of operation like transmit, receive, idle, and sleep mode. Further, when messages are transmitted in the network, various activities supported by different layers to transmit the messages from one layer to another uses different level of battery power. There is no external power supply to the sensors in the network. So, identifying suitable routing protocols and other models is very crucial to increase the lifetime of the network.

In this paper, we have analyzed a multipath routing protocol AOMDV and location-based routing protocol LAR1 to analyze their performance in terms of energy consumption. The individual nodes' energy consumption in receive and transmit mode is analyzed thoroughly with different simulation times. Because in various past studies, it shows that sensor nodes are using maximum power during receive and transmit mode operations. This study explores the insight understanding of the overall network life span.

The paper is organized as follows: Sect. 1 included the introduction of WSNs. In Sect. 2, we have included various works on energy consumption analysis for multipath as well as location-based routing. The experimental setup and assumptions made for analysis are discussed in Sect. 3. Finally, the paper is concluded in Sect. 4.

2 Literature Review

In this section, we have discussed various works related to the multipath and location-based routing used for data delivery in wireless sensor networks.

The application of multipath routing is increased day by day in supporting the various applications of WSNs [10]. Through the uses of multipath routing networks, the lifetime can be increased, and the network becomes more connected. In this survey, various concepts of the multipath routing protocols are discussed, along with fundamental challenges. Furthermore, extensive comparison among multipath routing techniques is discussed [10]. A survey on various multipath routing protocols is extensively discussed in [2].

In [5], authors have been proposed a novel mechanism to identify the best shortest path with minimum energy consumption. To achieve this goal, a genetic algorithm-based meta-heuristic optimization technique is used. The major advantages of this mechanism help in dynamically identifying the best possible path.

This mechanism is compared with other existing protocols, and simulation result achieves by the mechanism is better as compared to others methods.

In [7], the authors addressed the various design issues for wireless sensor networks and also thoroughly analyzed various design criteria for routing protocol for WSNs. The proposed technique is an energy-aware multipath routing algorithm. This protocol identifies the nodes with low energy levels and discards them when choosing an energy-efficient route. Further, the path is selected based on the remaining energy level, the number of hops in a route. The experiments show that energy consumption in multipath is lesser than the traditional routing protocols. This helps in prolonging the network operations and individual nodes' life.

The authors proposed an energy-efficient location-based routing protocol called Greedy Perimeter Stateless Routing for wireless sensor networks (GPSR-S) [1]. This protocol is an extended version of the well-known location-based GPSR protocol. The results show that GPSR-S is more energy-efficient than GPSR. The lifetime of the network is increased by 10% in GPSR-S.

A balanced routing technique is proposed in [4], where data is transmitted through multiple routes. The multiple paths depend on the weightage means the proportion of route utilization. This mechanism is compared with ETX routing scheme, and simulation shows that the balance scheme is outperformed than ETX technique.

In WSNs, energy consumption is a significant issue, and this is addressed in [8]. The single-path routing mechanisms fail to achieve better network lifetime and frequent disconnection in the network. Many node disjoint multipath between source and destination node is identified to overcome the problem of single path shortcomings. Authors have been proposed a distributed, scalable, as well as localized mechanism, are used. Through this process, a load-balancing algorithm is utilized to split the traffic into multiple paths. This proposed load-balancing mechanism is compared with various well-known mechanisms like directed diffusion, directed transmission, N-to-1 multipath routing, and the energy-aware routing protocols for evaluating the performance.

3 Methodology and Experimental Analysis

We have assumed that energy consumption by different activities other than receiving and transmit mode operation is negligible. We have observed different rounds of simulation that idle and sleep mode operations energy consumption is almost zero. Generally, if a node is always actively participate in communication, comparatively consumed higher energy. That is why nodes are operated through different modes to prolong the battery power as well as increase the lifetime of the network. We have set up a random topology of 30 sensor nodes and random CBR source and destination. This parameter is constant throughout all the experimental setup. The functioning and energy consumption of both the protocols are checked for 10, 20, and minutes of simulation. It is observed from studies and experiments that during transmit and receive mode operations, nodes are consuming maximum power, and in receive mode, operation consumes higher energy than transmit mode. The multipath routing protocol AOMDV [9] and location-based LAR1 routing protocols [6] are used for performance analysis done in Qualnet simulator. The simulation parameters considered for the analysis and their values are given in Table 1.

A. **Energy Consumption (in Transmit Mode)**

The total energy (power) consumed (in mWh) by radio interface in the transmission mode.

Figure 1 shows the overall consumption of energy during the transmit mode operation of sensor nodes. In terms of energy utilization, it is clearly visible from the graph that location-based routing protocol LAR-1 consumes more energy as compared to AOMDV. Some of the nodes are consuming energy quite similar in both the protocols.

Figure 2 shows the overall consumption of energy during the transmit mode operation of sensor nodes with increased simulation time 20 min. The data transmission begins at 1 s and terminates at 20 min. In terms of energy utilization, it is visible

Table 1 Simulation parameters and their values (Das et al. [3])

Parameter	Values
Simulation time	10, 20, 30 min
Nodes	30
Data rate	2 Mbps
Data item send	100
Packet reception model	PHY802.11b
Data size (Byte)	512
Data interval	1
Network area	1500 × 1500
Number of channel	1
Path loss model	Two ray
Battery model	Linear
Antenna model	Omni-directional
Radio type	802.11b
Full battery capacity	1200 mA.h
Routing protocol	AOMDV and LAR1
Energy model	Mica-Motes

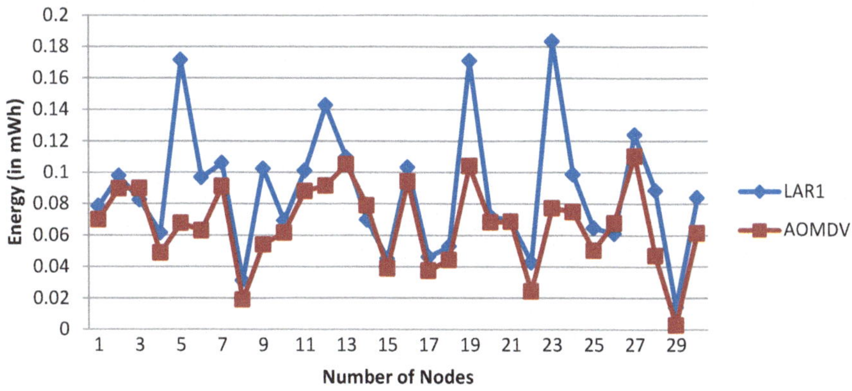

Fig. 1 Energy consumed LAR1 versus AOMDV (simulation time 10 min)

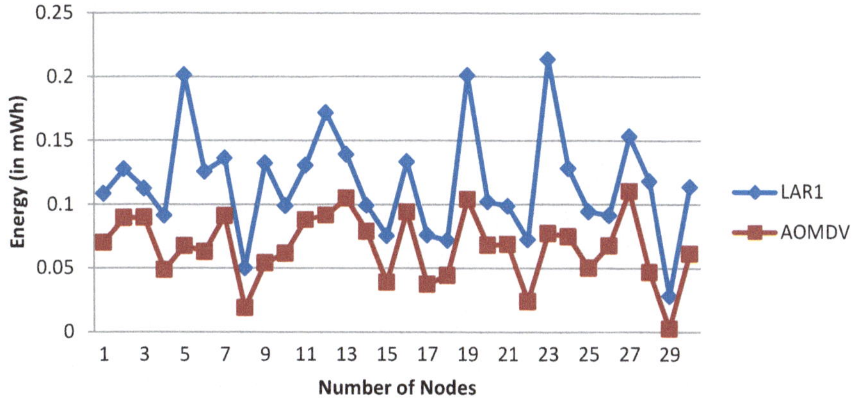

Fig. 2 Energy consumed LAR1 vs AOMDV (simulation time 20 min)

from the graph that location-based routing protocol LAR-1 consumes more energy as compared to AOMDV for all nodes.

Figure 3 shows the overall consumption of energy during transmit mode operation of sensor nodes with higher simulation time 30 min as compared to others mentioned in Figs. 1 and 2, respectively. The data transmission begins at 1 s and terminates at 30 min. All the nodes, while transmitting messages, consume lesser energy when transmitting packets with AOMDV protocol. During transmission, all the nodes consume higher energy in LAR1 operation.

B. **Energy Consumption (in Receive Mode)**

Total energy (power) consumed (in mWh) by radio interface in reception mode.

Figure 4 shows the overall consumption of energy during receive mode operation of sensor nodes. In terms of energy utilization, while receiving messages from various

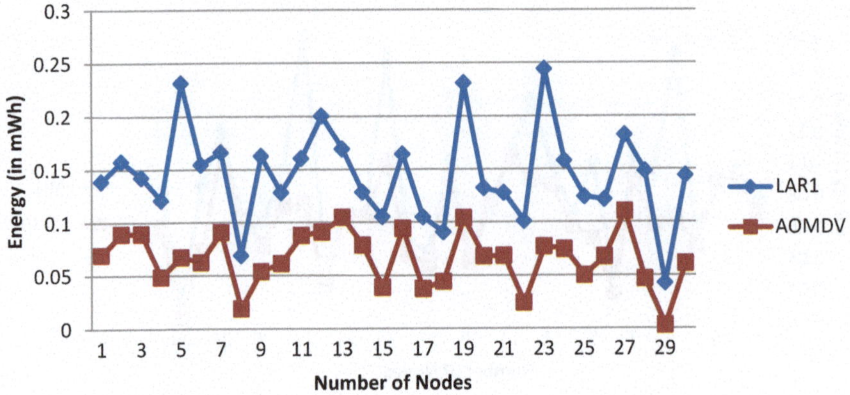

Fig. 3 Energy consumed LAR1 versus AOMDV (simulation time 30 min)

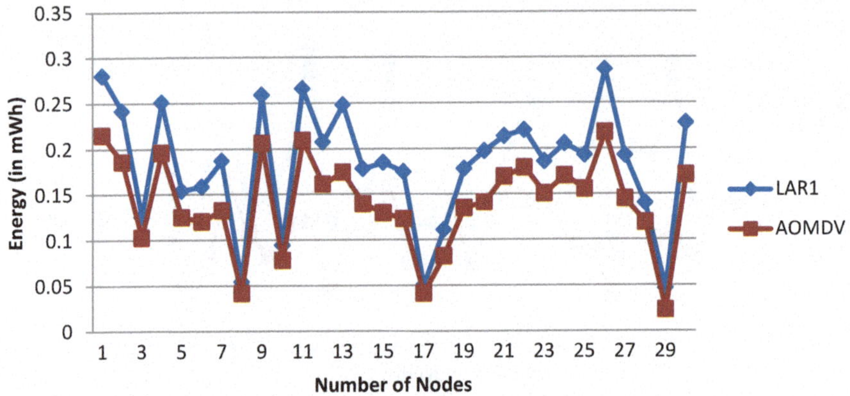

Fig. 4 Energy consumed LAR1 versus AOMDV (simulation time 10 min)

other nodes, from the graph, it is clear that location-based routing protocol LAR-1 consumes more energy as compared to AOMDV.

Figure 5 shows the overall consumption of energy during the receive mode operation of sensor nodes with increased simulation time 20 min. The data transmission by various sources starts at 1 s and terminates at 20 min. In terms of energy utilization, it is visible from the graph that location-based routing protocol LAR-1 consumes more energy as compared to AOMDV for all nodes.

Figure 6 shows the overall consumption of energy during the receive mode operation of sensor nodes with higher simulation time 30 min as compared to other simulation mentioned in Figs. 1 and 2. The data transmission in the network begins at 1 s and terminates at 30 min. All the nodes, while receiving messages from various

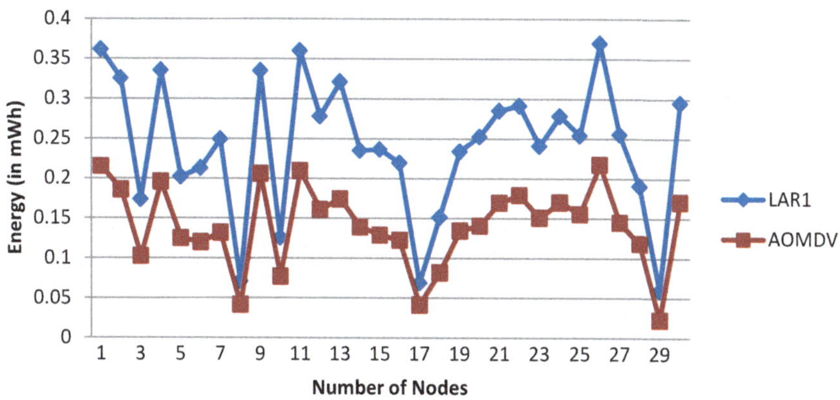

Fig. 5 Energy consumed LAR1 versus AOMDV (simulation time 20 min)

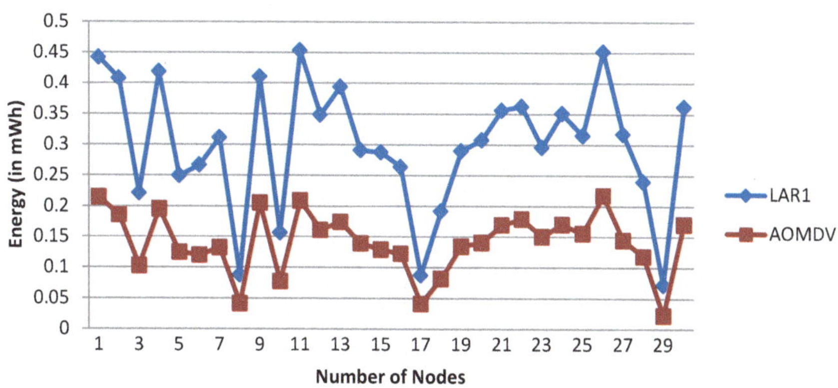

Fig. 6 Energy consumed LAR1 versus AOMDV (simulation time 30 min)

sources, consume lesser energy when working with AOMDV protocol. During the receiving of packets, all the nodes consume higher energy in LAR1 operation.

Overall, the LAR-1 routing protocol consumes higher energy in receive and transmit mode operation as compared to the AOMDV routing protocol. Therefore, the multipath routing protocol is outperformed than a location-based routing protocol. To increase the longevity of individual nodes as well as overall network lifetime, data transmission through multipath routing is preferable over location-based routing protocol.

4 Conclusions

In this paper, the energy consumption is analyzed for AOMDV, and LAR routing protocols with various energy models are analyzed. The obtained results show that the AOMDV routing protocol is outperformed than LAR1 in terms of energy consumption. The total energy consumes by all the nodes in AOMDV protocol 2.00194 mWh and LAR1 2.647416 mWh for simulation time 10 min during transmit mode operation. Further, for receive mode, operation LAR1 consumes 5.520279 and 4.251799 mWh. We have considered different simulation time to analyze the consumption of energy. From the result, it is easily concluded that if communication is done through AOMDV routing protocols, network lifetime will be increased, and nodes in the network may work for a longer period. In the future, evaluation of other types of routing protocols can be analyzed for a more insightful understanding of individual node performance.

References

1. Cho H, Baek Y (2005) Location-based routing protocol for energy efficiency in wireless sensor networks. In: Lecture notes in computer science (including subseries lecture notes in artificial intelligence and lecture notes in bioinformatics) 3823 LNCS: pp 622–631. https://doi.org/10.1007/11596042_65
2. Das Indrani (2019) Multipath routing protocols for ad hoc networks—a meticulous study and open issues. Compliance Eng J 10(8):256–275
3. Das I, Shaw RN, Das S (2020) Analysis of energy consumption in dynamic mobile ad hoc networks. In: Jain L, Tsihrintzis G, Balas V, Sharma D (eds) N data communication and networks, advances in intelligent systems and computing, Springer, pp 235–243
4. Dubal SS, Deshmukh A (2011) Energy efficient multipath routing for wireless sensor networks. In: 2011 Nirma university international conference on engineering: current trends in technology, NUiCONE 2011-conference proceedings, pp 8–10. https://doi.org/10.1109/NUiConE.2011.6153259
5. Genta Addisalem, Lobiyal DK, Abawajy Jemal H (2019) Energy efficient multipath routing algorithm for wireless multimedia sensor network. Sens (Switzerland) 19(17):1–21. https://doi.org/10.3390/s19173642
6. Ko Y, Vaidya NH (1998) Location-aided routing (LAR) in mobile ad hoc networks. In: Proceedings of the ACM/IEEE international conference on mobile computing and networking (MOBICOM'98), vol 3112, pp 1–16
7. Liping LV (2017) An energy aware multipath routing algorithm for wireless sensor networks. Int J Online Eng 13(4):45–56. https://doi.org/10.3991/ijoe.v13i04.7061
8. Ming Lu Y, WS Wong V (2006) An energy-efficient multipath routing protocol for wireless sensor networks. In: IEEE vehicular technology conference, pp 35–40. https://doi.org/10.1109/VTCF.2006.505
9. Marina Mahesh K, Das Samir R (2006) Ad Hoc on-demand multipath distance vector routing. Wirel Commun Mob Comput 6(7):969–988. https://doi.org/10.1002/wcm.432
10. Radi Marjan, Dezfouli Behnam, Bakar Kamalrulnizam Abu, Lee Malrey (2012) Multipath routing in wireless sensor networks: survey and research challenges. Sensors 12(1):650–685. https://doi.org/10.3390/s120100650

Correction to: On-Road Crime Detection Using Artificial Intelligence

Gopichand G, Vijayakumar, and Naga Swetha Pasupuleti

Correction to:
Chapter "On-Road Crime Detection Using Artificial Intelligence" in: M. N. Favorskaya et al. (eds.), *Innovations in Electrical and Electronic Engineering*, Lecture Notes in Electrical Engineering 661, https://doi.org/10.1007/978-981-15-4692-1_32

In the original version of the book, the following belated corrections have been incorporated: The author name "Gopinath" and his affiliation "School of Computer Science Vellore Institute of Technology, Vellore, India" has been changed to "Gopichand G" and "School of Computer Science and Engineering (SCOPE), Vellore Institute of Technology, Vellore - 632014, Tamilnadu, India" in the Frontmatter, Backmatter and in the chapter.

The updated version of this chapter can be found at
https://doi.org/10.1007/978-981-15-4692-1_32

M. N. Favorskaya et al. (eds.), *Innovations in Electrical and Electronic Engineering*, Lecture Notes in Electrical Engineering 661, https://doi.org/10.1007/978-981-15-4692-1_60

Author Index

M. N. Favorskaya et al. (eds.), *Innovations in Electrical and Electronic Engineering*,
Lecture Notes in Electrical Engineering 661, https://doi.org/10.1007/978-981-15-4692-1

The manufacturer's authorised representative in the EU is Springer Nature Customer Service Centre GmbH, Europaplatz 3, 69115 Heidelberg, Germany. If you have any concerns regarding our products, please contact ProductSafety@springernature.com

Printed and bound by CPI Group (UK) Ltd, Croydon, CR0 4YY
30/09/2026
02225503-0019